Science Monograph 49
September 1987

Wyoming Shrubland Ecology

A Chronologic Supplemented Multiple Source Annotated Bibliography

Herbert G. Fisser

Wyoming Agricultural Experiment Station
University of Wyoming
Laramie, Wyoming 82071

College of Agriculture
The University of Wyoming

Library of Congress Catalogue Card Number: 87-50211
ISBN: 0-941370-04-5

Computer programming assistance by Larry P. Germain

Colin Kaltenbach, Director, Agricultural Experiment Station, University of Wyoming, P.O. Box 3354, University Station, Laramie, WY 82071.

TABLE OF CONTENTS

ORGANIZATION AND GUIDE TO USE

As indicated in the Table of Contents, this publication comprises four major and interactive sections. The individual works of each of the 100 contributing authors is identified by six digit (year, source, item) code numbers. The Topic Outline comprises eleven major headings within which bibliographic items are given under individual subheadings and alphanumeric index codes (after Vallentine 1981). The Key Word Subject Index lists more than 800 individual headings under which are given the six digit code numbers and titles of related bibliography items. The Annotated Bibliography provides a chronological listing, by source category, from 1959 through 1985, and a listing of alphanumeric index codes for each citation and its individual six digit numeric code.

Copies of manuscripts, publications, data files, as well as permanent record photographs and color slides are on file at the University of Wyoming, Department of Range Management.

Vallentine, John F. 1981. U.S. — Canadian Range management, 1978-1980: A selected bibliography on ranges, pastures, wildlife, livestock, and ranching. Oryx Press, Phoenix, Arizona 166 pg.

HISTORICAL OVERVIEW

The arid rangelands of Wyoming were observed by many during 1850-1880, the settlement years. Aven Nelson, Ph.D., an early University of Wyoming botanist, traveled widely in the state and commented on the semidesert flora and apparent grazing management impacts during the 1890's. Robert L. Lang, Ph.D., began ecological investigations in 1936 with a range survey of Sweetwater County and soon thereafter Alan A. Beetle, Ph.D., developed complementary evaluation of the more mesic portions of the state. Their efforts during the decades of 1940-60 and the later development of grazing study and rangeland exclosure programs by Dixie R. Smith, Ph.D., from 1955 to 1961 provided a broad ecologically-oriented base from which were expanded the efforts reported in this publication. During the 1960's and 1970's the Range Management faculty, in addition to Lang, Beetle and Fisser, included Morten May, Ph.D., who initiated microclimatologically oriented descriptive ecological studies, ecology of disturbed lands, and range/wildlife integrated research. Several research associates and many graduate assistants developed programs for M.S. and Ph.D theses during these years. The primary thread of activities reported in this publication is oriented to research and study of the arid and semiarid rangelands of Wyoming, especially in the Big Horn and Wind River Basins.

PREFACE

This document fulfills the purpose and objective of preparing a consolidated record of activities and findings which have emanated from the semiarid shrubland ecology research program in the Range Management Department of the University of Wyoming. It also constitutes a major summary report to the Bureau of Land Management whose unbroken annual funding support to the University from 1955 to the present (1985) has provided the basis for continuity of research effort.

These studies of aridlands in Wyoming were begun by the senior editor during June 1959. Primary investigations were launched in the Big Horn and Wind River Basins with an intensive ecological condition class survey of rangelands near Ten Sleep, Thermopolis and Lander. Earlier studies initiated by range management personnel of the University of Wyoming included summer steer grazing of Gardner saltbush rangeland on the 15-mile drainage near Worland; winter sheep grazing also on Gardner saltbush, in the Dry Creek drainage northwest of Greybull; and ecological studies at 29 exclosures, most in the Big Horn and Wind River Basins, but also in Red Desert areas of Wamsutter and Rock Springs, and the Little Colorado Desert near Farson. These had been developed under the Halogeton Research Program of the Bureau of Land Management and have been incorporated into the aridland studies program.

The basic intent has been to obtain quantitative data from geographically and physiognomically diverse semiarid rangeland sites in Wyoming. Use of these data can and has resulted in solution of definitive management problems and identification of related ecosystem function. Biotic, edaphic and climatic relationships may be identified and extrapolated from this long-term, intensive monitoring program for comparison to similar rangelands of the U.S. and of other semiarid areas of the world. Desertification is recognized as being essentially a mismanagement induced phenomenon. Interpretation and synthesis of the Wyoming aridland shrub ecology records, with other regional and world distributed data sets, will significantly assist land managers and users. By application of proper management, reversal of rangeland deterioration can be achieved along with improved ecological condition class and increased products for human needs.

Research sites of the aridland shrub ecology program were located over much of Wyoming; extending from Cody-Deaver-Lovell in the north, 265 miles southward to Baggs and Burntfork; and from near Casper on the east, 235 miles westward to Sage. Research site locations were established in each of five physiographically defined regional areas within the state: 1. Big Horn Basin, 2. Wind River Basin, 3. Powder River Basin-Platte River Drainage, 4. Red Desert-Little Snake River Drainage, 5. Little Colorado Desert-Green and Bear River Drainages.

Intensive monitoring of vegetation, grazing, and environmental factors has been conducted on a variety of vegetation types including sagebrush-grass, saltbush, juniper-mahogany, wheatgrass, blue grama, greasewood, and mixed desert shrub complexes. Fenced exclosures were utilized on most sites and were one to eight acres in size. Annual reports, with detailed data summary requiring two volumes of 400 to 500 pages each during recent years, have been submitted to the BLM.

The most significant monitoring and research segments of the project are listed with brief characterization comment.

1. **Herbage Production and Cover** - Annually, 20 sites in 1962 and increased to 45 in 1965; 20 plot transects, cover estimate and clip, 0.3 m x 0.3 m quadrats in grass complexes, 0.3 m x 3.0 m quadrats in saltbush types; treatments include sagebrush control, grazing, pitting and seeding, within site vegetation type differentiation; shrub production all sites by weight estimate procedures.
2. **Environmental Data** - Quarterly seasonal precipitation recorded annually at 190 sites since 1961 and U.S. Weather Service records obtained at 60 locations. Soil moisture by neutron probe and gravimetric at several locations during the 1960's and 70's.
3. **Soils** - Detailed profile description obtained at 90 locations. Intensive soil-vegetation evaluation at 80 locations.
4. **Winter Sheep Grazing** - Four rates of use on saltbush, 8 pastures, 260 ha, 3 exclosures, 1956 through 1965.

5. **Summer Steer Grazing** - Two pastures, one of native saltbush rangeland and the other improved with waterspreader dikes and grass seedings. Each pasture 250 ha, 40 ha seeded dike area, 1960-1965.
6. **Permanent Quadrats Shrub Mapping** - Demographic and cover measures, 37 locations, 10 quadrats in exclosures. 10 outside, also shrub control treatments; 0.6 m x 6.0 m shrub mapping quadrats and 1.2 m x 1.2 m quadrats for understory point cover data. Data every four years from 1960.
7. **Transects of Permanent Quadrats** - Vegetation cover and composition, 20 quadrats 0.3 m x 0.3 m per line, 70 locations with combinations of grazing, chemical, pitting, seeding, burning, and dozing on many vegetation types. Transects from 1965 and biennially into 1970's and then occasionally as needed for associated research and management application.
8. **Erosion Transects** - Detailed point measurements to determine soil surface movement with 1.5 m transects on various treatments at 15 locations, from 1965 and irregularly after that.
9. **Harvester Ant Control** - Three levels of Mirex, grazing and non-grazing, 12 treatment sites 485 ha with 10,800 permanent 0.3 m x 0.3 m quadrats. Annual 1964 through 1968 and partial to present.
10. **Grazing Systems** - Three grazing systems with 1, 2, and 4 pasture units. Eight exclosures, 110 permanent quadrat transect lines for vegetation cover and composition on sagebrush-grass, juniper-browse, wheatgrass, and mixed desert shrub types. Browse utilization and density from seven 200 ft permanent point and strip transects in juniper-mahogany-bitterbrush type. Annually 1967-71, then 1975 and 1979. Remote sense vegetation type mapping with ERTS imagery, Skylab and high altitude photography.
11. **Phenology and Productivity** - Intensive study initiated in 1973 on prime species *Agropyron smithii, A. spicatum, Artemisia nova* and *A. tridentata* on 12 geographically and environmentally diverse sagebrush-grass sites from central to north central, south central and western Wyoming. Prime species phenology and life form measured on transects of 20 plants permanently located and identified by distance coordinates. Eleven transects of *Agropyron smithii,* 6 transects of *A. spicatum,* 3 transects of *Artemisia nova,* and 12 transects of *A. tridentata.* Also phenology of all species. Monitoring biweekly during growing season. Numerical systems established for separate vegetative and reproductive rankings. Plant height, leaf width and age class recorded for all permanent prime species plants. Environmental measures were ambient air temperature, maximum and minimum air temperatures since last sample date, soil moisture at 15, 30 and 46 cm, and soil temperature at 38 cm. Productivity clippings of prime species were initiated in 1979 and coordinated with entire phenodynamics program annually through 1982. The 20 years of continuous herbage production and precipitation records from 45 extremely diverse locations, with detailed soils information provides a data set unmatched. The additional intensive phenodynamics, productivity, and related environmental measures for the past 10 years also provides a data base from which can be obtained definitive evaluation of phenology relationships to environmental dynamics and management practices.

The initial predictive models of herbage production, phenodynamics, and environmental relationships, which will be the end product of the Wyoming data analyses, can be readily expanded for application to other states and regions, especially within the semiarid sagebrush, saltbush, juniper woodland, and greasewood ecosystems of western North America. Interactive discussion with other scientists, inspection of long-term data sets from similar ecosystems, and on-site verification trips would be valuable. Definition of the parameters characteristic of ecological systems, relationship to monitoring procedures, and interpretive applications on rangelands would provide updated information for assistance to land managers and users for minimization of land use impacts.

ACKNOWLEDGEMENTS

All aspects of this multi-phase research-management based program have been directed by Herb Fisser, Ph.D., senior editor. Through the years a variety of supplemental funding sources and personal contacts contributed to the continuity of the program. Both monetary and personnel support have been diverse and include:

USDI Bureau of Land Management Halogeton Research Program

USDI Bureau of Land Management Range Management Program

USDI Bureau of Land Management Soil, Water, Air Program

USDI Bureau of Land Management Grazing Management Program

USDI Bureau of Land Management Wyoming State Office Soils Program

USDI Bureau of Land Management Rawlins District Office Inventory Program

Wyoming Natural Resources Board

Wyoming Game and Fish Department

USDA Soil Conservation Service State Office, many district offices

USDA Forest Service, Big Horn N.F., Shoshone N.F., Bridger Teton N.F.

USDA Competitive Grants Program

USDA Hatch Program

USDA McIntyre-Stennis Program

USDA Extension Service

NSF IBP Grassland Ecosystem Program

Atlantic Richfield Company

El Paso Natural Gas Company

Colorado Interstate Pipeline Company

Especially notable for their assistance to the continuity of the Bureau of Land Management programs were Chris Vosler, Thane Johnson and Ron Clarke of the Denver Service Center, Roger Mertens of the Wyoming state office, and Rex Colton of the Worland District Office. The positive support from the Department of Interior Washington, D.C. offices and all levels of administration and management in the BLM has exhibited recognition for a continuing and positive need and search for greater knowledge of Wyoming's semiarid rangelands. More than 100 persons with the University of Wyoming have contributed to the materials in this report with presentations, annual reports, manuscripts, and more than 60 graduate theses and dissertations.

AUTHOR LIST AND INDEX

CHAPTER 1: AUTHOR LIST AND INDEX

The list of authors includes 100 persons who have been associated with the senior editor in research and publication of findings from the semiarid rangelands of Wyoming. Given with each author name are the six digit code numbers which identify each citation with which that author was associated. The code number such as 740407 consists of three segments. The first two digits represent year (74), the next two digits identify a source category in which the information is noted (04), and the last two digits identify the numerical occurrence of the item (07). Fisser's name, by intent, has not been included in the list since he is associated directly with most of the citations.

Source category is the term which has been developed to identify the major kinds of references to which the titles in the annotated bibliography apply.

Source Code Number	Source Category Description
S1	Research Activities
S2	Journal and Refereed Publications
S3	Presentations
S4	Published Proceedings/Abstracts
S5	Agricultural Experiment Station Publications
S6	Graduate Advisee Completion

AUTHOR INDEX

TOPIC OUTLINE

CHAPTER 2: BIBLIOGRAPHY BY TOPIC OUTLINE

Introduction

The topic outline used in this publication was taken from Vallentine, J., 1981. U.S.-Canadian range management, 1978-80: A selected bibliography on ranges, pastures, wildlife, livestock, and ranching. Oryx Press. Phoenix, Arizona. 166p. An additional major heading was derived based on recent work. The eleven major headings are listed below to give the reader a quick summary of the outline:

Topic Code Numbers	Topic Heading
1	Range Plants
2	Range Ecology
3	Range Resources
4	Grazing Management
5	Range and Ranch Development
6	Range Livestock
7	Range Wildlife
8	Range Watershed
9	Range Economics
10	Range Education and Research
11	Computer Storage, Programming, and Modeling

In most instances the outline subject words were taken directly from Vallentine (1981) and in all cases the alphanumeric categorization, such as 1B4 given in his publication, was utilized even though not all topical categories were encountered with citations of this publication. The topical headings such as 1B2 Cultivated Pasture and 1B3 Harvested Forages, not encountered in the bibliographic elements of this publication, were removed from the outline.

The Bibliography by Topic Outline is organized over each of the six citation source categories (S1, S2, S3, S4, S5, S6) as identified on page 1, by numerical and keyword topical code (example 1B4 Poisonous Plants) and then chronologically with the six digit code number given for each title (example 1D 650110 Harvester Ant Control Study). A table of contents format of the entire topical outline with pagination columns for each citation source follows.

Topical Outline

S1. RESEARCH ACTIVITIES

1. Range Plants

1B. FORAGE VALUE

1B4. Poisonous Plants

1B4 630106 Germination and longevity of halogeton seeds

1B4 650106 Germination and longevity of halogeton seeds

1D. PLANT INSECTS

1D 640106 Harvester ant control and related study of vegetation response and economic feasibility of the procedure

1D 650110 Harvester ant control study

1D 660107 Harvester ant ecology Big Horn Basin

1D 670105 Harvester ant ecology Big Horn and Wind River Basins

1E. PLANT PHYSIOLOGY

1E1. Reproduction and Germination

1E1 630106 Germination and longevity of halogeton seeds

1E1 650106 Germination and longevity of halogeton seeds

1E1 720103 In 1972 the Arid Land Ecology studies were amplified to include detailed phenological studies of native vegetation over a broad geographical and ecological site characterization.

1E1 730103 Intensive investigation of phenology and production relationships of vegetation as influenced by site characteristics of arid land sites representing diverse habitats of western Wyoming with specific attention given to big sagebrush, black sagebrush, western wheatgrass and bluebunch wheatgrass (initiated in 1973)

1E1 740102 Exclosure study: phenology, growth

1E1 750102 Exclosure study: phenology, growth

1E1 760102 Exclosure study: phenology, growth

1E1 760108 Grass seed sources at A.C. Hull 1940's plantings

1E1 770102 Exclosure study: phenology, growth

1E1 780102 Exclosure study: phenology, growth

1E1 790102 Exclosure study: phenology, growth

1E1 790103 Exclosure study: productivity/standing crop sampling

1E1 800102 Exclosure study: phenology, growth

1E1 800103 Exclosure study: productivity/standing crop sampling

1E1 810102 Exclosure study: phenology, growth

1E1 810103 Exclosure study: productivity/standing crop sampling

1E1 820102 Exclosure study: phenology, growth

1E1 820103 Exclosure study: productivity/standing crop sampling

1E2. Growth and Phenology

1E2 660110 Shrub production estimation study

1E2 660117 Saltbush root development and relation to grazing intensity

1E2 670108 Shrub production estimation study

1E2 680103 Study of moisture stress in woody species using pressure bomb

1E2 680115 Shrub production estimation trials

1E2 700106 Inspect SCS seeding along highway west of Moneta

1E2 710101 Determination of shrub production interrelated with herbaceous production and precipitation on some 50 sites in western Wyoming encompassing wide geographical and environmental variation (additional to Arid Land Ecology program) (initiated in 1971)

1E2 710104 Shortgrass prairie blue grama culm density study

1E2 710105 Shortgrass prairie seed production study

1E2 720102 Reseeding trials of grasses, forbs and shrubs on the bladed and disturbed area of a 120-mile long pipeline, extending across the Big Horn Basin. Funding by Colorado Interstate Pipeline Corp. Primary emphasis of trials for aesthetic improvement as well as erosion control and forage and browse production (initiated in 1972)

1E2 720103 In 1972 the Arid Land Ecology studies were amplified to include detailed phenological studies of native vegetation over a broad geographical and ecological site characterization

1E2 720108 Shortgrass prairie blue grama culm density study

1E2 720109 Shortgrass prairie seed production study

1E2 720111 Exclosure condition analyses

1E2 730103 Intensive investigation of phenology and production relationships of vegetation as influenced by site characteristics of arid land sites representing diverse habitats of western Wyoming with specific attention given to big sagebrush, black sagebrush, western wheatgrass and bluebunch wheatgrass (initiated in 1973)

1E2 730106 CIG pipeline seed plot study

1E2 730110 Shortgrass prairie blue grama culm density study

1E2 730111 Shortgrass prairie seed production study

1E2 740102 Exclosure study: phenology, growth

1E2 740106 CIG pipeline seed plot study

1E2 740107 Shortgrass prairie blue grama culm density study

1E2 740108 Shortgrass prairie seed production study

1E2 750102 Exclosure study: phenology, growth

1E2 750107 CIG pipeline seedplot study

1E2 750110 Shortgrass prairie blue grama culm density study

1E2 750111 Shortgrass prairie seed production study

1E2 760102 Exclosure study: phenology, growth

1E2 760107 CIG pipeline seed plot study

1E2 760108 Grass seed sources at A.C. Hull 1940's plantings

1E2 770102 Exclosure study: phenology, growth

1E2 780102 Exclosure study: phenology, growth

1E2 780105 Juniper GRI seeding

1E2 790102 Exclosure study: phenology, growth

1E2 790103 Exclosure study: productivity/standing crop sampling

1E2 800102 Exclosure study: phenology, growth

1E2 800103 Exclosure study: productivity/standing crop sampling

1E2 810102 Exclosure study: phenology, growth

1E2 810103 Exclosure study: productivity/standing crop sampling

1E2 810105 Woody plant density from 4 ft x 5 ft weight unit estimate plots

1E2 820102 Exclosure study: phenology, growth

1E2 820103 Exclosure study: productivity/standing crop sampling

1E2 820104 Exclosure study: woody plant density

1E4. Herbage Removal

1E4 590102 Summer steer grazing on saltbush rangeland with reseeding and water spreader treatments.

1E4 590103 Winter sheep grazing studies on saltbush rangeland.

1E4 600102 Summer steer grazing on saltbush rangeland with reseeding and water spreader treatments.

1E4 600103 Winter sheep grazing studies on saltbush rangeland.

1E4 610102 Summer steer grazing on saltbush rangeland with reseeding and water spreader treatments.

1E4 610103 Winter sheep grazing studies on saltbush rangeland.

1E4 620101 Winter sheep grazing on saltbush range

1E4 620102 Steer grazing study on saltbush range

1E4 630101 Winter sheep grazing on saltbush range

1E4 630102 Summer steer grazing on saltbush range

1E4 640101 Winter sheep grazing on saltbush range

1E4 640102 Summer steer grazing on saltbush range

1E4 650101 Winter sheep grazing on saltbush range

1E4 650102 Summer steer grazing on saltbush range

1E4 660101 Winter sheep grazing on saltbush rangeland

1E4 660109 Elk utilization and dietary preference study - Bull Elk Park

1E4 660112 Baggs grazing systems study

1E4 670101 Study of grazing systems and influence on vegetation and wildlife habitat (field work initiated in 1967)

1E4 670107 Elk utilization and dietary preference study - Bull Elk Park

1E4 690105 Baggs grazing systems study

1E4 700107 Baggs grazing systems study

1E4 750105 Baggs grazing systems study: impacts of grazing systems

1E4 790107 Baggs grazing systems study (BLM trend plots)

1E4 800105 Roadside intensive use site 7 miles east of Powder River

2. Range Ecology

2A. CLIMATIC FACTORS

2A 610105 Study of biotic communities: saltbush, greasewood, blue grama, sagebrush and limber pine; juniper community studies include soil-plant relationship, site-plant-mammal relationship, long term juniper invasion, juniper control by mechanical and chemical (Tordon) procedures with incorporated seeding studies, and forage production-precipitation relationship study under grazing and juniper control treatments

2A	620103	Exclosure study: production, cover, height, precipitation
2A	630103	Exclosure study: production, cover, height, precipitation
2A	630104	Soil moisture and temperature study
2A	630107	Precipitation pattern study
2A	640103	Exclosure study: production, cover, height, precipitation
2A	640104	Soil moisture and temperature study
2A	640105	Precipitation pattern study
2A	650103	Exclosure study: production, cover, height, precipitation
2A	650104	Soil moisture and temperature study
2A	650107	Precipitation pattern study
2A	650113	Microclimatic studies on desert shrub-types
2A	660105	Soil moisture and temperature
2A	660115	Sagebrush-wind-snow pattern relation study
2A	660118	Exclosure study: production, cover, precipitation
2A	670103	Soil moisture and temperatures
2A	670110	Exclosure study: production, cover, precipitation
2A	680105	Soil moisture and temperature
2A	680106	Exclosure study: production, cover, precipitation
2A	680116	Precipitation pattern study 1968
2A	690101	Exclosure study: production, cover, precipitation
2A	690103	Soil moisture and temperature
2A	700101	Exclosure study: production, cover, precipitation
2A	700104	Soil moisture and temperature
2A	710103	Soil moisture and temperature
2A	710111	Precipitation pattern study
2A	720103	In 1972 the Arid Land Ecology studies were amplified to include detailed phenological studies of native vegetation over a broad geographical and ecological site characterization
2A	720104	Exclosure study: production, cover, precipitation
2A	730102	Exclosure study: production, cover, precipitation
2A	730103	Intensive investigation of phenology and production relationships of vegetation as influenced by site characteristics of arid land sites representing diverse habitats of western Wyoming with specific attention given to big sagebrush, black sagebrush, western wheatgrass and bluebunch wheatgrass (initiated in 1973)
2A	730104	Precipitation pattern study
2A	740101	Exclosure study: production, cover, precipitation
2A	740102	Exclosure study: phenology, growth
2A	740103	Precipitation pattern study
2A	750101	Exclosure study: production, cover, precipitation
2A	750103	Precipitation pattern study
2A	760101	Exclosure study: production, cover, precipitation
2A	760102	Exclosure study: phenology, growth
2A	760103	Precipitation pattern study
2A	770101	Exclosure study: production, cover, precipitation
2A	770102	Exclosure study: phenology, growth
2A	770103	Precipitation pattern study
2A	770107	Snow catchment and revegetation on Jim Bridger mine spoil
2A	780101	Exclosure study: production, cover, precipitation
2A	780102	Exclosure study: phenology, growth
2A	780103	Precipitation pattern study
2A	790101	Exclosure study: production, cover, precipitation
2A	790102	Exclosure study: phenology, growth
2A	790103	Exclosure study: productivity/standing crop sampling
2A	790104	Precipitation pattern study
2A	800101	Exclosure study: production, cover, precipitation
2A	800103	Exclosure study: productivity/standing crop sampling

2A	800104	Precipitation pattern study
2A	810101	Exclosure study: production, cover, precipitation
2A	810103	Exclosure study: productivity/standing crop sampling
2A	810104	Precipitation pattern study
2A	820101	Exclosure study: production, cover, precipitation
2A	820103	Exclosure study: productivity/standing crop sampling
2A	820105	Precipitation pattern study
2A	830101	Precipitation pattern study
2A	830102	Computer program development for long-term weather record compilation
2A	830103	Computer program development for estimation of missing climate data
2A	830105	Summary of soil moisture and temperature data from Granite Mountain exclosure
2A	830108	Yield prediction from ERHYM model
2A	830109	Compilation of US Weather Service precipitation records - Wyoming
2A	840101	Precipitation pattern study
2A	840102	Computer program development for long-term weather record compilation
2A	840103	Computer program development for estimation of missing climatic data
2A	840105	Summary of soil moisture and temperature data from Granite Mountain exclosure
2A	840108	Yield prediction with ERHYM model
2A	840109	Compilation of U.S. Weather Service precipitation records - Wyoming

2B. BIOTIC FACTORS

2B	780101	Exclosure study: production, cover, precipitation
2B	790101	Exclosure study: production, cover, precipitation

2C. EDAPHIC FACTORS

2C	620103	Exclosure study: production, cover, height, precipitation
2C	630103	Exclosure study: production, cover, height, precipitation
2C	630104	Soil moisture and temperature study
2C	630105	Erosion transect study
2C	640103	Exclosure study: production, cover, height, precipitation
2C	640104	Soil moisture and temperature study
2C	650103	Exclosure study: production, cover, height, precipitation
2C	650104	Soil moisture and temperature study
2C	650105	Erosion transect study
2C	650113	Microclimatic studies on desert shrub-types
2C	660105	Soil moisture and temperature study
2C	660111	Limber pine ecology
2C	660113	Erosion transects for study of soil movement and relationship to climatic variation and grazing management
2C	660118	Exclosure study: production, cover, precipitation
2C	670103	Soil moisture and temperature study
2C	670109	Limber pine ecology
2C	670110	Exclosure study: production, cover, precipitation
2C	680105	Soil moisture and temperature study
2C	680106	Exclosure study: production, cover, precipitation

2C	690101	Exclosure study: production, cover, precipitation
2C	690103	Soil moisture and temperature study
2C	690111	Boars Tusk sand dune area tour
2C	700101	Exclosure study: production, cover, precipitation
2C	700104	Soil moisture and temperature study
2C	700105	Exclosure soil profile studies with SCS
2C	700108	Erosion transects
2C	710103	Soil moisture and temperature study
2C	710108	Exclosure soil profile studies with SCS
2C	720107	Exclosure soil profile studies with SCS
2C	730102	Exclosure study: production, cover, precipitation
2C	730107	Exclosure soil profile studies with SCS
2C	740101	Exclosure study: production, cover, precipitation
2C	750101	Exclosure study: production, cover, precipitation
2C	750108	Exclosure soil profile studies with SCS Dutch Nick
2C	760101	Exclosure study: production, cover, precipitation
2C	770101	Exclosure study: production, cover, precipitation
2C	770104	Exclosure soil profile studies with SCS
2C	770107	Snow catchment and revegetation on Jim Bridger mine spoil
2C	780101	Exclosure study: production, cover, precipitation
2C	780106	Soil-vegetation relation study
2C	780108	Exclosure soil profile studies with SCS
2C	790101	Exclosure study: production, cover, precipitation
2C	790105	Exclosure soil profile studies with SCS
2C	800101	Exclosure study: production, cover, precipitation
2C	810101	Exclosure study: production, cover, precipitation
2C	820101	Exclosure study: production, cover, precipitation
2C	830105	Summary of soil moisture and temperature data from Granite Mountain exclosure
2C	830106	Erosion transect data summarization and input to computer
2C	830108	Yield prediction from ERHYM model
2C	840105	Summary of soil moisture and temperature data from Granite Mountain exclosure
2C	840106	Erosion transect data summarization and input to computer
2C	840108	Yield prediction with ERHYM model

2D. PYRIC FACTORS

2D	610105	Study of biotic communities: saltbush, greasewood, blue grama, sagebrush and limber pine; juniper community studies include soil-plant relationship, site-plant-mammal relationship, long-term juniper invasion, juniper control by mechanical and chemical (Tordon) procedures with incorporated seeding studies, and forage production-precipitation relationship study under grazing and juniper control treatments
2D	790106	Juniper wildfire in Wind River Canyon 13 miles south of Thermopolis

2E. AUTECOLOGY

2E	610105	Study of biotic communities: saltbush, greasewood, blue grama, sagebrush and limber pine; juniper community studies include soil-plant relationship, site-plant-mammal relationship, long-term juniper invasion, juniper control by mechanical and chemical (Tordon) procedures with incorporated seeding studies, and forage production-precipitation relationship study under grazing and juniper control treatments
2E	630106	Germination and longevity of halogeton seeds
2E	640108	Ecology and control of juniper
2E	650106	Germination and longevity of halogeton seeds
2E	650112	Juniper control study on LU
2E	660104	Juniper control study on LU
2E	660108	Juniper ecology: permanent plot transects
2E	660111	Limber pine ecology
2E	670102	Juniper control study of LU
2E	670106	Juniper ecology: permanent plot transects and seeding trials
2E	670109	Limber pine ecology
2E	680103	Study of moisture stress in woody species using pressure bomb
2E	680110	Vegetation pattern on shortgrass-cactus range
2E	680111	Juniper ecology 20 ft x 50 ft permanent plots on GRI exclosures
2E	680114	Juniper GRI exclosure seedings
2E	690102	Juniper control study of LU
2E	690108	Vegetation pattern on shortgrass-cactus ranges
2E	690109	Juniper GRI exclosure seedings
2E	690112	Juniper GRI exclosure Tordon retention studies
2E	690112	Juniper GRI exclosure Tordon retention studies
2E	700102	Juniper control study
2E	700103	Juniper GRI seedings
2E	700110	Vegetation pattern on shortgrass-cactus range
2E	710102	Juniper GRI seedings
2E	710104	Shortgrass prairie blue grama culm density study
2E	710105	Shortgrass prairie seed production study
2E	720103	In 1972 the Arid Land Ecology studies were amplified to include detailed phenological studies of native vegetation over a broad geographical and ecological site characterization.
2E	720105	Juniper GRI seedings
2E	720108	Shortgrass prairie blue grama culm density study
2E	720109	Shortgrass prairie seed production study
2E	730103	Intensive investigation of phenology and production relationships of vegetation as influenced by site characteristics of arid land sites representing diverse habitats of western Wyoming with specific attention given to big sagebrush, black sagebrush, western wheatgrass and bluebunch wheatgrass (initiated in 1973)
2E	730110	Shortgrass prairie blue grama culm density study
2E	730111	Shortgrass prairie seed production study
2E	740102	Exclosure study: phenology, growth
2E	740107	Shortgrass prairie blue grama culm density study
2E	740108	Shortgrass prairie seed production study
2E	750102	Exclosure study: phenology, growth
2E	750110	Shortgrass prairie blue grama culm density study
2E	750111	Shortgrass prairie seed production study
2E	760102	Exclosure study: phenology, growth
2E	760108	Grass seed sources at A.C. Hull 1940's plantings
2E	770102	Exclosure study: phenology, growth

2E	780101	Exclosure study: production, cover, precipitation
2E	780102	Exclosure study: phenology, growth
2E	780104	Juniper GRI exclosure 20 ft x 50 ft permanent quadrats
2E	790101	Exclosure study: production, cover, precipitation
2E	790102	Exclosure study: phenology, growth
2E	790103	Exclosure study: productivity/standing crop sampling
2E	790108	Juniper and limber pine population dynamics and life history
2E	800102	Exclosure study: phenology, growth
2E	800103	Exclosure study: productivity/standing crop sampling
2E	800105	Roadside intensive use site 7 miles east of Powder River
2E	800106	Juniper and limber pine population dynamics and life history
2E	810102	Exclosure study: phenology, growth
2E	810103	Exclosure study: productivity/standing crop sampling
2E	810105	Woody plant density from 4 ft x 5 ft weight unit estimate plots
2E	820102	Exclosure study: phenology, growth
2E	820103	Exclosure study: productivity/standing crop sampling
2E	820104	Exclosure study: woody plant density
2E	840104	Compilation of herbage production and cover data from Wind River Basin exclosures
2E	840105	Summary of soil moisture and temperature data from Granite Mountain Exclosure

2F. SYNECOLOGY

2F	830107	Wyoming rangeland reference sites
2F	840107	Wyoming rangeland reference sites

2F1. Great Plains

2F1	690108	Vegetation pattern on shortgrass-cactus ranges
2F1	700110	Vegetation pattern on shortgrass-cactus range
2F1	710104	Shortgrass prairie blue grama culm density study
2F1	710105	Shortgrass prairie seed production study
2F1	720101	Study of biological and physical environmental characteristics as represented by remote imagery obtained from the N.A.S.A. Earth Resources Technology Satellite (ERTS) with emphasis on grazing impact of livestock and associated impact of browsing game animals on rangelands of southern Wyoming (initiated in 1972)
2F1	720108	Shortgrass prairie blue grama culm density study
2F1	720109	Shortgrass prairie seed production study
2F1	730101	On an area near Reno Junction, Wyoming, on which a strip mine operation is proposed, the ecology, vegetation and soils inventory studies were initiated in 1973. Continuing work is involved with the establishment of permanent monitoring locations for development and evaluation of best post-mining rehabilitation procedures
2F1	730105	Range inventory with remote sense analyses on Baggs grazing system
2F1	730110	Shortgrass prairie blue grama culm density study
2F1	730111	Shortgrass prairie seed production study

2F1	740104	Black Thunder pre-mining inventory
2F1	740107	Shortgrass prairie blue grama culm density study
2F1	740108	Shortgrass prairie seed production study
2F1	750104	Black Thunder pre-mining inventory
2F1	750110	Shortgrass prairie blue grama culm density study
2F1	750111	Shortgrass prairie seed production study
2F1	760105	Black Thunder pre-mining inventory
2F1	760108	Grass seed sources at A.C. Hull 1940's plantings
2F1	770106	Black Thunder pre-mining inventory

2F2. West

2F2	590101	Range condition class survey: Big Horn & Wind River Basin, Wyoming
2F2	600101	Range condition class survey: Big Horn & Wind River Basin, Wyoming.
2F2	610101	Range condition class survey: Big Horn & Wind River Basin, Wyoming
2F2	610105	Study of biotic communities: saltbush, greasewood, blue grama, sagebrush and limber pine; juniper community studies include soil-plant relationship, site-plant-mammal relationship, long-term juniper invasion, juniper control by mechanical and chemical (Tordon) procedures with incorporated seeding studies, and forage production-precipitation relationship study under grazing and juniper control treatments
2F2	630104	Soil moisture and temperature study
2F2	640104	Soil moisture and temperature study
2F2	640107	2 ft x 20 ft and 4 ft x 4 ft permanent plot sampling
2F2	650104	Soil moisture and temperature study
2F2	650109	1 ft x 1 ft permanent plot transects
2F2	660105	Soil moisture and temperature study
2F2	660106	1 ft x 1 ft permanent plot transects
2F2	660108	Juniper ecology: permanent plot transects
2F2	660111	Limber pine ecology
2F2	660114	2 ft x 20 ft and 4 ft x 4 ft permanent plot sampling
2F2	660116	Small mammal abundance and distribution in juniper
2F2	670101	Study of grazing systems and influence on vegetation and wildlife habitat (field work initiated in 1967)
2F2	670103	Soil moisture and temperature study
2F2	670104	1 ft x 1 ft permanent plot transects
2F2	670106	Juniper ecology: permanent plot transects and seeding trials
2F2	670109	Limber pine ecology
2F2	680105	Soil moisture and temperature study
2F2	680107	2 ft x 20 ft and 4 ft x 4 ft permanent plot sampling
2F2	680110	Vegetation pattern on shortgrass-cactus range
2F2	680113	1 ft x 1 ft permanent plot transects
2F2	690103	Soil moisture and temperature study
2F2	690104	1 ft x 1 ft permanent plot transects
2F2	690105	Baggs grazing systems study
2F2	690106	2 ft x 20 ft and 4 ft x 4 ft permanent plot sampling
2F2	690110	Block juniper site, 1 ft x 1 ft permanent plot studies
2F2	690111	Boars Tusk sand dune area tour
2F2	700104	Soil moisture and temperature study
2F2	700107	Baggs grazing systems study

2F2	700109	1 ft x 1 ft permanent plot transects
2F2	710103	Soil moisture and temperature study
2F2	710106	1 ft x 1 ft permanent plot transects
2F2	720103	In 1972 the Arid Land Ecology studies were amplified to include detailed phenological studies of native vegetation over a broad geographical and ecological site characterization
2F2	720106	2 ft x 20 ft and 4 ft x 4 ft permanent plot sampling
2F2	720111	Exclosure condition analyses
2F2	730103	Intensive investigation of phenology and production relationships of vegetation as influenced by site characteristics of arid land sites representing diverse habitats of western Wyoming with specific attention given to big sagebrush, black sagebrush, western wheatgrass and bluebunch wheatgrass (initiated in 1973)
2F2	730109	1 ft x 1 ft permanent plot transects
2F2	740102	Exclosure study: phenology, growth
2F2	750102	Exclosure study: phenology, growth
2F2	750105	1 Baggs grazing systems study: impacts of grazing systems
2F2	750105	2 Baggs grazing systems study: BLM trend plots
2F2	750106	1 ft x 1 ft permanent plot transects - Rawlins, Kemmerer, Pinedale
2F2	760102	Exclosure study: phenology, growth
2F2	760104	2 ft x 20 ft and 4 ft x 4 ft permanent plot sampling
2F2	760106	1 ft x 1 ft permanent plot transects - Kemmerer-Boulder Ridge exclosures
2F2	770102	Exclosure study: phenology, growth
2F2	770105	1 ft x 1 ft permanent plot transects
2F2	780102	Exclosure study: phenology, growth
2F2	780107	1 ft x 1 ft permanent plot transects
2F2	790102	Exclosure study: phenology, growth
2F2	790103	Exclosure study: productivity/standing crop sampling
2F2	790107	Baggs grazing systems study (BLM trend plots)
2F2	790108	Juniper and limber pine population dynamics and life history
2F2	800102	Exclosure study: phenology, growth
2F2	800103	Exclosure study: productivity/standing crop sampling
2F2	800105	Roadside intensive use site 7 miles east of Powder River
2F2	800106	Juniper and limber pine population dynamics and life history
2F2	810102	Exclosure study: phenology, growth
2F2	810103	Exclosure study: productivity/standing crop sampling
2F2	810105	Woody plant density from 4 ft x 5 ft weight unit estimate plots
2F2	820102	Exclosure study: phenology, growth
2F2	820103	Exclosure study: productivity/standing crop sampling
2F2	820104	Exclosure study: woody plant density
2F2	840104	Compilation of herbage production and cover data from Wind River Basin exclosures
2F2	840105	Summary of soil moisture and temperature data from Granite Mountain exclosure
2F2	840107	Wyoming rangeland reference sites
2F2	840108	Yield prediction with ERHYM model

3. Range Resources

3A. RANGE INVENTORY AND SURVEY

3A 610105 Study of biotic communities: saltbush, greasewood, blue grama, sagebrush and limber pine; juniper community studies include soil-plant relationship, site-plant-mammal relationship, long-term juniper invasion, juniper control by mechanical and chemical (Tordon) procedures with incorporated seeding studies, and forage production-precipitation relationship study under grazing and juniper control treatments

3A 630107 Precipitation pattern study

3A 640105 Precipitation pattern study

3A 650107 Precipitation pattern study

3A 680116 Precipitation pattern study

3A 710111 Precipitation pattern study

3A 730104 Precipitation pattern study

3A 740103 Precipitation pattern study

3A 750103 Precipitation pattern study

3A 760103 Precipitation pattern study

3A 770103 Precipitation pattern study

3A 780103 Precipitation pattern study

3A 790104 Precipitation pattern study

3A 790105 Exclosure soil profile studies with SCS

3A 800104 Precipitation pattern study

3A 810104 Precipitation pattern study

3A 820105 Precipitation pattern study

3A 830101 Precipitation pattern study

3A 830107 Wyoming rangeland reference sites

3A 840101 Precipitation pattern study

3A 840107 Wyoming rangeland reference sites

3A1. Site, Mapping, Remote Sensing

3A1 590101 Range condition class survey: Big Horn & Wind River Basin, Wyoming

3A1 600101 Range condition class survey: Big Horn & Wind River Basin, Wyoming.

3A1 610101 Range condition class survey: Big Horn & Wind River Basin, Wyoming

3A1 610104 Arid land ecology studies of vegetation, cover, composition, production, biochemical composition, precipitation, and site relationships under grazing, shrub control, seeding and pitting treatments

3A1 630108 Cumberland district exclosures, production, cover

3A1 650113 Microclimatic studies on desert shrub-types

3A1 660111 Limber pine ecology

3A1 670109 Limber pine ecology

3A1 710107 Wagon Wheel range survey

3A1 710109 Sheep Mountain deer habitat study

3A1 720101 Study of biological and physical environmental characteristics as represented by remote imagery obtained from the N.A.S.A. Earth Resources Technology Satellite (ERTS) with emphasis on grazing impact of livestock and associated impact of browsing game animals on rangelands of southern Wyoming (initiated in 1972)

3A1 720112 Wagon Wheel range survey

3A1 730101 On an area near Reno Junction, Wyoming, on which a strip mine operation is proposed, the ecology, vegetation and soils inventory studies were initiated in 1973. Continuing work is involved with the establishment of permanent monitoring locations for development and evaluation of best post-mining rehabilitation procedures

3A1 730105 Range inventory with remote sense analyses on Baggs grazing system

3A1	740104	Black Thunder pre-mining inventory
3A1	750104	Black Thunder pre-mining inventory
3A1	750109	Wagon Wheel range survey
3A1	760105	Black Thunder pre-mining inventory
3A1	770106	Black Thunder pre-mining inventory

3A2. Condition and Trend

3A2	590101	Range condition class survey: Big Horn & Wind River Basin, Wyoming
3A2	600101	Range condition class survey: Big Horn & Wind River Basin, Wyoming.
3A2	610101	Range condition class survey: Big Horn & Wind River Basin, Wyoming
3A2	610104	Arid land ecology studies of vegetation, cover, composition, production, biochemical composition, precipitation, and site relationships under grazing, shrub control, seeding and pitting treatments
3A2	640107	2 ft x 20 ft and 4 ft x 4 ft permanent plot sampling
3A2	650109	1 ft x 1 ft permanent plot transects
3A2	650111	Sagebrush cover estimation with angle gauge
3A2	660106	1 ft x 1 ft permanent plot transects
3A2	660112	Baggs grazing systems study
3A2	660114	2 ft x 20 ft and 4 ft x 4 ft permanent plot sampling
3A2	670101	Study of grazing systems and influence on vegetation and wildlife habitat (field work initiated in 1967)
3A2	670104	1 ft x 1 ft permanent plot transects
3A2	670106	Juniper ecology: permanent plot transects and seeding trials
3A2	680107	2 ft x 20 ft and 4 ft x 4 ft permanent plot sampling
3A2	680110	Vegetation pattern on shortgrass-cactus range
3A2	680113	1 ft x 1 ft permanent plot transects
3A2	690104	1 ft x 1 ft permanent plot transects
3A2	690105	Baggs grazing systems study
3A2	690106	2 ft x 20 ft and 4 ft x 4 ft permanent plot sampling
3A2	690108	Vegetation pattern on shortgrass-cactus ranges
3A2	690110	Block juniper site, 1 ft x 1 ft permanent plot transects
3A2	690111	Boars Tusk sand dune area tour
3A2	700107	Baggs grazing systems study
3A2	700109	1 ft x 1 ft permanent plot transects
3A2	700110	Vegetation pattern on shortgrass-cactus range
3A2	710106	1 ft x 1 ft permanent plot transects
3A2	710107	Wagon Wheel range survey
3A2	710110	Laramie Peak game range inventory
3A2	720103	In 1972 the Arid Land Ecology studies were amplified to include detailed phenological studies of native vegetation over a broad geographical and ecological site characterization
3A2	720106	2 ft x 20 ft and 4 ft x 4 ft permanent plot sampling
3A2	720110	Contour furrow and seedings study
3A2	720111	Exclosure condition analyses
3A2	720112	Wagon Wheel range survey
3A2	720113	Laramie Peak game range inventory
3A2	730101	On an area near Reno Junction, Wyoming, on which a strip mine operation is proposed, the ecology, vegetation and soils inventory studies were initiated in 1973. Continuing work is involved with the establishment of permanent monitoring locations for development and evaluation of best post-mining rehabilitation procedures

3A2	730103	Intensive investigation of phenology and production relationships of vegetation as influenced by site characteristics of arid land sites representing diverse habitats of western Wyoming with specific attention given to big sagebrush, black sagebrush, western wheatgrass and bluebunch wheatgrass (initiated in 1973)
3A2	730108	Mule deer winter range evaluation
3A2	730109	1 ft x 1 ft permanent plot transects
3A2	740102	Exclosure study: phenology, growth
3A2	740104	Black Thunder pre-mining inventory
3A2	740105	Mule deer winter range evaluation
3A2	750102	Exclosure study: phenology, growth
3A2	750104	Black Thunder pre-mining inventory
3A2	750105	1 Baggs grazing systems study: impacts of grazing systems
3A2	750106	1 ft x 1 ft permanent plot transects - Rawlins, Kemmerer, Pinedale
3A2	750109	Wagon Wheel range survey
3A2	760102	Exclosure study: phenology, growth
3A2	760104	2 ft x 20 ft and 4 ft x 4 ft permanent plot sampling
3A2	760105	Black Thunder pre-mining inventory
3A2	760106	1 ft x 1 ft permanent plot transects - Kemmerer-Boulder Ridge exclosures
3A2	770102	Exclosure study: phenology, growth
3A2	770105	1 ft x 1 ft permanent plot transects
3A2	770106	Black Thunder pre-mining inventory
3A2	780102	Exclosure study: phenology, growth
3A2	780104	Juniper GRI exclosure 20 ft x 50 ft permanent quadrats
3A2	780107	1 ft x 1 ft permanent plot transects
3A2	790102	Exclosure study: phenology, growth
3A2	790103	Exclosure study: productivity/standing crop sampling
3A2	790107	Baggs grazing systems study: BLM trend plots
3A2	790108	Juniper and limber pine population dynamics and life history
3A2	800102	Exclosure study: phenology, growth
3A2	800103	Exclosure study: productivity/standing crop sampling
3A2	800105	Roadside intensive use site 7 miles east of Powder River
3A2	800106	Juniper and limber pine population dynamics and life history
3A2	810102	Exclosure study: phenology, growth
3A2	810103	Exclosure study: productivity/standing crop sampling
3A2	820102	Exclosure study: phenology, growth
3A2	820103	Exclosure study: productivity/standing crop sampling
3A2	830108	Yield prediction from ERHYM model
3A2	840108	Yield prediction with ERHYM model

3A3. Production and Utilization

3A3	610104	Arid land ecology studies of vegetation, cover, composition, production, biochemical composition, precipitation, and site relationships under grazing, shrub control, seeding and pitting treatments

3A3	620103	Exclosure study: production, cover, height, precipitation
3A3	630103	Exclosure study: production, cover, height, precipitation
3A3	630108	Cumberland district exclosures: production, cover
3A3	640103	Exclosure study: production, cover, height, precipitation
3A3	650103	Exclosure study: production, cover, height, precipitation
3A3	660110	Shrub production estimation study
3A3	660118	Exclosure study: production, cover, precipitation
3A3	670108	Shrub production estimation study
3A3	670110	Exclosure study: production, cover, precipitation
3A3	680106	Exclosure study: production, cover, precipitation
3A3	680115	Shrub production estimation trials
3A3	690101	Exclosure study: production, cover, precipitation
3A3	700101	Exclosure study: production, cover, precipitation
3A3	710101	Determination of shrub production interrelated with herbaceous production and precipitation on some 50 sites in western Wyoming encompassing wide geographical and environmental variation (Additional to Arid Land Ecology program) (initiated in 1971)
3A3	710105	Shortgrass prairie seed production study
3A3	720104	Exclosure study: production, cover, precipitation.
3A3	720109	Shortgrass prairie seed production study
3A3	730102	Exclosure study: production, cover, precipitation
3A3	730111	Shortgrass prairie seed production study
3A3	740101	Exclosure study: production, cover, precipitation
3A3	740108	Shortgrass prairie seed production study
3A3	750101	Exclosure study: production, cover, precipitation
3A3	750105	2 Baggs grazing systems study: BLM trend plots
3A3	750111	Shortgrass prairie seed production study
3A3	760101	Exclosure study: production, cover, precipitation
3A3	760108	Grass seed sources at A.C. Hull 1940's plantings
3A3	770101	Exclosure study: production, cover, precipitation
3A3	780101	Exclosure study: production, cover, precipitation
3A3	790101	Exclosure study: production, cover, precipitation
3A3	800101	Exclosure study: production, cover, precipitation
3A3	810101	Exclosure study: production, cover, precipitation
3A3	820101	Exclosure study: production, cover, precipitation
3A3	830104	Compilation of herbage production and cover data from Wind River Basin exclosures
3A3	840104	Compilation of herbage production and cover data from Wind River Basin exclosures

3B. NATURAL RESOURCES PLANNING

3B	830107	Wyoming rangeland reference sites
3B	840107	Wyoming rangeland reference sites

3C. PUBLIC LAND ADMINISTRATION

3C	830107	Wyoming rangeland reference sites
3C	840107	Wyoming rangeland reference sites

4. Grazing Management

4 610104 Aridland ecology studies of vegetation, cover, composition, production, biochemical composition, precipitation, and site relationships under grazing, shrub control, seeding and pitting treatments

4A. STOCKING RATES

4A 590103 Winter sheep grazing on saltbush range

4A 600103 Winter sheep grazing on saltbush range

4A 610103 Winter sheep grazing on saltbush range

4A 620101 Winter sheep grazing on saltbush range

4A 630101 Winter sheep grazing on saltbush range

4A 640101 Winter sheep grazing on saltbush range

4A 650101 Winter sheep grazing on saltbush range

4A 660101 Winter sheep grazing on saltbush range

4A 660102 Grazing intensity influence on dietary intake by use of fistulated sheep

4A 660112 Baggs grazing systems study

4A 660117 Saltbush root development and relation to grazing intensity

4A 680108 Atrazine application to control annual weeds on saltbush ranges

4A 680109 Forb control with chemical on subalpine range Big Horn Mountains

4A 690107 Atrazine application to control annual weeds on saltbush range

4A 800105 Roadside intensive use site 7 miles east of Powder River

4B. GRAZING SEASONS AND SYSTEMS

4B 590102 Summer steer grazing on saltbush rangeland with reseeding and water spreader treatments.

4B 590103 Winter sheep grazing studies on saltbush rangeland.

4B 600102 Summer steer grazing on saltbush rangeland with reseeding and water spreader treatments.

4B 600103 Winter sheep grazing studies on saltbush rangeland.

4B 610102 Summer steer grazing on saltbush rangeland with reseeding and water spreader treatments.

4B 610103 Winter sheep grazing on saltbush range

4B 620101 Winter sheep grazing on saltbush range

4B 620102 Summer steer grazing on saltbush range

4B 630101 Winter sheep grazing on saltbush range

4B 630102 Summer steer grazing on saltbush range

4B 640101 Winter sheep grazing on saltbush range

4B 640102 Summer steer grazing on saltbush range

4B 650101 Winter sheep grazing on saltbush range

4B 650102 Summer steer grazing on saltbush range

4B 660101 Winter sheep grazing on saltbush range

4B 660112 Baggs grazing systems study

4B 670101 Study of grazing systems and influence on vegetation and wildlife habitat (field work initiated in 1967)

4B 680117 Baggs grazing system study

4B 690105 Baggs grazing system study

4B	700107	Baggs Grazing system study
4B	750105	1 Baggs grazing system study: impacts of grazing systems
4B	750105	2 Baggs grazing system study: BLM trend plots
4B	790107	Baggs grazing system study: BLM trend plots
4B	800105	Roadside intensive use site 7 miles east of Powder River

4C. GRAZING HABITS AND DISTRIBUTION

4C	670101	Study of grazing systems and influence on vegetation and wildlife habitat (field work initiated in 1967)
4C	680104	Study of elk-livestock competition in terms of utilization by class of animal and influence on vegetation
4C	690105	Baggs grazing system study
4C	700107	Baggs grazing system study
4C	750105	1 Baggs grazing system study: impacts of grazing systems
4C	750105	2 Baggs grazing system study: BLM trend plots
4C	790107	Baggs grazing system study: BLM trend plots
4C	800105	Roadside intensive use site 7 miles east of Powder River

5. Range and Ranch Development

5	610104	Aridland ecology studies of vegetation, cover, composition, production, biochemical composition, precipitation, and site relationships under grazing, shrub control, seeding and pitting treatments

5A. PLANT CONTROL

5A1. Methods of Control

5A1B. BIOLOGICAL

5A1B	640106	Harvester ant control and related study of vegetation response and economic feasibility of the procedure
5A1B	650110	Harvester ant control study
5A1B	660107	Harvester ant ecology Big Horn Basin

5A1C. MECHANICAL

5A1C	590102	Summer steer grazing on saltbush rangeland with reseeding and water spreader treatments.
5A1C	600102	Summer steer grazing on saltbush rangeland with reseeding and water spreader treatments.
5A1C	600104	Sagebrush control and seeding, 2 Mile, Demer, Smilo, Upper and Lower Govt.
5A1C	600104	2 Mechanical sagebrush control and seeding, 2 Mile, Demer, Smilo, Upper and Lower Govt.
5A1C	610102	Summer steer grazing on saltbush rangeland with reseeding and water spreader treatments
5A1C	620102	Summer steer grazing on saltbush range
5A1C	630102	Summer steer grazing on saltbush range
5A1C	640102	Summer steer grazing on saltbush range
5A1C	640108	Ecology and control of juniper
5A1C	650102	Summer steer grazing on saltbush range

5A1C650108 Seeding in juniper and juniper control

5A1C650112 Juniper control study on LU

5A1C660104 Juniper control study on LU

5A1C680101 Game range improvement program by control of juniper and seedling establishment studies of preferred browse species (initiated in 1968)

5A1C680102 Juniper control chemical, burn, mechanical on GRI exclosures.

5A1D. CHEMICAL

5A1D600104 Sagebrush control and seeding, 2 Mile, Demer, Smilo, Upper and Lower Govt.

5A1D600104 1 Chemical sagebrush control and seeding, 2 Mile, Demer, Smilo, Upper and Lower Govt.

5A1D620103 Exclosure study: production, cover, height, precipitation

5A1D630103 Exclosure study: production, cover, height, precipitation

5A1D630108 Cumberland district exclosures: production, cover

5A1D640103 Exclosure study: production, cover, height, precipitation

5A1D650103 Exclosure study: production, cover, height, precipitation

5A1D650108 Seeding in juniper and juniper control

5A1D660118 Exclosure study: production, cover, precipitation

5A1D670102 Juniper control study on LU

5A1D670105 Harvester ant ecology Big Horn and Wind River Basins

5A1D670110 Exclosure study: production, cover, precipitation

5A1D680101 Game range improvement program by control of juniper and seedling establishment studies of preferred browse species (initiated in 1968)

5A1D680102 Juniper control chemical, burn, mechanical on GRI exclosures.

5A1D680106 Exclosure study: production, cover, precipitation

5A1D680108 Atrazine application to control annual weeds on saltbush ranges

5A1D680109 Forb control with chemical on subalpine range Big Horn Mountains

5A1D680111 Juniper ecology 20 ft x 50 ft permanent plots on GRI exclosures

5A1D680112 Kane Deer exclosure seeding analysis

5A1D680114 Junipers GRI exclosure seedings

5A1D690101 Exclosure study: production, cover, precipitation

5A1D690102 Juniper control study of LU

5A1D690107 Atrazine application to control annual weeds on saltbush range

5A1D690109 Juniper GRI exclosure seedings

5A1D690112 Juniper GRI exclosure Tordon retention studies

5A1D700101 Exclosure study: production, cover, precipitation

5A1D700102 Juniper control study

5A1D700103 Juniper GRI exclosure seedings

5A1D710102 Juniper GRI exclosure seedings

5A1D720103 In 1972 the Arid Land Ecology studies were amplified to include detailed phenological studies of native vegetation over a broad geographical and ecological site characterization

5A1D720104 Exclosure study: production, cover, precipitation.

5A1D720105 Juniper GRI exclosure seedings

5A1D720110 Contour furrow and seedings study

5A1D730102 Exclosure study: production, cover, precipitation

5A1D730103 Intensive investigation of phenology and production relationships of vegetation as influenced by site characteristics of arid land sites representing diverse habitats of western Wyoming with specific attention given to big sagebrush, black sagebrush, western wheatgrass and bluebunch wheatgrass (initiated in 1973)

5A1D740101 Exclosure study: production, cover, precipitation

5A1D740102 Exclosure study: phenology, growth

5A1D750101 Exclosure study: production, cover, precipitation

5A1D760101 Exclosure study: production, cover, precipitation

5A1D760102 Exclosure study: phenology, growth

5A1D770101 Exclosure study: production, cover, precipitation

5A1D770102 Exclosure study: phenology, growth

5A1D780101 Exclosure study: production, cover, precipitation

5A1D780102 Exclosure study: phenology, growth

5A1D780104 Juniper GRI exclosure 20 ft x 50 ft permanent quadrats

5A1D790101 Exclosure study: production, cover, precipitation

5A1D790102 Exclosure study: phenology, growth

5A1D790103 Exclosure study: productivity/standing crop sampling

5A1D800101 Exclosure study: production, cover, precipitation

5A1D800103 Exclosure study: productivity/standing crop sampling

5A1D810101 Exclosure study: production, cover, precipitation

5A1D810103 Exclosure study: productivity/standing crop sampling

5A1D820101 Exclosure study: production, cover, precipitation

5A1D820103 Exclosure study: productivity/standing crop sampling

5A1E. PYRIC

5A1E680101 Game range improvement program by control of juniper and seedling establishment studies of preferred browse species (initiated in 1968)

5A1E680102 Juniper control chemical, burn, mechanical on GRI exclosures

5A1E680111 Juniper ecology 20 ft x 50 ft permanent plots on GRI exclosures

5A1E790106 Juniper wildfire in Wind River Canyon 13 miles south of Thermopolis

5A2. Regional Control

5A2D. INTERMOUNTAIN

5A2D630106 Germination and longevity of halogeton seeds

5A2D650106 Germination and longevity of halogeton seeds

5B. RANGE REVEGETATION

5B 620103 Exclosure study: production, cover, height, precipitation

5B 630103 Exclosure study: production, cover, height, precipitation

5B 640103 Exclosure study: production, cover, height, precipitation

5B 650103 Exclosure study: production, cover, height, precipitation

5B 660118 Exclosure study: production, cover, precipitation

5B 670110 Exclosure study: production, cover, precipitation

5B 680106 Exclosure study: production, cover, precipitation

5B 690101 Exclosure study: production, cover, precipitation

5B 700101 Exclosure study: production, cover, precipitation

5B 720104 Exclosure study: production, cover, precipitation

5B 720110 Contour furrow and seedings study

5B 730102 Exclosure study: production, cover, precipitation

5B 740101 Exclosure study: production, cover, precipitation

5B 750101 Exclosure study: production, cover, precipitation

5B 760101 Exclosure study: production, cover, precipitation

5B 770101 Exclosure study: production, cover, precipitation

5B 780101 Exclosure study: production, cover, precipitation

5B 790101 Exclosure study: production, cover, precipitation

5B 800101 Exclosure study: production, cover, precipitation

5B 810101 Exclosure study: production, cover, precipitation

5B 820101 Exclosure study: production, cover, precipitation

5B4. Intermountain

5B4 590102 Summer steer grazing on saltbush rangeland with reseeding and water spreader treatments.

5B4 600102 Summer steer grazing on saltbush rangeland with reseeding and water spreader treatments.

5B4 600104 Sagebrush control and seeding, 2 Mile, Demer, Smilo, Upper and Lower Govt.

5B4 600104 2 Mechanical sagebrush control and seeding, 2 Mile, Demer, Smilo, Upper and Lower Govt.

5B4 600104 3 Seeding results at 2 Mile, Demer, Smilo, Upper and Lower Govt.

5B4 610102 Summer steer grazing on saltbush rangeland with reseeding and water spreader treatments.

5B4 620102 Summer steer grazing on saltbush range

5B4 630102 Summer steer grazing on saltbush range

5B4 640102 Summer steer grazing on saltbush range

5B4 640109 Seeding and fertilizer studies

5B4 650102 Summer steer grazing on saltbush range

5B4 650108 Seeding in juniper and juniper control

5B4 660103 Seeding and fertilizer studies.

5B4 680112 Kane Deer exclosure seeding analysis

5B4 680114 Juniper GRI exclosure seedings

5B4 690109 Juniper GRI exclosure seedings

5B4 700103 Juniper GRI exclosure seedings

5B4 700106 Inspect SCS seeding along highway west of Moneta

5B4 710102 Juniper GRI exclosure seedings

5B4 720102 Reseeding trials of grasses, forbs and shrubs on the bladed and disturbed area of a 120-mile long pipeline, extending across the Big Horn Basin. Funding by Colorado Interstate Pipeline Corp. Primary emphasis of trials for aesthetic improvement as well as erosion control and forage and browse production. (initiated in 1972)

5B4 720105 Juniper GRI exclosure seedings

5B4 720109 Shortgrass prairie seed production study

5B4 730106 CIG pipeline seed plot study

5B4 730111 Shortgrass prairie seed production study

5B4 740106 CIG pipeline seed plot study

5B4 740108 Shortgrass prairie seed production study

5B4 750107 CIG pipeline seedplot study

5B4 750111 Shortgrass prairie seed production study

5B4 760107 CIG pipeline seed plot study

5B4 760108 Grass seed sources at A.C. Hull 1940's plantings

5B4 780105 Juniper GRI seeding

5C. WATER SPREADING, FURROWING, PITTING

5C	590102	Summer steer grazing on saltbush rangeland with reseeding and water spreader treatments.
5C	600102	Summer steer grazing on saltbush rangeland with reseeding and water spreader treatments.
5C	600104	2 Mechanical sagebrush control and seeding, 2 mile, Demer, Smilo, Upper and Lower Govt.
5C	610102	Summer steer grazing on saltbush rangeland with reseeding and water spreader treatments.
5C	620102	Summer steer grazing on saltbush range
5C	620103	Exclosure study: production, cover, height, precipitation
5C	630102	Summer steer grazing on saltbush range
5C	630103	Exclosure study: production, cover, height, precipitation
5C	640102	Summer steer grazing on saltbush range
5C	640103	Exclosure study: production, cover, height, precipitation
5C	640109	Seeding and fertilizer studies
5C	650102	Summer steer grazing on saltbush range
5C	650103	Exclosure study: production, cover, height, precipitation
5C	650109	1 ft x 1 ft permanent plot transects
5C	660103	Seeding and fertilizer studies.
5C	660106	1 ft x 1 ft permanent plot transects
5C	660118	Exclosure study: production, cover, precipitation
5C	670104	1 ft x 1 ft permanent plot transects
5C	670110	Exclosure study: production, cover, precipitation
5C	680106	Exclosure study: production, cover, precipitation
5C	680113	1 ft x 1 ft permanent plot transects
5C	690101	Exclosure study: production, cover, precipitation
5C	690104	1 ft x 1 ft permanent plot transects
5C	690110	Block juniper site, 1 ft x 1 ft permanent plot studies
5C	700101	Exclosure study: production, cover, precipitation
5C	700109	1 ft x 1 ft permanent plot transects
5C	710106	1 ft x 1 ft permanent plot transects
5C	710107	Wagon Wheel range survey
5C	720102	Reseeding trials of grasses, forbs and shrubs on the bladed and disturbed area of a 120-mile long pipeline, extending across the Big Horn Basin. Funding by Colorado Interstate Pipeline Corp. Primary emphasis of trials for aesthetic improvement as well as erosion control and forage and browse production. (initiated in 1972)
5C	720104	Exclosure study: production, cover, precipitation
5C	720110	Contour furrow and seedings study
5C	720112	Wagon Wheel range survey
5C	730102	Exclosure study: production, cover, precipitation
5C	730106	CIG pipeline seed plot study
5C	730109	1 ft x 1 ft permanent plot transects
5C	740101	Exclosure study: production, cover, precipitation
5C	740106	CIG pipeline seed plot study
5C	750101	Exclosure study: production, cover, precipitation
5C	750106	1 ft x 1 ft permanent plot transects - Rawlins, Kemmerer, Pinedale
5C	750107	CIG pipeline seedplot study
5C	750109	Wagon Wheel range survey
5C	760101	Exclosure study: production, cover, precipitation
5C	760106	1 ft x 1 ft permanent plot transects - Kemmerer-Boulder Ridge exclosures
5C	760107	CIG pipeline seed plot study

5C	770101	Exclosure study: production, cover, precipitation
5C	770105	1 ft x 1 ft permanent plot transects
5C	780101	Exclosure study: production, cover, precipitation
5C	780105	Juniper GRI exclosure seeding
5C	780107	1 ft x 1 ft permanent plot transects
5C	790101	Exclosure study: production, cover, precipitation
5C	800101	Exclosure study: production, cover, precipitation
5C	810101	Exclosure study: production, cover, precipitation
5C	820101	Exclosure study: production, cover, precipitation

6. Range Livestock

6A. GENERAL MANAGEMENT

6A	800105	Roadside intensive use site 7 miles east of Powder River

6A1. Cattle

6A1	590102	Summer steer grazing on saltbush rangeland with reseeding and water spreader treatments.
6A1	600102	Summer steer grazing on saltbush rangeland with reseeding and water spreader treatments.
6A1	610102	Summer steer grazing on saltbush rangeland with reseeding and water spreader treatments.
6A1	620102	Summer steer grazing on saltbush range
6A1	630102	Summer steer grazing on saltbush range
6A1	640102	Summer steer grazing on saltbush range
6A1	650102	Summer steer grazing on saltbush range

6A2. Sheep and Goats

6A2	590103	Winter sheep grazing on saltbush range
6A2	600103	Winter sheep grazing on saltbush range
6A2	610103	Winter sheep grazing on saltbush range
6A2	620101	Winter sheep grazing on saltbush range
6A2	630101	Winter sheep grazing on saltbush range
6A2	640101	Winter sheep grazing on saltbush range
6A2	650101	Winter sheep grazing on saltbush range
6A2	660101	Winter sheep grazing on saltbush range
6A2	660102	Grazing intensity influence on dietary intake by use of fistulated sheep

6A3. Horses

6A3	670101	Study of grazing systems and influence on vegetation and wildlife habitat (field work initiated in 1967)
6A3	690105	Baggs grazing system study
6A3	700107	Baggs grazing system study
6A3	750105	1 Baggs grazing system study: impacts of grazing systems
6A3	750105	2 Baggs grazing system study: BLM trend plots
6A3	790107	Baggs grazing system study: BLM trend plots

6B. NUTRITION AND FEEDING

6B 610104 Aridland ecology studies of vegetation, cover, composition, production, biochemical composition, precipitation, and site relationships under grazing, shrub control, seeding and pitting treatments

6B2. Nutrient Composition

6B2B. PASTURAGE - WEST

6B2B 660102 Grazing intensity influence on dietary intake by use of fistulated sheep.

7. Range Wildlife

7A. BIG GAME

7A 680101 Game range improvement program by control of juniper and seedling establishment studies of preferred browse species (initiated in 1968)

7A 680102 Juniper control chemical, burn, mechanical on GRI exclosures

7A 680111 Juniper ecology 20 ft x 50 ft permanent plots on GRI exclosures

7A 680112 Kane Deer exclosure seeding analysis

7A 680114 Juniper GRI exclosure seedings

7A 690109 Juniper GRI exclosure seedings

7A 690112 Juniper GRI exclosure Tordon retention studies

7A 700103 Juniper GRI exclosure seedings

7A 710102 Juniper GRI exclosure seedings

7A 720105 Juniper GRI exclosure seedings

7A1. Deer

7A1 670101 Study of grazing systems and influence on vegetation and wildlife habitat (field work initiated in 1967)

7A1 690105 Baggs grazing system study

7A1 700107 Baggs grazing system study

7A1 710109 Sheep Mountain deer habitat study

7A1 710110 Laramie Peak game range inventory

7A1 720113 Laramie Peak game range inventory

7A1 730108 Mule deer winter range evaluation

7A1 740105 Mule deer winter range evaluation

7A1 750105 1 Baggs grazing system study: impacts of grazing systems

7A1 750105 2 Baggs grazing system study: BLM trend plots

7A1 790107 Baggs grazing system study: BLM trend plots

7A2. Elk

7A2 660109 Elk utilization and dietary preference study - Bull Elk Park

7A2 670107 Elk utilization and dietary preference study - Bull Elk Park

7A2 680104 Study of elk-livestock competition in terms of utilization by class of animal and influence on vegetation

7A2 710110 Laramie Peak game range inventory

7A2 720113 Laramie Peak game range inventory

7A3. Antelope

7A3 670101 Study of grazing systems and influence on vegetation and wildlife habitat (field work initiated in 1967).

7A3 690105 Baggs grazing system study

7A3 700107 Baggs grazing system study

7A3 750105 1 Baggs grazing system study: impacts of grazing systems

7A3 790107 Baggs grazing system study: BLM trend plots

7D. NON-GAME

7D1. Rodents and Rabbits

7D1 660116 Small mammal abundance and distribution in juniper

8. Range Watersheds

8B. WATERSHED SOIL EROSION AND SEDIMENTATION

8B 630105 Erosion transect study

8B 650105 Erosion transect study

8B 650113 Microclimatic studies on desert shrub-types

8B 660113 Erosion transects for study of soil movement and relationship to climatic variation and grazing management

8B 690111 Boars Tusk sand dune area tour

8B 700105 Exclosure soil profile studies with SCS

8B 700108 Erosion transect study

8B 710108 Exclosure soil profile studies with SCS

8B 720107 Exclosure soil profile studies with SCS

8B 730107 Exclosure soil profile studies with SCS

8B 750108 Exclosure soil profile studies with SCS Dutch Nick

8B 760107 CIG pipeline seed plot study

8B 770104 Exclosure soil profile studies with SCS

8B 780106 Soil-vegetation relation study

8B 780108 Exclosure soil profile studies with SCS

8B 830106 Erosion transect data summarization and input to computer

8B 840106 Erosion transect data summarization and input to computer

8D. WATERSHED RESTORATION

8D 710107 Wagon Wheel range survey

8D 720102 Reseeding trials of grasses, forbs and shrubs on the bladed and disturbed area of a 120-mile long pipeline, extending across the Big Horn Basin. Funding by Colorado Interstate Pipeline Corp. Primary emphasis of trials for aesthetic improvement as well as erosion control and forage and browse production. (initiated in 1972)

8D 720112 Wagon Wheel range survey

8D 730101 On an area near Reno Junction, Wyoming, on which a strip mine operation is proposed, the ecology, vegetation and soils inventory studies were initiated in 1973. Continuing work is involved with the establishment of permanent monitoring locations for development and evaluation of best post-mining rehabilitation procedures

8D 730106 CIG pipeline seed plot study
8D 740104 Black Thunder pre-mining inventory
8D 740106 CIG pipeline seed plot study
8D 750104 Black Thunder pre-mining inventory
8D 750107 CIG pipeline seedplot study
8D 750109 Wagon Wheel range survey
8D 760105 Black Thunder pre-mining inventory
8D 770106 Black Thunder pre-mining inventory
8D 770107 Snow catchment and revegetation on Jim Bridger mine spoil
8D 780105 Juniper GRI exclosure seeding

10. Range Education and Research

10D. RANGE RESEARCH METHODS

10D 810105 Woody plant density from 4 ft x 5 ft weight unit estimate plots
10D 820104 Exclosure study: woody plant density

10D1. Plant-Soil Emphasis

10D1 620103 Exclosure study: production, cover, height, precipitation
10D1 630103 Exclosure study: production, cover, height, precipitation
10D1 630104 Soil moisture and temperature study
10D1 630105 Erosion transect study
10D1 640103 Exclosure study: production, cover, height, precipitation
10D1 640104 Soil moisture and temperature study
10D1 640106 Harvester ant control and related study of vegetation response and economic feasibility of the procedure
10D1 640107 2 ft x 20 ft and 4 ft x 4 ft permanent plot sampling
10D1 640108 Ecology and control of juniper
10D1 640109 Seeding and fertilizer studies
10D1 650103 Exclosure study: production, cover, height, precipitation
10D1 650104 Soil moisture and temperature study
10D1 650105 Erosion transect study
10D1 650109 1 ft x 1 ft permanent plot transects
10D1 650110 Harvester ant control study
10D1 650111 Sagebrush cover estimation with angle gauge
10D1 650112 Juniper control study on LU
10D1 650113 Microclimatic studies on desert shrub-types
10D1 660104 Juniper control study on LU
10D1 660108 Juniper ecology: permanent plot transects
10D1 660109 Elk utilization and dietary preference study - Bull Elk Park
10D1 660110 Shrub production estimation study
10D1 660111 Limber pine ecology
10D1 660115 Sagebrush-wind-snow pattern relation study
10D1 660116 Small mammal abundance and distribution in juniper
10D1 660117 Saltbush root development and relation to grazing intensity
10D1 670103 Soil moisture and temperature study
10D1 670106 Juniper ecology: permanent plot transects and seeding trials
10D1 670108 Shrub production estimation study
10D1 670109 Limber pine ecology
10D1 690108 Vegetation pattern on shortgrass-cactus ranges
10D1 690110 Block juniper site, 1 ft x 1 ft permanent plot studies
10D1 710104 Shortgrass prairie blue grama culm density study
10D1 710105 Shortgrass prairie seed production study
10D1 710107 Wagon Wheel range survey
10D1 750109 Wagon Wheel range survey

10D1 760106 1 ft x 1 ft permanent plot transects - Kemmerer-Boulder Ridge exclosures

10D1 770105 1 ft x 1 ft permanent plot transects

10D1 780108 Exclosure soil profile studies with SCS

10D1 790103 Exclosure study: productivity/standing crop sampling

10D1 800103 Exclosure study: productivity/standing crop sampling

10D1 810103 Exclosure study: productivity/standing crop sampling

10D1 820103 Exclosure study: productivity/standing crop sampling

11. Computer Storage, Programming and Modeling

11A. CLIMATIC

11A 830101 Precipitation pattern study

11A 830102 Computer program development for long-term weather record compilation

11A 830103 Computer program development for estimation of missing climate data

11A 830109 Compilation of US Weather Service precipitation records - Wyoming

11A 840101 Precipitation pattern study

11A 840102 Computer program development for long-term weather record compilation

11A 840103 Computer program development for estimation of missing climatic data

11A 840109 Compilation of U.S. Weather Service precipitation records - Wyoming

11C. EDAPHIC

11C 830105 Summary of soil moisture and temperature data from Granite Mountain exclosure

11C 830106 Erosion transect data summarization and input to computer

11C 840105 Summary of soil moisture and temperature data from Granite Mountain exclosure

11C 840106 Erosion transect data summarization and input to computer

11E. RANGELAND INVENTORY

11E1. Sites and Mapping

11E1 830107 Wyoming rangeland reference sites

11E1 840107 Wyoming rangeland reference sites

11E2. Condition and Trend

11E2 830107 Wyoming rangeland reference sites

11E2 830108 Yield prediction from ERHYM model

11E2 840107 Wyoming rangeland reference sites

11E2 840108 Yield prediction with ERHYM model

11E3. Production and Utilization

11E3	830104	Compilation of herbage production and cover data from Wind River Basin exclosures	11E3	840104	Compilation of herbage production and cover data from Wind River Basin exclosures
11E3	830107	Wyoming rangeland reference sites	11E3	840107	Wyoming rangeland reference sites
11E3	830108	Yield prediction from ERHYM model	11E3	840108	Yield prediction with ERHYM model

11F. PREDICTIVE MODELING

11F1. Herbage Yield

11F1	830104	Compilation of herbage production and cover data from Wind River Basin exclosures	11F1	840104	Compilation of herbage production and cover data from Wind River Basin exclosures
11F1	830108	Yield prediction from ERHYM model	11F1	840108	Yield prediction with ERHYM model

11F2. Phenology

11F2	830108	Yield prediction from ERHYM model	11F2	840108	Yield prediction with ERHYM model

S2. JOURNAL AND REFEREED PUBLICATIONS

1. Range Plants

1A. PLANT SYSTEMATICS

1A2. Floras and Plant Descriptions

1A2 620202 Fisser, H.G. 1962. List of plants observed on the Owl Creek and Wind River Ranges in Wyoming. Wyo. Range Manage. Issue 159. 10 p.

1A2 640202 Nichols, J.T. 1964. Plants of the 15-Mile Drainage, Washakie County, Wyoming. Wyo. Range Manage. 190:44-50.

1B. FORAGE VALUE

1B1. Range Plants

1B1 620202 Fisser, H.G. 1962. List of plants observed on the Owl Creek and Wind River Ranges in Wyoming. Wyo. Range Manage. Issue 159. 10 p.

1B1 640202 Nichols, J.T. 1964. Plants of the 15-Mile Drainage, Washakie County, Wyoming. Wyo. Range Manage. 190:44-50.

1B4. Poisonous Plants

1B4 620205 Vosler, L.C. 1962. Bibliography of *Halogeton glomeratus.* Wyo. Range Manage. Issue 158, 6 p.

1B4 790202 Cleaves, D.T. and J. Taylor. 1979. Competition as a control for *Halogeton glomeratus.* J. Colo.-Wyo. Acad. Sci. 11(1):18.

1D. PLANT INSECTS

1D 610202 Fisser, H.G., and R.J. Lavigne. 1961. *Trirhabda attenuata* (Say) (Coleoptera: Chrysomelidae), a leaf feeding beetle found on threetip sagebrush (*Artemisia tripartita* Rydb. subsp. *rupicola* Beetle). J. Range Manage. 14(4):278-279.

1D 660202 Lavigne, R.J. and H.G. Fisser. 1966. Controlling western harvester ants. Mtn. States Regional Public. 3. 4 p.

1D 660203 Wight, J.R. and J.T. Nichols. 1966. Effects of harvester ants on production of a saltbush community. J. Range Manage. 19(1):68-71.

1D 670203 Kirkham, D.R. 1967. Biological control as a tool in range management. Wyo. Range Manage. 246:254-262.

1D 700201 Fisser, H.G. and D.R. Kirkham. 1970. Reliability of harvester ant mound counts from aerial photographs. J. Colo.-Wyo. Acad. Sci. 7(1):42-43.

1D 720201 Kirkham, D.R. and H.G. Fisser. 1972. Rangeland relations and harvester ants in northcentral Wyoming. J. Range Manage. 25(1):55-60.

1E. PLANT PHYSIOLOGY

1E 620206 Vosler, L.C. 1962. Abstract of M.S. Thesis. Wyo. Range Manage. Issue 162. 1 p.

1E1. Reproduction and Germination

1E1 680202 Julian, T.D. 1968. *Atriplex nuttallii.* Wyo. Range Manage. 260:94-101.

1E1 760203 Kleinman, L.H., and H.G. Fisser. 1976. Phenology and environmental relationships of sagebrush-grass rangelands. J. Colo.-Wyo. Acad. Sci. 8(1):48-49.

1E2. Growth and Phenology

1E2 680205 Noller, G.L. 1968. The relationship of forage production to precipitation, cover, and soils in north central Wyoming. Wyo. Range Manage. 263:116-136.

1E2 700203 Lester, J.E. and H.G. Fisser. 1970. *Opuntia* clump size variation on the shortgrass plains, Pawnee Site, Nunn, Colorado. J. Colo.-Wyo. Acad. Sci. 7(1):37-38.

1E2 710201 Burrough, L.N. and H.G. Fisser. 1971. A greenhouse analysis of variance of picloram retention in soil and duff from juniper sites. J. Colo.-Wyo. Acad. Sci. Vol. 7(2):103.

1E2 730201 Wambolt, C.C. 1973. Conifer water potential as influenced by stand density and environmental factors. Canad. J. Bot. 51:2333-2337.

1E2 730202 Bedell, T.E. 1973. Revegetation of pipeline areas. J. Colo.-Wyo. Acad. Sci. 7(4):29-30.

1E2 760203 Kleinman, L.H., and H.G. Fisser. 1976. Phenology and environmental relationships of sagebrush-grass rangelands. J. Colo.-Wyo. Acad. Sci. 8(1):48-49.

1E2 760204 Taha, F.K., and H.G. Fisser. 1976. Production of a shortgrass prairie ecosystem as related to soils characteristics in eastern Wyoming. J. Colo.-Wyo. Acad. Sci. 8(1):50.

1E2 780203 Cox, J.R. and H.G. Fisser. 1978. The effects of water stress on phenology and carbohydrate storage in the shortgrass prairie. p. 529-535. In J.R. Goodin and D.K. Northington (eds.) 1979. Intern. Arid Lands Conference on Plant Resources. Intern. Center Arid and Semiarid Land Studies. Lubbock, Texas. October 8-15, 1978.

1E2 790201 Ries, R.E., and H.G. Fisser. 1979. Influence of environmental factors upon sagebrush and grass production in Wyoming. Agro-Ecosystems 5(1):41-55.

1E2 790202 Cleaves, D.T. and J. Taylor. 1979. Competition as a control for *Halogeton glomeratus.* J. Colo.-Wyo. Acad. Sci. 11(1):18.

1E2 800201 Waugh, W.J. and H.G. Fisser. 1980. Vegetation stagnation and decadence in a southeastern Wyoming big game exclosure: preliminary report. J. Colo.-Wyo. Acad. Sci. 12(1):29.

1E2 820201 Fisser, H.G. and N.E. Hargis. 1982. Numerical phenology of big and black sagebrush from visual description statements. J. Colo.-Wyo. Acad. Sci. 14(1):39.

1E2 820202 Taha, F.K., H.G. Fisser and R.E. Ries. 1982. A modified 100-point frame for repetition inventory. J. Range Manage. 36(1):124-125.

1E2 840201 Fisser, H.G. 1984. Culm weight dynamics as an index of sagebrush-grass community productivity. J. Colo.-Wyo. Acad. Sci. 16(1):25-26.

1E2 840203 Kinucan, R.J. and H.G. Fisser. 1984. Dynamics of vegetative plant part phenodynamics and related stress adaptations. Bull. Ecol. Soc. Amer. Program Issue. 65(2):253.

1E2 840204 Fisser, H.G. 1984. Productivity dynamics of sagebrush grass ecosystems as determined from biweekly sampling of a single grass and shrub species. Bull. Ecol. Soc. Amer. Program Issue. 65(2):258.

1E3. Photosynthesis and Plant Nutrients

1E3 760201 Cox, J.R., and H.G. Fisser. 1976. Methods of measuring environmental parameters on a pre-disturbed coal site. J. Colo.-Wyo. Acad. Sci. 8(1):45-46.

1E3 760205 Cox, J.R. 1976. A mixed method of carbohydrate analysis. J. Colo.-Wyo. Acad. Sci. 8(1):30.

1E3 780203 Cox, J.R. and H.G. Fisser. 1978. The effects of water stress on phenology and carbohydrate storage in the shortgrass prairie. p. 529-535. In J.R. Goodin and D.K. Northington (eds.) 1979. Intern. Arid Lands Conference on Plant Resources. Intern. Center Arid and Semiarid Land Studies. Lubbock, Texas. October 8-15, 1978.

1E3 800201 Waugh, W.J. and H.G. Fisser. 1980. Vegetation stagnation and decadence in a southeastern Wyoming big game exclosure: preliminary report. J. Colo.-Wyo. Acad. Sci. 12(1):29.

1E4. Herbage Removal

1E4 660203 Wight, J.R. and J.T. Nichols. 1966. Effects of harvester ants on production of a saltbush community. J. Range Manage. 19(1):68-71.

1E4 680203 Probasco, G.E. 1968. Dietary preference and utilization patterns of elk in the Big Horn Mountains, Wyoming. Wyo. Range Manage. 258:92.

1E4 760202 Gerhart, William, and H.G. Fisser. 1976. Some influences of grazing systems upon vegetation and wildlife. J. Colo.-Wyo. Acad. Sci. 8(1):68.

1E4 780203 Cox, J.R. and H.G. Fisser. 1978. The effects of water stress on phenology and carbohydrate storage in the shortgrass prairie. p. 529-535. In J.R. Goodin and D.K. Northington (eds.) 1979. Intern. Arid Lands Conference on Plant Resources. Intern. Center Arid and Semiarid Land Studies. Lubbock, Texas. October 8-15, 1978.

1E4 800201 Waugh, W.J. and H.G. Fisser. 1980. Vegetation stagnation and decadence in a southeastern Wyoming big game exclosure: preliminary report. J. Colo.-Wyo. Acad. Sci. 12(1):29.

1E4 800202 Rodgers, D.E. and H.G. Fisser. 1980. Vegetational response to grazing systems in Wyoming's Red Desert. J.Colo.-Wyo. Acad. Sci. 12(1):29-30.

2. Range Ecology

2A. CLIMATIC FACTORS

2A 620201 Fisser, H.G. 1962. An ecological study of the *Artemisia tripartita* subsp. *rupicola* and related shrub communities in Wyoming. Ph.D. Dissertation, University of Wyoming. 166 p.

2A 620204 Fisser, H.G. 1962. Abst. of Ph.D. Dissertation. Wyo. Range Manage. Issue 160. 2 p.

2A 620206 Vosler, L.C. 1962. Abstract of M.S. Thesis. Wyo. Range Manage. Issue 162. 1 p.

2A 670201 Fisser, H.G. 1967. Proper winter sheep grazing on saltbush rangeland. Wyo. Range Manage. 244:201-206.

2A 680201 Fisser, H.G. 1968. Soil moisture and temperature changes following sagebrush control. J. Range Manage. 21 (4):283-287.

2A 680204 Wambolt, C.L. 1968. Size, density, growth rate, and age, of Rocky Mountain Juniper and interrelationships with environmental factors. Wyo. Range Manage. 259:93.

2A 680205 Noller, G.L. 1968. The relationship of forage production to precipitation, cover, and soils in north central Wyoming. Wyo. Range Manage. 263:116-136.

2A 700202 Whysong, G.L. 1970. Predictive value of factors influencing perennial grass production on western Wyoming sagebrush-grass ranges. J. Colo.-Wyo. Acad. Sci. 7(1):36-37.

2A 730201 Wambolt, C.C. 1973. Conifer water potential as influenced by stand density and environmental factors. Canad. J. Bot. 51:2333-2337.

2A 760203 Kleinman, L.H., and H.G. Fisser. 1976. Phenology and environmental relationships of sagebrush-grass rangelands. J. Colo.-Wyo. Acad. Sci. 8(1):48-49.

2A 770201 Taha, F.K., and H.G. Fisser. 1977. Factors affecting growth and productivity of vegetation stratum in eastern Wyoming. J. Colo.-Wyo. Acad. Sci. 9(1):36.

2A 770202 Shawesh, O.M. and H.G. Fisser. 1977. Vegetation analysis and ecology of rangelands in northwest Libya, Africa. J. Colo.-Wyo. Acad. Sci. 9(1):35-36.

2A 780202 Shawesh, O.M. and H.G. Fisser. 1978. Ecology of rangeland communities in northwestern Libya, Africa. p. 190-192. In D.N. Hyder (ed.) Proc. 1st Intern. Rangeland Congress, Denver, Colorado. August 14-18, 1978.

2A 780203 Cox, J.R. and H.G. Fisser. 1978. The effects of water stress on phenology and carbohydrate storage in the shortgrass prairie. p. 529-535. In J.R. Goodin and D.K. Northington (eds.) 1979. Intern. Arid Lands Conference on Plant Resources. Intern. Center Arid and Semiarid Land Studies. Lubbock, Texas. October 8-15, 1978.

2A 790201 Ries, R.E., and H.G. Fisser. 1979. Influence of environmental factors upon sagebrush and grass production in Wyoming. Agro-Ecosystems 5(1):41-55.

2A 820201 Fisser, H.G. and N.E. Hargis. 1982. Numerical phenology of big and black sagebrush from visual description statements. J. Colo.-Wyo. Acad. Sci. 14(1):39.

2A 840201 Fisser, H.G. 1984. Culm weight dynamics as an index of sagebrush-grass community productivity. J. Colo.-Wyo. Acad. Sci. 16(1):25-26.

2A 840202 Germain, L.P. and H.G. Fisser. 1984. Application of kriging to ecological variables. J. Colo.-Wyo. Acad. Sci. 16(1):13-14.

2A 840203 Kinucan, R.J. and H.G. Fisser. 1984. Dynamics of vegetative plant part phenodynamics and related stress adaptations. Bull. Ecol. Soc. Amer. Program Issue. 65(2):253.

2A 840204 Fisser, H.G. 1984. Productivity dynamics of sagebrush grass ecosystems as determined from biweekly sampling of a single grass and shrub species. Bull. Ecol. Soc. Amer. Program Issue. 65(2):258.

2B. BIOTIC FACTORS

2B 670203 Kirkham, D.R. 1967. Biological control as a toll in range management. Wyo. Range Manage. 246:254-262.

2B 720201 Kirkham, D.R. and H.G. Fisser. 1972. Rangeland relations and harvester ants in northcentral Wyoming. J. Range Manage. 25(1):55-60.

2B 770201 Taha, F.K., and H.G. Fisser. 1977. Factors affecting growth and productivity of vegetation stratum in eastern Wyoming. J. Colo.-Wyo. Acad. Sci. 9(1):36.

2B 770202 Shawesh, O.M. and H.G. Fisser. 1977. Vegetation analysis and ecology of rangelands in northwest Libya, Africa. J. Colo.-Wyo. Acad. Sci. 9(1):35-36.

2B 780202 Shawesh, O.M. and H.G. Fisser. 1978. Ecology of rangeland communities in northwestern Libya, Africa. p. 190-192. In D.N. Hyder (ed.) Proc. 1st Intern. Rangeland Congress, Denver, Colorado. August 14-18, 1978.

2B 780203 Cox, J.R. and H.G. Fisser. 1978. The effects of water stress on phenology and carbohydrate storage in the shortgrass prairie. p. 529-535. In J.R. Goodin and D.K. Northington (eds.) 1979. Intern. Arid Lands Conference on Plant Resources. Intern. Center Arid and Semiarid Land Studies. Lubbock, Texas. October 8-15, 1978.

2B 790202 Cleaves, D.T. and J. Taylor. 1979. Competition as a control for *Halogeton glomeratus*. J. Colo.-Wyo. Acad. Sci. 11(1):18.

2B 840201 Fisser, H.G. 1984. Culm weight dynamics as an index of sagebrush-grass community productivity. J. Colo.-Wyo. Acad. Sci. 16(1):25-26.

2B 840204 Fisser, H.G. 1984. Productivity dynamics of sagebrush grass ecosystems as determined from biweekly sampling of a single grass and shrub species. Bull. Ecol. Soc. Amer. Program Issue. 65(2):258.

2C. EDAPHIC FACTORS

2C 620201 Fisser, H.G. 1962. An ecological study of the *Artemisia tripartita* subsp. *rupicola* and related shrub communities in Wyoming. Ph.D. Dissertation. University of Wyoming. 166 p.

2C 620204 Fisser, H.G. 1962. Abst. of Ph.D. Dissertation. Wyo. Range Manage. Issue 160. 2 p.

2C 620206 Vosler, L.C. 1962. Abstract of M.S. Thesis. Wyo. Range Manage. Issue 162. 1 p.

2C 660204 Wight, J.R. 1966. Juniper communities and range management in northwestern Wyoming. Wyo. Range Manage. 218:117-119.

2C 660205 Robinson, L.D. 1966. Abstract of M.S. Thesis. Wyo. Range Manage. 229:168.

2C 680201 Fisser, H.G. 1968. Soil moisture and temperature changes following sagebrush control. J. Range Manage. 21 (4): 283-287.

2C 720201 Kirkham, D.R. and H.G. Fisser. 1972. Rangeland relations and harvester ants in northcentral Wyoming. J. Range Manage. 25(1):55-60.

2C 730201 Wambolt, C.C. 1973. Conifer water potential as influenced by stand density and environmental factors. Canad. J. Bot. 51(12):2333-2337.

2C 760204 Taha, F.K., and H.G. Fisser. 1976. Production of a short-grass prairie ecosystem as related to soils characteristics in eastern Wyoming. J. Colo.-Wyo. Acad. Sci. 8(1):50.

2C 770201 Taha, F.K., and H.G. Fisser. 1977. Factors affecting growth and productivity of vegetation stratum in eastern Wyoming. J. Colo.-Wyo. Acad. Sci. 9(1):36.

2C 770202 Shawesh, O.M. and H.G. Fisser. 1977. Vegetation analysis and ecology of rangelands in northwest Libya, Africa. J. Colo.-Wyo. Acad. Sci. 9(1):35-36.

2C 780201 Fisser, H.G. 1978. Soil surface movement and relation to vegetation structure. p. 706-711. In Mohan K. Wali (ed.) Ecology and Coal Resource Development. Vol. 2, International Congress for Energy and the Ecosystem. University of North Dakota, Grand Forks, N.D. June 12-16, 1978. Pergamon Press, Elmsford, NY.

2C 780203 Cox, J.R. and H.G. Fisser. 1978. The effects of water stress on phenology and carbohydrate storage in the shortgrass prairie. p. 529-535. In J.R. Goodin and D.K. Northington (eds.) 1979. Intern. Arid Lands Conference on Plant Resources. Intern. Center Arid and Semiarid Land Studies. Lubbock, Texas. October 8-15, 1978.

2C 820201 Fisser, H.G. and N.E. Hargis. 1982. Numerical phenology of big and black sagebrush from visual description statements. J. Colo.-Wyo. Acad. Sci. 14(1):39.

2C 840203 Kinucan, R.J. and H.G. Fisser. 1984. Dynamics of vegetative plant part phenodynamics and related stress adaptations. Bull. Ecol. Soc. Amer. Program Issue. 65(2):253.

2E. AUTECOLOGY

2E 610202 Fisser, H.G., and R.J. Lavigne. 1961. *Trirhabda attenuata* (Say) (Coleoptera: Chrysomelidae), a leaf feeding beetle found on threetip sagebrush (*Artemisia tripartita* Rydb. subsp. *rupicola* Beetle). J. Range Manage. 14(4):278-279.

2E 620201 Fisser, H.G. 1962. An ecological study of the *Artemisia tripartita* subsp. *rupicola* and related shrub communities in Wyoming. Ph.D. Dissertation, University of Wyoming. 166 p.

2E 620204 Fisser, H.G. 1962. Abst. of Ph.D. Dissertation. Wyo. Range Manage. Issue 160. 2 p.

2E 620205 Vosler, L.C. 1962. Bibliography of *Halogeton glomeratus.* Wyo. Range Manage. Issue 158, 6 p.

2E 620206 Vosler, L.C. 1962. Abstract of M.S. Thesis. Wyo. Range Manage. Issue 162. 1 p.

2E 660204 Wight, J.R. 1966. Juniper communities and range management in northwestern Wyoming. Wyo. Range Manage. 218:117-119.

2E 660205 Robinson, L.D. 1966. Abstract of M.S. Thesis. Wyo. Range Manage. 229:168.

2E 680202 Julian, T.D. 1968. *Atriplex nuttallii.* Wyo. Range Manage. 260:94-101.

2E 680204 Wambolt, C.L. 1968. Size, density, growth rate and age of Rocky Mountain juniper and interrelationships with environmental factors. Wyo. Range Manage. 259:93.

2E 700203 Lester, J.E. and H.G. Fisser. 1970. *Opuntia* clump size variation on the shortgrass plains, Pawnee Site, Nunn, Colorado. J. Colo.-Wyo. Acad. Sci. 7(1):37-38.

2E 710201 Burrough, L.N. and H.G. Fisser. 1971. A greenhouse analysis of variance of picloram retention in soil and duff from juniper sites. J. Colo.-Wyo. Acad. Sci. Vol. 7(2):103.

2E 730201 Wambolt, C.C. 1973. Conifer water potential as influenced by stand density and environmental factors. Canad. J. Bot. 51:2333-2337.

2E 760203 Kleinman, L.H., and H.G. Fisser. 1976. Phenology and environmental relationships of sagebrush-grass rangelands. J. Colo.-Wyo. Acad. Sci. 8(1):48-49.

2E 760205 Cox, J.R. 1976. A mixed method of carbohydrate analysis. J. Colo.-Wyo. Acad. Sci. 8(1):30.

2E 770201 Taha, F.K., and H.G. Fisser. 1977. Factors affecting growth and productivity of vegetation stratum in eastern Wyoming. J. Colo.-Wyo. Acad. Sci. 9(1):36.

2E 780203 Cox, J.R. and H.G. Fisser. 1978. The effects of water stress on phenology and carbohydrate storage in the shortgrass prairie. p. 529-535. In J.R. Goodin and D.K. Northington (eds.) 1979. Intern. Arid Lands Conference on Plant Resources. Intern. Center Arid and Semiarid Land Studies. Lubbock, Texas. October 8-15, 1978.

2E 790202 Cleaves, D.T. and J. Taylor. 1979. Competition as a control for *Halogeton glomeratus*. J. Colo.-Wyo. Acad. Sci. 11(1):18.

2E 800201 Waugh, W.J. and H.G. Fisser. 1980. Vegetation stagnation and decadence in a southeastern Wyoming big game exclosure: preliminary report. J. Colo.-Wyo. Acad. Sci. 12(1):29.

2E 820201 Fisser, H.G. and N.E. Hargis. 1982. Numerical phenology of big and black sagebrush from visual description statements. J. Colo.-Wyo. Acad. Sci. 14(1):39.

2E 840201 Fisser, H.G. 1984. Culm weight dynamics as an index of sagebrush-grass community productivity. J. Colo.-Wyo. Acad. Sci. 16(1):25-26.

2E 840202 Germain, L.P. and H.G. Fisser. 1984. Application of kriging to ecological variables. J. Colo.-Wyo. Acad. Sci. 16(1):13-14.

2E 840204 Fisser, H.G. 1984. Productivity dynamics of sagebrush grass ecosystems as determined from biweekly sampling of a single grass and shrub species. Bull. Ecol. Soc. Amer. Program Issue. 65(2):258.

2F. SYNECOLOGY

2F1. Great Plains

2F1 700203 Lester, J.E. and H.G. Fisser. 1970. *Opuntia* clump size variation on the shortgrass plains, Pawnee Site, Nunn, Colorado. J. Colo.-Wyo. Acad. Sci. 7(1):37-38.

2F1 740202 Fisser, H.G. and J.C. Shaver. 1974. Remote sensing applied to the grassland biome. p. 293-295. In Proc. First International Congress of Ecology, The Hague, The Netherlands. September 10, 1974.

2F1 760201 Cox, J.R., and H.G. Fisser. 1976. Methods of measuring environmental parameters on a pre-disturbed coal site. J. Colo.-Wyo. Acad. Sci. 8(1):45-46.

2F1 760204 Taha, F.K., and H.G. Fisser. 1976. Production of a short-grass prairie ecosystem as related to soils characteristics in eastern Wyoming. J. Colo.-Wyo. Acad. Sci. 8(1):50.

2F1 760205 Cox, J.R. 1976. A mixed method of carbohydrate analysis. J. Colo.-Wyo. Acad. Sci. 8(1):30.

2F1 820202 Taha, F.K., H.G. Fisser and R.E. Ries. 1982. A modified 100-point frame for repetition inventory. J. Range Manage. 36(1):124-125.

2F2. West

2F2 600201 Fisser, H.G., A.A. Beetle, R.L. Lang, and D.R. Smith. 1960. A study of range condition classes and methods of range improvement in Wyoming. p. 133-134. In Range Research Progress in 17 Western States. State Exp. Sta. Div., Agr. Res. Serv., U.S.D.A. Wash. D.C.

2F2 620201 Fisser, H.G. 1962. An ecological study of the *Artemisia tripartita* subsp. *rupicola* and related shrub communities in Wyoming. Ph.D. Dissertation, University of Wyoming. 166 p.

2F2 620204 Fisser, H.G. 1962. Abst. of Ph.D. Dissertation. Wyo. Range Manage. Issue 160. 2 p.

2F2 660204 Wight, J.R. 1966. Juniper communities and range management in northwestern Wyoming. Wyo. Range Manage. 218:117-119.

2F2 660205 Robinson, L.D. 1966. Abstract of M.S. Thesis. Wyo. Range Manage. 229:168.

2F2 680201 Fisser, H.G. 1968. Soil moisture and temperature changes following sagebrush control. J. Range Manage. 21(4):283-287.

2F2 730201 Wambolt, C.C. 1973. Conifer water potential as influenced by stand density and environmental factors. Canad. J. Bot. 51(12):2333-2337.

2F2 760202 Gerhart, William, and H.G. Fisser. 1976. Some influences of grazing systems upon vegetation and wildlife. J. Colo.-Wyo. Acad. Sci. 8(1):68.

2F2 760203 Kleinman, L.H., and H.G. Fisser. 1976. Phenology and environmental relationships of sagebrush-grass rangelands. J. Colo.-Wyo. Acad. Sci. 8(1):48-49.

2F2 780201 Fisser, H.G. 1978. Soil surface movement and relation to vegetation structure. p. 706-711. In Mohan K. Wali (ed.) Ecology and Coal Resource Development. Vol. 2, International Congress for Energy and the Ecosystem. University of North Dakota, Grand Forks, N.D. June 12-16, 1978. Pergamon Press, Elmsford, NY.

2F2 780202 Shawesh, O.M. and H.G. Fisser. 1978. Ecology of rangeland communities in northwestern Libya, Africa. p. 190-192. In D.N. Hyder (ed.) Proc. 1st Intern. Rangeland Congress, Denver, Colorado. August 14-18, 1978.

2F2 790202 Cleaves, D.T. and J. Taylor. 1979. Competition as a control for *Halogeton glomeratus*. J. Colo.-Wyo. Acad. Sci. 11(1):18.

2F2 800201 Waugh, W.J. and H.G. Fisser. 1980. Vegetation stagnation and decadence in a southeastern Wyoming big game exclosure: preliminary report. J. Colo.-Wyo. Acad. Sci. 12(1):29.

2F2 800202 Rodgers, D.E. and H.G. Fisser. 1980. Vegetational response to grazing systems in Wyoming's Red Desert. J.Colo.-Wyo. Acad. Sci. 12(1):29-30.

2F2 820201 Fisser, H.G. and N.E. Hargis. 1982. Numerical phenology of big and black sagebrush from visual description statements. J. Colo.-Wyo. Acad. Sci. 14(1):39.

2F2 820202 Taha, F.K., H.G. Fisser and R.E. Ries. 1982. A modified 100-point frame for repetition inventory. J. Range Manage. 36(1):124-125.

2F2 840202 Germain, L.P. and H.G. Fisser. 1984. Application of kriging to ecological variables. J. Colo.-Wyo. Acad. Sci. 16(1):13-14.

3. Range Resources

3A. RANGE INVENTORY AND SURVEY

3A 620201 Fisser, H.G. 1962. An ecological study of the *Artemisia tripartita* subsp. *rupicola* and related shrub communities in Wyoming. Ph.D. Dissertation, University of Wyoming. 166 p.

3A 620202 Fisser, H.G. 1962. List of plants observed on the Owl Creek and Wind River Ranges in Wyoming. Wyo. Range Manage. Issue 159. 10 p.

3A 620204 Fisser, H.G. 1962. Abst. of Ph.D. Dissertation. Wyo. Range Manage. Issue 160. 2 p.

3A 660204 Wight, J.R. 1966. Juniper communities and range management in northwestern Wyoming. Wyo. Range Manage. 218:117-119.

3A1. Site, Mapping, Remote Sensing

3A1 600201 Fisser, H.G., A.A. Beetle, R.L. Lang, and D.R. Smith. 1960. A study of range condition classes and methods of range improvement in Wyoming. In Range Research Progress in 17 Western States. State Exp. Sta. Div., Agr. Res. Serv., U.S.D.A. p. 133-134.

3A1 680204 Wambolt, C.L. 1968. Size, density, growth rate, and age, of Rocky Mountain juniper and interrelationships with environmental factors. Wyo. Range Manage. 259:93.

3A1 740202 Fisser, H.G. and J.C. Shaver. 1974. Remote sensing applied to the grassland biome. p. 293-295. In Proc. First International Congress of Ecology, The Hague, The Netherlands. September 10, 1974.

3A1 760201 Cox, J.R., and H.G. Fisser. 1976. Methods of measuring environmental parameters on a pre-disturbed coal site. J. Colo.-Wyo. Acad. Sci. 8(1):45-46.

3A1 820202 Taha, F.K., H.G. Fisser and R.E. Ries. 1982. A modified 100-point frame for repetition inventory. J. Range Manage. 36(1):124-125.

3A2. Condition and Trend

3A2 600201 Fisser, H.G., A.A. Beetle, R.L. Lang, and D.R. Smith. 1960. A study of range condition classes and methods of range improvement in Wyoming. In Range Research Progress in 17 Western States. State Exp. Sta. Div., Agr. Res. Serv., U.S.D.A. p. 133-134.

3A2 620203 Fisser, H.G. 1962. A comparison of line intercept, inferred point contact, and weight measurement methods on foothill grassland vegetation. M.S. Thesis Abst. J. Range Manage. 15:124.

3A2 640201 Nichols. J.T. 1964. Cover, composition and production of contour-furrowed and seeded range as compared to native saltsage ranges. Wyo. Range Manage. 187:27-38.

3A2 660201 Fisser, H.G. and G.M. Van Dyne. 1966. Influence of number and spacing of points on accuracy and precision of basal cover estimates. J. Range Manage. 19(3):205-211.

3A2 660203 Wight, J.R. and J.T. Nichols. 1966. Effects of harvester ants on production of a saltbush community. J. Range Manage. 19(1):68-71.

3A2 660205 Robinson, L.D. 1966. Abstract of M.S. Thesis. Wyo. Range Manage. 229:168.

3A2 670201 Fisser, H.G. 1967. Proper winter sheep grazing on saltbush rangeland. Wyo. Range Manage. 244:201-206.

3A2 700203 Lester, J.E. and H.G. Fisser. 1970. *Opuntia* clump size variation on the shortgrass plains, Pawnee Site, Nunn, Colorado. J. Colo.-Wyo. Acad. Sci. 7(1):37-38.

3A2 740201 Fisser, H.G., M.H. Mackey, and J.T. Nichols. 1974. Contour-furrowing and seeding on Nuttall saltbush rangelands of Wyoming. J. Range Manage. 27(5):459-462.

3A2 740202 Fisser, H.G. and J.C. Shaver. 1974. Remote sensing applied to the grassland biome. p. 293-295. In Proc. First International Congress of Ecology, The Hague, The Netherlands. September 10, 1974.

3A2 760201 Cox, J.R., and H.G. Fisser. 1976. Methods of measuring environmental parameters on a pre-disturbed coal site. J. Colo.-Wyo. Acad. Sci. 8(1):45-46.

3A2 760203 Kleinman, L.H., and H.G. Fisser. 1976. Phenology and environmental relationships of sagebrush-grass rangelands. J. Colo.-Wyo. Acad. Sci. 8(1):48-49.

3A2 760204 Taha, F.K., and H.G. Fisser. 1976. Production of a short-grass prairie ecosystem as related to soils characteristics in eastern Wyoming. J. Colo.-Wyo. Acad. Sci. 8(1):50.

3A2 780201 Fisser, H.G. 1978. Soil surface movement and relation to vegetation structure. p. 706-711. In Mohan K. Wali (ed.) Ecology and Coal Resource Development. Vol. 2, International Congress for Energy and the Ecosystem. University of North Dakota, Grand Forks, N.D. June 12-16, 1978. Pergamon Press, Elmsford, NY.

3A2 820201 Fisser, H.G. and N.E. Hargis. 1982. Numerical phenology of big and black sagebrush from visual description statements. J. Colo.-Wyo. Acad. Sci. 14(1):39.

3A2 820202 Taha, F.K., H.G. Fisser and R.E. Ries. 1982. A modified 100-point frame for repetition inventory. J. Range Manage. 36(1):124-125.

3A3. Production and Utilization

3A3 610201 Fisser, H.G. 1961. Variable plot, square foot plot, and visual estimate for shrub crown cover measurements. J. Range Manage. 14(3):202-207.

3A3 630201 Van Dyne, G.M., W. Vogel and H.G. Fisser. 1963. Influence of plot size and shape on range herbage production estimates. Ecology 44(3):746-759.

3A3 640203 Nichols, J.T. 1964. Effect of protection from grazing on two vegetation types in the Big Horn Basin of Wyoming. Wyo. Range Manage. 192:50-69.

3A3 670204 Wight, J.R. 1967. The sample unit and its effect on saltbush yield estimates. J. Range Manage. 20(6):323-325.

3A3 680205 Noller, G.L. 1968. The relationship of forage production to precipitation, cover, and soils in north central Wyoming. Wyo. Range Manage. 263:116-136.

3A3 700202 Whysong, G.L. 1970. Predictive value of factors influencing perennial grass production on western Wyoming sagebrush-grass ranges. J. Colo.-Wyo. Acad. Sci. 7:(1)36-37.

3A3 790201 Ries, R.E., and H.G. Fisser. 1979. Influence of environmental factors upon sagebrush and grass production in Wyoming. Agro-Ecosystems 5(1):41-55.

3A3 840201 Fisser, H.G. 1984. Culm weight dynamics as an index of sagebrush-grass community productivity. J. Colo.-Wyo. Acad. Sci. 16(1):25-26.

3A3 840204 Fisser, H.G. 1984. Productivity dynamics of sagebrush grass ecosystems as determined from biweekly sampling of a single grass and shrub species. Bull. Ecol. Soc. Amer. Program Issue. 65(2):258.

4. Grazing Management

4A. STOCKING RATES

4A 640203 Nichols, J.T. 1964. Effect of protection from grazing on two vegetation types in the Big Horn Basin of Wyoming. Wyo. Range Manage. 192:50-69.

4A 670201 Fisser, H.G. 1967. Proper winter sheep grazing on saltbush rangeland. Wyo. Range Manage. 244:201-206.

4A 670202 Smith D.R., H.G. Fisser, Ned Jeffries, and P.O. Stratton. 1967. Rotation grazing on Wyoming's Big Horn Mountains. Wyo. Agr. Exp. Sta. Res. J. 13. 26 p.

4B. GRAZING SEASONS AND SYSTEMS

4B 670201 Fisser, H.G. 1967. Proper winter sheep grazing on saltbush rangeland. Wyo. Range Manage. 244:201-206.

4B 670202 Smith D.R., H.G. Fisser, Ned Jeffries, and P.O. Stratton. 1967. Rotation grazing on Wyoming's Big Horn Mountains. Wyo. Agr. Exp. Sta. Res. J. 13. 26 p.

4B 760202 Gerhart, William, and H.G. Fisser. 1976. Some influences of grazing systems upon vegetation and wildlife. J. Colo.-Wyo. Acad. Sci. 8(1):68.

4B 800202 Rodgers, D.E. and H.G. Fisser. 1980. Vegetational response to grazing systems in Wyoming's Red Desert. J.Colo.-Wyo. Acad. Sci. 12(1):29-30.

4C. GRAZING HABITS AND DISTRIBUTION

4C 760202 Gerhart, William, and H.G. Fisser. 1976. Some influences of grazing systems upon vegetation and wildlife. J. Colo.-Wyo. Acad. Sci. 8(1):68.

4C 800202 Rodgers, D.E. and H.G. Fisser. 1980. Vegetational response to grazing systems in Wyoming's Red Desert. J.Colo.-Wyo. Acad. Sci. 12(1):29-30.

5. Range and Ranch Development

5A. PLANT CONTROL

5A 620205 Vosler, L.C. 1962. Bibliography of *Halogeton glomeratus*. Wyo. Range Manage. Issue 158, 6 p.

5A1. Methods of Control

5A1B. BIOLOGICAL

5A1B610202 Fisser, H.G., and R.J. Lavigne. 1961. *Trirhabda attenuata* (Say) (Coleoptera: Chrysomelidae), a leaf feeding beetle found on threetip sagebrush (*Artemisia tripartita* Rydb. subsp. *rupicola* Beetle). J. Range Manage. 14(4):278-279.

5A1B660202 Lavigne, R.J. and H.G. Fisser. 1966. Controlling western harvester ants. Mtn. States Regional Public. 3. 4 p.

5A1B670203 Kirkham, D.R. 1967. Biological control as a tool in range management. Wyo. Range Manage. 246:254-262.

5A1B720201 Kirkham, D.R. and H.G. Fisser. 1972. Rangeland relations and harvester ants in north central Wyoming. J. Range Manage. 25(1):55-60.

5A1C. MECHANICAL

5A1C640201 Nichols. J.T. 1964. Cover, composition and production of contour-furrowed and seeded range as compared to native saltsage ranges. Wyo. Range Manage. 187:27-38.

5AID. CHEMICAL

5A1D660203 Wight, J.R. and J.T. Nichols. 1966. Effects of harvester ants on production of a saltbush community. J. Range Manage. 19(1):68-71.

5A1D710201 Burrough, L.N. and H.G. Fisser. 1971. A greenhouse analysis of variance of picloram retention in soil and duff from juniper sites. J. Colo.-Wyo. Acad. Sci. Vol. 7(2):103.

5A1D740201 Fisser, H.G., M.H. Mackey, and J.T. Nichols. 1974. Contour-furrowing and seeding on Nuttall saltbush rangelands of Wyoming. J. Range Manage. 27(5):459-462.

5A1D760203 Kleinman, L.H., and H.G. Fisser. 1976. Phenology and environmental relationships of sagebrush-grass rangelands. J. Colo.-Wyo. Acad. Sci. 8(1):48-49.

5A2. Regional Control

5A2D. INTERMOUNTAIN

5A2D680201 Fisser, H.G. 1968. Soil moisture and temperature changes following sagebrush control. J. Range Manage. 21(4):283-287.

5B. RANGE REVEGETATION

5B4. Intermountain

5B4 640201 Nichols. J.T. 1964. Cover, composition and production of contour-furrowed and seeded range as compared to native saltsage ranges. Wyo. Range Manage. 187:27-38.

5B4 730202 Bedell, T.E. 1973. Revegetation of pipeline areas. J. Colo.-Wyo. Acad. Sci. 7(4):29-30.

5B4 740201 Fisser, H.G., M.H. Mackey, and J.T. Nichols. 1974. Contour-furrowing and seeding on Nuttall saltbush rangelands of Wyoming. J. Range Manage. 27(5):459-462.

5C. WATER SPREADING, FURROWING, PITTING

5C 640201 Nichols. J.T. 1964. Cover, composition and production of contour-furrowed and seeded range as compared to native saltsage ranges. Wyo. Range Manage. 187:27-38.

5C 730202 Bedell, T.E. 1973. Revegetation of pipeline areas. J. Colo.-Wyo. Acad. Sci. 7(4):29-30.

5C 740201 Fisser, H.G., M.H. Mackey, and J.T. Nichols. 1974. Contour-furrowing and seeding on Nuttall saltbush rangelands of Wyoming. J. Range Manage. 27(5):459-462.

5C 780203 Cox, J.R. and H.G. Fisser. 1978. The effects of water stress on phenology and carbohydrate storage in the shortgrass prairie. p. 529-535. In J.R. Goodin and D.K. Northington (eds.) 1979. Intern. Arid Lands Conference on Plant Resources. Intern. Center Arid and Semiarid Land Studies. Lubbock, Texas. October 8-15, 1978.

6. Range Livestock

6A. GENERAL MANAGEMENT

6A 670202 Smith D.R., H.G. Fisser, Ned Jeffries, and P.O. Stratton. 1967. Rotation grazing on Wyoming's Big Horn Mountains. Wyo. Agr. Exp. Sta. Res. J. 13. 26 p.

6A 800201 Waugh, W.J. and H.G. Fisser. 1980. Vegetation stagnation and decadence in a southeastern Wyoming big game exclosure: preliminary report. J. Colo.-Wyo. Acad. Sci. 12(1):29.

6A2. Sheep and Goats

6A2 670201 Fisser, H.G. 1967. Proper winter sheep grazing on saltbush rangeland. Wyo. Range Manage. 244:201-206.

6A3. Horses

6A3 760202 Gerhart, William, and H.G. Fisser. 1976. Some influences of grazing systems upon vegetation and wildlife. J. Colo.-Wyo. Acad. Sci. 8(1):68.

6A3 800202 Rodgers, D.E. and H.G. Fisser. 1980. Vegetational response to grazing systems in Wyoming's Red Desert. J.Colo.-Wyo. Acad. Sci. 12(1):29-30.

7. Range Wildlife

7A. BIG GAME

7A 710201 Burrough, L.N. and H.G. Fisser. 1971. A greenhouse analysis of variance of picloram retention in soil and duff from juniper sites. J. Colo.-Wyo. Acad. Sci. Vol. 7(2):103.

7A1. Deer

7A1 760202 Gerhart, William, and H.G. Fisser. 1976. Some influences of grazing systems upon vegetation and wildlife. J. Colo.-Wyo. Acad. Sci. 8(1):68.

7A1 800201 Waugh, W.J. and H.G. Fisser. 1980. Vegetation stagnation and decadence in a southeastern Wyoming big game exclosure: preliminary report. J. Colo.-Wyo. Acad. Sci. 12(1):29.

7A1 800202 Rodgers, D.E. and H.G. Fisser. 1980. Vegetational response to grazing systems in Wyoming's Red Desert. J.Colo.-Wyo. Acad. Sci. 12(1):29-30.

7A2. Elk

7A2 680203 Probasco, G.E. 1968. Dietary preference and utilization patterns of elk in the Big Horn Mountains, Wyoming. Wyo. Range Manage. 258:92.

7A2 800201 Waugh, W.J. and H.G. Fisser. 1980. Vegetation stagnation and decadence in a southeastern Wyoming big game exclosure: preliminary report. J. Colo.-Wyo. Acad. Sci. 12(1):29.

7A3. Antelope

7A3 760202 Gerhart, William, and H.G. Fisser. 1976. Some influences of grazing systems upon vegetation and wildlife. J. Colo.-Wyo. Acad. Sci. 8(1):68.

7A3 800202 Rodgers, D.E. and H.G. Fisser. 1980. Vegetational response to grazing systems in Wyoming's Red Desert. J.Colo.-Wyo. Acad. Sci. 12(1):29-30.

7D. NON-GAME

7D1. Rodents and Rabbits

7D1 660205 Robinson, L.D. 1966. Abstract of M.S. Thesis. Wyo. Range Manage. 229:168.

8. Range Watersheds

8B. WATERSHED SOIL EROSION AND SEDIMENTATION

8B 760204 Taha, F.K., and H.G. Fisser. 1976. Production of a short-grass prairie ecosystem as related to soils characteristics in eastern Wyoming. J. Colo.-Wyo. Acad. Sci. 8(1):50.

8B 780201 Fisser, H.G. 1978. Soil surface movement and relation to vegetation structure. p. 706-711. In Mohan K. Wali (ed.) Ecology and Coal Resource Development. Vol. 2, International Congress for Energy and the Ecosystem. University of North Dakota, Grand Forks, N.D. June 12-16, 1978. Pergamon Press, Elmsford, NY.

8C. HYDROLOGY AND WATER YIELD

8C 770202 Shawesh, O.M. and H.G. Fisser. 1977. Vegetation analysis and ecology of rangelands in northwest Libya, Africa. J. Colo.-Wyo. Acad. Sci. 9(1):35-36.

8C 780202 Shawesh, O.M. and H.G. Fisser. 1978. Ecology of rangeland communities in northwestern Libya, Africa. p. 190-192. In D.N. Hyder (ed.) Proc. 1st Intern. Rangeland Congress, Denver, Colorado. August 14-18, 1978.

8D. WATERSHED RESTORATION

8D 730202 Bedell, T.E. 1973. Revegetation of pipeline areas. J. Colo.-Wyo. Acad. Sci. 7(4):29-30.

8D 760201 Cox, J.R., and H.G. Fisser. 1976. Methods of measuring environmental parameters on a pre-disturbed coal site. J. Colo.-Wyo. Acad. Sci. 8(1):45-46.

10. Range Education and Research

10D. RANGE RESEARCH METHODS

10D 610201 Fisser, H.G. 1961. Variable plot, square foot plot, and visual estimate for shrub crown cover measurements. J. Range Manage. 14(3):202-207.

10D 800201 Waugh, W.J. and H.G. Fisser. 1980. Vegetation stagnation and decadence in a southeastern Wyoming big game exclosure: preliminary report. J. Colo.-Wyo. Acad. Sci. 12(1):29.

10D 820202 Taha, F.K., H.G. Fisser and R.E. Ries. 1982. A modified 100-point frame for repetition inventory. J. Range Manage. 36(1):124-125.

10D1. Plant-Soil Emphasis

10D1 620201 Fisser, H.G. 1962. An ecological study of the *Artemisia tripartita* subsp. *rupicola* and related shrub communities in Wyoming. Ph.D. Dissertation, University of Wyoming. 166 p.

10D1 620203 Fisser, H.G. 1962. A comparison of line intercept, inferred point contact, and weight measurement methods on foothill grassland vegetation. M.S. Thesis Abst. J. Range Manage. 15:124.

10D1 630201 Van Dyne, G.M., W. Vogel and H.G. Fisser. 1963. Influence of plot size and shape on range herbage production estimates. Ecology 44(3):746-759.

10D1 640201 Nichols. J.T. 1964. Cover, composition and production of contour-furrowed and seeded range as compared to native saltsage ranges. Wyo. Range Manage. 187:27-38.

10D1 660201 Fisser, H.G. and G.M. Van Dyne. 1966. Influence of number and spacing of points on accuracy and precision of basal cover estimates. J. Range Manage. 19(3):205-211.

10D1 670204 Wight, J.R. 1967. The sample unit and its effect on saltbush yield estimates. J. Range Manage. 20(6):323-325.

10D1 680205 Noller, G.L. 1968. The relationship of forage production to precipitation, cover, and soils in north central Wyoming. Wyo. Range Manage. 263:116-136.

10D1 700201 Fisser, H.G. and D.R. Kirkham. 1970. Reliability of harvester ant mound counts from aerial photographs. J. Colo.-Wyo. Acad. Sci. 7(1):42-43.

10D1 700203 Lester, J.E. and H.G. Fisser. 1970. *Opuntia* clump size variation on the shortgrass plains, Pawnee Site, Nunn, Colorado. J. Colo.-Wyo. Acad. Sci. 7(1):37-38.

10D1 730201 Wambolt, C.C. 1973. Conifer water potential as influenced by stand density and environmental factors. Canad. J. Bot. 51(12):2333-2337.

10D1 760204 Taha, F.K., and H.G. Fisser. 1976. Production of a shortgrass prairie ecosystem as related to soils characteristics in eastern Wyoming. J. Colo.-Wyo. Acad. Sci. 8(1):50.

10D1 780203 Cox, J.R. and H.G. Fisser. 1978. The effects of water stress on phenology and carbohydrate storage in the shortgrass prairie. p. 529-535. In J.R. Goodin and D.K. Northington (eds.) 1979. Intern. Arid Lands Conference on Plant Resources. Intern. Center Arid and Semiarid Land Studies. Lubbock, Texas. October 8-15, 1978.

10D1 790201 Ries, R.E., and H.G. Fisser. 1979. Influence of environmental factors upon sagebrush and grass production in Wyoming. Agro-Ecosystems 5(1):41-55.

11. Computer Storage, Programming and Modeling

11A. CLIMATIC

11A 820201 Fisser, H.G. and N.E. Hargis. 1982. Numerical phenology of big and black sagebrush from visual description statements. J. Colo.-Wyo. Acad. Sci. 14(1):39.

11A 840202 Germain, L.P. and H.G. Fisser. 1984. Application of kriging to ecological variables. J. Colo.-Wyo. Acad. Sci. 16(1):13-14.

11B. BIOTIC

11B 820201 Fisser, H.G. and N.E. Hargis. 1982. Numerical phenology of big and black sagebrush from visual description statements. J. Colo.-Wyo. Acad. Sci. 14(1):39.

11C. EDAPHIC

11C 820201 Fisser, H.G. and N.E. Hargis. 1982. Numerical phenology of big and black sagebrush from visual description statements. J. Colo.-Wyo. Acad. Sci. 14(1):39.

11E. RANGELAND INVENTORY

11E 820201 Fisser, H.G. and N.E. Hargis. 1982. Numerical phenology of big and black sagebrush from visual description statements. J. Colo.-Wyo. Acad. Sci. 14(1):39.

11E2. Condition and Trend

11E2 700202 Whysong, G.L. 1970. Predictive value of factors influencing perennial grass production on western Wyoming sagebrush-grass ranges. J. Colo.-Wyo. Acad. Sci. 7:(1) 36-37.

11E2 840203 Kinucan, R.J. and H.G. Fisser. 1984. Dynamics of vegetative plant part phenodynamics and related stress adaptations. Bull. Ecol. Soc. Amer. Program Issue. 65(2):253.

11E3. Production and Utilization

11E3 790201 Ries, R.E., and H.G. Fisser. 1979. Influence of environmental factors upon sagebrush and grass production in Wyoming. Agro-Ecosystems 5(1):41-55.

11E3 840201 Fisser, H.G. 1984. Culm weight dynamics as an index of sagebrush-grass community productivity. J. Colo.-Wyo. Acad. Sci. 16(1):25-26.

11E3 840203 Kinucan, R.J. and H.G. Fisser. 1984. Dynamics of vegetative plant part phenodynamics and related stress adaptations. Bull. Ecol. Soc. Amer. Program Issue. 65(2):253.

11E3 840204 Fisser, H.G. 1984. Productivity dynamics of sagebrush grass ecosystems as determined from biweekly sampling of a single grass and shrub species. Bull. Ecol. Soc. Amer. Program Issue. 65(2):258.

11F. PREDICTIVE MODELING

11F1. Herbage Yield

11F1 840201 Fisser, H.G. 1984. Culm weight dynamics as an index of sagebrush-grass community productivity. J. Colo.-Wyo. Acad. Sci. 16(1):25-26.

11F1 840204 Fisser, H.G. 1984. Productivity dynamics of sagebrush grass ecosystems as determined from biweekly sampling of a single grass and shrub species. Bull. Ecol. Soc. Amer. Program Issue. 65(2):258.

11F2. Phenology

11F2 820201 Fisser, H.G. and N.E. Hargis. 1982. Numerical phenology of big and black sagebrush from visual description statements. J. Colo.-Wyo. Acad. Sci. 14(1):39.

S3. PRESENTATIONS

1. Range Plants

1A. PLANT SYSTEMATICS

1A1. Techniques and Nomenclature

1A1 760312 "Rangeland plants and plant identification." July 13, 1976. Wyoming OWRC/SRM Youth Range Camp. Tensleep, Wyoming. by H.G. Fisser.

1A1 770309 "Rangeland plants and plant identification." June 28, 1977. Wyoming OWRC/SRM Youth Range Camp. Tensleep, Wyoming. by H.G. Fisser.

1A2. Floras and Plant Description

1A2 770309 "Rangeland plants and plant identification." June 28, 1977. Wyoming OWRC/SRM Youth Range Camp. Tensleep, Wyoming. by H.G. Fisser.

1A2 790304 "Important rangeland plants of Wyoming and the West." September 5-November 21, 1979 Special 10-week lecture session course for BLM employees. University of Wyoming RgMgt 654M - Range Plant Identification and Ecology. Rock Springs, Wyoming. by H.G. Fisser.

1B. FORAGE VALUE

1B1. Range Plants

1B1 810304 "Esophageal fistulated steers for determination of dietary nutrient composition on large pasture native rangeland." February 11, 1981. 34th Ann. Meeting Soc. Range Manage. Tulsa, Oklahoma. by G.R. O'Connor.

1B4. Poisonous Plants

1B4 650303 "Halogeton abundance and distribution as influenced by degree of saltbush utilization." September 14, 1965. Range Weed Control Conf. Elko, Nevada. by H.G. Fisser.

1B4 790306 "Competition as a control for *Halogeton glomeratus*." April 30, 1979. 50th Ann. Meeting Colo.-Wyo. Acad. Sci. Pueblo, Colorado. by D.T. Cleaves.

1B4 830310 "*Atriplex*/grass and forb relationships under no grazing and shifting precipitation patterns in northcentral Wyoming." May 6, 1983. Wildland Shrub Symp. The biology of *Atriplex* and related chenopods. USDA Forest Service. Provo, Utah. by H.G. Fisser.

1D. PLANT INSECTS

1D 660303 "Effect of control of harvester ants on revegetation." November 11, 1966. Wyoming Woolgrowers Association Annual Meeting, Casper, Wyoming. by H.G. Fisser.

1D 670306 "Western harvester ant abundance in relation to grazing, vegetation, and soils in northcentral Wyoming." February 14, 1967. 20th Ann. Meeting Amer. Soc. Range Manage. Seattle, Washington. Feb.14, 1967. by D.R. Kirkham.

1D 700302 "Reliability of harvester ant mound counts from aerial photographs." May 1, 1970. Colo.-Wyo. Acad. Sci. Greeley, Colorado. by H.G. Fisser.

1E. PLANT PHYSIOLOGY

1E1. Reproduction and Germination

1E1 650304 "Exclosure studies: Responses to sagebrush spraying." Nov. 20, 1965. Wyoming Section American Society of Range Management Annual Winter Meeting. Worland, Wyoming. by H.G. Fisser.

1E1 700301 "Distribution of plant biomass to plant organs." January 3, 1970. I.B.P. Desert Biome Plant Specialists Meeting. Las Vegas, Nevada. by H.G. Fisser.

1E1 730301 "Methods of revegetation and restoration following a juniper wildfire." Feb. 6, 1973. 26th Ann. Meeting Soc. Range Manage. Boise, Idaho. by D.L. Layton.

1E1 750306 "Phenodynamics and environmental characteristics of sagebrush-grass rangelands." November 5, 1975. Wildland Shrubs Symp. Logan, Utah. by L.H. Kleinman.

1E1 760301 "Aridland ecology and phenology research in western Wyoming." March 3, 1976. Regional Rangeland Research Symposium. Bureau of Land Management. Denver, Colorado. by H.G. Fisser.

1E1 760311 "Phenology and environmental relationships of sagebrush-grass rangelands." April 22, 1976. 47th Ann. Meeting Colo.-Wyo. Acad Sci. Boulder, Colorado. by L.H. Kleinman.

1E1 770306 "Phenodynamics and ecology of sagebrush-grass rangelands." February 15, 1977. 30th Ann. Meeting Soc. Range Manage. Portland, Oregon. by L.H. Kleinman.

1E1 770307 "Factors affecting growth and productivity of vegetation stratum in eastern Wyoming." April 23, 1977. 48th Ann. Meeting Colo.-Wyo. Acad. Sci. Colorado Springs, Colorado. by F.K. Taha.

1E1 810302 "Phenological comparison of two sagebrush species in Wyoming." February 11, 1981. 34th Ann. Meeting Soc. Range Manage. Tulsa, Oklahoma. by N.E. Hargis.

1E1 810305 "Phenology and productivity comparisons of western wheatgrass (*Agropyron smithii*) and bluebunch wheatgrass (*A. spicatum*)." February 11, 1981. 34th Ann. Meeting Soc. Range Manage. Tulsa, Oklahoma. by M.K. Owens.

1E1 810315 "Phenology and culm weight variation between two wheatgrass species: *Agropyron smithii* and *A. spicatum*." June 17, 1981. 14th Intern. Grassl. Congress. Lexington, Kentucky. by M.K. Owens.

1E1 820301 "Numerical phenology of big and black sagebrush from visual interpretation statements." April 25, 1982. Colorado-Wyoming Academy of Science. Fort Collins, Colorado. by H.G. Fisser.

1E1 820302 "Vegetation productivity and phenology in the Wind River Basin." May 27, 1982. 11th Wyoming Shrub Ecology Workshop. Lander, Wyoming. by H.G. Fisser.

1E1 820305 "Numerical phenology scoring and use with quantitative variables." May 27, 1982. 11th Wyoming Shrub Ecology Workshop. Lander, Wyoming. by R.J. Kinucan.

1E1 830304 "Phenological description and numerical scoring of sagebrush." February 16, 1983. 36th Ann. Meeting Soc. Range Manage. Albuquerque, New Mexico. by H.G. Fisser.

1E1 830307 "Re-establishment rates of sagebrush following chemical control on grazed and ungrazed ranges in western Wyoming." February 17, 1983. 36th Ann. Meeting Soc. Range Manage. Albuquerque, New Mexico. by J.L. Mulholland.

1E1 830310 "*Atriplex*/grass and forb relationships under no grazing and shifting precipitation patterns in northcentral Wyoming." May 6, 1983. Wildland Shrub Symp. The biology of *Atriplex* and related chenopods. USDA Forest Service. Provo, Utah. by H.G. Fisser.

1E1 840301 "Detailed numerical phenology of sagebrush with visual description statements." February 14, 1984. 37th Ann. Meeting Soc. Range Manage. Rapid City, South Dakota. by N.E. Hargis and H.G. Fisser.

1E1 840304 "Detailed numerical phenology of sagebrush with visual description statements." February 16, 1984. 37th Ann. Meeting Soc. Range Manage. Rapid City, South Dakota. by H.G. Fisser.

1E1 840307 "Detailed numerical phenology of sagebrush with visual description statements." May 1, 1984. USDA-ARS/USDI-BLM Regional Workshop on Rangeland Hydrology, Modeling and Monitoring. Boise, Idaho. by H.G. Fisser.

1E1 840308 "Grazing management longevity and benefits of sagebrush control in western Wyoming." May 1, 1984. USDA-ARS/USDI-BLM Regional Workshop on Rangeland Hydrology, Modeling, and Monitoring. Boise, Idaho. by H.G. Fisser.

1E1 840311 "Reinvasion following sagebrush control." July 12, 1984. Shrub Research Consortium Third Ann. Wildland Shrub Symposium: the biology of *Artemisia* and *Chrysothamnus*. Provo, Utah. by H.G. Fisser.

1E1 840312 "Regional variation of phenology." July 12, 1984. Shrub Research Consortium Third Ann. Wildland Shrub Symposium: the biology of *Artemisia* and *Chrysothamnus*. Provo, Utah. by H.G. Fisser.

1E2. Growth and Phenology

1E2 690303 "Improvement studies on juniper sites in western Wyoming." November 21, 1969. American Society of Range Management, Wyoming Section, Annual Winter Meeting. Sheridan, Wyoming. by H.G. Fisser.

1E2 690305 "Size, density, growth rate and age of Rocky Mountain juniper and interrelationships with environmental parameters." Feb. 12, 1969. 22nd Ann. Meeting Soc. Range Manage. Calgary, Alberta. by C.L. Wambolt.

1E2 700301 "Distribution of plant biomass to plant organs." January 3, 1970. I.B.P. Desert Biome Plant Specialists Meeting. Las Vegas, Nevada. by H.G. Fisser.

1E2 700306 "Predictive value of factors influencing perennial grass production on western Wyoming sagebrush-grass ranges." May 1, 1970. Colo.-Wyo. Acad. Sci. Greeley, Colorado. by G.L. Whysong.

1E2 710301 "A greenhouse study of variance of picloram retention in soil and duff from juniper sites." April 30, 1971. Colo.-Wyo. Acad. Sci. Pueblo, Colorado. by H.G. Fisser.

1E2 710302 "Moisture stress in woody plants as influenced by environmental factors." Feb. 17, 1971. 24th Ann. Meeting Soc. Range Manage. Reno, Nevada. by C.L. Wambolt.

1E2 730301 "Methods of revegetation and restoration following a juniper wildfire." Feb. 6, 1973. 26th Ann. Meeting Soc. Range Manage. Boise, Idaho. by D.L. Layton.

1E2 730303 "Revegetation of pipeline areas." April 30, 1973. Colo.-Wyo. Acad. Sci. Ft. Collins, Colorado. by T.E. Bedell.

1E2 740301 "Productivity differences of semidesert grasslands in western Wyoming." February 5, 1974. 27th Ann. Meeting Soc. Range Manage. Tucson, Arizona. by H.G. Fisser.

1E2 740305 "Environmental factors affecting annual herbage production at 17 sagebrush-grass locations in Wyoming." Feb. 5, 1974. 27th Ann. Meeting Soc. Range Manage. Tucson, Ariz. by R.E. Ries.

1E2 750303 "Seed production ecology in the shortgrass prairie of the Central United States." Feb. 12, 1975. 28th Ann. Meeting Soc. Range Manage. Mexico City, Mexico. by J.A. Shrader.

1E2 750306 "Phenodynamics and environmental characteristics of sagebrush- grass rangelands." November 5, 1975. Wildland Shrubs Symp. Logan, Utah. by L.H. Kleinman.

1E2 760301 "Arid-land ecology and phenology research in western Wyoming." March 3, 1976. Regional Rangeland Research Symposium. Bureau of Land Management. Denver, Colorado. by H.G. Fisser.

1E2 760304 "Vegetative response of *Bouteloua gracilis* to four levels of utilization." February 18, 1976. 29th Ann. Meeting Soc. Range Manage. Omaha, Nebraska. by D.W. Stroud.

1E2 760305 "Physical and chemical analyses of range sites and inter-relationships with herbage production and forage value potentials." February 18, 1976. 29th Ann. Meeting Soc. Range Mange. Omaha, Nebraska. by F.K. Taha.

1E2 760306 "Vegetation re-establishment potential of strip-mining soils, Black Thunder Basin, Wyoming." February 19, 1976. 29th Ann. Meeting Soc. Range Manage. Omaha, Nebraska. by G.A. Lymbery.

1E2 760310 "Production of a shortgrass prairie ecosystem as related to soils characteristics in eastern Wyoming." April 22, 1976. 47th Ann. Meeting Colo.-Wyo. Acad. Sci. Boulder, Colorado. by F.K. Taha.

1E2 760311 "Phenology and environmental relationships of sagebrush-grass rangelands." April 22, 1976. 47th Ann. Meeting Colo.-Wyo. Acad Sci. Boulder, Colorado. by L.H. Kleinman.

1E2 770305 "The effect of environmental factors on carbohydrate storage and herbage production of five short-grass prairie plant species." February 16, 1977. 30th Ann. Meeting Soc. Range Manage. Portland, Oregon. by J.R. Cox.

1E2 770306 "Phenodynamics and ecology of sagebrush-grass rangelands." February 15, 1977. 30th Ann. Meeting Soc. Range Manage. Portland, Oregon. by L.H. Kleinman.

1E2 770307 "Factors affecting growth and productivity of vegetation stratum in eastern Wyoming." April 23, 1977. 48th Ann. Meeting Colo.-Wyo. Acad. Sci. Colorado Springs, Colorado. by F.K. Taha.

1E2 780309 "The effects of water stress on phenology and carbohydrate storage in the shortgrass prairie." October 11, 1978. Intern. Arid. Lands Conf. on Plant Resources. Texas Tech University. Lubbock, Texas. by J.R. Cox.

1E2 790305 "Soils - phenology relations on some arid rangelands of western Wyoming." February 15, 1979. 32nd Ann. Meeting Soc. Range Manage. Casper, Wyoming. by L.W. Young, III.

1E2 790306 "Competition as a control for *Halogeton glomeratus.*" April 30, 1979. 50th Ann. Meeting Colo.-Wyo. Acad. Sci. Pueblo, Colorado. by D.T. Cleaves.

1E2 800301 "Shrub establishment, dominance and ecology on the juniper and sagebrush-grass types in Wyoming." December 3, 1980. Shrub Establishment Symposium. Wyoming Game and Fish Department. Laramie, Wyo. by H.G. Fisser.

1E2 800303 "Vegetation stagnation and decadence in a southeastern Wyoming big game exclosure: a preliminary report." April 25, 1980. 51st Ann. Meeting Colo.-Wyo. Acad. Sci. Denver, Colorado. by W.J. Waugh.

1E2 810301 "Forage production and longevity of sagebrush control on semi-arid rangelands in western Wyoming." February 10, 1981. 34th Ann. Meeting Soc. Range Manage. Tulsa, Oklahoma. by W.E. Bensel.

1E2 810302 "Phenological comparison of two sagebrush species in Wyoming." February 11, 1981. 34th Ann. Meeting Soc. Range Manage. Tulsa, Oklahoma. by N.E. Hargis.

1E2 810303 "The effect of community changes on precipitation/ production relationships in exclosures of saltbush vegetation." February 11, 1981. 34th Ann. Meeting Soc. Range Manage. Tulsa, Oklahoma. by L.A. Joyce.

1E2 810305 "Phenology and productivity comparisons of western wheatgrass *(Agropyron smithii)* and bluebunch wheatgrass *(A. spicatum).*" February 11, 1981. 34th Ann. Meeting Soc. Range Manage. Tulsa, Oklahoma. by M.K. Owens.

1E2 810307 "Plant composition response to picloram application in the Wyoming juniper type." February 12, 1981. 34th Ann. Meeting Soc. Range Manage. Tulsa, Oklahoma. by H.G. Fisser.

1E2 810311 "Correlation of soil factors and environmental conditions with phenophase development of western wheatgrass *Agropyron smithii.*" February 10, 1981. 34th Ann. Meeting Soc. Range Manage. Tulsa, Oklahoma. by L.W. Young.

1E2 810314 "Stagnation and decadency of *Cercocarpus montanus* in a southeastern Wyoming big game exclosure." May 28, 1981. 10th Wyoming Shrub Ecology Workshop. Sybille Canyon, Wyoming. by W.J. Waugh.

1E2 810315 "Phenology and culm weight variation between two wheatgrass species: *Agropyron smithii* and *A. spicatum.*" June 17, 1981. 14th Intern. Grassl. Congress. Lexington, Kentucky. by M.K. Owens.

1E2 820301 "Numerical phenology of big and black sagebrush from visual interpretation statements." April 25, 1982. Colorado-Wyoming Academy of Science. Fort Collins, Colorado. by H.G. Fisser.

1E2 820302 "Vegetation productivity and phenology in the Wind River Basin." May 27, 1982. 11th Wyoming Shrub Ecology Workshop. Lander, Wyoming. by H.G. Fisser.

1E2 820305 "Numerical phenology scoring and use with quantitative variables." May 27, 1982. 11th Wyoming Shrub Ecology Workshop. Lander, Wyoming. by R.J. Kinucan.

1E2 830303 "Perennial grass production dynamics among five vegetation types in northcentral Wyoming." February 16, 1983. 36th Ann. Meeting Soc. Range Manage. Albuquerque, New Mexico. by H.G. Fisser.

1E2 830304 "Phenological description and numerical scoring of sagebrush." February 16, 1983. 36th Ann. Meeting Soc. Range Manage. Albuquerque, New Mexico. by H.G. Fisser.

1E2 830305 "Productivity differences of grasslands in semi-desert western Wyoming — A re-examination." February 16, 1983. 36th Ann. Meeting Soc. Range Manage. Albuquerque, New Mexico. by L.A. Joyce.

1E2 830306 "Dynamics of plant community development for diverse sagebrush sites in western Wyoming." February 15, 1983. 36th Ann. Meeting Soc. Range Manage. Albuquerque, New Mexico. by R.J. Kinucan.

1E2 830308 "Size and density parameters of western wheatgrass on diverse range sites in western Wyoming." February 15, 1983. 36th Ann. Meeting Soc. Range Manage. Albuquerque, New Mexico. by H.G. Fisser.

1E2 840301 "Detailed numerical phenology of sagebrush with visual description statements." February 14, 1984. 37th Ann. Meeting Soc. Range Manage. Rapid City, South Dakota. by N.E. Hargis and H.G. Fisser.

1E2 840302 "Productivity index procedures for sagebrush-grass rangeland." February 14, 1984. 37th Ann. Meeting Soc. Range Manage. Rapid City, South Dakota. by H.G. Fisser.

1E2 840304 "Detailed numerical phenology of sagebrush with visual description statements." February 16, 1984. 37th Ann. Meeting Soc. Range Manage. Rapid City, South Dakota. by H.G. Fisser.

1E2 840305 "Culm weight dynamics as an index of sagebrush-grass community productivity." April 27, 1984. 55th Ann. Meeting Colo.-Wyo. Acad. Sci. Gunnison,Colorado. by H.G. Fisser.

1E2 840307 "Detailed numerical phenology of sagebrush with visual description statements." May 1, 1984. USDA-ARS/USDI-BLM Regional Workshop on Rangeland Hydrology, Modeling and Monitoring. Boise, Idaho. by H.G. Fisser.

1E2 840308 "Grazing management longevity and benefits of sagebrush control in western Wyoming." May 1, 1984. USDA-ARS/USDI-BLM Regional Workshop on Rangeland Hydrology, Modeling and Monitoring. Boise, Idaho. by H.G. Fisser.

1E2 840311 "Reinvasion following sagebrush control." July 12, 1984. Shrub Research Consortium Third Ann. Wildland Shrub Symposium: the biology of *Artemisia* and *Chrysothamnus.* Provo, Utah. by H.G. Fisser.

1E2 840312 "Regional variation of phenology." July 12, 1984. Shrub Research Consortium Third Ann. Wildland Shrub Symposium: the biology of *Artemisia* and *Chrysothamnus.* Provo, Utah. by H.G. Fisser.

1E2 840314 "Herbage yield dynamics and seasonal precipitation relationships." July 12, 1984. Shrub Research Consortium Third Ann. Wildland Shrub Symposium: the biology of *Artemisia* and *Chrysothamnus*. Provo, Utah. by C.L. Hanson.

1E2 840315 "Productivity dynamics of sagebrush ecosystems as determined from biweekly sampling of a single grass and shrub species." August 9, 1984. Annual Meeting Ecological Society of America. Fort Collins, Colorado. by H.G. Fisser.

1E2 840316 "Dynamics of vegetative plant part phenodynamics and related stress adaptations." August 9, 1984. Annual Meeting Ecological Society of America. Fort Collins, Colorado. by H.G. Fisser.

1E3. Photosynthesis and Plant Nutrients

1E3 760307 "Methods of measuring environmental parameters on a pre-disturbed coal site." April 22, 1976. 47th Ann. Meeting Colo.-Wyo. Acad. Sci. Boulder, Colorado. by J.R. Cox.

1E3 760308 "A mixed method of carbohydrate analysis." April 22, 1976. 47th Ann. Meeting Colo.-Wyo. Acad. Sci. Boulder, Colorado. by J.R. Cox.

1E3 770305 "The effect of environmental factors on carbohydrate storage and herbage production of five short-grass prairie plant species." February 16, 1977. 30th Ann. Meeting Soc. Range Manage. Portland, Oregon. by J.R. Cox.

1E3 770307 "Factors affecting growth and productivity of vegetation stratum in eastern Wyoming." April 23, 1977. 48th Ann. Meeting Colo.-Wyo. Acad. Sci. Colorado Springs, Colorado. by F.K. Taha.

1E3 780309 "The effects of water stress on phenology and carbohydrate storage in the shortgrass prairie." October 11, 1978. Intern. Arid. Lands Conf. on Plant Resources. Texas Tech University. Lubbock, Texas. by J.R. Cox.

1E3 800303 "Vegetation stagnation and decadence in a southeastern Wyoming big game exclosure: a preliminary report." April 25, 1980. 51st Ann. Meeting Colo.-Wyo. Acad. Sci. Denver, Colorado. by W.J. Waugh.

1E3 810313 "Some recent research findings in the true mountain mahogany types." May 28, 1981. 10th Wyoming Shrub Ecology Workshop. Sybille Canyon, Wyoming. by H.G. Fisser.

1E3 810314 "Stagnation and decadency of *Cercocarpus montanus* in a southeastern Wyoming big game exclosure." May 28, 1981. 10th Wyoming Shrub Ecology Workshop. Sybille Canyon, Wyoming. by W.J. Waugh.

1E4. Herbage Removal

1E4 640301 "Winter sheep grazing studies on saltbush rangeland of Wyoming." February 14, 1964. 17th Ann. Meeting Amer. Soc. of Range Manage. Wichita, Kansas. by H.G. Fisser.

1E4 640303 "Summer steer grazing on saltbush rangeland." Nov. 20, 1964. Wyoming Section American Society of Range Management Annual Winter Meeting. Casper, Wyoming. by H.G. Fisser.

1E4 660302 "Sheep grazing intensity studies and vegetative changes on *Atriplex nuttallii*." August 4, 1966. Symposium on management of Salt Desert Shrub Vegetation. Cedar City, Utah. by H.G. Fisser.

1E4 660303 "Effect of control of harvester ants on revegetation." November 11, 1966. Wyoming Woolgrowers Association Annual Meeting, Casper, Wyoming. by H.G. Fisser.

1E4 660304 "Effect of sheep grazing intensities on vegetative change in Nuttall saltbush." December 3, 1966. Wyoming Section, American Society of Range Management Annual Winter Meeting. Riverton, Wyoming. by H.G. Fisser.

1E4 680302 "Dietary preferences and utilization of elk on spring range in the Big Horn Mountains of Wyoming." February 14, 1968. 21st Ann. Meeting Soc. Range Manage. Albuquerque, New Mexico. by G.E. Probasco.

1E4 760302 "Grazing systems research on northern desert shrub rangelands of Wyoming." March 11, 1976. GP-9 Committee Annual Meeting. Denver, Colorado. by H.G. Fisser.

1E4 760309 "Some influences of grazing systems upon vegetation and wildlife." April 22, 1976. 47th Ann. Meeting Colo.-Wyo. Acad Sci. Boulder, Colorado. by W.A. Gerhart.

1E4 770302 "An evaluation of three grazing management systems on northern desert shrub vegetation of southern Wyoming." February 17, 1977. 30th Ann Meeting Soc. Range Manage. Portland, Oregon. by H.G. Fisser.

1E4 770305 "The effect of environmental factors on carbohydrate storage and herbage production of five short-grass prairie plant species." February 16, 1977. 30th Ann. Meeting Soc. Range Manage. Portland, Oregon. by J.R. Cox.

1E4 780309 "The effects of water stress on phenology and carbohydrate storage in the shortgrass prairie." October 11, 1978. Intern. Arid. Lands Conf. on Plant Resources. Texas Tech University. Lubbock, Texas. by J.R. Cox.

1E4 800301 "Shrub establishment, dominance and ecology on the juniper and sagebrush-grass types in Wyoming." December 3, 1980. Shrub Establishment Symposium. Wyoming Game and Fish Department. Laramie, Wyo. by H.G. Fisser.

1E4 800303 "Vegetation stagnation and decadence in a southeastern Wyoming big game exclosure: a preliminary report." April 25, 1980. 51st Ann. Meeting Colo.-Wyo. Acad. Sci. Denver, Colorado. by W.J. Waugh.

1E4 800304 "Vegetational response to grazing systems in Wyoming's Red Desert." April 25, 1980. 51st Ann. Meeting Colo.-Wyo. Acad. Sci. Denver, Colorado. by D.E. Rodgers.

1E4 810306 "Grazing system impacts on semi-arid vegetation of Wyoming." February 12, 1981. 34th Ann. Meeting Soc. Range Manage. Tulsa, Oklahoma. by D.E. Rodgers.

1E4 810314 "Stagnation and decadency of *Cercocarpus montanus* in a southeastern Wyoming big game exclosure." May 28, 1981. 10th Wyoming Shrub Ecology Workshop. Sybille Canyon, Wyoming. by W.J. Waugh.

1E4 810316 "Vegetative response to grazing systems management in Wyoming's Red Desert." June 18, 1981. 14th Intern. Grassl. Congress. Lexington, Kentucky. by D.E. Rodgers.

1E4 830311 "Grazing, vegetation and soils interactions since 1967 in the Powder Rim grazing system." May 26, 1983. 12th Wyoming Shrub Ecology Workshop. Baggs, Wyoming. by H.G. Fisser.

1E4 830312 "Deer browse impacts on Utah juniper and true mountain mahogany in the Poison Butte unit west of Baggs, Wyoming." May 26, 1983. 12th Wyoming Shrub Ecology Workshop. Baggs, Wyoming. by H.G. Fisser.

2. Range Ecology

2A. CLIMATIC FACTORS

2A 630302 "Exclosure studies (Production, cover, height, and moisture phase)." March 19, 1963. Range Weed Control Conf. Portland, Oregon. by H.G. Fisser.

2A 630303 "Does weather influence rangeland potential?" September 10, 1963. Rangeland User Group. Worland, Wyoming. by H.G. Fisser.

2A 650301 "Vegetation changes from cold desert to subalpine communities." January 18, 1965. IX International Grassland Congress. Sao Paulo, Brazil. Presented in absentia by A. Ried.

2A 660301 "Soil moisture and temperature characteristics under grazed and non-grazed conditions." Accepted for presentation at X International Grassland Congress. Helsinki, Finland. July, 1966.

2A 670301 "Soil moisture and temperature characteristics as related to grass production following shrub control under grazed and non-grazed conditions." Feb. 14, 1967. 20th Ann. Meeting, Amer. Soc. of Range Manage. Seattle, Washington. by H.G. Fisser.

2A 670305 "Cooperative research with the University of Wyoming." March 10, 1967. Bureau of Land Management Wyoming State Advisory Board Meeting. Riverton, Wyoming. by H.G. Fisser.

2A 690304 "Soil moisture relationships of certain salt desert shrubs." Feb. 12, 1969. 22nd Ann. Meeting Soc. Range Manage. Calgary, Alberta. by R.E. Steger.

2A 690305 "Size, density, growth rate and age of Rocky Mountain juniper and interrelationships with environmental parameters." Feb. 12, 1969. 22nd Ann. Meeting Soc. Range Manage. Calgary, Alberta. by C.L. Wambolt.

2A 710302 "Moisture stress in woody plants as influenced by environmental factors." Feb. 17, 1971. 24th Ann. Meeting Soc. Range Manage. Reno, Nevada. by C.L. Wambolt.

2A 740301 "Productivity differences of semidesert grasslands in western Wyoming." February 5, 1974. 27th Ann. Meeting Soc. Range Manage. Tucson, Arizona. by H.G. Fisser.

2A 740303 "Relationship with soils and other environmental factors to rangeland production and rehabilitation." August 22, 1974. Bureau of Land Management Workshop Conference of Soils and Geology Specialists. Lander, Wyoming. by H.G. Fisser.

2A 740305 "Environmental factors affecting annual herbage production at 17 sagebrush-grass locations in Wyoming." Feb. 5, 1974. 27th Ann. Meeting Soc. Range Manage. Tucson, Ariz. by R.E. Ries.

2A 750303 "Seed production ecology in the shortgrass prairie of the Central United States." Feb. 12, 1975. 28th Ann. Meeting Soc. Range Manage. Mexico City, Mexico. by J.A. Shrader.

2A 760301 "Aridland ecology and phenology research in western Wyoming." March 3, 1976. Regional Rangeland Research Symposium. Bureau of Land Management. Denver, Colorado. by H.G. Fisser.

2A 760311 "Phenology and environmental relationships of sagebrush-grass rangelands." April 22, 1976. 47th Ann. Meeting Colo.-Wyo. Acad Sci. Boulder, Colorado. by L.H. Kleinman.

2A 770301 "Rangeland site ecology and associated plant physiology." January 19, 1977. Central High Plains Research Review. U.S.D.A.-A.R.S. Fort Collins, Colo. by H.G. Fisser.

2A 770303 "Development of snowfence systems on strip mine spoils to increase seedling establishment." February 16, 1977. 30th Ann. Meeting Soc. Range Manage. Portland, Oregon. by D.T. Cleaves.

2A 770304 "Long-term vegetation studies with permanent quadrats." February 16, 1972. 30th Ann. Meeting Soc. Range Manage. Portland, Oregon. by A.D. Gdara.

2A 770305 "The effect of environmental factors on carbohydrate storage and herbage production of five short-grass prairie plant species." February 16, 1977. 30th Ann. Meeting Soc. Range Manage. Portland, Oregon. by J.R. Cox.

2A 770306 "Phenodynamics and ecology of sagebrush-grass rangelands." February 15, 1977. 30th Ann. Meeting Soc. Range Manage. Portland, Oregon. by L.H. Kleinman.

2A 770308 "Vegetation analysis and ecology of rangelands in northwest Libya, Africa." April 23, 1977. 48th Ann. Meeting Colo.-Wyo. Acad. Sci. Colorado Springs, Colorado. by O.M. Shawesh.

2A 770309 "Rangeland plants and plant identification." June 28, 1977. Wyoming OWRC/SRM Youth Range Camp. Tensleep, Wyoming. by H.G. Fisser.

2A 780302 "Models for prediction of natural vegetation potential in eastern Wyoming." February 8, 1978. 31st Ann. Meeting Soc. Range Manage. San Antonio, Texas. by H.G. Fisser.

2A 780307 "Semiarid rangeland and grazing evaluation of shrubland from permanent quadrats." August 15, 1978. 1st Intern. Rangeland Congress. Denver, Colorado. by A.D. Gdara.

2A 780309 "The effects of water stress on phenology and carbohydrate storage in the shortgrass prairie." October 11, 1978. Intern. Arid. Lands Conf. on Plant Resources. Texas Tech University. Lubbock, Texas. by J.R. Cox.

2A 780311 "Range condition and trend review in the Lander Resource Area." December 11, 1978. Range survey report to Bureau Land Management. Cheyenne, Wyoming. by H.G. Fisser.

2A 790301 "Long-term herbage production response following chemical sagebrush control in western Wyoming." February 13, 1979. 32nd Ann. Meeting Soc. Range Manage. Casper, Wyoming. by H.G. Fisser.

2A 790305 "Soils - phenology relations on some arid rangelands of western Wyoming." February 15, 1979. 32nd Ann. Meeting Soc. Range Manage. Casper, Wyoming. by L.W. Young, III.

2A 800301 "Shrub establishment, dominance and ecology on the juniper and sagebrush-grass types in Wyoming." December 3, 1980. Shrub Establishment Symposium. Wyoming Game and Fish Department. Laramie, Wyo. by H.G. Fisser.

2A 810302 "Phenological comparison of two sagebrush species in Wyoming." February 11, 1981. 34th Ann. Meeting Soc. Range Manage. Tulsa, Oklahoma. by N.E. Hargis.

2A 810303 "The effect of community changes on precipitation/production relationships in exclosures of saltbush vegetation." February 11, 1981. 34th Ann. Meeting Soc. Range Manage. Tulsa, Oklahoma. by L.A. Joyce.

2A 810305 "Phenology and productivity comparisons of western wheatgrass *(Agropyron smithii)* and bluebunch wheatgrass *(A. spicatum)*." February 11, 1981. 34th Ann. Meeting Soc. Range Manage. Tulsa, Oklahoma. by M.K. Owens.

2A 810307 "Plant composition response to picloram application in the Wyoming juniper type." February 12, 1981. 34th Ann. Meeting Soc. Range Manage. Tulsa, Oklahoma. by H.G. Fisser.

2A 810309 "Vegetation dynamics of non-grazed semi-arid sagebrush grassland communities in the Big Horn Basin of Wyoming." February 11, 1981. 34th Ann. Meeting Soc. Range Manage. Tulsa, Oklahoma. by J.W. Uhlich.

2A 810311 "Correlation of soil factors and environmental conditions with phenophase development of western wheatgrass *Agropyron smithii*." February 10, 1981. 34th Ann. Meeting Soc. Range Manage. Tulsa, Oklahoma. by L.W. Young.

2A 810312 "Habitat and vegetation of semiarid rangelands, Libya, Africa." February 11, 1981. 34th Ann. Meeting Soc. Range Manage. Tulsa, Oklahoma. by O.M. Shawesh.

2A 810313 "Some recent research findings in the true mountain mahogany types." May 28, 1981. 10th Wyoming Shrub Ecology Workshop. Sybille Canyon, Wyoming. by H.G. Fisser.

2A 810315 "Phenology and culm weight variation between two wheatgrass species: *Agropyron smithii* and *A. spicatum*." June 17, 1981. 14th Intern. Grassl. Congress. Lexington, Kentucky. by M.K. Owens.

2A 820301 "Numerical phenology of big and black sagebrush from visual interpretation statements." April 25, 1982. Colorado-Wyoming Academy of Science. Fort Collins, Colorado. by H.G. Fisser.

2A 820302 "Vegetation productivity and phenology in the Wind River Basin." May 27, 1982. 11th Wyoming Shrub Ecology Workshop. Lander, Wyoming. by H.G. Fisser.

2A 820303 "Vegetation responses following strip chemical control of sagebrush at the Lower Government Draw site." May 28, 1982. 11th Wyoming Shrub Ecology Field Tour, Lander, Wyoming. by H.G. Fisser.

2A 820304 "Vegetation response to strip chemical control in the Government Draw area." May 27, 1982. 11th Wyoming Shrub Ecology Workshop. Lander, Wyoming. by S.K. Dean.

2A 820305 "Numerical phenology scoring and use with quantitative variables." May 27, 1982. 11th Wyoming Shrub Ecology Workshop. Lander, Wyoming. by R.J. Kinucan.

2A 820306 "Sagebrush and perennial grass dynamics at the Granite Mountain shrub control site." May 27, 1982. 11th Wyoming Shrub Ecology Workshop. Lander, Wyoming. by J.L. Mulholland.

2A 830301 "Long term vegetation responses to sagebrush control in western Wyoming." February 15, 1983. 36th Ann. Meeting Soc. Range Manage. Albuquerque, New Mexico. by S.K. Dean.

2A 830303 "Perennial grass production dynamics among five vegetation types in northcentral Wyoming." February 16, 1983. 36th Ann. Meeting Soc. Range Manage. Albuquerque, New Mexico. by H.G. Fisser.

2A 830304 "Phenological description and numerical scoring of sagebrush." February 16, 1983. 36th Ann. Meeting Soc. Range Manage. Albuquerque, New Mexico. by H.G. Fisser.

2A 830305 "Productivity differences of grasslands in semi-desert western Wyoming — A re-examination." February 16, 1983. 36th Ann. Meeting Soc. Range Manage. Albuquerque, New Mexico. by L.A. Joyce.

2A 830306 "Dynamics of plant community development for diverse sagebrush sites in western Wyoming." February 15, 1983. 36th Ann. Meeting Soc. Range Manage. Albuquerque, New Mexico. by R.J. Kinucan.

2A 830307 "Reestablishment rates of sagebrush following chemical control on grazed and ungrazed ranges in western Wyoming." February 17, 1983. 36th Ann. Meeting Soc. Range Manage. Albuquerque, New Mexico. by J.L. Mulholland.

2A 830308 "Size and density parameters of western wheatgrass on diverse range sites in western Wyoming." February 15, 1983. 36th Ann. Meeting Soc. Range Manage. Albuquerque, New Mexico. by H.G. Fisser.

2A 830310 "*Atriplex*/grass and forb relationships under no grazing and shifting precipitation patterns in northcentral Wyoming." May 6, 1983. Wildland Shrub Symp. The biology of *Atriplex* and related chenopods. USDA Forest Service. Provo, Utah. by H.G. Fisser.

2A 840301 "Detailed numerical phenology of sagebrush with visual description statements." February 14, 1984. 37th Ann. Meeting Soc. Range Manage. Rapid City, South Dakota. by N.E. Hargis and H.G. Fisser.

2A 840302 "Productivity index procedures for sagebrush-grass rangeland." February 14, 1984. 37th Ann. Meeting Soc. Range Manage. Rapid City, South Dakota. by H.G. Fisser.

2A 840303 "Estimation of missing precipitation data in Wyoming using traditional, kriging and best guess estimates." February 15, 1984. 37th Ann. Meeting Soc. Range Manage. Rapid City, South Dakota. by L.P. Germain.

2A 840304 "Detailed numerical phenology of sagebrush with visual description statements." February 16, 1984. 37th Ann. Meeting Soc. Range Manage. Rapid City, South Dakota. by H.G. Fisser.

2A 840305 "Culm weight dynamics as an index of sagebrush-grass community productivity." April 27, 1984. 55th Ann. Meeting Colo.-Wyo. Acad. Sci. Gunnison,Colorado. by H.G. Fisser.

2A 840306 "Application of kriging to ecological variables." April 27, 1984. 55th Ann. Meeting Colo.-Wyo. Acad. Sci. Gunnison, Colorado. by L.P. Germain.

2A 840307 "Detailed numerical phenology of sagebrush with visual description statements." May 1, 1984. USDA-ARS/USDI-BLM Regional Workshop on Rangeland Hydrology, Modeling and Monitoring. Boise, Idaho. by H.G. Fisser.

2A 840309 "Long term Wyoming rangeland monitoring and research cooperative studies of the USDI-BLM & University of Wyoming, Department of Range Management." May 1, 1984. USDA-ARS/USDI-BLM Regional Workshop on Rangeland Hydrology, Modeling and Monitoring. Boise, Idaho. by H.G. Fisser.

2A 840310 "Herbage production and community dynamics." July 12, 1984. Shrub Research Consortium Third Ann. Wildland Shrub Symposium: the biology of *Artemisia* and *Chrysothamnus*. Provo, Utah. by H.G. Fisser.

2A 840311 "Reinvasion following sagebrush control." July 12, 1984. Shrub Research Consortium Third Ann. Wildland Shrub Symposium: the biology of *Artemisia* and *Chrysothamnus*. Provo, Utah. by H.G. Fisser.

2A 840313 "Validation of a rangeland production model (ERHYM) for sagebrush sites." July 12, 1984. Shrub Research Consortium Third Ann. Wildland Shrub Symposium: the biology of *Artemisia* and *Chrysothamnus*. Provo, Utah. by J.R. Wight.

2A 840314 "Herbage yield dynamics and seasonal precipitation relationships." July 12, 1984. Shrub Research Consortium Third Ann. Wildland Shrub Symposium: the biology of *Artemisia* and *Chrysothamnus*. Provo, Utah. by C.L. Hanson.

2A 840315 "Productivity dynamics of sagebrush ecosystems as determined from biweekly sampling of a single grass and shrub species." August 9, 1984. Annual Meeting Ecological Society of America. Fort Collins, Colorado. by H.G. Fisser.

2A 840316 "Dynamics of vegetative plant part phenodynamics and related stress adaptations." August 9, 1984. Annual Meeting Ecological Society of America. Fort Collins, Colorado. by H.G. Fisser.

2B. BIOTIC FACTORS

2B 670303 "Plant control on salt desert shrub rangeland by sheep grazing." August 21, 1967. Range Weed Control Conf. Las Cruces, New Mexico. by H.G. Fisser.

2B 670306 "Western harvester ant abundance in relation to grazing, vegetation, and soils in northcentral Wyoming." February 14, 1967. 20th Ann. Meeting Amer. Soc. Range Manage. Seattle, Washington. Feb.14, 1967. by D.R. Kirkham.

2B 740301 "Productivity differences of semidesert grasslands in western Wyoming." February 5, 1974. 27th Ann. Meeting Soc. Range Manage. Tucson, Arizona. by H.G. Fisser.

2B 770301 "Rangeland site ecology and associated plant physiology." January 19, 1977. Central High Plains Research Review. U.S.D.A.-A.R.S. Fort Collins, Colo. by H.G. Fisser.

2B 770305 "The effect of environmental factors on carbohydrate storage and herbage production of five short-grass prairie plant species." February 16, 1977. 30th Ann. Meeting Soc. Range Manage. Portland, Oregon. by J.R. Cox.

2B 770308 "Vegetation analysis and ecology of rangelands in northwest Libya, Africa." April 23, 1977. 48th Ann. Meeting Colo.-Wyo. Acad. Sci. Colorado Springs, Colorado. by O.M. Shawesh.

2B 780308 "Ecology of rangeland communities in northwestern Libya, Africa." August 15, 1978. 1st Intern. Rangeland Congress. Denver, Colorado. by O.M. Shawesh.

2B 780309 "The effects of water stress on phenology and carbohydrate storage in the shortgrass prairie." October 11, 1978. Intern. Arid. Lands Conf. on Plant Resources. Texas Tech University. Lubbock, Texas. by J.R. Cox.

2B 780310 "Rangeland vegetation, improvement and management in Libya, Africa." October 12, 1978. Intern. Arid Lands Conf. on Plant Resources. Texas Tech University. Lubbock, Texas. by O.M. Shawesh.

2B 780311 "Range condition and trend review in the Lander Resource Area." December 11, 1978. Range survey report to Bureau Land Management. Cheyenne, Wyoming. by H.G. Fisser.

2B 790306 "Competition as a control for *Halogeton glomeratus.*" April 30, 1979. 50th Ann. Meeting Colo.-Wyo. Acad. Sci. Pueblo, Colorado. by D.T. Cleaves.

2B 810303 "The effect of community changes on precipitation/production relationships in exclosures of saltbush vegetation." February 11, 1981. 34th Ann. Meeting Soc. Range Manage. Tulsa, Oklahoma. by L.A. Joyce.

2B 810309 "Vegetation dynamics of non-grazed semiarid sagebrush grassland communities in the Big Horn Basin of Wyoming." February 11, 1981. 34th Ann. Meeting Soc. Range Manage. Tulsa, Oklahoma. by J.W. Uhlich.

2B 830305 "Productivity differences of grasslands in semi-desert western Wyoming — A re-examination." February 16, 1983. 36th Ann. Meeting Soc. Range Manage. Albuquerque, New Mexico. by L.A. Joyce.

2B 830307 "Reestablishment rates of sagebrush following chemical control on grazed and ungrazed ranges in western Wyoming." February 17, 1983. 36th Ann. Meeting Soc. Range Manage. Albuquerque, New Mexico. by J.L. Mulholland.

2B 840302 "Productivity index procedures for sagebrush-grass rangeland." February 14, 1984. 37th Ann. Meeting Soc. Range Manage. Rapid City, South Dakota. by H.G. Fisser.

2B 840305 "Culm weight dynamics as an index of sagebrush-grass community productivity." April 27, 1984. 55th Ann. Meeting Colo.-Wyo. Acad. Sci. Gunnison,Colorado. by H.G. Fisser.

2B 840315 "Productivity dynamics of sagebrush ecosystems as determined from biweekly sampling of a single grass and shrub species." August 9, 1984. Annual Meeting Ecological Society of America. Fort Collins, Colorado. by H.G. Fisser.

2C. EDAPHIC FACTORS

2C 630302 "Exclosure studies (production, cover, height, and moisture phase)." March 19, 1963. Range Weed Control Conf. Portland, Oregon. by H.G. Fisser.

2C 640304 "Soil-vegetation relations of the 15-Mile Drainage area in the Big Horn Basin of Wyoming." February 14, 1964. Amer. Soc. Range Manage. Ann. Meeting. Wichita, Kansas. by J.T. Nichols.

2C 650301 "Vegetation changes from cold desert to subalpine communities." January 18, 1965. IX International Grassland Congress. Sao Paulo, Brazil. Presented in absentia by A. Ried.

2C 660301 "Soil moisture and temperature characteristics under grazed and non-grazed conditions." Accepted for presentation at X International Grassland Congress. Helsinki, Finland. July, 1966.

2C 670301 "Soil moisture and temperature characteristics as related to grass production following shrub control under grazed and non-grazed conditions." Feb. 14, 1967. 20th Ann. Meeting, Amer. Soc. of Range Manage. Seattle, Washington. by H.G. Fisser.

2C 670306 "Western harvester ant abundance in relation to grazing, vegetation, and soils in northcentral Wyoming." February 14, 1967. 20th Ann. Meeting Amer. Soc. Range Manage. Seattle, Washington. Feb.14, 1967. by D.R. Kirkham.

2C 690304 "Soil moisture relationships of certain salt desert shrubs." Feb. 12, 1969. 22nd Ann. Meeting Soc. Range Manage. Calgary, Alberta. by R.E. Steger.

2C 690305 "Size, density, growth rate and age of Rocky Mountain juniper and interrelationships with environmental parameters." Feb. 12, 1969. 22nd Ann. Meeting Soc. Range Manage. Calgary, Alberta. by C.L. Wambolt.

2C 710302 "Moisture stress in woody plants as influenced by environmental factors." Feb. 17, 1971. 24th Ann. Meeting Soc. Range Manage. Reno, Nevada. by C.L. Wambolt.

2C 740303 "Relationship with soils and other environmental factors to rangeland production and rehabilitation." August 22, 1974. Bureau of Land Management Workshop Conference of Soils and Geology Specialists. Lander, Wyoming. by H.G. Fisser.

2C 740305 "Environmental factors affecting annual herbage production at 17 sagebrush-grass locations in Wyoming." Feb. 5, 1974. 27th Ann. Meeting Soc. Range Manage. Tucson, Ariz. by R.E. Ries.

2C 750303 "Seed production ecology in the shortgrass prairie of the Central United States." Feb. 12, 1975. 28th Ann. Meeting Soc. Range Manage. Mexico City, Mexico. by J.A. Shrader.

2C 750305 "A study of rangeland soil movement characteristics." August 12, 1975. Amer. Soc. Civil Engineer. Watershed Manage. Symp. Logan, Utah. by H.G. Fisser.

2C 760303 "Aridland ecological research activities in Wyoming and relationships to Libyan rangelands." May 15, 1976. College of Agriculture, University Al Fateh special seminar. Tripoli, Libya. by H.G. Fisser.

2C 760305 "Physical and chemical analyses of range sites and interrelationships with herbage production and forage value potentials." February 18, 1976. 29th Ann. Meeting Soc. Range Mange. Omaha, Nebraska. by F.K. Taha.

2C 760306 "Vegetation re-establishment potential of strip-mining soils, Black Thunder Basin, Wyoming." February 19, 1976. 29th Ann. Meeting Soc. Range Manage. Omaha, Nebraska. by G.A. Lymbery.

2C 760310 "Production of a shortgrass prairie ecosystem as related to soils characteristics in eastern Wyoming." April 22, 1976. 47th Ann. Meeting Colo.-Wyo. Acad. Sci. Boulder, Colorado. by F.K. Taha.

2C 770301 "Rangeland site ecology and associated plant physiology." January 19, 1977. Central High Plains Research Review. U.S.D.A.-A.R.S. Fort Collins, Colo. by H.G. Fisser.

2C 770303 "Development of snowfence systems on strip mine spoils to increase seedling establishment." February 16, 1977. 30th Ann. Meeting Soc. Range Manage. Portland, Oregon. by D.T. Cleaves.

2C 770305 "The effect of environmental factors on carbohydrate storage and herbage production of five shortgrass prairie plant species." February 16, 1977. 30th Ann. Meeting Soc. Range Manage. Portland, Oregon. by J.R. Cox.

2C 770308 "Vegetation analysis and ecology of rangelands in northwest Libya, Africa." April 23, 1977. 48th Ann. Meeting Colo.-Wyo. Acad. Sci. Colorado Springs, Colorado. by O.M. Shawesh.

2C 780301 "Soil erosion and movement studies in western Wyoming." February 8, 1978. 31st Ann. Meeting Soc. Range Manage. San Antonio, Texas. by H.G. Fisser.

2C 780302 "Models for prediction of natural vegetation potential in eastern Wyoming." February 8, 1978. 31st Ann. Meeting Soc. Range Manage. San Antonio, Texas. by H.G. Fisser.

2C 780304 "Soil surface movement and relation to vegetation structure." June 15, 1978. Intern. Congress for Energy and the Ecosystem. University of North Dakota, Grand Forks, North Dakota. by H.G. Fisser.

2C 780306 "Range management equals functional ecology." July 11, 1978. Wyoming OWRC/SRM Youth Range Camp. Tensleep, Wyoming. by H.G. Fisser.

2C 780308 "Ecology of rangeland communities in northwestern Libya, Africa." August 15, 1978. 1st Intern. Rangeland Congress. Denver, Colorado. by O.M. Shawesh.

2C 780309 "The effects of water stress on phenology and carbohydrate storage in the shortgrass prairie." October 11, 1978. Intern. Arid. Lands Conf. on Plant Resources. Texas Tech University. Lubbock, Texas. by J.R. Cox.

2C 780310 "Rangeland vegetation, improvement and management in Libya, Africa." October 12, 1978. Intern. Arid Lands Conf. on Plant Resources. Texas Tech University. Lubbock, Texas. by O.M. Shawesh.

2C 780311 "Range condition and trend review in the Lander Resource Area." December 11, 1978. Range survey report to Bureau Land Management. Cheyenne, Wyoming. by H.G. Fisser.

2C 790301 "Long-term herbage production response following chemical sagebrush control in western Wyoming." February 13, 1979. 32nd Ann. Meeting Soc. Range Manage. Casper, Wyoming. by H.G. Fisser.

2C 790302 "Soil-vegetation relationships on rangeland exclosures in the Grass Creek Planning Unit of north central Wyoming." September 5, 1979. Wyo. S.C.S. Soil Survey Work Planning Conference. Laramie, Wyoming. by H.G. Fisser.

2C 790305 "Soils-phenology relations on some arid rangelands of western Wyoming." February 15, 1979. 32nd Ann. Meeting Soc. Range Manage. Casper, Wyoming. by L.W. Young III.

2C 800301 "Shrub establishment, dominance and ecology on the juniper and sagebrush-grass types in Wyoming." December 3, 1980. Shrub Establishment Symposium. Wyoming Game and Fish Department. Laramie, Wyoming. by H.G. Fisser.

2C 800302 "Soil vegetation relationships on a semi-arid rangeland exclosure in north central Wyoming: long-term studies." February 14, 1980. 33rd Ann. Meeting Soc. Range Manage. San Diego, California. by D.C. Trueblood.

2C 810302 "Phenological comparison of two sagebrush species in Wyoming." February 11, 1981. 34th Ann. Meeting Soc. Range Manage. Tulsa, Oklahoma. by N.E. Hargis.

2C 810303 "The effect of community changes on precipitation/production relationships in exclosures of saltbush vegetation." February 11, 1981. 34th Ann. Meeting Soc. Range Manage. Tulsa, Oklahoma. by L.A. Joyce.

2C 810305 "Phenology and productivity comparisons of western wheatgrass *(Agropyron smithii)* and bluebunch wheatgrass *(A. spicatum)*." February 11, 1981. 34th Ann. Meeting Soc. Range Manage. Tulsa, Oklahoma. by M.K. Owens.

2C 810307 "Plant composition response to picloram application in the Wyoming juniper type." February 12, 1981. 34th Ann. Meeting Soc. Range Manage. Tulsa, Oklahoma. by H.G. Fisser.

2C 810308 "Soil-vegetation relationships in the Gardner saltbush vegetation types of Wyoming." February 10, 1981. 34th Ann. Meeting Soc. Range Manage. Tulsa, Oklahoma. by D.C. Trueblood.

2C 810309 "Vegetation dynamics of non-grazed semi-arid sagebrush grassland communities in the Big Horn Basin of Wyoming." February 11, 1981. 34th Ann. Meeting Soc. Range Manage. Tulsa, Oklahoma. by J.W. Uhlich.

2C 810310 "Population structure of Utah juniper in north central Wyoming." February 12, 1981. 34th Ann. Meeting Soc. Range Manage. Tulsa, Oklahoma. by W.J. Waugh.

2C 810311 "Correlation of soil factors and environmental conditions with phenophase development of western wheatgrass *Agropyron smithii*." February 10, 1981. 34th Ann. Meeting Soc. Range Manage. Tulsa, Oklahoma. by L.W. Young.

2C 810312 "Habitat and vegetation of semiarid rangelands, Libya, Africa." February 11, 1981. 34th Ann. Meeting Soc. Range Manage. Tulsa, Oklahoma. by O.M. Shawesh.

2C 810313 "Some recent research findings in the true mountain mahogany types." May 28, 1981. 10th Wyoming Shrub Ecology Workshop. Sybille Canyon, Wyoming. by H.G. Fisser.

2C 810315 "Phenology and culm weight variation between two wheatgrass species: *Agropyron smithii* and *A. spicatum*." June 17, 1981. 14th Intern. Grassl. Congress. Lexington, Kentucky. by M.K. Owens.

2C 820301 "Numerical phenology of big and black sagebrush from visual interpretation statements." April 25, 1982. Colorado-Wyoming Academy of Science. Fort Collins, Colorado. by H.G. Fisser.

2C 820303 "Vegetation responses following strip chemical control of sagebrush at the Lower Government Draw site." May 28, 1982. 11th Wyoming Shrub Ecology Field Tour. Lander, Wyoming. by H.G. Fisser.

2C 820304 "Vegetation response to strip chemical control in the Government Draw area." May 27, 1982. 11th Wyoming Shrub Ecology Workshop. Lander, Wyoming. by S.K. Dean.

2C 820305 "Numerical phenology scoring and use with quantitative variables." May 27, 1982. 11th Wyoming Shrub Ecology Workshop. Lander, Wyoming. by R.J. Kinucan.

2C 820306 "Sagebrush and perennial grass dynamics at the Granite Mountain shrub control site." May 27, 1982. 11th Wyoming Shrub Ecology Workshop. Lander, Wyoming. by J.L. Mulholland.

2C 830301 "Long term vegetation responses to sagebrush control in western Wyoming." February 15, 1983. 36th Ann. Meeting Soc. Range Manage. Albuquerque, New Mexico. by S.K. Dean.

2C 830303 "Perennial grass production dynamics among five vegetation types in north central Wyoming." February 16, 1983. 36th Ann. Meeting Soc. Range Manage. Albuquerque, New Mexico. by H.G. Fisser.

2C 830304 "Phenological description and numerical scoring of sagebrush." February 16, 1983. 36th Ann. Meeting Soc. Range Manage. Albuquerque, New Mexico. by H.G. Fisser.

2C 830306 "Dynamics of plant community development for diverse sagebrush sites in western Wyoming." February 15, 1983. 36th Ann. Meeting Soc. Range Manage. Albuquerque, New Mexico. by R.J. Kinucan.

2C 830307 "Reestablishment rates of sagebrush following chemical control on grazed and ungrazed ranges in western Wyoming." February 17, 1983. 36th Ann. Meeting Soc. Range Manage. Albuquerque, New Mexico. by J.L. Mulholland.

2C 830308 "Size and density parameters of western wheatgrass on diverse range sites in western Wyoming." February 15, 1983. 36th Ann. Meeting Soc. Range Manage. Albuquerque, New Mexico. by H.G. Fisser.

2C 830309 "Utah juniper invasion in north central Wyoming and historical climatic and management relationships." February 16, 1983. 36th Ann. Meeting Soc. Range Manage. Albuquerque, New Mexico. by H.G. Fisser.

2C 830310 "*Atriplex*/grass and forb relationships under no grazing and shifting precipitation patterns in northcentral Wyoming." May 6, 1983. Wildland Shrub Symp. The biology of *Atriplex* and related chenopods. USDA Forest Service. Provo, Utah. by H.G. Fisser.

2C 840301 "Detailed numerical phenology of sagebrush with visual description statements." February 14, 1984. 37th Ann. Meeting Soc. Range Manage. Rapid City, South Dakota. by N.E. Hargis and H.G. Fisser.

2C 840304 "Detailed numerical phenology of sagebrush with visual description statements." February 16, 1984. 37th Ann. Meeting Soc. Range Manage. Rapid City, South Dakota. by H.G. Fisser.

2C 840307 "Detailed numerical phenology of sagebrush with visual description statements." May 1, 1984. USDA-ARS/USDI-BLM Regional Workshop on Rangeland Hydrology, Modeling and Monitoring. Boise, Idaho. by H.G. Fisser.

2C 840309 "Long term Wyoming rangeland monitoring and research cooperative studies of the USDI-BLM & University of Wyoming, Department of Range Management." May 1, 1984. USDA-ARS/USDI-BLM Regional Workshop on Rangeland Hydrology, Modeling and Monitoring. Boise, Idaho. by H.G. Fisser.

2C 840311 "Reinvasion following sagebrush control." July 12, 1984. Shrub Research Consortium Third Ann. Wildland Shrub Symposium: the biology of *Artemisia* and *Chrysothamnus*. Provo, Utah. by H.G. Fisser.

2C 840313 "Validation of a rangeland production model (ERHYM) for sagebrush sites." July 12, 1984. Shrub Research Consortium Third Ann. Wildland Shrub Symposium: the biology of *Artemisia* and *Chrysothamnus*. Provo, Utah. by J.R. Wight.

2C 840316 "Dynamics of vegetative plant part phenodynamics and related stress adaptations." August 9, 1984. Annual Meeting Ecological Society of America. Fort Collins, Colorado. by H.G. Fisser.

2D. PYRIC FACTORS

2D 800301 "Shrub establishment, dominance and ecology on the juniper and sagebrush-grass types in Wyoming." December 3, 1980. Shrub Establishment Symposium. Wyoming Game and Fish Department. Laramie, Wyoming. by H.G. Fisser.

2D 810310 "Population structure of Utah juniper in north central Wyoming." February 12, 1981. 34th Ann. Meeting Soc. Range Manage. Tulsa, Oklahoma. by W.J. Waugh.

2D 830309 "Utah juniper invasion in north central Wyoming and historical climatic and management relationships." February 16, 1983. 36th Ann. Meeting Soc. Range Manage. Albuquerque, New Mexico. by H.G. Fisser.

2E. AUTECOLOGY

2E 640304 "Soil-vegetation relations of the 15-Mile Drainage area in the Big Horn Basin of Wyoming." February 14, 1964. Amer. Soc. Range Manage. Ann. Meeting. Wichita, Kansas. by J.T. Nichols.

2E 650304 "Exclosure studies-responses to sagebrush spraying." Nov. 20, 1965. Wyoming Section American Society of Range Management Annual Winter Meeting. Worland, Wyoming. by H.G. Fisser.

2E 670304 "Use of picloram for juniper control." August 21, 1967. Range Weed Control Conf. Las Cruces, New Mexico. by H.G. Fisser.

2E 670305 "Cooperative research with the University of Wyoming." March 10, 1967. Bureau of Land Management Wyoming State Advisory Board Meeting. Riverton, Wyoming. by H.G. Fisser.

2E 680301 "Patterns of live and dead vegetation in grassland ecosystems and their relationship to function." September 24, 1968. I.B.P. Grasslands Biome Synthesis Meeting. Halsey, Nebraska. by H.G. Fisser.

2E 690301 "Vegetation response to chemical control of low sagebrush *(Artemisia arbuscula)* in western Wyoming." August 26, 1969. Range Weed Control Conference. Reno, Nevada. by H.G. Fisser.

2E 690302 "Juniper game range improvement studies in western Wyoming." November 18, 1969. Wyoming Game and Fish Commission Biologists Meeting, Powell, Wyoming. by H.G. Fisser.

2E 690303 "Improvement studies on juniper sites in western Wyoming." November 21, 1969. American Society of Range Management, Wyoming Section, Annual Winter Meeting. Sheridan, Wyoming. by H.G. Fisser.

2E 690304 "Soil moisture relationships of certain salt desert shrubs." Feb. 12, 1969. 22nd Ann. Meeting Soc. Range Manage. Calgary, Alberta. by R.E. Steger.

2E 690305 "Size, density, growth rate and age of Rocky Mountain juniper and interrelationships with environmental parameters." Feb. 12, 1969. 22nd Ann. Meeting Soc. Range Manage. Calgary, Alberta. by C.L. Wambolt.

2E 700301 "Distribution of plant biomass to plant organs." January 3, 1970. I.B.P. Desert Biome Plant Specialists Meeting. Las Vegas, Nevada. by H.G. Fisser.

2E 700303 "*Opuntia* clump size, variation on the shortgrass plains-Pawnee Site, Nunn, Colorado." May 1, 1970. Colo.-Wyo. Acad. Sci., Greeley, Colorado. by H.G. Fisser.

2E 700305 "Significance of plant patterns in ecosystem functioning." November 14, 1970. I.B.P Grassland Biome Annual Review Meeting. Fort Collins, Colorado. by H.G. Fisser.

2E 710301 "A greenhouse study of variance of picloram retention in soil and duff from juniper sites." April 30, 1971. Colo.-Wyo. Acad. Sci. Pueblo, Colorado. by H.G. Fisser.

2E 710302 "Moisture stress in woody plants as influenced by environmental factors." Feb. 17, 1971. 24th Ann. Meeting Soc. Range Manage. Reno, Nevada. by C.L. Wambolt.

2E 720302 "Juniper ecology and response to protection from browsing." July 6, 1972. 1st Wyo. Shrub Ecology Workshop, Casper, Wyo. by H.G. Fisser.

2E 740301 "Productivity differences of semidesert grasslands in western Wyoming." February 5, 1974. 27th Ann. Meeting Soc. Range Manage. Tucson, Arizona. by H.G. Fisser.

2E 750303 "Seed production ecology in the shortgrass prairie of the central United States." February 12, 1975. 28th Ann. Meeting Soc. Range Manage. Mexico City, Mexico. by J.A. Shrader.

2E 760301 "Aridland ecology and phenology research in western Wyoming." March 3, 1976. Regional Rangeland Research Symposium. Bureau of Land Management. Denver, Colorado. by H.G. Fisser.

2E 760304 "Vegetative response of *Bouteloua gracilis* to four levels of utilization." February 18, 1976. 29th Ann. Meeting Soc. Range Manage. Omaha, Nebraska. by D.W. Stroud.

2E 760306 "Vegetation re-establishment potential of strip-mining soils, Black Thunder Basin, Wyoming." February 19, 1976. 29th Ann. Meeting Soc. Range Manage. Omaha, Nebraska. by G.A. Lymbery.

2E 760308 "A mixed method of carbohydrate analysis." April 22, 1976. 47th Ann. Meeting Colo.-Wyo. Acad. Sci. Boulder, Colorado. by J.R. Cox.

2E 760311 "Phenology and environmental relationships of sagebrush-grass rangelands." April 22, 1976. 47th Ann. Meeting Colo.-Wyo. Acad Sci. Boulder, Colorado. by L.H. Kleinman.

2E 770301 "Rangeland site ecology and associated plant physiology." January 19, 1977. Central High Plains Research Review. U.S.D.A.-A.R.S. Fort Collins, Colo. by H.G. Fisser.

2E 770304 "Long-term vegetation studies with permanent quadrats." February 16, 1972. 30th Ann. Meeting Soc. Range Manage. Portland, Oregon. by A.D. Gdara.

2E 770305 "The effect of environmental factors on carbohydrate storage and herbage production of five short-grass prairie plant species." February 16, 1977. 30th Ann. Meeting Soc. Range Manage. Portland, Oregon. by J.R. Cox.

2E 770306 "Phenodynamics and ecology of sagebrush-grass rangelands." February 15, 1977. 30th Ann. Meeting Soc. Range Manage. Portland, Oregon. by L.H. Kleinman.

2E 780302 "Models for prediction of natural vegetation potential in eastern Wyoming." February 8, 1978. 31st Ann. Meeting Soc. Range Manage. San Antonio, Texas. by H.G. Fisser.

2E 780305 "Plant succession on plains pocket gopher mounds." February 7, 1978. 31st Ann. Meeting Soc. Range Manage. San Antonio, Texas. by M.A. Foster.

2E 780307 "Semiarid rangeland and grazing evaluation of shrubland from permanent quadrats." August 15, 1978. 1st Intern. Rangeland Congress. Denver, Colorado. by A.D. Gdara.

2E 780309 "The effects of water stress on phenology and carbohydrate storage in the shortgrass prairie." October 11, 1978. Intern. Arid. Lands Conf. on Plant Resources. Texas Tech University. Lubbock, Texas. by J.R. Cox.

2E 780311 "Range condition and trend review in the Lander Resource Area." December 11, 1978. Range survey report to Bureau Land Management. Cheyenne, Wyoming. by H.G. Fisser.

2E 790301 "Long-term herbage production response following chemical sagebrush control in western Wyoming." February 13, 1979. 32nd Ann. Meeting Soc. Range Manage. Casper, Wyoming. by H.G. Fisser.

2E 790303 "Arid rangeland sampling." September 27, 1979. University of Wyoming Zoo. Biol. Sampling Seminar. Laramie, Wyoming. by H.G. Fisser.

2E 790304 "Important rangeland plants of Wyoming and the West." September 5-November 21. 1979. Special 10-week lecture session course for BLM employees. University of Wyoming RgMgt 654M - Range Plant Identification and Ecology. Rock Springs, Wyoming. by H.G. Fisser.

2E 790305 "Soils-phenology relations on some arid rangelands of western Wyoming." February 15, 1979. 32nd Ann. Meeting Soc. Range Manage. Casper, Wyoming. by L.W. Young, III.

2E 790306 "Competition as a control for *Halogeton glomeratus.*" April 30, 1979. 50th Ann. Meeting Colo.-Wyo. Acad. Sci. Pueblo, Colorado. by D.T. Cleaves.

2E 800301 "Shrub establishment, dominance and ecology on the juniper and sagebrush-grass types in Wyoming." December 3, 1980. Shrub Establishment Symposium. Wyoming Game and Fish Department. Laramie, Wyoming. by H.G. Fisser.

2E 800303 "Vegetation stagnation and decadence in a southeastern Wyoming big game exclosure: a preliminary report." April 25, 1980. 51st Ann. Meeting Colo.-Wyo. Acad. Sci. Denver, Colorado. by W.J. Waugh.

2E 810301 "Forage production and longevity of sagebrush control on semiarid rangelands in western Wyoming." February 10, 1981. 34th Ann. Meeting Soc. Range Manage. Tulsa, Oklahoma. by W.E. Bensel.

2E 810302 "Phenological comparison of two sagebrush species in Wyoming." February 11, 1981. 34th Ann. Meeting Soc. Range Manage. Tulsa, Oklahoma. by N.E. Hargis.

2E 810303 "The effect of community changes on precipitation/production relationships in exclosures of saltbush vegetation." February 11, 1981. 34th Ann. Meeting Soc. Range Manage. Tulsa, Oklahoma. by L.A. Joyce.

2E 810305 "Phenology and productivity comparisons of western wheatgrass *(Agropyron smithii)* and bluebunch wheatgrass *(A. spicatum).*" February 11, 1981. 34th Ann. Meeting Soc. Range Manage. Tulsa, Oklahoma. by M.K. Owens.

2E 810307 "Plant composition response to picloram application in the Wyoming juniper type." February 12, 1981. 34th Ann. Meeting Soc. Range Manage. Tulsa, Oklahoma. by H.G. Fisser.

2E 810309 "Vegetation dynamics of non-grazed semiarid sagebrush grassland communities in the Big Horn Basin of Wyoming." February 11, 1981. 34th Ann. Meeting Soc. Range Manage. Tulsa, Oklahoma. by J.W. Uhlich.

2E 810310 "Population structure of Utah juniper in north central Wyoming." February 12, 1981. 34th Ann. Meeting Soc. Range Manage. Tulsa, Oklahoma. by W.J. Waugh.

2E 810311 "Correlation of soil factors and environmental conditions with phenophase development of western wheatgrass *(Agropyron smithii).*" February 10, 1981. 34th Ann. Meeting Soc. Range Manage. Tulsa, Oklahoma. by L.W. Young.

2E 810312 "Habitat and vegetation of semiarid rangelands, Libya, Africa." February 11, 1981. 34th Ann. Meeting Soc. Range Manage. Tulsa, Oklahoma. by O.M. Shawesh.

2E 810313 "Some recent research findings in the true mountain mahogany types." May 28, 1981. 10th Wyoming Shrub Ecology Workshop. Sybille Canyon, Wyoming. by H.G. Fisser.

2E 810314 "Stagnation and decadency of *Cercocarpus montanus* in a southeastern Wyoming big game exclosure." May 28, 1981. 10th Wyoming Shrub Ecology Workshop. Sybille Canyon, Wyoming. by W.J. Waugh.

2E 810315 "Phenology and culm weight variation between two wheatgrass species: *Agropyron smithii* and *A. spicatum.*" June 17, 1981. 14th Intern. Grassl. Congress. Lexington, Kentucky. by M.K. Owens.

2E 820301 "Numerical phenology of big and black sagebrush from visual interpretation statements." April 25, 1982. Colorado-Wyoming Academy of Science. Fort Collins, Colorado. by H.G. Fisser.

2E 820302 "Vegetation productivity and phenology in the Wind River Basin." May 27, 1982. 11th Wyoming Shrub Ecology Workshop. Lander, Wyoming. by H.G. Fisser.

2E 820303 "Vegetation responses following strip chemical control of sagebrush at the Lower Government Draw site." May 28, 1982. 11th Wyoming Shrub Ecology Field Tour. Lander, Wyoming. by H.G. Fisser.

2E 820304 "Vegetation response to strip chemical control in the Government Draw area." May 27, 1982. 11th Wyoming Shrub Ecology Workshop. Lander, Wyoming. by S.K. Dean.

2E 820305 "Numerical phenology scoring and use with quantitative variables." May 27, 1982. 11th Wyoming Shrub Ecology Workshop. Lander, Wyoming. by R.J. Kinucan.

2E 820306 "Sagebrush and perennial grass dynamics at the Granite Mountain shrub control site." May 27, 1982. 11th Wyoming Shrub Ecology Workshop. Lander, Wyoming. by J.L. Mulholland.

2E 830301 "Long term vegetation responses to sagebrush control in western Wyoming." February 15, 1983. 36th Ann. Meeting Soc. Range Manage. Albuquerque, New Mexico. by S.K. Dean.

2E 830303 "Perennial grass production dynamics among five vegetation types in north central Wyoming." February 16, 1983. 36th Ann. Meeting Soc. Range Manage. Albuquerque, New Mexico. by H.G. Fisser.

2E 830304 "Phenological description and numerical scoring of sagebrush." February 16, 1983. 36th Ann. Meeting Soc. Range Manage. Albuquerque, New Mexico. by H.G. Fisser.

2E 830305 "Productivity differences of grasslands in semi-desert western Wyoming — A re-examination." February 16, 1983. 36th Ann. Meeting Soc. Range Manage. Albuquerque, New Mexico. by L.A. Joyce.

2E 830307 "Reestablishment rates of sagebrush following chemical control on grazed and ungrazed ranges in western Wyoming." February 17, 1983. 36th Ann. Meeting Soc. Range Manage. Albuquerque, New Mexico. by J.L. Mulholland.

2E 830308 "Size and density parameters of western wheatgrass on diverse range sites in western Wyoming." February 15, 1983. 36th Ann. Meeting Soc. Range Manage. Albuquerque, New Mexico. by H.G. Fisser.

2E 830309 "Utah juniper invasion in north central Wyoming and historical climatic and management relationships." February 16, 1983. 36th Ann. Meeting Soc. Range Manage. Albuquerque, New Mexico. by H.G. Fisser.

2E 830310 "*Atriplex*/grass and forb relationships under no grazing and shifting precipitation patterns in north central Wyoming." May 6, 1983. Wildland Shrub Symp. The biology of *Atriplex* and related chenopods. USDA Forest Service. Provo, Utah. by H.G. Fisser.

2E 840301 "Detailed numerical phenology of sagebrush with visual description statements." February 14, 1984. 37th Ann. Meeting Soc. Range Manage. Rapid City, South Dakota. by N.E. Hargis and H.G. Fisser.

2E 840302 "Productivity index procedures for sagebrush-grass rangeland." February 14, 1984. 37th Ann. Meeting Soc. Range Manage. Rapid City, South Dakota. by H.G. Fisser.

2E 840303 "Estimation of missing precipitation data in Wyoming using traditional, kriging and best guess estimates." February 15, 1984. 37th Ann. Meeting Soc. Range Manage. Rapid City, South Dakota. by L.P. Germain.

2E 840304 "Detailed numerical phenology of sagebrush with visual description statements." February 16, 1984. 37th Ann. Meeting Soc. Range Manage. Rapid City, South Dakota. by H.G. Fisser.

2E 840305 "Culm weight dynamics as an index of sagebrush-grass community productivity." April 27, 1984. 55th Ann. Meeting Colo.-Wyo. Acad. Sci. Gunnison,Colorado. by H.G. Fisser.

2E 840306 "Application of kriging to ecological variables." April 27, 1984. 55th Ann. Meeting Colo.-Wyo. Acad. Sci. Gunnison, Colorado. by L.P. Germain.

2E 840307 "Detailed numerical phenology of sagebrush with visual description statements." May 1, 1984. USDA-ARS/USDI-BLM Regional Workshop on Rangeland Hydrology, Modeling and Monitoring. Boise, Idaho. by H.G. Fisser.

2E 840308 "Grazing management longevity and benefits of sagebrush control in western Wyoming." May 1, 1984. USDA-ARS/USDI-BLM Regional Workshop on Rangeland Hydrology, Modeling and Monitoring. Boise, Idaho. by H.G. Fisser.

2E 840309 "Long-term Wyoming rangeland monitoring and research cooperative studies of the USDI-BLM & University of Wyoming, Department of Range Management." May 1, 1984. USDA-ARS/USDI-BLM Regional Workshop on Rangeland Hydrology, Modeling and Monitoring. Boise, Idaho. by H.G. Fisser.

2E 840310 "Herbage production and community dynamics." July 12, 1984. Shrub Research Consortium Third Ann. Wildland Shrub Symposium: the biology of *Artemisia* and *Chrysothamnus*. Provo, Utah. by H.G. Fisser.

2E 840311 "Reinvasion following sagebrush control." July 12, 1984. Shrub Research Consortium Third Ann. Wildland Shrub Symposium: the biology of *Artemisia* and *Chrysothamnus*. Provo, Utah. by H.G. Fisser.

2E 840312 "Regional variation of phenology." July 12, 1984. Shrub Research Consortium Third Ann. Wildland Shrub Symposium: the biology of *Artemisia* and *Chrysothamnus*. Provo, Utah. by H.G. Fisser.

2E 840315 "Productivity dynamics of sagebrush ecosystems as determined from biweekly sampling of a single grass and shrub species." August 9, 1984. Annual Meeting Ecological Society of America. Fort Collins, Colorado. by H.G. Fisser.

2F. SYNECOLOGY

2F 680301 "Patterns of live and dead vegetation in grassland ecosystems and their relationship to function." September 24, 1968. I.B.P. Grasslands Biome Synthesis Meeting. Halsey, Nebraska. by H.G. Fisser.

2F 750306 "Phenodynamics and environmental characteristics of sagebrush-grass rangelands." November 5, 1975. Wildland Shrubs Symp. Logan, Utah. by L.H. Kleinman.

2F 780308 "Ecology of rangeland communities in northwestern Libya, Africa." August 15, 1978. 1st Intern. Rangeland Congress. Denver, Colorado. by O.M. Shawesh.

2F 780310 "Rangeland vegetation, improvement and management in Libya, Africa." October 12, 1978. Intern. Arid Lands Conf. on Plant Resources. Texas Tech University. Lubbock, Texas. by O.M. Shawesh.

2F 810312 "Habitat and vegetation of semiarid rangelands, Libya, Africa." February 11, 1981. 34th Ann. Meeting Soc. Range Manage. Tulsa, Oklahoma. by O.M. Shawesh.

2F1. Great Plains

2F1 700303 "*Opuntia* clump size, variation on the shortgrass plains - Pawnee Site, Nunn, Colorado." May 1, 1970. Colo.-Wyo. Acad. Sci., Greeley, Colorado. by H.G. Fisser.

2F1 700304 "The I.B.P. and its implications for natural resource management." June 24, 1970. Wyoming Game and Fish Commission. Biologists Meeting. Lander, Wyoming. by H.G. Fisser.

2F1 700305 "Significance of plant patterns in ecosystem functioning." November 14, 1970. I.B.P Grassland Biome Annual Review Meeting. Fort Collins, Colorado. by H.G. Fisser.

2F1 720301 "Panchromatic, color, and infrared photography for detailed vegetation mapping." February 8, 1972. 25th Ann. Meeting Soc. Range Manage. Washington, D.C. by H.G. Fisser.

2F1 740302 "Remote sensing applied to the grassland biome." September 19, 1974. First International Congress of Ecology, The Hague, The Netherlands. by H.G. Fisser.

2F1 740304 "Relationships of range production and rehabilitation to soils and other environmental factors." Aug. 22, 1974. Workshop of BLM Soils and Geology Specialists. Lander, Wyo. by H.G. Fisser.

2F1 750301 "Pre-disturbance vegetation inventory and ecological monitoring procedures for rangeland strip-mining." February 12, 1975. 28th Ann. Meeting Soc. Range Manage. Mexico City, Mexico. by H.G. Fisser.

2F1 750302 "Vegetation response to seismic motion from underground nuclear detonation." Feb. 13, 1975. 28th Ann. Meeting Soc. Range Manage. Mexico City, Mexico. by H.G. Fisser.

2F1 750304 "Pre-disturbance ecological studies improve and define potential for surface mine reclamation." October 22, 1975. 3rd NCA/BCR Coal Conference and Expo II. Third Symposium on Surface Mining and Reclamation. Louisville, Kentucky. by H.G. Fisser.

2F1 760304 Vegetative response of *Bouteloua gracilis* to four levels of utilization." February 18, 1976. 29th Ann. Meeting Soc. Range Manage. Omaha, Nebraska. by D.W. Stroud.

2F1 760305 "Physical and chemical analyses of range sites and interrelationships with herbage production and forage value potentials." February 18, 1976. 29th Ann. Meeting Soc. Range Mange. Omaha, Nebraska. by F.K. Taha.

2F1 760306 "Vegetation re-establishment potential of strip-mining soils, Black Thunder Basin, Wyoming." February 19, 1976. 29th Ann. Meeting Soc. Range Manage. Omaha, Nebraska. by G.A. Lymbery.

2F1 760307 "Methods of measuring environmental parameters on a pre-disturbed coal site." April 22, 1976. 47th Ann. Meeting Colo.-Wyo. Acad. Sci. Boulder, Colorado. by J.R. Cox.

2F1 760308 "A mixed method of carbohydrate analysis." April 22, 1976. 47th Ann. Meeting Colo.-Wyo. Acad. Sci. Boulder, Colorado. by J.R. Cox.

2F1 760310 "Production of a shortgrass prairie ecosystem as related to soils characteristics in eastern Wyoming." April 22, 1976. 47th Ann. Meeting Colo.-Wyo. Acad. Sci. Boulder, Colorado. by F.K. Taha.

2F1 770307 "Factors affecting growth and productivity of vegetation stratum in eastern Wyoming." April 23, 1977. 48th Ann. Meeting Colo.-Wyo. Acad. Sci. Colorado Springs, Colorado. by F.K. Taha.

2F1 780305 "Plant succession on plains pocket gopher mounds." February 7, 1978. 31st Ann. Meeting Soc. Range Manage. San Antonio, Texas. by M.A. Foster.

2F2. West

2F2 610301 "Range condition class survey in the Big Horn Basin of Wyoming." Sept. 14, 1961. Range Weed Control Conf. Worland, Wyoming. by H.G. Fisser.

2F2 620301 "Concepts of range condition class survey." Nov. 30, 1962. Wyoming Wool Growers Association Convention. Worland, Wyoming. by H.G. Fisser.

2F2 640304 "Soil-vegetation relations of the 15-Mile Drainage area in the Big Horn Basin of Wyoming." February 14, 1964. Amer. Soc. Range Manage. Ann. Meeting. Wichita, Kansas. by J.T. Nichols.

2F2 650301 "Vegetation changes from cold desert to subalpine communities." January 18, 1965. IX International Grassland Congress. Sao Paulo, Brazil. Presented in absentia by A. Ried.

2F2 660301 "Soil moisture and temperature characteristics under grazed and non-grazed conditions." Accepted for presentation at X International Grassland Congress. Helsinki, Finland. July, 1966.

2F2 670301 "Soil moisture and temperature characteristics as related to grass production following shrub control under grazed and non-grazed conditions." Feb. 14, 1967. 20th Ann. Meeting, Amer. Soc. of Range Manage. Seattle, Washington. by H.G. Fisser.

2F2 690301 "Vegetation response to chemical control of low sagebrush *(Artemisia arbuscula)* in western Wyoming." August 26, 1969. Range Weed Control Conference. Reno, Nevada. by H.G. Fisser.

2F2 710302 "Moisture stress in woody plants as influenced by environmental factors." Feb. 17, 1971. 24th Ann. Meeting Soc. Range Manage. Reno, Nevada. by C.L. Wambolt.

2F2 720302 "Juniper ecology and response to protection from browsing." July 6, 1972. 1st Wyo. Shrub Ecology Workshop. Casper, Wyo. by H.G. Fisser.

2F2 740305 "Environmental factors affecting annual herbage production at 17 sagebrush-grass locations in Wyoming." Feb. 5, 1974. 27th Ann. Meeting Soc. Range Manage. Tucson, Ariz. by R.E. Ries.

2F2 750303 "Seed production ecology in the shortgrass prairie of the Central United States." Feb. 12, 1975. 28th Ann. Meeting Soc. Range Manage. Mexico City, Mexico. by J.A. Shrader.

2F2 760301 "Aridland ecology and phenology research in western Wyoming." March 3, 1976. Regional Rangeland Research Symposium. Bureau of Land Management. Denver, Colorado. by H.G. Fisser.

2F2 760302 "Grazing systems research on northern desert shrub rangelands of Wyoming." March 11, 1976. GP-9 Committee Annual Meeting. Denver, Colorado. by H.G. Fisser.

2F2 760303 "Aridland ecological research activities in Wyoming and relationships to Libyan rangelands." May 15, 1976. College of Agriculture, University Al Fateh special seminar. Tripoli, Libya. by H.G. Fisser.

2F2 760309 "Some influences of grazing systems upon vegetation and wildlife." April 22, 1976. 47th Ann. Meeting Colo.-Wyo. Acad Sci. Boulder, Colorado. by W.A. Gerhart.

2F2 760311 "Phenology and environmental relationships of sagebrush-grass rangelands." April 22, 1976. 47th Ann. Meeting Colo.-Wyo. Acad Sci. Boulder, Colorado. by L.H. Kleinman.

2F2 770302 "An evaluation of three grazing management systems on northern desert shrub vegetation of southern Wyoming." February 17, 1977. 30th Ann Meeting Soc. Range Manage. Portland, Oregon. by H.G. Fisser.

2F2 770304 "Long-term vegetation studies with permanent quadrats." February 16, 1972. 30th Ann. Meeting Soc. Range Manage. Portland, Oregon. by A.D. Gdara.

2F2 770306 "Phenodynamics and ecology of sagebrush-grass rangelands." February 15, 1977. 30th Ann. Meeting Soc. Range Manage. Portland, Oregon. by L.H. Kleinman.

2F2 770309 "Rangeland plants and plant identification." June 28, 1977. Wyoming OWRC/SRM Youth Range Camp. Tensleep, Wyoming. by H.G. Fisser.

2F2 780301 "Soil erosion and movement studies in western Wyoming." February 8, 1978. 31st Ann. Meeting Soc. Range Manage. San Antonio, Texas. by H.G. Fisser.

2F2 780302 "Models for prediction of natural vegetation potential in eastern Wyoming." February 8, 1978. 31st Ann. Meeting Soc. Range Manage. San Antonio, Texas. by H.G. Fisser.

2F2 780304 "Soil surface movement and relation to vegetation structure." June 15, 1978. Intern. Congress for Energy and the Ecosystem. University of North Dakota, Grand Forks, North Dakota. by H.G. Fisser.

2F2 780306 "Range management equals functional ecology." July 11, 1978. Wyoming OWRC/SRM Youth Range Camp. Tensleep, Wyoming. by H.G. Fisser.

2F2 780307 "Semiarid rangeland and grazing evaluation of shrubland from permanent quadrats." August 15, 1978. 1st Intern. Rangeland Congress. Denver, Colorado. by A.D. Gdara.

2F2 780311 "Range condition and trend review in the Lander Resource Area." December 11, 1978. Range survey report to Bureau of Land Management. Cheyenne, Wyoming. by H.G. Fisser.

2F2 790301 "Long-term herbage production response following chemical sagebrush control in western Wyoming." February 13, 1979. 32nd Ann. Meeting Soc. Range Manage. Casper, Wyoming. by H.G. Fisser.

2F2 790302 "Soil-vegetation relationships on rangeland exclosures in the Grass Creek Planning Unit of northcentral Wyoming." September 5, 1979. Wyo. S.C.S. Soil Survey Work Planning Conference. Laramie, Wyoming. by H.G. Fisser.

2F2 790304 "Important rangeland plants of Wyoming and the West." September 5-November 21. 1979. Special 10-week lecture session course for BLM employees. University of Wyoming RgMgt 654M - Range Plant Identification and Ecology. Rock Springs, Wyoming. by H.G. Fisser.

2F2 790305 "Soils-phenology relations on some arid rangelands of western Wyoming." February 15, 1979. 32nd Ann. Meeting Soc. Range Manage. Casper, Wyoming. by L.W. Young III.

2F2 790306 "Competition as a control for *Halogeton glomeratus.*" April 30, 1979. 50th Ann. Meeting Colo.-Wyo. Acad. Sci. Pueblo, Colorado. by D.T. Cleaves.

2F2 800301 "Shrub establishment, dominance and ecology on the juniper and sagebrush-grass types in Wyoming." December 3, 1980. Shrub Establishment Symposium. Wyoming Game and Fish Department. Laramie, Wyoming. by H.G. Fisser.

2F2 800302 "Soil vegetation relationships on a semiarid rangeland exclosure in north central Wyoming: long-term studies." February 14, 1980. 33rd Ann. Meeting Soc. Range Manage. San Diego, California. by D.C. Trueblood.

2F2 800303 "Vegetation stagnation and decadence in a southeastern Wyoming big game exclosure: a preliminary report." April 25, 1980. 51st Ann. Meeting Colo.-Wyo. Acad. Sci. Denver, Colorado. by W.J. Waugh.

2F2 800304 "Vegetational response to grazing systems in Wyoming's Red Desert." April 25, 1980. 51st Ann. Meeting Colo.-Wyo. Acad. Sci. Denver, Colorado. by D.E. Rodgers.

2F2 810301 "Forage production and longevity of sagebrush control on semi-arid rangelands in western Wyoming." February 10, 1981. 34th Ann. Meeting Soc. Range Manage. Tulsa, Oklahoma. by W.E. Bensel.

2F2 810302 "Phenological comparison of two sagebrush species in Wyoming." February 11, 1981. 34th Ann. Meeting Soc. Range Manage. Tulsa, Oklahoma. by N.E. Hargis.

2F2 810303 "The effect of community changes on precipitation/production relationships in exclosures of saltbush vegetation." February 11, 1981. 34th Ann. Meeting Soc. Range Manage. Tulsa, Oklahoma. by L.A. Joyce.

2F2 810306 "Grazing system impacts on semiarid vegetation of Wyoming." February 12, 1981. 34th Ann. Meeting Soc. Range Manage. Tulsa, Oklahoma. by D.E. Rodgers.

2F2 810307 "Plant composition response to picloram application in the Wyoming juniper type." February 12, 1981. 34th Ann. Meeting Soc. Range Manage. Tulsa, Oklahoma. by H.G. Fisser.

2F2 810308 "Soil-vegetation relationships in the Gardner saltbush vegetation types of Wyoming." February 10, 1981. 34th Ann. Meeting Soc. Range Manage. Tulsa, Oklahoma. by D.C. Trueblood.

2F2 810309 "Vegetation dynamics of non-grazed semiarid sagebrush grassland communities in the Big Horn Basin of Wyoming." February 11, 1981. 34th Ann. Meeting Soc. Range Manage. Tulsa, Oklahoma. by J.W. Uhlich.

2F2 810310 "Population structure of Utah juniper in north central Wyoming." February 12, 1981. 34th Ann. Meeting Soc. Range Manage. Tulsa, Oklahoma. by W.J. Waugh.

2F2 810311 "Correlation of soil factors and environmental conditions with phenophase development of western wheatgrass *Agropyron smithii.*" February 10, 1981. 34th Ann. Meeting Soc. Range Manage. Tulsa, Oklahoma. by L.W. Young.

2F2 810313 "Some recent research findings in the true mountain mahogany types." May 28, 1981. 10th Wyoming Shrub Ecology Workshop. Sybille Canyon, Wyoming. by H.G. Fisser.

2F2 810314 "Stagnation and decadency of *Cercocarpus montanus* in a southeastern Wyoming big game exclosure." May 28, 1981. 10th Wyoming Shrub Ecology Workshop. Sybille Canyon, Wyoming. by W.J. Waugh.

2F2 810316 "Vegetative response to grazing systems management in Wyoming's Red Desert." June 18, 1981. 14th Intern. Grassl. Congress. Lexington, Kentucky. by D.E. Rodgers.

2F2 820301 "Numerical phenology of big and black sagebrush from visual interpretation statements." April 25, 1982. Colorado-Wyoming Academy of Science. Fort Collins, Colorado. by H.G. Fisser.

2F2 820302 "Vegetation productivity and phenology in the Wind River Basin." May 27, 1982. 11th Wyoming Shrub Ecology Workshop. Lander, Wyoming. by H.G. Fisser.

2F2 820303 "Vegetation responses following strip chemical control of sagebrush at the Lower Government Draw site." May 28, 1982. 11th Wyoming Shrub Ecology Field Tour, Lander, Wyoming. by H.G. Fisser.

2F2 820304 "Vegetation response to strip chemical control in the Government Draw area." May 27, 1982. 11th Wyoming Shrub Ecology Workshop. Lander, Wyoming. by S.K. Dean.

2F2 820305 "Numerical phenology scoring and use with quantitative variables." May 27, 1982. 11th Wyoming Shrub Ecology Workshop. Lander, Wyoming. by R.J. Kinucan.

2F2 820306 "Sagebrush and perennial grass dynamics at the Granite Mountain shrub control site." May 27, 1982. 11th Wyoming Shrub Ecology Workshop. Lander, Wyoming. by J.L. Mulholland.

2F2 830301 "Long-term vegetation responses to sagebrush control in western Wyoming." February 15, 1983. 36th Ann. Meeting Soc. Range Manage. Albuquerque, New Mexico. by S.K. Dean.

2F2 830303 "Perennial grass production dynamics among five vegetation types in north central Wyoming." February 16, 1983. 36th Ann. Meeting Soc. Range Manage. Albuquerque, New Mexico. by H.G. Fisser.

2F2 830304 "Phenological description and numerical scoring of sagebrush." February 16, 1983. 36th Ann. Meeting Soc. Range Manage. Albuquerque, New Mexico. by H.G. Fisser.

2F2 830307 "Reestablishment rates of sagebrush following chemical control on grazed and ungrazed ranges in western Wyoming." February 17, 1983. 36th Ann. Meeting Soc. Range Manage. Albuquerque, New Mexico. by J.L. Mulholland.

2F2 830308 "Size and density parameters of western wheatgrass on diverse range sites in western Wyoming." February 15, 1983. 36th Ann. Meeting Soc. Range Manage. Albuquerque, New Mexico. by H.G. Fisser.

2F2 830309 "Utah juniper invasion in north central Wyoming and historical climatic and management relationships." February 16, 1983. 36th Ann. Meeting Soc. Range Manage. Albuquerque, New Mexico. by H.G. Fisser.

2F2 830310 "*Atriplex*/grass and forb relationships under no grazing and shifting precipitation patterns in northcentral Wyoming." May 6, 1983. Wildland Shrub Symp. The biology of *Atriplex* and related chenopods. USDA Forest Service. Provo, Utah. by H.G. Fisser.

2F2 830311 "Grazing, vegetation and soils interactions since 1967 in the Powder Rim grazing system." May 26, 1983. 12th Wyoming Shrub Ecology Workshop. Baggs, Wyoming. by H.G. Fisser.

2F2 830312 "Deer browse impacts on Utah juniper and true mountain mahogany in the Poison Butte unit west of Baggs, Wyoming." May 26, 1983. 12th Wyoming Shrub Ecology Workshop. Baggs, Wyoming. by H.G. Fisser.

2F2 840301 "Detailed numerical phenology of sagebrush with visual description statements." February 14, 1984. 37th Ann. Meeting Soc. Range Manage. Rapid City, South Dakota. by N.E. Hargis and H.G. Fisser.

2F2 840303 "Estimation of missing precipitation data in Wyoming using traditional, kriging and best guess estimates." February 15, 1984. 37th Ann. Meeting Soc. Range Manage. Rapid City, South Dakota. by L.P. Germain.

2F2 840304 "Detailed numerical phenology of sagebrush with visual description statements." February 16, 1984. 37th Ann. Meeting Soc. Range Manage. Rapid City, South Dakota. by H.G. Fisser.

2F2 840306 "Application of kriging to ecological variables." April 27, 1984. 55th Ann. Meeting Colo.-Wyo. Acad. Sci. Gunnison, Colorado. by L.P. Germain.

2F2 840307 "Detailed numerical phenology of sagebrush with visual description statements." May 1, 1984. USDA-ARS/USDI-BLM Regional Workshop on Rangeland Hydrology, Modeling and Monitoring. Boise, Idaho. by H.G. Fisser.

2F2 840308 "Grazing management longevity and benefits of sagebrush control in western Wyoming." May 1, 1984. USDA-ARS/USDI-BLM Regional Workshop on Rangeland Hydrology, Modeling and Monitoring. Boise, Idaho. by H.G. Fisser.

2F2 840309 "Long term Wyoming rangeland monitoring and research cooperative studies of the USDI-BLM & University of Wyoming, Department of Range Management." May 1, 1984. USDA-ARS/USDI-BLM Regional Workshop on Rangeland Hydrology, Modeling and Monitoring. Boise, Idaho. by H.G. Fisser.

2F2 840310 "Herbage production and community dynamics." July 12, 1984. Shrub Research Consortium Third Ann. Wildland Shrub Symposium: the biology of *Artemisia* and *Chrysothamnus*. Provo, Utah. by H.G. Fisser.

2F2 840311 "Reinvasion following sagebrush control." July 12, 1984. Shrub Research Consortium Third Ann. Wildland Shrub Symposium: the biology of *Artemisia* and *Chrysothamnus*. Provo, Utah. by H.G. Fisser.

2F2 840312 "Regional variation of phenology." July 12, 1984. Shrub Research Consortium Third Ann. Wildland Shrub Symposium: the biology of *Artemisia* and *Chrysothamnus*. Provo, Utah. by H.G. Fisser.

2F2 840313 "Validation of a rangeland production model (ERHYM) for sagebrush sites." July 12, 1984. Shrub Research Consortium Third Ann. Wildland Shrub Symposium: the biology of *Artemisia* and *Chrysothamnus*. Provo, Utah. by J.R. Wight.

2F2 840314 "Herbage yield dynamics and seasonal precipitation relationships." July 12, 1984. Shrub Research Consortium Third Ann. Wildland Shrub Symposium: the biology of *Artemisia* and *Chrysothamnus*. Provo, Utah. by C.L. Hanson.

3. Range Resources

3A. RANGE INVENTORY AND SURVEY

3A 610302 "Theories and methods of range surveys." December 4, 1961. Wyo. Agr. Exp. Sta. Superintendents Conference. Laramie, Wyoming. by H.G. Fisser.

3A 630303 "Does weather influence rangeland potential?" September 10, 1963. Rangeland User Group. Worland, Wyoming. by H.G. Fisser.

3A 790302 "Soil-vegetation relationships on rangeland exclosures in the Grass Creek Planning Unit of north central Wyoming." September 5, 1979. Wyo. S.C.S. Soil Survey Work Planning Conference. Laramie, Wyoming. by H.G. Fisser.

3A 790303 "Arid rangeland sampling." September 27, 1979. University of Wyoming Zoo. Biol. Sampling Seminar. Laramie, Wyoming. by H.G. Fisser.

3A1. Site, Mapping, Remote Sensing

3A1 610301 "Range condition class survey in the Big Horn Basin of Wyoming." September 14, 1961. Range Weed Control Conf. Worland, Wyoming. by H.G. Fisser.

3A1 620301 "Concepts of range condition class survey." November 30, 1962. Wyoming Wool Growers Association Convention. Worland, Wyoming. by H.G. Fisser.

3A1 650301 "Vegetation changes from cold desert to subalpine communities." January 18, 1965. IX International Grassland Congress. Sao Paulo, Brazil. Presented in absentia by A. Ried.

3A1 720301 "Panchromatic, color, and infrared photography for detailed vegetation mapping." February 8, 1972. 25th Ann. Meeting Soc. Range Manage. Washington, D.C. by H.G. Fisser.

3A1 740302 "Remote sensing applied to the grassland biome." September 19, 1974. First International Congress of Ecology, The Hague, The Netherlands. by H.G. Fisser.

3A1 740304 "Relationships of range production and rehabilitation to soils and other environmental factors." August 22, 1974. Workshop of BLM Soils and Geology Specialists. Lander, Wyo. by H.G. Fisser.

3A1 750301 "Pre-disturbance vegetation inventory and ecological monitoring procedures for rangeland strip-mining." February 12, 1975. 28th Ann. Meeting Soc. Range Manage. Mexico City, Mexico. by H.G. Fisser.

3A1 750302 "Vegetation response to seismic motion from underground nuclear detonation." February 13, 1975. 28th Ann. Meeting Soc. Range Manage. Mexico City, Mexico. by H.G. Fisser.

3A1 750304 "Pre-disturbance ecological studies improve and define potential for surface mine reclamation." October 22, 1975. 3rd NCA/BCR Coal Conference and Expo II. Third Symposium on Surface Mining and Reclamation. Louisville, Kentucky. by H.G. Fisser.

3A1 760306 "Vegetation re-establishment potential of strip-mining soils, Black Thunder Basin, Wyoming." February 19, 1976. 29th Ann. Meeting Soc. Range Manage. Omaha, Nebraska. by G.A. Lymbery.

3A1 760307 "Methods of measuring environmental parameters on a pre-disturbed coal site." April 22, 1976. 47th Ann. Meeting Colo.-Wyo. Acad. Sci. Boulder, Colorado. by J.R. Cox.

3A1 760312 "Rangeland plants and plant identification." July 13, 1976. Wyoming OWRC/SRM Youth Range Camp. Tensleep, Wyoming. by H.G. Fisser.

3A1 780303 "Status of remote sensing in Wyoming." March 9, 1978. Annual Meeting GPC-7 Committee. Brownsville, Texas. by H.G. Fisser.

3A1 780305 "Plant succession on plains pocket gopher mounds." February 7, 1978. 31st Ann. Meeting Soc. Range Manage. San Antonio, Texas. by M.A. Foster.

3A1 780308 "Ecology of rangeland communities in northwestern Libya, Africa." August 15, 1978. 1st Intern. Rangeland Congress. Denver, Colorado. by O.M. Shawesh.

3A1 780310 "Rangeland vegetation, improvement and management in Libya, Africa." October 12, 1978. Intern. Arid Lands Conf. on Plant Resources. Texas Tech University. Lubbock, Texas. by O.M. Shawesh.

3A1 780311 "Range condition and trend review in the Lander Resource Area." December 11, 1978. Range survey report to Bureau Land Management. Cheyenne, Wyoming. by H.G. Fisser.

3A1 840309 "Long-term Wyoming rangeland monitoring and research cooperative studies of the USDI-BLM & University of Wyoming, Department of Range Management." May 1, 1984. USDA-ARS/USDI-BLM Regional Workshop on Rangeland Hydrology, Modeling and Monitoring. Boise, Idaho. by H.G. Fisser.

3A2. Condition and Trend

3A2 610301 "Range condition class survey in the Big Horn Basin of Wyoming." September 14, 1961. Range Weed Control Conf. Worland, Wyoming. by H.G. Fisser.

3A2 620301 "Concepts of range condition class survey." November 30, 1962. Wyoming Wool Growers Association Convention. Worland, Wyoming. by H.G. Fisser.

3A2 670303 "Plant control on salt desert shrub rangeland by sheep grazing." August 21, 1967. Range Weed Control Conf. Las Cruces, New Mexico. by H.G. Fisser.

3A2 680301 "Patterns of live and dead vegetation in grassland ecosystems and their relationship to function." September 24, 1968. I.B.P. Grasslands Biome Synthesis Meeting. Halsey, Nebraska. by H.G. Fisser.

3A2 690301 "Vegetation response to chemical control of low sagebrush *(Artemisia arbuscula)* in western Wyoming." August 26, 1969. Range Weed Control Conference. Reno, Nevada. by H.G. Fisser.

3A2 690304 "Soil moisture relationships of certain salt desert shrubs." February 12, 1969. 22nd Ann. Meeting Soc. Range Manage. Calgary, Alberta. by R.E. Steger.

3A2 700303 "*Opuntia* clump size, variation on the shortgrass plains-Pawnee Site, Nunn, Colorado." May 1, 1970. Colo.-Wyo. Acad. Sci., Greeley, Colorado. by H.G. Fisser.

3A2 700304 "The I.B.P. and its implications for natural resource management." June 24, 1970. Wyoming Game and Fish Commission. Biologists Meeting. Lander, Wyoming. by H.G. Fisser.

3A2 700305 "Significance of plant patterns in ecosystem functioning." November 14, 1970. I.B.P Grassland Biome Annual Review Meeting. Fort Collins, Colorado. by H.G. Fisser.

3A2 720301 "Panchromatic, color, and infrared photography for detailed vegetation mapping." February 8, 1972. 25th Ann. Meeting Soc. Range Manage. Washington, D.C. by H.G. Fisser.

3A2 730302 "Evaluation of furrowing and seeding in a north central Wyoming Nuttall saltbush area." February 6, 1973. 26th Ann. Meeting Soc. Range Manage. Boise, Idaho. by M.H. Mackey.

3A2 740302 "Remote sensing applied to the grassland biome." September 19, 1974. First International Congress of Ecology, The Hague, The Netherlands. by H.G. Fisser.

3A2 740304 "Relationships of range production and rehabilitation to soils and other environmental factors." August 22, 1974. Workshop of BLM Soils and Geology Specialists. Lander, Wyoming. by H.G. Fisser.

3A2 750301 "Pre-disturbance vegetation inventory and ecological monitoring procedures for rangeland strip mining." February 12, 1975. 28th Ann. Meeting Soc. Range Manage. Mexico City, Mexico. by H.G. Fisser.

3A2 750302 "Vegetation response to seismic motion from underground nuclear detonation." February 13, 1975. 28th Ann. Meeting Soc. Range Manage. Mexico City, Mexico. by H.G. Fisser.

3A2 750304 "Pre-disturbance ecological studies improve and define potential for surface mine reclamation." October 22, 1975. 3rd NCA/BCR Coal Conference and Expo II. Third Symposium on Surface Mining and Reclamation. Louisville, Kentucky. by H.G. Fisser.

3A2 750306 "Phenodynamics and environmental characteristics of sagebrush-grass rangelands." November 5, 1975. Wildland Shrubs Symp. Logan, Utah. by L.H. Kleinman.

3A2 760301 "Aridland ecology and phenology research in western Wyoming." March 3, 1976. Regional Rangeland Research Symposium. Bureau of Land Management. Denver, Colorado. by H.G. Fisser.

3A2 760303 "Aridland ecological research activities in Wyoming and relationships to Libyan rangelands." May 15, 1976. College of Agriculture, University Al Fateh special seminar. Tripoli, Libya. by H.G. Fisser.

3A2 760305 "Physical and chemical analyses of range sites and interrelationships with herbage production and forage value potentials." February 18, 1976. 29th Ann. Meeting Soc. Range Mange. Omaha, Nebraska. by F.K. Taha.

3A2 760306 "Vegetation re-establishment potential of strip-mining soils, Black Thunder Basin, Wyoming." February 19, 1976. 29th Ann. Meeting Soc. Range Manage. Omaha, Nebraska. by G.A. Lymbery.

3A2 760307 "Methods of measuring environmental parameters on a pre-disturbed coal site." April 22, 1976. 47th Ann. Meeting Colo.-Wyo. Acad. Sci. Boulder, Colorado. by J.R. Cox.

3A2 760310 "Production of a shortgrass prairie ecosystem as related to soils characteristics in eastern Wyoming." April 22, 1976. 47th Ann. Meeting Colo.-Wyo. Acad. Sci. Boulder, Colorado. by F.K. Taha.

3A2 760311 "Phenology and environmental relationships of sagebrush-grass rangelands." April 22, 1976. 47th Ann. Meeting Colo.-Wyo. Acad Sci. Boulder, Colorado. by L.H. Kleinman.

3A2 770304 "Long-term vegetation studies with permanent quadrats." February 16, 1972. 30th Ann. Meeting Soc. Range Manage. Portland, Oregon. by A.D. Gdara.

3A2 770306 "Phenodynamics and ecology of sagebrush-grass rangelands." February 15, 1977. 30th Ann. Meeting Soc. Range Manage. Portland, Oregon. by L.H. Kleinman.

3A2 770307 "Factors affecting growth and productivity of vegetation stratum in eastern Wyoming." April 23, 1977. 48th Ann. Meeting Colo.-Wyo. Acad. Sci. Colorado Springs, Colorado. by F.K. Taha.

3A2 780301 "Soil erosion and movement studies in western Wyoming." February 8, 1978. 31st Ann. Meeting Soc. Range Manage. San Antonio, Texas. by H.G. Fisser.

3A2 780302 "Models for prediction of natural vegetation potential in eastern Wyoming." February 8, 1978. 31st Ann. Meeting Soc. Range Manage. San Antonio, Texas. by H.G. Fisser.

3A2 780304 "Soil surface movement and relation to vegetation structure." June 15, 1978. Intern. Congress for Energy and the Ecosystem. University of North Dakota. Grand Forks, North Dakota. by H.G. Fisser.

3A2 780306 "Range management equals functional ecology." July 11, 1978. Wyoming OWRC/SRM Youth Range Camp. Tensleep, Wyoming. by H.G. Fisser.

3A2 780307 "Semiarid rangeland and grazing evaluation of shrubland from permanent quadrats." August 15, 1978. 1st Intern. Rangeland Congress. Denver, Colorado. by A.D. Gdara.

3A2 780311 "Range condition and trend review in the Lander Resource Area." December 11, 1978. Range survey report to Bureau Land Management. Cheyenne, Wyoming. by H.G. Fisser.

3A2 790301 "Long-term herbage production response following chemical sagebrush control in western Wyoming." February 13, 1979. 32nd Ann. Meeting Soc. Range Manage. Casper, Wyoming. by H.G. Fisser.

3A2 790302 "Soil-vegetation relationships on rangeland exclosures in the Grass Creek Planning Unit of north central Wyoming." September 5, 1979. Wyo. S.C.S. Soil Survey Work Planning Conference. Laramie, Wyoming. by H.G. Fisser.

3A2 790304 "Important rangeland plants of Wyoming and the West." September 5-November 21. 1979. Special 10-week lecture session course for BLM employees. University of Wyoming RgMgt 654M - Range Plant Identification and Ecology. Rock Springs, Wyoming. by H.G. Fisser.

3A2 790305 "Soils-phenology relations on some arid rangelands of western Wyoming." February 15, 1979. 32nd Ann. Meeting Soc. Range Manage. Casper, Wyoming. by L.W. Young III.

3A2 800301 "Shrub establishment, dominance and ecology on the juniper and sagebrush-grass types in Wyoming." December 3, 1980. Shrub Establishment Symposium. Wyoming Game and Fish Department. Laramie, Wyoming by H.G. Fisser.

3A2 800302 "Soil vegetation relationships on a semiarid rangeland exclosure in north central Wyoming: long term studies." February 14, 1980. 33rd Ann. Meeting Soc. Range Manage. San Diego, California. by D.C. Trueblood.

3A2 810301 "Forage production and longevity of sagebrush control on semiarid rangelands in western Wyoming." February 10, 1981. 34th Ann. Meeting Soc. Range Manage. Tulsa, Oklahoma. by W.E. Bensel.

3A2 810302 "Phenological comparison of two sagebrush species in Wyoming." February 11, 1981. 34th Ann. Meeting Soc. Range Manage. Tulsa, Oklahoma. by N.E. Hargis.

3A2 810303 "The effect of community changes on precipitation/production relationships in exclosures of saltbush vegetation." February 11, 1981. 34th Ann. Meeting Soc. Range Manage. Tulsa, Oklahoma. by L.A. Joyce.

3A2 810305 "Phenology and productivity comparisons of western wheatgrass *(Agropyron smithii)* and bluebunch wheatgrass *(A. spicatum).*" February 11, 1981. 34th Ann. Meeting Soc. Range Manage. Tulsa, Oklahoma. by M.K. Owens.

3A2 810307 "Plant composition response to picloram application in the Wyoming juniper type." February 12, 1981. 34th Ann. Meeting Soc. Range Manage. Tulsa, Oklahoma. by H.G. Fisser.

3A2 810308 "Soil-vegetation relationships in the Gardner saltbush vegetation types of Wyoming." February 10, 1981. 34th Ann. Meeting Soc. Range Manage. Tulsa, Oklahoma. by D.C. Trueblood.

3A2 810309 "Vegetation dynamics of non-grazed semiarid sagebrush grassland communities in the Big Horn Basin of Wyoming." February 11, 1981. 34th Ann. Meeting Soc. Range Manage. Tulsa, Oklahoma. by J.W. Uhlich.

3A2 810310 "Population structure of Utah juniper in north central Wyoming." February 12, 1981. 34th Ann. Meeting Soc. Range Manage. Tulsa, Oklahoma. by W.J. Waugh.

3A2 810315 "Phenology and culm weight variation between two wheatgrass species: *Agropyron smithii* and *A. spicatum.*" June 17, 1981. 14th Intern. Grassl. Congress. Lexington, Kentucky. by M.K. Owens.

3A2 820301 "Numerical phenology of big and black sagebrush from visual interpretation statements." April 25, 1982. Colorado-Wyoming Academy of Science. Fort Collins, Colorado. by H.G. Fisser.

3A2 820302 "Vegetation productivity and phenology in the Wind River Basin." May 27, 1982. 11th Wyoming Shrub Ecology Workshop. Lander, Wyoming. by H.G. Fisser.

3A2 820303 "Vegetation responses following strip chemical control of sagebrush at the Lower Government Draw site." May 28, 1982. 11th Wyoming Shrub Ecology Field Tour, Lander, Wyoming. by H.G. Fisser.

3A2 820304 "Vegetation response to strip chemical control in the Government Draw area." May 27, 1982. 11th Wyoming Shrub Ecology Workshop. Lander, Wyoming. by S.K. Dean.

3A2 820305 "Numerical phenology scoring and use with quantitative variables." May 27, 1982. 11th Wyoming Shrub Ecology Workshop. Lander, Wyoming. by R.J. Kinucan.

3A2 820306 "Sagebrush and perennial grass dynamics at the Granite Mountain shrub control site." May 27, 1982. 11th Wyoming Shrub Ecology Workshop. Lander, Wyoming. by J.L. Mulholland.

3A2 830301 "Long-term vegetation responses to sagebrush control in western Wyoming." February 15, 1983. 36th Ann. Meeting Soc. Range Manage. Albuquerque, New Mexico. by S.K. Dean.

3A2 830302 "Economic feasibility of resting rangeland to improve forage production in central Wyoming." February 16, 1983. 36th Ann. Meeting Soc. Range Manage. Albuquerque, New Mexico. by T.L. Deboodt.

3A2 830303 "Perennial grass production dynamics among five vegetation types in north central Wyoming." February 16, 1983. 36th Ann. Meeting Soc. Range Manage. Albuquerque, New Mexico. by H.G. Fisser.

3A2 830304 "Phenological description and numerical scoring of sagebrush." February 16, 1983. 36th Ann. Meeting Soc. Range Manage. Albuquerque, New Mexico. by H.G. Fisser.

3A2 830309 "Utah juniper invasion in north central Wyoming and historical climatic and management relationships." February 16, 1983. 36th Ann. Meeting Soc. Range Manage. Albuquerque, New Mexico. by H.G. Fisser.

3A2 840301 "Detailed numerical phenology of sagebrush with visual description statements." February 14, 1984. 37th Ann. Meeting Soc. Range Manage. Rapid City, South Dakota. by N.E. Hargis and H.G. Fisser.

3A2 840304 "Detailed numerical phenology of sagebrush with visual description statements." February 16, 1984. 37th Ann. Meeting Soc. Range Manage. Rapid City, South Dakota. by H.G. Fisser.

3A2 840307 "Detailed numerical phenology of sagebrush with visual description statements." May 1, 1984. USDA-ARS/USDI-BLM Regional Workshop on Rangeland Hydrology, Modeling and Monitoring. Boise, Idaho. by H.G. Fisser.

3A2 840309 "Long term Wyoming rangeland monitoring and research cooperative studies of the USDI-BLM & University of Wyoming, Department of Range Management." May 1, 1984. USDA-ARS/USDI-BLM Regional Workshop on Rangeland Hydrology, Modeling and Monitoring. Boise, Idaho. by H.G. Fisser.

3A2 840312 "Regional variation of phenology." July 12, 1984. Shrub Research Consortium Third Ann. Wildland Shrub Symposium: the biology of *Artemisia* and *Chrysothamnus.* Provo, Utah. by H.G. Fisser.

3A2 840313 "Validation of a rangeland production model (ERHYM) for sagebrush sites." July 12, 1984. Shrub Research Consortium Third Ann. Wildland Shrub Symposium: the biology of *Artemisia* and *Chrysothamnus.* Provo, Utah. by J.R. Wight.

3A3. Production and Utilization

3A3 630302 "Exclosure studies (Production, cover, height, and moisture phase)." March 19, 1963. Range Weed Control Conf. Portland, Oregon. by H.G. Fisser.

3A3 640302 "The Dry Creek Halogeton Pastures and response to winter sheep use." July 7, 1964. BLM Interagency Tour, Greybull, Wyoming. by H.G. Fisser.

3A3 650303 "Halogeton abundance and distribution as influenced by degree of saltbush utilization." September 14, 1965. Range Weed Control Conf. Elko, Nevada. by H.G. Fisser.

3A3 670305 "Cooperative research with the University of Wyoming." March 10, 1967. Bureau of Land Management Wyoming State Advisory Board Meeting. Riverton, Wyoming. by H.G. Fisser.

3A3 700306 "Predictive value of factors influencing perennial grass production on western Wyoming sagebrush-grass ranges." May 1, 1970. Colo.-Wyo. Acad. Sci. Greeley, Colorado. by G.L. Whysong.

3A3 720302 "Juniper ecology and response to protection from browsing." July 6, 1972. 1st Wyo. Shrub Ecology Workshop, Casper, Wyoming. by H.G. Fisser.

3A3 740301 "Productivity differences of semidesert grasslands in western Wyoming." February 5, 1974. 27th Ann. Meeting Soc. Range Manage. Tucson, Arizona. by H.G. Fisser.

3A3 740303 "Relationship with soils and other environmental factors to rangeland production and rehabilitation." August 22, 1974. Bureau of Land Management Workshop Conference of Soils and Geology Specialists. Lander, Wyoming. by H.G. Fisser.

3A3 790301 "Long-term herbage production response following chemical sagebrush control in western Wyoming." February 13, 1979. 32nd Ann. Meeting Soc. Range Manage. Casper, Wyoming. by H.G. Fisser.

3A3 800301 "Shrub establishment, dominance and ecology on the juniper and sagebrush-grass types in Wyoming." December 3, 1980. Shrub Establishment Symposium. Wyoming Game and Fish Department. Laramie, Wyoming. by H.G. Fisser.

3A3 810301 "Forage production and longevity of sagebrush control on semiarid rangelands in western Wyoming." February 10, 1981. 34th Ann. Meeting Soc. Range Manage. Tulsa, Oklahoma. by W.E. Bensel.

3A3 810303 "The effect of community changes on precipitation/ production relationships in exclosures of saltbush vegetation." February 11, 1981. 34th Ann. Meeting Soc. Range Manage. Tulsa, Oklahoma. by L.A. Joyce.

3A3 810311 "Correlation of soil factors and environmental conditions with phenophase development of western wheatgrass (*Agropyron smithii*)." February 10, 1981. 34th Ann. Meeting Soc. Range Manage. Tulsa, Oklahoma. by L.W. Young.

3A3 820303 "Vegetation responses following strip chemical control of sagebrush at the Lower Government Draw site." May 28, 1982. 11th Wyoming Shrub Ecology Field Tour. Lander, Wyoming. by H.G. Fisser.

3A3 820304 "Vegetation response to strip chemical control in the Government Draw area." May 27, 1982. 11th Wyoming Shrub Ecology Workshop. Lander, Wyoming. by S.K. Dean.

3A3 820306 "Sagebrush and perennial grass dynamics at the Granite Mountain shrub control site." May 27, 1982. 11th Wyoming Shrub Ecology Workshop. Lander, Wyoming. by J.L. Mulholland.

3A3 830301 "Long-term vegetation responses to sagebrush control in western Wyoming." February 15, 1983. 36th Ann. Meeting Soc. Range Manage. Albuquerque, New Mexico. by S.K. Dean.

3A3 830302 "Economic feasibility of resting rangeland to improve forage production in central Wyoming." February 16, 1983. 36th Ann. Meeting Soc. Range Manage. Albuquerque, New Mexico. by T.L. Deboodt.

3A3 830303 "Perennial grass production dynamics among five vegetation types in northcentral Wyoming." February 16, 1983. 36th Ann. Meeting Soc. Range Manage. Albuquerque, New Mexico. by H.G. Fisser.

3A3 830305 "Productivity differences of grasslands in semi-desert western Wyoming — A re-examination." February 16, 1983. 36th Ann. Meeting Soc. Range Manage. Albuquerque, New Mexico. by L.A. Joyce.

3A3 840301 "Detailed numerical phenology of sagebrush with visual description statements." February 14, 1984. 37th Ann. Meeting Soc. Range Manage. Rapid City, South Dakota. by N.E. Hargis and H.G. Fisser.

3A3 840302 "Productivity index procedures for sagebrush-grass rangeland." February 14, 1984. 37th Ann. Meeting Soc. Range Manage. Rapid City, South Dakota. by H.G. Fisser.

3A3 840305 "Culm weight dynamics as an index of sagebrush-grass community productivity." April 27, 1984. 55th Ann. Meeting Colo.-Wyo. Acad. Sci. Gunnison, Colorado. by H.G. Fisser.

3A3 840308 "Grazing management longevity and benefits of sagebrush control in western Wyoming." May 1, 1984. USDA-ARS/USDI-BLM Regional Workshop on Rangeland Hydrology, Modeling and Monitoring. Boise, Idaho. by H.G. Fisser.

3A3 840309 "Long term Wyoming rangeland monitoring and research cooperative studies of the USDI-BLM & University of Wyoming, Department of Range Management." May 1, 1984. USDA-ARS/USDI-BLM Regional Workshop on Rangeland Hydrology, Modeling and Monitoring. Boise, Idaho. by H.G. Fisser.

3A3 840310 "Herbage production and community dynamics." July 12, 1984. Shrub Research Consortium Third Ann. Wildland Shrub Symposium: the biology of *Artemisia* and *Chrysothamnus*. Provo, Utah. by H.G. Fisser.

3A3 840311 "Reinvasion following sagebrush control." July 12, 1984. Shrub Research Consortium Third Ann. Wildland Shrub Symposium: the biology of *Artemisia* and *Chrysothamnus*. Provo, Utah. by H.G. Fisser.

3A3 840314 "Herbage yield dynamics and seasonal precipitation relationships." July 12, 1984. Shrub Research Consortium Third Ann. Wildland Shrub Symposium: the biology of *Artemisia* and *Chrysothamnus*. Provo, Utah. by C.L. Hanson.

3A3 840315 "Productivity dynamics of sagebrush ecosystems as determined from biweekly sampling of a single grass and shrub species." August 9, 1984. Annual Meeting Ecological Society of America. Fort Collins, Colorado. by H.G. Fisser.

3B. NATURAL RESOURCES PLANNING

3B 780303 "Status of remote sensing in Wyoming." March 9, 1978. Annual Meeting GPC-7 Committee. Brownsville, Texas. by H.G. Fisser.

3C. PUBLIC LAND ADMINISTRATION

3C 780303 "Status of remote sensing in Wyoming." March 9, 1978. Annual Meeting GPC-7 Committee. Brownsville, Texas. by H.G. Fisser.

4. Grazing Management

4A. STOCKING RATES

4A 630301 "Progress of cattle and sheep grazing trials on halogeton infested ranges." March 19, 1963. Range Weed Control Conf. Portland, Oregon. by H.G. Fisser.

4A 640301 "Winter sheep grazing studies on saltbush rangeland of Wyoming." February 14, 1964. 17th Ann. Meeting Amer. Soc. of Range Manage. Wichita, Kansas. by H.G. Fisser.

4A 650302 "Comparison of rotation and season-long grazing on the Big Horn Mountains." July 22, 1965. Soil Conservation Society Annual Meeting, Wyoming Chapter. Big Horn Mountains. by H.G. Fisser.

4A 650303 "Halogeton abundance and distribution as influenced by degree of saltbush utilization." September 14, 1965. Range Weed Control Conf. Elko, Nevada. by H.G. Fisser.

4A 650305 "Grass seeded spreader systems can increase carrying capacity." May 26, 1965. Washakie County Agr. Ext. Service Range Improvement Conf. Worland, Wyoming. by H.G. Fisser.

4A 660302 "Sheep grazing intensity studies and vegetative changes on *Atriplex nuttallii.*" August 4, 1966. Symposium on management of Salt Desert Shrub Vegetation. Cedar City, Utah. by H.G. Fisser.

4A 660304 "Effect of sheep grazing intensities on vegetative change in Nuttall saltbush." December 3, 1966. Wyoming Section, Amer. Soc. of Range Manage. Annual Winter Meeting. Riverton, Wyoming. by H.G. Fisser.

4A 670302 "Influence of summer steer use under rotation and season-long grazing systems." July 29, 1967. Wyoming Section, Amer. Soc. of Range Manage. Annual Meeting. Big Horn Mountains. by H.G. Fisser.

4A 670306 "Western harvester ant abundance in relation to grazing, vegetation, and soils in north central Wyoming." February 14, 1967. 20th Ann. Meeting Amer. Soc. of Range Manage. Seattle, Washington. Feb.14, 1967. by D.R. Kirkham.

4A 830310 "*Atriplex*/grass and forb relationships under no grazing and shifting precipitation patterns in north central Wyoming." May 6, 1983. Wildland Shrub Symp. The biology of *Atriplex* and related chenopods. USDA Forest Service. Provo, Utah. by H.G. Fisser.

4B. GRAZING SEASONS AND SYSTEMS

4B 630301 "Progress of cattle and sheep grazing trials on halogeton infested ranges." March 19, 1963. Range Weed Control Conf. Portland, Oregon. by H.G. Fisser.

4B 640301 "Winter sheep grazing studies on saltbush rangeland of Wyoming." February 14, 1964. 17th Ann. Meeting Amer. Soc. of Range Manage. Wichita, Kansas. by H.G. Fisser.

4B 640302 "The Dry Creek Halogeton Pastures and response to winter sheep use." July 7, 1964. BLM Interagency Tour, Greybull, Wyoming. by H.G. Fisser.

4B 640303 "Summer steer grazing on saltbush rangeland" November 20, 1964. Wyoming Section American Society of Range Management Annual Winter Meeting. Casper, Wyoming. by H.G. Fisser.

4B 650302 "Comparison of rotation and season-long grazing on the Big Horn Mountains." July 22, 1965. Soil Conservation Society Annual Meeting, Wyoming Chapter. Big Horn Mountains. by H.G. Fisser.

4B 660302 "Sheep grazing intensity studies and vegetative changes on *Atriplex nuttallii.*" August 4, 1966. Symposium on management of Salt Desert Shrub Vegetation. Cedar City, Utah. by H.G. Fisser.

4B 660304 "Effect of sheep grazing intensities on vegetative change in Nuttall saltbush." December 3, 1966. Wyoming Section, American Society of Range Management Annual Winter Meeting. Riverton, Wyoming. by H.G. Fisser.

4B 670302 "Influence of summer steer use under rotation and season-long grazing systems." July 29, 1967. Wyoming Section, American Society of Range Management Annual Meeting. Big Horn Mountains. by H.G. Fisser.

4B 760302 "Grazing systems research on northern desert shrub rangelands of Wyoming." March 11, 1976. GP-9 Committee Annual Meeting. Denver, Colorado. by H.G. Fisser.

4B 760309 "Some influences of grazing systems upon vegetation and wildlife." April 22, 1976. 47th Ann. Meeting Colo.-Wyo. Acad Sci. Boulder, Colorado. by W.A. Gerhart.

4B 770302 "An evaluation of three grazing management systems on northern desert shrub vegetation of southern Wyoming." February 17, 1977. 30th Ann Meeting Soc. Range Manage. Portland, Oregon. by H.G. Fisser.

4B 800301 "Shrub establishment, dominance and ecology on the juniper and sagebrush-grass types in Wyoming." December 3, 1980. Shrub Establishment Symposium. Wyoming Game and Fish Department. Laramie, Wyoming. by H.G. Fisser.

4B 800304 "Vegetational response to grazing systems in Wyoming's Red Desert." April 25, 1980. 51st Ann. Meeting Colo.-Wyo. Acad. Sci. Denver, Colorado. by D.E. Rodgers.

4B 810306 "Grazing system impacts on semiarid vegetation of Wyoming." February 12, 1981. 34th Ann. Meeting Soc. Range Manage. Tulsa, Oklahoma. by D.E. Rodgers.

4B 810316 "Vegetative response to grazing systems management in Wyoming's Red Desert." June 18, 1981. 14th Intern. Grassl. Congress. Lexington, Kentucky. by D.E. Rodgers.

4B 830302 "Economic feasibility of resting rangeland to improve forage production in central Wyoming." February 16, 1983. 36th Ann. Meeting Soc. Range Manage. Albuquerque, New Mexico. by T.L. Deboodt.

4B 830310 "*Atriplex*/grass and forb relationships under no grazing and shifting precipitation patterns in north central Wyoming." May 6, 1983. Wildland Shrub Symp. The biology of *Atriplex* and related chenopods. USDA Forest Service. Provo, Utah. by H.G. Fisser.

4B 830311 "Grazing, vegetation and soils interactions since 1967 in the Powder Rim grazing system." May 26, 1983. 12th Wyoming Shrub Ecology Workshop. Baggs, Wyoming. by H.G. Fisser.

4B 840308 "Grazing management longevity and benefits of sagebrush control in western Wyoming." May 1, 1984. USDA-ARS/USDI-BLM Regional Workshop on Rangeland Hydrology, Modeling and Monitoring. Boise, Idaho. by H.G. Fisser.

4C. GRAZING HABITS AND DISTRIBUTION

4C 760302 "Grazing systems research on northern desert shrub rangelands of Wyoming." March 11, 1976. GP-9 Committee Annual Meeting. Denver, Colorado. by H.G. Fisser.

4C 760309 "Some influences of grazing systems upon vegetation and wildlife." April 22, 1976. 47th Ann. Meeting Colo.-Wyo. Acad Sci. Boulder, Colorado. by W.A. Gerhart.

4C 770302 "An evaluation of three grazing management systems on northern desert shrub vegetation of southern Wyoming." February 17, 1977. 30th Ann Meeting Soc. Range Manage. Portland, Oregon. by H.G. Fisser.

4C 800301 "Shrub establishment, dominance and ecology on the juniper and sagebrush-grass types in Wyoming." December 3, 1980. Shrub Establishment Symposium. Wyoming Game and Fish Department. Laramie, Wyo. by H.G. Fisser.

4C 800304 "Vegetational response to grazing systems in Wyoming's Red Desert." April 25, 1980. 51st Ann. Meeting Colo.-Wyo. Acad. Sci. Denver, Colorado. by D.E. Rodgers.

4C 810306 "Grazing system impacts on semiarid vegetation of Wyoming." February 12, 1981. 34th Ann. Meeting Soc. Range Manage. Tulsa, Oklahoma. by D.E. Rodgers.

4C 810316 "Vegetative response to grazing systems management in Wyoming's Red Desert." June 18, 1981. 14th Intern. Grassl. Congress. Lexington, Kentucky. by D.E. Rodgers.

4C 830311 "Grazing, vegetation and soils interactions since 1967 in the Powder Rim grazing system." May 26, 1983. 12th Wyoming Shrub Ecology Workshop. Baggs, Wyoming. by H.G. Fisser.

4C 830312 "Deer browse impacts on Utah juniper and true mountain mahogany in the Poison Butte unit west of Baggs, Wyoming." May 26, 1983. 12th Wyoming Shrub Ecology Workshop. Baggs, Wyoming. by H.G. Fisser.

5. Range and Ranch Development

5A. PLANT CONTROL

5A1. Methods of Control

5A1 760301 "Aridland ecology and phenology research in western Wyoming." March 3, 1976. Regional Rangeland Research Symposium. Bureau of Land Management. Denver, Colorado. by H.G. Fisser.

5A1B. BIOLOGICAL

5A1B 660303 "Effect of control of harvester ants on revegetation." November 11, 1966. Wyoming Woolgrowers Association Annual Meeting. Casper, Wyoming. by H.G. Fisser.

5A1B 670303 "Plant control on salt desert shrub rangeland by sheep grazing." August 21, 1967. Range Weed Control Conf. Las Cruces, New Mexico. by H.G. Fisser.

5A1C. MECHANICAL

5A1C 640303 "Summer steer grazing on saltbush rangeland." Nov. 20, 1964. Wyoming Section American Society of Range Management Annual Winter Meeting. Casper, Wyoming. by H.G. Fisser.

5A1C800301 "Shrub establishment, dominance and ecology on the juniper and sagebrush-grass types in Wyoming." December 3, 1980. Shrub Establishment Symposium. Wyoming Game and Fish Department. Laramie, Wyoming. by H.G. Fisser.

5A1D. CHEMICAL

5A1D630302 "Exclosure studies (production, cover, height, and moisture phase)." March 19, 1963. Range Weed Control Conf. Portland, Oregon. by H.G. Fisser.

5A1D650304 "Exclosure studies - Responses to sagebrush spraying." November 20, 1965. Wyoming Section, American Society of Range Management Annual Winter Meeting. Worland, Wyoming. by H.G. Fisser.

5A1D670304 "Use of picloram for juniper control." August 21, 1967. Range Weed Control Conf. Las Cruces, New Mexico. by H.G. Fisser.

5A1D690301 "Vegetation response to chemical control of low sagebrush *(Artemisia arbuscula)* in western Wyoming." August 26, 1969. Range Weed Control Conference. Reno, Nevada. by H.G. Fisser.

5A1D690302 "Juniper game range improvement studies in western Wyoming." November 18, 1969. Wyoming Game and Fish Commission Biologists Meeting. Powell, Wyoming. by H.G. Fisser.

5A1D690303 "Improvement studies on juniper sites in western Wyoming." November 21, 1969. American Society of Range Management, Wyoming Section, Annual Winter Meeting. Sheridan, Wyoming. by H.G. Fisser.

5A1D710301 "A greenhouse study of variance of picloram retention in soil and duff from juniper sites." April 30, 1971. Colo.-Wyo. Acad. Sci. Pueblo, Colorado. by H.G. Fisser.

5A1D730302 "Evaluation of furrowing and seeding in a north central Wyoming Nuttall saltbush area." February 6, 1973. 26th Ann. Meeting Soc. Range Manage. Boise, Idaho. by M.H. Mackey.

5A1D760311 "Phenology and environmental relationships of sagebrush-grass rangelands." April 22, 1976. 47th Ann. Meeting Colo.-Wyo. Acad Sci. Boulder, Colorado. by L.H. Kleinman.

5A1D770306 "Phenodynamics and ecology of sagebrush-grass rangelands." February 15, 1977. 30th Ann. Meeting Soc. Range Manage. Portland, Oregon. by L.H. Kleinman.

5A1D790301 "Long-term herbage production response following chemical sagebrush control in western Wyoming." February 13, 1979. 32nd Ann. Meeting Soc. Range Manage. Casper, Wyoming. by H.G. Fisser.

5A1D800301 "Shrub establishment, dominance and ecology on the juniper and sagebrush-grass types in Wyoming." December 3, 1980. Shrub Establishment Symposium. Wyoming Game and Fish Department. Laramie, Wyoming. by H.G. Fisser.

5A1D810301 "Forage production and longevity of sagebrush control on semiarid rangelands in western Wyoming." February 10, 1981. 34th Ann. Meeting Soc. Range Manage. Tulsa, Oklahoma. by W.E. Bensel.

5A1D810307 "Plant composition response to picloram application in the Wyoming juniper type." February 12, 1981. 34th Ann. Meeting Soc. Range Manage. Tulsa, Oklahoma. by H.G. Fisser.

5A1D820302 "Vegetation productivity and phenology in the Wind River Basin." May 27, 1982. 11th Wyoming Shrub Ecology Workshop. Lander, Wyoming. by H.G. Fisser.

5A1D820303 "Vegetation responses following strip chemical control of sagebrush at the Lower Government Draw site." May 28, 1982. 11th Wyoming Shrub Ecology Field Tour. Lander, Wyoming. by H.G. Fisser.

5A1D820304 "Vegetation response to strip chemical control in the Government Draw area." May 27, 1982. 11th Wyoming Shrub Ecology Workshop. Lander, Wyoming. by S.K. Dean.

5A1D820306 "Sagebrush and perennial grass dynamics at the Granite Mountain shrub control site." May 27, 1982. 11th Wyoming Shrub Ecology Workshop. Lander, Wyoming. by J.L. Mulholland.

5A1D830301 "Long term vegetation responses to sagebrush control in western Wyoming." February 15, 1983. 36th Ann. Meeting Soc. Range Manage. Albuquerque, New Mexico. by S.K. Dean.

5A1D830302 "Economic feasibility of resting rangeland to improve forage production in central Wyoming." February 16, 1983. 36th Ann. Meeting Soc. Range Manage. Albuquerque, New Mexico. by T.L. Deboodt.

5A1D830307 "Reestablishment rates of sagebrush following chemical control on grazed and ungrazed ranges in western Wyoming." February 17, 1983. 36th Ann. Meeting Soc. Range Manage. Albuquerque, New Mexico. by J.L. Mulholland.

5A1D840308 "Grazing management longevity and benefits of sagebrush control in western Wyoming." May 1, 1984. USDA-ARS/USDI-BLM Regional Workshop on Rangeland Hydrology, Modeling and Monitoring. Boise, Idaho. by H.G. Fisser.

5A1D840311 "Reinvasion following sagebrush control." July 12, 1984. Shrub Research Consortium Third Ann. Wildland Shrub Symposium: the biology of *Artemisia* and *Chrysothamnus*. Provo, Utah. by H.G. Fisser.

5A1E. PYRIC

5A1E730301 "Methods of revegetation and restoration following a juniper wildfire." February 6, 1973. 26th Ann. Meeting Soc. Range Manage. Boise, Idaho. by D.L. Layton.

5A1E800301 "Shrub establishment, dominance and ecology on the juniper and sagebrush-grass types in Wyoming." December 3, 1980. Shrub Establishment Symposium. Wyoming Game and Fish Department. Laramie, Wyo. by H.G. Fisser.

5A2. Regional Control

5A2D. INTERMOUNTAIN

5A2D630301 "Progress of cattle and sheep grazing trials on halogeton infested ranges." March 19, 1963. Range Weed Control Conf. Portland, Oregon. by H.G. Fisser.

5A2D810301 "Forage production and longevity of sagebrush control on semiarid rangelands in western Wyoming." February 10, 1981. 34th Ann. Meeting Soc. Range Manage. Tulsa, Oklahoma. by W.E. Bensel.

5B. RANGE REVEGETATION

5B 630302 "Exclosure studies (production, cover, height, and moisture phase)." March 19, 1963. Range Weed Control Conf. Portland, Oregon. by H.G. Fisser.

5B 780310 "Rangeland vegetation, improvement and management in Libya, Africa." October 12, 1978. Intern. Arid Lands Conf. on Plant Resources. Texas Tech University. Lubbock, Texas. by O.M. Shawesh.

5B4. Intermountain

5B4 640303 "Summer steer grazing on saltbush rangeland" Nov. 20, 1964. Wyoming Section American Society of Range Management Annual Winter Meeting. Casper, Wyoming. by H.G. Fisser.

5B4 650305 "Grass seeded spreader systems can increase carrying capacity." May 26, 1965. Washakie County Agr. Ext. Service Range Improvement Conf. Worland, Wyoming. by H.G. Fisser.

5B4 730301 "Methods of revegetation and restoration following a juniper wildfire." February 6, 1973. 26th Ann. Meeting Soc. Range Manage. Boise, Idaho. by D.L. Layton.

5B4 730302 "Evaluation of furrowing and seeding in a north central Wyoming Nuttall saltbush area." February 6, 1973. 26th Ann. Meeting Soc. Range Manage. Boise, Idaho. by M.H. Mackey.

5B4 730303 "Revegetation of pipeline areas." April 30, 1973. Colo.-Wyo. Acad. Sci. Ft. Collins, Colorado. by T.E. Bedell.

5B4 740303 "Relationship with soils and other environmental factors to rangeland production and rehabilitation." August 22, 1974. Bureau of Land Management Workshop Conference of Soils and Geology Specialists. Lander, Wyoming. by H.G. Fisser.

5B4 790301 "Long-term herbage production response following chemical sagebrush control in western Wyoming." February 13, 1979. 32nd Ann. Meeting Soc. Range Manage. Casper, Wyoming. by H.G. Fisser.

5B4 800301 "Shrub establishment, dominance and ecology on the juniper and sagebrush-grass types in Wyoming." December 3, 1980. Shrub Establishment Symposium. Wyoming Game and Fish Department. Laramie, Wyoming. by H.G. Fisser.

5B4 820303 "Vegetation responses following strip chemical control of sagebrush at the Lower Government Draw site." May 28, 1982. 11th Wyoming Shrub Ecology Field Tour. Lander, Wyoming. by H.G. Fisser.

5B4 820304 "Vegetation response to strip chemical control in the Government Draw area." May 27, 1982. 11th Wyoming Shrub Ecology Workshop. Lander, Wyoming. by S.K. Dean.

5B4 820306 "Sagebrush and perennial grass dynamics at the Granite Mountain shrub control site." May 27, 1982. 11th Wyoming Shrub Ecology Workshop. Lander, Wyoming. by J.L. Mulholland.

5B4 830301 "Long-term vegetation responses to sagebrush control in western Wyoming." February 15, 1983. 36th Ann. Meeting Soc. Range Manage. Albuquerque, New Mexico. by S.K. Dean.

5C. WATER SPREADING, FURROWING, PITTING

5C 630302 "Exclosure studies (production, cover, height, and moisture phase)." March 19, 1963. Range Weed Control Conf. Portland, Oregon. by H.G. Fisser.

5C 640303 "Summer steer grazing on saltbush rangeland" Nov. 20, 1964. Wyoming Section American Society of Range Management Annual Winter Meeting. Casper, Wyoming. by H.G. Fisser.

5C 650305 "Grass seeded spreader systems can increase carrying capacity." May 26, 1965. Washakie County Agr. Ext. Service Range Improvement Conf. Worland, Wyoming. by H.G. Fisser.

5C 730301 "Methods of revegetation and restoration following a juniper wildfire." February 6, 1973. 26th Ann. Meeting Soc. Range Manage. Boise, Idaho. by D.L. Layton.

5C 730302 "Evaluation of furrowing and seeding in a northcentral Wyoming Nuttall saltbush area." February 6, 1973. 26th Ann. Meeting Soc. Range Manage. Boise, Idaho. by M.H. Mackey.

5C 730303 "Revegetation of pipeline areas." April 30, 1973. Colo.-Wyo. Acad. Sci. Ft. Collins, Colorado. by T.E. Bedell.

5C 760303 "Aridland ecological research activities in Wyoming and relationships to Libyan rangelands." May 15, 1976. College of Agriculture, University Al Fateh special seminar. Tripoli, Libya. by H.G. Fisser.

5C 770305 "The effect of environmental factors on carbohydrate storage and herbage production of five shortgrass prairie plant species." February 16, 1977. 30th Ann. Meeting Soc. Range Manage. Portland, Oregon. by J.R. Cox.

5C 780309 "The effects of water stress on phenology and carbohydrate storage in the shortgrass prairie." October 11, 1978. Intern. Arid. Lands Conf. on Plant Resources. Texas Tech University. Lubbock, Texas. by J.R. Cox.

5C 800301 "Shrub establishment, dominance and ecology on the juniper and sagebrush-grass types in Wyoming." December 3, 1980. Shrub Establishment Symposium. Wyoming Game and Fish Department. Laramie, Wyoming. by H.G. Fisser.

6. Range Livestock

6A. GENERAL MANAGEMENT

6A 800303 "Vegetation stagnation and decadence in a southeastern Wyoming big game exclosure: a preliminary report." April 25, 1980. 51st Ann. Meeting Colo.-Wyo. Acad. Sci. Denver, Colorado. by W.J. Waugh.

6A 810314 "Stagnation and decadency of *Cercocarpus montanus* in a southeastern Wyoming big game exclosure." May 28, 1981. 10th Wyoming Shrub Ecology Workshop. Sybille Canyon, Wyoming. by W.J. Waugh.

6A 830302 "Economic feasibility of resting rangeland to improve forage production in central Wyoming." February 16, 1983. 36th Ann. Meeting Soc. Range Manage. Albuquerque, New Mexico. by T.L. Deboodt.

6A1. Cattle

6A1 630301 "Progress of cattle and sheep grazing trials on halogeton infested ranges." March 19, 1963. Range Weed Control Conf. Portland, Oregon. by H.G. Fisser.

6A1 640303 "Summer steer grazing on saltbush rangeland." November 20, 1964. Wyoming Section American Society of Range Management Annual Winter Meeting. Casper, Wyoming. by H.G. Fisser.

6A1 670302 "Influence of summer steer use under rotation and season-long grazing systems." July 29, 1967. Wyoming Section, American Society of Range Management Annual Meeting. Big Horn Mountains. by H.G. Fisser.

6A1 670305 "Cooperative research with the University of Wyoming." March 10, 1967. Bureau of Land Management Wyoming State Advisory Board Meeting. Riverton, Wyoming. by H.G. Fisser.

6A1 810304 "Esophageal fistulated steers for determination of dietary nutrient composition on large pasture native rangeland." February 11, 1981. 34th Ann. Meeting Soc. Range Manage. Tulsa, Oklahoma. by G.R. O'Connor.

6A2. Sheep and Goats

6A2 630301 "Progress of cattle and sheep grazing trials on halogeton infested ranges." March 19, 1963. Range Weed Control Conf. Portland, Oregon. by H.G. Fisser.

6A2 640301 "Winter sheep grazing studies on saltbush rangeland of Wyoming." February 14, 1964. 17th Ann. Meeting Amer. Soc. of Range Manage. Wichita, Kansas. by H.G. Fisser.

6A2 640302 "The Dry Creek Halogeton Pastures and response to winter sheep use." July 7, 1964. BLM Interagency Tour, Greybull, Wyoming. by H.G. Fisser.

6A2 660302 "Sheep grazing intensity studies and vegetative changes on *Atriplex nuttallii.*" August 4, 1966. Symposium on management of Salt Desert Shrub Vegetation. Cedar City, Utah. by H.G. Fisser.

6A2 660304 "Effect of sheep grazing intensities on vegetative change in Nuttall saltbush." December 3, 1966. Wyoming Section, American Society of Range Management Annual Winter Meeting. Riverton, Wyoming. by H.G. Fisser.

6A2 670305 "Cooperative research with the University of Wyoming." March 10, 1967. Bureau of Land Management Wyoming State Advisory Board Meeting. Riverton, Wyoming. by H.G. Fisser.

6A2 810304 "Esophageal fistulated steers for determination of dietary nutrient composition on large pasture native rangeland." February 11, 1981. 34th Ann. Meeting Soc. Range Manage. Tulsa, Oklahoma. by G.R. O'Connor.

6A2 830310 "*Atriplex*/grass and forb relationships under no grazing and shifting precipitation patterns in north central Wyoming." May 6, 1983. Wildland Shrub Symp. The biology of *Atriplex* and related chenopods. USDA Forest Service. Provo, Utah. by H.G. Fisser.

6A3. Horses

6A3 760302 "Grazing systems research on northern desert shrub rangelands of Wyoming." March 11, 1976. GP-9 Committee Annual Meeting. Denver, Colorado. by H.G. Fisser.

6A3 760304 "Vegetative response of *Bouteloua gracilis* to four levels of utilization." February 18, 1976. 29th Ann. Meeting Soc. Range Manage. Omaha, Nebraska. by D.W. Stroud.

6A3 760309 "Some influences of grazing systems upon vegetation and wildlife." April 22, 1976. 47th Ann. Meeting Colo.-Wyo. Acad Sci. Boulder, Colorado. by W.A. Gerhart.

6A3 770302 "An evaluation of three grazing management systems on northern desert shrub vegetation of southern Wyoming." February 17, 1977. 30th Ann Meeting Soc. Range Manage. Portland, Oregon. by H.G. Fisser.

6A3 800304 "Vegetational response to grazing systems in Wyoming's Red Desert." April 25, 1980. 51st Ann. Meeting Colo.-Wyo. Acad. Sci. Denver, Colorado. by D.E. Rodgers.

6A3 810306 "Grazing system impacts on semiarid vegetation of Wyoming." February 12, 1981. 34th Ann. Meeting Soc. Range Manage. Tulsa, Oklahoma. by D.E. Rodgers.

6A3 810316 "Vegetative response to grazing systems management in Wyoming's Red Desert." June 18, 1981. 14th Intern. Grassl. Congress. Lexington, Kentucky. by D.E. Rodgers.

6A3 830311 "Grazing, vegetation and soils interactions since 1967 in the Powder Rim grazing system." May 26, 1983. 12th Wyoming Shrub Ecology Workshop. Baggs, Wyoming. by H.G. Fisser.

6A3 830312 "Deer browse impacts on Utah juniper and true mountain mahogany in the Poison Butte unit west of Baggs, Wyoming." May 26, 1983. 12th Wyoming Shrub Ecology Workshop. Baggs, Wyoming. by H.G. Fisser.

6B. NUTRITION AND FEEDING

6B2. Nutrient Composition

6B2 810304 "Esophageal fistulated steers for determination of dietary nutrient composition on large pasture native rangeland." February 11, 1981. 34th Ann. Meeting Soc. Range Manage. Tulsa, Oklahoma. by G.R. O'Connor.

7. Range Wildlife

7A. BIG GAME

7A 690302 "Juniper game range improvement studies in western Wyoming." November 18, 1969. Wyoming Game and Fish Commission Biologists Meeting, Powell, Wyoming. by H.G. Fisser.

7A 690303 "Improvement studies on juniper sites in western Wyoming." November 21, 1969. American Society of Range Management, Wyoming Section, Annual Winter Meeting. Sheridan, Wyoming. by H.G. Fisser.

7A 710301 "A greenhouse study of variance of picloram retention in soil and duff from juniper sites." April 30, 1971. Colo.-Wyo. Acad. Sci. Pueblo, Colorado. by H.G. Fisser.

7A 830312 "Deer browse impacts on Utah juniper and true mountain mahogany in the Poison Butte unit west of Baggs, Wyoming." May 26, 1983. 12th Wyoming Shrub Ecology Workshop. Baggs, Wyoming. by H.G. Fisser.

7A1. Deer

7A1 760302 "Grazing systems research on northern desert shrub rangelands of Wyoming." March 11, 1976. GP-9 Committee Annual Meeting. Denver, Colorado. by H.G. Fisser.

7A1 760309 "Some influences of grazing systems upon vegetation and wildlife." April 22, 1976. 47th Ann. Meeting Colo.-Wyo. Acad Sci. Boulder, Colorado. by W.A. Gerhart.

7A1 770302 "An evaluation of three grazing management systems on northern desert shrub vegetation of southern Wyoming." February 17, 1977. 30th Ann Meeting Soc. Range Manage. Portland, Oregon. by H.G. Fisser.

7A1 800301 "Shrub establishment, dominance and ecology on the juniper and sagebrush-grass types in Wyoming." December 3, 1980. Shrub Establishment Symposium. Wyoming Game and Fish Department. Laramie, Wyoming. by H.G. Fisser.

7A1 800303 "Vegetation stagnation and decadence in a southeastern Wyoming big game exclosure: a preliminary report." April 25, 1980. 51st Ann. Meeting Colo.-Wyo. Acad. Sci. Denver, Colorado. by W.J. Waugh.

7A1 800304 "Vegetational response to grazing systems in Wyoming's Red Desert." April 25, 1980. 51st Ann. Meeting Colo.-Wyo. Acad. Sci. Denver, Colorado. by D.E. Rodgers.

7A1 810306 "Grazing system impacts on semiarid vegetation of Wyoming." February 12, 1981. 34th Ann. Meeting Soc. Range Manage. Tulsa, Oklahoma. by D.E. Rodgers.

7A1 810313 "Some recent research findings in the true mountain mahogany types." May 28, 1981. 10th Wyoming Shrub Ecology Workshop. Sybille Canyon, Wyoming. by H.G. Fisser.

7A1 810316 "Vegetative response to grazing systems management in Wyoming's Red Desert." June 18, 1981. 14th Intern. Grassl. Congress. Lexington, Kentucky. by D.E. Rodgers.

7A1 830311 "Grazing, vegetation and soils interactions since 1967 in the Powder Rim grazing system." May 26, 1983. 12th Wyoming Shrub Ecology Workshop. Baggs, Wyoming. by H.G. Fisser.

7A1 830312 "Deer browse impacts on Utah juniper and true mountain mahogany in the Poison Butte unit west of Baggs, Wyoming." May 26, 1983. 12th Wyoming Shrub Ecology Workshop. Baggs, Wyoming. by H.G. Fisser.

7A2. Elk

7A2 680302 "Dietary preferences and utilization of elk on spring range in the Big Horn Mountains of Wyoming." February 14, 1968. 21st Ann. Meeting Soc. Range Manage. Albuquerque, New Mexico. by G.E. Probasco.

7A2 800303 "Vegetation stagnation and decadence in a southeastern Wyoming big game exclosure: a preliminary report." April 25, 1980. 51st Ann. Meeting Colo.-Wyo. Acad. Sci. Denver, Colorado. by W.J. Waugh.

7A3. Antelope

7A3 760302 "Grazing systems research on northern desert shrub rangelands of Wyoming." March 11, 1976. GP-9 Committee Annual Meeting. Denver, Colorado. by H.G. Fisser.

7A3 760309 "Some influences of grazing systems upon vegetation and wildlife." April 22, 1976. 47th Ann. Meeting Colo.-Wyo. Acad Sci. Boulder, Colorado. by W.A. Gerhart.

7A3 770302 "An evaluation of three grazing management systems on northern desert shrub vegetation of southern Wyoming." February 17, 1977. 30th Ann Meeting Soc. Range Manage. Portland, Oregon. by H.G. Fisser.

7A3 800304 "Vegetational response to grazing systems in Wyoming's Red Desert." April 25, 1980. 51st Ann. Meeting Colo.-Wyo. Acad. Sci. Denver, Colorado. by D.E. Rodgers.

7A3 810306 "Grazing system impacts on semi-arid vegetation of Wyoming." February 12, 1981. 34th Ann. Meeting Soc. Range Manage. Tulsa, Oklahoma. by D.E. Rodgers.

7A3 810316 "Vegetative response to grazing systems management in Wyoming's Red Desert." June 18, 1981. 14th Intern. Grassl. Congress. Lexington, Kentucky. by D.E. Rodgers.

7A3 830311 "Grazing, vegetation and soils interactions since 1967 in the Powder Rim grazing system." May 26, 1983. 12th Wyoming Shrub Ecology Workshop. Baggs, Wyoming. by H.G. Fisser.

7A3 830312 "Deer browse impacts on Utah juniper and true mountain mahogany in the Poison Butte unit west of Baggs, Wyoming." May 26, 1983. 12th Wyoming Shrub Ecology Workshop. Baggs, Wyoming. by H.G. Fisser.

7D. NON-GAME

7D1. Rodents and Rabbits

7D1 780305 "Plant succession on plains pocket gopher mounds." February 7, 1978. 31st Ann. Meeting Soc. Range Manage. San Antonio, Texas. by M.A. Foster.

8. Range Watersheds

8B. WATERSHED SOIL EROSION AND SEDIMENTATION

8B 640304 "Soil-vegetation relations of the 15-Mile Drainage area in the Big Horn Basin of Wyoming." February 14, 1964. Amer. Soc. Range Manage. Ann. Meeting. Wichita, Kansas. by J.T. Nichols.

8B 750305 "A study of rangeland soil movement characteristics." August 12, 1975. Amer. Soc. Civil Engineer. Watershed Manage. Symp. Logan, Utah. by H.G. Fisser.

8B 760305 "Physical and chemical analyses of range sites and interrelationships with herbage production and forage value potentials." February 18, 1976. 29th Ann. Meeting Soc. Range Mange. Omaha, Nebraska. by F.K. Taha.

8B 760310 "Production of a shortgrass prairie ecosystem as related to soils characteristics in eastern Wyoming." April 22, 1976. 47th Ann. Meeting Colo.-Wyo. Acad. Sci. Boulder, Colorado. by F.K. Taha.

8B 780301 "Soil erosion and movement studies in western Wyoming." February 8, 1978. 31st Ann. Meeting Soc. Range Manage. San Antonio, Texas. by H.G. Fisser.

8B 780304 "Soil surface movement and relation to vegetation structure." June 15, 1978. Intern. Congress for Energy and the Ecosystem. University of North Dakota. Grand Forks, North Dakota. by H.G. Fisser.

8B 790302 "Soil-vegetation relationships on rangeland exclosures in the Grass Creek Planning Unit of northcentral Wyoming." September 5, 1979. Wyo. S.C.S. Soil Survey Work Planning Conference. Laramie, Wyoming. by H.G. Fisser.

8B 800302 "Soil vegetation relationships on a semiarid rangeland exclosure in north central Wyoming: long term studies." February 14, 1980. 33rd Ann. Meeting Soc. Range Manage. San Diego, California. by D.C. Trueblood.

8B 810308 "Soil-vegetation relationships in the Gardner saltbush vegetation types of Wyoming." February 10, 1981. 34th Ann. Meeting Soc. Range Manage. Tulsa, Oklahoma. by D.C. Trueblood.

8B 840309 "Long term Wyoming rangeland monitoring and research cooperative studies of the USDI-BLM & University of Wyoming, Department of Range Management." May 1, 1984. USDA-ARS/USDI-BLM Regional Workshop on Rangeland Hydrology, Modeling and Monitoring. Boise, Idaho. by H.G. Fisser.

8C. HYDROLOGY AND WATER YIELD

8C 770308 "Vegetation analysis and ecology of rangelands in northwest Libya, Africa." April 23, 1977. 48th Ann. Meeting Colo.-Wyo. Acad. Sci. Colorado Springs, Colorado. by O.M. Shawesh.

8C 780308 "Ecology of rangeland communities in northwestern Libya, Africa." August 15, 1978. 1st Intern. Rangeland Congress. Denver, Colorado. by O.M. Shawesh.

8C 780310 "Rangeland vegetation, improvement and management in Libya, Africa." October 12, 1978. Intern. Arid Lands Conf. on Plant Resources. Texas Tech University. Lubbock, Texas. by O.M. Shawesh.

8D. WATERSHED RESTORATION

8D 730303 "Revegetation of pipeline areas." April 30, 1973. Colo.-Wyo. Acad. Sci. Ft. Collins, Colorado. by T.E. Bedell.

8D 740304 "Relationships of range production and rehabilitation to soils and other environmental factors." Aug. 22, 1974. Workshop of BLM Soils and Geology Specialists. Lander, Wyoming. by H.G. Fisser.

8D 750301 "Pre-disturbance vegetation inventory and ecological monitoring procedures for rangeland strip-mining." February 12, 1975. 28th Ann. Meeting Soc. Range Manage. Mexico City, Mexico. by H.G. Fisser.

8D 750302 "Vegetation response to seismic motion from underground nuclear detonation." February 13, 1975. 28th Ann. Meeting Soc. Range Manage. Mexico City, Mexico. by H.G. Fisser.

8D 750304 "Pre-disturbance ecological studies improve and define potential for surface mine reclamation." October 22, 1975. 3rd NCA/BCR Coal Conference and Expo II. Third Symposium on Surface Mining and Reclamation. Louisville, Kentucky. by H.G. Fisser.

8D 760306 "Vegetation re-establishment potential of strip-mining soils, Black Thunder Basin, Wyoming." February 19, 1976. 29th Ann. Meeting Soc. Range Manage. Omaha, Nebraska. by G.A. Lymbery.

8D 760307 "Methods of measuring environmental parameters on a pre-disturbed coal site." April 22, 1976. 47th Ann. Meeting Colo.-Wyo. Acad. Sci. Boulder, Colorado. by J.R. Cox.

8D 770303 "Development of snowfence systems on strip mine spoils to increase seedling establishment." February 16, 1977. 30th Ann. Meeting Soc. Range Manage. Portland, Oregon. by D.T. Cleaves.

8D 770307 "Factors affecting growth and productivity of vegetation stratum in eastern Wyoming." April 23, 1977. 48th Ann. Meeting Colo.-Wyo. Acad. Sci. Colorado Springs, Colorado. by F.K. Taha.

9. Range Economics

9A. RANCH PLANNING

9A 830302 "Economic feasibility of resting rangeland to improve forage production in central Wyoming." February 16, 1983. 36th Ann. Meeting Soc. Range Manage. Albuquerque, New Mexico. by T.L. Deboodt.

10. Range Education and Research

10A. RANGE PHILOSOPHY

10A 780306 "Range management equals functional ecology." July 11, 1978. Wyoming OWRC/SRM Youth Range Camp. Tensleep, Wyoming. by H.G. Fisser.

10D. RANGE RESEARCH METHODS

10D 780309 "The effects of water stress on phenology and carbohydrate storage in the shortgrass prairie." October 11, 1978. Intern. Arid. Lands Conf. on Plant Resources. Texas Tech University. Lubbock, Texas. by J.R. Cox.

10D 790303 "Arid rangeland sampling." September 27, 1979. University of Wyoming Zoo. Biol. Sampling Seminar. Laramie, Wyoming. by H.G. Fisser.

10D 800303 "Vegetation stagnation and decadence in a southeastern Wyoming big game exclosure: a preliminary report." April 25, 1980. 51st Ann. Meeting Colo.-Wyo. Acad. Sci. Denver, Colorado. by W.J. Waugh.

10D 810304 "Esophageal fistulated steers for determination of dietary nutrient composition on large pasture native rangeland." February 11, 1981. 34th Ann. Meeting Soc. Range Manage. Tulsa, Oklahoma. by G.R. O'Connor.

10D 810313 "Some recent research findings in the true mountain mahogany types." May 28, 1981. 10th Wyoming Shrub Ecology Workshop. Sybille Canyon, Wyoming. by H.G. Fisser.

10D 820302 "Vegetation productivity and phenology in the Wind River Basin." May 27, 1982. 11th Wyoming Shrub Ecology Workshop. Lander, Wyoming. by H.G. Fisser.

10D 820303 "Vegetation responses following strip chemical control of sagebrush at the Lower Government Draw site." May 28, 1982. 11th Wyoming Shrub Ecology Field Tour. Lander, Wyoming. by H.G. Fisser.

10D 820304 "Vegetation response to strip chemical control in the Government Draw area." May 27, 1982. 11th Wyoming Shrub Ecology Workshop. Lander, Wyoming. by S.K. Dean.

10D 820306 "Sagebrush and perennial grass dynamics at the Granite Mountain shrub control site." May 27, 1982. 11th Wyoming Shrub Ecology Workshop. Lander, Wyoming. by J.L. Mulholland.

10D 830301 "Long term vegetation responses to sagebrush control in western Wyoming." February 15, 1983. 36th Ann. Meeting Soc. Range Manage. Albuquerque, New Mexico. by S.K. Dean.

10D 830308 "Size and density parameters of western wheatgrass on diverse range sites in western Wyoming." February 15, 1983. 36th Ann. Meeting Soc. Range Manage. Albuquerque, New Mexico. by H.G. Fisser.

10D 840309 "Long term Wyoming rangeland monitoring and research cooperative studies of the USDI-BLM & University of Wyoming, Department of Range Management." May 1, 1984. USDA-ARS/USDI-BLM Regional Workshop on Rangeland Hydrology, Modeling and Monitoring. Boise, Idaho. by H.G. Fisser.

10D 840310 "Herbage production and community dynamics." July 12, 1984. Shrub Research Consortium Third Ann. Wildland Shrub Symposium: the biology of *Artemisia* and *Chrysothamnus*. Provo, Utah. by H.G. Fisser.

10D1. Plant-Soil Emphasis

10D1 610302 "Theories and methods of range surveys." December 4, 1961. Wyo. Agr. Exp. Sta. Superintendents Conference. Laramie, Wyoming. by H.G. Fisser.

10D1 620301 "Concepts of range condition class survey." November 30, 1962. Wyoming Wool Growers Association Convention. Worland, Wyoming. by H.G. Fisser.

10D1 630302 "Exclosure studies (production, cover, height, and moisture phase)." March 19, 1963. Range Weed Control Conf. Portland, Oregon. by H.G. Fisser.

10D1 650305 "Grass seeded spreader systems can increase carrying capacity." May 26, 1965. Washakie County Agr. Ext. Service Range Improvement Conf. Worland, Wyoming. by H.G. Fisser.

10D1 670305 "Cooperative research with the University of Wyoming." March 10, 1967. Bureau of Land Management Wyoming State Advisory Board Meeting. Riverton, Wyoming. by H.G. Fisser.

10D1 720301 "Panchromatic, color, and infrared photography for detailed vegetation mapping." February 8, 1972. 25th Ann. Meeting Soc. Range Manage. Washington, D.C. by H.G. Fisser.

10D1 750305 "A study of rangeland soil movement characteristics." August 12, 1975. Amer. Soc. Civil Engineer. Watershed Manage. Symp. Logan, Utah. by H.G. Fisser.

10D1 760301 "Aridland ecology and phenology research in western Wyoming." March 3, 1976. Regional Rangeland Research Symposium. Bureau of Land Management. Denver, Colorado. by H.G. Fisser.

10D1 760303 "Aridland ecological research activities in Wyoming and relationships to Libyan rangelands." May 15, 1976. College of Agriculture, University Al Fateh special seminar. Tripoli, Libya. by H.G. Fisser.

10D1 760305 "Physical and chemical analyses of range sites and interrelationships with herbage production and forage value potentials." February 18, 1976. 29th Ann. Meeting Soc. Range Mange. Omaha, Nebraska. by F.K. Taha.

10D1 770304 "Long-term vegetation studies with permanent quadrats." February 16, 1972. 30th Ann. Meeting Soc. Range Manage. Portland, Oregon. by A.D. Gdara.

10D1 770305 "The effect of environmental factors on carbohydrate storage and herbage production of five shortgrass prairie plant species." February 16, 1977. 30th Ann. Meeting Soc. Range Manage. Portland, Oregon. by J.R. Cox.

10D1 780307 "Semiarid rangeland and grazing evaluation of shrubland from permanent quadrats." August 15, 1978. 1st Intern. Rangeland Congress. Denver, Colorado. by A.D. Gdara.

10D1 780309 "The effects of water stress on phenology and carbohydrate storage in the shortgrass prairie." October 11, 1978. Intern. Arid. Lands Conf. on Plant Resources. Texas Tech University. Lubbock, Texas. by J.R. Cox.

10D1 790302 "Soil-vegetation relationships on rangeland exclosures in the Grass Creek Planning Unit of northcentral Wyoming." September 5, 1979. Wyo. S.C.S. Soil Survey Work Planning Conference. Laramie, Wyoming. by H.G. Fisser.

10D1 790305 "Soils-phenology relations on some arid rangelands of western Wyoming." February 15, 1979. 32nd Ann. Meeting Soc. Range Manage. Casper, Wyoming. by L.W. Young, III.

10D1 800302 "Soil vegetation relationships on a semiarid rangeland exclosure in north central Wyoming: long term studies." February 14, 1980. 33rd Ann. Meeting Soc. Range Manage. San Diego, California. by D.C. Trueblood.

10D1 810308 "Soil-vegetation relationships in the Gardner saltbush vegetation types of Wyoming." February 10, 1981. 34th Ann. Meeting Soc. Range Manage. Tulsa, Oklahoma. by D.C. Trueblood.

10D1 830311 "Grazing, vegetation and soils interactions since 1967 in the Powder Rim grazing system." May 26, 1983. 12th Wyoming Shrub Ecology Workshop. Baggs, Wyoming. by H.G. Fisser.

10D1 830312 "Deer browse impacts on Utah juniper and true mountain mahogany in the Poison Butte unit west of Baggs, Wyoming." May 26, 1983. 12th Wyoming Shrub Ecology Workshop. Baggs, Wyoming. by H.G. Fisser.

11. Computer Storage, Programming and Modeling

11A. CLIMATIC

11A 810302 "Phenological comparison of two sagebrush species in Wyoming." February 11, 1981. 34th Ann. Meeting Soc. Range Manage. Tulsa, Oklahoma. by N.E. Hargis.

11A 810308 "Soil-vegetation relationships in the Gardner saltbush vegetation types of Wyoming." February 10, 1981. 34th Ann. Meeting Soc. Range Manage. Tulsa, Oklahoma. by D.C. Trueblood.

11A 820301 "Numerical phenology of big and black sagebrush from visual interpretation statements." April 25, 1982. Colorado-Wyoming Academy of Science. Fort Collins, Colorado. by H.G. Fisser.

11A 820305 "Numerical phenology scoring and use with quantitative variables." May 27, 1982. 11th Wyoming Shrub Ecology Workshop. Lander, Wyoming. by R.J. Kinucan.

11A 830304 "Phenological description and numerical scoring of sagebrush." February 16, 1983. 36th Ann. Meeting Soc. Range Manage. Albuquerque, New Mexico. by H.G. Fisser.

11A 840301 "Detailed numerical phenology of sagebrush with visual description statements." February 14, 1984. 37th Ann. Meeting Soc. Range Manage. Rapid City, South Dakota. by N.E. Hargis and H.G. Fisser.

11A 840303 "Estimation of missing precipitation data in Wyoming using traditional, kriging and best guess estimates." February 15, 1984. 37th Ann. Meeting Soc. Range Manage. Rapid City, South Dakota. by L.P. Germain.

11A 840304 "Detailed numerical phenology of sagebrush with visual description statements." February 16, 1984. 37th Ann. Meeting Soc. Range Manage. Rapid City, South Dakota. by H.G. Fisser.

11A 840306 "Application of kriging to ecological variables." April 27, 1984. 55th Ann. Meeting Colo.-Wyo. Acad. Sci. Gunnison, Colorado. by L.P. Germain.

11A 840307 "Detailed numerical phenology of sagebrush with visual description statements." May 1, 1984. USDA-ARS/USDI-BLM Regional Workshop on Rangeland Hydrology, Modeling and Monitoring. Boise, Idaho. by H.G. Fisser.

11A 840312 "Regional variation of phenology." July 12, 1984. Shrub Research Consortium Third Ann. Wildland Shrub Symposium: the biology of *Artemisia* and *Chrysothamnus.* Provo, Utah. by H.G. Fisser.

11B. BIOTIC

11B 810302 "Phenological comparison of two sagebrush species in Wyoming." February 11, 1981. 34th Ann. Meeting Soc. Range Manage. Tulsa, Oklahoma. by N.E. Hargis.

11B 820301 "Numerical phenology of big and black sagebrush from visual interpretation statements." April 25, 1982. Colorado-Wyoming Academy of Science. Fort Collins, Colorado. by H.G. Fisser.

11B 820305 "Numerical phenology scoring and use with quantitative variables." May 27, 1982. 11th Wyoming Shrub Ecology Workshop. Lander, Wyoming. by R.J. Kinucan.

11B 830304 "Phenological description and numerical scoring of sagebrush." February 16, 1983. 36th Ann. Meeting Soc. Range Manage. Albuquerque, New Mexico. by H.G. Fisser.

11B 840301 "Detailed numerical phenology of sagebrush with visual description statements." February 14, 1984. 37th Ann. Meeting Soc. Range Manage. Rapid City, South Dakota. by N.E. Hargis and H.G. Fisser.

11B 840304 "Detailed numerical phenology of sagebrush with visual description statements." February 16, 1984. 37th Ann. Meeting Soc. Range Manage. Rapid City, South Dakota. by H.G. Fisser.

11B 840307 "Detailed numerical phenology of sagebrush with visual description statements." May 1, 1984. USDA-ARS/USDI-BLM Regional Workshop on Rangeland Hydrology, Modeling and Monitoring. Boise, Idaho. by H.G. Fisser.

11B 840312 "Regional variation of phenology." July 12, 1984. Shrub Research Consortium Third Ann. Wildland Shrub Symposium: the biology of *Artemisia* and *Chrysothamnus.* Provo, Utah. by H.G. Fisser.

11C. EDAPHIC

11C 810302 "Phenological comparison of two sagebrush species in Wyoming." February 11, 1981. 34th Ann. Meeting Soc. Range Manage. Tulsa, Oklahoma. by N.E. Hargis.

11C 810308 "Soil-vegetation relationships in the Gardner saltbush vegetation types of Wyoming." February 10, 1981. 34th Ann. Meeting Soc. Range Manage. Tulsa, Oklahoma. by D.C. Trueblood.

11C 820301 "Numerical phenology of big and black sagebrush from visual interpretation statements." April 25, 1982. Colorado-Wyoming Academy of Science. Fort Collins, Colorado. by H.G. Fisser.

11C 820305 "Numerical phenology scoring and use with quantitative variables." May 27, 1982. 11th Wyoming Shrub Ecology Workshop. Lander, Wyoming. by R.J. Kinucan.

11C 830304 "Phenological description and numerical scoring of sagebrush." February 16, 1983. 36th Ann. Meeting Soc. Range Manage. Albuquerque, New Mexico. by H.G. Fisser.

11C 840301 "Detailed numerical phenology of sagebrush with visual description statements." February 14, 1984. 37th Ann. Meeting Soc. Range Manage. Rapid City, South Dakota. by N.E. Hargis and H.G. Fisser.

11C 840304 "Detailed numerical phenology of sagebrush with visual description statements." February 16, 1984. 37th Ann. Meeting Soc. Range Manage. Rapid City, South Dakota. by H.G. Fisser.

11C 840307 "Detailed numerical phenology of sagebrush with visual description statements." May 1, 1984. USDA-ARS/USDI-BLM Regional Workshop on Rangeland Hydrology, Modeling and Monitoring. Boise, Idaho. by H.G. Fisser.

11C 840312 "Regional variation of phenology." July 12, 1984. Shrub Research Consortium Third Ann. Wildland Shrub Symposium: the biology of *Artemisia* and *Chrysothamnus*. Provo, Utah. by H.G. Fisser.

11E. RANGELAND INVENTORY

11E 810302 "Phenological comparison of two sagebrush species in Wyoming." February 11, 1981. 34th Ann. Meeting Soc. Range Manage. Tulsa, Oklahoma. by N.E. Hargis.

11E 810306 "Grazing system impacts on semiarid vegetation of Wyoming." February 12, 1981. 34th Ann. Meeting Soc. Range Manage. Tulsa, Oklahoma. by D.E. Rodgers.

11E 810308 "Soil-vegetation relationships in the Gardner saltbush vegetation types of Wyoming." February 10, 1981. 34th Ann. Meeting Soc. Range Manage. Tulsa, Oklahoma. by D.C. Trueblood.

11E 810312 "Habitat and vegetation of semiarid rangelands, Libya, Africa." February 11, 1981. 34th Ann. Meeting Soc. Range Manage. Tulsa, Oklahoma. by O.M. Shawesh.

11E 820301 "Numerical phenology of big and black sagebrush from visual interpretation statements." April 25, 1982. Colorado-Wyoming Academy of Science. Fort Collins, Colorado. by H.G. Fisser.

11E 820305 "Numerical phenology scoring and use with quantitative variables." May 27, 1982. 11th Wyoming Shrub Ecology Workshop. Lander, Wyoming. by R.J. Kinucan.

11E 830304 "Phenological description and numerical scoring of sagebrush." February 16, 1983. 36th Ann. Meeting Soc. Range Manage. Albuquerque, New Mexico. by H.G. Fisser.

11E 830307 "Reestablishment rates of sagebrush following chemical control on grazed and ungrazed ranges in western Wyoming." February 17, 1983. 36th Ann. Meeting Soc. Range Manage. Albuquerque, New Mexico. by J.L. Mulholland.

11E 840301 "Detailed numerical phenology of sagebrush with visual description statements." February 14, 1984. 37th Ann. Meeting Soc. Range Manage. Rapid City, South Dakota. by N.E. Hargis and H.G. Fisser.

11E 840307 "Detailed numerical phenology of sagebrush with visual description statements." May 1, 1984. USDA-ARS/USDI-BLM Regional Workshop on Rangeland Hydrology, Modeling and Monitoring. Boise, Idaho. by H.G. Fisser.

11E1. Sites and Mapping

11E1 840304 "Detailed numerical phenology of sagebrush with visual description statements." February 16, 1984. 37th Ann. Meeting Soc. Range Manage. Rapid City, South Dakota. by H.G. Fisser.

11E2. Condition and Trend

11E2 740305 "Environmental factors affecting annual herbage production at 17 sagebrush-grass locations in Wyoming." Feb. 5, 1974. 27th Ann. Meeting Soc. Range Manage. Tucson, Ariz. by R.E. Ries.

11E2 830302 "Economic feasibility of resting rangeland to improve forage production in central Wyoming." February 16, 1983. 36th Ann. Meeting Soc. Range Manage. Albuquerque, New Mexico. by T.L. Deboodt.

11E2 830308 "Size and density parameters of western wheatgrass on diverse range sites in western Wyoming." February 15, 1983. 36th Ann. Meeting Soc. Range Manage. Albuquerque, New Mexico. by H.G. Fisser.

11E2 840309 "Long term Wyoming rangeland monitoring and research cooperative studies of the USDI-BLM & University of Wyoming, Department of Range Management." May 1, 1984. USDA-ARS/USDI-BLM Regional Workshop on Rangeland Hydrology, Modeling and Monitoring. Boise, Idaho. by H.G. Fisser.

11E2 840312 "Regional variation of phenology." July 12, 1984. Shrub Research Consortium Third Ann. Wildland Shrub Symposium: the biology of *Artemisia* and *Chrysothamnus.* Provo, Utah. by H.G. Fisser.

11E2 840313 "Validation of a rangeland production model (ERHYM) for sagebrush sites." July 12, 1984. Shrub Research Consortium Third Ann. Wildland Shrub Symposium: the biology of *Artemisia* and *Chrysothamnus.* Provo, Utah. by J.R. Wight.

11E2 840316 "Dynamics of vegetative plant part phenodynamics and related stress adaptations." August 9, 1984. Annual Meeting Ecological Society of America. Fort Collins, Colorado. by H.G. Fisser.

11E3. Production and Utilization

11E3 700306 "Predictive value of factors influencing perennial grass production on western Wyoming sagebrush-grass ranges." May 1, 1970. Colo.-Wyo. Acad. Sci. Greeley, Colorado. by G.L. Whysong.

11E3 810305 "Phenology and productivity comparisons of western wheatgrass *(Agropyron smithii)* and bluebunch wheatgrass *(A. spicatum).*" February 11, 1981. 34th Ann. Meeting Soc. Range Manage. Tulsa, Oklahoma. by M.K. Owens.

11E3 810315 "Phenology and culm weight variation between two wheatgrass species: *Agropyron smithii* and *A. spicatum.*" June 17, 1981. 14th Intern. Grassl. Congress. Lexington, Kentucky. by M.K. Owens.

11E3 840302 "Productivity index procedures for sagebrush-grass rangeland." February 14, 1984. 37th Ann. Meeting Soc. Range Manage. Rapid City, South Dakota. by H.G. Fisser.

11E3 840305 "Culm weight dynamics as an index of sagebrush-grass community productivity." April 27, 1984. 55th Ann. Meeting Colo.-Wyo. Acad. Sci. Gunnison, Colorado. by H.G. Fisser.

11E3 840309 "Long term Wyoming rangeland monitoring and research cooperative studies of the USDI-BLM & University of Wyoming, Department of Range Management." May 1, 1984. USDA-ARS/USDI-BLM Regional Workshop on Rangeland Hydrology, Modeling and Monitoring. Boise, Idaho. by H.G. Fisser.

11E3 840313 "Validation of a rangeland production model (ERHYM) for sagebrush sites." July 12, 1984. Shrub Research Consortium Third Ann. Wildland Shrub Symposium: the biology of *Artemisia* and *Chrysothamnus.* Provo, Utah. by J.R. Wight.

11E3 840314 "Herbage yield dynamics and seasonal precipitation relationships." July 12, 1984. Shrub Research Consortium Third Ann. Wildland Shrub Symposium: the biology of *Artemisia* and *Chrysothamnus.* Provo, Utah. by C.L. Hanson.

11E3 840315 "Productivity dynamics of sagebrush ecosystems as determined from biweekly sampling of a single grass and shrub species." August 9, 1984. Annual Meeting Ecological Society of America. Fort Collins, Colorado. by H.G. Fisser.

11E3 840316 "Dynamics of vegetative plant part phenodynamics and related stress adaptations." August 9, 1984. Annual Meeting Ecological Society of America. Fort Collins, Colorado. by H.G. Fisser.

11F. PREDICTIVE MODELING

11F1. Herbage Yield

11F1 780302 "Models for prediction of natural vegetation potential in eastern Wyoming." February 8, 1978. 31st Ann. Meeting Soc. Range Manage. San Antonio, Texas. by H.G. Fisser.

11F1 840301 "Detailed numerical phenology of sagebrush with visual description statements." February 14, 1984. 37th Ann. Meeting Soc. Range Manage. Rapid City, South Dakota. by N.E. Hargis and H.G. Fisser.

11F1 840302 "Productivity index procedures for sagebrush-grass rangeland." February 14, 1984. 37th Ann. Meeting Soc. Range Manage. Rapid City, South Dakota. by H.G. Fisser.

11F1 840305 "Culm weight dynamics as an index of sagebrush-grass community productivity." April 27, 1984. 55th Ann. Meeting Colo.-Wyo. Acad. Sci. Gunnison, Colorado. by H.G. Fisser.

11F1 840307 "Detailed numerical phenology of sagebrush with visual description statements." May 1, 1984. USDA-ARS/USDI-BLM Regional Workshop on Rangeland Hydrology, Modeling and Monitoring. Boise, Idaho. by H.G. Fisser.

11F1 840309 "Long term Wyoming rangeland monitoring and research cooperative studies of the USDI-BLM & University of Wyoming, Department of Range Management." May 1, 1984. USDA-ARS/USDI-BLM Regional Workshop on Rangeland Hydrology, Modeling and Monitoring. Boise, Idaho. by H.G. Fisser.

11F1 840310 "Herbage production and community dynamics." July 12, 1984. Shrub Research Consortium Third Ann. Wildland Shrub Symposium: the biology of *Artemisia* and *Chrysothamnus.* Provo, Utah. by H.G. Fisser.

11F1 840313 "Validation of a rangeland production model (ERHYM) for sagebrush sites." July 12, 1984. Shrub Research Consortium Third Ann. Wildland Shrub Symposium: the biology of *Artemisia* and *Chrysothamnus.* Provo, Utah. by J.R. Wight.

11F1 840314 "Herbage yield dynamics and seasonal precipitation relationships." July 12, 1984. Shrub Research Consortium Third Ann. Wildland Shrub Symposium: the biology of *Artemisia* and *Chrysothamnus.* Provo, Utah. by C.L. Hanson.

11F1 840315 "Productivity dynamics of sagebrush ecosystems as determined from biweekly sampling of a single grass and shrub species." August 9, 1984. Annual Meeting Ecological Society of America. Fort Collins, Colorado. by H.G. Fisser.

11F2. Phenology

11F2 750306 "Phenodynamics and environmental characteristics of sagebrush-grass rangelands." November 5, 1975. Wildland Shrubs Symp. Logan, Utah. by L.H. Kleinman.

11F2 810302 "Phenological comparison of two sagebrush species in Wyoming." February 11, 1981. 34th Ann. Meeting Soc. Range Manage. Tulsa, Oklahoma. by N.E. Hargis.

11F2 810305 "Phenology and productivity comparisons of western wheatgrass *(Agropyron smithii)* and bluebunch wheatgrass *(A. spicatum).*" February 11, 1981. 34th Ann. Meeting Soc. Range Manage. Tulsa, Oklahoma. by M.K. Owens.

11F2 810315 "Phenology and culm weight variation between two wheatgrass species: *Agropyron smithii* and *A. spicatum.*" June 17, 1981. 14th Intern. Grassl. Congress. Lexington, Kentucky. by M.K. Owens.

11F2 820301 "Numerical phenology of big and black sagebrush from visual interpretation statements." April 25, 1982. Colorado-Wyoming Academy of Science. Fort Collins, Colorado. by H.G. Fisser.

11F2 820305 "Numerical phenology scoring and use with quantitative variables." May 27, 1982. 11th Wyoming Shrub Ecology Workshop. Lander, Wyoming. by R.J. Kinucan.

11F2 830304 "Phenological description and numerical scoring of sagebrush." February 16, 1983. 36th Ann. Meeting Soc. Range Manage. Albuquerque, New Mexico. by H.G. Fisser.

11F2 840304 "Detailed numerical phenology of sagebrush with visual description statements." February 16, 1984. 37th Ann. Meeting Soc. Range Manage. Rapid City, South Dakota. by H.G. Fisser.

11F2 840309 "Long term Wyoming rangeland monitoring and research cooperative studies of the USDI-BLM & University of Wyoming, Department of Range Management." May 1, 1984. USDA-ARS/USDI-BLM Regional Workshop on Rangeland Hydrology, Modeling and Monitoring. Boise, Idaho. by H.G. Fisser.

11F2 840312 "Regional variation of phenology." July 12, 1984. Shrub Research Consortium Third Ann. Wildland Shrub Symposium: the biology of *Artemisia* and *Chrysothamnus.* Provo, Utah. by H.G. Fisser.

11F2 840313 "Validation of a rangeland production model (ERHYM) for sagebrush sites." July 12, 1984. Shrub Research Consortium Third Ann. Wildland Shrub Symposium: the biology of *Artemisia* and *Chrysothamnus.* Provo, Utah. by J.R. Wight.

S4. PUBLISHED PROCEEDINGS/ABSTRACTS

1B. FORAGE VALUE

1B1. Range Plants

1B1 810404 O'Connor, G.R., H.G. Fisser and J.W. Waggoner. 1981. Esophageal-fistulated steers for determination of dietary nutrient composition on large pasture native rangeland. p. 26. In Absts. Soc. Range Manage. 34th Ann. Meeting. Tulsa, Okla. Feb. 8-13, 1981.

1B1 830402 Fisser, H.G. and K.L. Johnson (eds.) 1983. Wyoming Shrublands: Proc. 9th Wyoming Shrub Ecology Workshop. June 5-6, 1980. Lander, Wyoming. University of Wyoming, Laramie. 40 p.

1B1 830403 Fisser, H.G. and K.L. Johnson (eds.) 1983. Wyoming Shrublands: Proc. 10th Wyoming Shrub Ecology Workshop. May 28-29, 1981. Sybille Canyon, Wyoming. University of Wyoming, Laramie. 40 p.

1B4. Poisonous Plants

1B4 610401 Vosler, L.C. 1961. Halogeton seed burial study. p. 1. In Proc. Range Weed Research Tour. Worland, Wyoming. September 14-15, 1961.

1B4 650402 Fisser, H.G. 1965. Halogeton abundance and distribution as influenced by degree of saltbush utilization. p. 11-13. In Proc. Range Weed Control Conf. Elko, Nevada. Aug. 2, 1965.

1B4 840407 Fisser, H.G. and L.A. Joyce. 1984. Saltbush community changes with and without grazing over a 20-year period in north central Wyoming. Submitted to Proc. 2nd Intern. Rangeland Congress. Adelaide, Australia. May 13-18, 1984.

1B4 840413 Fisser, H.G. and L.A. Joyce. 1984. *Atriplex*/grass and forb relationships under no grazing and shifting precipitation patterns in north central Wyoming. p. 87-96. In Tiedeman, A.R., E.D. McArthur, H.C. Stutz, R. Stevens, and K.L. Johnson (eds.). Proc. Wildland Shrub Symp. "Biology of *Atriplex* and related chenopods." Provo, Utah. May 2-6, 1983. USDA Forest Service General Technical Report 172.

1D. PLANT INSECTS

1D 670402 Kirkham, D.R. 1967. Western Harvester Ant abundance in relation to grazing, vegetation and soil in northcentral Wyoming. p. 28. In 20th Ann. Meeting Proc. Am. Soc. Range Manage. Seattle, Washington. Feb. 14, 1967.

1E. PLANT PHYSIOLOGY

1E1. Reproduction and Germination

1E1 610401 Vosler, L.C. 1961. Halogeton seed burial study. p. 1. In Proc. Range Weed Research Tour. Worland, Wyoming. September 14-15, 1961.

1E1 700401 Fisser, H.G. 1970. Distribution of plant biomass to plant organs. p. 34-42 and 110-115. In Proc. Intern. Biol. Prog. Desert Biome Plant Specialists Meeting. Jan. 3-4, 1970. Las Vegas, Nevada. January 3-4, 1970.

1E1 730402 Layton, D.L. 1973. Methods of revegetation and restoration following a juniper wildfire. p. 12. In Proc. 26th Ann. Meeting Soc. Range Manage. Boise, Idaho. February 4-9, 1973.

1E1 750403 Kleinman, L.H. and H.G. Fisser. 1975. Phenodynamics and environmental characteristics of sagebrush-grass rangelands. p. 161. In H.C. Stutz, (ed.) Proc. Wildland Shrubs Symp. U.S. Forest Service Shrub Sciences Lab. Provo, Utah. November 4-6, 1975.

1E1 770402 Kleinman, Larry H. and H.G. Fisser. 1977. Phenodynamics and ecology of sagebrush-grass rangelands. p. 11. In Proc. Soc. Range Manage. 30th Ann. Meeting. Portland, Oregon. February 14-17, 1977.

1E1 810402 Hargis, N.E. and H.G. Fisser. 1981. Phenological comparison of two sagebrush species in Wyoming. p. 21. In Absts. Soc. Range Manage. 34th Ann. Meeting. Tulsa, Oklahoma. Feb. 8-13, 1981.

1E1 810405 Owens, M.K. and H.G. Fisser. 1981. Phenology and productivity comparisons of western wheatgrass *(Agropyron smithii)* and bluebunch wheatgrass *(A. spicatum)*. p. 21. In Absts. Soc. Range Manage. 34th Ann. Meeting. Tulsa, Oklahoma. Feb. 8-13, 1981.

1E1 810415 Owens, M.K. and H.G. Fisser. 1981. Phenology and culm weight variation between two wheatgrass species: *Agropyron smithii* and *A. spicatum*. p. 205. In Summaries of Papers. J. Allan Smith (ed.). XIV International Grassland Congress. Lexington, Kentucky. June 14-23, 1981.

1E1 820403 Kinucan, R.J. 1982. Numerical phenology scoring and use with quantitative environmental variables. p. 36-39. In: Fisser, H.G. and K.L. Johnson eds. 1984. Proceedings of the 11th Wyoming Shrub Ecology Workshop. Lander, Wyoming. May 27-28, 1982. University of Wyoming, Laramie.

1E1 820404 Fisser, H.G. 1982. Vegetation productivity and phenology in the Wind River Basin. p. 42-45. In: Fisser, H.G. and K.L. Johnson, eds. 1984. Proceedings of the 11th Wyoming Shrub Ecology Workshop. Lander, Wyoming. May 27-28, 1982. University of Wyoming, Laramie.

1E1 830408 Fisser, H.G. 1983. Phenological description and numerical scoring of sagebrush. abst. 136. In Absts. 36th Ann. Meeting. Soc. Range Manage. Albuquerque, New Mexico. February 14-17, 1983.

1E1 830411 Mulholland, J.L. and H.G. Fisser. 1983. Reestablishment rates of sagebrush following chemical control on grazed and ungrazed ranges in western Wyoming. abst. 295. In Absts. 36th Ann. Meeting. Soc. Range Manage. Albuquerque, New Mexico. February 14-17, 1983.

1E1 840402 Fisser, H.G. and N.E. Hargis. 1984. Detailed numerical phenology of sagebrush with visual description of statements. p. 44 . In Proc 37th Ann. Meeting Soc. Range Manage. Rapid City, South Dakota. February 12-17, 1984.

1E1 840407 Fisser, H.G. and L.A. Joyce. 1984. Saltbush community changes with and without grazing over a 20-year period in north central Wyoming. Submitted to Proc. 2nd Intern. Rangeland Congress. Adelaide, Australia. May 13-18, 1984.

1E1 840408 Mulholland, J.L. and H.G. Fisser. 1984. Big sagebrush response and reestablishment following chemical control in Wyoming, U.S.A. Submitted to Proc. 2nd Intern. Rangeland Congress. Adelaide, Australia. May 13-18, 1984.

1E1 840413 Fisser, H.G. and L.A. Joyce. 1984. *Atriplex*/grass and forb relationships under no grazing and shifting precipitation patterns in north central Wyoming. p. 87-96. In Tiedeman, A.R., E.D. McArthur, H.C. Stutz, R. Stevens, and K.L. Johnson (eds.). Proc. Wildland Shrub Symp. "Biology of *Atriplex* and related chenopods." Provo, Utah. May 2-6, 1983. USDA Forest Service General Technical Report 172.

1E1 840415 Fisser, H.G. 1984. Biology and ecology of sagebrush in Wyoming: II. Reinvasion following control. p. 21. In Abst. Shrub Research Consortium Third Annual Wildland Shrub Symposium: the biology of *Artemisia* and *Chrysothamnus*. Provo, Utah. July 9-13, 1984.

1E1 840416 Fisser, H.G. 1984. Biology and ecology of sagebrush in Wyoming: III. Regional variation of phenology. p. 22. In Abst. Shrub Research Consortium Third Annual Wildland Shrub Symposium: the biology of *Artemisia* and *Chrysothamnus*. Provo, Utah. July 9-13, 1984.

1E1 840419 Fisser, H.G. and K.L. Johnson, eds. 1984. Wyoming Shrublands Proc. 11th Wyoming Shrub Ecology Workshop. Lander, Wyoming. May 27-28, 1982. University of Wyoming, Laramie.

1E2. Growth and Phenology

1E2 690403 Wambolt, C.L. 1969. Size, density, growth rate and age of Rocky Mountain juniper and interrelationships with environmental factors. p. 57. In Proc. 22nd Ann. Meeting Soc. Range Manage. Calgary, Alberta. Feb. 10-13, 1969.

1E2 700401 Fisser, H.G. 1970. Distribution of plant biomass to plant organs. p. 34-42 and 110-115. In Proc. Intern. Biol. Prog. Desert Biome Plant Specialists Meeting. Jan. 3-4, 1970. Las Vegas, Nevada. January 3-4, 1970.

1E2 710401 Wambolt, C.L. 1971. Moisture stress in woody plants as influenced by environmental factors. p. 25-26. In Proc. 24th Ann. Meeting. Soc. Range Manage. Reno, Nevada. February 14-18, 1971.

1E2 730402 Layton, D.L. 1973. Methods of revegetation and restoration following a juniper wildfire. p. 12. In Proc. 26th Ann. Meeting Soc. Range Manage. Boise, Idaho. February 4-9, 1973.

1E2 740401 Fisser, H.G. 1974. Productivity differences of semidesert grasslands in western Wyoming. p. 22-23. In Proc. 27th Ann. Meeting, Soc. Range Manage. Tuscon, Arizona. Feb. 5, 1974.

1E2 740403 Stroud, D.W. and H.G. Fisser. 1974. Influence of livestock grazing on growth patterns, plant density and soil microrelief of a *Bouteloua gracilis* prairie, p. 347-351. In V.G. Iglovikov, A.P. Movsissyants, F.F. Sarganov, and A.A. Stepanenko (Compilers). Biological and physiological aspects of the intensification of grassland utilization. XII International Grassland Congress, Moscow, USSR. June 12-16, 1974.

1E2 740404 Fisser, H.G. 1974. Productivity potential differences of grasslands of semidesert Western Wyoming p. 116-124. In V.G. Iglovikov, A.P. Movsissyants, F.F. Sarganov, and A.A. Stepanenko (Compilers). Improvement of natural and production of seeded meadows and pastures. XII International Grassland Congress, Moscow, USSR. June 12-16, 1974.

1E2 750403 Kleinman, L.H. and H.G. Fisser. 1975. Phenodynamics and environmental characteristics of sagebrush-grass rangelands. p. 161. In H.C. Stutz, (ed.) Proc. Wildland Shrubs Symp. U.S. Forest Service Shrub Sciences Lab. Provo, Utah. November 4-6, 1975.

1E2 750406 Shrader, J.A. and H.G. Fisser. 1975. Seed production ecology in the shortgrass prairie of the Central United States. p. 29. In Proc. 28th Ann. Meeting Soc. Range Mange. Mexico City, Mexico. February 10-14, 1975.

1E2 760401 Stroud, D.W., and H.G. Fisser. 1976. Vegetative response of *Bouteloua gracilis* to four levels of utilization. p. 11. In Proc. Soc. Range Manage. 29th Ann. Meeting. Omaha, Nebraska. February 16-20, 1976.

1E2 760402 Taha, Faisal K., and H.G. Fisser. 1976. Physical and chemical analyses of range sites and interrelationships with herbage production and forage value potentials. p. 13. In Proc. Soc. Range Manage. 29th Ann. Meeting. Omaha, Nebraska. February 16-20, 1976.

1E2 760403 Lymbery, Gordon A., and H.G. Fisser. 1976. Vegetation re-establishment potential of strip-mined soils, Black Thunder Basin, Wyoming. p. 49. In Proc. Soc. Range Manage. 29th Ann. Meeting. Omaha, Nebraska. February 16-20, 1976.

1E2 770402 Kleinman, Larry H. and H.G. Fisser. 1977. Phenodynamics and ecology of sagebrush-grass rangelands. p. 11. In Proc. Soc. Range Manage. 30th Ann. Meeting. Portland, Oregon. February 14-17, 1977.

1E2 770403 Cox, Jerry R., and H.G. Fisser. 1977. The effect of environmental factors on carbohydrate storage and herbage production of five short-grass prairie plant species. p. 41. In Proc. Soc. Range Manage. 30th Ann. Meeting. Portland, Oregon. February 14-17, 1977.

1E2 790401 Cox J.R., and H.G. Fisser. 1979. The effects of water stress on phenology and carbohydrate. p. 529-535. In J.R. Goodin and D.K. Northington (eds.) Proc. of the International Arid Lands Conference. International Center for Arid and Semi-Arid Land Studies. Texas Tech. University, Lubbock, Texas. October 8-15, 1978.

1E2 790402 Young, L.W., and H.G. Fisser. 1979. Soils-phenology relations on some arid rangelands of western Wyoming. p. 53-54. In Proc. 32nd Ann. Meeting, Soc. Range Manage. Casper, Wyoming. February 12-15, 1979.

1E2 810401 Bensel, W.E. and H.G. Fisser. 1981. Forage production and longevity of sagebrush control on semiarid rangelands in western Wyoming. p. 9. In Absts. Soc. Range Manage. 34th Ann. Meeting. Tulsa, Oklahoma. February 8-13, 1981.

1E2 810402 Hargis, N.E. and H.G. Fisser. 1981. Phenological comparison of two sagebrush species in Wyoming. p. 21. In Absts. Soc. Range Manage. 34th Ann. Meeting. Tulsa, Oklahoma. February 8-13, 1981.

1E2 810403 Joyce, L.A. and H.G. Fisser. 1981. The effect of community changes on precipitation/ production relationships in exclosures of saltbush vegetation. p. 22. In Absts. Papers Soc. Range Manage. 34th Ann. Meeting. Tulsa, Oklahoma. February 8-13, 1981.

1E2 810405 Owens, M.K. and H.G. Fisser. 1981. Phenology and productivity comparisons of western wheatgrass *(Agropyron smithii)* and bluebunch wheatgrass *(A. spicatum)*. p. 21. In Absts. Soc. Range Manage. 34th Ann. Meeting. Tulsa, Oklahoma. February 8-13, 1981.

1E2 810407 Spaeth, K.E. and H.G. Fisser. 1981. Plant composition response to picloram application in the Wyoming juniper type. p. 35. In Absts. Soc. Range Manage. 34th Ann. Meeting. Tulsa, Oklahoma. February 8-13, 1981.

1E2 810411 Young, L.W. and H.G. Fisser. 1981. Correlation of soil factors and environmental conditions with phenophase development of western wheatgrass *(Agropyron smithii)*. p. 12. In Absts. Soc. Range Manage. 34th Ann. Meeting. Tulsa, Oklahoma. February 8-13, 1981.

1E2 810413 Fisser, H.G. 1981. Shrub establishment, dominance, and ecology on the juniper and sagebrush-grass types in Wyoming. p. 23-27. In L.S. Stelter, E.J. DePuit, S.A. Mikol (eds.). Proc. Symposium on Shrub Establishment on Disturbed Arid and Semiarid Lands. December 3, 1980. Wyoming Game and Fish Dept. Publication.

1E2 810415 Owens, M.K. and H.G. Fisser. 1981. Phenology and culm weight variation between two wheatgrass species: *Agropyron smithii* and *A. spicatum*. p. 205. In Summaries of Papers. J. Allan Smith (ed.). XIV International Grassland Congress. Lexington, Kentucky. June 14-23, 1981.

1E2 820403 Kinucan, R.J. 1982. Numerical phenology scoring and use with quantitative environmental variables. p. 36-39. In: Fisser, H.G. and K.L. Johnson eds. 1984. Proceedings of the 11th Wyoming Shrub Ecology Workshop. Lander, Wyoming. May 27-28, 1982. University of Wyoming, Laramie.

1E2 820404 Fisser, H.G. 1982. Vegetation productivity and phenology in the Wind River Basin. p. 42-45. In: Fisser, H.G. and K.L. Johnson, eds. 1984. Proceedings of the 11th Wyoming Shrub Ecology Workshop. Lander, Wyoming. May 27-28, 1982. University of Wyoming, Laramie.

1E2 830402 Fisser, H.G. and K.L. Johnson (eds.) 1983. Wyoming Shrublands: Proc. 9th Wyoming Shrub Ecology Workshop. June 5-6, 1980. Lander, Wyoming. University of Wyoming, Laramie. 40 p.

1E2 830403 Fisser, H.G. and K.L. Johnson (eds.) 1983. Wyoming Shrublands: Proc. 10th Wyoming Shrub Ecology Workshop. May 28-29, 1981. Sybille Canyon, Wyoming. University of Wyoming, Laramie. 40 p.

1E2 830404 Waugh, W.J. 1983. Stagnation and decadence of *Cercocarpus montanus* in a southeastern Wyoming big game exclosure, p. 11-14. In H.G. Fisser and K.L. Johnson (eds.). Wyoming Shrublands: Proc. 10th Wyoming Shrub Ecology Workshop, Sybille Canyon, Wyoming, May 28-29, 1981. University of Wyoming, Laramie.

1E2 830407 Elwakeel, A.S. and H.G. Fisser. 1983. Perennial grass production dynamics among five vegetation types in north central Wyoming. abst. 208. In Absts. 36th Ann. Meeting. Soc. Range Manage. Albuquerque, New Mexico. February 14-17, 1983.

1E2 830408 Fisser, H.G. 1983. Phenological description and numerical scoring of sagebrush. abst. 136. In Absts. 36th Ann. Meeting. Soc. Range Manage. Albuquerque, New Mexico. February 14-17, 1983.

1E2 830409 Joyce, L.A. and H.G. Fisser. 1983. Productivity differences of grasslands in semidesert western Wyoming — A re-examination. abst. 207. In Absts. 36th Ann. Meeting. Soc. Range Manage. Albuquerque, New Mexico. February 14-17, 1983.

1E2 830410 Kinucan, R.J. and H.G. Fisser. 1983. Dynamics of plant community development for diverse sagebrush sites in western Wyoming. abst. 65. In Absts. 36th Ann. Meeting. Soc. Range Manage. Albuquerque, New Mexico. February 14-17, 1983.

1E2 830412 Owens, M.K. and H.G. Fisser. 1983. Size and density parameters of western wheatgrass on diverse range sites in western Wyoming. abst. 128. In Absts. 36th Ann. Meeting. Soc. Range Manage. Albuquerque, New Mexico. February 14-17, 1983.

1E2 840401 Fisser, H.G. 1984. Productivity index procedures for sagebrush-grass rangeland. abst. 010. In Abst. 37th Ann. Meeting. Soc. Range Manage. Rapid City, South Dakota. February 12-17, 1984.

1E2 840402 Fisser, H.G. and N.E. Hargis. 1984. Detailed numerical phenology of sagebrush with visual description of statements. p. 44 . In Proc 37th Ann. Meeting Soc. Range Manage. Rapid City, South Dakota. February 12-17, 1984.

1E2 840406 Kinucan, R.J. and H.G. Fisser. 1984. Vegetative phenology patterns of semiarid plant communities in Wyoming, U.S.A. Submitted to Proc. 2nd Intern. Rangeland Congress. Adelaide, Australia. May 13-18, 1984.

1E2 840410 Fisser, H.G. 1984. Culm weight dynamics as an index of sagebrush-grass productivity. p. 17. In Proc. 55th Ann. Meeting Colo.-Wyo. Acad. Sci., Western State College, Gunnison, Colorado. April 27-28, 1984.

1E2 840411 Fisser, H.G. 1984. Detailed numerical phenology of sagebrush with visual description statements. Posterboard display at USDA-ARS/USDI-BLM Regional Workshop on Rangeland Hydrology, Modeling and Monitoring. Boise, Idaho. April 31-May 2, 1984.

1E2 840415 Fisser, H.G. 1984. Biology and ecology of sagebrush in Wyoming: II. Reinvasion following control. p. 21. In Abst. Shrub Research Consortium Third Annual Wildland Shrub Symposium: the biology of *Artemisia* and *Chrysothamnus*. Provo, Utah. July 9-13, 1984.

1E2 840416 Fisser, H.G. 1984. Biology and ecology of sagebrush in Wyoming: III. Regional variation of phenology. p. 22. In Abst. Shrub Research Consortium Third Annual Wildland Shrub Symposium: the biology of *Artemisia* and *Chrysothamnus*. Provo, Utah. July 9-13, 1984.

1E2 840418 Fisser, H.G., C.W. Hanson and J.R. Wight. 1984. Biology and ecology of sagebrush in Wyoming: V. Herbage yield dynamics and seasonal precipitation relationships. p. 13. In Abst. Shrub Research Consortium Third Annual Wildland Shrub Symposium: the biology of *Artemisia* and *Chrysothamnus*. Provo, Utah. July 9-13, 1984.

1E2 840419 Fisser, H.G. and K.L. Johnson, eds. 1984. Wyoming Shrublands Proc. 11th Wyoming Shrub Ecology Workshop. Lander, Wyoming. May 27-28, 1982. University of Wyoming, Laramie.

1E3. Photosynthesis and Plant Nutrients

1E3 770403 Cox, Jerry R., and H.G. Fisser. 1977. The effect of environmental factors on carbohydrate storage and herbage production of five short-grass prairie plant species. p. 41. In Proc. Soc. Range Manage. 30th Ann. Meeting. Portland, Oregon. February 14-17, 1977.

1E3 790401 Cox, J.R., and H.G. Fisser. 1979. The effects of water stress on phenology and carbohydrate. p. 529-535. In J.R. Goodin and D.K. Northington (eds.) Proc. of the International Arid Lands Conference. International Center for Arid and Semiarid Land Studies. Texas Tech. University, Lubbock, Texas. October 8-15, 1978.

1E3 830401 Fisser, Herbert G. 1983. Some recent research findings in the true mountain mahogany types. p. 7-10. In H.G. Fisser and K.L. Johnson (eds.) Wyoming Shrublands: Proc. 10th Wyo. Shrub Ecology Workshop. Sybille Canyon, Wyoming. May 28-29, 1981. University of Wyoming, Laramie.

1E3 830404 Waugh, W.J. 1983. Stagnation and decadence of *Cercocarpus montanus* in a southeastern Wyoming big game exclosure, p. 11-14. In H.G. Fisser and K.L. Johnson (eds.). Wyoming Shrublands: Proc. 10th Wyoming Shrub Ecology Workshop, Sybille Canyon, Wyoming, May 28-29, 1981. University of Wyoming, Laramie.

1E4. Herbage Removal

1E4 640401 Fisser, H.G. 1964. Winter sheep grazing studies on saltbush rangeland of Wyoming. p. 53. In Proc. 17th Ann. Meeting, Am. Soc. Range Manage. Wichita, Kansas. February 14, 1964.

1E4 640403 Fisser, H.G. 1964. Summer steer grazing on saltbush rangeland. p. 7. In Proc. Wyo. Sec. Am. Soc. Range Manage. Casper, Wyoming. November 19-20, 1964.

1E4 650402 Fisser, H.G. 1965. Halogeton abundance and distribution as influenced by degree of saltbush utilization. p. 11-13. In Proc. Range Weed Control Conf. Elko, Nevada. August 2, 1965.

1E4 660401 Fisser, H.G. 1966. Sheep grazing intensity studies and vegetative changes on *Atriplex nuttallii.* p. 226-237. In Proc. Salt Desert Shrub Symposium. Cedar City, Utah. August 1-3, 1966.

1E4 680401 Probasco, G.E. 1968. Dietary preferences and utilization of elk on spring range in the northern Big Horn Mountains of Wyoming. p. 22. In Proc. 21st Ann. Meeting Soc. Range Manage. Albuquerque, New Mexico. February 14, 1968.

1E4 770401 Gerhart, William A., and H.G. Fisser. 1977. An evaluation of three grazing management systems on northern desert shrub vegetation of southern Wyoming. p. 52. In Proc. Soc. Range Manage. 30th Ann. Meeting. Portland, Oregon. February 17, 1977.

1E4 770403 Cox, Jerry R., and H.G. Fisser. 1977. The effect of environmental factors on carbohydrate storage and herbage production of five shortgrass prairie plant species. p. 41. In Proc. Soc. Range Manage. 30th Ann. Meeting. Portland, Oregon. February 14-17, 1977.

1E4 790401 Cox, J.R., and H.G. Fisser. 1979. The effects of water stress on phenology and carbohydrate. p. 529-535. In J.R. Goodin and D.K. Northington (eds.) Proc. of the International Arid Lands Conference. International Center for Arid and Semiarid Land Studies. Texas Tech. University, Lubbock, Texas. October 8-15, 1978.

1E4 810406 Rodgers, D.E. and H.G. Fisser. 1981. Grazing system impacts on semiarid vegetation of Wyoming. p. 34. In Absts. Soc. Range Manage. 34th Ann. Meeting. Tulsa, Oklahoma. February 8-13, 1981.

1E4 810413 Fisser, H.G. 1981. Shrub establishment, dominance, and ecology on the juniper and sagebrush-grass types in Wyoming. p. 23-27. In L.S. Stelter, E.J. DePuit, S.A. Mikol (eds.). Proc. Symposium on Shrub Establishment on Disturbed Arid and Semiarid Lands. December 3, 1980. Wyo. Game and Fish Dept. Publication.

1E4 810414 Rodgers, D.E. and H.G. Fisser. 1981. Vegetative response to grazing systems management in Wyoming's Red Desert. p. 451. In Summaries of Papers. J. Allan Smith (ed.). XIV International Grassland Congress. Lexington, Kentucky. June 14-23, 1981.

1E4 820401 Fisser, H.G. 1982. Vegetation response to grazing management strategies on semiarid shrublands of Wyoming. p. 162-165. In D.D. Briske and M.M. Kothman (eds.). Proc. Natl. Conf. on Grazing Techniques. Dept. Range Sci. Texas A & M University. College Station, Texas. November 10-13, 1982.

1E4 830402 Fisser, H.G. and K.L. Johnson (eds.) 1983. Wyoming Shrublands: Proc. 9th Wyoming Shrub Ecology Workshop. June 5-6, 1980. Lander, Wyoming. University of Wyoming, Laramie. 40 p.

1E4 830403 Fisser, H.G. and K.L. Johnson (eds.) 1983. Wyoming Shrublands: Proc. 10th Wyoming Shrub Ecology Workshop. May 28-29, 1981. Sybille Canyon, Wyoming. University of Wyoming, Laramie. 40 p.

1E4 830404 Waugh, W.J. 1983. Stagnation and decadence of *Cercocarpus montanus* in a southeastern Wyoming big game exclosure, p. 11-14. In H.G. Fisser and K.L. Johnson (eds.). Wyoming Shrublands: Proc. 10th Wyoming Shrub Ecology Workshop, Sybille Canyon, Wyoming, May 28-29, 1981. University of Wyoming, Laramie.

2. Range Ecology

2 630402 Fisser, H.G. 1963. Exclosure studies (production, cover, height, and moisture phases). p. 23-24. In Proc. Range Weed Control Conf. Portland, Oregon. March 19, 1963.

2A. CLIMATIC FACTORS

2A 610404 Hull, A.C. Jr. 1961. Seeding salt desert shrub lands in Wyoming. p. 7-11. In Proc. Range Weed Research Tour. Worland, Wyoming. September 14-15, 1961.

2A 650401 Fisser, H.G. 1965. Vegetation changes from cold desert to subalpine communities. p. 425-427. In Proc. IX International Grassland Congress. Sao Paulo, Brazil. January 18, 1965.

2A 660402 Fisser, H.G. 1966. Soil moisture and temperature characteristics as related to grass production following shrub control under grazed and ungrazed conditions. In Proc. X Intern. Grassl. Congress. Helsinki, Finland. July 1966.

2A 670401 Fisser, H.G. 1967. Soil moisture and temperature characteristics as related to grass production following shrub control under grazed and non-grazed conditions. p. 23-25. In 20th Ann. Meeting Proc. Am. Soc. Range Manage. Seattle, Washington. February 14, 1967.

2A 690402 Steger, R.E. 1969. Soil moisture relationships of certain salt desert shrubs. p. 50. In Proc. 22nd Ann. Meeting Soc. Range Manage. Calgary, Alberta. February 10-13, 1969.

2A 690403 Wambolt, C.L. 1969. Size, density, growth rate and age of Rocky Mountain juniper and interrelationships with environmental factors. p. 57. In Proc. 22nd Ann. Meeting Soc. Range Manage. Calgary, Alberta. February 10-13, 1969.

2A 710401 Wambolt, C.L. 1971. Moisture stress in woody plants as influenced by environmental factors. p. 25-26. In Proc. 24th Ann. Meeting. Soc. Range Manage. Reno, Nevada. February 14-18, 1971.

2A 740401 Fisser, H.G. 1974. Productivity differences of semidesert grasslands in western Wyoming. p. 22-23. In Proc. 27th Ann. Meeting, Soc. Range Manage. Tucson, Arizona. February 5, 1974.

2A 740404 Fisser, H.G. 1974. Productivity potential differences of grasslands of semidesert Western Wyoming p. 116-124. In V.G. Iglovikov, A.P. Movsissyants, F.F. Sarganov, and A.A. Stepanenko (Compilers). Improvement of natural and production of seeded meadows and pastures. XII International Grassland Congress, Moscow, USSR. June 12-16, 1974.

2A 740406 Ries, R.E. 1974. Environmental factors affecting annual herbage production at 17 sagebrush-grass locations in Wyoming. p. 21-22. In Proc. 27th Ann. Meeting Soc. Range Manage. Tucson, Arizona. February 3-8, 1974.

2A 760401 Stroud, D.W., and H.G. Fisser. 1976. Vegetative response of *Bouteloua gracilis* to four levels of utilization. p. 11. In Proc. Soc. Range Manage. 29th Ann. Meeting. Omaha, Nebraska. February 16-20, 1976.

2A 770402 Kleinman, Larry H. and H.G. Fisser. 1977. Phenodynamics and ecology of sagebrush-grass rangelands. p. 11. In Proc. Soc. Range Manage. 30th Ann. Meeting. Portland, Oregon. February 14-17, 1977.

2A 770403 Cox, Jerry R., and H.G. Fisser. 1977. The effect of environmental factors on carbohydrate storage and herbage production of five short-grass prairie plant species. p. 41 In Proc. Soc. Range Manage. 30th Ann. Meeting. Portland, Oregon. February 14-17, 1977.

2A 770404 Cleaves, D.T. 1977. Development of snowfence systems on strip mine spoils to increase seedling establishment. p. 32. In Proc. Soc. Range Manage. 30th Ann. Meeting. Portland, Oregon. February 14-17, 1977.

2A 770405 Gdara, A.O. 1977. Long-term vegetation studies with permanent quadrats. p. 8. In Proc. Soc. Range Manage. 30th Ann. Meeting. Portland, Oregon. February 14-17, 1977.

2A 780402 Taha, Faisal K. and H.G. Fisser. 1978. Models for prediction of natural vegetation potential in eastern Wyoming. p. 19. In Proc. Soc. Range Manage. 31st Ann. Meeting. San Antonio, Texas. February 7-9, 1978.

2A 780404 Gdara, A.O. and H.G. Fisser. 1978. Semiarid rangeland and grazing evaluation of shrubland from permanent quadrats. Manuscript submitted to 1st Intern. Rangeland Congress, Denver, Colorado. August 14-18, 1978.

2A 780406 Fisser, H.G. 1978. A structural definition of semiarid *Artemisia* and *Atriplex* rangeland ecosystems. Abstract for 2nd Intern. Congress of Ecology. Jerusalem, Israel. September 10-16, 1978.

2A 790401 Cox, J.R., and H.G. Fisser. 1979. The effects of water stress on phenology and carbohydrate. p. 529-535. In J.R. Goodin and D.K. Northington (eds.) Proc. of the International Arid Lands Conference. International Center for Arid and Semiarid Land Studies. Texas Tech. University, Lubbock, Texas. October 8-15, 1978.

2A 790402 Young, L.W., and H.G. Fisser. 1979. Soils-phenology relations on some arid rangelands of western Wyoming. p. 53-54. In Proc. 32nd Ann. Meeting, Soc. Range Manage. Casper, Wyoming. February 12-15, 1979.

2A 810402 Hargis, N.E. and H.G. Fisser. 1981. Phenological comparison of two sagebrush species in Wyoming. p. 21. In Absts. Soc. Range Manage. 34th Ann. Meeting. Tulsa, Oklahoma. February 8-13, 1981.

2A 810403 Joyce, L.A. and H.G. Fisser. 1981. The effect of community changes on precipitation/production relationships in exclosures of saltbush vegetation. p. 22. In Absts. Papers Soc. Range Manage. 34th Ann. Meeting. Tulsa, Oklahoma. February 8-13, 1981.

2A 810405 Owens, M.K. and H.G. Fisser. 1981. Phenology and productivity comparisons of western wheatgrass *(Agropyron smithii)* and bluebunch wheatgrass *(A. spicatum)*. p. 21. In Absts. Soc. Range Manage. 34th Ann. Meeting. Tulsa, Oklahoma. February 8-13, 1981.

2A 810407 Spaeth, K.E. and H.G. Fisser. 1981. Plant composition response to picloram application in the Wyoming juniper type. p. 35. In Absts. Soc. Range Manage. 34th Ann. Meeting. Tulsa, Oklahoma. February 8-13, 1981.

2A 810409 Uhlich, J.W. and H.G. Fisser. 1981. Vegetation dynamics of non-grazed semiarid sagebrush grassland communities in the Big Horn Basin of Wyoming. p. 22. In Absts. Soc. Range Manage. 34th Ann. Meeting. Tulsa, Oklahoma. February 8-13, 1981.

2A 810411 Young, L.W. and H.G. Fisser. 1981. Correlation of soil factors and environmental conditions with phenophase development of western wheatgrass *(Agropyron smithii)*. p. 12. In Absts. Soc. Range Manage. 34th Ann. Meeting. Tulsa, Oklahoma. February 8-13, 1981.

2A 810412 Shawesh, O.M. and H.G. Fisser. 1981. Habitat and vegetation of semiarid rangelands, Libya, Africa. p. 23. In Absts. Soc. Range Manage. 34th Ann. Meeting. Tulsa, Oklahoma. February 8-13, 1981.

2A 810413 Fisser, H.G. 1981. Shrub establishment, dominance, and ecology on the juniper and sagebrush-grass types in Wyoming. p. 23-27. In L.S. Stelter, E.J. DePuit, S.A. Mikol (eds.). Proc. Symposium on Shrub Establishment on Disturbed Arid and Semiarid Lands. December 3, 1980. Wyoming Game and Fish Dept. Publication.

2A 810415 Owens, M.K. and H.G. Fisser. 1981. Phenology and culm weight variation between two wheatgrass species: *Agropyron smithii* and *A. spicatum*. p. 205. In Summaries of Papers. J. Allan Smith (ed.). XIV International Grassland Congress. Lexington, Kentucky. June 14-23, 1981.

2A 820402 Dean, S.K. 1982. Vegetation response to strip chemical control in the Government Draw area. p. 35-36. In: Fisser, H.G. and K.L. Johnson, eds. 1984. Proceedings of the 11th Wyoming Shrub Ecology Workshop. Lander, Wyoming. May 27-28, 1982. University of Wyoming, Laramie.

2A 820403 Kinucan, R.J. 1982. Numerical phenology scoring and use with quantitative environmental variables. p. 36-39. In: Fisser, H.G. and K.L. Johnson eds.1984. Proceedings of the 11th Wyoming Shrub Ecology Workshop. Lander, Wyoming. May 27-28, 1982. University of Wyoming, Laramie.

2A 820404 Fisser, H.G. 1982. Vegetation productivity and phenology in the Wind River Basin. p. 42-45. In: Fisser, H.G. and K.L. Johnson, eds. 1984. Proceedings of the 11th Wyoming Shrub Ecology Workshop. Lander, Wyoming. May 27-28, 1982. University of Wyoming, Laramie.

2A 820405 Mulholland, J.L. 1982. Sagebrush and perennial grass dynamics at the Granite Mountain shrub control site. p. 39-42. In: Fisser, H.G. and K.L. Johnson, eds. 1984. Proceedings of the 11th Wyoming Shrub Ecology Workshop. Lander, Wyoming. May 27-28, 1982. University of Wyoming, Laramie.

2A 830401 Fisser, Herbert G. 1983. Some recent research findings in the true mountain mahogany types. p. 7-10. In H.G. Fisser and K.L. Johnson (eds.) Wyoming Shrublands: Proc. 10th Wyo. Shrub Ecology Workshop. Sybille Canyon, Wyoming. May 28-29, 1981. University of Wyoming, Laramie.

2A 830402 Fisser, H.G. and K.L. Johnson (eds.) 1983. Wyoming Shrublands: Proc. 9th Wyoming Shrub Ecology Workshop. June 5-6, 1980. Lander, Wyoming. University of Wyoming, Laramie. 40 p.

2A 830403 Fisser, H.G. and K.L. Johnson (eds.) 1983. Wyoming Shrublands: Proc. 10th Wyoming Shrub Ecology Workshop. May 28-29, 1981. Sybille Canyon, Wyoming. University of Wyoming, Laramie. 40 p.

2A 830405 Dean, S.K. and H.G. Fisser. 1983. Long-term vegetation responses to sagebrush control in western Wyoming. abst. 29. In Absts. 36th Ann. Meeting. Soc. Range Manage. Albuquerque, New Mexico. February 14-17, 1983.

2A 830407 Elwakeel, A.S. and H.G. Fisser. 1983. Perennial grass production dynamics among five vegetation types in north central Wyoming. abst. 208. In Absts. 36th Ann. Meeting. Soc. Range Manage. Albuquerque, New Mexico. February 14-17, 1983.

2A 830408 Fisser, H.G. 1983. Phenological description and numerical scoring of sagebrush. abst. 136. In Absts. 36th Ann. Meeting. Soc. Range Manage. Albuquerque, New Mexico. February 14-17, 1983.

2A 830409 Joyce, L.A. and H.G. Fisser. 1983. Productivity differences of grasslands in semi-desert western Wyoming — A re-examination. abst. 207. In Absts. 36th Ann. Meeting. Soc. Range Manage. Albuquerque, New Mexico. February 14-17, 1983.

2A 830410 Kinucan, R.J. and H.G. Fisser. 1983. Dynamics of plant community development for diverse sagebrush sites in western Wyoming. abst. 65. In Absts. 36th Ann. Meeting. Soc. Range Manage. Albuquerque, New Mexico. February 14-17, 1983.

2A 830411 Mulholland, J.L. and H.G. Fisser. 1983. Reestablishment rates of sagebrush following chemical control on grazed and ungrazed ranges in western Wyoming. abst. 295. In Absts. 36th Ann. Meeting. Soc. Range Manage. Albuquerque, New Mexico. February 14-17, 1983.

2A 830412 Owens, M.K. and H.G. Fisser. 1983. Size and density parameters of western wheatgrass on diverse range sites in western Wyoming. abst. 128. In Absts. 36th Ann. Meeting. Soc. Range Manage. Albuquerque, New Mexico. February 14-17, 1983.

2A 840401 Fisser, H.G. 1984. Productivity index procedures for sagebrush-grass rangeland. abst. 010. In Abst. 37th Ann. Meeting. Soc. Range Manage. Rapid City, South Dakota. February 12-17, 1984.

2A 840402 Fisser, H.G. and N.E. Hargis. 1984. Detailed numerical phenology of sagebrush with visual description statements. p. 44. In Proc. 37th Ann. Meeting Soc. Range Manage. Rapid City, South Dakota. February 12-17, 1984.

2A 840403 Germain, L.P. and H.G. Fisser. 1984. Estimation of missing precipitation data using traditional, kriging, and subjective techniques. abst. 160. In Abst. 37th Ann. Meeting Soc. Range Manage. Rapid City, South Dakota. February 12-17, 1984.

2A 840405 Dean, S.K. and H.G. Fisser. 1984. Grazing management, longevity, and benefits of sagebrush control on semiarid rangelands. Submitted to Proc. 2nd Intern. Rangeland Congress. Adelaide, Australia. May 13-18, 1984.

2A 840406 Kinucan, R.J. and H.G. Fisser. 1984. Vegetative phenology patterns of semiarid plant communities in Wyoming, U.S.A. Submitted to Proc. 2nd Intern. Rangeland Congress. Adelaide, Australia. May 13-18, 1984.

2A 840407 Fisser, H.G. and L.A. Joyce. 1984. Saltbush community changes with and without grazing over a 20-year period in north central Wyoming. Submitted to Proc. 2nd Intern. Rangeland Congress. Adelaide, Australia. May 13-18, 1984.

2A 840408 Mulholland, J.L. and H.G. Fisser. 1984. Big sagebrush response and reestablishment following chemical control in Wyoming, U.S.A. Submitted to Proc. 2nd Intern. Rangeland Congress. Adelaide, Australia. May 13-18, 1984.

2A 840409 Germain, L.P. and H.G. Fisser. 1984. Application of kriging to ecological variables. p. 12. In Proc. 55th Ann. Meeting Colo.-Wyo. Acad. Sci., Western State College, Gunnison, Colorado. April 27-28, 1984.

2A 840410 Fisser, H.G. 1984. Culm weight dynamics as an index of sagebrush-grass community productivity. p. 17. In Proc. 55th Ann. Meeting Colo.-Wyo. Acad. Sci., Western State College, Gunnison, Colorado. April 27-28, 1984.

2A 840411 Fisser, H.G. 1984. Detailed numerical phenology of sagebrush with visual description statements. Posterboard display at USDA-ARS/USDI-BLM Regional Workshop on Rangeland Hydrology, Modeling and Monitoring. Boise, Idaho. April 31-May 2, 1984.

2A 840412 Dean, S.K. and H.G. Fisser. 1984. Grazing management longevity and benefits of sagebrush control in western Wyoming, U.S.A. Posterboard display at USDA-ARS/USDI-BLM Regional Workshop on Rangeland Hydrology, Modeling and Monitoring. Boise, Idaho. April 30 through May 2, 1984.

2A 840413 Fisser, H.G. and L.A. Joyce. 1984. *Atriplex*/grass and forb relationships under no grazing and shifting precipitation patterns in north central Wyoming. p. 87-96. In Tiedeman, A.R., E.D. McArthur, H.C. Stutz, R. Stevens, and K.L. Johnson (eds.). Proc. Wildland Shrub Symp. "Biology of *Atriplex* and related chenopods." Provo, Utah. May 2-6, 1983. USDA Forest Service General Technical Report 172.

2A 840414 Fisser, H.G. 1984. Biology and ecology of sagebrush in Wyoming: I. Herbage production and community dynamics. p. 21. In Abst. Shrub Research Consortium Third Annual Wildland Shrub Symposium: the biology of *Artemisia* and *Chrysothamnus.* Provo, Utah. July 9-13, 1984.

2A 840415 Fisser, H.G. 1984. Biology and ecology of sagebrush in Wyoming: II. Reinvasion following control. p. 21. In Abst. Shrub Research Consortium Third Annual Wildland Shrub Symposium: the biology of *Artemisia* and *Chrysothamnus.* Provo, Utah. July 9-13, 1984.

2A 840417 Fisser, H.G., J.R. Wight and C.W. Hanson. 1984. Biology and ecology of sagebrush in Wyoming: IV. Validation of a rangeland production model (ERHYM) for sagebrush sites. p. 13. In Abst. Shrub Research Consortium Third Annual Wildland Shrub Symposium: the biology of *Artemisia* and *Chrysothamnus.* Provo, Utah. July 9-13, 1984.

2A 840418 Fisser, H.G., C.W. Hanson and J.R. Wight. 1984. Biology and ecology of sagebrush in Wyoming: V. Herbage yield dynamics and seasonal precipitation relationships. p. 13. In Abst. Shrub Research Consortium Third Annual Wildland Shrub Symposium: the biology of *Artemisia* and *Chrysothamnus.* Provo, Utah. July 9-13, 1984.

2A 840419 Fisser, H.G. and K.L. Johnson, eds. 1984. Wyoming Shrublands Proc. 11th Wyoming Shrub Ecology Workshop. Lander, Wyoming. May 27-28, 1982. University of Wyoming, Laramie.

2B. BIOTIC FACTORS

2B 670402 Kirkham, D.R. 1967. Western harvester ant abundance in relation to grazing, vegetation and soil in north central Wyoming. p. 28. In Proc. 20th Ann. Meeting Am. Soc. Range Manage. Seattle, Washington. February 14, 1967.

2B 670403 Fisser, H.G. 1967. Plant control on salt desert shrub rangeland by sheep grazing. p. 48-54. In Proc. Range Weed Control Conf. Las Cruces, New Mexico. August 21, 1967. p. 48-54.

2B 740401 Fisser, H.G. 1974. Productivity differences of semi-desert grasslands in western Wyoming. p. 22-23. In Proc. 27th Ann. Meeting, Soc. Range Manage. Tucson, Arizona. February 5, 1974.

2B 740404 Fisser, H.G. 1974. Productivity potential differences of grasslands of semi-desert Western Wyoming p. 116-124. In V.G. Iglovikov, A.P. Movsissyants, F.F. Sarganov, and A.A. Stepanenko (Compilers). Improvement of natural and production of seeded meadows and pastures. XII International Grassland Congress, Moscow, USSR. June 12-16, 1974.

2B 770403 Cox, Jerry R., and H.G. Fisser. 1977. The effect of environmental factors on carbohydrate storage and herbage production of five shortgrass prairie plant species. p. 41. In Proc. Soc. Range Manage. 30th Ann. Meeting. Portland, Oregon. February 14-17, 1977.

2B 780405 Shawesh, O.M. and H.G. Fisser. 1978. Rangeland development and improvement in Libya, Africa. p. 65. In J.R. Goodin and D.K. Northington (eds.) Abstracts Intern. Arid Lands Conf. Intern. Center for Arid and Semiarid Land Studies. Texas Tech University, Lubbock, Texas. October 8-15, 1978.

2B 790401 Cox, J.R., and H.G. Fisser. 1979. The effects of water stress on phenology and carbohydrate. p. 529-535. In J.R. Goodin and D.K. Northington (eds.) Proc. of the International Arid Lands Conference. International Center for Arid and Semiarid Land Studies. Texas Tech. University. Lubbock, Texas. October 8-15, 1978.

2B 810403 Joyce, L.A. and H.G. Fisser. 1981. The effect of community changes on precipitation/production relationships in exclosures of saltbush vegetation. p. 22. In Absts. Papers Soc. Range Manage. 34th Ann. Meeting. Tulsa, Oklahoma. February 8-13, 1981.

2B 810409 Uhlich, J.W. and H.G. Fisser. 1981. Vegetation dynamics of non-grazed semiarid sagebrush grassland communities in the Big Horn Basin of Wyoming. p. 22. In Absts. Soc. Range Manage. 34th Ann. Meeting. Tulsa, Oklahoma. February 8-13, 1981.

2B 830409 Joyce, L.A. and H.G. Fisser. 1983. Productivity differences of grasslands in semi-desert western Wyoming — A re-examination. abst. 207. In Absts. 36th Ann. Meeting. Soc. Range Manage. Albuquerque, New Mexico. February 14-17, 1983.

2B 830411 Mulholland, J.L. and H.G. Fisser. 1983. Reestablishment rates of sagebrush following chemical control on grazed and ungrazed ranges in western Wyoming. abst. 295. In Absts. 36th Ann. Meeting. Soc. Range Manage. Albuquerque, New Mexico. February 14-17, 1983.

2B 840401 Fisser, H.G. 1984. Productivity index procedures for sagebrush-grass rangeland. abst. 010. In Abst. 37th Ann. Meeting. Soc. Range Manage. Rapid City, South Dakota. February 12-17, 1984.

2B 840408 Mulholland, J.L. and H.G. Fisser. 1984. Big sagebrush response and reestablishment following chemical control in Wyoming, U.S.A. Submitted to Proc. 2nd Intern. Rangeland Congress. Adelaide, Australia. May 13-18, 1984.

2B 840410 Fisser, H.G. 1984. Culm weight dynamics as an index of sagebrush-grass community productivity. p. 17. In Proc. 55th Ann. Meeting Colo.-Wyo. Acad. Sci., Western State College, Gunnison, Colorado. April 27-28, 1984.

2C. EDAPHIC FACTORS

2C 640402 Nichols, J.R. 1964. Soil-vegetation relations of the 15-Mile Drainage area in the Big Horn Basin of Wyoming. p. 25. In Proc. 17th Ann. Meeting Am. Soc. Range Manage. Wichita, Kansas. February 14, 1964.

2C 650401 Fisser, H.G. 1965. Vegetation changes from cold desert to subalpine communities. p. 425-427. In Proc. IX International Grassland Congress. Sao Paulo, Brazil. January 18, 1965.

2C 660402 Fisser, H.G. 1966. Soil moisture and temperature characteristics as related to grass production following shrub control under grazed and ungrazed conditions. In Proc. X Intern. Grassl. Congress. Helsinki, Finland. July, 1966.

2C 670401 Fisser, H.G. 1967. Soil moisture and temperature characteristics as related to grass production following shrub control under grazed and non-grazed conditions. p. 23-25. In 20th Ann. Meeting Proc. Am. Soc. Range Manage. Seattle, Washington. February 14, 1967.

2C 670402 Kirkham, D.R. 1967. Western harvester ant abundance in relation to grazing, vegetation and soil in northcentral Wyoming. p. 28. In 20th Ann. Meeting Proc. Am. Soc. Range Manage. Seattle, Washington. February 14, 1967.

2C 690402 Steger, R.E. Soil moisture relationships of certain salt desert shrubs. p. 50. In Proc. 22nd Ann. Meeting Soc. Range Manage. Calgary, Alberta. February 10-13, 1969.

2C 690403 Wambolt, C.L. 1969. Size, density, growth rate and age of Rocky Mountain juniper and interrelationships with environmental factors. p. 57. In Proc. 22nd Ann. Meeting Soc. Range Manage. Calgary, Alberta. February 10-13, 1969.

2C 710401 Wambolt, C.L. 1971. Moisture stress in woody plants as influenced by environmental factors. p. 25-26. In Proc. 24th Ann. Meeting. Soc. Range Manage. Reno, Nevada. February 14-18, 1971.

2C 740403 Stroud, D.W. and H.G. Fisser. 1974. Influence of livestock grazing on growth patterns, plant density and soil microrelief of a *Bouteloua gracilis* prairie, p. 347-351. In V.G. Iglovikov, A.P. Movsissyants, F.F. Sarganov, and A.A. Stepanenko (Compilers) Biological and physiological aspects of the intensification of grassland utilization. XII International Grassland Congress, Moscow, USSR. June 12-16, 1974.

2C 740406 Ries, R.E. 1974. Environmental factors affecting annual herbage production at 17 sagebrush-grass locations in Wyoming. p. 21-22. In Proc. 27th Ann. Meeting Soc. Range Manage. Tucson, Arizona. February 3-8, 1974.

2C 750401 Fisser, H.G. 1975. A study of rangeland soil movement characteristics. p. 421-422. In Proc. Amer. Soc. Civil Engineers Watershed Manage. Symp. Logan, Utah. August 11-13, 1975.

2C 760402 Taha, Faisal K., and H.G. Fisser. 1976. Physical and chemical analyses of range sites and interrelationships with herbage production and forage value potentials. p. 13. In Proc. Soc. Range Manage. 29th Ann. Meeting. Omaha, Nebraska. February 16-20, 1976.

2C 760403 Lymbery, Gordon A., and H.G. Fisser. 1976. Vegetation re-establishment potential of strip-mined soils, Black Thunder Basin, Wyoming. p. 49. In Proc. Soc. Range Manage. 29th Ann. Meeting. Omaha, Nebraska. February 16-20, 1976.

2C 770403 Cox, Jerry R., and H.G. Fisser. 1977. The effect of environmental factors on carbohydrate storage and herbage production of five shortgrass prairie plant species. p. 41. In Proc. Soc. Range Manage. 30th Ann. Meeting. Portland, Oregon. February 14-17, 1977.

2C 770404 Cleaves, D.T. 1977. Development of snowfence systems on strip mine spoils to increase seedling establishment. p. 32. In Proc. Soc. Range Manage. 30th Ann. Meeting. Portland, Oregon. February 14-17, 1977.

2C 780401 Fisser, Herbert G. 1978. Soil erosion and movement studies in western Wyoming. p. 16. In Proc. Soc. Range Manage. 31st Ann. Meeting, San Antonio, Texas. February 7-9, 1978.

2C 780402 Taha, Faisal K. and H.G. Fisser. 1978. Models for prediction of natural vegetation potential in eastern Wyoming. p. 19. In Proc. Soc. Range Manage. 31st Ann. Meeting. San Antonio, Texas. February 7-9, 1978.

2C 780405 Shawesh, O.M. and H.G. Fisser. 1978. Rangeland development and improvement in Libya, Africa. p. 65. In J.R. Goodin and D.K. Northington (eds.) Abstracts Intern. Arid Lands Conf. Intern. Center for Arid and Semiarid Land Studies. Texas Tech University. Lubbock, Texas. October 8-15, 1978.

2C 780406 Fisser, H.G. 1978. A structural definition of semiarid *Artemisia* and *Atriplex* rangeland ecosystems. Abstract for 2nd Intern. Congress of Ecology. Jerusalem, Israel. September 10-16, 1978.

2C 790401 Cox J.R., and H.G. Fisser. 1979. The effects of water stress on phenology and carbohydrate. p. 529-535. In J.R. Goodin and D.K. Northington (eds.) Proc. of the International Arid Lands Conference. International Center for Arid and Semiarid Land Studies. Texas Tech. University. Lubbock, Texas. October 8-15, 1978.

2C 790402 Young, L.W., and H.G. Fisser. 1979. Soils-phenology relations on some arid rangelands of western Wyoming. p. 53-54. In Proc. 32nd Ann. Meeting, Soc. Range Manage. Casper, Wyoming. February 12-15, 1979.

2C 800401 Trueblood, D.C. and H.G. Fisser. 1980. Soil vegetation relationships at semiarid rangeland exclosures in north central Wyoming — long term studies. p. 31. In Proc. 33rd Ann. Meeting Soc. Range Manage. San Diego, California. February 11-14, 1980.

2C 800402 Hajjar, S., G. Spitzer, G. Meyer, and H.G. Fisser. 1980. An evaluation of water and soil samples for forage and other agricultural development in the Tabuk region. A report submitted to A.M. Aziz, Governor, Tabuk Province, Saudi Arabia. Amer. Res. Design Scientists, Inc. 49 p.

2C 810402 Hargis, N.E. and H.G. Fisser. 1981. Phenological comparison of two sagebrush species in Wyoming. p. 21. In Absts. Soc. Range Manage. 34th Ann. Meeting. Tulsa, Oklahoma. February 8-13, 1981.

2C 810403 Joyce, L.A. and H.G. Fisser. 1981. The effect of community changes on precipitation/production relationships in exclosures of saltbush vegetation. p. 22. In Absts. Papers Soc. Range Manage. 34th Ann. Meeting. Tulsa, Oklahoma. February 8-13, 1981.

2C 810405 Owens, M.K. and H.G. Fisser. 1981. Phenology and productivity comparisons of western wheatgrass *(Agropyron smithii)* and bluebunch wheatgrass *(A. spicatum)*. p. 21. In Absts. Soc. Range Manage. 34th Ann. Meeting. Tulsa, Oklahoma. February 8-13, 1981.

2C 810407 Spaeth, K.E. and H.G. Fisser. 1981. Plant composition response to picloram application in the Wyoming juniper type. p. 35. In Absts. Soc. Range Manage. 34th Ann. Meeting. Tulsa, Oklahoma. February 8-13, 1981.

2C 810408 Trueblood, D.C. and H.G. Fisser. 1981. Soil-vegetation relationships in the Gardner saltbush vegetation type of Wyoming. p. 12. In Absts. Soc. Range Manage. 34th Ann. Meeting. Tulsa, Oklahoma. February 8-13, 1981.

2C 810409 Uhlich, J.W. and H.G. Fisser. 1981. Vegetation dynamics of non-grazed semiarid sagebrush grassland communities in the Big Horn Basin of Wyoming. p.22. In Absts. Soc. Range Manage. 34th Ann. Meeting. Tulsa, Oklahoma. February 8-13, 1981.

2C 810410 Waugh, W.J. and H.G. Fisser. 1981. Population structure of Utah juniper in northcentral Wyoming. p. 40. In Absts. Soc. Range Manage. 34th Ann. Meeting. Tulsa, Oklahoma. February 8-13, 1981.

2C 810411 Young, L.W. and H.G. Fisser. 1981. Correlation of soil factors and environmental conditions with phenophase development of western wheatgrass *(Agropyron smithii).* p. 12. In Absts. Soc. Range Manage. 34th Ann. Meeting. Tulsa, Oklahoma. February 8-13, 1981.

2C 810412 Shawesh, O.M. and H.G. Fisser. 1981. Habitat and vegetation of semiarid rangelands, Libya, Africa. p. 23. In Absts. Soc. Range Manage. 34th Ann. Meeting. Tulsa, Oklahoma. February 8-13, 1981.

2C 810413 Fisser, H.G. 1981. Shrub establishment, dominance, and ecology on the juniper and sagebrush-grass types in Wyoming. p. 23-27. In L.S. Stelter, E.J. DePuit, S.A. Mikol (eds.). Proc. Symposium on Shrub Establishment on Disturbed Arid and Semiarid Lands. December 3, 1980. Wyoming Game and Fish Dept. Publication.

2C 810415 Owens, M.K. and H.G. Fisser. 1981. Phenology and culm weight variation between two wheatgrass species: *Agropyron smithii* and *A. spicatum.* p. 205. In Summaries of Papers. J. Allan Smith (ed.). XIV International Grassland Congress. Lexington, Kentucky. June 14-23, 1981.

2C 820402 Dean, S.K. 1982. Vegetation response to strip chemical control in the Government Draw area. p. 35-36. In: Fisser, H.G. and K.L. Johnson, eds. 1984. Proceedings of the 11th Wyoming Shrub Ecology Workshop. Lander, Wyoming. May 27-28, 1982. University of Wyoming, Laramie.

2C 820403 Kinucan, R.J. 1982. Numerical phenology scoring and use with quantitative environmental variables. p. 36-39. In: Fisser, H.G. and K.L. Johnson eds. 1984. Proceedings of the 11th Wyoming Shrub Ecology Workshop. Lander, Wyoming. May 27-28, 1982. University of Wyoming, Laramie.

2C 820405 Mulholland, J.L. 1982. Sagebrush and perennial grass dynamics at the Granite Mountain shrub control site. p. 39-42. In: Fisser, H.G. and K.L. Johnson, eds. 1984. Proceedings of the 11th Wyoming Shrub Ecology Workshop. Lander, Wyoming. May 27-28, 1982. University of Wyoming, Laramie.

2C 830401 Fisser, Herbert G. 1983. Some recent research findings in the true mountain mahogany types. p. 7-10. In H.G. Fisser and K.L. Johnson (eds.) Wyoming Shrublands: Proc. 10th Wyo. Shrub Ecology Workshop. Sybille Canyon, Wyoming. May 28-29, 1981. University of Wyoming, Laramie.

2C 830402 Fisser, H.G. and K.L. Johnson (eds.) 1983. Wyoming Shrublands: Proc. 9th Wyoming Shrub Ecology Workshop. June 5-6, 1980. Lander, Wyoming. University of Wyoming, Laramie. 40 p.

2C 830403 Fisser, H.G. and K.L. Johnson (eds.) 1983. Wyoming Shrublands: Proc. 10th Wyoming Shrub Ecology Workshop. May 28-29, 1981. Sybille Canyon, Wyoming. University of Wyoming, Laramie. 40 p.

2C 830405 Dean, S.K. and H.G. Fisser. 1983. Long term vegetation responses to sagebrush control in western Wyoming. abst. 29. In Absts. 36th Ann. Meeting. Soc. Range Manage. Albuquerque, New Mexico. February 14-17, 1983.

2C 830407 Elwakeel, A.S. and H.G. Fisser. 1983. Perennial grass production dynamics among five vegetation types in north central Wyoming. abst. 208. In Absts. 36th Ann. Meeting. Soc. Range Manage. Albuquerque, New Mexico. February 14-17, 1983.

2C 830408 Fisser, H.G. 1983. Phenological description and numerical scoring of sagebrush. abst. 136. In Absts. 36th Ann. Meeting. Soc. Range Manage. Albuquerque, New Mexico. February 14-17, 1983.

2C 830410 Kinucan, R.J. and H.G. Fisser. 1983. Dynamics of plant community development for diverse sagebrush sites in western Wyoming. abst. 65. In Absts. 36th Ann. Meeting. Soc. Range Manage. Albuquerque, New Mexico. February 14-17, 1983.

2C 830411 Mulholland, J.L. and H.G. Fisser. 1983. Reestablishment rates of sagebrush following chemical control on grazed and ungrazed ranges in western Wyoming. abst. 295. In Absts. 36th Ann. Meeting. Soc. Range Manage. Albuquerque, New Mexico. February 14-17, 1983.

2C 830412 Owens, M.K. and H.G. Fisser. 1983. Size and density parameters of western wheatgrass on diverse range sites in western Wyoming. abst. 128. In Absts. 36th Ann. Meeting. Soc. Range Manage. Albuquerque, New Mexico. February 14-17, 1983.

2C 830413 Waugh, W.J. and H.G. Fisser. 1983. Utah juniper invasion in north central Wyoming and historical climatic and management relationships. abst. 202. In Absts. 36th Ann. Meeting. Soc. Range Manage. Albuquerque, New Mexico. February 14-17, 1983.

2C 840402 Fisser, H.G. and N.E. Hargis. 1984. Detailed numerical phenology of sagebrush with visual description statements. p. 44. In Proc. 37th Ann. Meeting Soc. Range Manage. Rapid City, South Dakota. February 12-17, 1984.

2C 840404 Fisser, H.G. and W.J. Waugh. 1984. Population structure anomalies of *Juniperus osteosperma* invasion in north central Wyoming, U.S.A. Submitted to Proc. 2nd Intern. Rangeland Congress. Adelaide, Australia. May 13-18, 1984.

2C 840405 Dean, S.K. and H.G. Fisser. 1984. Grazing management, longevity, and benefits of sagebrush control on semiarid rangelands. Submitted to Proc. 2nd Intern. Rangeland Congress. Adelaide, Australia. May 13-18, 1984.

2C 840406 Kinucan, R.J. and H.G. Fisser. 1984. Vegetative phenology patterns of semiarid plant communities in Wyoming, U.S.A. Submitted to Proc. 2nd Intern. Rangeland Congress. Adelaide, Australia. May 13-18, 1984.

2C 840407 Fisser, H.G. and L.A. Joyce. 1984. Saltbush community changes with and without grazing over a 20-year period in north central Wyoming. Submitted to Proc. 2nd Intern. Rangeland Congress. Adelaide, Australia. May 13-18, 1984.

2C 840408 Mulholland, J.L. and H.G. Fisser. 1984. Big sagebrush response and reestablishment following chemical control in Wyoming, U.S.A. Submitted to Proc. 2nd Intern. Rangeland Congress. Adelaide, Australia. May 13-18, 1984.

2C 840411 Fisser, H.G. 1984. Detailed numerical phenology of sagebrush with visual description statements. Posterboard display at USDA-ARS/USDI-BLM Regional Workshop on Rangeland Hydrology, Modeling and Monitoring. Boise, Idaho. April 30 through May 2, 1984.

2C 840412 Dean, S.K. and H.G. Fisser. 1984. Grazing management, longevity, and benefits of sagebrush control in western Wyoming, U.S.A. Posterboard display at USDA-ARS/USDI-BLM Regional Workshop on Rangeland Hydrology, Modeling and Monitoring. Boise, Idaho. April 30 through May 2, 1984.

2C 840413 Fisser, H.G. and L.A. Joyce. 1984. *Atriplex*/grass and forb relationships under no grazing and shifting precipitation patterns in north central Wyoming. p. 87-96. In Tiedeman, A.R., E.D. McArthur, H.C. Stutz, R. Stevens, and K.L. Johnson (eds.). Proc. Wildland Shrub Symp. "Biology of *Atriplex* and related chenopods." Provo, Utah. May 2-6, 1983. USDA Forest Service General Technical Report 172.

2C 840415 Fisser, H.G. 1984. Biology and ecology of sagebrush in Wyoming: II. Reinvasion following control. p. 21. In Abst. Shrub Research Consortium Third Annual Wildland Shrub Symposium: the biology of *Artemisia* and *Chrysothamnus*. Provo, Utah. July 9-13, 1984.

2C 840417 Fisser, H.G., J.R. Wight and C.W. Hanson. 1984. Biology and ecology of sagebrush in Wyoming: IV. Validation of a rangeland production model (ERHYM) for sagebrush sites. p. 13. In Abst. Shrub Research Consortium Third Annual Wildland Shrub Symposium: the biology of *Artemisia* and *Chrysothamnus*. Provo, Utah. July 9-13, 1984.

2D. PYRIC FACTORS

2D 810410 Waugh, W.J. and H.G. Fisser. 1981. Population structure of Utah juniper in north central Wyoming. p. 40. In Absts. Soc. Range Manage. 34th Ann. Meeting. Tulsa, Oklahoma. February 8-13, 1981.

2D 810413 Fisser, H.G. 1981. Shrub establishment, dominance, and ecology on the juniper and sagebrush-grass types in Wyoming. p. 23-27. In L.S. Stelter, E.J. DePuit, S.A. Mikol (eds.). Proc. Symposium on Shrub Establishment on Disturbed Arid and Semiarid Lands. December 3, 1980. Wyoming Game and Fish Dept. Publication.

2D 830402 Fisser, H.G. and K.L. Johnson (eds.) 1983. Wyoming Shrublands: Proc. 9th Wyoming Shrub Ecology Workshop. June 5-6, 1980. Lander, Wyoming. University of Wyoming, Laramie. 40 p.

2D 830403 Fisser, H.G. and K.L. Johnson (eds.) 1983. Wyoming Shrublands: Proc. 10th Wyoming Shrub Ecology Workshop. May 28-29, 1981. Sybille Canyon, Wyoming. University of Wyoming, Laramie. 40 p.

2D 830413 Waugh, W.J. and H.G. Fisser. 1983. Utah juniper invasion in north central Wyoming and historical climatic and management relationships. abst. 202. In Absts. 36th Ann. Meeting. Soc. Range Manage. Albuquerque, New Mexico. February 14-17, 1983.

2D 840404 Fisser, H.G. and W.J. Waugh. 1984. Population structure anomalies of *Juniperus osteosperma* invasion in north central Wyoming, U.S.A. Submitted to Proc. 2nd Intern. Rangeland Congress. Adelaide, Australia. May 13-18, 1984.

2E. AUTECOLOGY

2E 610401 Vosler, L.C. 1961. Halogeton seed burial study. p. 1. In Proc. Range Weed Research Tour. Worland, Wyoming. September 14-15, 1961.

2E 640402 Nichols, J.R. 1964. Soil-vegetation relations of the 15-Mile Drainage area in the Big Horn Basin of Wyoming. p. 25. In Proc. 17th Ann. Meeting Am. Soc. Range Manage. Wichita, Kansas. February 14, 1964.

2E 670404 Fisser, H.G. 1967. Use of picloram for juniper control. p. 37-39. In Proc. Range Weed Control Conf. Las Cruces, New Mexico. August 21, 1967.

2E 690401 Fisser, H.G. 1969. Plant pattern and distribution in ecosystems and relationships to function. p. 183-196. In R.L. Dix and R.G. Beidleman (eds.) Proc. Information Synthesis Project Grassland Biome. U.S. I.B.P. Range Science Series No. 2. Colo. State Univ. Ft. Collins, Colorado.

2E 690402 Steger, R.E. Soil moisture relationships of certain salt desert shrubs. p. 50. In Proc. 22nd Ann. Meeting Soc. Range Manage. Calgary, Alberta. February 10-13, 1969.

2E 690403 Wambolt, C.L. 1969. Size, density, growth rate and age of Rocky Mountain juniper and interrelationships with environmental factors. p. 57. In Proc. 22nd Ann. Meeting Soc. Range Manage. Calgary, Alberta. February 10-13, 1969.

2E 700401 Fisser, H.G. 1970. Distribution of plant biomass to plant organs. p. 34-42 and 110-115. In Proc. Intern. Biol. Prog. Desert Biome Plant Specialists Meeting. Las Vegas, Nevada. January 3-4, 1970.

2E 710401 Wambolt, C.L. 1971. Moisture stress in woody plants as influenced by environmental factors. p. 25-26. In Proc. 24th Ann. Meeting. Soc. Range Manage. Reno, Nevada. February 14-18, 1971.

2E 740401 Fisser, H.G. 1974. Productivity differences of semi-desert grasslands in western Wyoming. p. 22-23. In Proc. 27th Ann. Meeting, Soc. Range Manage. Tuscon, Arizona. February 5, 1974.

2E 740403 Stroud, D.W. and H.G. Fisser. 1974. Influence of livestock grazing on growth patterns, plant density and soil microrelief of a *Bouteloua gracilis* prairie, p. 347-351. In V.G. Iglovikov, A.P. Movsissyants, F.F. Sarganov, and A.A. Stepanenko (Compilers) Biological and physiological aspects of the intensification of grassland utilization. XII International Grassland Congress, Moscow, USSR. June 12-16, 1974.

2E 740404 Fisser, H.G. 1974. Productivity potential differences of grasslands of semidesert Western Wyoming p. 116-124. In V.G. Iglovikov, A.P. Movsissyants, F.F. Sarganov, and A.A. Stepanenko (Compilers) Improvement of natural and production of seeded meadows and pastures. XII International Grassland Congress, Moscow, USSR. June 12-16, 1974.

2E 750406 Shrader, J.A. and H.G. Fisser. 1975. Seed production ecology in the shortgrass prairie of the Central United States. p. 29. In Proc. 28th Ann. Meeting Soc. Range Mange. Mexico City, Mexico. February 10-14, 1975.

2E 760403 Lymbery, Gordon A., and H.G. Fisser. 1976. Vegetation re-establishment potential of strip-mined soils, Black Thunder Basin, Wyoming. p. 49. In Proc. Soc. Range Manage. 29th Ann. Meeting. Omaha, Nebraska. February 16-20, 1976.

2E 770402 Kleinman, Larry H. and H.G. Fisser. 1977. Phenodynamics and ecology of sagebrush-grass rangelands. p. 11. In Proc. Soc. Range Manage. 30th Ann. Meeting. Portland, Oregon. February 14-17, 1977.

2E 770403 Cox, Jerry R., and H.G. Fisser. 1977. The effect of environmental factors on carbohydrate storage and herbage production of five shortgrass prairie plant species. p. 41. In Proc. Soc. Range Manage. 30th Ann. Meeting. Portland, Oregon. February 14-17, 1977.

2E 770405 Gdara, A.O. 1977. Long-term vegetation studies with permanent quadrats. p. 8. In Proc. Soc. Range Manage. 30th Ann. Meeting. Portland, Oregon. February 14-17, 1977.

2E 780402 Taha, Faisal K. and H.G. Fisser. 1978. Models for prediction of natural vegetation potential in eastern Wyoming. p. 19. In Proc. Soc. Range Manage. 31st Ann. Meeting, San Antonio, Texas. February 7-9, 1978.

2E 780403 Foster, M.A., J. Stubbendieck, and H.G. Fisser. 1978. Plant succession on plains pocket gopher mounds. p. 8. In Proc. Soc. Range Manage. 31st Ann. Meeting, San Antonio, Texas. February 7-9, 1978.

2E 780404 Gdara, A.O. and H.G. Fisser. 1978. Semiarid rangeland and grazing evaluation of shrubland from permanent quadrats. Manuscript submitted to 1st Intern. Rangeland Congress, Denver, Colorado. August 14-18, 1978.

2E 780406 Fisser, H.G. 1978. A structural definition of semiarid *Artemisia* and *Atriplex* rangeland ecosystems. Abstract for 2nd Intern. Congress of Ecology. Jerusalem, Israel. September 10-16, 1978.

2E 790401 Cox J.R., and H.G. Fisser. 1979. The effects of water stress on phenology and carbohydrate. p. 529-535. In J.R. Goodin and D.K. Northington (eds.) Proc. of the International Arid Lands Conference. International Center for Arid and Semi-Arid Land Studies. Texas Tech. University, Lubbock, Texas. October 8-15, 1978.

2E 790402 Young, L.W., and H.G. Fisser. 1979. Soils-phenology relations on some arid rangelands of western Wyoming. p. 53-54. In Proc. 32nd Ann. Meeting, Soc. Range Manage. Casper, Wyoming. February 12-15, 1979.

2E 810401 Bensel, W.E. and H.G. Fisser. 1981. Forage production and longevity of sagebrush control on semiarid rangelands in western Wyoming. p. 9. In Absts. Soc. Range Manage. 34th Ann. Meeting. Tulsa, Oklahoma. February 8-13, 1981.

2E 810402 Hargis, N.E. and H.G. Fisser. 1981. Phenological comparison of two sagebrush species in Wyoming. p. 21. In Absts. Soc. Range Manage. 34th Ann. Meeting. Tulsa, Oklahoma. February 8-13, 1981.

2E 810403 Joyce, L.A. and H.G. Fisser. 1981. The effect of community changes on precipitation/production relationships in exclosures of saltbush vegetation. p. 22. In Absts. Papers Soc. Range Manage. 34th Ann. Meeting. Tulsa, Oklahoma. February 8-13, 1981.

2E 810405 Owens, M.K. and H.G. Fisser. 1981. Phenology and productivity comparisons of western wheatgrass *(Agropyron smithii)* and bluebunch wheatgrass *(A. spicatum)*. p. 21. In Absts. Soc. Range Manage. 34th Ann. Meeting. Tulsa, Oklahoma. February 8-13, 1981.

2E 810407 Spaeth, K.E. and H.G. Fisser. 1981. Plant composition response to picloram application in the Wyoming juniper type. p. 35. In Absts. Soc. Range Manage. 34th Ann. Meeting. Tulsa, Oklahoma. February 8-13, 1981.

2E 810409 Uhlich, J.W. and H.G. Fisser. 1981. Vegetation dynamics of non-grazed semiarid sagebrush grassland communities in the Big Horn Basin of Wyoming. p.22. In Absts. Soc. Range Manage. 34th Ann. Meeting. Tulsa, Oklahoma. February 8-13, 1981.

2E 810410 Waugh, W.J. and H.G. Fisser. 1981. Population structure of Utah juniper in north central Wyoming. p. 40. In Absts. Soc. Range Manage. 34th Ann. Meeting. Tulsa, Oklahoma. February 8-13, 1981.

2E 810411 Young, L.W. and H.G. Fisser. 1981. Correlation of soil factors and environmental conditions with phenophase development of western wheatgrass *(Agropyron smithii)*. p. 12. In Absts. Soc. Range Manage. 34th Ann. Meeting. Tulsa, Oklahoma. February 8-13, 1981.

2E 810412 Shawesh, O.M. and H.G. Fisser. 1981. Habitat and vegetation of semiarid rangelands, Libya, Africa. p. 23. In Absts. Soc. Range Manage. 34th Ann. Meeting. Tulsa, Oklahoma. February 8-13, 1981.

2E 810413 Fisser, H.G. 1981. Shrub establishment, dominance, and ecology on the juniper and sagebrush-grass types in Wyoming. p. 23-27. In L.S. Stelter, E.J. DePuit, S.A. Mikol (eds.). Proc. Symposium on Shrub Establishment on Disturbed Arid and Semiarid Lands. December 3, 1980. Wyoming Game and Fish Dept. Publication.

2E 810415 Owens, M.K. and H.G. Fisser. 1981. Phenology and culm weight variation between two wheatgrass species: *Agropyron smithii* and *A. spicatum*. p. 205. In Summaries of Papers. J. Allan Smith (ed.). XIV International Grassland Congress.

2E 820402 Dean, S.K. 1982. Vegetation response to strip chemical control in the Government Draw area. p. 35-36. In: Fisser, H.G. and K.L. Johnson, eds. 1984. Proceedings of the 11th Wyoming Shrub Ecology Workshop. Lander, Wyoming. May 27-28, 1982. University of Wyoming, Laramie.

2E 820403 Kinucan, R.J. 1982. Numerical phenology scoring and use with quantitative environmental variables. p. 36-39. In: Fisser, H.G. and K.L. Johnson eds.1984. Proceedings of the 11th Wyoming Shrub Ecology Workshop. Lander, Wyoming. May 27-28, 1982. University of Wyoming, Laramie.

2E 820404 Fisser, H.G. 1982. Vegetation productivity and phenology in the Wind River Basin. p. 42-45. In: Fisser, H.G. and K.L. Johnson, eds. 1984. Proceedings of the 11th Wyoming Shrub Ecology Workshop. Lander, Wyoming. May 27-28, 1982. University of Wyoming, Laramie.

2E 820405 Mulholland, J.L. 1982. Sagebrush and perennial grass dynamics at the Granite Mountain shrub control site. p. 39-42. In: Fisser, H.G. and K.L. Johnson, eds. 1984. Proceedings of the 11th Wyoming Shrub Ecology Workshop. Lander, Wyoming. May 27-28, 1982. University of Wyoming, Laramie.

2E 830401 Fisser, H.G. 1983. Some recent research findings in the true mountain mahogany types. p. 7-10. In H.G. Fisser and K.L. Johnson (eds.) Wyoming Shrublands: Proc. 10th Wyo. Shrub Ecology Workshop. Sybille Canyon, Wyoming. May 28-29, 1981. University of Wyoming, Laramie.

2E 830402 Fisser, H.G. and K.L. Johnson (eds.) 1983. Wyoming Shrublands: Proc. 9th Wyoming Shrub Ecology Workshop. June 5-6, 1980. Lander, Wyoming. University of Wyoming, Laramie. 40 p.

2E 830403 Fisser, H.G. and K.L. Johnson (eds.) 1983. Wyoming Shrublands: Proc. 10th Wyoming Shrub Ecology Workshop. May 28-29, 1981. Sybille Canyon, Wyoming. University of Wyoming, Laramie. 40 p.

2E 830404 Waugh, W.J. 1983. Stagnation and decadence of *Cercocarpus montanus* in a southeastern Wyoming big game exclosure. p. 11-14. In H.G. Fisser and K.L. Johnson (eds.). Wyoming Shrublands: Proc. 10th Wyoming Shrub Ecology Workshop, Sybille Canyon, Wyoming, May 28-29, 1981. University of Wyoming, Laramie.

2E 830405 Dean, S.K. and H.G. Fisser. 1983. Long-term vegetation responses to sagebrush control in western Wyoming. abst. 29. In Absts. 36th Ann. Meeting. Soc. Range Manage. Albuquerque, New Mexico. February 14-17, 1983.

2E 830407 Elwakeel, A.S. and H.G. Fisser. 1983. Perennial grass production dynamics among five vegetation types in north-central Wyoming. abst. 208. In Absts. 36th Ann. Meeting. Soc. Range Manage. Albuquerque, New Mexico. February 14-17, 1983.

2E 830408 Fisser, H.G. 1983. Phenological description and numerical scoring of sagebrush. abst. 136. In Absts. 36th Ann. Meeting. Soc. Range Manage. Albuquerque, New Mexico. February 14-17, 1983.

2E 830409 Joyce, L.A. and H.G. Fisser. 1983. Productivity differences of grasslands in semidesert western Wyoming — A re-examination. abst. 207. In Absts. 36th Ann. Meeting. Soc. Range Manage. Albuquerque, New Mexico. February 14-17, 1983.

2E 830410 Kinucan, R.J. and H.G. Fisser. 1983. Dynamics of plant community development for diverse sagebrush sites in western Wyoming. abst. 65. In Absts. 36th Ann. Meeting. Soc. Range Manage. Albuquerque, New Mexico. February 14-17, 1983.

2E 830411 Mulholland, J.L. and H.G. Fisser. 1983. Reestablishment rates of sagebrush following chemical control on grazed and ungrazed ranges in western Wyoming. abst. 295. In Absts. 36th Ann. Meeting. Soc. Range Manage. Albuquerque, New Mexico. February 14-17, 1983.

2E 830412 Owens, M.K. and H.G. Fisser. 1983. Size and density parameters of western wheatgrass on diverse range sites in western Wyoming. abst. 128. In Absts. 36th Ann. Meeting. Soc. Range Manage. Albuquerque, New Mexico. February 14-17, 1983.

2E 830413 Waugh, W.J. and H.G. Fisser. 1983. Utah juniper invasion in north central Wyoming and historical climatic and management relationships. abst. 202. In Absts. 36th Ann. Meeting. Soc. Range Manage. Albuquerque, New Mexico. February 14-17, 1983.

2E 840401 Fisser, H.G. 1984. Productivity index procedures for sagebrush-grass rangeland. abst. 010. In Abst. 37th Ann. Meeting. Soc. Range Manage. Rapid City, South Dakota. February 12-17, 1984.

2E 840402 Fisser, H.G. and N.E. Hargis. 1984. Detailed numerical phenology of sagebrush with visual description statements. p. 44 . In Proc 37th Ann. Meeting Soc. Range Manage. Rapid City, South Dakota. February 12-17, 1984.

2E 840403 Germain, L.P. and H.G. Fisser. 1984. Estimation of missing precipitation data using traditional, kriging, and subjective techniques. abst. 160. In Abst. 37th Ann. Meeting Soc. Range Manage. Rapid City, South Dakota. February 12-17, 1984.

2E 840404 Fisser, H.G. and W.J. Waugh. 1984. Population structure anomalies of *Juniperus osteosperma* invasion in northcentral Wyoming, U.S.A. Submitted to Proc. 2nd Intern. Rangeland Congress. Adelaide, Australia. May 13-18, 1984.

2E 840405 Dean, S.K. and H.G. Fisser. 1984. Grazing management, longevity, and benefits of sagebrush control on semiarid rangelands. Submitted to Proc. 2nd Intern. Rangeland Congress. Adelaide, Australia. May 13-18, 1984.

2E 840406 Kinucan, R.J. and H.G. Fisser. 1984. Vegetative phenology patterns of semi-arid plant communities in Wyoming, U.S.A. Submitted to Proc. 2nd Intern. Rangeland Congress. Adelaide, Australia. May 13-18, 1984.

2E 840407 Fisser, H.G. and L.A. Joyce. 1984. Saltbush community changes with and without grazing over a 20-year period in north central Wyoming. Submitted to Proc. 2nd Intern. Rangeland Congress. Adelaide, Australia. May 13-18, 1984.

2E 840408 Mulholland, J.L. and H.G. Fisser. 1984. Big sagebrush response and reestablishment following chemical control in Wyoming, U.S.A. Submitted to Proc. 2nd Intern. Rangeland Congress. Adelaide, Australia. May 13-18, 1984.

2E 840409 Germain, L.P. and H.G. Fisser. 1984. Application of kriging to ecological variables. p. 12. In Proc. 55th Ann. Meeting Colo.-Wyo. Acad. Sci., Western State College. Gunnison, Colorado. April 27-28, 1984.

2E 840410 Fisser, H.G. 1984. Culm weight dynamics as an index of sagebrush-grass community productivity. p. 17. In Proc. 55th Ann. Meeting Colo.-Wyo. Acad. Sci., Western State College. Gunnison, Colorado. Ap[ril 27-28, 1984.

2E 840411 Fisser, H.G. 1984. Detailed numerical phenology of sagebrush with visual description statements. Posterboard display at USDA-ARS/USDI-BLM Regional Workshop on Rangeland Hydrology, Modeling and Monitoring. Boise, Idaho. April 31 through May 2, 1984.

2E 840412 Dean, S.K. and H.G. Fisser. 1984. Grazing management, longevity, and benefits of sagebrush control in western Wyoming, U.S.A. Posterboard display at USDA-ARS/USDI-BLM Regional Workshop on Rangeland Hydrology, Modeling and Monitoring. Boise, Idaho. April 31 through May 2, 1984.

2E 840413 Fisser, H.G. and L.A. Joyce. 1984. *Atriplex*/grass and forb relationships under no grazing and shifting precipitation patterns in north central Wyoming. p. 87-96. In Tiedeman, A.R., E.D. McArthur, H.C. Stutz, R. Stevens, and K.L. Johnson (eds.). Proc. Wildland Shrub Symp. "Biology of *Atriplex* and related chenopods." Provo, Utah. May 2-6, 1983. USDA Forest Service General Technical Report 172.

2E 840414 Fisser, H.G. 1984. Biology and ecology of sagebrush in Wyoming: I. Herbage production and community dynamics. p. 21. In Abst. Shrub Research Consortium Third Annual Wildland Shrub Symposium: the biology of *Artemisia* and *Chrysothamnus*. Provo, Utah. July 9-13, 1984.

2E 840415 Fisser, H.G. 1984. Biology and ecology of sagebrush in Wyoming: II. Reinvasion following control. p. 21. In Abst. Shrub Research Consortium Third Annual Wildland Shrub Symposium: the biology of *Artemisia* and *Chrysothamnus*. Provo, Utah. July 9-13, 1984.

2E 840416 Fisser, H.G. 1984. Biology and ecology of sagebrush in Wyoming: III. Regional variation of phenology. p. 22. In Abst. Shrub Research Consortium Third Annual Wildland Shrub Symposium: the biology of *Artemisia* and *Chrysothamnus*. Provo, Utah. July 9-13, 1984.

2E 840419 Fisser, H.G. and K.L. Johnson, eds. 1984. Wyoming Shrublands Proc. 11th Wyoming Shrub Ecology Workshop. Lander, Wyoming. May 27-28, 1982. University of Wyoming, Laramie.

2F. SYNECOLOGY

2F 750403 Kleinman, L.H. and H.G. Fisser. 1975. Phenodynamics and environmental characteristics of sagebrush-grass rangelands. p. 161. In H.C. Stutz, (ed.) Proc. Wildland Shrubs Symp. U.S. Forest Service Shrub Sciences Lab. Provo, Utah. November 4-6, 1975.

2F 780405 Shawesh, O.M. and H.G. Fisser. 1978. Rangeland development and improvement in Libya, Africa. p. 65. In J.R. Goodin and D.K. Northington (eds.) abstracts Intern. Arid Lands Conf. Intern. Center for Arid and Semiarid Land Studies. Texas Tech University. Lubbock, Texas. October 8-15, 1978.

2F 810412 Shawesh, O.M. and H.G. Fisser. 1981. Habitat and vegetation of semiarid rangelands, Libya, Africa. p. 23. In Absts. Soc. Range Manage. 34th Ann. Meeting. Tulsa, Oklahoma. February 8-13, 1981.

2F1. Great Plains

2F1 720401 Shaver, Jens C. and H.G. Fisser. 1972. Panchromatic, color and infrared photography for detailed vegetation mapping. p. 33. In Proc. 25th Ann. Meeting Soc. Range Manage. Washington, D.C. February 8, 1972.

2F1 740402 Shaver, J.C. and H.G. Fisser. 1974. Detailed grassland type delineation with panchromatic, color and infrared aerial photographs. p. 369-373. In V.G. Iglovikov, A.P. Movsissyants, F.F. Sarganov, A.A. Stepanenko (Compilers). Improvement of natural and production of seeded meadows and pastures. XII International Grassland Congress, Moscow, USSR. June 12-16, 1974.

2F1 740403 Stroud, D.W. and H.G. Fisser. 1974. Influence of livestock grazing on growth patterns, plant density and soil microrelief of a *Bouteloua gracilis* prairie, p. 347-351. In V.G. Iglovikov, A.P. Movsissyants, F.F. Sarganov, and A.A. Stepanenko (Compilers). Biological and physiological aspects of the intensification of grassland utilization. XII International Grassland Congress, Moscow, USSR. June 12-16, 1974.

2F1 750402 Fisser, H.G. and R.E. Ries. 1975. Pre-disturbance ecological studies improve and define potential for surface mine reclamation. p. 128-134. In Proc. 3rd NCA-BCR Coal Conference and Expo II. Vol. I. Louisville, Kentucky. October 21-23, 1975.

2F1 750405 Ries, R.E. and H.G. Fisser. 1975. Pre-disturbance vegetation inventory and ecological monitoring procedures for rangeland strip mining. p. 40. In Proc. 28th Ann. Meeting Soc. Range Manage. Mexico City, Mexico. February 10-14, 1975.

2F1 750406 Shrader, J.A. and H.G. Fisser. 1975. Seed production ecology in the shortgrass prairie of the Central United States. p. 29. In Proc. 28th Ann. Meeting Soc. Range Mange. Mexico City, Mexico. February 10-14, 1975.

2F1 760401 Stroud, D.W., and H.G. Fisser. 1976. Vegetative response of *Bouteloua gracilis* to four levels of utilization. p. 11. In Proc. Soc. Range Manage. 29th Ann. Meeting. Omaha, Nebraska. February 16-20, 1976.

2F1 760402 Taha, Faisal K., and H.G. Fisser. 1976. Physical and chemical analyses of range sites and interrelationships with herbage production and forage value potentials. p. 13. In Proc. Soc. Range Manage. 29th Ann. Meeting. Omaha, Nebraska. February 16-20, 1976.

2F1 760403 Lymbery, Gordon A., and H.G. Fisser. 1976. Vegetation re-establishment potential of strip-mined soils, Black Thunder Basin, Wyoming. p. 49. In Proc. Soc. Range Manage. 29th Ann. Meeting. Omaha, Nebraska. February 16-20, 1976.

2F1 780403 Foster, M.A., J. Stubbendieck, and H.G. Fisser. 1978. Plant succession on plains pocket gopher mounds. p. 8. In Proc. Soc. Range Manage. 31st Ann. Meeting, San Antonio, Texas. February 7-9, 1978.

2F2. West

2F2 640402 Nichols, J.R. 1964. Soil-vegetation relations of the 15-Mile Drainage area in the Big Horn Basin of Wyoming. p. 25. In Proc. 17th Ann. Meeting Am. Soc. Range Manage. Wichita, Kansas. February 14, 1964.

2F2 650401 Fisser, H.G. 1965. Vegetation changes from cold desert to subalpine communities. p. 425-427. In Proc. IX International Grassland Congress. Sao Paulo, Brazil. January 18, 1965.

2F2 660402 Fisser, H.G. 1966. Soil moisture and temperature characteristics as related to grass production following shrub control under grazed and ungrazed conditions. In Proc. X Intern. Grassl. Congress. Helsinki, Finland. July 1966.

2F2 670401 Fisser, H.G. 1967. Soil moisture and temperature characteristics as related to grass production following shrub control under grazed and non-grazed conditions. p. 23-25. In 20th Ann. Meeting Proc. Am. Soc. Range Manage. Seattle, Washington. February 14, 1967.

2F2 690401 Fisser, H.G. 1969. Plant pattern and distribution in ecosystems and relationships to function. p. 183-196. In R.L. Dix and R.G. Beidleman (eds.) Proc. Information Synthesis Project Grassland Biome. U.S. I.B.P. Range Science Series No. 2. Colo. State Univ. Ft. Collins, Colorado.

2F2 710401 Wambolt, C.L. 1971. Moisture stress in woody plants as influenced by environmental factors. p. 25-26. In Proc. 24th Ann. Meeting. Soc. Range Manage. Reno, Nevada. February 14-18, 1971.

2F2 750403 Kleinman, L.H. and H.G. Fisser. 1975. Phenodynamics and environmental characteristics of sagebrush-grass rangelands. p. 161. In H.C. Stutz, (ed.) Proc. Wildland Shrubs Symp. U.S. Forest Service Shrub Sciences Lab. Provo, Utah. November 4-6, 1975.

2F2 770401 Gerhart, William A., and H.G. Fisser. 1977. An evaluation of three grazing management systems on northern desert shrub vegetation of southern Wyoming. p. 52. In Proc. Soc. Range Manage. 30th Ann. Meeting. Portland, Oregon. February 17, 1977.

2F2 770402 Kleinman, Larry H. and H.G. Fisser. 1977. Phenodynamics and ecology of sagebrush-grass rangelands. p. 11. In Proc. Soc. Range Manage. 30th Ann. Meeting. Portland, Oregon. February 14-17, 1977.

2F2 770405 Gdara, A.O. 1977. Long-term vegetation studies with permanent quadrats. p. 8. In Proc. Soc. Range Manage. 30th Ann. Meeting. Portland, Oregon. February 14-17, 1977.

2F2 780401 Fisser, Herbert G. 1978. Soil erosion and movement studies in western Wyoming. p. 16. In Proc. Soc. Range Manage. 31st Ann. Meeting, San Antonio, Texas. February 7-9, 1978.

2F2 780402 Taha, Faisal K. and H.G. Fisser. 1978. Models for prediction of natural vegetation potential in eastern Wyoming. p. 19. In Proc. Soc. Range Manage. 31st Ann. Meeting, San Antonio, Texas. February 7-9, 1978.

2F2 780404 Gdara, A.O. and H.G. Fisser. 1978. Semiarid rangeland and grazing evaluation of shrubland from permanent quadrats. Manuscript submitted to 1st Intern. Rangeland Congress, Denver, Colorado. August 14-18, 1978.

2F2 780406 Fisser, H.G. 1978. A structural definition of semiarid *Artemisia* and *Atriplex* rangeland ecosystems. Abstract for 2nd Intern. Congress of Ecology. Jerusalem, Israel. September 10-16, 1978.

2F2 790402 Young, L.W., and H.G. Fisser. 1979. Soils-phenology relations on some arid rangelands of western Wyoming. p. 53-54. In Proc. 32nd Ann. Meeting, Soc. Range Manage. Casper, Wyoming. February 12-15, 1979.

2F2 800401 Trueblood, D.C. and H.G. Fisser. 1980. Soil vegetation relationships at semiarid rangeland exclosures in north central Wyoming: long term studies. p. 31. In Proc. 33rd Ann. Meeting Soc. Range Manage. San Diego, Calif. February 11-14, 1980.

2F2 810401 Bensel, W.E. and H.G. Fisser. 1981. Forage production and longevity of sagebrush control on semiarid rangelands in western Wyoming. p. 9. In Absts. Soc. Range Manage. 34th Ann. Meeting. Tulsa, Oklahoma. February 8-13, 1981.

2F2 810402 Hargis, N.E. and H.G. Fisser. 1981. Phenological comparison of two sagebrush species in Wyoming. p. 21. In Absts. Soc. Range Manage. 34th Ann. Meeting. Tulsa, Oklahoma. February 8-13, 1981.

2F2 810403 Joyce, L.A. and H.G. Fisser. 1981. The effect of community changes on precipitation/ production relationships in exclosures of saltbush vegetation. p. 22. In Absts. Papers Soc. Range Manage. 34th Ann. Meeting. Tulsa, Oklahoma. February 8-13, 1981.

2F2 810405 Owens, M.K. and H.G. Fisser. 1981. Phenology and productivity comparisons of western wheatgrass *(Agropyron smithii)* and bluebunch wheatgrass *(A. spicatum)*. p. 21. In Absts. Soc. Range Manage. 34th Ann. Meeting. Tulsa, Oklahoma. February 8-13, 1981.

2F2 810406 Rodgers, D.E. and H.G. Fisser. 1981. Grazing system impacts on semiarid vegetation of Wyoming. p. 34. In Absts. Soc. Range Manage. 34th Ann. Meeting. Tulsa, Oklahoma. February 8-13, 1981.

2F2 810407 Spaeth, K.E. and H.G. Fisser. 1981. Plant composition response to picloram application in the Wyoming juniper type. p. 35. In Absts. Soc. Range Manage. 34th Ann. Meeting. Tulsa, Oklahoma. February 8-13, 1981.

2F2 810408 Trueblood, D.C. and H.G. Fisser. 1981. Soil-vegetation relationships in the Gardner saltbush vegetation type of Wyoming. p. 12. In Absts. Soc. Range Manage. 34th Ann. Meeting. Tulsa, Oklahoma. February 8-13, 1981.

2F2 810409 Uhlich, J.W. and H.G. Fisser. 1981. Vegetation dynamics of non-grazed semiarid sagebrush grassland communities in the Big Horn Basin of Wyoming. p. 22. In Absts. Soc. Range Manage. 34th Ann. Meeting. Tulsa, Oklahoma. February 8-13, 1981.

2F2 810410 Waugh, W.J. and H.G. Fisser. 1981. Population structure of Utah juniper in north central Wyoming. p. 40. In Absts. Soc. Range Manage. 34th Ann. Meeting. Tulsa, Oklahoma. February 8-13, 1981.

2F2 810411 Young, L.W. and H.G. Fisser. 1981. Correlation of soil factors and environmental conditions with phenophase development of western wheatgrass *(Agropyron smithii)*. p. 12. In Absts. Soc. Range Manage. 34th Ann. Meeting. Tulsa, Oklahoma. February 8-13, 1981.

2F2 810413 Fisser, H.G. 1981. Shrub establishment, dominance, and ecology on the juniper and sagebrush-grass types in Wyoming. p. 23-27. In L.S. Stelter, E.J. DePuit, S.A. Mikol (eds.). Proc. Symposium on Shrub Establishment on Disturbed Arid and Semiarid Lands. December 3, 1980. Wyoming Game and Fish Dept. Publication.

2F2 810414 Rodgers, D.E. and H.G. Fisser. 1981. Vegetative response to grazing systems management in Wyoming's Red Desert. p. 451. In Summaries of Papers. J. Allan Smith (ed.). XIV International Grassland Congress. Lexington, Kentucky. June 14-23, 1981.

2F2 810415 Owens, M.K. and H.G. Fisser. 1981. Phenology and culm weight variation between two wheatgrass species: *Agropyron smithii* and *A. spicatum*. p. 205. In Summaries of Papers. J. Allan Smith (ed.). XIV International Grassland Congress. Lexington, Kentucky. June 14-23, 1981.

2F2 820401 Fisser, H.G. 1982. Vegetation response to grazing management strategies on semiarid shrublands of Wyoming. p. 162-165. In D.D. Briske and M.M. Kothman (eds.). Proc. Natl. Conf. on Grazing Techniques. Dept. Range Sci. Texas A & M University. College Station, Texas. November 10-13, 1982.

2F2 820402 Dean, S.K. 1982. Vegetation response to strip chemical control in the Government Draw area. p. 35-36. In: Fisser, H.G. and K.L. Johnson, eds. 1984. Proceedings of the 11th Wyoming Shrub Ecology Workshop. Lander, Wyoming. May 27-28, 1982. University of Wyoming, Laramie.

2F2 820403 Kinucan, R.J. 1982. Numerical phenology scoring and use with quantitative environmental variables. p. 36-39. In: Fisser, H.G. and K.L. Johnson eds.1984. Proceedings of the 11th Wyoming Shrub Ecology Workshop. Lander, Wyoming. May 27-28, 1982. University of Wyoming, Laramie.

2F2 820404 Fisser, H.G. 1982. Vegetation productivity and phenology in the Wind River Basin. p. 42-45. In: Fisser, H.G. and K.L. Johnson, eds. 1984. Proceedings of the 11th Wyoming Shrub Ecology Workshop. Lander, Wyoming. May 27-28, 1982. University of Wyoming, Laramie.

2F2 820405 Mulholland, J.L. 1982. Sagebrush and perennial grass dynamics at the Granite Mountain shrub control site. p. 39-42. In: Fisser, H.G. and K.L. Johnson, eds. 1984. Proceedings of the 11th Wyoming Shrub Ecology Workshop. Lander, Wyoming. May 27-28, 1982. University of Wyoming, Laramie.

2F2 830401 Fisser, Herbert G. 1983. Some recent research findings in the true mountain mahogany types. p. 7-10. In H.G. Fisser and K.L. Johnson (eds.) Wyoming Shrublands: Proc. 10th Wyoming Shrub Ecology Workshop. Sybille Canyon, Wyoming. May 28-29, 1981. University of Wyoming, Laramie.

2F2 830402 Fisser, H.G. and K.L. Johnson (eds.) 1983. Wyoming Shrublands: Proc. 9th Wyoming Shrub Ecology Workshop. June 5-6, 1980. Lander, Wyoming. University of Wyoming, Laramie. 40 p.

2F2 830403 Fisser, H.G. and K.L. Johnson (eds.) 1983. Wyoming Shrublands: Proc. 10th Wyoming Shrub Ecology Workshop. May 28-29, 1981. Sybille Canyon, Wyoming. University of Wyoming, Laramie. 40 p.

2F2 830404 Waugh, W.J. 1983. Stagnation and decadence of *Cercocarpus montanus* in a southeastern Wyoming big game exclosure, p. 11-14. In H.G. Fisser and K.L. Johnson (eds.). Wyoming Shrublands: Proc. 10th Wyoming Shrub Ecology Workshop, Sybille Canyon, Wyoming, May 28-29, 1981. University of Wyoming, Laramie.

2F2 830405 Dean, S.K. and H.G. Fisser. 1983. Long-term vegetation responses to sagebrush control in western Wyoming. abst. 29. In Absts. 36th Ann. Meeting. Soc. Range Manage. Albuquerque, New Mexico. February 14-17, 1983.

2F2 830407 Elwakeel, A.S. and H.G. Fisser. 1983. Perennial grass production dynamics among five vegetation types in north central Wyoming. abst. 208. In Absts. 36th Ann. Meeting. Soc. Range Manage. Albuquerque, New Mexico. February 14-17, 1983.

2F2 830408 Fisser, H.G. 1983. Phenological description and numerical scoring of sagebrush. abst. 136. In Absts. 36th Ann. Meeting. Soc. Range Manage. Albuquerque, New Mexico. February 14-17, 1983.

2F2 830410 Kinucan, R.J. and H.G. Fisser. 1983. Dynamics of plant community development for diverse sagebrush sites in western Wyoming. abst. 65. In Absts. 36th Ann. Meeting. Soc. Range Manage. Albuquerque, New Mexico. February 14-17, 1983.

2F2 830411 Mulholland, J.L. and H.G. Fisser. 1983. Reestablishment rates of sagebrush following chemical control on grazed and ungrazed ranges in western Wyoming. abst. 295. In Absts. 36th Ann. Meeting. Soc. Range Manage. Albuquerque, New Mexico. February 14-17, 1983.

2F2 830412 Owens, M.K. and H.G. Fisser. 1983. Size and density parameters of western wheatgrass on diverse range sites in western Wyoming. abst. 128. In Absts. 36th Ann. Meeting. Soc. Range Manage. Albuquerque, New Mexico. February 14-17, 1983.

2F2 830413 Waugh, W.J. and H.G. Fisser. 1983. Utah juniper invasion in north central Wyoming and historical climatic and management relationships. abst. 202. In Absts. 36th Ann. Meeting. Soc. Range Manage. Albuquerque, New Mexico. February 14-17, 1983.

2F2 840402 Fisser, H.G. and N.E. Hargis. 1984. Detailed numerical phenology of sagebrush with visual description of statements. p. 44 . In Proc 37th Ann. Meeting Soc. Range Manage. Rapid City, South Dakota. February 12-17, 1984.

2F2 840403 Germain, L.P. and H.G. Fisser. 1984. Estimation of missing precipitation data using traditional, kridging, and subjective techniques. abst. 160. In Abst. 37th Ann. Meeting Soc. Range Manage. Rapid City, South Dakota. February 12-17, 1984.

2F2 840404 Fisser, H.G. and W.J. Waugh. 1984. Population structure anomalies of *Juniperus osteosperma* invasion in northcentral Wyoming, U.S.A. Submitted to Proc. 2nd Intern. Rangeland Congress. Adelaide, Australia. May 13-18, 1984.

2F2 840405 Dean, S.K. and H.G. Fisser. 1984. Grazing management, longevity, and benefits of sagebrush control on semiarid rangelands. Submitted to Proc. 2nd Intern. Rangeland Congress. Adelaide, Australia. May 13-18, 1984.

2F2 840406 Kinucan, R.J. and H.G. Fisser. 1984. Vegetative phenology patterns of semiarid plant communities in Wyoming, U.S.A. Submitted to Proc. 2nd Intern. Rangeland Congress. Adelaide, Australia. May 13-18, 1984.

2F2 840407 Fisser, H.G. and L.A. Joyce. 1984. Saltbush community changes with and without grazing over a 20-year period in north central Wyoming. Submitted to Proc. 2nd Intern. Rangeland Congress. Adelaide, Australia. May 13-18, 1984.

2F2 840408 Mulholland, J.L. and H.G. Fisser. 1984. Big sagebrush response and reestablishment following chemical control in Wyoming, U.S.A. Submitted to Proc. 2nd Intern. Rangeland Congress. Adelaide, Australia. May 13-18, 1984.

2F2 840409 Germain, L.P. and H.G. Fisser. 1984. Application of kriging to ecological variables. p. 12. In Proc. 55th Ann. Meeting Colo.-Wyo. Acad. Sci., Western State College. Gunnison, Colorado. April 27-28, 1984.

2F2 840411 Fisser, H.G. 1984. Detailed numerical phenology of sagebrush with visual description statements. Posterboard display at USDA-ARS/USDI-BLM Regional Workshop on Rangeland Hydrology, Modeling and Monitoring. Boise, Idaho. April 30 through May 2, 1984.

2F2 840412 Dean, S.K. and H.G. Fisser. 1984. Grazing management longevity and benefits of sagebrush control in western Wyoming, U.S.A. Posterboard display at USDA-ARS/USDI-BLM Regional Workshop on Rangeland Hydrology, Modeling and Monitoring. Boise, Idaho. April 30 through May 2, 1984.

2F2 840413 Fisser, H.G. and L.A. Joyce. 1984. *Atriplex*/grass and forb relationships under no grazing and shifting precipitation patterns in north central Wyoming. p. 87-96. In Tiedeman, A.R., E.D. McArthur, H.C. Stutz, R. Stevens, and K.L. Johnson (eds.). Proc. Wildland Shrub Symp. "Biology of *Atriplex* and related chenopods." Provo, Utah. May 2-6, 1983. USDA Forest Service General Technical Report 172.

2F2 840414 Fisser, H.G. 1984. Biology and ecology of sagebrush in Wyoming: I. Herbage production and community dynamics. p. 21. In Abst. Shrub Research Consortium Third Annual Wildland Shrub Symposium: the biology of *Artemisia* and *Chrysothamnus*. Provo, Utah. July 9-13, 1984.

2F2 840415 Fisser, H.G. 1984. Biology and ecology of sagebrush in Wyoming: II. Reinvasion following control. p. 21. In Abst. Shrub Research Consortium Third Annual Wildland Shrub Symposium: the biology of *Artemisia* and *Chrysothamnus*. Provo, Utah. July 9-13, 1984.

2F2 840416 Fisser, H.G. 1984. Biology and ecology of sagebrush in Wyoming: III. Regional variation of phenology. p. 22. In Abst. Shrub Research Consortium Third Annual Wildland Shrub Symposium: the biology of *Artemisia* and *Chrysothamnus*. Provo, Utah. July 9-13, 1984.

2F2 840417 Fisser, H.G., J.R. Wight and C.W. Hanson. 1984. Biology and ecology of sagebrush in Wyoming: IV. Validation of a rangeland production model (ERHYM) for sagebrush sites. p. 13. In Abst. Shrub Research Consortium Third Annual Wildland Shrub Symposium: the biology of *Artemisia* and *Chrysothamnus*. Provo, Utah. July 9-13, 1984.

2F2 840418 Fisser, H.G., C.W. Hanson and J.R. Wight. 1984. Biology and ecology of sagebrush in Wyoming: V. Herbage yield dynamics and seasonal precipitation relationships. p. 13. In Abst. Shrub Research Consortium Third Annual Wildland Shrub Symposium: the biology of *Artemisia* and *Chrysothamnus*. Provo, Utah. July 9-13, 1984.

2F2 840419 Fisser, H.G. and K.L. Johnson, eds. 1984. Wyoming Shrublands Proc. 11th Wyoming Shrub Ecology Workshop. Lander, Wyoming. May 27-28, 1982. University of Wyoming, Laramie.

3. Range Resources

3A. RANGE INVENTORY AND SURVEY

3A 740407 Whicker, F.W., A.W. Alldredge, H.G. Fisser, W.C. Hanson, J.E. Johnson & G. Post. 1974. Ecological considerations for Project Wagon Wheel. Phase I: Pre-event biospheric inventory, impact analysis and mitigation, and baseline documentation. A study conducted for El Paso Natural Gas Co. PNE-WW-35. 280 p.

3A 750407 Whicker, F.W., A.W. Alldredge, H.G. Fisser, and G. Post. 1975. Ecological Considerations for Project Wagon Wheel and hydraulic fracturing activities. Phase II(a). Annual summary report for 1974. A study conducted for El Paso Natural Gas Co. PNE-WW-40. 33 p.

3A 760404 Whicker, F.W., A.W. Alldredge, H.G. Fisser, W.C. Hanson, and G. Post. 1976. Ecological considerations for Project Wagon Wheel and hydraulic fracturing activities. Phase II(b): Annual summary for 1975 and project completion report. A study conducted for El Paso Natural Gas Company PNE-WW-85. 53 p.

3A1. Site, Mapping, Remote Sensing

3A1 650401 Fisser, H.G. 1965. Vegetation changes from cold desert to subalpine communities. p. 425-427. In Proc. IX International Grassland Congress. Sao Paulo, Brazil. January 18, 1965.

3A1 720401 Shaver, Jens C. and H.G. Fisser. 1972. Panchromatic, color and infrared photography for detailed vegetation mapping. p. 33. In Proc. 25th Ann. Meeting Soc. Range Manage. Washington, D.C. February 8, 1972.

3A1 720402 Fisser, H.G. 1972. An ecological description of biotic communities of the Wagon Wheel environs. p. I-1 thru I-48. In Rogers L.A. (ed.). Project Wagon Wheel Study Report Number Two. El Paso Natural Gas Co., PME-WW-13.

3A1 740402 Shaver, J.C. and H.G. Fisser. 1974. Detailed grassland type delineation with panchromatic, color and infrared aerial photographs, p. 369-373. In V.G. Iglovikov, A.P. Movsissyants, F.F. Sarganov, A.A. Stepanenko (Compilers). Improvement of natural and production of seeded meadows and pastures. XII International Grassland Congress, Moscow, USSR. June 12-16, 1974.

3A1 740407 Whicker, F.W., A.W. Alldredge, H.G. Fisser, W.C. Hanson, J.E. Johnson & G. Post. 1974. Ecological considerations for Project Wagon Wheel. Phase I: Pre-event biospheric inventory, impact analysis and mitigation, and baseline documentation. A study conducted for El Paso Natural Gas Co. PNE-WW-35. 280 p.

3A1 750402 Fisser, H.G. and R.E. Ries. 1975. Pre-disturbance ecological studies improve and define potential for surface mine reclamation. p. 128-134. In Proc. 3rd NCA-BCR Coal Conference and Expo II. Vol. I. Louisville, Kentucky. October 21-23, 1975.

3A1 750404 Fisser, H.G. 1975. Vegetation response to seismic motion from underground nuclear detonation. p. 37. In Proc. Soc. Range Manage. 28th Ann. Meeting. Mexico City, Mexico. February 10-14, 1975.

3A1 750405 Ries, R.E. and H.G. Fisser. 1975. Pre-disturbance vegetation inventory and ecological monitoring procedures for rangeland strip mining. p. 40. In Proc. 28th Ann. Meeting Soc. Range Manage. Mexico City, Mexico. February 10-14, 1975.

3A1 750407 Whicker, F.W., A.W. Alldredge, H.G. Fisser, and G. Post. 1975. Ecological Considerations for Project Wagon Wheel and hydraulic fracturing activities. Phase II(a). Annual summary report for 1974. A study conducted for El Paso Natural Gas Co. PNE-WW-40. 33 p.

3A1 760404 Whicker, F.W., A.W. Alldredge, H.G. Fisser, W.C. Hanson, and G. Post. 1976. Ecological considerations for Project Wagon Wheel and hydraulic fracturing activities. Phase II(b): Annual summary for 1975 and project completion report. A study conducted for El Paso Natural Gas Company PNE-WW-85. 53 p.

3A1 760403 Lymbery, Gordon A., and H.G. Fisser. 1976. Vegetation re-establishment potential of strip-mined soils, Black Thunder Basin, Wyoming. p. 49. In Proc. Soc. Range Manage. 29th Ann. Meeting. Omaha, Nebraska. February 16-20, 1976.

3A1 780403 Foster, M.A., J. Stubbendieck, and H.G. Fisser. 1978. Plant succession on plains pocket gopher mounds. p. 8. In Proc. Soc. Range Manage. 31st Ann. Meeting, San Antonio, Texas. February 7-9, 1978.

3A1 780405 Shawesh, O.M. and H.G. Fisser. 1978. Rangeland development and improvement in Libya, Africa. p. 65. In J.R. Goodin and D.K. Northington (eds.) abstracts Intern. Arid Lands Conf. Intern. Center for Arid and Semiarid Land Studies. Texas Tech University. Lubbock, Texas. October 8-15, 1978.

3A1 810402 Hargis, N.E. and H.G. Fisser. 1981. Phenological comparison of two sagebrush species in Wyoming. p. 21. In Absts. Soc. Range Manage. 34th Ann. Meeting. Tulsa, Oklahoma. February 8-13, 1981.

3A2. Condition and Trend

3A2 670403 Fisser, H.G. 1967. Plant control on salt desert shrub rangeland by sheep grazing. p. 48-54. In Proc. Range Weed Control Conf. Las Cruces, New Mexico. August 21, 1967.

3A2 690401 Fisser, H.G. 1969. Plant pattern and distribution in ecosystems and relationships to function. p. 183-196. In R.L. Dix and R.G. Beidleman (eds.) Proc. Information Synthesis Project Grassland Biome. U.S. I.B.P. Range Science Series No. 2. Colo. State Univ. Ft. Collins, Colorado.

3A2 690402 Steger, R.E. Soil moisture relationships of certain salt desert shrubs. p. 50. In Proc. 22nd Ann. Meeting Soc. Range Manage. Calgary, Alberta. February 10-13, 1969.

3A2 720401 Shaver, Jens C. and H.G. Fisser. 1972. Panchromatic, color and infrared photography for detailed vegetation mapping. p. 33. In Proc. 25th Ann. Meeting Soc. Range Manage. Washington, D.C. February 8, 1972.

3A2 720402 Fisser, H.G. 1972. An ecological description of biotic communities of the Wagon Wheel environs. p. I-1 thru I-48. In Rogers L.A. (ed.). Project Wagon Wheel Study Report Number Two. El Paso Natural Gas Co., PME-WW-13.

3A2 730401 Mackey, M.H. 1973. Evaluation of furrowing and seeding in a north central Wyoming Nuttall saltbush area. p. 13. In Proc. 26th Ann. Meeting. Soc. Range Manage. Boise, Idaho. February 4-9, 1973.

3A2 740402 Shaver, J.C. and H.G. Fisser. 1974. Detailed grassland type delineation with panchromatic, color and infrared aerial photographs, p. 369-373. In V.G. Iglovikov, A.P. Movsissyants, F.F. Sarganov, A.A. Stepanenko (Compilers). Improvement of natural and production of seeded meadows and pastures. XII International Grassland Congress, Moscow, USSR. June 12-16, 1974.

3A2 740406 Ries, R.E. 1974. Environmental factors affecting annual herbage production at 17 sagebrush-grass locations in Wyoming. p. 21-22. In Proc. 27th Ann. Meeting Soc. Range Manage. Tucson, Arizona. February 3-8, 1974.

3A2 740407 Whicker, F.W., A.W. Alldredge, H.G. Fisser, W.C. Hanson, J.E. Johnson & G. Post. 1974. Ecological considerations for Project Wagon Wheel. Phase I: Pre-event biospheric inventory, impact analysis and mitigation, and baseline documentation. A study conducted for El Paso Natural Gas Co. PNE-WW-35. 280 p.

3A2 750402 Fisser, H.G. and R.E. Ries. 1975. Pre-disturbance ecological studies improve and define potential for surface mine reclamation. p. 128-134. In Proc. 3rd NCA-BCR Coal Conference and Expo II. Vol. I. Louisville, Kentucky. October 21-23, 1975.

3A2 750403 Kleinman, L.H. and H.G. Fisser. 1975. Phenodynamics and environmental characteristics of sagebrush-grass rangelands. p. 161. In H.C. Stutz, (ed.) Proc. Wildland Shrubs Symp. U.S. Forest Service Shrub Sciences Lab. Provo, Utah. November 4-6, 1975.

3A2 750404 Fisser, H.G. 1975. Vegetation response to seismic motion from underground nuclear detonation. p. 37. In Proc. Soc. Range Manage. 28th Ann. Meeting. Mexico City, Mexico. February 10-14, 1975.

3A2 750405 Ries, R.E. and H.G. Fisser. 1975. Pre-disturbance vegetation inventory and ecological monitoring procedures for rangeland strip mining. p. 40. In Proc. 28th Ann. Meeting Soc. Range Manage. Mexico City, Mexico. February 10-14, 1975.

3A2 750407 Whicker, F.W., A.W. Alldredge, H.G. Fisser, and G. Post. 1975. Ecological Considerations for Project Wagon Wheel and hydraulic fracturing activities. Phase II(a). Annual summary report for 1974. A study conducted for El Paso Natural Gas Co. PNE-WW-40. 33 p.

3A2 760404 Whicker, F.W., A.W. Alldredge, H.G. Fisser, W.C. Hanson, and G. Post. 1976. Ecological considerations for Project Wagon Wheel and hydraulic fracturing activities. Phase II(b): Annual summary for 1975 and project completion report. A study conducted for El Paso Natural Gas Company PNE-WW-85. 53 p.

3A2 760402 Taha, Faisal K., and H.G. Fisser. 1976. Physical and chemical analyses of range sites and interrelationships with herbage production and forage value potentials. p. 13. In Proc. Soc. Range Manage. 29th Ann. Meeting. Omaha, Nebraska. February 16-20, 1976.

3A2 760403 Lymbery, Gordon A., and H.G. Fisser. 1976. Vegetation re-establishment potential of strip-mined soils, Black Thunder Basin, Wyoming. p. 49. In Proc. Soc. Range Manage. 29th Ann. Meeting. Omaha, Nebraska. February 16-20, 1976.

3A2 770402 Kleinman, Larry H. and H.G. Fisser. 1977. Phenodynamics and ecology of sagebrush-grass rangelands. p. 11. In Proc. Soc. Range Manage. 30th Ann. Meeting. Portland, Oregon. February 14-17, 1977.

3A2 770405 Gdara, A.O. 1977. Long-term vegetation studies with permanent quadrats. p. 8. In Proc. Soc. Range Manage. 30th Ann. Meeting. Portland, Oregon. February 14-17, 1977.

3A2 780401 Fisser, Herbert G. 1978. Soil erosion and movement studies in western Wyoming. p. 16. In Proc. Soc. Range Manage. 31st Ann. Meeting. San Antonio, Texas. February 7-9, 1978.

3A2 780402 Taha, Faisal K. and H.G. Fisser. 1978. Models for prediction of natural vegetation potential in eastern Wyoming. p. 19. In Proc. Soc. Range Manage. 31st Ann. Meeting, San Antonio, Texas. February 7-9, 1978.

3A2 780404 Gdara, A.O. and H.G. Fisser. 1978. Semiarid rangeland and grazing evaluation of shrubland from permanent quadrats. Manuscript submitted to 1st Intern. Rangeland Congress, Denver, Colorado. August 14-18, 1978.

3A2 790402 Young, L.W., and H.G. Fisser. 1979. Soils-phenology relations on some arid rangelands of western Wyoming. p. 53-54. In Proc. 32nd Ann. Meeting, Soc. Range Manage. Casper, Wyoming. February 12-15, 1979.

3A2 800401 Trueblood, D.C. and H.G. Fisser. 1980. Soil vegetation relationships at semiarid rangeland exclosures in north central Wyoming: long term studies. p. 31. In Proc. 33rd Ann. Meeting Soc. Range Manage. San Diego, California. February 11-14, 1980.

3A2 810401 Bensel, W.E. and H.G. Fisser. 1981. Forage production and longevity of sagebrush control on semiarid rangelands in western Wyoming. p. 9. In Absts. Soc. Range Manage. 34th Ann. Meeting. Tulsa, Oklahoma. February 8-13, 1981.

3A2 810402 Hargis, N.E. and H.G. Fisser. 1981. Phenological comparison of two sagebrush species in Wyoming. p. 21. In Absts. Soc. Range Manage. 34th Ann. Meeting. Tulsa, Oklahoma. February 8-13, 1981.

3A2 810403 Joyce, L.A. and H.G. Fisser. 1981. The effect of community changes on precipitation/production relationships in exclosures of saltbush vegetation. p. 22. In Absts. Papers Soc. Range Manage. 34th Ann. Meeting. Tulsa, Oklahoma. February 8-13, 1981.

3A2 810405 Owens, M.K. and H.G. Fisser. 1981. Phenology and productivity comparisons of western wheatgrass *(Agropyron smithii)* and bluebunch wheatgrass *(A. spicatum).* p. 21. In Absts. Soc. Range Manage. 34th Ann. Meeting. Tulsa, Oklahoma. February 8-13, 1981.

3A2 810407 Spaeth, K.E. and H.G. Fisser. 1981. Plant composition response to picloram application in the Wyoming juniper type. p. 35. In Absts. Soc. Range Manage. 34th Ann. Meeting. Tulsa, Oklahma. February 8-13, 1981.

3A2 810408 Trueblood, D.C. and H.G. Fisser. 1981. Soil-vegetation relationships in the Gardner saltbush vegetation type of Wyoming. p. 12. In Absts. Soc. Range Manage. 34th Ann. Meeting. Tulsa, Oklahoma. February 8-13, 1981.

3A2 810409 Uhlich, J.W. and H.G. Fisser. 1981. Vegetation dynamics of non-grazed semiarid sagebrush grassland communities in the Big Horn Basin of Wyoming. p.22. In Absts. Soc. Range Manage. 34th Ann. Meeting. Tulsa, Oklahoma. February 8-13, 1981.

3A2 810410 Waugh, W.J. and H.G. Fisser. 1981. Population structure of Utah juniper in north central Wyoming. p. 40. In Absts. Soc. Range Manage. 34th Ann. Meeting. Tulsa, Oklahoma. February 8-13, 1981.

3A2 810413 Fisser, H.G. 1981. Shrub establishment, dominance, and ecology on the juniper and sagebrush-grass types in Wyoming. p. 23-27. In L.S. Stelter, E.J. DePuit, S.A. Mikol (eds.). Proc. Symposium on Shrub Establishment on Disturbed Arid and Semiarid Lands. December 3, 1980. Wyoming Game and Fish Dept. Publication.

3A2 810415 Owens, M.K. and H.G. Fisser. 1981. Phenology and culm weight variation between two wheatgrass species: *Agropyron smithii* and *A. spicatum.* p. 205. In Summaries of Papers. J. Allan Smith (ed.). XIV International Grassland Congress.

3A2 820402 Dean, S.K. 1982. Vegetation response to strip chemical control in the Government Draw area. p. 35-36. In: Fisser, H.G. and K.L. Johnson, eds. 1984. Proceedings of the 11th Wyoming Shrub Ecology Workshop. Lander, Wyoming. May 27-28, 1982. University of Wyoming, Laramie.

3A2 820403 Kinucan, R.J. 1982. Numerical phenology scoring and use with quantitative environmental variables. p. 36-39. In: Fisser, H.G. and K.L. Johnson eds. 1984. Proceedings of the 11th Wyoming Shrub Ecology Workshop. Lander, Wyoming. May 27-28, 1982. University of Wyoming, Laramie.

3A2 820405 Mulholland, J.L. 1982. Sagebrush and perennial grass dynamics at the Granite Mountain shrub control site. p. 39-42. In: Fisser, H.G. and K.L. Johnson, eds. 1984. Proceedings of the 11th Wyoming Shrub Ecology Workshop. Lander, Wyoming. May 27-28, 1982. University of Wyoming, Laramie.

3A2 830402 Fisser, H.G. and K.L. Johnson (eds.) 1983. Wyoming Shrublands: Proc. 9th Wyoming Shrub Ecology Workshop. June 5-6, 1980. Lander, Wyoming. University of Wyoming, Laramie. 40 p.

3A2 830403 Fisser, H.G. and K.L. Johnson (eds.) 1983. Wyoming Shrublands: Proc. 10th Wyoming Shrub Ecology Workshop. May 28-29, 1981. Sybille Canyon, Wyoming. University of Wyoming, Laramie. 40 p.

3A2 830405 Dean, S.K. and H.G. Fisser. 1983. Long-term vegetation responses to sagebrush control in western Wyoming. abst. 29. In Absts. 36th Ann. Meeting. Soc. Range Manage. Albuquerque, New Mexico. February 14-17, 1983.

3A2 830406 Deboodt, T.L., E.B. Bradley, and H.G. Fisser. 1983. Economic feasibility of resting rangeland to improve forage production in central Wyoming. abst. 215. In Absts. 36th Ann. Meeting. Soc. Range Manage. Albuquerque, New Mexico. February 14-17, 1983.

3A2 830407 Elwakeel, A.S. and H.G. Fisser. 1983. Perennial grass production dynamics among five vegetation types in north central Wyoming. abst. 208. In Absts. 36th Ann. Meeting. Soc. Range Manage. Albuquerque, New Mexico. February 14-17, 1983.

3A2 830408 Fisser, H.G. 1983. Phenological description and numerical scoring of sagebrush. abst. 136. In Absts. 36th Ann. Meeting. Soc. Range Manage. Albuquerque, New Mexico. February 14-17, 1983.

3A2 830413 Waugh, W.J. and H.G. Fisser. 1983. Utah juniper invasion in north central Wyoming and historical climatic and management relationships. abst. 202. In Absts. 36th Ann. Meeting. Soc. Range Manage. Albuquerque, New Mexico. February 14-17, 1983.

3A2 840402 Fisser, H.G. and N.E. Hargis. 1984. Detailed numerical phenology of sagebrush with visual description statements. p. 44. In Proc 37th Ann. Meeting Soc. Range Manage. Rapid City, South Dakota. February 12-17, 1984.

3A2 840404 Fisser, H.G. and W.J. Waugh. 1984. Population structure anomalies of *Juniperus osteosperma* invasion in north central Wyoming, U.S.A. Submitted to Proc. 2nd Intern. Rangeland Congress. Adelaide, Australia. May 13-18, 1984.

3A2 840405 Dean, S.K. and H.G. Fisser. 1984. Grazing management, longevity, and benefits of sagebrush control on semiarid rangelands. Submitted to Proc. 2nd Intern. Rangeland Congress. Adelaide, Australia. May 13-18, 1984.

3A2 840411 Fisser, H.G. 1984. Detailed numerical phenology of sagebrush with visual description statements. Posterboard display at USDA-ARS/USDI-BLM Regional Workshop on Rangeland Hydrology, Modeling and Monitoring. Boise, Idaho. April 31 through May 2, 1984.

3A2 840412 Dean, S.K. and H.G. Fisser. 1984. Grazing management longevity and benefits of sagebrush control in western Wyoming, U.S.A. Posterboard display at USDA-ARS/USDI-BLM Regional Workshop on Rangeland Hydrology, Modeling and Monitoring. Boise, Idaho. April 31 through May 2, 1984.

3A2 840416 Fisser, H.G. 1984. Biology and ecology of sagebrush in Wyoming: III. Regional variation of phenology. p. 22. In Abst. Shrub Research Consortium Third Annual Wildland Shrub Symposium: the biology of *Artemisia* and *Chrysothamnus*. Provo, Utah. July 9-13, 1984.

3A2 840417 Fisser, H.G., J.R. Wight and C.W. Hanson. 1984. Biology and ecology of sagebrush in Wyoming: IV. Validation of a rangeland production model (ERHYM) for sagebrush sites. p. 13. In Abst. Shrub Research Consortium Third Annual Wildland Shrub Symposium: the biology of *Artemisia* and *Chrysothamnus*. Provo, Utah. July 9-13, 1984.

3A2 840419 Fisser, H.G. and K.L. Johnson, eds. 1984. Wyoming Shrublands Proc. 11th Wyoming Shrub Ecology Workshop. Lander, Wyoming. May 27-28, 1982. University of Wyoming, Laramie.

3A3. Production and Utilization

3A3 630402 Fisser, H.G. 1963. Exclosure studies (production, cover, height, and moisture phases). p. 23-24. In Proc. Range Weed Control Conf. Portland, Oregon. March 19, 1963.

3A3 740401 Fisser, H.G. 1974. Productivity differences of semidesert grasslands in western Wyoming. p. 22-23. In Proc. 27th Ann. Meeting, Soc. Range Manage. Tuscon, Arizona. February 5, 1974.

3A3 740404 Fisser, H.G. 1974. Productivity potential differences of grasslands of semidesert Western Wyoming p. 116-124. In V.G. Iglovikov, A.P. Movsissyants, F.F. Sarganov, and A.A. Stepanenko (Compilers). Improvement of natural and production of seeded meadows and pastures. XII International Grassland Congress. Moscow, USSR. June 12-16, 1974.

3A3 740405 Whysong, G.L. and H.G. Fisser. 1974. Statistical models for description and prediction of perennial grass production. p. 408-413. In V.G. Iglovikov, A.P. Movsissyants, F.F. Sarganov, and A.A. Stepanenko (Compilers). Biological and physiological aspects of the intensification of grassland utilization. XII International Grassland Congress. Moscow, USSR. June 12-16, 1974.

3A3 750406 Shrader, J.A. and H.G. Fisser. 1975. Seed production ecology in the shortgrass prairie of the Central United States. p. 29. In Proc. 28th Ann. Meeting Soc. Range Mange. Mexico City, Mexico. February 10-14, 1975.

3A3 780406 Fisser, H.G. 1978. A structural definition of semiarid *Artemisia* and *Atriplex* rangeland ecosystems. Abstract for 2nd Intern. Congress of Ecology. Jerusalem, Israel. September 10-16, 1978.

3A3 810401 Bensel, W.E. and H.G. Fisser. 1981. Forage production and longevity of sagebrush control on semiarid rangelands in western Wyoming. p. 9. In Absts. Soc. Range Manage. 34th Ann. Meeting. Tulsa, Oklahoma. February 8-13, 1981.

3A3 810403 Joyce, L.A. and H.G. Fisser. 1981. The effect of community changes on precipitation/ production relationships in exclosures of saltbush vegetation. p. 22. In Absts. Papers Soc. Range Manage. 34th Ann. Meeting. Tulsa, Oklahoma. February 8-13, 1981.

3A3 810411 Young, L.W. and H.G. Fisser. 1981. Correlation of soil factors and environmental conditions with phenophase development of western wheatgrass *(Agropyron smithii).* p. 12. In Absts. Soc. Range Manage. 34th Ann. Meeting. Tulsa, Oklahoma. February 8-13, 1981.

3A3 810413 Fisser, H.G. 1981. Shrub establishment, dominance, and ecology on the juniper and sagebrush-grass types in Wyoming. p. 23-27. In L.S. Stelter, E.J. DePuit, S.A. Mikol (eds.). Proc. Symposium on Shrub Establishment on Disturbed Arid and Semiarid Lands. December 3, 1980. Wyoming Game and Fish Dept. Publication

3A3 820402 Dean, S.K. 1982. Vegetation response to strip chemical control in the Government Draw area. p. 35-36. In: Fisser, H.G. and K.L. Johnson, eds. 1984. Proceedings of the 11th Wyoming Shrub Ecology Workshop. Lander, Wyoming. May 27-28, 1982. University of Wyoming, Laramie.

3A3 820405 Mulholland, J.L. 1982. Sagebrush and perennial grass dynamics at the Granite Mountain shrub control site. p. 39-42. In: Fisser, H.G. and K.L. Johnson, eds. 1984. Proceedings of the 11th Wyoming Shrub Ecology Workshop. Lander, Wyoming. May 27-28, 1982. University of Wyoming, Laramie.

3A3 830402 Fisser, H.G. and K.L. Johnson (eds.) 1983. Wyoming Shrublands: Proc. 9th Wyoming Shrub Ecology Workshop. June 5-6, 1980. Lander, Wyoming. University of Wyoming, Laramie. 40 p.

3A3 830403 Fisser, H.G. and K.L. Johnson (eds.) 1983. Wyoming Shrublands: Proc. 10th Wyoming Shrub Ecology Workshop. May 28-29, 1981. Sybille Canyon, Wyoming. University of Wyoming, Laramie. 40 p.

3A3 830405 Dean, S.K. and H.G. Fisser. 1983. Long-term vegetation responses to sagebrush control in western Wyoming. abst. 29. In Absts. 36th Ann. Meeting. Soc. Range Manage. Albuquerque, New Mexico. February 14-17, 1983.

3A3 830406 Deboodt, T.L., E.B. Bradley, and H.G. Fisser. 1983. Economic feasibility of resting rangeland to improve forage production in central Wyoming. abst. 215. In Absts. 36th Ann. Meeting. Soc. Range Manage. Albuquerque, New Mexico. February 14-17, 1983.

3A3 830407 Elwakeel, A.S. and H.G. Fisser. 1983. Perennial grass production dynamics among five vegetation types in north central Wyoming. abst. 208. In Absts. 36th Ann. Meeting. Soc. Range Manage. Albuquerque, New Mexico. February 14-17, 1983.

3A3 830409 Joyce, L.A. and H.G. Fisser. 1983. Productivity differences of grasslands in semidesert western Wyoming — A re-examination. abst. 207. In Absts. 36th Ann. Meeting. Soc. Range Manage. Albuquerque, New Mexico. February 14-17, 1983.

3A3 840401 Fisser, H.G. 1984. Productivity index procedures for sagebrush-grass rangeland. abst. 010. In Abst. 37th Ann. Meeting. Soc. Range Manage. Rapid City, South Dakota. February 12-17, 1984.

3A3 840405 Dean, S.K. and H.G. Fisser. 1984. Grazing management, longevity, and benefits of sagebrush control on semiarid rangelands. Submitted to Proc. 2nd Intern. Rangeland Congress. Adelaide, Australia. May 13-18, 1984.

3A3 840410 Fisser, H.G. 1984. Culm weight dynamics as an index of sagebrush-grass community productivity. p. 17. In Proc. 55th Ann. Meeting Colo.-Wyo. Acad. Sci., Western State College. Gunnison, Colorado. April 27-28, 1984.

3A3 840411 Fisser, H.G. 1984. Detailed numerical phenology of sagebrush with visual description statements. Posterboard display at USDA-ARS/USDI-BLM Regional Workshop on Rangeland Hydrology, Modeling, and Monitoring. Boise, Idaho. April 31 - May 2, 1984.

3A3 840412 Dean, S.K. and H.G. Fisser. 1984. Grazing management longevity and benefits of sagebrush control in western Wyoming, U.S.A. Posterboard display at USDA-ARS/USDI-BLM Regional Workshop on Rangeland Hydrology, Modeling and Monitoring. Boise, Idaho. April 30 through May 2, 1984.

3A3 840414 Fisser, H.G. 1984. Biology and ecology of sagebrush in Wyoming: I. Herbage production and community dynamics. p. 21. In Abst. Shrub Research Consortium Third Annual Wildland Shrub Symposium: the biology of *Artemisia* and *Chrysothamnus.* Provo, Utah. July 9-13, 1984.

3A3 840415 Fisser, H.G. 1984. Biology and ecology of sagebrush in Wyoming: II. Reinvasion following control. p. 21. In Abst. Shrub Research Consortium Third Annual Wildland Shrub Symposium: the biology of *Artemisia* and *Chrysothamnus.* Provo, Utah. July 9-13, 1984.

3A3 840418 Fisser, H.G., C.W. Hanson and J.R. Wight. 1984. Biology and ecology of sagebrush in Wyoming: V. Herbage yield dynamics and seasonal precipitation relationships. p. 13. In Abst. Shrub Research Consortium Third Annual Wildland Shrub Symposium: the biology of *Artemisia* and *Chrysothamnus.* Provo, Utah. July 9-13, 1984.

3B. NATURAL RESOURCES PLANNING

3B 770406 Fisser, H.G. 1977. Rangeland and wildlife resources of the South Pass Historic Area. p. 77-88. In D.S. Warder. Management suggestions: South Pass Mining Area. Prepared for the Bureau of Land Management contract YA-510-PH-110. March 31, 1977. Parks and Recreation Department. University of Wyoming, Laramie.

4. Grazing Management

4 610405 Colton, R.D. 1961. Sagebrush spraying tour. p. 12-14. In Proc. Range Weed Research Tour. Worland, Wyoming. September 14-15, 1961.

4A. STOCKING RATES

4A 610403 Smith, D.R. 1961. Intensity of grazing study. p. 4-6. In Proc. Range Weed Research Tour. Worland, Wyoming. September 14-15, 1961.

4A 630401 Fisser, H.G. 1963. Progress of cattle and sheep grazing trials on halogeton infested ranges. p. 7-8. In Proc. Range Weed Control Conf. Portland, Oregon. March 19, 1963.

4A 640401 Fisser, H.G. 1964. Winter sheep grazing studies on saltbush rangeland of Wyoming. p. 53. In Proc. 17th Ann. Meeting, Am. Soc. Range Manage. Wichita, Kansas. February 14, 1964.

4A 650402 Fisser, H.G. 1965. Halogeton abundance and distribution as influenced by degree of saltbush utilization. p. 11-13. In Proc. Range Weed Control Conf. Elko, Nevada. August 2, 1965.

4A 660401 Fisser, H.G. 1966. Sheep grazing intensity studies and vegetative changes on *Atriplex nuttallii.* p. 226-237. In Proc. Salt Desert Shrub Symposium. Cedar City, Utah. August 1-3, 1966.

4A 740403 Stroud, D.W. and H.G. Fisser. 1974. Influence of livestock grazing on growth patterns, plant density and soil microrelief of a *Bouteloua gracilis* prairie, p. 347-351. In V.G. Iglovikov, A.P. Movsissyants, F.F. Sarganov, and A.A. Stepanenko (Compilers). Biological and physiological aspects of the intensification of grassland utilization. XII International Grassland Congress. Moscow, USSR. June 12-16, 1974.

4A 840407 Fisser, H.G. and L.A. Joyce. 1984. Saltbush community changes with and without grazing over a 20-year period in north central Wyoming. Submitted to Proc. 2nd Intern. Rangeland Congress. Adelaide, Australia. May 13-18, 1984.

4A 840413 Fisser, H.G. and L.A. Joyce. 1984. *Atriplex*/grass and forb relationships under no grazing and shifting precipitation patterns in north central Wyoming. p. 87-96. In Tiedeman, A.R., E.D. McArthur, H.C. Stutz, R. Stevens, and K.L. Johnson (eds.). Proc. Wildland Shrub Symp. "Biology of *Atriplex* and related chenopods." Provo, Utah. May 2-6, 1983. USDA Forest Service General Technical Report 172.

4B. GRAZING SEASONS AND SYSTEMS

4B 610403 Smith, D.R. 1961. Intensity of grazing study. p. 4-6. In Proc. Range Weed Research Tour. Worland, Wyoming. September 14-15, 1961.

4B 630401 Fisser, H.G. 1963. Progress of cattle and sheep grazing trials on halogeton infested ranges. p. 7-8. In Proc. Range Weed Control Conf. Portland, Oregon. March 19, 1963.

4B 640401 Fisser, H.G. 1964. Winter sheep grazing studies on saltbush rangeland of Wyoming. p. 53. In Proc. 17th Ann. Meeting, Am. Soc. Range Manage. Wichita, Kansas. February 14, 1964.

4B 640403 Fisser, H.G. 1964. Summer steer grazing on saltbush rangeland. p. 7. In Proc. Wyoming Sec. Am. Soc. Range Manage. Casper, Wyoming. November 19-20, 1964.

4B 660401 Fisser, H.G. 1966. Sheep grazing intensity studies and vegetative changes on *Atriplex nuttallii.* p. 226-237. In Proc. Salt Desert Shrub Symposium. Cedar City, Utah. August 1-3, 1966.

4B 770401 Gerhart, William A., and H.G. Fisser. 1977. An evaluation of three grazing management systems on northern desert shrub vegetation of southern Wyoming. p. 52. In Proc. Soc. Range Manage. 30th Ann. Meeting. Portland, Oregon. February 17, 1977.

4B 770406 Fisser, H.G. 1977. Rangeland and wildlife resources of the South Pass Historic Area. p. 77-88. In D.S. Warder. Management suggestions: South Pass Mining Area. Prepared for the Bureau of Land Management contract YA-510-PH-110. March 31, 1977. Parks and Recreation Department. University of Wyoming, Laramie.

4B 810406 Rodgers, D.E. and H.G. Fisser. 1981. Grazing system impacts on semiarid vegetation of Wyoming. p. 34. In Absts. Soc. Range Manage. 30th Ann. Meeting. Tulsa, Oklahoma. February 8-13, 1981.

4B 810413 Fisser, H.G. 1981. Shrub establishment, dominance, and ecology on the juniper and sagebrush-grass types in Wyoming. p. 23-27. In L.S. Stelter, E.J. DePuit, S.A. Mikol (eds.). Proc. Symposium on Shrub Establishment on Disturbed Arid and Semiarid Lands. December 3, 1980. Wyoming Game and Fish Dept. Publication.

4B 810414 Rodgers, D.E. and H.G. Fisser. 1981. Vegetative response to grazing systems management in Wyoming's Red Desert. p. 451. In Summaries of Papers. J. Allan Smith (ed.). XIV International Grassland Congress. Lexington, Kentucky. June 14-23, 1981.

4B 820401 Fisser, H.G. 1982. Vegetation response to grazing management strategies on semiarid shrublands of Wyoming. p. 162-165. In D.D. Briske and M.M. Kothman (eds.). Proc. Natl. Conf. on Grazing Techniques. Dept. Range Sci. Texas A & M University. College Station, Texas. November 10-13, 1982.

4B 830402 Fisser, H.G. and K.L. Johnson (eds.) 1983. Wyoming Shrublands: Proc. 9th Wyoming Shrub Ecology Workshop. June 5-6, 1980. Lander, Wyoming. University of Wyoming, Laramie. 40 p.

4B 830403 Fisser, H.G. and K.L. Johnson (eds.) 1983. Wyoming Shrublands: Proc. 10th Wyoming Shrub Ecology Workshop. May 28-29, 1981. Sybille Canyon, Wyoming. University of Wyoming, Laramie. 40 p.

4B 830406 Deboodt, T.L., E.B. Bradley, and H.G. Fisser. 1983. Economic feasibility of resting rangeland to improve forage production in central Wyoming. abst. 215. In Absts. 36th Ann. Meeting. Soc. Range Manage. Albuquerque, New Mexico. February 14-17, 1983.

4B 840407 Fisser, H.G. and L.A. Joyce. 1984. Saltbush community changes with and without grazing over a 20-year period in north central Wyoming. Submitted to Proc. 2nd Intern. Rangeland Congress. Adelaide, Australia. May 13-18, 1984.

4B 840413 Fisser, H.G. and L.A. Joyce. 1984. *Atriplex*/grass and forb relationships under no grazing and shifting precipitation patterns in north central Wyoming. p. 87-96. In Tiedeman, A.R., E.D. McArthur, H.C. Stutz, R. Stevens, and K.L. Johnson (eds.). Proc. Wildland Shrub Symp. "Biology of *Atriplex* and related chenopods." Provo, Utah. May 2-6, 1983. USDA Forest Service General Technical Report 172.

4C. GRAZING HABITS AND DISTRIBUTION

4C 770401 Gerhart, William A., and H.G. Fisser. 1977. An evaluation of three grazing management systems on northern desert shrub vegetation of southern Wyoming. p. 52. In Proc. Soc. Range Manage. 30th Ann. Meeting. Portland, Oregon. February 17, 1977.

4C 810406 Rodgers, D.E. and H.G. Fisser. 1981. Grazing system impacts on semiarid vegetation of Wyoming. p. 34. In Absts. Soc. Range Manage. 34th Ann. Meeting. Tulsa, Oklahoma. February 8-13, 1981.

4C 810413 Fisser, H.G. 1981. Shrub establishment, dominance, and ecology on the juniper and sagebrush-grass types in Wyoming. p. 23-27. In L.S. Stelter, E.J. DePuit, S.A. Mikol (eds.). Proc. Symposium on Shrub Establishment on Disturbed Arid and Semiarid Lands. December 3, 1980. Wyoming Game and Fish Dept. Publication.

4C 810414 Rodgers, D.E. and H.G. Fisser. 1981. Vegetative response to grazing systems management in Wyoming's Red Desert. p. 451. In Summaries of Papers. J. Allan Smith (ed.). XIV International Grassland Congress. Lexington, Kentucky. June 14-23, 1981.

4C 820401 Fisser, H.G. 1982. Vegetation response to grazing management strategies on semiarid shrublands of Wyoming. p. 162-165. In D.D. Briske and M.M. Kothman (eds.). Proc. Natl. Conf. on Grazing Techniques. Dept. Range Sci. Texas A & M University. College Station, Texas. November 10-13, 1982.

4C 830402 Fisser, H.G. and K.L. Johnson (eds.) 1983. Wyoming Shrublands: Proc. 9th Wyoming Shrub Ecology Workshop. June 5-6, 1980. Lander, Wyoming. University of Wyoming, Laramie. 40 p.

4C 830403 Fisser, H.G. and K.L. Johnson (eds.) 1983. Wyoming Shrublands: Proc. 10th Wyoming Shrub Ecology Workshop. May 28-29, 1981. Sybille Canyon, Wyoming. University of Wyoming, Laramie. 40 p.

5. Range and Ranch Development

5A. PLANT CONTROL

5A1. Methods of Control

5A1B. BIOLOGICAL

5A1B670403 Fisser, H.G. 1967. Plant control on salt desert shrub rangeland by sheep grazing. p. 48-54. In Proc. Range Weed Control Conf. Las Cruces, New Mexico. August 21, 1967. p. 48-54.

5A1C. MECHANICAL

5A1C 610402 Vosler, L.C. 1961. Studies of mechanically treated watersheds. p. 1-3. In Proc. Range Weed Research Tour. Worland, Wyoming. September 14-15, 1961.

5A1C 640403 Fisser, H.G. 1964. Summer steer grazing on saltbush rangeland. p. 7. In Proc. Wyo. Sec. Am. Soc. Range Manage. Casper, Wyoming. November 19-20, 1964.

5A1C 810413 Fisser, H.G. 1981. Shrub establishment, dominance, and ecology on the juniper and sagebrush-grass types in Wyoming. p. 23-27. In L.S. Stelter, E.J. DePuit, S.A. Mikol (eds.). Proc. Symposium on Shrub Establishment on Disturbed Arid and Semiarid Lands. December 3, 1980. Wyoming Game and Fish Dept. Publication.

5A1C 830402 Fisser, H.G. and K.L. Johnson (eds.) 1983. Wyoming Shrublands: Proc. 9th Wyoming Shrub Ecology Workshop. June 5-6, 1980. Lander, Wyoming. University of Wyoming, Laramie. 40 p.

5A1C 830403 Fisser, H.G. and K.L. Johnson (eds.) 1983. Wyoming Shrublands: Proc. 10th Wyoming Shrub Ecology Workshop. May 28-29, 1981. Sybille Canyon, Wyoming. University of Wyoming, Laramie. 40 p.

5A1D. CHEMICAL.

5A1D610405 Colton, R.D. 1961. Sagebrush spraying tour. p. 12-14. In Proc. Range Weed Research Tour. Worland, Wyoming. September 14-15, 1961.

5A1D630402 Fisser, H.G. 1963. Exclosure studies (production, cover, height, and moisture phases). p. 23-24. In Proc. Range Weed Control Conf. Portland, Oregon. March 19, 1963.

5A1D670404 Fisser, H.G. 1967. Use of picloram for juniper control. p. 37-39. In Proc. Range Weed Control Conf. Las Cruces, New Mexico. August 21, 1967.

5A1D730401 Mackey, M.H. 1973. Evaluation of furrowing and seeding in a north central Wyoming Nuttall saltbush area. p. 13. In Proc. 26th Ann. Meeting. Soc. Range Manage. Boise, Idaho. February 4-9, 1973.

5A1D770402 Kleinman, Larry H. and H.G. Fisser. 1977. Phenodynamics and ecology of sagebrush-grass rangelands. p. 11. In Proc. Soc. Range Manage. 30th Ann. Meeting. Portland, Oregon. February 14-17, 1977.

5A1D790403 Orpet, J.E., and H.G. Fisser. 1979. Long-term herbage production response following the chemical control of big sagebrush. p. 29. In Proc. 32nd Ann. Meeting, Soc. Range Manage. Casper, Wyoming. February 12-15, 1979.

5A1D810401 Bensel, W.E. and H.G. Fisser. 1981. Forage production and longevity of sagebrush control on semiarid rangelands in western Wyoming. p. 9. In Absts. Soc. Range Manage. 34th Ann. Meeting. Tulsa, Oklahoma. February 8-13, 1981.

5A1D810407 Spaeth, K.E. and H.G. Fisser. 1981. Plant composition response to picloram application in the Wyoming juniper type. p. 35. In Absts. Soc. Range Manage. 34th Ann. Meeting. Tulsa, Oklahoma. February 8-13, 1981.

5A1D810413 Fisser, H.G. 1981. Shrub establishment, dominance, and ecology on the juniper and sagebrush-grass types in Wyoming. p. 23-27. In L.S. Stelter, E.J. DePuit, S.A. Mikol (eds.). Proc. Symposium on Shrub Establishment on Disturbed Arid and Semiarid Lands. December 3, 1980. Wyoming Game and Fish Dept.

5A1D820402 Dean, S.K. 1982. Vegetation response to strip chemical control in the Government Draw area. p. 35-36. In: Fisser, H.G. and K.L. Johnson, eds. 1984. Proceedings of the 11th Wyoming Shrub Ecology Workshop. Lander, Wyoming. May 27-28, 1982. University of Wyoming, Laramie.

5A1D820404 Fisser, H.G. 1982. Vegetation productivity and phenology in the Wind River Basin. p. 42-45. In: Fisser, H.G. and K.L. Johnson, eds. 1984. Proceedings of the 11th Wyoming Shrub Ecology Workshop. Lander, Wyoming. May 27-28, 1982. University of Wyoming, Laramie.

5A1D820405 Mulholland, J.L. 1982. Sagebrush and perennial grass dynamics at the Granite Mountain shrub control site. p. 39-42. In: Fisser, H.G. and K.L. Johnson, eds. 1984. Proceedings of the 11th Wyoming Shrub Ecology Workshop. Lander, Wyoming. May 27-28, 1982. University of Wyoming, Laramie. Publication.

5A1D830402 Fisser, H.G. and K.L. Johnson (eds.) 1983. Wyoming Shrublands: Proc. 9th Wyoming Shrub Ecology Workshop. June 5-6, 1980. Lander, Wyoming. University of Wyoming, Laramie. 40 p.

5A1D830403 Fisser, H.G. and K.L. Johnson (eds.) 1983. Wyoming Shrublands: Proc. 10th Wyoming Shrub Ecology Workshop. May 28-29, 1981. Sybille Canyon, Wyoming. University of Wyoming, Laramie. 40 p.

5A1D830405 Dean, S.K. and H.G. Fisser. 1983. Long-term vegetation responses to sagebrush control in western Wyoming. abst. 29. In Absts. 36th Ann. Meeting. Soc. Range Manage. Albuquerque, New Mexico. February 14-17, 1983.

5A1D830406 Deboodt, T.L., E.B. Bradley, and H.G. Fisser. 1983. Economic feasibility of resting rangeland to improve forage production in central Wyoming. abst. 215. In Absts. 36th Ann. Meeting. Soc. Range Manage. Albuquerque, New Mexico. February 14-17, 1983.

5A1D830411 Mulholland, J.L. and H.G. Fisser. 1983. Reestablishment rates of sagebrush following chemical control on grazed and ungrazed ranges in western Wyoming. abst. 295. In Absts. 36th Ann. Meeting. Soc. Range Manage. Albuquerque, New Mexico. February 14-17, 1983.

5A1D840405 Dean, S.K. and H.G. Fisser. 1984. Grazing management, longevity, and benefits of sagebrush control on semiarid rangelands. Submitted to Proc. 2nd Intern. Rangeland Congress. Adelaide, Australia. May 13-18, 1984.

5A1D840408 Mulholland, J.L. and H.G. Fisser. 1984. Big sagebrush response and reestablishment following chemical control in Wyoming, U.S.A. Submitted to Proc. 2nd Intern. Rangeland Congress. Adelaide, Australia. May 13-18, 1984.

5A1D840412 Dean, S.K. and H.G. Fisser. 1984. Grazing management longevity and benefits of sagebrush control in western Wyoming, U.S.A. Posterboard display at USDA-ARS/USDI-BLM Regional Workshop on Rangeland Hydrology, Modeling and Monitoring. Boise, Idaho. April 30 through May 2, 1984.

5A1D840415 Fisser, H.G. 1984. Biology and ecology of sagebrush in Wyoming: II. Reinvasion following control. p. 21. In Abst. Shrub Research Consortium Third Annual Wildland Shrub Symposium: the biology of *Artemisia* and *Chrysothamnus*. Provo, Utah. July 9-13, 1984.

5A1D840419 Fisser, H.G. and K.L. Johnson, eds. 1984. Wyoming Shrublands Proc. 11th Wyoming Shrub Ecology Workshop. Lander, Wyoming. May 27-28, 1982. University of Wyoming, Laramie.

5A1E. PYRIC

5A1E730402 Layton, D.L. 1973. Methods of revegetation and restoration following a juniper wildfire. p. 12. In Proc. 26th Ann. Meeting Soc. Range Mange. Boise, Idaho. February 4-9, 1973.

5A1E810413 Fisser, H.G. 1981. Shrub establishment, dominance, and ecology on the juniper and sagebrush-grass types in Wyoming. p. 23-27. In L.S. Stelter, E.J. DePuit, S.A. Mikol (eds.). Proc. Symposium on Shrub Establishment on Disturbed Arid and Semiarid Lands. December 3, 1980. Wyoming Game and Fish Dept. Publication.

5A1E830402 Fisser, H.G. and K.L. Johnson (eds.) 1983. Wyoming Shrublands: Proc. 9th Wyoming Shrub Ecology Workshop. June 5-6, 1980. Lander, Wyoming. University of Wyoming, Laramie. 40 p. Proc. 10th Wyoming Shrub Ecology Workshop. May 28-29, 1981. Sybille Canyon, Wyoming. University of Wyoming, Laramie. 40 p.

5A2. Regional Control

5A2D. INTERMOUNTAIN

5A2D610401 Vosler, L.C. 1961. Halogeton seed burial study. p. 1. In Proc. Range Weed Research Tour. Worland, Wyoming. September 14-15, 1961.

5A2D630401 Fisser, H.G. 1963. Progress of cattle and sheep grazing trials on halogeton infested ranges. p. 7-8. In Proc. Range Weed Control Conf. Portland, Oregon. March 19, 1963.

5A2D810401 Bensel, W.E. and H.G. Fisser. 1981. Forage production and longevity of sagebrush control on semiarid rangelands in western Wyoming. p. 9. In Absts. Soc. Range Manage. 34th Ann. Meeting. Tulsa, Oklahoma. February 8-13, 1981.

5B. RANGE REVEGETATION

5B 630402 Fisser, H.G. 1963. Exclosure studies (production, cover, height, and moisture phases). p. 23-24. In Proc. Range Weed Control Conf. Portland, Oregon. March 19, 1963.

5B 780405 Shawesh, O.M. and H.G. Fisser. 1978. Rangeland development and improvement in Libya, Africa. p. 65. In J.R. Goodin and D.K. Northington (eds.) Abstracts Intern. Arid Lands Conf. Intern. Center for Arid and Semiarid Land Studies. Texas Tech University. Lubbock, Texas. October 8-15, 1978.

5B4. Intermountain

5B4 610402 Vosler, L.C. 1961. Studies of mechanically treated watersheds. p. 1-3. In Proc. Range Weed Research Tour. Worland, Wyoming. September 14-15, 1961.

5B4 610404 Hull, A.C. Jr. 1961. Seeding salt desert shrub lands in Wyoming. p. 7-11. In Proc. Range Weed Research Tour. Worland, Wyoming. September 14-15, 1961.

5B4 640403 Fisser, H.G. 1964. Summer steer grazing on saltbush rangeland. p. 7. In Proc. Wyo. Sec. Am. Soc. Range Manage. Casper, Wyoming. Nov. 19-20, 1964.

5B4 730401 Mackey, M.H. 1973. Evaluation of furrowing and seeding in a north central Wyoming Nuttall saltbush area. p. 13. In Proc. 26th Ann. Meeting. Soc. Range Manage. Boise, Idaho. February 4-9, 1973.

5B4 730402 Layton, D.L. 1973. Methods of revegetation and restoration following a juniper wildfire. p. 12. In Proc. 26th Ann. Meeting Soc. Range Mange. Boise, Idaho. February 4-9, 1973.

5B4 750406 Shrader, J.A. and H.G. Fisser. 1975. Seed production ecology in the shortgrass prairie of the Central United States. p. 29. In Proc. 28th Ann. Meeting Soc. Range Mange. Mexico City, Mexico. February 10-14, 1975.

5B4 810413 Fisser, H.G. 1981. Shrub establishment, dominance, and ecology on the juniper and sagebrush-grass types in Wyoming. p. 23-27. In L.S. Stelter, E.J. DePuit, S.A. Mikol (eds.). Proc. Symposium on Shrub Establishment on Disturbed Arid and Semiarid Lands. December 3, 1980. Wyoming Game and Fish Dept. Publication

5B4 820402 Dean, S.K. 1982. Vegetation response to strip chemical control in the Government Draw area. p. 35-36. In: Fisser, H.G. and K.L. Johnson, eds. 1984. Proceedings of the 11th Wyoming Shrub Ecology Workshop. Lander, Wyoming. May 27-28, 1982. University of Wyoming, Laramie.

5B4 820405 Mulholland, J.L. 1982. Sagebrush and perennial grass dynamics at the Granite Mountain shrub control site. p. 39-42. In: Fisser, H.G. and K.L. Johnson, eds. 1984. Proceedings of the 11th Wyoming Shrub Ecology Workshop. Lander, Wyoming. May 27-28, 1982. University of Wyoming, Laramie.

5B4 830402 Fisser, H.G. and K.L. Johnson (eds.) 1983. Wyoming Shrublands: Proc. 9th Wyoming Shrub Ecology Workshop. June 5-6, 1980. Lander, Wyoming. University of Wyoming, Laramie. 40 p.

5B4 830403 Fisser, H.G. and K.L. Johnson (eds.) 1983. Wyoming Shrublands: Proc. 10th Wyoming Shrub Ecology Workshop. May 28-29, 1981. Sybille Canyon, Wyoming. University of Wyoming, Laramie. 40 p.

5B4 830405 Dean, S.K. and H.G. Fisser. 1983. Long-term vegetation responses to sagebrush control in western Wyoming. abst. 29. In Absts. 36th Ann. Meeting. Soc. Range Manage. Albuquerque, New Mexico. February 14-17, 1983.

5B4 840405 Dean, S.K. and H.G. Fisser. 1984. Grazing management, longevity, and benefits of sagebrush control on semiarid rangelands. Submitted to Proc. 2nd Intern. Rangeland Congress. Adelaide, Australia. May 13-18, 1984.

5B4 840412 Dean, S.K. and H.G. Fisser. 1984. Grazing management longevity and benefits of sagebrush control in western Wyoming, U.S.A. Posterboard display at USDA-ARS/USDI-BLM Regional Workshop on Rangeland Hydrology, Modeling and Monitoring. Boise, Idaho. April 30 through May 2, 1984.

5C. WATER SPREADING, FURROWING, PITTING

5C 610402 Vosler, L.C. 1961. Studies of mechanically treated watersheds. p. 1-3. In Proc. Range Weed Research Tour. Worland, Wyoming. September 14-15, 1961.

5C 610404 Hull, A.C. Jr. 1961. Seeding salt desert shrub lands in Wyoming. p. 7-11. In Proc. Range Weed Research Tour. Worland, Wyoming. September 14-15, 1961.

5C 630402 Fisser, H.G. 1963. Exclosure studies (production, cover, height, and moisture phases). p. 23-24. In Proc. Range Weed Control Conf. Portland, Oregon. March 19, 1963.

5C 640403 Fisser, H.G. 1964. Summer steer grazing on saltbush rangeland. p. 7. In Proc. Wyo. Sec. Am. Soc. Range Manage. Casper, Wyoming. November 19-20, 1964.

5C 720402 Fisser, H.G. 1972. An ecological description of biotic communities of the Wagon Wheel environs. p. I-1 thru I-48. In Rogers L.A. (ed.). Project Wagon Wheel Study Report Number Two. El Paso Natural Gas Co., PME-WW-13.

5C 730401 Mackey, M.H. 1973. Evaluation of furrowing and seeding in a north central Wyoming Nuttall saltbush area. p. 13. In Proc. 26th Ann. Meeting. Soc. Range Manage. Boise, Idaho. February 4-9, 1973.

5C 730402 Layton, D.L. 1973. Methods of revegetation and restoration following a juniper wildfire. p. 12. In Proc. 26th Ann. Meeting Soc. Range Mange. Boise, Idaho. February 4-9, 1973.

5C 740407 Whicker, F.W., A.W. Alldredge, H.G. Fisser, W.C. Hanson, J.E. Johnson & G. Post. 1974. Ecological considerations for Project Wagon Wheel. Phase I: Pre-event biospheric inventory, impact analysis and mitigation, and baseline documentation. A study conducted for El Paso Natural Gas Co. PNE-WW-35. 280 p.

5C 750404 Fisser, H.G. 1975. Vegetation response to seismic motion from underground nuclear detonation. p. 37. In Proc. Soc. Range Manage. 28th Ann. Meeting. Mexico City, Mexico. February 10-14, 1975.

5C 750407 Whicker, F.W., A.W. Alldredge, H.G. Fisser, and G. Post. 1975. Ecological Considerations for Project Wagon Wheel and hydraulic fracturing activities. Phase II(a). Annual summary report for 1974. A study conducted for El Paso Natural Gas Co. PNE-WW-40. 33 p.

5C 760404 Whicker, F.W., A.W. Alldredge, H.G. Fisser, W.C. Hanson, and G. Post. 1976. Ecological considerations for Project Wagon Wheel and hydraulic fracturing activities. Phase II(b): Annual summary for 1975 and project completion report. A study conducted for El Paso Natural Gas Company PNE-WW. 53 p.

5C 770403 Cox, Jerry R., and H.G. Fisser. 1977. The effect of environmental factors on carbohydrate storage and herbage production of five shortgrass prairie plant species. p. 41 In Proc. Soc. Range Manage. 30th Ann. Meeting. Portland, Oregon. February 14-17, 1977.

5C 790401 Cox J.R., and H.G. Fisser. 1979. The effects of water stress on phenology and carbohydrate. p. 529-535. In J.R. Goodin and D.K. Northington (eds.) Proc. of the International Arid Lands Conference. International Center for Arid and Semiarid Land Studies. Texas Tech University. Lubbock, Texas. October 8-15, 1978.

5C 800402 Hajjar, S., G. Spitzer, G. Meyer, and H.G. Fisser. 1980. An evaluation of water and soil samples for forage and other agricultural development in the Tabuk region. A report submitted to A.M. Aziz, Governor, Tabuk Province, Saudi Arabia. Amer. Res. Design Scientists, Inc. 49 p.

5C 810413 Fisser, H.G. 1981. Shrub establishment, dominance, and ecology on the juniper and sagebrush-grass types in Wyoming. p. 23-27. In L.S. Stelter, E.J. DePuit, S.A. Mikol (eds.). Proc. Symposium on Shrub Establishment on Disturbed Arid and Semiarid Lands. December 3, 1980. Wyoming Game and Fish Dept. Publication.

5C 830402 Fisser, H.G. and K.L. Johnson (eds.) 1983. Wyoming Shrublands: Proc. 9th Wyoming Shrub Ecology Workshop. June 5-6, 1980. Lander, Wyoming. University of Wyoming, Laramie. 40 p.

5C 830403 Fisser, H.G. and K.L. Johnson (eds.) 1983. Wyoming Shrublands: Proc. 10th Wyoming Shrub Ecology Workshop. May 28-29, 1981. Sybille Canyon, Wyoming. University of Wyoming, Laramie. 40 p.

5D. FERTILIZATION

5D 800402 Hajjar, S., G. Spitzer, G. Meyer, and H.G. Fisser. 1980. An evaluation of water and soil samples for forage and other agricultural development in the Tabuk region. A report submitted to A.M. Aziz, Governor, Tabuk Province, Saudi Arabia. Amer. Res. Design Scientists, Inc. 49 p.

6. Range Livestock

6 610403 Smith, D.R. 1961. Intensity of grazing study. p. 4-6. In Proc. Range Weed Research Tour. Worland, Wyoming. September 14-15, 1961.

6A. GENERAL MANAGEMENT

6A 830404 Waugh, W.J. 1983. Stagnation and decadence of *Cercocarpus montanus* in a southeastern Wyoming big game exclosure, p. 11-14. In H.G. Fisser and K.L. Johnson (eds.). Wyoming Shrublands: Proc. 10th Wyoming Shrub Ecology Workshop, Sybille Canyon, Wyoming, May 28-29, 1981. University of Wyoming, Laramie.

6A 830406 Deboodt, T.L., E.B. Bradley, and H.G. Fisser. 1983. Economic feasibility of resting rangeland to improve forage production in central Wyoming. abst. 215. In Absts. 36th Ann. Meeting. Soc. Range Manage. Albuquerque, New Mexico. February 14-17, 1983.

6A1. Cattle

6A1 630401 Fisser, H.G. 1963. Progress of cattle and sheep grazing trials on halogeton infested ranges. p. 7-8. In Proc. Range Weed Control Conf. Portland, Oregon. March 19, 1963.

6A1 640403 Fisser, H.G. 1964. Summer steer grazing on saltbush rangeland. p. 7. In Proc. Wyoming Sec. Am. Soc. Range Manage. Casper, Wyoming. November 19-20, 1964.

6A1 770406 Fisser, H.G. 1977. Rangeland and wildlife resources of the South Pass Historic Area. p. 77-88. In D.S. Warder. Management suggestions: South Pass Mining Area. Prepared for the Bureau of Land Management contract YA-510-PH-110. March 31, 1977. Parks and Recreation Department. University of Wyoming, Laramie.

6A1 810404 O'Connor, G.R., H.G. Fisser and J.W. Waggoner. 1981. Esophageal-fistulated steers for determination of dietary nutrient composition on large pasture native rangeland. p. 26. In Absts. Soc. Range Manage. 34th Ann. Meeting. Tulsa, Oklahoma. February 8-13, 1981.

6A2. Sheep and Goats

6A2 630401 Fisser, H.G. 1963. Progress of cattle and sheep grazing trials on halogeton infested ranges. p. 7-8. In Proc. Range Weed Control Conf. Portland, Oregon. March 19, 1963.

6A2 640401 Fisser, H.G. 1964. Winter sheep grazing studies on saltbush rangeland of Wyoming. p. 53. In Proc. 17th Ann. Meeting, Am. Soc. Range Manage. Wichita, Kansas. February 14, 1964.

6A2 660401 Fisser, H.G. 1966. Sheep grazing intensity studies and vegetative changes on *Atriplex nuttallii.* p. 226-237. In Proc. Salt Desert Shrub Symposium. Cedar City, Utah. August 1-3, 1966.

6A2 840407 Fisser, H.G. and L.A. Joyce. 1984. Saltbush community changes with and without grazing over a 20-year period in north central Wyoming. Submitted to Proc. 2nd Intern. Rangeland Congress. Adelaide, Australia. May 13-18, 1984.

6A2 840413 Fisser, H.G. and L.A. Joyce. 1984. *Atriplex*/grass and forb relationships under no grazing and shifting precipitation patterns in north central Wyoming. p. 87-96. In Tiedeman, A.R., E.D. McArthur, H.C. Stutz, R. Stevens, and K.L. Johnson (eds.). Proc. Wildland Shrub Symp. "Biology of *Atriplex* and related chenopods." Provo, Utah. May 2-6, 1983. USDA Forest Service General Technical Report 172.

6A3. Horses

6A3 760401 Stroud, D.W., and H.G. Fisser. 1976. Vegetative response of *Bouteloua gracilis* to four levels of utilization. p. 11. In Proc. Soc. Range Manage. 29th Ann. Meeting. Omaha, Nebraska. February 16-20, 1976.

6A3 770401 Gerhart, William A., and H.G. Fisser. 1977. An evaluation of three grazing management systems on northern desert shrub vegetation of southern Wyoming. p. 52. In Proc. Soc. Range Manage. 30th Ann. Meeting. Portland, Oregon. February 17, 1977.

6A3 810406 Rodgers, D.E. and H.G. Fisser. 1981. Grazing system impacts on semiarid vegetation of Wyoming. p. 34. In Absts. Soc. Range Manage. 34th Ann. Meeting. Tulsa, Oklahoma. February 8-13, 1981.

6A3 810414 Rodgers, D.E. and H.G. Fisser. 1981. Vegetative response to grazing systems management in Wyoming's Red Desert. p. 451. In Summaries of Papers. J. Allan Smith (ed.). XIV International Grassland Congress. Lexington, Kentucky. June 14-23, 1981.

6A3 820401 Fisser, H.G. 1982. Vegetation response to grazing management strategies on semiarid shrublands of Wyoming. p. 162-165. In D.D. Briske and M.M. Kothman (eds.). Proc. Natl. Conf. on Grazing Techniques. Dept. Range Sci. Texas A & M University. College Station, Texas. November 10-13, 1982.

6B. NUTRITION AND FEEDING

6B2. Nutrient Composition

6B2 810404 O'Connor, G.R., H.G. Fisser and J.W. Waggoner. 1981. Esophageal-fistulated steers for determination of dietary nutrient composition on large pasture native rangeland. p. 26. In Absts. Soc. Range Manage. 34th Ann. Meeting. Tulsa, Oklahoma. February 8-13, 1981.

7. Range Wildlife

7A. BIG GAME

7A1. Deer

7A1 770401 Gerhart, William A., and H.G. Fisser. 1977. An evaluation of three grazing management systems on northern desert shrub vegetation of southern Wyoming. p. 52. In Proc. Soc. Range Manage. 30th Ann. Meeting. Portland, Oregon. February 17, 1977.

7A1 770406 Fisser, H.G. 1977. Rangeland and wildlife resources of the South Pass Historic Area. p. 77-88. In D.S. Warder. Management suggestions: South Pass Mining Area. Prepared for the Bureau of Land Management contract YA-510-PH-110. March 31, 1977. Parks and Recreation Department. University of Wyoming, Laramie.

7A1 810406 Rodgers, D.E. and H.G. Fisser. 1981. Grazing system impacts on semiarid vegetation of Wyoming. p. 34. In Absts. Soc. Range Manage. 34th Ann. Meeting. Tulsa, Oklahoma. February 8-13, 1981.

7A1 810413 Fisser, H.G. 1981. Shrub establishment, dominance, and ecology on the juniper and sagebrush-grass types in Wyoming. p. 23-27. In L.S. Stelter, E.J. DePuit, S.A. Mikol (eds.). Proc. Symposium on Shrub Establishment on Disturbed Arid and Semiarid Lands. December 3, 1980. Wyoming Game and Fish Dept. Publication.

7A1 810414 Rodgers, D.E. and H.G. Fisser. 1981. Vegetative response to grazing systems management in Wyoming's Red Desert. p. 451. In Summaries of Papers. J. Allan Smith (ed.). XIV International Grassland Congress. Lexington, Kentucky. June 14-23, 1981.

7A1 820401 Fisser, H.G. 1982. Vegetation response to grazing management strategies on semiarid shrublands of Wyoming. p. 162-165. In D.D. Briske and M.M. Kothman (eds.). Proc. Natl. Conf. on Grazing Techniques. Dept. Range Sci. Texas A & M University. College Station, Texas. November 10-13, 1982.

7A1 830401 Fisser, Herbert G. 1983. Some recent research findings in the true mountain mahogany types. p. 7-10. In H.G. Fisser and K.L. Johnson (eds.) Wyoming Shrublands: Proc. 10th Wyo. Shrub Ecology Workshop. Sybille Canyon, Wyoming. May 28-29, 1981. University of Wyoming, Laramie.

7A1 830402 Fisser, H.G. and K.L. Johnson (eds.) 1983. Wyoming Shrublands: Proc. 9th Wyoming Shrub Ecology Workshop. June 5-6, 1980. Lander, Wyoming. University of Wyoming, Laramie. 40 p.

7A1 830403 Fisser, H.G. and K.L. Johnson (eds.) 1983. Wyoming Shrublands: Proc. 10th Wyoming Shrub Ecology Workshop. May 28-29, 1981. Sybille Canyon, Wyoming. University of Wyoming, Laramie. 40 p.

7A2. Elk

7A2 680401 Probasco, G.E. 1968. Dietary preferences and utilization of elk on spring range in the northern Big Horn Mountains of Wyoming. p. 22. In Proc. 21st Ann. Meeting Soc. Range Manage. Albuquerque, New Mexico. February 14, 1968.

7A2 770406 Fisser, H.G. 1977. Rangeland and wildlife resources of the South Pass Historic Area. p. 77-88. In D.S. Warder. Management suggestions: South Pass Mining Area. Prepared for the Bureau of Land Management contract YA-510-PH-110. March 31, 1977. Parks and Recreation Department. University of Wyoming, Laramie.

7A2 830403 Fisser, H.G. and K.L. Johnson (eds.) 1983. Wyoming Shrublands: Proc. 10th Wyoming Shrub Ecology Workshop. May 28-29, 1981. Sybille Canyon, Wyoming. University of Wyoming, Laramie. 40 p.

7A3. Antelope

7A3 770401 Gerhart, William A., and H.G. Fisser. 1977. An evaluation of three grazing management systems on northern desert shrub vegetation of southern Wyoming. p. 52. In Proc. Soc. Range Manage. 30th Ann. Meeting. Portland, Oregon. February 17, 1977.

7A3 770406 Fisser, H.G. 1977. Rangeland and wildlife resources of the South Pass Historic Area. p. 77-88. In D.S. Warder. Management suggestions: South Pass Mining Area. Prepared for the Bureau of Land Management contract YA-510-PH-110. March 31, 1977. Parks and Recreation Department. University of Wyoming, Laramie.

7A3 810406 Rodgers, D.E. and H.G. Fisser. 1981. Grazing system impacts on semiarid vegetation of Wyoming. p. 34. In Absts. Soc. Range Manage. 34th Ann. Meeting. Tulsa, Okla. Feb. 8-13, 1981.

7A3 810414 Rodgers, D.E. and H.G. Fisser. 1981. Vegetative response to grazing systems management in Wyoming's Red Desert. p. 451. In Summaries of Papers. J. Allan Smith (ed.). XIV International Grassland Congress. Lexington, Ky. June 14-23, 1981.

7A3 820401 Fisser, H.G. 1982. Vegetation response to grazing management strategies on semiarid shrublands of Wyoming. p. 162-165. In D.D. Briske and M.M. Kothman (eds.). Proc. Natl. Conf. on Grazing Techniques. Dept. Range Sci. Texas A & M University. College Station, Texas. November 10-13, 1982.

7A3 830403 Fisser, H.G. and K.L. Johnson (eds.) 1983. Wyoming Shrublands: Proc. 10th Wyoming Shrub Ecology Workshop. May 28-29, 1981. Sybille Canyon, Wyoming. University of Wyoming, Laramie. 40 p.

7A5. Moose

7A5 770406 Fisser, H.G. 1977. Rangeland and wildlife resources of the South Pass Historic Area. p. 77-88. In D.S. Warder. Management suggestions: South Pass Mining Area. Prepared for the Bureau of Land Management contract YA-510-PH-110. March 31, 1977. Parks and Recreation Department. University of Wyoming, Laramie.

7D. NON-GAME

7D1. Rodents and Rabbits

7D1 780403 Foster, M.A., J. Stubbendieck, and H.G. Fisser. 1978. Plant succession on plains pocket gopher mounds. p. 8. In Proc. Soc. Range Manage. 31st Ann. Meeting, San Antonio, Texas. February 7-9, 1978.

8. Range Watersheds

8A. WATER RESOURCES

8A 770406 Fisser, H.G. 1977. Rangeland and wildlife resources of the South Pass Historic Area. p. 77-88. In D.S. Warder. Management suggestions: South Pass Mining Area. Prepared for the Bureau of Land Management contract YA-510-PH-110. March 31, 1977. Parks and Recreation Department. University of Wyoming, Laramie.

8A 800402 Hajjar, S., G. Spitzer, G. Meyer, and H.G. Fisser. 1980. An evaluation of water and soil samples for forage and other agricultural development in the Tabuk region. A report submitted to A.M. Aziz, Governor, Tabuk Province, Saudi Arabia. Amer. Res. Design Scientists, Inc. 49 p.

8B. WATERSHED SOIL EROSION AND SEDIMENTATION

8B 640402 Nichols, J.R. 1964. Soil-vegetation relations of the 15-Mile Drainage area in the Big Horn Basin of Wyoming. p. 25. In Proc. 17th Ann. Meeting Am. Soc. Range Manage. Wichita, Kansas. February 14, 1964.

8B 750401 Fisser, H.G. 1975. A study of rangeland soil movement characteristics. p. 421-422. In Proc. Amer. Soc. Civil Engineers Watershed Manage. Symp. Logan, Utah. August 11-13, 1975.

8B 760402 Taha, Faisal K., and H.G. Fisser. 1976. Physical and chemical analyses of range sites and interrelationships with herbage production and forage value potentials. p. 13. In Proc. Soc. Range Manage. 29th Ann. Meeting. Omaha, Nebraska. February 16-20, 1976.

8B 780401 Fisser, Herbert G. 1978. Soil erosion and movement studies in western Wyoming. p. 16. In Proc. Soc. Range Manage. 31st Ann. Meeting, San Antonio, Texas. February 7-9, 1978.

8B 800401 Trueblood, D.C. and H.G. Fisser. 1980. Soil vegetation relationships at semiarid rangeland exclosures in north central Wyoming: long term studies. p. 31. In Proc. 33rd Ann. Meeting Soc. Range Manage. San Diego, California. February 11-14, 1980.

8B 810408 Trueblood, D.C. and H.G. Fisser. 1981. Soil-vegetation relationships in the Gardner saltbush vegetation type of Wyoming. p. 12. In Absts. Soc. Range Manage. 34th Ann. Meeting. Tulsa, Oklahoma. February 8-13, 1981.

8C. HYDROLOGY AND WATER YIELD

8C 780405 Shawesh, O.M. and H.G. Fisser. 1978. Rangeland development and improvement in Libya, Africa. p. 65. In J.R. Goodin and D.K. Northington (eds.) abstracts Intern. Arid Lands Conf. Intern. Center for Arid and Semiarid Land Studies. Texas Tech University. Lubbock, Texas. October 8-15, 1978.

8C 800402 Hajjar, S., G. Spitzer, G. Meyer, and H.G. Fisser. 1980. An evaluation of water and soil samples for forage and other agricultural development in the Tabuk region. A report submitted to A.M. Aziz, Governor, Tabuk Province, Saudi Arabia. Amer. Res. Design Scientists, Inc. 49 p.

8C 830402 Fisser, H.G. and K.L. Johnson (eds.) 1983. Wyoming Shrublands: Proc. 9th Wyoming Shrub Ecology Workshop. June 5-6, 1980. Lander, Wyoming. University of Wyoming, Laramie. 40 p.

8C 830403 Fisser, H.G. and K.L. Johnson (eds.) 1983. Wyoming Shrublands: Proc. 10th Wyoming Shrub Ecology Workshop. May 28-29, 1981. Sybille Canyon, Wyoming. University of Wyoming, Laramie. 40 p.

8D. WATERSHED RESTORATION

8D 720402 Fisser, H.G. 1972. An ecological description of biotic communities of the Wagon Wheel environs. p. I-1 thru I-48. In Rogers L.A. (ed.). Project Wagon Wheel Study Report Number Two. El Paso Natural Gas Co., PME-WW-13.

8D 740407 Whicker, F.W., A.W. Alldredge, H.G. Fisser, W.C. Hanson, J.E. Johnson & G. Post. 1974. Ecological considerations for Project Wagon Wheel. Phase I: Pre-event biospheric inventory, impact analysis and mitigation, and baseline documentation. A study conducted for El Paso Natural Gas Co. PNE-WW-35. 280 p.

8D 750402 Fisser, H.G. and R.E. Ries. 1975. Pre-disturbance ecological studies improve and define potential for surface mine reclamation. p. 128-134. In Proc. 3rd NCA-BCR Coal Conference and Expo II. Vol. I. Louisville, Kentucky. October 21-23, 1975.

8D 750404 Fisser, H.G. 1975. Vegetation response to seismic motion from underground nuclear detonation. p. 37. In Proc. Soc. Range Manage. 28th Ann. Meeting. Mexico City, Mexico. February 10-14, 1975.

8D 750405 Ries, R.E. and H.G. Fisser. 1975. Pre-disturbance vegetation inventory and ecological monitoring procedures for rangeland strip-mining. p. 40. In Proc. 28th Ann. Meeting Soc. Range Manage. Mexico City, Mexico. February 10-14, 1975.

8D 750407 Whicker, F.W., A.W. Alldredge, H.G. Fisser, and G. Post. 1975. Ecological Considerations for Project Wagon Wheel and Hydraulic Fracturing Activities. Phase II(a). Annual summary report for 1974. A study conducted for El Paso Natural Gas Co. PNE-WW-40. 33 p.

8D 760403 Lymbery, Gordon A., and H.G. Fisser. 1976. Vegetation re-establishment potential of strip-mined soils, Black Thunder Basin, Wyoming. p. 49. In Proc. Soc. Range Manage. 29th Ann. Meeting. Omaha, Nebraska. February 16-20, 1976.

8D 760404 Whicker, F.W., A.W. Alldredge, H.G. Fisser, W.C. Hanson, and G. Post. 1976. Ecological considerations for Project Wagon Wheel and hydraulic fracturing activities. Phase II(b): Annual summary for 1975 and project completion report. A study conducted for El Paso Natural Gas Company PNE-WW-85. 53 p.

8D 770404 Cleaves, D.T. 1977. Development of snowfence systems on strip mine spoils to increase seedling establishment. p. 32. In Proc. Soc. Range Manage. 30th Ann. Meeting. Portland, Oregon. February 14-17, 1977.

8D 800402 Hajjar, S., G. Spitzer, G. Meyer, and H.G. Fisser. 1980. An evaluation of water and soil samples for forage and other agricultural development in the Tabuk region. A report submitted to A.M. Aziz, Governor, Tabuk Province, Saudi Arabia. Amer. Res. Design Scientists, Inc. 49 p.

8D 830402 Fisser, H.G. and K.L. Johnson (eds.) 1983. Wyoming Shrublands: Proc. 9th Wyoming Shrub Ecology Workshop. June 5-6, 1980. Lander, Wyoming. University of Wyoming, Laramie. 40 p.

8D 830403 Fisser, H.G. and K.L. Johnson (eds.) 1983. Wyoming Shrublands: Proc. 10th Wyoming Shrub Ecology Workshop. May 28-29, 1981. Sybille Canyon, Wyoming. University of Wyoming, Laramie. 40 p.

9. Range Economics

9A. RANCH PLANNING

9A 770406 Fisser, H.G. 1977. Rangeland and wildlife resources of the South Pass Historic Area. p. 77-88. In D.S. Warder. Management suggestions: South Pass Mining Area. Prepared for the Bureau of Land Management contract YA-510-PH-110. March 31, 1977. Parks and Recreation Department. University of Wyoming, Laramie.

9A 830406 Deboodt, T.L., E.B. Bradley, and H.G. Fisser. 1983. Economic feasibility of resting rangeland to improve forage production in central Wyoming. abst. 215. In Absts. 36th Ann. Meeting. Soc. Range Manage. Albuquerque, New Mexico. February 14-17, 1983.

10. Range Education and Research

10D. RANGE RESEARCH METHODS

10D 800402 Hajjar, S., G. Spitzer, G. Meyer, and H.G. Fisser. 1980. An evaluation of water and soil samples for forage and other agricultural development in the Tabuk region. A report submitted to A.M. Aziz, Governor, Tabuk Province, Saudi Arabia. Amer. Res. Design Scientists, Inc. 49 p.

10D 810404 O'Connor, G.R., H.G. Fisser and J.W. Waggoner. 1981. Esophageal-fistulated steers for determination of dietary nutrient composition on large pasture native rangeland. p. 26. In Absts. Soc. Range Manage. 34th Ann. Meeting. Tulsa, Oklahoma. February 8-13, 1981.

10D 820402 Dean, S.K. 1982. Vegetation response to strip chemical control in the Government Draw area. p. 35-36. In: Fisser, H.G. and K.L. Johnson, eds. 1984. Proceedings of the 11th Wyoming Shrub Ecology Workshop. Lander, Wyoming. May 27-28, 1982. University of Wyoming, Laramie.

10D 820404 Fisser, H.G. 1982. Vegetation productivity and phenology in the Wind River Basin. p. 42-45. In: Fisser, H.G. and K.L. Johnson, eds. 1984. Proceedings of the 11th Wyoming Shrub Ecology Workshop. Lander, Wyoming. May 27-28, 1982. University of Wyoming, Laramie.

10D 820405 Mulholland, J.L. 1982. Sagebrush and perennial grass dynamics at the Granite Mountain shrub control site. p. 39-42. In: Fisser, H.G. and K.L. Johnson, eds. 1984. Proceedings of the 11th Wyoming Shrub Ecology Workshop. Lander, Wyoming. May 27-28, 1982. University of Wyoming, Laramie.

10D 830401 Fisser, Herbert G. 1983. Some recent research findings in the true mountain mahogany types. p. 7-10. In H.G. Fisser and K.L. Johnson (eds.) Wyoming Shrublands: Proc. 10th Wyo. Shrub Ecology Workshop. Sybille Canyon, Wyoming. May 28-29, 1981. University of Wyoming, Laramie.

10D 830405 Dean, S.K. and H.G. Fisser. 1983. Long-term vegetation responses to sagebrush control in western Wyoming. abst. 29. In Absts. 36th Ann. Meeting. Soc. Range Manage. Albuquerque, New Mexico. February 14-17, 1983.

10D 830412 Owens, M.K. and H.G. Fisser. 1983. Size and density parameters of western wheatgrass on diverse range sites in western Wyoming. abst. 128. In Absts. 36th Ann. Meeting. Soc. Range Manage. Albuquerque, New Mexico. February 14-17, 1983.

10D 840405 Dean, S.K. and H.G. Fisser. 1984. Grazing management, longevity, and benefits of sagebrush control on semiarid rangelands. Submitted to Proc. 2nd Intern. Rangeland Congress. Adelaide, Australia. May 13-18, 1984.

10D 840412 Dean, S.K. and H.G. Fisser. 1984. Grazing management longevity and benefits of sagebrush control in western Wyoming, U.S.A. Posterboard display at USDA-ARS/USDI-BLM Regional Workshop on Rangeland Hydrology, Modeling and Monitoring. Boise, Idaho, April 30 through May 2, 1984.

10D 840414 Fisser, H.G. 1984. Biology and ecology of sagebrush in Wyoming: I. Herbage production and community dynamics. p. 21. In Abst. Shrub Research Consortium Third Annual Wildland Shrub Symposium: the biology of *Artemisia* and *Chrysothamnus*. Provo, Utah. July 9-13, 1984.

10D 840419 Fisser, H.G. and K.L. Johnson, eds. 1984. Wyoming Shrublands Proc. 11th Wyoming Shrub Ecology Workshop. Lander, Wyoming. May 27-28, 1982. University of Wyoming, Laramie.

10D1. Plant-Soil Emphasis

10D1 690401 Fisser, H.G. 1969. Plant pattern and distribution in ecosystems and relationships to function. p. 183-196. In R.L. Dix and R.G. Beidleman (eds.) Proc. Information Synthesis Project Grassland Biome. U.S. I.B.P. Range Science Series No. 2. Colo. State Univ. Ft. Collins, Colorado.

10D1 770403 Cox, Jerry R., and H.G. Fisser. 1977. The effect of environmental factors on carbohydrate storage and herbage production of five shortgrass prairie plant species. p. 41 In Proc. Soc. Range Manage. 30th Ann. Meeting. Portland, Oregon. February 14-17, 1977.

10D1 770405 Gdara, A.O. 1977. Long-term vegetation studies with permanent quadrats. p. 8. In Proc. Soc. Range Manage. 30th Ann. Meeting. Portland, Oregon. February 14-17, 1977.

10D1 780404 Gdara, A.O. and H.G. Fisser. 1978. Semiarid rangeland and grazing evaluation of shrubland from permanent quadrats. Manuscript submitted to 1st Intern. Rangeland Congress, Denver, Colorado. August 14-18, 1978.

10D1 790401 Cox J.R., and H.G. Fisser. 1979. The effects of water stress on phenology and carbohydrate. p. 529-535. In J.R. Goodin and D.K. Northington (eds.) Proc. of the International Arid Lands Conference. International Center for Arid and Semiarid Land Studies. Texas Tech University. Lubbock, Texas. October 8-15, 1978.

10D1 790402 Young, L.W., and H.G. Fisser. 1979. Soils-phenology relations on some arid rangelands of western Wyoming. p. 53-54. In Proc. 32nd Ann. Meeting, Soc. Range Manage. Casper, Wyoming. February 12-15, 1979.

10D1 800401 Trueblood, D.C. and H.G. Fisser. 1980. Soil vegetation relationships at semiarid rangeland exclosures in north central Wyoming: long term studies. p. 31. In Proc. 33rd Ann. Meeting Soc. Range Manage. San Diego, California. February 11-14, 1980.

10D1 810408 Trueblood, D.C. and H.G. Fisser. 1981. Soil-vegetation relationships in the Gardner saltbush vegetation type of Wyoming. p. 12. In Absts. Soc. Range Manage. 34th Ann. Meeting. Tulsa, Oklahoma. February 8-13, 1981.

11. Computer Storage, Programming and Modeling

11A. CLIMATIC

11A 810402 Hargis, N.E. and H.G. Fisser. 1981. Phenological comparison of two sagebrush species in Wyoming. p. 21. In Absts. Soc. Range Manage. 34th Ann. Meeting. Tulsa, Oklahoma. February 8-13, 1981.

11A 810408 Trueblood, D.C. and H.G. Fisser. 1981. Soil-vegetation relationships in the Gardner saltbush vegetation type of Wyoming. p. 12. In Absts. Soc. Range Manage. 34th Ann.

11A 820403 Kinucan, R.J. 1982. Numerical phenology scoring and use with quantitative environmental variables. p. 36-39. In: Fisser, H.G. and K.L. Johnson eds.1984. Proceedings of the 11th Wyoming Shrub Ecology Workshop. Lander, Wyoming. May 27-28, 1982. University of Wyoming, Laramie.

11A 830408 Fisser, H.G. 1983. Phenological description and numerical scoring of sagebrush. abst. 136. In Absts. 36th Ann. Meeting. Soc. Range Manage. Albuquerque, New Mexico. February 14-17, 1983.

11A 840402 Fisser, H.G. and N.E. Hargis. 1984. Detailed numerical phenology of sagebrush with visual description of statements. p. 44 . In Proc 37th Ann. Meeting Soc. Range Manage. Rapid City, South Dakota. February 12-17, 1984.

11A 840403 Germain, L.P. and H.G. Fisser. 1984. Estimation of missing precipitation data using traditional, kriging, and subjective techniques. abst. 160. In Abst. 37th Ann. Meeting Soc. Range Manage. Rapid City, South Dakota. February 12-17, 1984.

11A 840409 Germain, L.P. and H.G. Fisser. 1984. Application of kriging to ecological variables. p. 12. In Proc. 55th Ann. Meeting Colo.-Wyo. Acad. Sci., Western State College. Gunnison, Colorado. April 27-28, 1984.

11A 840411 Fisser, H.G. 1984. Detailed numerical phenology of sagebrush with visual description statements. Posterboard display at USDA-ARS/USDI-BLM Regional Workshop on Rangeland Hydrology, Modeling and Monitoring. Boise, Idaho. April 31 through May 2, 1984.

11A 840413 Fisser, H.G. and L.A. Joyce. 1984. *Atriplex*/grass and forb relationships under no grazing and shifting precipitation patterns in north central Wyoming. p. 87-96. In Tiedeman, A.R., E.D. McArthur, H.C. Stutz, R. Stevens, and K.L. Johnson (eds.). Proc. Wildland Shrub Symp. "Biology of *Atriplex* and related chenopods." Provo, Utah. May 2-6, 1983. USDA Forest Service General Technical Report 172.

11A 840416 Fisser, H.G. 1984. Biology and ecology of sagebrush in Wyoming: III. Regional variation of phenology. p. 22. In Abst. Shrub Research Consortium Third Annual Wildland Shrub Symposium: the biology of *Artemisia* and *Chrysothamnus*. Provo, Utah. July 9-13, 1984.

11B. BIOTIC

11B 810402 Hargis, N.E. and H.G. Fisser. 1981. Phenological comparison of two sagebrush species in Wyoming. p. 21. In Absts. Soc. Range Manage. 34th Ann. Meeting. Tulsa, Oklahoma. February 8-13, 1981.

11B 820403 Kinucan, R.J. 1982. Numerical phenology scoring and use with quantitative environmental variables. p. 36-39. In: Fisser, H.G. and K.L. Johnson eds. 1984. Proceedings of the 11th Wyoming Shrub Ecology Workshop. Lander, Wyoming. May 27-28, 1982. University of Wyoming, Laramie.

11B 830408 Fisser, H.G. 1983. Phenological description and numerical scoring of sagebrush. abst. 136. In Absts. 36th Ann. Meeting. Soc. Range Manage. Albuquerque, New Mexico. February 14-17, 1983.

11B 840402 Fisser, H.G. and N.E. Hargis. 1984. Detailed numerical phenology of sagebrush with visual description statements. p. 44. In Proc 37th Ann. Meeting Soc. Range Manage. Rapid City, South Dakota. February 12-17, 1984.

11B 840411 Fisser, H.G. 1984. Detailed numerical phenology of sagebrush with visual description statements. Posterboard display at USDA-ARS/USDI-BLM Regional Workshop on Rangeland Hydrology, Modeling and Monitoring. Boise, Idaho. April 31 through May 2, 1984.

11B 840416 Fisser, H.G. 1984. Biology and ecology of sagebrush in Wyoming: III. Regional variation of phenology. p. 22. In Abst. Shrub Research Consortium Third Annual Wildland Shrub Symposium: the biology of *Artemisia* and *Chrysothamnus*. Provo, Utah. July 9-13, 1984.

11C. EDAPHIC

11C 810402 Hargis, N.E. and H.G. Fisser. 1981. Phenological comparison of two sagebrush species in Wyoming. p. 21. In Absts. Soc. Range Manage. 34th Ann. Meeting. Tulsa, Oklahoma. February 8-13, 1981.

11C 810408 Trueblood, D.C. and H.G. Fisser. 1981. Soil-vegetation relationships in the Gardner saltbush vegetation type of Wyoming. p. 12. In Absts. Soc. Range Manage. 34th Ann. Meeting. Tulsa, Oklahoma. February 8-13, 1981.

11C 820403 Kinucan, R.J. 1982. Numerical phenology scoring and use with quantitative environmental variables. p. 36-39. In: Fisser, H.G. and K.L. Johnson eds. 1984. Proceedings of the 11th Wyoming Shrub Ecology Workshop. Lander, Wyoming. May 27-28, 1982. University of Wyoming, Laramie.

11C 830408 Fisser, H.G. 1983. Phenological description and numerical scoring of sagebrush. abst. 136. In Absts. 36th Ann. Meeting. Soc. Range Manage. Albuquerque, New Mexico. February 14-17, 1983.

11C 840402 Fisser, H.G. and N.E. Hargis. 1984. Detailed numerical phenology of sagebrush with visual description statements. p. 44. In Proc 37th Ann. Meeting Soc. Range Manage. Rapid City, South Dakota. February 12-17, 1984.

11C 840411 Fisser, H.G. 1984. Detailed numerical phenology of sagebrush with visual description statements. Posterboard display at USDA-ARS/USDI-BLM Regional Workshop on Rangeland Hydrology, Modeling and Monitoring. Boise, Idaho. April 31 through May 2, 1984.

11C 840416 Fisser, H.G. 1984. Biology and ecology of sagebrush in Wyoming: III. Regional variation of phenology. p. 22. In Abst. Shrub Research Consortium Third Annual Wildland Shrub Symposium: the biology of *Artemisia* and *Chrysothamnus*. Provo, Utah. July 9-13, 1984.

11E. RANGELAND INVENTORY

11E 810402 Hargis, N.E. and H.G. Fisser. 1981. Phenological comparison of two sagebrush species in Wyoming. p. 21. In Absts. Soc. Range Manage. 34th Ann. Meeting. Tulsa, Oklahoma. February 8-13, 1981.

11E 810408 Trueblood, D.C. and H.G. Fisser. 1981. Soil-vegetation relationships in the Gardner saltbush vegetation type of Wyoming. p. 12. In Absts. Soc. Range Manage. 34th Ann. Meeting. Tulsa, Oklahoma. February 8-13, 1981.

11E 810412 Shawesh, O.M. and H.G. Fisser. 1981. Habitat and vegetation of semiarid rangelands, Libya, Africa. p. 23. In Absts. Soc. Range Manage. 34th Ann. Meeting. Tulsa, Oklahoma. February 8-13, 1981.

11E 810414 Rodgers, D.E. and H.G. Fisser. 1981. Vegetative response to grazing systems management in Wyoming's Red Desert. p. 451. In Summaries of Papers. J. Allan Smith (ed.). XIV International Grassland Congress. Lexington, Kentucky. June 14-23, 1981.

11E 820403 Kinucan, R.J. 1982. Numerical phenology scoring and use with quantitative environmental variables. p. 36-39. In: Fisser, H.G. and K.L. Johnson eds. 1984. Proceedings of the 11th Wyoming Shrub Ecology Workshop. Lander, Wyoming. May 27-28, 1982. University of Wyoming, Laramie.

11E 830408 Fisser, H.G. 1983. Phenological description and numerical scoring of sagebrush. abst. 136. In Absts. 36th Ann. Meeting. Soc. Range Manage. Albuquerque, New Mexico. February 14-17, 1983.

11E 830411 Mulholland, J.L. and H.G. Fisser. 1983. Reestablishment rates of sagebrush following chemical control on grazed and ungrazed ranges in western Wyoming. abst. 295. In Absts. 36th Ann. Meeting. Soc. Range Manage. Albuquerque, New Mexico. February 14-17, 19823.

11E 840402 Fisser, H.G. and N.E. Hargis. 1984. Detailed numerical phenology of sagebrush with visual description statements. p. 44. In Proc 37th Ann. Meeting Soc. Range Manage. Rapid City, South Dakota. February 12-17, 1984.

11E 840411 Fisser, H.G. 1984. Detailed numerical phenology of sagebrush with visual description statements. Posterboard display at USDA-ARS/USDI-BLM Regional Workshop on Rangeland Hydrology, Modeling and Monitoring. Boise, Idaho. April 31 through May 2, 1984.

11E2. Condition and Trend

11E2 830406 Deboodt, T.L., E.B. Bradley, and H.G. Fisser. 1983. Economic feasibility of resting rangeland to improve forage production in central Wyoming. abst. 215. In Absts. 36th Ann. Meeting. Soc. Range Manage. Albuquerque, New Mexico. February 14-17, 1983.

11E2 830412 Owens, M.K. and H.G. Fisser. 1983. Size and density parameters of western wheatgrass on diverse range sites in western Wyoming. abst. 128. In Absts. 36th Ann. Meeting. Soc. Range Manage. Albuquerque, New Mexico. February 14-17, 1983.

11E2 840408 Mulholland, J.L. and H.G. Fisser. 1984. Big sagebrush response and reestablishment following chemical control in Wyoming, U.S.A. Submitted to Proc. 2nd Intern. Rangeland Congress. Adelaide, Australia. May 13-18, 1984.

11E2 840413 Fisser, H.G. and L.A. Joyce. 1984. *Atriplex*/grass and forb relationships under no grazing and shifting precipitation patterns in north central Wyoming. p. 87-96. In Tiedeman, A.R., E.D. McArthur, H.C. Stutz, R. Stevens, and K.L. Johnson (eds.). Proc. Wildland Shrub Symp. "Biology of *Atriplex* and related chenopods." Provo, Utah. May 2-6, 1983. USDA Forest Service General Technical Report 172.

11E2 840416 Fisser, H.G. 1984. Biology and ecology of sagebrush in Wyoming: III. Regional variation of phenology. p. 22. In Abst. Shrub Research Consortium Third Annual Wildland Shrub Symposium: the biology of *Artemisia* and *Chrysothamnus.* Provo, Utah. July 9-13, 1984.

11E2 840417 Fisser, H.G., J.R. Wight and C.W. Hanson. 1984. Biology and ecology of sagebrush in Wyoming: IV. Validation of a rangeland production model (ERHYM) for sagebrush sites. p. 13. In Abst. Shrub Research Consortium Third Annual Wildland Shrub Symposium: the biology of *Artemisia* and *Chrysothamnus.* Provo, Utah. July 9-13, 1984.

11E3. Production and Utilization

11E3 740405 Whysong, G.L. and H.G. Fisser. 1974. Statistical models for description and prediction of perennial grass production. p. 408-413. In V.G. Iglovikov, A.P. Movsissyants, F.F. Sarganov, and A.A. Stepanenko (Compilers) Biological and physiological aspects of the intensification of grassland utilization. XII International Grassland Congress. Moscow, USSR. June 12-16, 1974.

11E3 810405 Owens, M.K. and H.G. Fisser. 1981. Phenology and productivity comparisons of western wheatgrass *(Agropyron smithii)* and bluebunch wheatgrass *(A. spicatum).* p. 21. In Absts. Soc. Range Manage. 34th Ann. Meeting. Tulsa, Oklahoma. February 8-13, 1981.

11E3 810415 Owens, M.K. and H.G. Fisser. 1981. Phenology and culm weight variation between two wheatgrass species: *Agropyron smithii* and *A. spicatum.* p. 205. In Summaries of Papers. J. Allan Smith (ed.). XIV International Grassland Congress. Lexington, Kentucky. June 14-23, 1981.

11E3 840401 Fisser, H.G. 1984. Productivity index procedures for sagebrush-grass rangeland. abst. 010. In Abst. 37th Ann. Meeting. Soc. Range Manage. Rapid City, South Dakota. February 12-17, 1984.

11E3 840408 Mulholland, J.L. and H.G. Fisser. 1984. Big sagebrush response and reestablishment following chemical control in Wyoming, U.S.A. Submitted to Proc. 2nd Intern. Rangeland Congress. Adelaide, Australia. May 13-18, 1984. (In press.)

11E3 840410 Fisser, H.G. 1984. Culm weight dynamics as an index of sagebrush-grass community productivity. p. 17. In Proc. 55th Ann. Meeting Colo.-Wyo. Acad. Sci., Western State College. Gunnison, Colorado. April 27-28, 1984.

11E3 840413 Fisser, H.G. and L.A. Joyce. 1984. *Atriplex*/grass and forb relationships under no grazing and shifting precipitation patterns in north central Wyoming. p. 87-96. In Tiedeman, A.R., E.D. McArthur, H.C. Stutz, R. Stevens, and K.L. Johnson (eds.). Proc. Wildland Shrub Symp. "Biology of *Atriplex* and related chenopods." Provo, Utah. May 2-6, 1983. USDA Forest Service General Technical Report 172.

11E3 840417 Fisser, H.G., J.R. Wight and C.W. Hanson. 1984. Biology and ecology of sagebrush in Wyoming: IV. Validation of a rangeland production model (ERHYM) for sagebrush sites. p. 13. In Abst. Shrub Research Consortium Third Annual Wildland Shrub Symposium: the biology of *Artemisia* and *Chrysothamnus*. Provo, Utah. July 9-13, 1984.

11E3 840418 Fisser, H.G., C.W. Hanson and J.R. Wight. 1984. Biology and ecology of sagebrush in Wyoming: V. Herbage yield dynamics and seasonal precipitation relationships. p. 13. In Abst. Shrub Research Consortium Third Annual Wildland Shrub Symposium: the biology of *Artemisia* and *Chrysothamnus*. Provo, Utah. July 9-13, 1984.

11F. PREDICTIVE MODELING

11F1. Herbage Yield

11F1 780402 Taha, Faisal K. and H.G. Fisser. 1978. Models for prediction of natural vegetation potential in eastern Wyoming. p. 19. In Proc. Soc. Range Manage. 31st Ann. Meeting, San Antonio, Texas. February 7-9, 1978.

11F1 840401 Fisser, H.G. 1984. Productivity index procedures for sagebrush-grass rangeland. abst. 010. In Abst. 37th Ann. Meeting. Soc. Range Manage. Rapid City, South Dakota. February 12-17, 1984.

11F1 840402 Fisser, H.G. and N.E. Hargis. 1984. Detailed numerical phenology of sagebrush with visual description statements. p. 44. In Proc 37th Ann. Meeting Soc. Range Manage. Rapid City, South Dakota. February 12-17, 1984.

11F1 840410 Fisser, H.G. 1984. Culm weight dynamics as an index of sagebrush-grass community productivity. p. 17. In Proc. 55th Ann. Meeting Colo.-Wyo. Acad. Sci., Western State College. Gunnison, Colorado. April 27-28, 1984.

11F1 840411 Fisser, H.G. 1984. Detailed numerical phenology of sagebrush with visual description statements. Posterboard display at USDA-ARS/USDI-BLM Regional Workshop on Rangeland Hydrology, Modeling and Monitoring. Boise, Idaho. April 31 through May 2, 1984.

11F1 840413 Fisser, H.G. and L.A. Joyce. 1984. Atriplex/grass and forb relationships under no grazing and shifting precipitation patterns in north central Wyoming. p. 87-96. In Tiedeman, A.R., E.D. McArthur, H.C. Stutz, R. Stevens, and K.L. Johnson (eds.). Proc. Wildland Shrub Symp. "Biology of *Atriplex* and related chenopods." Provo, Utah. May 2-6, 1983. USDA Forest Service General Technical Report 172.

11F1 840414 Fisser, H.G. 1984. Biology and ecology of sagebrush in Wyoming: I. Herbage production and community dynamics. p. 21. In Abst. Shrub Research Consortium Third Annual Wildland Shrub Symposium: the biology of *Artemisia* and *Chrysothamnus*. Provo, Utah. July 9-13, 1984.

11F1 840417 Fisser, H.G., J.R. Wight and C.W. Hanson. 1984. Biology and ecology of sagebrush in Wyoming: IV. Validation of a rangeland production model (ERHYM) for sagebrush sites. p. 13. In Abst. Shrub Research Consortium Third Annual Wildland Shrub Symposium: the biology of *Artemisia* and *Chrysothamnus*. Provo, Utah. July 9-13, 1984.

11F1 840418 Fisser, H.G., C.W. Hanson and J.R. Wight. 1984. Biology and ecology of sagebrush in Wyoming: V. Herbage yield dynamics and seasonal precipitation relationships. p. 13. In Abst. Shrub Research Consortium Third Annual Wildland Shrub Symposium: the biology of *Artemisia* and *Chrysothamnus*. Provo, Utah. July 9-13, 1984.

11F2. Phenology

11F2 820403 Kinucan, R.J. 1982. Numerical phenology scoring and use with quantitative environmental variables. p. 36-39. In: Fisser, H.G. and K.L. Johnson eds.1984. Proceedings of the 11th Wyoming Shrub Ecology Workshop. Lander, Wyoming. May 27-28, 1982. University of Wyoming, Laramie.

11F2 840416 Fisser, H.G. 1984. Biology and ecology of sagebrush in Wyoming: III. Regional variation of phenology. p. 22. In Abst. Shrub Research Consortium Third Annual Wildland Shrub Symposium: the biology of *Artemisia* and *Chrysothamnus*. Provo, Utah. July 9-13, 1984.

11F2 840417 Fisser, H.G., J.R. Wight and C.W. Hanson. 1984. Biology and ecology of sagebrush in Wyoming: IV. Validation of a rangeland production model (ERHYM) for sagebrush sites. p. 13. In Abst. Shrub Research Consortium Third Annual Wildland Shrub Symposium: the biology of *Artemisia* and *Chrysothamnus*. Provo, Utah. July 9-13, 1984.

S5. WYOMING AGRICULTURAL EXPERIMENT STATION PUBLICATIONS

1. Range Plants

1B. FORAGE VALUE

1B4. Poisonous Plants

1B4 640501 6 Fisser, H.G., J.T. Nichols, R.W. Hamner, and J.R. Wight. 1964. Halogeton Research. 1963 Results. Univ. of Wyo. Coop. Res. Rep. to the BLM. Sections I - VIII. Wyo. Agr. Exp. Sta. Mimeo. p. 1-61. Section VI - H.G. Fisser, R.W. Hamner. Germination and longevity of halogeton seeds. 1 p.

1D. PLANT INSECTS

1D 660502 Lavigne, R.J. 1966. Summary of research during 1964 and 1965 for the control of the western harvester ants, *Pogonomyrmex occidentalis* (Cresson). Wyo. Agr. Exp. Sta. Mimeo. 12 p.

1E. PLANT PHYSIOLOGY

1E1. Reproduction and Germination

1E1 640501 6 Fisser, H.G., J.T. Nichols, R.W. Hamner, and J.R. Wight. 1964. Halogeton Research. 1963 Results. Univ. of Wyo. Coop. Res. Rep. to the BLM. Sections I - VIII. Wyo. Agr. Exp. Sta. Mimeo. p. 1-61. Section VI - H.G. Fisser, R.W. Hamner. Germination and longevity of halogeton seeds. 1 p.

1E1 740502 1 Fisser, H.G. and L.H. Kleinman. 1974. Phenology and production on semiarid shrub types. 1973 results. Univ. of Wyo. Coop. Res. Rep. to the BLM. Sections I, II and III. Wyo. Agr. Exp. Sta. Sci. Rep. 607. 256 p. Section I - Phenology study.

1E1 750503 1 Fisser, H.G., and L.H. Kleinman. 1975. Phenology and production on semiarid shrub-types. 1974 results. Univ. Wyo. Coop. Res. Rep. to the BLM. Sections I, II, and III. Wyo. Agr. Exp. Sta. Sci. Rep. 729. 399 p. Section I - Phenology study.

1E1 760501 1 Fisser, H.G., and L.H. Kleinman. 1976. Phenology and production studies on semiarid shrub-types. 1975 results. Univ. of Wyo. Coop. Res. Rep. to the BLM. Sections I, II and III. Wyo. Agr. Exp. Sta. Sci. Rep. 831. 410 p. Section I - Phenology study.

1E1 770502 1 Fisser, H.G., L.W. Young, J.E. Orpet and N.E. Hargis. 1977. Phenology and production studies on semiarid shrub-types. 1976 results. Univ. of Wyo. Coop. Res. Rep. to the BLM. Sections I, II, III and IV. Wyo. Agr. Exp. Sta. Sci. Rep. 930. 438 p. Section I - Phenology study.

1E1 770502 2 Fisser, H.G., L.W. Young, J.E. Orpet and N.E. Hargis. 1977. Phenology and production studies on semiarid shrub-types. 1976 results. Univ. of Wyo. Coop. Res. Rep. to the BLM. Sections I, II, III and IV. Wyo. Agr. Exp. Sta. Sci. Rep. 930. 438 p. Section II - Phenology step-wise regression for predictive modeling.

1E1 780503 1 Fisser, H.G., L.W. Young, J.E. Orpet and N.E. Hargis. 1978. Phenology and production studies on semiarid shrub-types. 1977 results. Univ. of Wyo. Coop. Res. Rep. to the BLM. Sections I, II, III and IV. Wyo. Agr. Exp. Sta. Sci. Rep. 970. 439 p. Section I - Phenology study.

1E1 780503 2 Fisser, H.G., L.W. Young, J.E. Orpet and N.E. Hargis. 1978. Phenology and production studies on semiarid shrub-types. 1977 results. Univ. of Wyo. Coop. Res. Rep. to the BLM. Sections I, II, III and IV. Wyo. Agr. Exp. Sta. Sci. Rep. 970. 439 p. Section II - Phenology step-wise regression for predictive modeling.

1E1 790502 1 Fisser, H.G., L.W. Young, N.E. Hargis and K.E. Spaeth. 1979. Phenology and production studies on semiarid shrub-types. 1978 results. Univ. of Wyo. Coop. Res. Rep. to the BLM. Sections I, II, III and IV. Wyo. Agr. Exp. Sta. Sci. Rep. 985. 489 p. Section I - Phenology study.

1E1 790502 2 Fisser, H.G., L.W. Young, N.E. Hargis and K.E. Spaeth. 1979. Phenology and production studies on semiarid shrub-types. 1978 results. Univ. of Wyo. Coop. Res. Rep. to the BLM. Sections I, II, III and IV. Wyo. Agr. Exp. Sta. Sci. Rep. 985. 489 p. Section II - Phenology step-wise regression for predictive modeling.

1E1 800501 1 Fisser, H.G., M.K. Owens, W.J. Waugh, N.E. Hargis and J. Uhlich. 1980. Phenology and production studies on semiarid shrub-types. 1979 Results. Univ. of Wyo. Coop. Res. Rep. to the BLM. Vol. I: Phenology Study. Sections I and II. Wyo. Agr. Exp. Sta. Sci. Rep. 1044. 387 p. Section I - Phenology study.

1E1 800501 2 Fisser, H.G., M.K. Owens, W.J. Waugh, N.E. Hargis and J. Uhlich. 1980. Phenology and production studies on semiarid shrub-types. 1979 Results. Univ. of Wyo. Coop. Res. Rep. to the BLM. Vol. I: Phenology Study. Sections I and II. Wyo. Agr. Exp. Sta. Sci. Rep. 1044. 387 p. Section II - Phenology step-wise regression for predictive modeling.

1E1 800502 4 Fisser, H.G., M.K. Owens, J.W. Uhlich, W.J. Waugh, D.E. Rodgers, and N.E. Hargis. 1980. Phenology and production studies on semiarid shrub-types. 1979 Results. Univ. of Wyo. Coop. Res. Rep. to the BLM. Vol. II: Precipitation, Productivity and Production Section. Sections III, IV, and V. Wyo. Agr. Exp. Sta. Sci. Rep. 1106. 219 p. Section IV - Productivity - phenology relationships.

1E1 820501 2 Fisser, H.G., S.K. Dean, T.L. Deboodt, R.J. Kinucan. 1982. Phenology and production studies on semiarid shrub-types. 1982 Results. Univ. of Wyo. Coop. Res. Rep. to the BLM - Vol. I: Phenology study sections I and II. Wyo. Agr. Exp. Sta. Sci. Rep. 1230. 415 p. Section II - Phenology multiple regression for predictive modeling.

1E1 820502 4 Fisser, H.G., S.K. Dean, T.L. Deboodt, R.J. Kinucan. 1982. Phenology and production studies on semiarid shrub-types. 1982 results. Univ. of Wyo. Coop. Res. Rep. to the BLM. Vol. II: Precipitation, productivity, and production sections III, IV, and V. Wyo. Agr. Exp. Sta. Sci. Rep. 1231. 256 p. Section IV - Productivity-phenology interrelation study.

1E1 830501 1 Fisser, H.G., S.K. Dean, T.L. Deboodt, R.S. Kinucan and J.L. Mulholland. 1983. Phenology and production studies on semiarid shrub-types. 1982 results. Univ. of Wyo. Coop. Res. Rep. to the BLM. Vol. I: Phenology Study, Sections I and II. Wyo. Agr. Exp. Sta. Sci. Rep. 1230. 430 p. Section I - Phenology study.

1E1 830501 2 Fisser, H.G., S.K. Dean, T.L. Deboodt, R.S. Kinucan and J.L. Mulholland. 1983. Phenology and production studies on semiarid shrub-types. 1982 results. Univ. of Wyo. Coop. Res. Rep. to the BLM. Vol. I: Phenology Study, Sections I and II. Wyo. Agr. Exp. Sta. Sci. Rep. 1230. 430 p. Section II - Phenology multiple regression for predictive modeling.

1E1 830502 4 Fisser, H.G., S.K. Dean, T.L. Deboodt, R.S. Kinucan and J.L. Mulholland. 1983. Phenology and production studies on semiarid shrub-types. 1982 results. Univ. of Wyo. Coop. Res. Rep. to the BLM. Vol. II: Precipitation, Productivity and Production, Sections III, IV, and V. Wyo. Agr. Exp. Sta. Sci. Rep. 1231. 256 p. Section IV - Productivity - phenology interrelation study.

1E1 830503 Fisser, H.G. 1983. Aridland shrub ecology and phenology research program. Wyo. Agr. Exp. Sta. Mimeo. 3 p.

1E2. Growth and Phenology

1E2 710503 Fisser, H.G. 1971. Significance of plant pattern in ecosystem functioning. Wyo. Agr. Exp. Sta. Mimeo. 2 p.

1E2 710504 Fisser, H.G. 1971. Evaluation of Pawnee Site frequency data. Wyo. Agr. Exp. Sta. Mimeo. 6 p.

1E2 720504 Stroud, D.W. and H.G. Fisser. 1972. Plant pattern micro-relief and *Bouteloua gracilis* culm density studies on the Pawnee Site: 1970-1971. U.S. I.B.P. Grassland Biome Technical Report 186. 26 p.

1E2 740502 1 Fisser, H.G. and L.H. Kleinman. 1974. Phenology and production on semiarid shrub-types. 1973 results. Univ. of Wyo. Coop. Res. Rep. to the BLM. Sections I, II and III. Wyo. Agr. Exp. Sta. Sci. Rep. 607. 256 p. Section I - Phenology study.

1E2 740503 Shrader, J.A. and H.G. Fisser. 1974. Seed production ecology of the shortgrass prairie of the central United States. Wyo. Agr. Exp. Sta. Mimeo. 2 p.

1E2 750503 1 Fisser, H.G., and L.H. Kleinman. 1975. Phenology and production on semiarid shrub-types. 1974 results. Univ. Wyo. Coop. Res. Rep. to the BLM. Sections I, II, and III. Wyo. Agr. Exp. Sta. Sci. Rep. 729. 399 p. Section I - Phenology study.

1E2 760501 1 Fisser, H.G., and L.H. Kleinman. 1976. Phenology and production studies on semiarid shrub-types. 1975 results. Univ. of Wyo. Coop. Res. Rep. to the BLM. Sections I, II and III. Wyo. Agr. Exp. Sta. Sci. Rep. 831. 410 p. Section I - Phenology study.

1E2 770502 1 Fisser, H.G., L.W. Young, J.E. Orpet and N.E. Hargis. 1977. Phenology and production studies on semiarid shrub-types. 1976 results. Univ. of Wyo. Coop. Res. Rep. to the BLM. Sections I, II, III and IV. Wyo. Agr. Exp. Sta. Sci. Rep. 930. 438 p. Section I - Phenology study.

1E2 780501 Fisser, H.G., J.R. Cox and M. Mecke. 1978. Herbage structure, production and phenology on Black Thunder. Annual Progress Report to Atlantic-Richfield Company. 1977 results. Wyo. Agr. Exp. Sta. Sci. Rep. 856. 157 p.

1E2 780502 Fisser, H.G., and J.R. Cox. 1978. Herbage structure, production and phenology on Black Thunder. Final Summary Report to Atlantic-Richfield Company. Wyo. Agr. Exp. Sta. Sci. Rep. 968. 243 p.

1E2 780503 1 Fisser, H.G., L.W. Young, J.E. Orpet and N.E. Hargis. 1978. Phenology and production studies on semiarid shrub-types. 1977 results. Univ. of Wyo. Coop. Res. Rep. to the BLM. Sections I, II, III and IV. Wyo. Agr. Exp. Sta. Sci. Rep. 970. 439 p. Section I - Phenology study.

1E2 790502 1 Fisser, H.G., L.W. Young, N.E. Hargis and K.E. Spaeth. 1979. Phenology and production studies on semiarid shrub-types. 1978 results. Univ. of Wyo. Coop. Res. Rep. to the BLM. Sections I, II, III and IV. Wyo. Agr. Exp. Sta. Sci. Rep. 985. 489 p. Section I - Phenology study.

1E2 800501 1 Fisser, H.G., M.K. Owens, W.J. Waugh, N.E. Hargis and J. Uhlich. 1980. Phenology and production studies on semiarid shrub-types. 1979 Results. Univ. of Wyo. Coop. Res. Rep. to the BLM. Vol. I: Phenology Study. Sections I and II. Wyo. Agr. Exp. Sta. Sci. Rep. 1044. 387 p. Section I - Phenology study.

1E2 800502 4 Fisser, H.G., M.K. Owens, J.W. Uhlich, W.J. Waugh, D.E. Rodgers, and N.E. Hargis. 1980. Phenology and production studies on semiarid shrub-types. 1979 Results. Univ. of Wyo. Coop. Res. Rep. to the BLM. Vol. II: Precipitation, Productivity and Production Section. Sections III, IV, and V. Wyo. Agr. Exp. Sta. Sci. Rep. 1106. 219 p. Section IV - Productivity - phenology relationships.

1E2 820501 1 Fisser, H.G., S.K. Dean, T.L. Deboodt, R.J. Kinucan. 1982. Phenology and production studies on semiarid shrub-types. 1982 Results. Univ. of Wyo. Coop. Res. Rep. to the BLM - Vol. I: Phenology study sections I and II. Wyo. Agr. Exp. Sta. Sci. Rep. 1230. 415 p. Section I - Phenology study.

1E2 820502 4 Fisser, H.G., S.K. Dean, T.L. Deboodt, R.J. Kinucan. 1982. Phenology and production studies on semiarid shrub-types. 1982 results. Univ. of Wyo. Coop. Res. Rep. to the BLM. Vol. II: Precipitation, productivity, and production sections III, IV, and V. Wyo. Agr. Exp. Sta. Sci. Rep. 1231. 256 p. Section IV - Productivity-phenology interrelation study.

1E2 830501 1 Fisser, H.G., S.K. Dean, T.L. Deboodt, R.S. Kinucan and J.L. Mulholland, 1983. Phenology and production studies on semiarid shrub-types. 1982 results. Univ. of Wyo. Coop. Res. Rep. to the BLM. Vol. I: Phenology Study, Sections I and II. Wyo. Agr. Exp. Sta. Sci. Rep. 1230. 430 p. Section I - Phenology study.

1E2 830502 4 Fisser, H.G., S.K. Dean, T.L. Deboodt, R.S. Kinucan and J.L. Mulholland. 1983. Phenology and production studies on semiarid shrub-types. 1982 results. Univ. of Wyo. Coop. Res. Rep. to the BLM. Vol. II: Precipitation, Productivity and Production, Sections III, IV, and V. Wyo. Agr. Exp. Sta. Sci. Rep. 1231. 256 p. Section IV - Productivity - phenology interrelation study.

1E2 840501 Fisser, H.G. and L.P. Germain. 1984. Phenology and production studies on semiarid shrub-types. Statistical tests and computer programs. Wyo. Agr. Exp. Sta. Sci. Rep. 15 p.

1E2 840502 Fisser, H.G. 1984. Statistics and interpretation. Wyo. Agr. Exp. Sta. Sci. Rep. 5 p.

1E4. *Herbage Removal*

1E4 630503 1 Fisser, H.G., J.T. Nichols, and R.W. Hamner. 1963. Halogeton Research. 1962 Results. Univ. of Wyo. Coop. Res. Report to the BLM. Sections I - III. Wyo. Agr. Exp. Sta. Mimeo. p. 1-38. Section I - H.G. Fisser and J.T. Nichols. Winter sheep grazing on saltbush range (Greybull). p. 1-6.

1E4 630503 2 Fisser, H.G., J.T. Nichols, and R.W. Hamner. 1963. Halogeton Research. 1962 Results. Univ. of Wyo. Coop. Res. Rep. to the BLM. Sections I - III. Wyo. Agr. Exp. Sta. Mimeo. p. 1-38. Section II - H.G. Fisser and J.T. Nichols. Summer steer grazing on saltbush range and seeded water spreaders (Worland). p. 7-11.

1E4 640501 1 Fisser, H.G., J.T. Nichols, R.W. Hamner, and J.R. Wight. 1964. Halogeton Research. 1963 Results. Univ. of Wyo. Coop. Res. Rep. to the BLM. Sections I - VIII. Wyo. Agr. Exp. Sta. Mimeo. p. 1-61. Section I - H.G. Fisser and R.W. Hamner. Winter sheep grazing on saltbush range (Greybull). 8 p.

1E4 640501 2 Fisser, H.G., J.T. Nichols, R.W. Hamner, and J.R. Wight. 1964. Halogeton Research. 1963 Results. Univ. of Wyo. Coop. Res. Rep. to the BLM. Sections I - VIII. Wyo. Agr. Exp. Sta. Mimeo. p. 1-61. Section II - H.G. Fisser and R.W. Hamner. Summer steer grazing on saltbush range and seeded water spreaders (Worland). 6 p.

1E4 650501 1 Fisser, H.G., J.R. Wight, J.R. Flesland, L.D. Robinson. 1965. Halogeton Research. 1964 Results. Univ. of Wyo. Coop. Res. Rep. to the BLM. Section I - VI. Wyo. Agr. Exp. Sta. Mimeo. p. 1-82. Section I - H.G. Fisser and J.R. Wight. Winter sheep grazing on saltbush range (Greybull). 13 p.

1E4 650501 2 Fisser, H.G., J.R. Wight, J.R. Flesland, L.D. Robinson. 1965. Halogeton Research. 1964 Results. Univ. of Wyo. Coop. Res. Rep. to the BLM. Section I - VI. Wyo. Agr. Exp. Sta. Mimeo. p. 1-82. Section II - H.G. Fisser and J.R. Wight. Spring steer grazing on saltbush range and seeded water spreaders (Worland). 7 p.

1E4 660501 1 Fisser, H.G., L.D. Robinson, D.R. Kirkam, R. Steger, and G. Noller. 1966. Halogeton Research. 1965 Results. Univ. of Wyo. Coop. Res. Rep. to the BLM. Sections I - V. Wyo. Agr. Exp. Sta. Mimeo. 134 p. Section I - Winter sheep grazing on saltbush range (Greybull). 11 p.

1E4 670501 Fisser, H.G. and R.E. Steger. 1967. Winter sheep grazing study on saltbush rangeland. 1966 Results. Univ. of Wyo. Coop. Res. Rep. to the BLM. Wyo. Agr. Exp. Sta. Sci. Rep. 36. 9 p.

1E4 680501 Gibbens, R.P., H.G. Fisser, and M. May. 1968. Influence of grazing management systems on vegetation and wildlife habitat. Annual report of progress, July, 1967 - July, 1968. Univ. of Wyo. Coop. Res. Rep. to the BLM. Wyo. Agr. Exp. Sta. Sci. Rep. 102. 9 p.

1E4 690502 Gibbens, R.P., H.G. Fisser and M. May. 1969. Influence of grazing management systems on vegetation and wildlife habitat. Annual report of progress, July, 1968 - July, 1969. Univ. of Wyo. Coop. Res. Rep. to the BLM. Wyo. Agr. Exp. Sta. Sci. Rep. 179. 132 p.

1E4 700502 Gibbens, R.P. and H.G. Fisser. 1970. Influence of grazing management systems on vegetation and wildlife habitat. Annual report of progress, July, 1969 - July, 1970. Univ. of Wyo. Coop. Res. Rep. to the BLM. Wyo. Agr. Exp. Sta. Sci. Rep. 241. 32 p.

1E4 700505 Gibbens, R.P., H.G. Fisser and M. May. 1970. Influence of grazing management systems on vegetation and wildlife habitat. Semiann. Rep. to the BLM. Wyo. Agr. Exp. Sta. Mimeo. 2 p.

1E4 710501 Fisser, H.G. and R.P. Gibbens. 1971. Influence of grazing management systems on vegetation and wildlife habitat. Annual report of progress, July, 1970 - July, 1971. Univ. of Wyo. Coop. Res. Rep. to the BLM. Wyo. Agr. Exp. Sta. Sci. Rep. 336. 92 p.

1E4 720501 Gibbens, R.P., and H.G. Fisser. 1972. Influences of grazing management systems on vegetation and wildlife habitat. Annual Progress Report. Wyo. Agr. Exp. Sta. Sci. Rep. 414. 69 p.

1E4 750501 Gibbens, R.P., and H.G. Fisser. 1975. Influence of grazing management systems on vegetation and wildlife habitat in the Red Desert Region of Wyoming. Wyo. Agr. Exp. Sta. Sci. Mono. 29. 23 p.

1E4 790504 Fisser, H.G. 1979. Grazing management system studies of northern desert shrub vegetation of southern Wyoming. Wyo. Agr. Exp. Sta. Mimeo. 3 p.

1E4 800505 Fisser, H.G., D.E. Rodgers and W.E. Gerhart. 1980. Influences of grazing management systems on vegetation and wildlife habitat. 1975 and 1979 Results and Long-Term Summary. Univ. of Wyo. Coop. Res. Rep. to the BLM. Wyo. Agr. Exp. Sta. Sci. Rep. 1047. 188 p.

1E4 820503 Fisser, H.G. 1982. Vegetation response to grazing management strategies on semiarid shrublands of Wyoming. Wyo. Agr. Exp. Sta. Mimeo. prepared for Grazing Systems Conference, Texas A & M University. November 1982. 4 p.

2. Range Ecology

2A. CLIMATIC FACTORS

2A 600501 Fisser, H.G. 1960. The writings of Lincoln Ellison. Wyo. Range Manage. Issue 136.

2A 600502 Fisser, H.G., A.A. Beetle, and R.L. Lang. 1960. A study of range conditions in the Big Horn and Wind River Basins. Prel. Report II. Wyo. Agr. Exp. Sta. Mimeo. 10 p.

2A 620503 Fisser, H.G. 1962. An ecological study of the *Artemisia tripartita* subsp. *rupicola* and related shrub communities in Wyoming. In Crop Care in Wyoming. p. 1-3. June.

2A 630502 Fisser, H.G. 1963. Ecological studies of the *Artemisia tripartita* subsp. *rupicola* and related shrub communities in Wyoming. In 12-year report to the Wyo. Natural Res. Board. Wyo. Agr. Exp. Sta. Mimeo Part VI. 2 p.

2A 630503 3 Fisser, H.G., J.T. Nichols, and R.W. Hamner. 1963. Halogeton Research. 1962 Results. Univ. of Wyo. Coop. Res. Rep. to the BLM. Sections I - III. Wyo. Agr. Exp. Sta. Mimeo. p. 1-38. Section III - H.G. Fisser and R.W. Hamner. Exclosure studies (production, cover, height, and precipitation). p. 12-38.

2A 640501 3 Fisser, H.G., J.T. Nichols, R.W. Hamner, and J.R. Wight. 1964. Halogeton Research. 1963 Results. Univ. of Wyo. Coop. Res. Rep. to the BLM. Sections I - VIII. Wyo. Agr. Exp. Sta. Mimeo. p. 1-61. Section III - H.G. Fisser, J.T. Nichols, and J.R. Wight. Exclosure studies (production, cover, height and precipitation). 33 p.

2A 640501 4 Fisser, H.G., J.T. Nichols, R.W. Hamner, and J.R. Wight. 1964. Halogeton Research. 1963 Results. Univ. of Wyo. Coop. Res. Rep. to the BLM. Sections I - VIII. Wyo. Agr. Exp. Sta. Mimeo. p. 1-61. Section IV - H.G. Fisser, J.T. Nichols, and J.R. Wight. Soil moisture and temperature study. 3 p.

2A 640501 7 Fisser, H.G., J.T. Nichols, R.W. Hamner, and J.R. Wight. 1964. Halogeton Research. 1963 Results. Univ. of Wyo. Coop. Res. Rep. to the BLM. Sections I - VIII. Wyo. Agr. Exp. Sta. Mimeo. p. 1-61. Section VII - H.G. Fisser. Precipitation pattern study. 5 p.

2A 650501 3 Fisser, H.G., J.R. Wight, J.R. Flesland, L.D. Robinson. 1965. Halogeton Research. 1964 Results. Univ. of Wyo. Coop. Res. Rep. to the BLM. Section I - VI. Wyo. Agr. Exp. Sta. Mimeo. p. 1-82. Section III - H.G. Fisser, J.R. Wight, J. R. Flesland, and L.D. Robinson. Exclosure studies (production, cover, height and precipitation). 49 p.

2A 650501 4 Fisser, H.G., J.R. Wight, J.R. Flesland, L.D. Robinson. 1965. Halogeton Research. 1964 Results. Univ. of Wyo. Coop. Res. Rep. to the BLM. Section I - VI. Wyo. Agr. Exp. Sta. Mimeo. p. 1-82. Section IV - H.G. Fisser, J.R. Wight, and J.R. Flesland. Soil moisture and temperature study. 6 p.

2A 650501 5 Fisser, H.G., J.R. Wight, J.R. Flesland, L.D. Robinson. 1965. Halogeton Research. 1964 Results. Univ. of Wyo. Coop. Res. Rep. to the BLM. Section I - VI. Wyo. Agr. Exp. Sta. Mimeo. p. 1-82. Section V - H.G. Fisser. Precipitation pattern study. 5 p.

2A 660501 3 Fisser, H.G., L.D. Robinson, D.R. Kirkam, R. Steger, and G. Noller. 1966. Halogeton Research. 1965 Results. Univ. of Wyo. Coop. Res. Rep. to the BLM. Sections I - V. Wyo. Agr. Exp. Sta. Mimeo. 134 p. Section III - Exclosure studies (production, cover, height and precipitation). 105 p.

2A 660501 4 Fisser, H.G., L.D. Robinson, D.R. Kirkam, R. Steger, and G. Noller. 1966. Halogeton Research. 1965 Results. Univ. of Wyo. Coop. Res. Rep. to the BLM. Sections I - V. Wyo. Agr. Exp. Sta. Mimeo. 134 p. Section IV - Soil moisture and temperature study. 7 p.

2A 660501 5 Fisser, H.G., L.D. Robinson, D.R. Kirkam, R. Steger, and G. Noller. 1966. Halogeton Research. 1965 Results. Univ. of Wyo. Coop. Res. Rep. to the BLM. Sections I - V. Wyo. Agr. Exp. Sta. Mimeo. 134 p. Section V - Precipitation pattern study. 4 p.

2A 660503 Fisser, H.G. and R.E. Steger. 1966. Microclimate studies in desert shrub types. Wyo. Agr. Exp. Sta. Mimeo. 14 p.

2A 670502 1 Fisser, H.G., G. Noller, R.E. Steger, D.R. Kirkham and T. King. 1967. Arid Land Ecology Research. 1966 Results. Univ. of Wyo. Coop. Res. Rep. to the BLM. Section I - IV. Wyo. Agr. Exp. Sta. Sci. Rep. 41. 137 p. Section I - Soil moisture and temperature study.

2A 670502 2 Fisser, H.G., G. Noller, R.E. Steger, D.R. Kirkham and T. King. 1967. Arid Land Ecology Research. 1966 Results. Univ. of Wyo. Coop. Res. Rep. to the BLM. Section I - IV. Wyo. Agr. Exp. Sta. Sci. Rep. 41. 137 p. Section II - Microclimate studies in desert shrub communities.

2A 670502 3 Fisser, H.G., G. Noller, R.E. Steger, D.R. Kirkham and T. King. 1967. Arid Land Ecology Research. 1966 Results. Univ. of Wyo. Coop. Res. Rep. to the BLM. Section I - IV. Wyo. Agr. Exp. Sta. Sci. Rep. 41. 137 p. Section III - Precipitation pattern study.

2A 680504 1 Fisser, H.G. and G.L. Noller. 1968. Aridland Ecology. 1967 results. Univ. of Wyo. Coop. Res. Rep. to the BLM. Sections I, II, and III, Wyo. Agr. Exp. Sta. Sci. Rep. 132. 93 p. Section I - Soil moisture and temperature study.

2A 680504 2 Fisser, H.G. and G.L. Noller. 1968. Aridland Ecology. 1967 results. Univ. of Wyo. Coop. Res. Rep. to the BLM. Sections I, II, and III, Wyo. Agr. Exp. Sta. Sci. Rep. 132. 93 p. Section II - Precipitation pattern study.

2A 690501 1 Fisser, H.G. and G.L. Whysong. 1969. Aridland Ecology Research. 1968 results. Univ. of Wyo. Coop. Res. Rep. to the BLM. Sections I - IV. Wyo. Agr. Exp. Sta. Sci. Rep. 178. 114 p. Section I - Soil moisture and temperature.

2A 690501 2 Fisser, H.G. and G.L. Whysong. 1969. Aridland Ecology Research. 1968 results. Univ. of Wyo. Coop. Res. Rep. to the BLM. Sections I - IV. Wyo. Agr. Exp. Sta. Sci. Rep. 178. 114 p. Section II - Microclimate studies in desert shrub communities.

2A 690501 3 Fisser, H.G. and G.L. Whysong. 1969. Aridland Ecology Research. 1968 results. Univ. of Wyo. Coop. Res. Rep. to the BLM. Sections I - IV. Wyo. Agr. Exp. Sta. Sci. Rep. 178. 114 p. Section III - Precipitation pattern study.

2A 700501 1 Fisser, H.G. and G.L. Whysong. 1970. Aridland Ecology Research. 1969 results. Univ. of Wyo. Coop. Res. Rep. to the BLM. Sections I - IV. Wyo. Agr. Exp. Sta. Sci. Rep. 240. 128 p. Section I - Soil moisture and temperature.

2A 700501 2 Fisser, H.G. and G.L. Whysong. 1970. Aridland Ecology Research. 1969 results. Univ. of Wyo. Coop. Res. Rep. to the BLM. Sections I - IV. Wyo. Agr. Exp. Sta. Sci. Rep. 240. 128 p. Section II - Microclimate studies.

2A 700501 3 Fisser, H.G. and G.L. Whysong. 1970. Aridland Ecology Research. 1969 results. Univ. of Wyo. Coop. Res. Rep. to the BLM. Sections I - IV. Wyo. Agr. Exp. Sta. Sci. Rep. 240. 128 p. Section III - Precipitation pattern study.

2A 710502 1 Fisser, H.G., and R.E. Ries. 1971. Aridland Ecology Research. 1970 results. Univ. of Wyo. Coop. Res. Rep. to the BLM. Sections I - IV. Wyo. Agr. Exp. Sta. Sci. Rep. 350. 133 p. Section I - Soil moisture and temperature studies.

2A 710502 2 Fisser, H.G., and R.E. Ries. 1971. Aridland Ecology Research. 1970 results. Univ. of Wyo. Coop. Res. Rep. to the BLM. Sections I - IV. Wyo. Agr. Exp. Sta. Sci. Rep. 350. 133 p. Section II - Soil moisture and temperature studies on microclimate study sites in the Big Horn Basin.

2A 710502 3 Fisser, H.G., and R.E. Ries. 1971. Aridland Ecology Research. 1970 results. Univ. of Wyo. Coop. Res. Rep. to the BLM. Sections I - IV. Wyo. Agr. Exp. Sta. Sci. Rep. 350. 133 p. Section III - Precipitation pattern study.

2A 720502 1 Fisser, H.G., and R.E. Ries. 1972. Aridland Ecology Research. 1971 results. Univ. of Wyo. Coop. Res. Rep. to the BLM. Sections I and II. Wyo. Agr. Exp. Sta. Sci. Rep. 428. 130 p. Section I - Precipitation pattern study.

2A 730501 1 Fisser, H.G., and R.E. Ries. 1973. Aridland Ecology Research. 1972 results. Univ. of Wyo. Coop. Res. Rep. to the BLM. Sections I and II. Wyo. Agr. Exp. Sta. Sci. Rep. 521. 128 p. Section I - Precipitation pattern study.

2A 740501 Fisser, H.G. and R.E. Ries. 1974. Summary of data, 1962-1970. Aridland Ecology Research. Univ. of Wyo. Coop. Res. Rep. to the BLM. Wyo. Agr. Exp. Sta. Sci. Rep. 556. 132 p.

2A 740502 1 Fisser, H.G. and L.H. Kleinman. 1974. Phenology and production on semiarid shrub-types. 1973 results. Univ. of Wyo. Coop. Res. Rep. to the BLM. Sections I, II and III. Wyo. Agr. Exp. Sta. Sci. Rep. 607. 256 p. Section I - Phenology study.

2A 740502 2 Fisser, H.G. and L.H. Kleinman. 1974. Phenology and production on semiarid shrub-types. 1973 results. Univ. of Wyo. Coop. Res. Rep. to the BLM. Sections I, II and III. Wyo. Agr. Exp. Sta. Sci. Rep. 607. 256 p. Section II - Precipitation pattern study.

2A 750503 2 Fisser, H.G., and L.H. Kleinman. 1975. Phenology and production on semiarid shrub-types. 1974 results. Univ. Wyo. Coop. Res. Rep. to the BLM. Sections I, II, and III. Wyo. Agr. Exp. Sta. Sci. Rep. 729. 399 p. Section II - Precipitation pattern study.

2A 750503 3 Fisser, H.G., and L.H. Kleinman. 1975. Phenology and production on semiarid shrub-types. 1974 results. Univ. Wyo. Coop. Res. Rep. to the BLM. Sections I, II, and III. Wyo. Agr. Exp. Sta. Sci. Rep. 729. 399 p. Section III - Exclosure studies (production, cover and precipitation).

2A 760501 2 Fisser, H.G., and L.H. Kleinman. 1976. Phenology and production studies on semiarid shrub-types. 1975 results. Univ. of Wyo. Coop. Res. Rep. to the BLM. Sections I, II and III. Wyo. Agr. Exp. Sta. Sci. Rep. 831. 410 p. Section II - Precipitation pattern study.

2A 760501 3 Fisser, H.G., and L.H. Kleinman. 1976. Phenology and production studies on semiarid shrub-types. 1975 results. Univ. of Wyo. Coop. Res. Rep. to the BLM. Sections I, II and III. Wyo. Agr. Exp. Sta. Sci. Rep. 831. 410 p. Section III - Exclosure studies (production, cover, precipitation).

2A 770502 1 Fisser, H.G., L.W. Young, J.E. Orpet and N.E. Hargis. 1977. Phenology and production studies on semiarid shrub-types. 1976 results. Univ. of Wyo. Coop. Res. Rep. to the BLM. Sections I, II, III and IV. Wyo. Agr. Exp. Sta. Sci. Rep. 930. 438 p. Section I - Phenology study.

2A 770502 2 Fisser, H.G., L.W. Young, J.E. Orpet and N.E. Hargis. 1977. Phenology and production studies on semiarid shrub-types. 1976 results. Univ. of Wyo. Coop. Res. Rep. to the BLM. Sections I, II, III and IV. Wyo. Agr. Exp. Sta. Sci. Rep. 930. 438 p. Section II - Phenology step-wise regression for predictive modeling.

2A 770502 3 Fisser, H.G., L.W. Young, J.E. Orpet and N.E. Hargis. 1977. Phenology and production studies on semiarid shrub-types. 1976 results. Univ. of Wyo. Coop. Res. Rep. to the BLM. Sections I, II, III and IV. Wyo. Agr. Exp. Sta. Sci. Rep. 930. 438 p. Section III - Precipitation pattern study.

2A 770502 4 Fisser, H.G., L.W. Young, J.E. Orpet and N.E. Hargis. 1977. Phenology and production studies on semiarid shrub-types. 1976 results. Univ. of Wyo. Coop. Res. Rep. to the BLM. Sections I, II, III and IV. Wyo. Agr. Exp. Sta. Sci. Rep. 930. 438 p. Section IV - Exclosure studies (production, cover and precipitation).

2A 780501 Fisser, H.G., J.R. Cox and M. Mecke. 1978. Herbage structure, production and phenology on Black Thunder. Annual Progress Report to Atlantic-Richfield Company. 1977 results. Wyo. Agr. Exp. Sta. Sci. Rep. 856. 157 p.

2A 780502 Fisser, H.G., and J.R. Cox. 1978. Herbage structure, production and phenology on Black Thunder. Final Summary Report to Atlantic-Richfield Company. Wyo. Agr. Exp. Sta. Sci. Rep. 968. 243 p.

2A 780503 1 Fisser, H.G., L.W. Young, J.E. Orpet and N.E. Hargis. 1978. Phenology and production studies on semiarid shrub-types. 1977 results. Univ. of Wyo. Coop. Res. Rep. to the BLM. Sections I, II, III and IV. Wyo. Agr. Exp. Sta. Sci. Rep. 970. 439 p. Section I - Phenology study.

2A 780503 2 Fisser, H.G., L.W. Young, J.E. Orpet and N.E. Hargis. 1978. Phenology and production studies on semiarid shrub-types. 1977 results. Univ. of Wyo. Coop. Res. Rep. to the BLM. Sections I, II, III and IV. Wyo. Agr. Exp. Sta. Sci. Rep. 970. 439 p. Section II - Phenology step-wise regression for predictive modeling.

2A 780503 3 Fisser, H.G., L.W. Young, J.E. Orpet and N.E. Hargis. 1978. Phenology and production studies on semiarid shrub-types. 1977 results. Univ. of Wyo. Coop. Res. Rep. to the BLM. Sections I, II, III and IV. Wyo. Agr. Exp. Sta. Sci. Rep. 970. 439 p. Section III - Precipitation pattern study.

2A 780503 4 Fisser, H.G., L.W. Young, J.E. Orpet and N.E. Hargis. 1978. Phenology and production studies on semiarid shrub-types. 1977 results. Univ. of Wyo. Coop. Res. Rep. to the BLM. Sections I, II, III and IV. Wyo. Agr. Exp. Sta. Sci. Rep. 970. 439 p. Section IV - Exclosure studies (production, cover and precipitation).

2A 790501 Fisser, H.G., D.C. Trueblood and D.D. Samuelson. 1979. Soil vegetation relationships on rangeland exclosures on the Grass Creek Planning Unit of north central Wyoming. Univ. of Wyo. Coop. Res. Rep. to the BLM. Wyo. Agr. Exp. Sta. Sci. Rep. 969. 279 p.

2A 790502 1 Fisser, H.G., L.W. Young, N.E. Hargis and K.E. Spaeth. 1979. Phenology and production studies on semiarid shrub-types. 1978 results. Univ. of Wyo. Coop. Res. Rep. to the BLM. Sections I, II, III and IV. Wyo. Agr. Exp. Sta. Sci. Rep. 985. 489 p. Section I - Phenology study.

2A 790502 2 Fisser, H.G., L.W. Young, N.E. Hargis and K.E. Spaeth. 1979. Phenology and production studies on semiarid shrub-types. 1978 results. Univ. of Wyo. Coop. Res. Rep. to the BLM. Sections I, II, III and IV. Wyo. Agr. Exp. Sta. Sci. Rep. 985. 489 p. Section II - Phenology step-wise regression for predictive modeling.

2A 790502 3 Fisser, H.G., L.W. Young, N.E. Hargis and K.E. Spaeth. 1979. Phenology and production studies on semiarid shrub-types. 1978 results. Univ. of Wyo. Coop. Res. Rep. to the BLM. Sections I, II, III and IV. Wyo. Agr. Exp. Sta. Sci. Rep. 985. 489 p. Section III - Precipitation pattern study.

2A 790502 4 Fisser, H.G., L.W. Young, N.E. Hargis and K.E. Spaeth. 1979. Phenology and production studies on semiarid shrub-types. 1978 results. Univ. of Wyo. Coop. Res. Rep. to the BLM. Sections I, II, III and IV. Wyo. Agr. Exp. Sta. Sci. Rep. 985. 489 p. Section IV - Exclosure studies (production, cover and precipitation).

2A 800501 1 Fisser, H.G., M.K. Owens, W.J. Waugh, N.E. Hargis and J. Uhlich. 1980. Phenology and production studies on semiarid shrub-types. 1979 Results. Univ. of Wyo. Coop. Res. Rep. to the BLM. Vol. I: Phenology Study. Sections I and II. Wyo. Agr. Exp. Sta. Sci. Rep. 1044. 387 p. Section I - Phenology study.

2A 800501 2 Fisser, H.G., M.K. Owens, W.J. Waugh, N.E. Hargis and J. Uhlich. 1980. Phenology and production studies on semiarid shrub-types. 1979 Results. Univ. of Wyo. Coop. Res. Rep. to the BLM. Vol. I: Phenology Study. Sections I and II. Wyo. Agr. Exp. Sta. Sci. Rep. 1044. 387 p. Section II - Phenology step-wise regression for predictive modeling.

2A 800502 3 Fisser, H.G., M.K. Owens, J.W. Uhlich, W.J. Waugh, D.E. Rodgers, and N.E. Hargis. 1980. Phenology and production studies on semiarid shrub-types. 1979 Results. Univ. of Wyo. Coop. Res. Rep. to the BLM. Vol. II: Precipitation, Productivity and Production Section. Sections III, IV, and V. Wyo. Agr. Exp. Sta. Sci. Rep. 1106. 219 p. Section III - Precipitation pattern study.

2A 800502 4 Fisser, H.G., M.K. Owens, J.W. Uhlich, W.J. Waugh, D.E. Rodgers, and N.E. Hargis. 1980. Phenology and production studies on semiarid shrub-types. 1979 Results. Univ. of Wyo. Coop. Res. Rep. to the BLM. Vol. II: Precipitation, Productivity and Production Section. Sections III, IV, and V. Wyo. Agr. Exp. Sta. Sci. Rep. 1106. 219 p. Section IV - Productivity-phenology relationships.

2A 800502 5 Fisser, H.G., M.K. Owens, J.W. Uhlich, W.J. Waugh, D.E. Rodgers, and N.E. Hargis. 1980. Phenology and production studies on semiarid shrub-types. 1979 Results. Univ. of Wyo. Coop. Res. Rep. to the BLM. Vol. II: Precipitation, Productivity and Production Section. Sections III, IV, and V. Wyo. Agr. Exp. Sta. Sci. Rep. 1106. 219 p. Section V - Exclosure studies (production, cover and precipitation).

2A 820501 2 Fisser, H.G., S.K. Dean, T.L. Deboodt, R.J. Kinucan. 1982. Phenology and production studies on semiarid shrub-types. 1982 Results. Univ. of Wyo. Coop. Res. Rep. to the BLM - Vol. I: Phenology study sections I and II. Wyo. Agr. Exp. Sta. Sci. Rep. 1230. 415 p. Section II - Phenology multiple regression for predictive modeling.

2A 820502 3 Fisser, H.G., S.K. Dean, T.L. Deboodt, R.J. Kinucan. 1982. Phenology and production studies on semiarid shrub-types. 1982 results. Univ. of Wyo. Coop. Res. Rep. to the BLM. Vol. II: Precipitation, productivity, and production sections III, IV, and V. Wyo. Agr. Exp. Sta. Sci. Rep. 1231. 256 p. Section III - Precipitation pattern study.

2A 820502 4 Fisser, H.G., S.K. Dean, T.L. Deboodt, R.J. Kinucan. 1982. Phenology and production studies on semiarid shrub-types. 1982 results. Univ. of Wyo. Coop. Res. Rep. to the BLM. Vol. II: Precipitation, productivity, and production sections III, IV, and V. Wyo. Agr. Exp. Sta. Sci. Rep. 1231. 256 p. Section IV - Productivity-phenology interrelation study.

2A 820502 5 Fisser, H.G., S.K. Dean, T.L. Deboodt, R.J. Kinucan. 1982. Phenology and production studies on semiarid shrub-types. 1982 results. Univ. of Wyo. Coop. Res. Rep. to the BLM. Vol. II: Precipitation, productivity, and production sections III, IV, and V. Wyo. Agr. Exp. Sta. Sci. Rep. 1231. 256 p. Section V - Exclosure studies (production, cover and precipitation studies).

2A 830501 2 Fisser, H.G., S.K. Dean, T.L. Deboodt, R.S. Kinucan and J.L. Mulholland, 1983. Phenology and production studies on semiarid shrub-types. 1982 results. Univ. of Wyo. Coop. Res. Rep. to the BLM. Vol. I: Phenology Study, Sections I and II. Wyo. Agr. Exp. Sta. Sci. Rep. 1230. 430 p. Section II - Phenology multiple regression for predictive modeling.

2A 830502 3 Fisser, H.G., S.K. Dean, T.L. Deboodt, R.S. Kinucan and J.L. Mulholland. 1983. Phenology and production studies on semiarid shrub-types. 1982 results. Univ. of Wyo. Coop. Res. Rep. to the BLM. Vol. II: Precipitation, Productivity and Production, Sections III, IV, and V. Wyo. Agr. Exp. Sta. Sci. Rep. 1231. 256 p. Section III - Precipitation pattern study.

2A 830502 4 Fisser, H.G., S.K. Dean, T.L. Deboodt, R.S. Kinucan and J.L. Mulholland. 1983. Phenology and production studies on semiarid shrub-types. 1982 results. Univ. of Wyo. Coop. Res. Rep. to the BLM. Vol. II: Precipitation, Productivity and Production, Sections III, IV, and V. Wyo. Agr. Exp. Sta. Sci. Rep. 1231. 256 p. Section IV - Productivity - phenology interrelation study.

2A 830502 5 Fisser, H.G., S.K. Dean, T.L. Deboodt, R.S. Kinucan and J.L. Mulholland. 1983. Phenology and production studies on semiarid shrub-types. 1982 results. Univ. of Wyo. Coop. Res. Rep. to the BLM. Vol. II: Precipitation, Productivity and Production, Sections III, IV, and V. Wyo. Agr. Exp. Sta. Sci. Rep. 1231. 256 p. Section V - Exclosure studies (production, cover and precipitation).

2A 830503 Fisser, H.G. 1983. Aridland shrub ecology and phenology research program. Wyo. Agr. Exp. Sta. Mimeo. 3 p.

2A 840501 Fisser, H.G. and L.P. Germain. 1984. Phenology and production studies on semiarid shrub-types. Statistical tests and computer programs. Wyo. Agr. Exp. Sta. Sci. Rep. 15 p.

2A 840502 Fisser, H.G. 1984. Statistics and interpretation. Wyo. Agr. Exp. Sta. Sci. Rep. 5 p.

2C. EDAPHIC FACTORS

2C 620503 Fisser, H.G. 1962. An ecological study of the *Artemisia tripartita* subsp. *rupicola* and related shrub communities in Wyoming. In Crop Care in Wyoming. p. 1-3. June.

2C 630502 Fisser, H.G. 1963. Ecological studies of the *Artemisia tripartita* subsp. *rupicola* and related shrub communities in Wyoming. In 12-year report to the Wyo. Natural Res. Board. Wyo. Agr. Exp. Sta. Mimeo Part VI. 2 pp.

2C 630503 3 Fisser, H.G., J.T. Nichols, and R.W. Hamner. 1963. Halogeton Research. 1962 Results. Univ. of Wyo. Coop. Res. Rep. to the BLM. Sections I - III. Wyo. Agr. Exp. Sta. Mimeo. p. 1-38. Section III - H.G. Fisser and R.W. Hamner. Exclosure studies (production, cover, height, and precipitation). p. 12-38.

2C 640501 3 Fisser, H.G., J.T. Nichols, R.W. Hamner, and J.R. Wight. 1964. Halogeton Research. 1963 Results. Univ. of Wyo. Coop. Res. Rep. to the BLM. Sections I - VIII. Wyo. Agr. Exp. Sta. Mimeo. p. 1-61. Section III - H.G. Fisser, J.T. Nichols, and J.R. Wight. Exclosure studies (production, cover, height and precipitation). 33 p.

2C 640501 4 Fisser, H.G., J.T. Nichols, R.W. Hamner, and J.R. Wight. 1964. Halogeton Research. 1963 Results. Univ. of Wyo. Coop. Res. Rep. to the BLM. Sections I - VIII. Wyo. Agr. Exp. Sta. Mimeo. p. 1-61. Section IV - H.G. Fisser, J.T. Nichols, and J.R. Wight. Soil moisture and temperature study. 3 p.

2C 640501 5 Fisser, H.G., J.T. Nichols, R.W. Hamner, and J.R. Wight. 1964. Halogeton Research. 1963 Results. Univ. of Wyo. Coop. Res. Rep. to the BLM. Sections I - VIII. Wyo. Agr. Exp. Sta. Mimeo. p. 1-61. Section V - H.G. Fisser, J.T. Nichols, and J.R. Wight. Erosion transect study. 2 p.

2C 650501 3 Fisser, H.G., J.R. Wight, J.R. Flesland, L.D. Robinson. 1965. Halogeton Research. 1964 Results. Univ. of Wyo. Coop. Res. Rep. to the BLM. Section I - VI. Wyo. Agr. Exp. Sta. Mimeo. p. 1-82. Section III - H.G. Fisser, J.R. Wight, J. R. Flesland, and L.D. Robinson. Exclosure studies (production, cover, height and precipitation). 49 p.

2C 650501 4 Fisser, H.G., J.R. Wight, J.R. Flesland, L.D. Robinson. 1965. Halogeton Research. 1964 Results. Univ. of Wyo. Coop. Res. Rep. to the BLM. Section I - VI. Wyo. Agr. Exp. Sta. Mimeo. p. 1-82. Section IV - H.G. Fisser, J.R. Wight, and J.R. Flesland. Soil moisture and temperature study. 6 p.

2C 660501 4 Fisser, H.G., L.D. Robinson, D.R. Kirkam, R. Steger, and G. Noller. 1966. Halogeton Research. 1965 Results. Univ. of Wyo. Coop. Res. Rep. to the BLM. Sections I - V. Wyo. Agr. Exp. Sta. Mimeo. 134 p. Section IV - Soil moisture and temperature study. 7 p.

2C 660503 Fisser, H.G. and R.E. Steger. 1966. Microclimate studies in desert shrub types. Wyo. Agr. Exp. Sta. Mimeo. 14 p.

2C 670502 1 Fisser, H.G., G. Noller, R.E. Steger, D.R. Kirkham and T. King. 1967. Aridland Ecology Research. 1966 Results. Univ. of Wyo. Coop. Res. Rep. to the BLM. Section I - IV. Wyo. Agr. Exp. Sta. Sci. Rep. 41. 137 p. Section I - Soil moisture and temperature study.

2C 670502 2 Fisser, H.G., G. Noller, R.E. Steger, D.R. Kirkham and T. King. 1967. Aridland Ecology Research. 1966 Results. Univ. of Wyo. Coop. Res. Rep. to the BLM. Section I - IV. Wyo. Agr. Exp. Sta. Sci. Rep. 41. 137 p. Section II - Microclimate studies in desert shrub communities.

2C 670502 4 Fisser, H.G., G. Noller, R.E. Steger, D.R. Kirkham and T. King. 1967. Aridland Ecology Research. 1966 Results. Univ. of Wyo. Coop. Res. Rep. to the BLM. Section I - IV. Wyo. Agr. Exp. Sta. Sci. Rep. 41. 137 p. Section IV - Exclosure studies (production, cover and precipitation).

2C 680504 1 Fisser, H.G. and G.L. Noller. 1968. Aridland Ecology. 1967 results. Univ. of Wyo. Coop. Res. Rep. to the BLM. Sections I, II, and III, Wyo. Agr. Exp. Sta. Sci. Rep. 132. 93 p. Section I - Soil moisture and temperature study.

2C 680504 3 Fisser, H.G. and G.L. Noller. 1968. Aridland Ecology. 1967 results. Univ. of Wyo. Coop. Res. Rep. to the BLM. Sections I, II, and III, Wyo. Agr. Exp. Sta. Sci. Rep. 132. 93 p. Section III - Exclosure studies (production, cover and precipitation).

2C 680505 Wight, J.R. and H.G. Fisser. 1968. *Juniperus osteosperma* in northwestern Wyoming - Their distribution and ecology. Wyo. Agr. Exp. Sta. Sci. Mono. 7. 31 p.

2C 690501 1 Fisser, H.G. and G.L. Whysong. 1969. Aridland Ecology Research. 1968 results. Univ. of Wyo. Coop. Res. Rep. to the BLM. Sections I - IV. Wyo. Agr. Exp. Sta. Sci. Rep. 178. 114 p. Section I - Soil moisture and temperature.

2C 690501 2 Fisser, H.G. and G.L. Whysong. 1969. Aridland Ecology Research. 1968 results. Univ. of Wyo. Coop. Res. Rep. to the BLM. Sections I - IV. Wyo. Agr. Exp. Sta. Sci. Rep. 178. 114 p. Section II - Microclimate studies in desert shrub communities.

2C 690501 4 Fisser, H.G. and G.L. Whysong. 1969. Aridland Ecology Research. 1968 results. Univ. of Wyo. Coop. Res. Rep. to the BLM. Sections I - IV. Wyo. Agr. Exp. Sta. Sci. Rep. 178. 114 p. Section IV - Exclosure studies (production, cover and precipitation).

2C 700501 1 Fisser, H.G. and G.L. Whysong. 1970. Aridland Ecology Research. 1969 results. Univ. of Wyo. Coop. Res. Rep. to the BLM. Sections I - IV. Wyo. Agr. Exp. Sta. Sci. Rep. 240. 128 p. Section I - Soil moisture and temperature.

2C 700501 2 Fisser, H.G. and G.L. Whysong. 1970. Aridland Ecology Research. 1969 results. Univ. of Wyo. Coop. Res. Rep. to the BLM. Sections I - IV. Wyo. Agr. Exp. Sta. Sci. Rep. 240. 128 p. Section II - Microclimate studies.

2C 700501 4 Fisser, H.G. and G.L. Whysong. 1970. Aridland Ecology Research. 1969 results. Univ. of Wyo. Coop. Res. Rep. to the BLM. Sections I - IV. Wyo. Agr. Exp. Sta. Sci. Rep. 240. 128 p. Section IV - Exclosure studies (production, cover and precipitation).

2C 710502 1 Fisser, H.G., and R.E. Ries. 1971. Aridland Ecology Research. 1970 results. Univ. of Wyo. Coop. Res. Rep. to the BLM. Sections I - IV. Wyo. Agr. Exp. Sta. Sci. Rep. 350. 133 p. Section I - Soil moisture and temperature studies.

2C 710502 2 Fisser, H.G., and R.E. Ries. 1971. Aridland Ecology Research. 1970 results. Univ. of Wyo. Coop. Res. Rep. to the BLM. Sections I - IV. Wyo. Agr. Exp. Sta. Sci. Rep. 350. 133 p. Section II - Soil moisture and temperature studies on microclimate study sites in the Big Horn Basin.

2C 710502 4 Fisser, H.G., and R.E. Ries. 1971. Aridland Ecology Research. 1970 results. Univ. of Wyo. Coop. Res. Rep. to the BLM. Sections I - IV. Wyo. Agr. Exp. Sta. Sci. Rep. 350. 133 p. Section IV - exclosure studies (production, cover, and precipitation).

2C 710505 Fisser, H.G. and R.E. Ries. 1971. Semi-annual progress report to the Bureau of Land Management, January 1971. Wyo. Agr. Exp. Sta. Mimeo. 1 p.

2C 720502 2 Fisser, H.G., and R.E. Ries. 1972. Aridland Ecology Research. 1971 results. Univ. of Wyo. Coop. Res. Rep. to the BLM. Sections I and II. Wyo. Agr. Exp. Sta. Sci. Rep. 428. 130 pp. Section II - Exclosure studies (production, cover and precipitation).

2C 730502 Shaver, J.C. and H.G. Fisser. 1973. Evaluation of vegetation types differentiated by aerial photo interpretation of panchromatic, color, and infrared imagery. U.S. I.B.P. Biome Technical Report 238. 86 p.

2C 740501 Fisser, H.G. and R.E. Ries. 1974. Summary of data, 1962-1970. Aridland Ecology Research. Univ. of Wyo. Coop. Res. Rep. to the BLM. Wyo. Agr. Exp. Sta. Sci. Rep. 556. 132 p.

2C 740502 3 Fisser, H.G. and L.H. Kleinman. 1974. Phenology and production on semiarid shrub-types. 1973 results. Univ. of Wyo. Coop. Res. Rep. to the BLM. Sections I, II and III. Wyo. Agr. Exp. Sta. Sci. Rep. 607. 256 p. Section III - Exclosure studies (production, cover and precipitation).

2C 750503 3 Fisser, H.G., and L.H. Kleinman. 1975. Phenology and production on semiarid shrub-types. 1974 results. Univ. Wyo. Coop. Res. Rep. to the BLM. Sections I, II, and III. Wyo. Agr. Exp. Sta. Sci. Rep. 729. 399 p. Section III - Exclosure studies (production, cover and precipitation).

2C 760501 3 Fisser, H.G., and L.H. Kleinman. 1976. Phenology and production studies on semiarid shrub-types. 1975 results. Univ. of Wyo. Coop. Res. Rep. to the BLM. Sections I, II and III. Wyo. Agr. Exp. Sta. Sci. Rep. 831. 410 p. Section III - Exclosure studies (production, cover, precipitation).

2C 770502 4 Fisser, H.G., L.W. Young, J.E. Orpet and N.E. Hargis. 1977. Phenology and production studies on semiarid shrub-types. 1976 results. Univ. of Wyo. Coop. Res. Rep. to the BLM. Sections I, II, III and IV. Wyo. Agr. Exp. Sta. Sci. Rep. 930. 438 p. Section IV - Exclosure studies (production, cover and precipitation).

2C 780503 4 Fisser, H.G., L.W. Young, J.E. Orpet and N.E. Hargis. 1978. Phenology and production studies on semiarid shrub-types. 1977 results. Univ. of Wyo. Coop. Res. Rep. to the BLM. Sections I, II, III and IV. Wyo. Agr. Exp. Sta. Sci. Rep. 970. 439 p. Section IV - Exclosure studies (production, cover and precipitation).

2C 790501 Fisser, H.G., D.C. Trueblood and D.D. Samuelson. 1979. Soil vegetation relationships on rangeland exclosures on the Grass Creek Planning Unit of north central Wyoming. Univ. of Wyo. Coop. Res. Rep. to the BLM. Wyo. Agr. Exp. Sta. Sci. Rep. 969. 279 p.

2C 790502 4 Fisser, H.G., L.W. Young, N.E. Hargis and K.E. Spaeth. 1979. Phenology and production studies on semiarid shrub-types. 1978 results. Univ. of Wyo. Coop. Res. Rep. to the BLM. Sections I, II, III and IV. Wyo. Agr. Exp. Sta. Sci. Rep. 985. 489 p. Section IV - Exclosure studies (production, cover and precipitation).

2C 800502 5 Fisser, H.G., M.K. Owens, J.W. Uhlich, W.J. Waugh, D.E. Rodgers, and N.E. Hargis. 1980. Phenology and production studies on semiarid shrub-types. 1979 Results. Univ. of Wyo. Coop. Res. Rep. to the BLM. Vol. II: Precipitation, Productivity and Production Section. Sections III, IV, and V. Wyo. Agr. Exp. Sta. Sci. Rep. 1106. 219 p. Section V - Exclosure studies (production, cover and precipitation).

2C 800503 Fisser, H.G., D.C. Trueblood and C.H. Butterfield. 1980. Soil vegetation relationships on rangeland exclosures in the Overland Planning Unit of south central Wyoming. Univ. of Wyo. Coop. Res. Rep. to the BLM. Wyo. Agr. Exp. Sta. Sci. Rep. 1042. 75 p.

2C 800504 Fisser, H.G., and D.C. Trueblood. 1980. Soil vegetation relationships on rangeland exclosures on the Salt Wells Planning Unit of southwestern Wyoming. Univ. of Wyo. Coop. Res. Rep. to the BLM. Wyo. Agr. Exp. Sta. Sci. Rep. 1041. 68 p.

2C 820502 5 Fisser, H.G., S.K. Dean, T.L. Deboodt, R.J. Kinucan. 1982. Phenology and production studies on semiarid and shrub-types. 1982 results. Univ. of Wyo. Coop. Res. Rep. to the BLM. Vol. II: Precipitation, productivity, and production sections III, IV, and V. Wyo. Agr. Exp. Sta. Sci. Rep. 1231. 256 p. Section V - Exclosure studies (production, cover and precipitation).

2C 830502 5 Fisser, H.G., S.K. Dean, T.L. Deboodt, R.S. Kinucan and J.L. Mulholland. 1983. Phenology and production studies on semiarid shrub-types. 1982 results. Univ. of Wyo. Coop. Res. Rep. to the BLM. Vol. II: Precipitation, Productivity and Production, Sections III, IV, and V. Wyo. Agr. Exp. Sta. Sci. Rep. 1231. 256 p.

2E. AUTECOLOGY

2E 620503 Fisser, H.G. 1962. An ecological study of the *Artemisia tripartita* subsp. *rupicola* and related shrub communities in Wyoming. In Crop Care in Wyoming. p. 1-3. June.

2E 630502 Fisser, H.G. 1963. Ecological studies of the *Artemisia tripartita* subsp. *rupicola* and related shrub communities in Wyoming. In 12-year report to the Wyo. Natural Res. Board. Wyo. Agr. Exp. Sta. Mimeo Part VI. 2 p.

2E 640501 6 Fisser, H.G., J.T. Nichols, R.W. Hamner, and J.R. Wight. 1964. Halogeton Research. 1963 Results. Univ. of Wyo. Coop. Res. Rep. to the BLM. Sections I - VIII. Wyo. Agr. Exp. Sta. Mimeo. p. 1-61. Section VI - H.G. Fisser, R.W. Hamner. Germination and longevity of halogeton seeds. 1 p.

2E 670502 2 Fisser, H.G., G. Noller, R.E. Steger, D.R. Kirkham and T. King. 1967. Aridland Ecology Research. 1966 Results. Univ. of Wyo. Coop. Res. Rep. to the BLM. Section I - IV. Wyo. Agr. Exp. Sta. Sci. Rep. 41. 137 p. Section II - Microclimate studies in desert shrub communities.

2E 680505 Wight, J.R. and H.G. Fisser. 1968. *Juniperus osteosperma* in northwestern Wyoming - Their distribution and ecology. Wyo. Agr. Exp. Sta. Sci. Mono. 7. 31 p.

2E 690501 2 Fisser, H.G. and G.L. Whysong. 1969. Aridland Ecology Research. 1968 results. Univ. of Wyo. Coop. Res. Rep. to the BLM. Sections I - IV. Wyo. Agr. Exp. Sta. Sci. Rep. 178. 114 p. Section II - Microclimate studies in desert shrub communities.

2E 690503 Fisser, H.G., J.E. Lester, G.C. Mitchell, and L.J. Bledsoe. 1969. Preliminary report of methodology and results for analyses of plant pattern project research on the Pawnee Site. U.S. I.B.P. Grassland Biome Technical Report 9. 68 p.

2E 700501 2 Fisser, H.G. and G.L. Whysong. 1970. Aridland Ecology Research. 1969 results. Univ. of Wyo. Coop. Res. Rep. to the BLM. Sections I - IV. Wyo. Agr. Exp. Sta. Sci. Rep. 240. 128 p. Section II - Microclimate studies.

2E 710502 2 Fisser, H.G., and R.E. Ries. 1971. Aridland Ecology Research. 1970 results. Univ. of Wyo. Coop. Res. Rep. to the BLM. Sections I - IV. Wyo. Agr. Exp. Sta. Sci. Rep. 350. 133 p. Section II - Soil moisture and temperature studies on microclimate study sites in the Big Horn Basin.

2E 710503 Fisser, H.G. 1971. Significance of plant pattern in ecosystem functioning. Wyo. Agr. Exp. Sta. Mimeo. 2 p.

2E 710504 Fisser, H.G. 1971. Evaluation of Pawnee Site frequency data. Wyo. Agr. Exp. Sta. Mimeo. 6 p.

2E 710505 Fisser, H.G. and R.E. Ries. 1971. Semi-annual progress report to the Bureau of Land Management, January 1971. Wyo. Agr. Exp. Sta. Mimeo. 1 p.

2E 720504 Stroud, D.W. and H.G. Fisser. 1972. Plant pattern micro-relief and *Bouteloua gracilis* culm density studies on the Pawnee Site: 1970-1971. U.S. I.B.P. Grassland Biome Technical Report 186. 26 p.

2E 740502 1 Fisser, H.G. and L.H. Kleinman. 1974. Phenology and production on semiarid shrub-types. 1973 results. Univ. of Wyo. Coop. Res. Rep. to the BLM. Sections I, II and III. Wyo. Agr. Exp. Sta. Sci. Rep. 607. 256 p. Section I - Phenology study.

2E 740503 Shrader, J.A. and H.G. Fisser. 1874. Seed production ecology of the shortgrass prairie of the central United States. Wyo. Agr. Exp. Sta. Mimeo. 2 p.

2E 750503 1 Fisser, H.G., and L.H. Kleinman. 1975. Phenology and production on semiarid shrub-types. 1974 results. Univ. Wyo. Coop. Res. Rep. to the BLM. Sections I, II, and III. Wyo. Agr. Exp. Sta. Sci. Rep. 729. 399 p. Section I - Phenology study.

2E 760501 1 Fisser, H.G., and L.H. Kleinman. 1976. Phenology and production studies on semiarid shrub-types. 1975 results. Univ. of Wyo. Coop. Res. Rep. to the BLM. Sections I, II and III. Wyo. Agr. Exp. Sta. Sci. Rep. 831. 410 p. Section I - Phenology study.

2E 770502 1 Fisser, H.G., L.W. Young, J.E. Orpet and N.E. Hargis. 1977. Phenology and production studies on semiarid shrub-types. 1976 results. Univ. of Wyo. Coop. Res. Rep. to the BLM. Sections I, II, III and IV. Wyo. Agr. Exp. Sta. Sci. Rep. 930. 438 p. Section I - Phenology study.

2E 770502 2 Fisser, H.G., L.W. Young, J.E. Orpet and N.E. Hargis. 1977. Phenology and production studies on semiarid shrub-types. 1976 results. Univ. of Wyo. Coop. Res. Rep. to the BLM. Sections I, II, III and IV. Wyo. Agr. Exp. Sta. Sci. Rep. 930. 438 p. Section II - Phenology step-wise regression for predictive modeling.

2E 780503 1 Fisser, H.G., L.W. Young, J.E. Orpet and N.E. Hargis. 1978. Phenology and production studies on semiarid shrub-types. 1977 results. Univ. of Wyo. Coop. Res. Rep. to the BLM. Sections I, II, III and IV. Wyo. Agr. Exp. Sta. Sci. Rep. 970. 439 p. Section I - Phenology study.

2E 780503 2 Fisser, H.G., L.W. Young, J.E. Orpet and N.E. Hargis. 1978. Phenology and production studies on semiarid shrub-types. 1977 results. Univ. of Wyo. Coop. Res. Rep. to the BLM. Sections I, II, III and IV. Wyo. Agr. Exp. Sta. Sci. Rep. 970. 439 p. Section II - Phenology step-wise regression for predictive modeling.

2E 780503 4 Fisser, H.G., L.W. Young, J.E. Orpet and N.E. Hargis. 1978. Phenology and production studies on semiarid shrub-types. 1977 results. Univ. of Wyo. Coop. Res. Rep. to the BLM. Sections I, II, III and IV. Wyo. Agr. Exp. Sta. Sci. Rep. 970. 439 p. Section IV - Exclosure studies (production, cover and precipitation).

2E 790501 Fisser, H.G., D.C. Trueblood and D.D. Samuelson. 1979. Soil vegetation relationships on rangeland exclosures on the Grass Creek Planning Unit of north central Wyoming. Univ. of Wyo. Coop. Res. Rep. to the BLM. Wyo. Agr. Exp. Sta. Sci. Rep. 969. 279 p.

2E 790502 1 Fisser, H.G., L.W. Young, N.E. Hargis and K.E. Spaeth. 1979. Phenology and production studies on semiarid shrub-types. 1978 results. Univ. of Wyo. Coop. Res. Rep. to the BLM. Sections I, II, III and IV. Wyo. Agr. Exp. Sta. Sci. Rep. 985. 489 p. Section I - Phenology study.

2E 790502 2 Fisser, H.G., L.W. Young, N.E. Hargis and K.E. Spaeth. 1979. Phenology and production studies on semiarid shrub-types. 1978 results. Univ. of Wyo. Coop. Res. Rep. to the BLM. Sections I, II, III and IV. Wyo. Agr. Exp. Sta. Sci. Rep. 985. 489 p. Section II - Phenology step-wise regression for predictive modeling.

2E 790502 4 Fisser, H.G., L.W. Young, N.E. Hargis and K.E. Spaeth. 1979. Phenology and production studies on semiarid shrub-types. 1978 results. Univ. of Wyo. Coop. Res. Rep. to the BLM. Sections I, II, III and IV. Wyo. Agr. Exp. Sta. Sci. Rep. 985. 489 p. Section IV - Exclosure studies (production, cover and precipitation).

2E 800501 1 Fisser, H.G., M.K. Owens, W.J. Waugh, N.E. Hargis and J. Uhlich. 1980. Phenology and production studies on semiarid shrub-types. 1979 Results. Univ. of Wyo. Coop. Res. Rep. to the BLM. Vol. I: Phenology Study. Sections I and II. Wyo. Agr. Exp. Sta. Sci. Rep. 1044. 387 p. Section I - Phenology study.

2E 800501 2 Fisser, H.G., M.K. Owens, W.J. Waugh, N.E. Hargis and J. Uhlich. 1980. Phenology and production studies on semiarid shrub-types. 1979 Results. Univ. of Wyo. Coop. Res. Rep. to the BLM. Vol. I: Phenology Study. Sections I and II. Wyo. Agr. Exp. Sta. Sci. Rep. 1044. 387 p. Section II - Phenology step-wise regression for predictive modeling.

2E 800502 4 Fisser, H.G., M.K. Owens, J.W. Uhlich, W.J. Waugh, D.E. Rodgers, and N.E. Hargis. 1980. Phenology and production studies on semiarid shrub-types. 1979 Results. Univ. of Wyo. Coop. Res. Rep. to the BLM. Vol. II: Precipitation, Productivity and Production Section. Sections III, IV, and V. Wyo. Agr. Exp. Sta. Sci. Rep. 1106. 219 p. Section IV - Productivity - phenology relationships.

2E 800502 5 Fisser, H.G., M.K. Owens, J.W. Uhlich, W.J. Waugh, D.E. Rodgers, and N.E. Hargis. 1980. Phenology and production studies on semiarid shrub-types. 1979 Results. Univ. of Wyo. Coop. Res. Rep. to the BLM. Vol. II: Precipitation, Productivity and Production Section. Sections III, IV, and V. Wyo. Agr. Exp. Sta. Sci. Rep. 1106. 219 p. Section V - Exclosure studies (production, cover and precipitation).

2E 820501 1 Fisser, H.G., S.K. Dean, T.L. Deboodt, R.J. Kinucan. 1982. Phenology and production studies on semiarid shrub-types. 1982 Results. Univ. of Wyo. Coop. Res. Rep. to the BLM - Vol. I: Phenology study sections I and II. Wyo. Agr. Exp. Sta. Sci. Rep. 1230. 415 p. Section I - Phenology study.

2E 820501 2 Fisser, H.G., S.K. Dean, T.L. Deboodt, R.J. Kinucan. 1982. Phenology and production studies on semiarid shrub-types. 1982 Results. Univ. of Wyo. Coop. Res. Rep. to the BLM - Vol. I: Phenology study sections I and II. Wyo. Agr. Exp. Sta. Sci. Rep. 1230. 415 p. Section II - Phenology multiple regression for predictive modeling.

2E 820502 4 Fisser, H.G., S.K. Dean, T.L. Deboodt, R.J. Kinucan. 1982. Phenology and production studies on semiarid and shrub-types. 1982 results. Univ. of Wyo. Coop. Res. Rep. to the BLM. Vol. II: Precipitation, productivity, and production sections III, IV, and V. Wyo. Agr. Exp. Sta. Sci. Rep. 1231. 256 p. Section IV - Productivity-phenology interrelation study.

2E 830501 1 Fisser, H.G., S.K. Dean, T.L. Deboodt, R.S. Kinucan and J.L. Mulholland, 1983. Phenology and production studies on semiarid shrub-types. 1982 results. Univ. of Wyo. Coop. Res. Rep. to the BLM. Vol. I: Phenology Study, Sections I and II. Wyo. Agr. Exp. Sta. Sci. Rep. 1230. 430 p. Section I - Phenology study.

2E 830501 2 Fisser, H.G., S.K. Dean, T.L. Deboodt, R.S. Kinucan and J.L. Mulholland, 1983. Phenology and production studies on semiarid shrub-types. 1982 results. Univ. of Wyo. Coop. Res. Rep. to the BLM. Vol. I: Phenology Study, Sections I and II. Wyo. Agr. Exp. Sta. Sci. Rep. 1230. 430 p. Section II - Phenology multiple regression for predictive modeling.

2E 830502 4 Fisser, H.G., S.K. Dean, T.L. Deboodt, R.S. Kinucan and J.L. Mulholland. 1983. Phenology and production studies on semiarid shrub-types. 1982 results. Univ. of Wyo. Coop. Res. Rep. to the BLM. Vol. II: Precipitation, Productivity and Production, Sections III, IV, and V. Wyo. Agr. Exp. Sta. Sci. Rep. 1231. 256 p. Section IV - Productivity-phenology interrelation study.

2E 830503 Fisser, H.G. 1983. Aridland shrub ecology and phenology research program. Wyo. Agr. Exp. Sta. Mimeo. 3 p.

2F. SYNECOLOGY

2F 830503 Fisser, H.G. 1983. Aridland shrub ecology and phenology research program. Wyo. Agr. Exp. Sta. Mimeo. 3 p.

2F1. Great Plains

2F1 690503 Fisser, H.G., J.E. Lester, G.C. Mitchell, and L.J. Bledsoe. 1969. Preliminary report of methodology and results for analyses of plant pattern project research on the Pawnee Site. U.S. I.B.P. Grassland Biome Technical Report 9. 68 p.

2F1 710503 Fisser, H.G. 1971. Significance of plant pattern in ecosystem functioning. Wyo. Agr. Exp. Sta. Mimeo. 2 p.

2F1 710504 Fisser, H.G. 1971. Evaluation of Pawnee Site frequency data. Wyo. Agr. Exp. Sta. Mimeo. 6 p.

2F1 720503 Shaver, J.C. and H.G. Fisser. 1972. Detailed vegetation mapping of the I.B.P. Pawnee Site. U.S. I.B.P. Grassland Biome Technical Report 150. 44 p.

2F1 720504 Stroud, D.W. and H.G. Fisser. 1972. Plant pattern micro-relief and *Bouteloua gracilis* culm density studies on the Pawnee Site: 1970-1971. U.S. I.B.P. Grassland Biome Technical Report 186. 26 p.

2F1 740503 Shrader, J.A. and H.G. Fisser. 1874. Seed production ecology of the shortgrass prairie of the central United States. Wyo. Agr. Exp. Sta. Mimeo. 2 p.

2F1 750502 Fisser, H.G., F.K. Taha, G.A. Lymbery and J.R. Cox. 1975. Herbage structure, productivity and phenology on Black Thunder. Annual Progress Report to Atlantic-Richfield Company. 1974. results. Wyo. Agr. Exp. Sta. Sci. Rep. 662. 118 p.

2F1 760502 Fisser, H.G., J.R. Cox and F.K. Taha. 1976. Herbage structure, productivity and phenology on Black Thunder. Annual Progress Report to Atlantic-Richfield Company. 1975 results. Wyo. Agr. Exp. Sta. Sci. Rep. 779. 161 p.

2F1 760503 Ries, R.E., H.G. Fisser, and A.T. Harrison. 1976. Vegetation, Section IX. p. 1-85. In Bergman, H.L. and M.D. Marcus (eds.). Final Report Environmental Assessment of the Black Thunder Mine Site, Campbell, Colorado. University of Wyoming, Laramie, Wyoming.

2F1 770501 Fisser, H.G., J.R. Cox and F.K. Taha. 1977. Herbage structure, production and phenology on Black Thunder. Annual Progress Report to Atlantic-Richfield Company. 1976 results. Wyo. Agr. Exp. Sta. Sci. Rep. 832. 208 p.

2F1 780501 Fisser, H.G., J.R. Cox and M. Mecke. 1978. Herbage structure, production and phenology on Black Thunder. Annual Progress Report to Atlantic-Richfield Company. 1977 results. Wyo. Agr. Exp. Sta. Sci. Rep. 856. 157 p.

2F1 780502 Fisser, H.G., and J.R. Cox. 1978. Herbage structure, production and phenology on Black Thunder. Final Summary Report to Atlantic-Richfield Company. Wyo. Agr. Exp. Sta. Sci. Rep. 968. 243 p.

2F2. West

2F2 590501 Fisser, H.G., R.L. Lang, and A.A. Beetle. 1959. A study of range conditions in the Big Horn and Wind River Basins. Pre. Report I. Wyo. Agr. Exp. Sta. Mimeo Circ. 119. 5 p.

2F2 590502 Fisser, H.G. 1959. Range survey. In Crop Care in Wyoming. p. 22, December.

2F2 600502 Fisser, H.G., A.A. Beetle, and R.L. Lang. 1960. A study of range conditions in the Big Horn and Wind River Basins. Prel. Report II. Wyo. Agr. Exp. Sta. Mimeo. 10 p.

2F2 610501 Fisser, H.G. 1961. The range condition survey of the Big Horn and Wind River Basins. University of Wyoming. Plant Sci. Div., Range Manage. Notes 4:1.

2F2 620501 Fisser, H.G., C. Williams, A.A. Beetle, and R.L. Lang. 1962. A study of range conditions in the Big Horn and Wind River Basins. Prel. Report III. Wyo. Agr. Exp. Sta. Mimeo. 7 p.

2F2 620502 Nichols, J.T., L.C. Vosler, D.R. Smith, and H.G. Fisser. 1962. Halogeton studies. In Coop. Res. Rep. to the BLM. 1961 Results. Wyo. Agr. Exp. Sta. Mimeo. p. 1-17.

2F2 620503 Fisser, H.G. 1962. An ecological study of the *Artemisia tripartita* subsp. *rupicola* and related shrub communities in Wyoming. In Crop Care in Wyoming. p. 1-3. June.

2F2 630501 Fisser, H.G. 1963. A study of range conditions in the Big Horn and Wind River Basins. Prel. Report IV. Wyo. Agr. Exp. Sta. Mimeo Report. 13 p.

2F2 630502 Fisser, H.G. 1963. Ecological studies of the *Artemisia tripartita* subsp. *rupicola* and related shrub communities in Wyoming. In 12-year report to the Wyo. Natural Res. Board. Wyo. Agr. Exp. Sta. Mimeo Part VI. 2 p.

2F2 640501 4 Fisser, H.G., J.T. Nichols, R.W. Hamner, and J.R. Wight. 1964. Halogeton Research. 1963 Results. Univ. of Wyo. Coop. Res. Rep. to the BLM. Sections I - VIII. Wyo. Agr. Exp. Sta. Mimeo. p. 1-61. Section IV - H.G. Fisser, J.T. Nichols, and J.R. Wight. Soil moisture and temperature study. 3 p.

2F2 640502 Fisser, H.G. 1964. Range survey in Wyoming's Big Horn Basin. Wyo. Agr. Exp. Sta. Bull. 424. 44 p.

2F2 650501 4 Fisser, H.G., J.R. Wight, J.R. Flesland, L.D. Robinson. 1965. Halogeton Research. 1964 Results. Univ. of Wyo. Coop. Res. Rep. to the BLM. Section I - VI. Wyo. Agr. Exp. Sta. Mimeo. p. 1-82. Section IV - H.G. Fisser, J.R. Wight, and J.R. Flesland. Soil moisture and temperature study. 6 p.

2F2 650502 Fisser, H.G. 1965. University of Wyoming and Bureau of Land Management Cooperative research in the Wind River Basin. Wyo. Agr. Exp. Sta. Mimeo. 3 p.

2F2 660501 4 Fisser, H.G., L.D. Robinson, D.R. Kirkam, R. Steger, and G. Noller. 1966. Halogeton Research. 1965 Results. Univ. of Wyo. Coop. Res. Rep. to the BLM. Sections I - V. Wyo. Agr. Exp. Sta. Mimeo. 134 p. Section IV - Soil moisture and temperature study. 7 p.

2F2 670502 1 Fisser, H.G., G. Noller, R.E. Steger, D.R. Kirkham and T. King. 1967. Aridland Ecology Research. 1966 Results. Univ. of Wyo. Coop. Res. Rep. to the BLM. Section I - IV. Wyo. Agr. Exp. Sta. Sci. Rep. 41. 137 p. Section I - Soil moisture and temperature study.

2F2 670503 Fisser, H.G. 1967. Exclosure study with transects of permanent plots. Univ. of Wyo. Coop. Res. Rep. to the BLM. Wyo. Agr. Exp. Sta. Sci. Rep. 69. 94 p.

2F2 670504 Fisser, H.G. 1967. Range condition class survey - concepts and application. In Crop Care in Wyoming. p. 2-5. March.

2F2 680501 Gibbens, R.P., H.G. Fisser, and M. May. 1968. Influence of grazing management systems on vegetation and wildlife habitat. Annual report of progress, July, 1967 - July, 1968. Univ. of Wyo. Coop. Res. Rep. to the BLM. Wyo. Agr. Exp. Sta. Sci. Rep. 102. 9 p.

2F2 680502 Fisser, H.G. 1968. Exclosure studies with transect of permanent plots. 1967 results. Univ. of Wyo. Coop. Res. Rep. to the BLM. Wyo. Agr. Exp. Sta. Sci. Rep. 118. 52 p.

2F2 680503 Fisser, H.G. 1968. Long-term exclosure studies with 2 ft x 20 ft and 4 ft x 4 ft permanent plots. 1959-1964 results. Univ. Wyo. Coop. Res. Rep. to the BLM. Wyo. Agr. Exp. Sta. Sci. Rep. 131. 78 p.

2F2 680504 1 Fisser, H.G. and G.L. Noller. 1968. Aridland Ecology. 1967 results. Univ. of Wyo. Coop. Res. Rep. to the BLM. Sections I, II, and III, Wyo. Agr. Exp. Sta. Sci. Rep. 132. 93 p. Section I - Soil moisture and temperature study.

2F2 680505 Wight, J.R. and H.G. Fisser. 1968. *Juniperus osteosperma* in northwestern Wyoming - Their distribution and ecology. Wyo. Agr. Exp. Sta. Sci. Mono. 7. 31 p.

2F2 680506 Fisser, H.G. 1968. Synthetic study area nomination and related analytic research proposals to the Desert Biome Ecosystem Analysis Program of the International Biological Program. Wyo. Agr. Exp. Sta. Univ. of Wyo. Mimeo. 82 p.

2F2 690501 1 Fisser, H.G. and G.L. Whysong. 1969. Aridland Ecology Research. 1968 results. Univ. of Wyo. Coop. Res. Rep. to the BLM. Sections I - IV. Wyo. Agr. Exp. Sta. Sci. Rep. No. 178. 114 p. Section I - Soil moisture and temperature.

2F2 690502 Gibbens, R.P., H.G. Fisser and M. May. 1969. Influence of grazing management systems on vegetation and wildlife habitat. Annual report of progress, July, 1968 - July, 1969. Univ. of Wyo. Coop. Res. Rep. to the BLM. Wyo. Agr. Exp. Sta. Sci. Rep. 179. 132 p.

2F2 700501 1 Fisser, H.G. and G.L. Whysong. 1970. Aridland Ecology Research. 1969 results. Univ. of Wyo. Coop. Res. Rep. to the BLM. Sections I - IV. Wyo. Agr. Exp. Sta. Sci. Rep. 240. 128 p. Section I - Soil moisture and temperature.

2F2 700502 Gibbens, R.P. and H.G. Fisser. 1970. Influence of grazing management systems on vegetation and wildlife habitat. Annual report of progress, July, 1969 - July, 1970. Univ. of Wyo. Coop. Res. Rep. to the BLM. Wyo. Agr. Exp. Sta. Sci. Rep. 241. 32 p.

2F2 700503 Whysong, G.L. and H.G. Fisser. 1970. Frequency, density and basal area measures on sagebrush-grass ranges. 1969 results. Univ. of Wyo. Coop. Res. Rep. to the BLM. Wyo. Agr. Exp. Sta. Sci. Rep. 265. 35 p.

2F2 700504 Fisser, H.G. 1970. Exclosure studies with transects of permanent plots. 1969 results. Univ. of Wyo. Coop. Res. Rep. to the BLM. Wyo. Agr. Exp. Sta. Sci. Rep. 275. 84 p.

2F2 700505 Gibbens, R.P., H.G. Fisser and M. May. 1970. Influence of grazing management systems on vegetation and wildlife habitat. Semiann. Rep. to the BLM. Wyo. Agr. Exp. Sta. Mimeo. 2 p.

2F2 700506 Fisser, H.G. 1970. Aridland Ecology, semi-annual progress report to the Bureau of Land Management, March 1970. Wyo. Agr. Ext. Sta. Mimeo. 1 p.

2F2 710501 Fisser, H.G. and R.P. Gibbens. 1971. Influence of grazing management systems on vegetation and wildlife habitat. Annual report of progress, July, 1970 - July, 1971. Univ. of Wyo. Coop. Res. Rep. to the BLM. Wyo. Agr. Exp. Sta. Sci. Rep. 336. 92 p.

2F2 710502 1 Fisser, H.G., and R.E. Ries. 1971. Aridland Ecology Research. 1970 results. Univ. of Wyo. Coop. Res. Rep. to the BLM. Sections I - IV. Wyo. Agr. Exp. Sta. Sci. Rep. 350. 133 p. Section I - Soil moisture and temperature studies.

2F2 710505 Fisser, H.G. and R.E. Ries. 1971. Semi-annual progress report to the Bureau of Land Management, January 1971. Wyo. Agr. Exp. Sta. Mimeo. 1 p.

2F2 720501 Gibbens, R.P., and H.G. Fisser. 1972. Influences of grazing management systems on vegetation and wildlife habitat. Annual Progress Report. Wyo. Agr. Exp. Sta. Sci. Rep. 414. 69 p.

2F2 740502 1 Fisser, H.G. and L.H. Kleinman. 1974. Phenology and production on semiarid shrub-types. 1973 results. Univ. of Wyo. Coop. Res. Rep. to the BLM. Sections I, II and III. Wyo. Agr. Exp. Sta. Sci. Rep. 607. 256 p. Section I - Phenology study.

2F2 750501 Gibbens, R.P., and H.G. Fisser. 1975. Influence of grazing management systems on vegetation and wildlife habitat in the Red Desert Region of Wyoming. Wyo. Agr. Exp. Sta. Sci. Mono. 29. 23 p.

2F2 750503 1 Fisser, H.G., and L.H. Kleinman. 1975. Phenology and production on semiarid shrub-types. 1974 results. Univ. Wyo. Coop. Res. Rep. to the BLM. Sections I, II, and III. Wyo. Agr. Exp. Sta. Sci. Rep. 729. 399 p. Section I - Phenology study.

2F2 760501 1 Fisser, H.G., and L.H. Kleinman. 1976. Phenology and production studies on semiarid shrub-types. 1975 results. Univ. of Wyo. Coop. Res. Rep. to the BLM. Sections I, II and III. Wyo. Agr. Exp. Sta. Sci. Rep. 831. 410 p. Section I - Phenology study.

2F2 770502 1 Fisser, H.G., L.W. Young, J.E. Orpet and N.E. Hargis. 1977. Phenology and production studies on semiarid shrub-types. 1976 results. Univ. of Wyo. Coop. Res. Rep. to the BLM. Sections I, II, III and IV. Wyo. Agr. Exp. Sta. Sci. Rep. 930. 438 p. Section I - Phenology study.

2F2 770502 2 Fisser, H.G., L.W. Young, J.E. Orpet and N.E. Hargis. 1977. Phenology and production studies on semiarid shrub-types. 1976 results. Univ. of Wyo. Coop. Res. Rep. to the BLM. Sections I, II, III and IV. Wyo. Agr. Exp. Sta. Sci. Rep. 930. 438 p. Section II - Phenology step-wise regression for predictive modeling.

2F2 780503 1 Fisser, H.G., L.W. Young, J.E. Orpet and N.E. Hargis. 1978. Phenology and production studies on semiarid shrub-types. 1977 results. Univ. of Wyo. Coop. Res. Rep. to the BLM. Sections I, II, III and IV. Wyo. Agr. Exp. Sta. Sci. Rep. 970. 439 p. Section I - Phenology study.

2F2 780503 2 Fisser, H.G., L.W. Young, J.E. Orpet and N.E. Hargis. 1978. Phenology and production studies on semiarid shrub-types. 1977 results. Univ. of Wyo. Coop. Res. Rep. to the BLM. Sections I, II, III and IV. Wyo. Agr. Exp. Sta. Sci. Rep. 970. 439 p. Section II - Phenology step-wise regression for predictive modeling.

2F2 780503 4 Fisser, H.G., L.W. Young, J.E. Orpet and N.E. Hargis. 1978. Phenology and production studies on semiarid shrub-types. 1977 results. Univ. of Wyo. Coop. Res. Rep. to the BLM. Sections I, II, III and IV. Wyo. Agr. Exp. Sta. Sci. Rep. 970. 439 p. Section IV - Exclosure studies (production, cover and precipitation).

2F2 790501 Fisser, H.G., D.C. Trueblood and D.D. Samuelson. 1979. Soil vegetation relationships on rangeland exclosures on the Grass Creek Planning Unit of north central Wyoming. Univ. of Wyo. Coop. Res. Rep. to the BLM. Wyo. Agr. Exp. Sta. Sci. Rep. 969. 279 p.

2F2 790502 1 Fisser, H.G., L.W. Young, N.E. Hargis and K.E. Spaeth. 1979. Phenology and production studies on semiarid shrub-types. 1978 results. Univ. of Wyo. Coop. Res. Rep. to the BLM. Sections I, II, III and IV. Wyo. Agr. Exp. Sta. Sci. Rep. 985. 489 p. Section I - Phenology study.

2F2 790502 2 Fisser, H.G., L.W. Young, N.E. Hargis and K.E. Spaeth. 1979. Phenology and production studies on semiarid shrub-types. 1978 results. Univ. of Wyo. Coop. Res. Rep. to the BLM. Sections I, II, III and IV. Wyo. Agr. Exp. Sta. Sci. Rep. 985. 489 p. Section II - Phenology step-wise regression for predictive modeling.

2F2 790502 4 Fisser, H.G., L.W. Young, N.E. Hargis and K.E. Spaeth. 1979. Phenology and production studies on semiarid shrub-types. 1978 results. Univ. of Wyo. Coop. Res. Rep. to the BLM. Sections I, II, III and IV. Wyo. Agr. Exp. Sta. Sci. Rep. 985. 489 p. Section IV - Exclosure studies (production, cover and precipitation).

2F2 790504 Fisser, H.G. 1979. Grazing management system studies of northern desert shrub vegetation of southern Wyoming. Wyo. Agr. Exp. Sta. Mimeo. 3 p.

2F2 800501 1 Fisser, H.G., M.K. Owens, W.J. Waugh, N.E. Hargis and J. Uhlich. 1980. Phenology and production studies on semiarid shrub-types. 1979 Results. Univ. of Wyo. Coop. Res. Rep. to the BLM. Vol. I: Phenology Study. Sections I and II. Wyo. Agr. Exp. Sta. Sci. Rep. 1044. 387 p. Section I - Phenology study.

2F2 800501 2 Fisser, H.G., M.K. Owens, W.J. Waugh, N.E. Hargis and J. Uhlich. 1980. Phenology and production studies on semiarid shrub-types. 1979 Results. Univ. of Wyo. Coop. Res. Rep. to the BLM. Vol. I: Phenology Study. Sections I and II. Wyo. Agr. Exp. Sta. Sci. Rep. 1044. 387 p. Section II - Phenology step-wise regression for predictive modeling.

2F2 800502 4 Fisser, H.G., M.K. Owens, J.W. Uhlich, W.J. Waugh, D.E. Rodgers, and N.E. Hargis. 1980. Phenology and production studies on semiarid shrub-types. 1979 Results. Univ. of Wyo. Coop. Res. Rep. to the BLM. Vol. II: Precipitation, Productivity and Production Section. Sections III, IV, and V. Wyo. Agr. Exp. Sta. Sci. Rep. 1106. 219 p. Section IV - Productivity-phenology relationships.

2F2 800502 5 Fisser, H.G., M.K. Owens, J.W. Uhlich, W.J. Waugh, D.E. Rodgers, and N.E. Hargis. 1980. Phenology and production studies on semiarid shrub-types. 1979 Results. Univ. of Wyo. Coop. Res. Rep. to the BLM. Vol. II: Precipitation, Productivity and Production Section. Sections III, IV, and V. Wyo. Agr. Exp. Sta. Sci. Rep. 1106. 219 p. Section V - Exclosure studies (production, cover and precipitation).

2F2 800503 Fisser, H.G., D.C. Trueblood and C.H. Butterfield. 1980. Soil vegetation relationships on rangeland exclosures in the Overland Planning Unit of south central Wyoming. Univ. of Wyo. Coop. Res. Rep. to the BLM. Wyo. Agr. Exp. Sta. Sci. Rep. 1042. 75 p.

2F2 800504 Fisser, H.G., and D.C. Trueblood. 1980. Soil vegetation relationships on rangeland exclosures on the Salt Wells Planning Unit of southwestern Wyoming. Univ. of Wyo. Coop. Res. Rep. to the BLM. Wyo. Agr. Exp. Sta. Sci. Rep. 1041. 68 p.

2F2 800505 Fisser, H.G., D.E. Rodgers and W.E. Gerhart. 1980. Influences of grazing management systems on vegetation and wildlife habitat. 1975 and 1979 Results and Long-Term Summary. Univ. of Wyo. Coop. Res. Rep. to the BLM. Wyo. Agr. Exp. Sta. Sci. Rep. 1047. 188 p.

2F2 820501 1 Fisser, H.G., S.K. Dean, T.L. Deboodt, R.J. Kinucan. 1982. Phenology and production studies on semiarid shrub-types. 1982 Results. Univ. of Wyo. Coop. Res. Rep. to the BLM - Vol. I: Phenology study sections I and II. Wyo. Agr. Exp. Sta. Sci. Rep. 1230. 415 p. Section I - Phenology study.

2F2 820501 2 Fisser, H.G., S.K. Dean, T.L. Deboodt, R.J. Kinucan. 1982. Phenology and production studies on semiarid shrub-types. 1982 Results. Univ. of Wyo. Coop. Res. Rep. to the BLM - Vol. I: Phenology study sections I and II. Wyo. Agr. Exp. Sta. Sci. Rep. 1230. 415 p. Section II - Phenology multiple regression for predictive modeling.

2F2 820502 4 Fisser, H.G., S.K. Dean, T.L. Deboodt, R.J. Kinucan. 1982. Phenology and production studies on semiarid shrub-types. 1982 results. Univ. of Wyo. Coop. Res. Rep. to the BLM. Vol. II: Precipitation, productivity, and production sections III, IV, and V. Wyo. Agr. Exp. Sta. Sci. Rep. 1231. 256 p. Section IV - Productivity-phenology interrelation study.

2F2 820503 Fisser, H.G. 1982. Vegetation response to grazing management strategies on semiarid shrublands of Wyoming. Wyo. Agr. Exp. Sta. Mimeo. prepared for Grazing Systems Conference, Texas A & M University. November 1982. 4 p.

2F2 830501 1 Fisser, H.G., S.K. Dean, T.L. Deboodt, R.S. Kinucan and J.L. Mulholland, 1983. Phenology and production studies on semiarid shrub-types. 1982 results. Univ. of Wyo. Coop. Res. Rep. to the BLM. Vol. I: Phenology Study, Sections I and II. Wyo. Agr. Exp. Sta. Sci. Rep. 1230. 430 p. Section I - Phenology study.

2F2 830501 2 Fisser, H.G., S.K. Dean, T.L. Deboodt, R.S. Kinucan and J.L. Mulholland, 1983. Phenology and production studies on semiarid shrub-types. 1982 results. Univ. of Wyo. Coop. Res. Rep. to the BLM. Vol. I: Phenology Study, Sections I and II. Wyo. Agr. Exp. Sta. Sci. Rep. 1230. 430 p. Section II - Phenology multiple regression for predictive modeling.

2F2 830502 4 Fisser, H.G., S.K. Dean, T.L. Deboodt, R.S. Kinucan and J.L. Mulholland. 1983. Phenology and production studies on semiarid shrub-types. 1982 results. Univ. of Wyoming Coop. Res. Rep. to the BLM. Vol. II: Precipitation, Productivity and Production, Sections III, IV, and V. Wyo. Agr. Exp. Sta. Sci. Rep. 1231. 256 p. Section IV - Productivity-phenology interrelation study.

2F2 830503 Fisser, H.G. 1983. Aridland shrub ecology and phenology research program. Wyo. Agr. Exp. Sta. Mimeo. 3 p.

3. Range Resources

3A. RANGE INVENTORY AND SURVEY

3A 620503 Fisser, H.G. 1962. An ecological study of the *Artemisia tripartita* subsp. *rupicola* and related shrub communities in Wyoming. In Crop Care in Wyoming. p. 1-3. June.

3A 630502 Fisser, H.G. 1963. Ecological studies of the *Artemisia tripartita* subsp. *rupicola* and related shrub communities in Wyoming. In 12-year report to the Wyo. Natural Res. Board. Wyo. Agr. Exp. Sta. Mimeo Part VI. 2 p.

3A 640501 7 Fisser, H.G., J.T. Nichols, R.W. Hamner, and J.R. Wight. 1964. Halogeton Research. 1963 Results. Univ. of Wyo. Coop. Res. Rep. to the BLM. Sections I - VIII. Wyo. Agr. Exp. Sta. Mimeo. p. 1-61. Section VII - H.G. Fisser. Precipitation pattern study. 5 p.

3A 650501 5 Fisser, H.G., J.R. Wight, J.R. Flesland, L.D. Robinson. 1965. Halogeton Research. 1964 Results. Univ. of Wyo. Coop. Res. Rep. to the BLM. Section I - VI. Wyo. Agr. Exp. Sta. Mimeo. p. 1-82. Section V - H.G. Fisser. Precipitation pattern study. 5 p.

3A 660501 5 Fisser, H.G., L.D. Robinson, D.R. Kirkam, R. Steger, and G. Noller. 1966. Halogeton Research. 1965 Results. Univ. of Wyo. Coop. Res. Rep. to the BLM. Sections I - V. Wyo. Agr. Exp. Sta. Mimeo. 134 p. Section V - Precipitation pattern study. 4 p.

3A 670502 3 Fisser, H.G., G. Noller, R.E. Steger, D.R. Kirkham and T. King. 1967. Aridland Ecology Research. 1966 Results. Univ. of Wyo. Coop. Res. Rep. to the BLM. Section I - IV. Wyo. Agr. Exp. Sta. Sci. Rep. 41. 137 p. Section III - Precipitation pattern study.

3A 680504 2 Fisser, H.G. and G.L. Noller. 1968. Aridland Ecology. 1967 results. Univ. of Wyo. Coop. Res. Rep. to the BLM. Sections I, II, and III, Wyo. Agr. Exp. Sta. Sci. Rep. 132. 93 p. Section II - Precipitation pattern study.

3A 690501 3 Fisser, H.G. and G.L. Whysong. 1969. Aridland Ecology Research. 1968 results. Univ. of Wyo. Coop. Res. Rep. to the BLM. Sections I - IV. Wyo. Agr. Exp. Sta. Sci. Rep. 178. 114 p. Section III - Precipitation pattern study.

3A 690503 Fisser, H.G., J.E. Lester, G.C. Mitchell, and L.J. Bledsoe. 1969. Preliminary report of methodology and results for analyses of plant pattern project research on the Pawnee Site. U.S. I.B.P. Grassland Biome Technical Report 9. 68 p.

3A 700501 3 Fisser, H.G. and G.L. Whysong. 1970. Aridland Ecology Research. 1969 results. Univ. of Wyo. Coop. Res. Rep. to the BLM. Sections I - IV. Wyo. Agr. Exp. Sta. Sci. Rep. 240. 128 p. Section III - Precipitation pattern study.

3A 710502 3 Fisser, H.G., and R.E. Ries. 1971. Aridland Ecology Research. 1970 results. Univ. of Wyo. Coop. Res. Rep. to the BLM. Sections I - IV. Wyo. Agr. Exp. Sta. Sci. Rep. 350. 133 p. Section III - Precipitation pattern study.

3A 720502 1 Fisser, H.G., and R.E. Ries. 1972. Aridland Ecology Research. 1971 results. Univ. of Wyo. Coop. Res. Rep. to the BLM. Sections I and II. Wyo. Agr. Exp. Sta. Sci. Rep. 428. 130 p. Section I - Precipitation pattern study.

3A 730501 1 Fisser, H.G., and R.E. Ries. 1973. Aridland Ecology Research. 1972 results. Univ. of Wyo. Coop. Res. Rep. to the BLM. sections I and II. Wyo. Agr. Exp. Sta. Sci. Rep. 521. 128 p. Section I - Precipitation pattern study.

3A 740501 Fisser, H.G. and R.E. Ries. 1974. Summary of data, 1962-1970. Aridland Ecology Research. Univ. of Wyo. Coop. Res. Rep. to the BLM. Wyo. Agr. Exp. Sta. Sci. Rep. 556. 132 p.

3A 740502 2 Fisser, H.G. and L.H. Kleinman. 1974. Phenology and production on semiarid shrub-types. 1973 results. Univ. of Wyo. Coop. Res. Rep. to the BLM. Sections I, II and III. Wyo. Agr. Exp. Sta. Sci. Rep. 607. 256 p. Section II - Precipitation pattern study.

3A 750503 2 Fisser, H.G., and L.H. Kleinman. 1975. Phenology and production on semiarid shrub-types. 1974 results. Univ. Wyo. Coop. Res. Rep. to the BLM. Sections I, II, and III. Wyo. Agr. Exp. Sta. Sci. Rep. 729. 399 p. Section II - Precipitation pattern study.

3A 760501 2 Fisser, H.G., and L.H. Kleinman. 1976. Phenology and production studies on semiarid shrub-types. 1975 results. Univ. of Wyo. Coop. Res. Rep. to the BLM. Sections I, II and III. Wyo. Agr. Exp. Sta. Sci. Rep. 831. 410 p. Section II - Precipitation pattern study.

3A 770502 3 Fisser, H.G., L.W. Young, J.E. Orpet and N.E. Hargis. 1977. Phenology and production studies on semiarid shrub-types. 1976 results. Univ. of Wyo. Coop. Res. Rep. to the BLM. Sections I, II, III and IV. Wyo. Agr. Exp. Sta. Sci. Rep. 930. 438 p. Section III - Precipitation pattern study.

3A 780503 3 Fisser, H.G., L.W. Young, J.E. Orpet and N.E. Hargis. 1978. Phenology and production studies on semiarid shrub-types. 1977 results. Univ. of Wyo. Coop. Res. Rep. to the BLM. Sections I, II, III and IV. Wyo. Agr. Exp. Sta. Sci. Rep. 970. 439 p. Section III - Precipitation pattern study.

3A 790502 3 Fisser, H.G., L.W. Young, N.E. Hargis and K.E. Spaeth. 1979. Phenology and production studies on semiarid shrub-types. 1978 results. Univ. of Wyo. Coop. Res. Rep. to the BLM. Sections I, II, III and IV. Wyo. Agr. Exp. Sta. Sci. Rep. 985. 489 p. Section III - Precipitation pattern study.

3A 800502 3 Fisser, H.G., M.K. Owens, J.W. Uhlich, W.J. Waugh, D.E. Rodgers, and N.E. Hargis. 1980. Phenology and production studies on semiarid shrub-types. 1979 Results. Univ. of Wyo. Coop. Res. Rep. to the BLM. Vol. II: Precipitation, Productivity and Production Section. Sections III, IV, and V. Wyo. Agr. Exp. Sta. Sci. Rep. 1106. 219 p. Section III - Precipitation pattern study.

3A 820502 3 Fisser, H.G., S.K. Dean, T.L. Deboodt, R.J. Kinucan. 1982. Phenology and production studies on semiarid shrub-types. 1982 results. Univ. of Wyo. Coop. Res. Rep. to the BLM. Vol. II: Precipitation, productivity, and production sections III, IV, and V. Wyo. Agr. Exp. Sta. Sci. Rep. 1231. 256 p. Section III - Precipitation pattern study.

3A 830502 3 Fisser, H.G., S.K. Dean, T.L. Deboodt, R.S. Kinucan and J.L. Mulholland. 1983. Phenology and production studies on semiarid shrub-types. 1982 results. Univ. of Wyo. Coop. Res. Rep. to the BLM. Vol. II: Precipitation, Productivity and Production, Sections III, IV, and V. Wyo. Agr. Exp. Sta. Sci. Rep. 1231. 256 p. Section III - Precipitation pattern study.

3A1. Site, Mapping, Remote Sensing

3A1 590501 Fisser, H.G., R.L. Lang, and A.A. Beetle. 1959. A study of range conditions in the Big Horn and Wind River Basins. Pre. Report I. Wyo. Agr. Exp. Sta. Mimeo Circ. 119. 5 p.

3A1 600503 Fisser, H.G., A.A. Beetle, R.L. Lang, and D.R. Smith. 1960. Range Management survey studies. In Crop Care in Wyoming. p. 20-22, December.

3A1 610501 Fisser, H.G. 1961. The range condition survey of the Big Horn and Wind River Basins. University of Wyoming. Plant Sci. Div., Range Manage. Notes 4:1.

3A1 620501 Fisser, H.G., C. Williams, A.A. Beetle, and R.L. Lang. 1962. A study of range conditions in the Big Horn and Wind River Basins. Prel. Report III. Wyo. Agr. Exp. Sta. Mimeo. 7 p.

3A1 630501 Fisser, H.G. 1963. A study of range conditions in the Big Horn and Wind River Basins. Prel. Report IV. Wyo. Agr. Exp. Sta. Mimeo Report. 13 p.

3A1 640501 8 Fisser, H.G., J.T. Nichols, R.W. Hamner, and J.R. Wight. 1964. Halogeton Research. 1963 Results. Univ. of Wyo. Coop. Res. Rep. to the BLM. Sections I - VIII. Wyo. Agr. Exp. Sta. Mimeo. p. 1-61. Section VIII - H.G. Fisser, R.W. Hamner, and J.T. Nichols. Cumberland district exclosures. 3 p.

3A1 640502 Fisser, H.G. 1964. Range survey in Wyoming's Big Horn Basin. Wyo. Agr. Exp. Sta. Bull. 424. 44 p.

3A1 650502 Fisser, H.G. 1965. University of Wyoming and Bureau of Land Management Cooperative research in the Wind River Basin. Wyo. Agr. Exp. Sta. Mimeo. 3 p.

3A1 660503 Fisser, H.G. and R.E. Steger. 1966. Microclimate studies in desert shrub-types. Wyo. Agr. Exp. Sta. Mimeo. 14 p.

3A1 670504 Fisser, H.G. 1967. Range condition class survey - concepts and application. In Crop Care in Wyoming. p. 2-5. March.

3A1 720503 Shaver, J.C. and H.G. Fisser. 1972. Detailed vegetation mapping of the I.B.P. Pawnee Site. U.S. I.B.P. Grassland Biome Technical Report 150. 44 p.

3A1 730502 Shaver, J.C. and H.G. Fisser. 1973. Evaluation of vegetation types differentiated by aerial photo interpretation of panchromatic, color, and infrared imagery. U.S. I.B.P. Biome Technical Report 238. 86 p.

3A1 750502 Fisser, H.G., F.K. Taha, G.A. Lymbery and J.R. Cox. 1975. Herbage structure, productivity and phenology on Black Thunder. Annual Progress Report to Atlantic-Richfield Company. 1974. results. Wyo. Agr. Exp. Sta. Sci. Rep. 662. 118 p.

3A1 760502 Fisser, H.G., J.R. Cox and F.K. Taha. 1976. Herbage structure, productivity and phenology on Black Thunder. Annual Progress Report to Atlantic-Richfield Company. 1975 results. Wyo. Agr. Exp. Sta. Sci. Rep. 779. 161 p.

3A1 760503 Ries, R.E., H.G. Fisser, and A.T. Harrison. 1976. Vegetation, Section IX. p. 1-85. In Bergman, H.L. and M.D. Marcus (eds.). Final Report Environmental Assessment of the Black Thunder Mine Site, Campbell, Colorado. University of Wyoming, Laramie, Wyoming.

3A1 770501 Fisser, H.G., J.R. Cox and F.K. Taha. 1977. Herbage structure, production and phenology on Black Thunder. Annual Progress Report to Atlantic-Richfield Company. 1976 results. Wyo. Agr. Exp. Sta. Sci. Rep. 832. 208 p.

3A1 780501 Fisser, H.G., J.R. Cox and M. Mecke. 1978. Herbage structure, production and phenology on Black Thunder. Annual Progress Report to Atlantic-Richfield Company. 1977 results. Wyo. Agr. Exp. Sta. Sci. Rep. 856. 157 p.

3A1 780502 Fisser, H.G., and J.R. Cox. 1978. Herbage structure, production and phenology on Black Thunder. Final Summary Report to Atlantic-Richfield Company. Wyo. Agr. Exp. Sta. Sci. Rep. 968. 243 p.

3A2. Condition and Trend

3A2 590501 Fisser, H.G., R.L. Lang, and A.A. Beetle. 1959. A study of range conditions in the Big Horn and Wind River Basins. Pre. Report I. Wyo. Agr. Exp. Sta. Mimeo Circ. 119. 5 p.

3A2 590502 Fisser, H.G. 1959. Range survey. In Crop Care in Wyoming. p. 22, December.

3A2 600502 Fisser, H.G., A.A. Beetle, and R.L. Lang. 1960. A study of range conditions in the Big Horn and Wind River Basins. Prel. Report II. Wyo. Agr. Exp. Sta. Mimeo. 10 p.

3A2 600503 Fisser, H.G., A.A. Beetle, R.L. Lang, and D.R. Smith. 1960. Range Management survey studies. In Crop Care in Wyoming. p. 20-22, December.

3A2 610501 Fisser, H.G. 1961. The range condition survey of the Big Horn and Wind River Basins. University of Wyoming. Plant Sci. Div., Range Manage. Notes 4:1.

3A2 620501 Fisser, H.G., C. Williams, A.A. Beetle, and R.L. Lang. 1962. A study of range conditions in the Big Horn and Wind River Basins. Prel. Report III. Wyo. Agr. Exp. Sta. Mimeo. 7 p.

3A2 620505 Fisser, H.G. 1962. Information to be obtained from the range management vegetation measurement transects. Wyo. Agr. Exp. Sta. Mimeo. 2 p.

3A2 630501 Fisser, H.G. 1963. A study of range conditions in the Big Horn and Wind River Basins. Prel. Report IV. Wyo. Agr. Exp. Sta. Mimeo Report. 13 p.

3A2 640502 Fisser, H.G. 1964. Range survey in Wyoming's Big Horn Basin. Wyo. Agr. Exp. Sta. Bull. 424. 44 p.

3A2 670502 2 Fisser, H.G., G. Noller, R.E. Steger, D.R. Kirkham and T. King. 1967. Aridland Ecology Research. 1966 Results. Univ. of Wyo. Coop. Res. Rep. to the BLM. Section I - IV. Wyo. Agr. Exp. Sta. Sci. Rep. 41. 137 p. Section II - Microclimate studies in desert shrub communities.

3A2 670503 Fisser, H.G. 1967. Exclosure study with transects of permanent plots. Univ. of Wyo. Coop. Res. Rep. to the BLM. Wyo. Agr. Exp Sta. Sci. Rep. 69. 94 p.

3A2 670504 Fisser, H.G. 1967. Range condition class survey - concepts and application. In Crop Care in Wyoming. p. 2-5. March.

3A2 680501 Gibbens, R.P., H.G. Fisser, and M. May. 1968. Influence of grazing management systems on vegetation and wildlife habitat. Annual report of progress, July, 1967 - July, 1968. Univ. of Wyo. Coop. Res. Rep. to the BLM. Wyo. Agr. Exp. Sta. Sci. Rep. 102. 9 p.

3A2 680502 Fisser, H.G. 1968. Exclosure studies with transects of permanent plots. 1967 results. Univ. of Wyo. Coop. Res. Rep. to the BLM. Wyo. Agr. Exp. Sta. Sci. Rep. 118. 52 p.

3A2 680503 Fisser, H.G. 1968. Long-term exclosure studies with 2 ft x 20 ft and 4 ft x 4 ft permanent plots. 1959-1964 results. Univ. Wyo. Coop. Res. Rep. to the BLM. Wyo. Agr. Exp. Sta. Sci. Rep. 131. 78 p.

3A2 680505 Wight, J.R. and H.G. Fisser. 1968. *Juniperus osteosperma* in northwestern Wyoming - Their distribution and ecology. Wyo. Agr. Exp. Sta. Sci. Mono. 7. 31 p.

3A2 690501 2 Fisser, H.G. and G.L. Whysong. 1969. Aridland Ecology Research. 1968 results. Univ. of Wyo. Coop. Res. Rep. to the BLM. Sections I - IV. Wyo. Agr. Exp. Sta. Sci. Rep. 178. 114 p. Section II - Microclimate studies in desert shrub communities.

3A2 690502 Gibbens, R.P., H.G. Fisser and M. May. 1969. Influence of grazing management systems on vegetation and wildlife habitat. Annual report of progress, July, 1968 - July, 1969. Univ. of Wyo. Coop. Res. Rep. to the BLM. Wyo. Agr. Exp. Sta. Sci. Rep. 179. 132 p.

3A2 700501 2 Fisser, H.G. and G.L. Whysong. 1970. Aridland Ecology Research. 1969 results. Univ. of Wyo. Coop. Res. Rep. to the BLM. Sections I - IV. Wyo. Agr. Exp. Sta. Sci. Rep. 240. 128 p. Section II - Microclimate studies.

3A2 700502 Gibbens, R.P. and H.G. Fisser. 1970. Influence of grazing management systems on vegetation and wildlife habitat. Annual report of progress, July, 1969 - July, 1970. Univ. of Wyo. Coop. Res. Rep. to the BLM. Wyo. Agr. Exp. Sta. Sci. Rep. 241. 32 p.

3A2 700503 Whysong, G.L. and H.G. Fisser. 1970. Frequency, density and basal area measures on sagebrush-grass ranges. 1969 results. Univ. of Wyo. Coop. Res. Rep. to the BLM. Wyo. Agr. Exp. Sta. Sci. Rep. 265. 35 p.

3A2 700504 Fisser, H.G. 1970. Exclosure studies with transects of permanent plots. 1969 results. Univ. of Wyo. Coop. Res. Rep. to the BLM. Wyo. Agr. Exp. Sta. Sci. Rep. 275. 84 p.

3A2 700505 Gibbens, R.P., H.G. Fisser and M. May. 1970. Influence of grazing management systems on vegetation and wildlife habitat. Semiann. Rep. to the BLM. Wyo. Agr. Exp. Sta. Mimeo. 2 p.

3A2 700506 Fisser, H.G. 1970. Aridland Ecology, semi-annual progress report to the Bureau of Land Management, March 1970. Wyo. Agr. Ext. Sta. Mimeo. 1p.

3A2 710501 Fisser, H.G. and R.P. Gibbens. 1971. Influence of grazing management systems on vegetation and wildlife habitat. Annual report of progress, July, 1970 - July, 1971. Univ. of Wyo. Coop. Res. Rep. to the BLM. Wyo. Agr. Exp. Sta. Sci. Rep. 336. 92 p.

3A2 710502 2 Fisser, H.G., and R.E. Ries. 1971. Aridland Ecology Research. 1970 results. Univ. of Wyo. Coop. Res. Rep. to the BLM. Sections I - IV. Wyo. Agr. Exp. Sta. Sci. Rep. 350. 133 p. Section II - Soil moisture and temperature studies on microclimate study sites in the Big Horn Basin.

3A2 710503 Fisser, H.G. 1971. Significance of plant pattern in ecosystem functioning. Wyo. Agr. Exp. Sta. Mimeo. 2 p.

3A2 710504 Fisser, H.G. 1971. Evaluation of Pawnee Site frequency data. Wyo. Agr. Exp. Sta. Mimeo. 6 p.

3A2 710505 Fisser, H.G. and R.E. Ries. 1971. Semi-annual progress report to the Bureau of Land Management, January 1971. Wyo. Agr. Exp. Sta. Mimeo. 1 p.

3A2 720501 Gibbens, R.P., and H.G. Fisser. 1972. Influences of grazing management systems on vegetation and wildlife habitat. Annual Progress Report. Wyo. Agr. Exp. Sta. Sci. Rep. 414. 69 p.

3A2 720503 Shaver, J.C. and H.G. Fisser. 1972. Detailed vegetation mapping of the I.B.P. Pawnee Site. U.S. I.B.P. Grassland Biome Technical Report 150. 44 p.

3A2 750502 Fisser, H.G., F.K. Taha, G.A. Lymbery and J.R. Cox. 1975. Herbage structure, productivity and phenology on Black Thunder. Annual Progress Report to Atlantic-Richfield Company. 1974. results. Wyo. Agr. Exp. Sta. Sci. Rep. 662. 118 p.

3A2 750503 1 Fisser, H.G., and L.H. Kleinman. 1975. Phenology and production on semiarid shrub-types. 1974 results. Univ. Wyo. Coop. Res. Rep. to the BLM. Sections I, II, and III. Wyo. Agr. Exp. Sta. Sci. Rep. 729. 399 p. Section I - Phenology study.

3A2 760501 1 Fisser, H.G., and L.H. Kleinman. 1976. Phenology and production studies on semiarid shrub-types. 1975 results. Univ. of Wyo. Coop. Res. Rep. to the BLM. Sections I, II and III. Wyo. Agr. Exp. Sta. Sci. Rep. 831. 410 p. Section I - Phenology study.

3A2 760502 Fisser, H.G., J.R. Cox and F.K. Taha. 1976. Herbage structure, productivity and phenology on Black Thunder. Annual Progress Report to Atlantic-Richfield Company. 1975 results. Wyo. Agr. Exp. Sta. Sci. Rep. 779. 161 p.

3A2 760503 Ries, R.E., H.G. Fisser, and A.T. Harrison. 1976. Vegetation, Section IX. p. 1-85. In Bergman, H.L. and M.D. Marcus (eds.). Final Report Environmental Assessment of the Black Thunder Mine Site, Campbell, Colorado. University of Wyoming, Laramie, Wyoming.

3A2 770501 Fisser, H.G., J.R. Cox and F.K. Taha. 1977. Herbage structure, production and phenology on Black Thunder. Annual Progress Report to Atlantic-Richfield Company. 1976 results. Wyo. Agr. Exp. Sta. Sci. Rep. 832. 208 p.

3A2 770502 1 Fisser, H.G., L.W. Young, J.E. Orpet and N.E. Hargis. 1977. Phenology and production studies on semiarid shrub-types. 1976 results. Univ. of Wyo. Coop. Res. Rep. to the BLM. Sections I, II, III and IV. Wyo. Agr. Exp. Sta. Sci. Rep. 930. 438 p. Section I - Phenology study.

3A2 770502 2 Fisser, H.G., L.W. Young, J.E. Orpet and N.E. Hargis. 1977. Phenology and production studies on semiarid shrub-types. 1976 results. Univ. of Wyo. Coop. Res. Rep. to the BLM. Sections I, II, III and IV. Wyo. Agr. Exp. Sta. Sci. Rep. 930. 438 p. Section II - Phenology step-wise regression for predictive modeling.

3A2 780501 Fisser, H.G., J.R. Cox and M. Mecke. 1978. Herbage structure, production and phenology on Black Thunder. Annual Progress Report to Atlantic-Richfield Company. 1977 results. Wyo. Agr. Exp. Sta. Sci. Rep. 856. 157 p.

3A2 780502 Fisser, H.G., and J.R. Cox. 1978. Herbage structure, production and phenology on Black Thunder. Final Summary Report to Atlantic-Richfield Company. Wyo. Agr. Exp. Sta. Sci. Rep. 968. 243 p.

3A2 780503 1 Fisser, H.G., L.W. Young, J.E. Orpet and N.E. Hargis. 1978. Phenology and production studies on semiarid shrub-types. 1977 results. Univ. of Wyo. Coop. Res. Rep. to the BLM. Sections I, II, III and IV. Wyo. Agr. Exp. Sta. Sci. Rep. 970. 439 p. Section I - Phenology study.

3A2 780503 2 Fisser, H.G., L.W. Young, J.E. Orpet and N.E. Hargis. 1978. Phenology and production studies on semiarid shrub-types. 1977 results. Univ. of Wyo. Coop. Res. Rep. to the BLM. Sections I, II, III and IV. Wyo. Agr. Exp. Sta. Sci. Rep. 970. 439 p. Section II - Phenology step-wise regression for predictive modeling.

3A2 790501 Fisser, H.G., D.C. Trueblood and D.D. Samuelson. 1979. Soil vegetation relationships on rangeland exclosures on the Grass Creek Planning Unit of north central Wyoming. Univ. of Wyo. Coop. Res. Rep. to the BLM. Wyo. Agr. Exp. Sta. Sci. Rep. 969. 279 p.

3A2 790502 1 Fisser, H.G., L.W. Young, N.E. Hargis and K.E. Spaeth. 1979. Phenology and production studies on semiarid shrub-types. 1978 results. Univ. of Wyo. Coop. Res. Rep. to the BLM. Sections I, II, III and IV. Wyo. Agr. Exp. Sta. Sci. Rep. 985. 489 p. Section I - Phenology study

3A2 790502 2 Fisser, H.G., L.W. Young, N.E. Hargis and K.E. Spaeth. 1979. Phenology and production studies on semiarid shrub-types. 1978 results. Univ. of Wyo. Coop. Res. Rep. to the BLM. Sections I, II, III and IV. Wyo. Agr. Exp. Sta. Sci. Rep. 985. 489 p. Section II - Phenology step-wise regression for predictive modeling.

3A2 800501 1 Fisser, H.G., M.K. Owens, W.J. Waugh, N.E. Hargis and J. Uhlich. 1980. Phenology and production studies on semiarid shrub-types. 1979 Results. Univ. of Wyo. Coop. Res. Rep. to the BLM. Vol. I: Phenology Study. Sections I and II. Wyo. Agr. Exp. Sta. Sci. Rep. 1044. 387 p. Section I - Phenology study.

3A2 800501 2 Fisser, H.G., M.K. Owens, W.J. Waugh, N.E. Hargis and J. Uhlich. 1980. Phenology and production studies on semiarid shrub-types. 1979 Results. Univ. of Wyo. Coop. Res. Rep. to the BLM. Vol. I: Phenology Study. Sections I and II. Wyo. Agr. Exp. Sta. Sci. Rep. 1044. 387 p. Section II - Phenology step-wise regression for predictive modeling.

3A2 800502 4 Fisser, H.G., M.K. Owens, J.W. Uhlich, W.J. Waugh, D.E. Rodgers, and N.E. Hargis. 1980. Phenology and production studies on semiarid shrub-types. 1979 Results. Univ. of Wyo. Coop. Res. Rep. to the BLM. Vol. II: Precipitation, Productivity and Production Section. Sections III, IV, and V. Wyo. Agr. Exp. Sta. Sci. Rep. 1106. 219 p. Section IV - Productivity-phenology relationships.

3A2 800503 Fisser, H.G., D.C. Trueblood and C.H. Butterfield. 1980. Soil vegetation relationships on rangeland exclosures in the Overland Planning Unit of south central Wyoming. Univ. of Wyo. Coop. Res. Rep. to the BLM. Wyo. Agr. Exp. Sta. Sci. Rep. 1042. 75 p.

3A2 800504 Fisser, H.G., and D.C. Trueblood. 1980. Soil vegetation relationships on rangeland exclosures on the Salt Wells Planning Unit of southwestern Wyoming. Univ. of Wyo. Coop. Res. Rep. to the BLM. Wyo. Agr. Exp. Sta. Sci. Rep. 1041. 68 p.

3A2 820501 1 Fisser, H.G., S.K. Dean, T.L. Deboodt, R.J. Kinucan. 1982. Phenology and production studies on semiarid shrub-types. 1982 Results. Univ. of Wyo. Coop. Res. Rep. to the BLM - Vol. I: Phenology study sections I and II. Wyo. Agr. Exp. Sta. Sci. Rep. 1230. 415 p. Section I - Phenology study.

3A2 820501 2 Fisser, H.G., S.K. Dean, T.L. Deboodt, R.J. Kinucan. 1982. Phenology and production studies on semiarid shrub-types. 1982 Results. Univ. of Wyo. Coop. Res. Rep. to the BLM - Vol. I: Phenology study sections I and II. Wyo. Agr. Exp. Sta. Sci. Rep. 1230. 415 p. Section II - Phenology multiple regression for predictive modeling.

3A2 820502 4 Fisser, H.G., S.K. Dean, T.L. Deboodt, R.J. Kinucan. 1982. Phenology and production studies on semiarid shrub-types. 1982 results. Univ. of Wyo. Coop. Res. Rep. to the BLM. Vol. II: Precipitation, productivity, and production sections III, IV, and V. Wyo. Agr. Exp. Sta. Sci. Rep. 1231. 256 p. Section IV - Productivity-phenology interrelation study.

3A2 830501 1 Fisser, H.G., S.K. Dean, T.L. Deboodt, R.S. Kinucan and J.L. Mulholland, 1983. Phenology and production studies on semiarid shrub-types. 1982 results. Univ. of Wyo. Coop. Res. Rep. to the BLM. Vol. I: Phenology Study, Sections I and II. Wyo. Agr. Exp. Sta. Sci. Rep. 1230. 430 p. Section I - Phenology study

3A2 830501 2 Fisser, H.G., S.K. Dean, T.L. Deboodt, R.S. Kinucan and J.L. Mulholland, 1983. Phenology and production studies on semiarid shrub-types. 1982 results. University of Wyoming Coop. Res. Rep. to the BLM. Vol. I: Phenology Study, Sections I and II. Wyo. Agr. Exp. Sta. Sci. Rep. 1230. 430 p. Section II - Phenology multiple regression for predictive modeling.

3A2 830502 4 Fisser, H.G., S.K. Dean, T.L. Deboodt, R.S. Kinucan and J.L. Mulholland. 1983. Phenology and production studies on semiarid shrub-types. 1982 results. Univ. of Wyo. Coop. Res. Rep. to the BLM. Vol. II: Precipitation, Productivity and Production, Sections III, IV, and V. Wyo. Agr. Exp. Sta. Sci. Rep. 1231. 256 p. Section IV - Productivity-phenology interrelation study.

3A2 830503 Fisser, H.G. 1983. Aridland shrub ecology and phenology research program. Wyo. Agr. Exp. Sta. Mimeo. 3 p.

3A3. Production and Utilization

3A3 620502 Nichols, J.T., L.C. Vosler, D.R. Smith, and H.G. Fisser. 1962. Halogeton studies. In Coop. Res. Rep. to the BLM. 1961 Results. Wyo. Agr. Exp. Sta. Mimeo. p. 1-17.

3A3 630503 3 Fisser, H.G., J.T. Nichols, and R.W. Hamner. 1963. Halogeton Research. 1962 Results. Univ. of Wyo. Coop. Res. Rep. to the BLM. Sections I - III. Wyo. Agr. Exp. Sta. Mimeo. p. 1-38. Section III - H.G. Fisser and R.W. Hamner. Exclosure studies (production, cover, height, and precipitation). p. 12-38.

3A3 640501 3 Fisser, H.G., J.T. Nichols, R.W. Hamner, and J.R. Wight. 1964. Halogeton Research. 1963 Results. Univ. of Wyo. Coop. Res. Rep. to the BLM. Sections I - VIII. Wyo. Agr. Exp. Sta. Mimeo. p. 1-61. Section III - H.G. Fisser, J.T. Nichols, and J.R. Wight. Exclosure studies (production, cover, height and precipitation). 33 p.

3A3 640501 8 Fisser, H.G., J.T. Nichols, R.W. Hamner, and J.R. Wight. 1964. Halogeton Research. 1963 Results. Univ. of Wyo. Coop. Res. Rep. to the BLM. Sections I - VIII. Wyo. Agr. Exp. Sta. Mimeo. p. 1-61. Section VIII - H.G. Fisser, R.W. Hamner, and J.T. Nichols. Cumberland district exclosures. 3 p.

3A3 650501 3 Fisser, H.G., J.R. Wight, J.R. Flesland, L.D. Robinson. 1965. Halogeton Research. 1964 Results. Univ. of Wyo. Coop. Res. Report to the BLM. Section I - VI. Wyo. Agr. Exp. Sta. Mimeo. p. 1-82. Section III - H.G. Fisser, J.R. Wight, J. R. Flesland, and L.D. Robinson. Exclosure studies (production, cover, height and precipitation). 49 p.

3A3 660501 3 Fisser, H.G., L.D. Robinson, D.R. Kirkam, R. Steger, and G. Noller. 1966. Halogeton Research. 1965 Results. Univ. of Wyo. Coop. Res. Rep. to the BLM. Sections I - V. Wyo. Agr. Exp. Sta. Mimeo. 134 p. Section III - Exclosure studies (production, cover, height and precipitation). 105 p.

3A3 670502 4 Fisser, H.G., G. Noller, R.E. Steger, D.R. Kirkham and T. King. 1967. Aridland Ecology Research. 1966 Results. Univ. of Wyo. Coop. Res. Rep. to the BLM. Section I - IV. Wyo. Agr. Exp. Sta. Sci. Rep. 41. 137 p. Section IV - Exclosure studies (production, cover and precipitation).

3A3 680504 3 Fisser, H.G. and G.L. Noller. 1968. Aridland Ecology. 1967 results. Univ. of Wyo. Coop. Res. Rep. to the BLM. Sections I, II, and III, Wyo. Agr. Exp. Sta. Sci. Rep. 132. 93 p. Section III - Exclosure studies (production, cover and precipitation).

3A3 690501 4 Fisser, H.G. and G.L. Whysong. 1969. Aridland Ecology Research. 1968 results. Univ. of Wyo. Coop. Res. Rep. to the BLM. Sections I - IV. Wyo. Agr. Exp. Sta. Sci. Rep. 178. 114 p. Section IV - Exclosure studies (production, cover and precipitation).

3A3 700501 4 Fisser, H.G. and G.L. Whysong. 1970. Aridland Ecology Research. 1969 results. Univ. of Wyo. Coop. Res. Rep. to the BLM. Sections I - IV. Wyo. Agr. Exp. Sta. Sci. Rep. 240. 128 p. Section IV - Exclosure studies (production, cover and precipitation).

3A3 700506 Fisser, H.G. 1970. Aridland Ecology, semi-annual progress report to the Bureau of Land Management, March 1970. Wyo. Agr. Ext. Sta. Mimeo. 1 p.

3A3 710502 4 Fisser, H.G., and R.E. Ries. 1971. Aridland Ecology Research. 1970 results. Univ. of Wyo. Coop. Res. Rep. to the BLM. Sections I - IV. Wyo. Agr. Exp. Sta. Sci. Rep. 350. 133 p. Section IV - exclosure studies (production, cover, and precipitation).

3A3 710505 Fisser, H.G. and R.E. Ries. 1971. Semi-annual progress report to the Bureau of Land Management, January 1971. Wyo. Agr. Exp. Sta. Mimeo. 1 p.

3A3 720502 2 Fisser, H.G., and R.E. Ries. 1972. Aridland Ecology Research. 1971 results. Univ. of Wyo. Coop. Res. Rep. to the BLM. Sections I and II. Wyo. Agr. Exp. Sta. Sci. Rep. 428. 130 p. Section II - Exclosure studies (production, cover and precipitation).

3A3 730501 2 Fisser, H.G., and R.E. Ries. 1973. Aridland Ecology Research. 1972 results. Univ. of Wyo. Coop. Res. Rep. to the BLM. Sections I and II. Wyo. Agr. Exp. Sta. Sci. Rep. 521. 128 p. Section II - Exclosure studies (production, cover and precipitation).

3A3 740501 Fisser, H.G. and R.E. Ries. 1974. Summary of data, 1962-1970. Aridland Ecology Research. Univ. of Wyo. Coop. Res. Rep. to the BLM. Wyo. Agr. Exp. Sta. Sci. Rep. 556. 132 p.

3A3 740502 1 Fisser, H.G. and L.H. Kleinman. 1974. Phenology and production on semiarid shrub-types. 1973 results. Univ. of Wyo. Coop. Res. Rep. to the BLM. Sections I, II and III. Wyo. Agr. Exp. Sta. Sci. Rep. 607. 256 p. Section I - Phenology study.

3A3 740502 3 Fisser, H.G. and L.H. Kleinman. 1974. Phenology and production on semiarid shrub-types. 1973 results. Univ. of Wyo. Coop. Res. Rep. to the BLM. Sections I, II and III. Wyo. Agr. Exp. Sta. Sci. Rep. 607. 256 p. Section III - Exclosure studies (production, cover and precipitation).

3A3 740503 Shrader, J.A. and H.G. Fisser. 1874. Seed production ecology of the shortgrass prairie of the central United States. Wyo. Agr. Exp. Sta. Mimeo. 2 p.

3A3 750503 3 Fisser, H.G., and L.H. Kleinman. 1975. Phenology and production on semiarid shrub-types. 1974 results. Univ. Wyo. Coop. Res. Rep. to the BLM. Sections I, II, and III. Wyo. Agr. Exp. Sta. Sci. Rep. 729. 399 p. Section III - Exclosure studies (production, cover and precipitation).

3A3 760501 3 Fisser, H.G., and L.H. Kleinman. 1976. Phenology and production studies on semiarid shrub-types. 1975 results. Univ. of Wyo. Coop. Res. Rep. to the BLM. Sections I, II and III. Wyo. Agr. Exp. Sta. Sci. Rep. 831. 410 p. Section III - Exclosure studies (production, cover, precipitation).

3A3 770502 4 Fisser, H.G., L.W. Young, J.E. Orpet and N.E. Hargis. 1977. Phenology and production studies on semiarid shrub-types. 1976 results. Univ. of Wyo. Coop. Res. Rep. to the BLM. Sections I, II, III and IV. Wyo. Agr. Exp. Sta. Sci. Rep. 930. 438 p. Section IV - Exclosure studies (production, cover and precipitation).

3A3 780503 4 Fisser, H.G., L.W. Young, J.E. Orpet and N.E. Hargis. 1978. Phenology and production studies on semiarid shrub-types. 1977 results. Univ. of Wyo. Coop. Res. Rep. to the BLM. Sections I, II, III and IV. Wyo. Agr. Exp. Sta. Sci. Rep. 970. 439 p. Section IV - Exclosure studies (production, cover and precipitation).

3A3 790501 Fisser, H.G., D.C. Trueblood and D.D. Samuelson. 1979. Soil vegetation relationships on rangeland exclosures on the Grass Creek Planning Unit of north central Wyoming. Univ. of Wyo. Coop. Res. Rep. to the BLM. Wyo. Agr. Exp. Sta. Sci. Rep. 969. 279 p.

3A3 790502 4 Fisser, H.G., L.W. Young, N.E. Hargis and K.E. Spaeth. 1979. Phenology and production studies on semiarid shrub-types. 1978 results. Univ. of Wyo. Coop. Res. Rep. to the BLM. Sections I, II, III and IV. Wyo. Agr. Exp. Sta. Sci. Rep. 985. 489 p. Section IV - Exclosure studies (production, cover and precipitation).

3A3 800502 5 Fisser, H.G., M.K. Owens, J.W. Uhlich, W.J. Waugh, D.E. Rodgers, and N.E. Hargis. 1980. Phenology and production studies on semiarid shrub-types. 1979 Results. Univ. of Wyo. Coop. Res. Rep. to the BLM. Vol. II: Precipitation, Productivity and Production Section. Sections III, IV, and V. Wyo. Agr. Exp. Sta. Sci. Rep. 1106. 219 p. Section V - Exclosure (production, cover and precipitation).

3A3 820502 5 Fisser, H.G., S.K. Dean, T.L. Deboodt, R.J. Kinucan. 1982. Phenology and production studies on semiarid shrub-types. 1982 results. Univ. of Wyo. Coop. Res. Rep. to the BLM. Vol. II: Precipitation, productivity, and production sections III, IV, and V. Wyo. Agr. Exp. Sta. Sci. Rep. 1231. 256 p. Section V - Exclosure studies (production, cover and precipitation).

3A3 830502 5 Fisser, H.G., S.K. Dean, T.L. Deboodt, R.S. Kinucan and J.L. Mulholland. 1983. Phenology and production studies on semiarid shrub-types. 1982 results. Univ. of Wyo. Coop. Res. Rep. to the BLM. Vol. II: Precipitation, Productivity and Production, Sections III, IV, and V. Wyo. Agr. Exp. Sta. Sci. Rep. 1231. 256 p. Section V - Exclosure (production, cover and precipitation).

3A3 840501 Fisser, H.G. and L.P. Germain. 1984. Phenology and production studies on semiarid shrub-types. Statistical tests and computer programs. Wyo. Agr. Exp. Sta. Sci. Rep. 15 p.

3A3 840502 Fisser, H.G. 1984. Statistics and interpretation. Wyo. Agr. Exp. Sta. Sci. Rep. 5 p.

3B. NATURAL RESOURCES PLANNING

3B 790503 Fisser, H.G. 1979. Rangeland and wildlife resources of the South Pass historic mining area. Wyo. Agr. Exp. Sta. Mimeo. 8 p.

4. Grazing Management

4A. STOCKING RATES

4A 620504 Fisser, H.G., M. May, W.M. Johnson, P. Stratton, and D.R. Smith. 1962. Comparison of rotation and season-long summer grazing on subalpine range in Wyoming. Wyo. Agr. Exp. Sta. Mimeo Circ. 177. 3 p.

4A 630503 1 Fisser, H.G., J.T. Nichols, and R.W. Hamner. 1963. Halogeton Research. 1962 Results. Univ. of Wyo. Coop. Res. Rep. to the BLM. Sections I - III. Wyo. Agr. Exp. Sta. Mimeo. p. 1-38. Section I - H.G. Fisser and J.T. Nichols. Winter sheep grazing on saltbush range (Greybull). p. 1-6.

4A 640501 1 Fisser, H.G., J.T. Nichols, R.W. Hamner, and J.R. Wight. 1964. Halogeton Research. 1963 Results. Univ. of Wyo. Coop. Res. Rep. to the BLM. Sections I - VIII. Wyo. Agr. Exp. Sta. Mimeo. p. 1-61. Section I - H.G. Fisser and R.W. Hamner. Winter sheep grazing on saltbush range (Greybull). 8 p.

4A 650501 1 Fisser, H.G., J.R. Wight, J.R. Flesland, L.D. Robinson. 1965. Halogeton Research. 1964 Results. Univ. of Wyo. Coop. Res. Rep. to the BLM. Section I - VI. Wyo. Agr. Exp. Sta. Mimeo. p. 1-82. Section I - H.G. Fisser and J.R. Wight. Winter sheep grazing on saltbush range (Greybull). 13 p.

4A 660501 1 Fisser, H.G., L.D. Robinson, D.R. Kirkam, R. Steger, and G. Noller. 1966. Halogeton Research. 1965 Results. Univ. of Wyo. Coop. Res. Rep. to the BLM. Sections I - V. Wyo. Agr. Exp. Sta. Mimeo. 134 p. Section I - Winter sheep grazing on saltbush range (Greybull). 11 p.

4A 670501 Fisser, H.G. and R.E. Steger. 1967. Winter sheep grazing study on saltbush rangeland. 1966 Results. Univ. of Wyo. Coop. Res. Rep. to the BLM. Wyo. Agr. Exp. Sta. Sci. Rep. 36. 9 p.

4B. GRAZING SEASONS AND SYSTEMS

4B 620504 Fisser, H.G., M. May, W.M. Johnson, P. Stratton, and D.R. Smith. 1962. Comparison of rotation and season-long summer grazing on subalpine range in Wyoming. Wyo. Agr. Exp. Sta. Mimeo Circ. 177. 3 p.

4B 630503 1 Fisser, H.G., J.T. Nichols, and R.W. Hamner. 1963. Halogeton Research. 1962 Results. Univ. of Wyo. Coop. Res. Rep. to the BLM. Sections I - III. Wyo. Agr. Exp. Sta. Mimeo. p. 1-38. Section I - H.G. Fisser and J.T. Nichols. Winter sheep grazing on saltbush range (Greybull). p. 1-6.

4B 630503 2 Fisser, H.G., J.T. Nichols, and R.W. Hamner. 1963. Halogeton Research. 1962 Results. Univ. of Wyo. Coop. Res. Rep. to the BLM. Sections I - III. Wyo. Agr. Exp. Sta. Mimeo. p. 1-38. Section II - H.G. Fisser and J.T. Nichols. Summer steer grazing on saltbush range and seeded water spreaders (Worland). p. 7-11.

4B 640501 1 Fisser, H.G., J.T. Nichols, R.W. Hamner, and J.R. Wight. 1964. Halogeton Research. 1963 Results. Univ. of Wyo. Coop. Res. Rep. to the BLM. Sections I - VIII. Wyo. Agr. Exp. Sta. Mimeo. p. 1-61. Section I - H.G. Fisser and R.W. Hamner. Winter sheep grazing on saltbush range (Greybull). 8 p.

4B 640501 2 Fisser, H.G., J.T. Nichols, R.W. Hamner, and J.R. Wight. 1964. Halogeton Research. 1963 Results. Univ. of Wyo. Coop. Res. Rep. to the BLM. Sections I - VIII. Wyo. Agr. Exp. Sta. Mimeo. p. 1-61. Section II - H.G. Fisser and R.W. Hamner. Summer steer grazing on saltbush range and seeded water spreaders (Worland). 6 p.

4B 650501 1 Fisser, H.G., J.R. Wight, J.R. Flesland, L.D. Robinson. 1965. Halogeton Research. 1964 Results. Univ. of Wyo. Coop. Res. Rep. to the BLM. Section I - VI. Wyo. Agr. Exp. Sta. Mimeo. p. 1-82. Section I - H.G. Fisser and J.R. Wight. Winter sheep grazing on saltbush range (Greybull). 13 p.

4B 650501 2 Fisser, H.G., J.R. Wight, J.R. Flesland, L.D. Robinson. 1965. Halogeton Research. 1964 Results. Univ. of Wyo. Coop. Res. Rep. to the BLM. Section I - VI. Wyo. Agr. Exp. Sta. Mimeo. p. 1-82. Section II - H.G. Fisser and J.R. Wight. Spring steer grazing on saltbush range and seeded water spreaders (Worland). 7 p.

4B 660501 1 Fisser, H.G., L.D. Robinson, D.R. Kirkam, R. Steger, and G. Noller. 1966. Halogeton Research. 1965 Results. Univ. of Wyo. Coop. Res. Rep. to the BLM. Sections I - V. Wyo. Agr. Exp. Sta. Mimeo. 134 p. Section I - Winter sheep grazing on saltbush range (Greybull). 11 p.

4B 660501 2 Fisser, H.G., L.D. Robinson, D.R. Kirkam, R. Steger, and G. Noller. 1966. Halogeton Research. 1965 Results. Univ. of Wyo. Coop. Res. Rep. to the BLM. Sections I - V. Wyo. Agr. Exp. Sta. Mimeo. 134 p. Section II - Steer grazing studies on saltbush range and seeded water spreaders (Worland). 7 p.

4B 670501 Fisser, H.G. and R.E. Steger. 1967. Winter sheep grazing study on saltbush rangeland. 1966 Results. Univ. of Wyo. Coop. Res. Rep. to the BLM. Wyo. Agr. Exp. Sta. Sci. Rep. 36. 9 p.

4B 680501 Gibbens, R.P., H.G. Fisser, and M. May. 1968. Influence of grazing management systems on vegetation and wildlife habitat. Annual report of progress, July, 1967 - July, 1968. Univ. of Wyo. Coop. Res. Rep. to the BLM. Wyo. Agr. Exp. Sta. Sci. Rep. 102. 9 p.

4B 690502 Gibbens, R.P., H.G. Fisser and M. May. 1969. Influence of grazing management systems on vegetation and wildlife habitat. Annual report of progress, July, 1968 - July, 1969. Univ. of Wyo. Coop. Res. Rep. to the BLM. Wyo. Agr. Exp. Sta. Sci. Rep. 179. 132 p.

4B 700502 Gibbens, R.P. and H.G. Fisser. 1970. Influence of grazing management systems on vegetation and wildlife habitat. Annual report of progress, July, 1969 - July, 1970. Univ. of Wyo. Coop. Res. Rep. to the BLM. Wyo. Agr. Exp. Sta. Sci. Rep. 241. 32 p.

4B 700505 Gibbens, R.P., H.G. Fisser and M. May. 1970. Influence of grazing management systems on vegetation and wildlife habitat. Semiann. Rep. to the BLM. Wyo. Agr. Exp. Sta. Mimeo. 2 p.

4B 710501 Fisser, H.G. and R.P. Gibbens. 1971. Influence of grazing management systems on vegetation and wildlife habitat. Annual report of progress, July, 1970 - July, 1971. Univ. of Wyo. Coop. Res. Rep. to the BLM. Wyo. Agr. Exp. Sta. Sci. Rep. 336. 92 p.

4B 720501 Gibbens, R.P., and H.G. Fisser. 1972. Influences of grazing management systems on vegetation and wildlife habitat. Annual Progress Report. Wyo. Agr. Exp. Sta. Sci. Rep. 414. 69 p.

4B 750501 Gibbens, R.P., and H.G. Fisser. 1975. Influence of grazing management systems on vegetation and wildlife habitat in the Red Desert Region of Wyoming. Wyo. Agr. Exp. Sta. Sci. Mono. 29. 23 p.

4B 790503 Fisser, H.G. 1979. Rangeland and wildlife resources of the South Pass historic mining area. Wyo. Agr. Exp. Sta. Mimeo. 8 p.

4B 790504 Fisser, H.G. 1979. Grazing management system studies of northern desert shrub vegetation of southern Wyoming. Wyo. Agr. Exp. Sta. Mimeo. 3 p.

4B 800505 Fisser, H.G., D.E. Rodgers and W.E. Gerhart. 1980. Influences of grazing management systems on vegetation and wildlife habitat. 1975 and 1979 Results and Long-Term Summary. Univ. of Wyo. Coop. Res. Rep. to the BLM. Wyo. Agr. Exp. Sta. Sci. Rep. 1047. 188 p.

4B 820503 Fisser, H.G. 1982. Vegetation response to grazing management strategies on semiarid shrublands of Wyoming. Wyo. Agr. Exp. Sta. Mimeo. prepared for Grazing Systems Conference, Texas A & M University. November 1982. 4 p.

4C. GRAZING HABITS AND DISTRIBUTION

4C 680501 Gibbens, R.P., H.G. Fisser, and M. May. 1968. Influence of grazing management systems on vegetation and wildlife habitat. Annual report of progress, July, 1967 - July, 1968. Univ. of Wyo. Coop. Res. Rep. to the BLM. Wyo. Agr. Exp. Sta. Sci. Rep. 102. 9 p.

4C 690502 Gibbens, R.P., H.G. Fisser and M. May. 1969. Influence of grazing management systems on vegetation and wildlife habitat. Annual report of progress, July, 1968 - July, 1969. Univ. of Wyo. Coop. Res. Rep. to the BLM. Wyo. Agr. Exp. Sta. Sci. Rep. 179. 132 p.

4C 700502 Gibbens, R.P. and H.G. Fisser. 1970. Influence of grazing management systems on vegetation and wildlife habitat. Annual report of progress, July, 1969 - July, 1970. Univ. of Wyo. Coop. Res. Rep. to the BLM. Wyo. Agr. Exp. Sta. Sci. Rep. 241. 32 p.

4C 700505 Gibbens, R.P., H.G. Fisser and M. May. 1970. Influence of grazing management systems on vegetation and wildlife habitat. Semiann. Rep. to the BLM. Wyo. Agr. Exp. Sta. Mimeo. 2 p.

4C 710501 Fisser, H.G. and R.P. Gibbens. 1971. Influence of grazing management systems on vegetation and wildlife habitat. Annual report of progress, July, 1970 - July, 1971. Univ. of Wyo. Coop. Res. Rep. to the BLM. Wyo. Agr. Exp. Sta. Sci. Rep. 336. 92 p.

4C 720501 Gibbens, R.P., and H.G. Fisser. 1972. Influences of grazing management systems on vegetation and wildlife habitat. Annual Progress Report. Wyo. Agr. Exp. Sta. Sci. Rep. 414. 69 p.

4C 750501 Gibbens, R.P., and H.G. Fisser. 1975. Influence of grazing management systems on vegetation and wildlife habitat in the Red Desert Region of Wyoming. Wyo. Agr. Exp. Sta. Sci. Mono. 29. 23 p.

4C 790504 Fisser, H.G. 1979. Grazing management system studies of northern desert shrub vegetation of southern Wyoming. Wyo. Agr. Exp. Sta. Mimeo. 3 p.

4C 800505 Fisser, H.G., D.E. Rodgers and W.E. Gerhart. 1980. Influences of grazing management systems on vegetation and wildlife habitat. 1975 and 1979 Results and Long-Term Summary. Univ. of Wyo. Coop. Res. Rep. to the BLM. Wyo. Agr. Exp. Sta. Sci. Rep. 1047. 188 p.

4C 820503 Fisser, H.G. 1982. Vegetation response to grazing management strategies on semiarid shrublands of Wyoming. Wyo. Agr. Exp. Sta. Mimeo. prepared for Grazing Systems Conference, Texas A & M University. November 1982. 4 p.

5. Range and Ranch Development

5A. PLANT CONTROL

5A 740502 3 Fisser, H.G. and L.H. Kleinman. 1974. Phenology and production on semiarid shrub-types. 1973 results. Univ. of Wyo. Coop. Res. Rep. to the BLM. Sections I, II and III. Wyo. Agr. Exp. Sta. Sci. Rep. 607. 256 p. Section III - Exclosure studies (production, cover and precipitation).

5A1. Methods of Control

5A1B. BIOLOGICAL

5A1B660502 Lavigne, R.J. 1966. Summary of research during 1964 and 1965 for the control of the western harvester ants, *Pogonomyrmex owyheei* (Cole). Wyo. Agr. Exp. Sta. Mimeo. 12 p.

5A1C. MECHANICAL

5A1C 630503 2 Fisser, H.G., J.T. Nichols, and R.W. Hamner. 1963. Halogeton Research. 1962 Results. Univ. of Wyo. Coop. Res. Rep. to the BLM. Sections I - III. Wyo. Agr. Exp. Sta. Mimeo. p. 1-38. Section II - H.G. Fisser and J.T. Nichols. Summer steer grazing on saltbush range and seeded water spreaders (Worland). p. 7-11.

5A1C 640501 2 Fisser, H.G., J.T. Nichols, R.W. Hamner, and J.R. Wight. 1964. Halogeton Research. 1963 Results. Univ. of Wyo. Coop. Res. Rep. to the BLM. Sections I - VIII. Wyo. Agr. Exp. Sta. Mimeo. p. 1-61. Section II - H.G. Fisser and R.W. Hamner. Summer steer grazing on saltbush range and seeded water spreaders (Worland). 6 p.

5A1C 650501 2 Fisser, H.G., J.R. Wight, J.R. Flesland, L.D. Robinson. 1965. Halogeton Research. 1964 Results. Univ. of Wyo. Coop. Res. Rep. to the BLM. Section I - VI. Wyo. Agr. Exp. Sta. Mimeo. p. 1-82. Section II - H.G. Fisser and J.R. Wight. Spring steer grazing on saltbush range and seeded water spreaders (Worland). 7 p.

5A1C 660501 2 Fisser, H.G., L.D. Robinson, D.R. Kirkam, R. Steger, and G. Noller. 1966. Halogeton Research. 1965 Results. Univ. of Wyo. Coop. Res. Rep. to the BLM. Sections I - V. Wyo. Agr. Exp. Sta. Mimeo. 134 p. Section II - Steer grazing studies on saltbush range and seeded water spreaders (Worland). 7 p.

5A1D. CHEMICAL

5A1D630503 3 Fisser, H.G., J.T. Nichols, and R.W. Hamner. 1963. Halogeton Research. 1962 Results. Univ. of Wyo. Coop. Res. Rep. to the BLM. Sections I - III. Wyo. Agr. Exp. Sta. Mimeo. p. 1-38. Section III - H.G. Fisser and R.W. Hamner. Exclosure studies (production, cover, height, and precipitation). p. 12-38.

5A1D640501 3 Fisser, H.G., J.T. Nichols, R.W. Hamner, and J.R. Wight. 1964. Halogeton Research. 1963 Results. Univ. of Wyo. Coop. Res. Rep. to the BLM. Sections I - VIII. Wyo. Agr. Exp. Sta. Mimeo. p. 1-61. Section III - H.G. Fisser, J.T. Nichols, and J.R. Wight. Exclosure studies (production, cover, height and precipitation). 33 p.

5A1D640501 8 Fisser, H.G., J.T. Nichols, R.W. Hamner, and J.R. Wight. 1964. Halogeton Research. 1963 Results. Univ. of Wyo. Coop. Res. Rep. to the BLM. Sections I - VIII. Wyo. Agr. Exp. Sta. Mimeo. p. 1-61. Section VIII - H.G. Fisser, R.W. Hamner, and J.T. Nichols. Cumberland district exclosures. 3 p.

5A1D650501 3 Fisser, H.G., J.R. Wight, J.R. Flesland, L.D. Robinson. 1965. Halogeton Research. 1964 Results. Univ. of Wyo. Coop. Res. Rep. to the BLM. Section I - VI. Wyo. Agr. Exp. Sta. Mimeo. p. 1-82. Section III - H.G. Fisser, J.R. Wight, J.R. Flesland, and L.D. Robinson. Exclosure studies (production, cover, height and precipitation). 49 p.

5A1D660501 3 Fisser, H.G., L.D. Robinson, D.R. Kirkam, R. Steger, and G. Noller. 1966. Halogeton Research. 1965 Results. Univ. of Wyo. Coop. Res. Rep. to the BLM. Sections I - V. Wyo. Agr. Exp. Sta. Mimeo. 134 p. Section III - Exclosure studies (production, cover, height and precipitation). 105 p.

5A1D670502 4 Fisser, H.G., G. Noller, R.E. Steger, D.R. Kirkham and T. King. 1967. Aridland Ecology Research. 1966 Results. Univ. of Wyo. Coop. Res. Rep. to the BLM. Section I - IV. Wyo. Agr. Exp. Sta. Sci. Rep. 41. 137 p. Section IV - Exclosure studies (production, cover and precipitation).

5A1D680504 3 Fisser, H.G. and G.L. Noller. 1968. Aridland Ecology. 1967 results. Univ. of Wyo. Coop. Res. Rep. to the BLM. Sections I, II, and III, Wyo. Agr. Exp. Sta. Sci. Rep. 132. 93 p. Section III - Exclosure studies (production, cover and precipitation).

5A1D690501 4 Fisser, H.G. and G.L. Whysong. 1969. Aridland Ecology Research. 1968 results. Univ. of Wyo. Coop. Res. Rep. to the BLM. Sections I - IV. Wyo. Agr. Exp. Sta. Sci. Rep. 178. 114 p. Section IV - Exclosure studies (production, cover and precipitation).

5A1D700501 4 Fisser, H.G. and G.L. Whysong. 1970. Aridland Ecology Research. 1969 results. Univ. of Wyo. Coop. Res. Rep. to the BLM. Sections I - IV. Wyo. Agr. Exp. Sta. Sci. Rep. 240. 128 p. Section IV - Exclosure studies (production, cover and precipitation).

5A1D710502 4 Fisser, H.G., and R.E. Ries. 1971. Aridland Ecology Research. 1970 results. Univ. of Wyo. Coop. Res. Rep. to the BLM. Sections I - IV. Wyo. Agr. Exp. Sta. Sci. Rep. 350. 133 p. Section IV - exclosure studies (production, cover and precipitation).

5A1D710505 Fisser, H.G. and R.E. Ries. 1971. Semi-annual progress report to the Bureau of Land Management, January 1971. Wyo. Agr. Exp. Sta. Mimeo. 1 p.

5A1D720502 2 Fisser, H.G., and R.E. Ries. 1972. Aridland Ecology Research. 1971 results. Univ. of Wyo. Coop. Res. Rep. to the BLM. Sections I and II. Wyo. Agr. Exp. Sta. Sci. Rep. 428. 130 p. Section II - Exclosure studies (production, cover and precipitation).

5A1D730502 Shaver, J.C. and H.G. Fisser. 1973. Evaluation of vegetation types differentiated by aerial photo interpretation of panchromatic, color, and infrared imagery. U.S. I.B.P. Biome Technical Report 238. 86 p.

5A1D740501 Fisser, H.G. and R.E. Ries. 1974. Summary of data, 1962-1970. Aridland Ecology Research. Univ. of Wyo. Coop. Res. Rep. to the BLM. Wyo. Agr. Exp. Sta. Sci. Rep. 556. 132 p.

5A1D740502 1 Fisser, H.G. and L.H. Kleinman. 1974. Phenology and production on semiarid shrub-types. 1973 results. University of Wyo. Coop. Res. Rep. to the BLM. Sections I, II and III. Wyo. Agr. Exp. Sta. Sci. Rep. 607. 256 p. Section I - Phenology study.

5A1D740502 3 Fisser, H.G. and L.H. Kleinman. 1974. Phenology and production on semiarid shrub-types. 1973 results. Univ. of Wyo. Coop. Res. Rep. to the BLM. Sections I, II and III. Wyo. Agr. Exp. Sta. Sci. Rep. 607. 256 p. Section III - Exclosure studies (production, cover and precipitation).

5A1D750503 3 Fisser, H.G., and L.H. Kleinman. 1975. Phenology and production on semiarid shrub-types. 1974 results. Univ. Wyo. Coop. Res. Rep. to the BLM. Sections I, II, and III. Wyo. Agr. Exp. Sta. Sci. Rep. 729. 399 p. Section III - Exclosure studies (production, cover and precipitation).

5A1D760501 3 Fisser, H.G., and L.H. Kleinman. 1976. Phenology and production studies on semiarid shrub-types. 1975 results. Univ. of Wyo. Coop. Res. Rep. to the BLM. Sections I, II and III. Wyo. Agr. Exp. Sta. Sci. Rep. 831. 410 p. Section III - Exclosure studies (production, cover and precipitation).

5A1D770502 1 Fisser, H.G., L.W. Young, J.E. Orpet and N.E. Hargis. 1977. Phenology and production studies on semiarid shrub-types. 1976 results. Univ. of Wyo. Coop. Res. Rep. to the BLM. Sections I, II, III and IV. Wyo. Agr. Exp. Sta. Sci. Rep. 930. 438 p. Section I - Phenology study.

5A1D770502 2 Fisser, H.G., L.W. Young, J.E. Orpet and N.E. Hargis. 1977. Phenology and production studies on semiarid shrub-types. 1976 results. Univ. of Wyo. Coop. Res. Rep. to the BLM. Sections I, II, III and IV. Wyo. Agr. Exp. Sta. Sci. Rep. 930. 438 p. Section II - Phenology step-wise regression for predictive modeling.

5A1D770502 4 Fisser, H.G., L.W. Young, J.E. Orpet and N.E. Hargis. 1977. Phenology and production studies on semiarid shrub-types. 1976 results. Univ. of Wyo. Coop. Res. Rep. to the BLM. Sections I, II, III and IV. Wyo. Agr. Exp. Sta. Sci. Rep. 930. 438 p. Section IV - Exclosure studies (production, cover and precipitation).

5A1D780503 1 Fisser, H.G., L.W. Young, J.E. Orpet and N.E. Hargis. 1978. Phenology and production studies on semiarid shrub-types. 1977 results. Univ. of Wyo. Coop. Res. Rep. to the BLM. Sections I, II, III and IV. Wyo. Agr. Exp. Sta. Sci. Rep. 970. 439 p. Section I - Phenology study.

5A1D780503 4 Fisser, H.G., L.W. Young, J.E. Orpet and N.E. Hargis. 1978. Phenology and production studies on semiarid shrub-types. 1977 results. Univ. of Wyo. Coop. Res. Rep. to the BLM. Sections I, II, III and IV. Wyo. Agr. Exp. Sta. Sci. Rep. 970. 439 p. Section IV - Exclosure studies (production, cover and precipitation).

5A1D790501 Fisser, H.G., D.C. Trueblood and D.D. Samuelson. 1979. Soil vegetation relationships on rangeland exclosures on the Grass Creek Planning Unit of north central Wyoming. Univ. of Wyo. Coop. Res. Rep. to the BLM. Wyo. Agr. Exp. Sta. Sci. Rep. 969. 279 p.

5A1D790502 1 Fisser, H.G., L.W. Young, N.E. Hargis and K.E. Spaeth. 1979. Phenology and production studies on semiarid shrub-types. 1978 results. Univ. of Wyo. Coop. Res. Rep. to the BLM. Sections I, II, III and IV. Wyo. Agr. Exp. Sta. Sci. Rep. 985. 489 p. Section I - Phenology study.

5A1D790502 4 Fisser, H.G., L.W. Young, N.E. Hargis and K.E. Spaeth. 1979. Phenology and production studies on semiarid shrub-types. 1978 results. Univ. of Wyo. Coop. Res. Rep. to the BLM. Sections I, II, III and IV. Wyo. Agr. Exp. Sta. Sci. Rep. 985. 489 p. Section IV - Exclosure studies (production, cover and precipitation).

5A1D800501 1 Fisser, H.G., M.K. Owens, W.J. Waugh, N.E. Hargis and J. Uhlich. 1980. Phenology and production studies on semiarid shrub-types. 1979 Results. Univ. of Wyo. Coop. Res. Rep. to the BLM. Vol. I: Phenology Study. Sections I and II. Wyo. Agr. Exp. Sta. Sci. Rep. 1044. 387 p. Section I - Phenology study.

5A1D800502 4 Fisser, H.G., M.K. Owens, J.W. Uhlich, W.J. Waugh, D.E. Rodgers, and N.E. Hargis. 1980. Phenology and production studies on semiarid shrub-types. 1979 Results. Univ. of Wyo. Coop. Res. Rep. to the BLM. Vol. II: Precipitation, Productivity and Production Section. Sections III, IV, and V. Wyo. Agr. Exp. Sta. Sci. Rep. 1106. 219 p. Section IV - Productivity-phenology relationships.

5A1D800502 5 Fisser, H.G., M.K. Owens, J.W. Uhlich, W.J. Waugh, D.E. Rodgers, and N.E. Hargis. 1980. Phenology and production studies on semiarid shrub-types. 1979 Results. Univ. of Wyo. Coop. Res. Rep. to the BLM. Vol. II: Precipitation, Productivity and Production Section. Sections III, IV, and V. Wyo. Agr. Exp. Sta. Sci. Rep. 1106. 219 p. Section V - Exclosure studies (production, cover and precipitation).

5A1D820502 5 Fisser, H.G., S.K. Dean, T.L. Deboodt, R.J. Kinucan. 1982. Phenology and production studies on semiarid shrub-types. 1982 results. Univ. of Wyo. Coop. Res. Rep. to the BLM. Vol. II: Precipitation, productivity, and production sections III, IV, and V. Wyo. Agr. Exp. Sta. Sci. Rep. 1231. 256 p. Section V - Exclosure studies (production, cover and precipitation).

5A1D830502 4 Fisser, H.G., S.K. Dean, T.L. Deboodt, R.S. Kinucan and J.L. Mulholland. 1983. Phenology and production studies on semiarid shrub-types. 1982 results. Univ. of Wyo. Coop. Res. Rep. to the BLM. Vol. II: Precipitation, Productivity and Production, Sections III, IV, and V. Wyo. Agr. Exp. Sta. Sci. Rep. 1231. 256 p. Section IV - Productivity-phenology interrelation study.

5A1D830502 5 Fisser, H.G., S.K. Dean, T.L. Deboodt, R.S. Kinucan and J.L. Mulholland. 1983. Phenology and production studies on semiarid shrub-types. 1982 results. Univ. of Wyo. Coop. Res. Rep. to the BLM. Vol. II: Precipitation, Productivity and Production, Sections III, IV, and V. Wyo. Agr. Exp. Sta. Sci. Rep. 1231. 256 p. Section V - Exclosure studies (production, cover and precipitation).

5A1D830503 Fisser, H.G. 1983. Aridland shrub ecology and phenology research program. Wyo. Agr. Exp. Sta. Mimeo. 3 p.

5A2. Regional Control

5A2D. INTERMOUNTAIN

5A2D640501 6 Fisser, H.G., J.T. Nichols, R.W. Hamner, and J.R. Wight. 1964. Halogeton Research. 1963 Results. Univ. of Wyo. Coop. Res. Rep. to the BLM. Sections I - VIII. Wyo. Agr. Exp. Sta. Mimeo. p. 1-61. Section VI - H.G. Fisser, R.W. Hamner. Germination and longevity of halogeton seeds. 1 p.

5B. RANGE REVEGETATION

5B 630503 3 Fisser, H.G., J.T. Nichols, and R.W. Hamner. 1963. Halogeton Research. 1962 Results. Univ. of Wyo. Coop. Res. Rep. to the BLM. Sections I - III. Wyo. Agr. Exp. Sta. Mimeo. p. 1-38. Section III - H.G. Fisser and R.W. Hamner. Exclosure studies (production, cover, height, and precipitation). p. 12-38.

5B 640501 3 Fisser, H.G., J.T. Nichols, R.W. Hamner, and J.R. Wight. 1964. Halogeton Research. 1963 Results. Univ. of Wyo. Coop. Res. Rep. to the BLM. Sections I - VIII. Wyo. Agr. Exp. Sta. Mimeo. p. 1-61. Section III - H.G. Fisser, J.T. Nichols, and J.R. Wight. Exclosure studies (production, cover, height and precipitation). 33 p.

5B 650501 3 Fisser, H.G., J.R. Wight, J.R. Flesland, L.D. Robinson. 1965. Halogeton Research. 1964 Results. Univ. of Wyo. Coop. Res. Rep. to the BLM. Section I - VI. Wyo. Agr. Exp. Sta. Mimeo. p. 1-82. Section III - H.G. Fisser, J.R. Wight, J. R. Flesland, and L.D. Robinson. Exclosure studies (production, cover, height and precipitation). 49 p.

5B 660501 3 Fisser, H.G., L.D. Robinson, D.R. Kirkam, R. Steger, and G. Noller. 1966. Halogeton Research. 1965 Results. Univ. of Wyo. Coop. Res. Rep. to the BLM. Sections I - V. Wyo. Agr. Exp. Sta. Mimeo. 134 p. Section III - Exclosure studies (production, cover, height and precipitation). 105 p.

5B 670502 4 Fisser, H.G., G. Noller, R.E. Steger, D.R. Kirkham and T. King. 1967. Aridland Ecology Research. 1966 Results. Univ. of Wyo. Coop. Res. Rep. to the BLM. Section I - IV. Wyo. Agr. Exp. Sta. Sci. Rep. 41. 137 p. Section IV - Exclosure studies (production, cover and precipitation).

5B 680504 3 Fisser, H.G. and G.L. Noller. 1968. Aridland Ecology. 1967 results. Univ. of Wyo. Coop. Res. Rep. to the BLM. Sections I, II, and III, Wyo. Agr. Exp. Sta. Sci. Rep. 132. 93 p. Section III - Exclosure studies (production, cover and precipitation).

5B 690501 4 Fisser, H.G. and G.L. Whysong. 1969. Aridland Ecology Research. 1968 results. Univ. of Wyo. Coop. Res. Rep. to the BLM. Sections I - IV. Wyo. Agr. Exp. Sta. Sci. Rep. 178. 114 p. Section IV - Exclosure studies (production, cover and precipitation).

5B 700501 4 Fisser, H.G. and G.L. Whysong. 1970. Aridland Ecology Research. 1969 results. Univ. of Wyo. Coop. Res. Rep. to the BLM. Sections I - IV. Wyo. Agr. Exp. Sta. Sci. Rep. 240. 128 p. Section IV - Exclosure studies (production, cover and precipitation).

5B 710502 4 Fisser, H.G., and R.E. Ries. 1971. Aridland Ecology Research. 1970 results. Univ. of Wyo. Coop. Res. Rep. to the BLM. Sections I - IV. Wyo. Agr. Exp. Sta. Sci. Rep. 350. 133 p. Section IV - Exclosure studies (production, cover and precipitation).

5B 710505 Fisser, H.G. and R.E. Ries. 1971. Semi-annual progress report to the Bureau of Land Management, January 1971. Wyo. Agr. Exp. Sta. Mimeo. 1 p.

5B 720502 2 Fisser, H.G., and R.E. Ries. 1972. Aridland Ecology Research. 1971 results. Univ. of Wyo. Coop. Res. Rep. to the BLM. Sections I and II. Wyo. Agr. Exp. Sta. Sci. Rep. 428. 130 p. Section II - Exclosure studies (production, cover and precipitation).

5B 730502 Shaver, J.C. and H.G. Fisser. 1973. Evaluation of vegetation types differentiated by aerial photo interpretation of panchromatic, color, and infrared imagery. U.S. I.B.P. Biome Technical Report 238. 86 p.

5B 740501 Fisser, H.G. and R.E. Ries. 1974. Summary of data, 1962-1970. Aridland Ecology Research. Univ. of Wyo. Coop. Res. Rep. to the BLM. Wyo. Agr. Exp. Sta. Sci. Rep. 556. 132 p.

5B 740502 3 Fisser, H.G. and L.H. Kleinman. 1974. Phenology and production on semiarid shrub-types. 1973 results. Univ. of Wyo. Coop. Res. Rep. to the BLM. Sections I, II and III. Wyo. Agr. Exp. Sta. Sci. Rep. 607. 256 p. Section III - Exclosure studies (production, cover and precipitation).

5B 750503 3 Fisser, H.G., and L.H. Kleinman. 1975. Phenology and production on semiarid shrub-types. 1974 results. Univ. Wyo. Coop. Res. Rep. to the BLM. Sections I, II, and III. Wyo. Agr. Exp. Sta. Sci. Rep. 729. 399 p. Section III - Exclosure studies (production, cover and precipitation).

5B 760501 3 Fisser, H.G., and L.H. Kleinman. 1976. Phenology and production studies on semiarid shrub-types. 1975 results. Univ. of Wyo. Coop. Res. Rep. to the BLM. Sections I, II and III. Wyo. Agr. Exp. Sta. Sci. Rep. 831. 410 p. Section III - Exclosure studies (production, cover, precipitation).

5B 770502 4 Fisser, H.G., L.W. Young, J.E. Orpet and N.E. Hargis. 1977. Phenology and production studies on semiarid shrub-types. 1976 results. Univ. of Wyo. Coop. Res. Rep. to the BLM. Sections I, II, III and IV. Wyo. Agr. Exp. Sta. Sci. Rep. 930. 438 p. Section IV - Exclosure studies (production, cover and precipitation).

5B 780503 4 Fisser, H.G., L.W. Young, J.E. Orpet and N.E. Hargis. 1978. Phenology and production studies on semiarid shrub-types. 1977 results. Univ. of Wyo. Coop. Res. Rep. to the BLM. Sections I, II, III and IV. Wyo. Agr. Exp. Sta. Sci. Rep. 970. 439 p. Section IV - Exclosure studies (production, cover and precipitation).

5B 790502 4 Fisser, H.G., L.W. Young, N.E. Hargis and K.E. Spaeth. 1979. Phenology and production studies on semiarid shrub-types. 1978 results. Univ. of Wyo. Coop. Res. Rep. to the BLM. Sections I, II, III and IV. Wyo. Agr. Exp. Sta. Sci. Rep. 985. 489 p. Section IV - Exclosure studies (production, cover and precipitation).

5B 800502 5 Fisser, H.G., M.K. Owens, J.W. Uhlich, W.J. Waugh, D.E. Rodgers, and N.E. Hargis. 1980. Phenology and production studies on semiarid shrub-types. 1979 Results. Univ. of Wyo. Coop. Res. Rep. to the BLM. Vol. II: Precipitation, Productivity and Production Section. Sections III, IV, and V. Wyo. Agr. Exp. Sta. Sci. Rep. 1106. 219 p. Section V - Exclosure studies (production, cover and precipitation).

5B 820502 5 Fisser, H.G., S.K. Dean, T.L. Deboodt, R.J. Kinucan. 1982. Phenology and production studies on semiarid shrub-types. 1982 results. Univ. of Wyo. Coop. Res. Rep. to the BLM. Vol. II: Precipitation, productivity, and production sections III, IV, and V. Wyo. Agr. Exp. Sta. Sci. Rep. 1231. 256 p. Section V - Exclosure studies (production, cover and precipitation).

5B 830502 5 Fisser, H.G., S.K. Dean, T.L. Deboodt, R.S. Kinucan and J.L. Mulholland. 1983. Phenology and production studies on semiarid shrub-types. 1982 results. Univ. of Wyo. Coop. Res. Rep. to the BLM. Vol. II: Precipitation, Productivity and Production, Sections III, IV, and V. Wyo. Agr. Exp. Sta. Sci. Rep. 1231. 256 p. Section V - Exclosure studies (production, cover and precipitation).

5B4. Intermountain

5B4 630503 2 Fisser, H.G., J.T. Nichols, and R.W. Hamner. 1963. Halogeton Research. 1962 Results. Univ. of Wyo. Coop. Res. Rep. to the BLM. Sections I - III. Wyo. Agr. Exp. Sta. Mimeo. p. 1-38. Section II - H.G. Fisser and J.T. Nichols. Summer steer grazing on saltbush range and seeded water spreaders (Worland). p. 7-11.

5B4 640501 2 Fisser, H.G., J.T. Nichols, R.W. Hamner, and J.R. Wight. 1964. Halogeton Research. 1963 Results. Univ. of Wyo. Coop. Res. Rep. to the BLM. Sections I - VIII. Wyo. Agr. Exp. Sta. Mimeo. p. 1-61. Section II - H.G. Fisser and R.W. Hamner. Summer steer grazing on saltbush range and seeded water spreaders (Worland). 6 p.

5B4 650501 2 Fisser, H.G., J.R. Wight, J.R. Flesland, L.D. Robinson. 1965. Halogeton Research. 1964 Results. Univ. of Wyo. Coop. Res. Rep. to the BLM. Section I - VI. Wyo. Agr. Exp. Sta. Mimeo. p. 1-82. Section II - H.G. Fisser and J.R. Wight. Spring steer grazing on saltbush range and seeded water spreaders (Worland). 7 p.

5B4 650501 6 Fisser, H.G., J.R. Wight, J.R. Flesland, L.D. Robinson. 1965. Halogeton Research. 1964 Results. Univ. of Wyo. Coop. Res. Rep. to the BLM. Section I - VI. Wyo. Agr. Exp. Sta. Mimeo. p. 1-82. Section VI - H.G. Fisser. Seeding and fertilizer trials in saltbush range. 2 p.

5B4 660501 2 Fisser, H.G., L.D. Robinson, D.R. Kirkam, R. Steger, and G. Noller. 1966. Halogeton Research. 1965 Results. Univ. of Wyo. Coop. Res. Rep. to the BLM. Sections I - V. Wyo. Agr. Exp. Sta. Mimeo. 134 p. Section II - Steer grazing studies on saltbush range and seeded water spreaders (Worland). 7 p.

5B4 740503 Shrader, J.A. and H.G. Fisser. 1874. Seed production ecology of the shortgrass prairie of the central United States. Wyo. Agr. Exp. Sta. Mimeo. 2 p.

5B4 790501 Fisser, H.G., D.C. Trueblood and D.D. Samuelson. 1979. Soil vegetation relationships on rangeland exclosures on the Grass Creek Planning Unit of north central Wyoming. Univ. of Wyo. Coop. Res. Rep. to the BLM. Wyo. Agr. Exp. Sta. Sci. Rep. 969. 279 p.

5C. WATER SPREADING, FURROWING, PITTING

5C 630503 2 Fisser, H.G., J.T. Nichols, and R.W. Hamner. 1963. Halogeton Research. 1962 Results. Univ. of Wyo. Coop. Res. Rep. to the BLM. Sections I - III. Wyo. Agr. Exp. Sta. Mimeo. p. 1-38. Section II - H.G. Fisser and J.T. Nichols. Summer steer grazing on saltbush range and seeded water spreaders (Worland). p. 7-11.

5C 630503 3 Fisser, H.G., J.T. Nichols, and R.W. Hamner. 1963. Halogeton Research. 1962 Results. Univ. of Wyo. Coop. Res. Rep. to the BLM. Sections I - III. Wyo. Agr. Exp. Sta. Mimeo. p. 1-38. Section III - H.G. Fisser and R.W. Hamner. Exclosure studies (production, cover, height, and precipitation). p. 12-38.

5C 640501 2 Fisser, H.G., J.T. Nichols, R.W. Hamner, and J.R. Wight. 1964. Halogeton Research. 1963 Results. Univ. of Wyo. Coop. Res. Rep. to the BLM. Sections I - VIII. Wyo. Agr. Exp. Sta. Mimeo. p. 1-61. Section II - H.G. Fisser and R.W. Hamner. Summer steer grazing on saltbush range and seeded water spreaders (Worland). 6 p.

5C 640501 3 Fisser, H.G., J.T. Nichols, R.W. Hamner, and J.R. Wight. 1964. Halogeton Research. 1963 Results. Univ. of Wyo. Coop. Res. Rep. to the BLM. Sections I - VIII. Wyo. Agr. Exp. Sta. Mimeo. p. 1-61. Section III - H.G. Fisser, J.T. Nichols, and J.R. Wight. Exclosure studies (production, cover, height and precipitation). 33 p.

5C 650501 2 Fisser, H.G., J.R. Wight, J.R. Flesland, L.D. Robinson. 1965. Halogeton Research. 1964 Results. Univ. of Wyo. Coop. Res. Rep. to the BLM. Section I - VI. Wyo. Agr. Exp. Sta. Mimeo. p. 1-82. Section II - H.G. Fisser and J.R. Wight. Spring steer grazing on saltbush range and seeded water spreaders (Worland). 7 p.

5C 650501 3 Fisser, H.G., J.R. Wight, J.R. Flesland, L.D. Robinson. 1965. Halogeton Research. 1964 Results. Univ. of Wyo. Coop. Res. Rep. to the BLM. Section I - VI. Wyo. Agr. Exp. Sta. Mimeo. p. 1-82. Section III - H.G. Fisser, J.R. Wight, J. R. Flesland, and L.D. Robinson. Exclosure studies (production, cover, height and precipitation). 49 p.

5C 650501 6 Fisser, H.G., J.R. Wight, J.R. Flesland, L.D. Robinson. 1965. Halogeton Research. 1964 Results. Univ. of Wyo. Coop. Res. Rep. to the BLM. Section I - VI. Wyo. Agr. Exp. Sta. Mimeo. p. 1-82. Section VI - H.G. Fisser. Seeding and fertilizer trials in saltbush range. 2 p.

5C 660501 2 Fisser, H.G., L.D. Robinson, D.R. Kirkam, R. Steger, and G. Noller. 1966. Halogeton Research. 1965 Results. Univ. of Wyo. Coop. Res. Rep. to the BLM. Sections I - V. Wyo. Agr. Exp. Sta. Mimeo. 134 p. Section II - Steer grazing studies on saltbush range and seeded water spreaders (Worland). 7 p.

5C 660501 3 Fisser, H.G., L.D. Robinson, D.R. Kirkam, R. Steger, and G. Noller. 1966. Halogeton Research. 1965 Results. Univ. of Wyo. Coop. Res. Rep. to the BLM. Sections I - V. Wyo. Agr. Exp. Sta. Mimeo. 134 p. Section III - Exclosure studies (production, cover, height and precipitation). 105 p.

5C 670502 4 Fisser, H.G., G. Noller, R.E. Steger, D.R. Kirkham and T. King. 1967. Aridland Ecology Research. 1966 Results. Univ. of Wyo. Coop. Res. Rep. to the BLM. Section I - IV. Wyo. Agr. Exp. Sta. Sci. Rep. 41. 137 p. Section IV - Exclosure studies (production, cover and precipitation).

5C 670503 Fisser, H.G. 1967. Exclosure study with transects of permanent plots. Univ. of Wyo. Coop. Res. Rep. to the BLM. Wyo. Agr. Exp. Sta. Sci. Rep. 69. 94 p.

5C 680502 Fisser, H.G. 1968. Exclosure studies with transects of permanent plots. 1967 results. Univ. of Wyo. Coop. Res. Rep. to the BLM. Wyo. Agr. Exp. Sta. Sci. Rep. 118. 52 p.

5C 680504 3 Fisser, H.G. and G.L. Noller. 1968. Aridland Ecology. 1967 results. Univ. of Wyo. Coop. Res. Rep. to the BLM. Sections I, II, and III, Wyo. Agr. Exp. Sta. Sci. Rep. 132. 93 p. Section III - Exclosure studies (production, cover and precipitation).

5C 690501 4 Fisser, H.G. and G.L. Whysong. 1969. Aridland Ecology Research. 1968 results. Univ. of Wyo. Coop. Res. Rep. to the BLM. Sections I - IV. Wyo. Agr. Exp. Sta. Sci. Rep. 178. 114 p. Section IV - Exclosure studies (production, cover and precipitation).

5C 700501 4 Fisser, H.G. and G.L. Whysong. 1970. Aridland Ecology Research. 1969 results. Univ. of Wyo. Coop. Res. Rep. to the BLM. Sections I - IV. Wyo. Agr. Exp. Sta. Sci. Rep. 240. 128 p. Section IV - Exclosure studies (production, cover and precipitation).

5C 700504 Fisser, H.G. 1970. Exclosure studies with transects of permanent plots. 1969 results. Univ. of Wyo. Coop. Res. Rep. to the BLM. Wyo. Agr. Exp. Sta. Sci. Rep. 275. 84 p.

5C 710502 4 Fisser, H.G., and R.E. Ries. 1971. Aridland Ecology Research. 1970 results. Univ. of Wyo. Coop. Res. Rep. to the BLM. Sections I - IV. Wyo. Agr. Exp. Sta. Sci. Rep. 350. 133 p. Section IV - Exclosure studies (production, cover, and precipitation).

5C 710505 Fisser, H.G. and R.E. Ries. 1971. Semi-annual progress report to the Bureau of Land Management, January 1971. Wyo. Agr. Exp. Sta. Mimeo. 1 p.

5C 720502 2 Fisser, H.G., and R.E. Ries. 1972. Aridland Ecology Research. 1971 results. Univ. of Wyo. Coop. Res. Rep. to the BLM. Sections I and II. Wyo. Agr. Exp. Sta. Sci. Rep. 428. 130 p. Section II - Exclosure studies (production, cover and precipitation phases).

5C 730502 Shaver, J.C. and H.G. Fisser. 1973. Evaluation of vegetation types differentiated by aerial photo interpretation of panchromatic, color, and infrared imagery. U.S. I.B.P. Biome Technical Report 238. 86 p.

5C 740501 Fisser, H.G. and R.E. Ries. 1974. Summary of data, 1962-1970. Aridland Ecology Research. Univ. of Wyo. Coop. Res. Rep. to the BLM. Wyo. Agr. Exp. Sta. Sci. Rep. 556. 132 p.

5C 740502 3 Fisser, H.G. and L.H. Kleinman. 1974. Phenology and production on semiarid shrub-types. 1973 results. University of Wyo. Coop. Res. Rep. to the BLM. Sections I, II and III. Wyo. Agr. Exp. Sta. Sci. Rep. 607. 256 p. Section III - Exclosure studies (production, cover and precipitation phase).

5C 750503 3 Fisser, H.G., and L.H. Kleinman. 1975. Phenology and production on semiarid shrub-types. 1974 results. Univ. Wyo. Coop. Res. Rep. to the BLM. Sections I, II, and III. Wyo. Agr. Exp. Sta. Sci. Rep. 729. 399 p. Section III - Exclosure studies (production, cover and precipitation).

5C 760501 3 Fisser, H.G., and L.H. Kleinman. 1976. Phenology and production studies on semiarid shrub-types. 1975 results. Univ. of Wyo. Coop. Res. Rep. to the BLM. Sections I, II and III. Wyo. Agr. Exp. Sta. Sci. Rep. 831. 410 p. Section III - Exclosure studies (production, cover, precipitation).

5C 770502 4 Fisser, H.G., L.W. Young, J.E. Orpet and N.E. Hargis. 1977. Phenology and production studies on semiarid shrub-types. 1976 results. Univ. of Wyo. Coop. Res. Rep. to the BLM. Sections I, II, III and IV. Wyo. Agr. Exp. Sta. Sci. Rep. 930. 438 p. Section IV - Exclosure studies (production, cover and precipitation).

5C 780503 4 Fisser, H.G., L.W. Young, J.E. Orpet and N.E. Hargis. 1978. Phenology and production studies on semiarid shrub-types. 1977 results. Univ. of Wyo. Coop. Res. Rep. to the BLM. Sections I, II, III and IV. Wyo. Agr. Exp. Sta. Sci. Rep. 970. 439 p. Section IV - Exclosure studies (production, cover and precipitation).

5C 790502 4 Fisser, H.G., L.W. Young, N.E. Hargis and K.E. Spaeth. 1979. Phenology and production studies on semiarid shrub-types. 1978 results. Univ. of Wyo. Coop. Res. Rep. to the BLM. Sections I, II, III and IV. Wyo. Agr. Exp. Sta. Sci. Rep. 985. 489 p. Section IV - Exclosure studies (production, cover and precipitation).

5C 800502 5 Fisser, H.G., M.K. Owens, J.W. Uhlich, W.J. Waugh, D.E. Rodgers, and N.E. Hargis. 1980. Phenology and production studies on semiarid shrub-types. 1979 Results. Univ. of Wyo. Coop. Res. Rep. to the BLM. Vol. II: Precipitation, Productivity and Production Section. Sections III, IV, and V. Wyo. Agr. Exp. Sta. Sci. Rep. 1106. 219 p. Section V - Exclosure studies (production,cover and precipitation).

5C 820502 5 Fisser, H.G., S.K. Dean, T.L. Deboodt, R.J. Kinucan. 1982. Phenology and production studies on semiarid shrub-types. 1982 results. Univ. of Wyo. Coop. Res. Rep. to the BLM. Vol. II: Precipitation, productivity, and production sections III, IV, and V. Wyo. Agr. Exp. Sta. Sci. Rep. 1231. 256 p. Section V - Exclosure studies (production, cover and precipitation).

5C 830502 5 Fisser, H.G., S.K. Dean, T.L. Deboodt, R.S. Kinucan and J.L. Mulholland. 1983. Phenology and production studies on semiarid shrub-types. 1982 results. Univ. of Wyo. Coop. Res. Rep. to the BLM. Vol. II: Precipitation, Productivity and Production, Sections III, IV, and V. Wyo. Agr. Exp. Sta. Sci. Rep. 1231. 256 p. Section V - Exclosure studies (production,cover and precipitation).

6. Range Livestock

6A. GENERAL MANAGEMENT

6A1. Cattle

6A1 630503 2 Fisser, H.G., J.T. Nichols, and R.W. Hamner. 1963. Halogeton Research. 1962 Results. Univ. of Wyo. Coop. Res. Rep. to the BLM. Sections I - III. Wyo. Agr. Exp. Sta. Mimeo. p. 1-38. Section II - H.G. Fisser and J.T. Nichols. Summer steer grazing on saltbush range and seeded water spreaders (Worland). p. 7-11.

6A1 640501 2 Fisser, H.G., J.T. Nichols, R.W. Hamner, and J.R. Wight. 1964. Halogeton Research. 1963 Results. Univ. of Wyo. Coop. Res. Rep. to the BLM. Sections I - VIII. Wyo. Agr. Exp. Sta. Mimeo. p. 1-61. Section II - H.G. Fisser and R.W. Hamner. Summer steer grazing on saltbush range and seeded water spreaders (Worland). 6 p.

6A1 650501 2 Fisser, H.G., J.R. Wight, J.R. Flesland, L.D. Robinson. 1965. Halogeton Research. 1964 Results. Univ. of Wyo. Coop. Res. Rep. to the BLM. Section I - VI. Wyo. Agr. Exp. Sta. Mimeo. p. 1-82. Section II - H.G. Fisser and J.R. Wight. Spring steer grazing on saltbush range and seeded water spreaders (Worland). 7 p.

6A1 660501 2 Fisser, H.G., L.D. Robinson, D.R. Kirkam, R. Steger, and G. Noller. 1966. Halogeton Research. 1965 Results. Univ. of Wyo. Coop. Res. Rep. to the BLM. Sections I - V. Wyo. Agr. Exp. Sta. Mimeo. 134 p. Section II - Steer grazing studies on saltbush range and seeded water spreaders (Worland). 7 p.

6A1 790503 Fisser, H.G. 1979. Rangeland and wildlife resources of the South Pass historic mining area. Wyo. Agr. Exp. Sta. Mimeo. 8 p.

6A2. Sheep and Goats

6A2 620502 Nichols, J.T., L.C. Vosler, D.R. Smith, and H.G. Fisser. 1962. Halogeton studies. In Coop. Res. Rep. to the BLM. 1961 Results. Wyo. Agr. Exp. Sta. Mimeo. p. 1-17.

6A2 630503 1 Fisser, H.G., J.T. Nichols, and R.W. Hamner. 1963. Halogeton Research. 1962 Results. Univ. of Wyo. Coop. Res. Rep. to the BLM. Sections I - III. Wyo. Agr. Exp. Sta. Mimeo. p. 1-38. Section I - H.G. Fisser and J.T. Nichols. Winter sheep grazing on saltbush range (Greybull). p. 1-6.

6A2 640501 1 Fisser, H.G., J.T. Nichols, R.W. Hamner, and J.R. Wight. 1964. Halogeton Research. 1963 Results. Univ. of Wyo. Coop. Res. Rep. to the BLM. Sections I - VIII. Wyo. Agr. Exp. Sta. Mimeo. p. 1-61. Section I - H.G. Fisser and R.W. Hamner. Winter sheep grazing on saltbush range (Greybull). 8 p.

6A2 650501 1 Fisser, H.G., J.R. Wight, J.R. Flesland, L.D. Robinson. 1965. Halogeton Research. 1964 Results. Univ. of Wyo. Coop. Res. Rep. to the BLM. Section I - VI. Wyo. Agr. Exp. Sta. Mimeo. p. 1-82. Section I - H.G. Fisser and J.R. Wight. Winter sheep grazing on saltbush range (Greybull). 13 p.

6A2 660501 1 Fisser, H.G., L.D. Robinson, D.R. Kirkam, R. Steger, and G. Noller. 1966. Halogeton Research. 1965 Results. Univ. of Wyo. Coop. Res. Rep. to the BLM. Sections I - V. Wyo. Agr. Exp. Sta. Mimeo. 134 p. Section I - Winter sheep grazing on saltbush range (Greybull). 11 p.

6A2 670501 Fisser, H.G. and R.E. Steger. 1967. Winter sheep grazing study on saltbush rangeland. 1966 Results. Univ. of Wyo. Coop. Res. Rep. to the BLM. Wyo. Agr. Exp. Sta. Sci. Rep. 36. 9 p.

6A3. Horses

6A3 680501 Gibbens, R.P., H.G. Fisser, and M. May. 1968. Influence of grazing management systems on vegetation and wildlife habitat. Annual report of progress, July, 1967 - July, 1968. Univ. of Wyo. Coop. Res. Rep. to the BLM. Wyo. Agr. Exp. Sta. Sci. Rep. 102. 9 p.

6A3 690502 Gibbens, R.P., H.G. Fisser and M. May. 1969. Influence of grazing management systems on vegetation and wildlife habitat. Annual report of progress, July, 1968 - July, 1969. Univ. of Wyo. Coop. Res. Rep. to the BLM. Wyo. Agr. Exp. Sta. Sci. Rep. 179. 132 p.

6A3 700502 Gibbens, R.P. and H.G. Fisser. 1970. Influence of grazing management systems on vegetation and wildlife habitat. Annual report of progress, July, 1969 - July, 1970. Univ. of Wyo. Coop. Res. Rep. to the BLM. Wyo. Agr. Exp. Sta. Sci. Rep. 241. 32 p.

6A3 700505 Gibbens, R.P., H.G. Fisser and M. May. 1970. Influence of grazing management systems on vegetation and wildlife habitat. Semiann. Rep. to the BLM. Wyo. Agr. Exp. Sta. Mimeo. 2 p.

6A3 710501 Fisser, H.G. and R.P. Gibbens. 1971. Influence of grazing management systems on vegetation and wildlife habitat. Annual report of progress, July, 1970 - July, 1971. Univ. of Wyo. Coop. Res. Rep. to the BLM. Wyo. Agr. Exp. Sta. Sci. Rep. 336. 92 p.

6A3 720501 Gibbens, R.P., and H.G. Fisser. 1972. Influences of grazing management systems on vegetation and wildlife habitat. Annual Progress Report. Wyo. Agr. Exp. Sta. Scientific Rep. 414. 69 p.

6A3 750501 Gibbens, R.P., and H.G. Fisser. 1975. Influence of grazing management systems on vegetation and wildlife habitat in the Red Desert Region of Wyoming. Wyo. Agr. Exp. Sta. Sci. Mono. 29. 23 p.

6A3 790504 Fisser, H.G. 1979. Grazing management system studies of northern desert shrub vegetation of southern Wyoming. Wyo. Agr. Exp. Sta. Mimeo. 3 p.

6A3 800505 Fisser, H.G., D.E. Rodgers and W.E. Gerhart. 1980. Influences of grazing management systems on vegetation and wildlife habitat. 1975 and 1979 Results and Long-Term Summary. Univ. of Wyo. Coop. Res. Rep. to the BLM. Wyo. Agr. Exp. Sta. Sci. Rep. 1047. 188 p.

6A3 820503 Fisser, H.G. 1982. Vegetation response to grazing management strategies on semiarid shrublands of Wyoming. Wyo. Agr. Exp. Sta. Mimeo. prepared for Grazing Systems Conference, Texas A & M University. November 1982. 4 p.

7. Range Wildlife

7A. BIG GAME

7A1. Deer

7A1 680501 Gibbens, R.P., H.G. Fisser, and M. May. 1968. Influence of grazing management systems on vegetation and wildlife habitat. Annual report of progress, July, 1967 - July, 1968. Univ. of Wyo. Coop. Res. Rep. to the BLM. Wyo. Agr. Exp. Sta. Sci. Rep. 102. 9 p.

7A1 690502 Gibbens, R.P., H.G. Fisser and M. May. 1969. Influence of grazing management systems on vegetation and wildlife habitat. Annual report of progress, July, 1968 - July, 1969. Univ. of Wyo. Coop. Res. Rep. to the BLM. Wyo. Agr. Exp. Sta. Sci. Rep. 179. 132 p.

7A1 700502 Gibbens, R.P. and H.G. Fisser. 1970. Influence of grazing management systems on vegetation and wildlife habitat. Annual report of progress, July, 1969 - July, 1970. Univ. of Wyo. Coop. Res. Rep. to the BLM. Wyo. Agr. Exp. Sta. Sci. Rep. 241. 32 p.

7A1 700505 Gibbens, R.P., H.G. Fisser and M. May. 1970. Influence of grazing management systems on vegetation and wildlife habitat. Semiann. Rep. to the BLM. Wyo. Agr. Exp. Sta. Mimeo. 2 p.

7A1 710501 Fisser, H.G. and R.P. Gibbens. 1971. Influence of grazing management systems on vegetation and wildlife habitat. Annual report of progress, July, 1970 - July, 1971. Univ. of Wyo. Coop. Res. Rep. to the BLM. Wyo. Agr. Exp. Sta. Sci. Rep. 336. 92 p.

7A1 720501 Gibbens, R.P., and H.G. Fisser. 1972. Influences of grazing management systems on vegetation and wildlife habitat. Annual Progress Report. Wyo. Agr. Exp. Sta. Sci. Rep. 414. 69 p.

7A1 750501 Gibbens, R.P., and H.G. Fisser. 1975. Influence of grazing management systems on vegetation and wildlife habitat in the Red Desert Region of Wyoming. Wyo. Agr. Exp. Sta. Sci. Mono. 29. 23 p.

7A1 790503 Fisser, H.G. 1979. Rangeland and wildlife resources of the South Pass historic mining area. Wyo. Agr. Exp. Sta. Mimeo. 8 p.

7A1 790504 Fisser, H.G. 1979. Grazing management system studies of northern desert shrub vegetation of southern Wyoming. Wyo. Agr. Exp. Sta. Mimeo. 3 p.

7A1 800505 Fisser, H.G., D.E. Rodgers and W.E. Gerhart. 1980. Influences of grazing management systems on vegetation and wildlife habitat. 1975 and 1979 Results and Long-Term Summary. Univ. of Wyo. Coop. Res. Rep. to the BLM. Wyo. Agr. Exp. Sta. Sci. Rep. 1047. 188 p.

7A1 820503 Fisser, H.G. 1982. Vegetation response to grazing management strategies on semiarid shrublands of Wyoming. Wyo. Agr. Exp. Sta. Mimeo. prepared for Grazing Systems Conference, Texas A & M University. November 1982. 4 p.

7A2. Elk

7A2 790503 Fisser, H.G. 1979. Rangeland and wildlife resources of the South Pass historic mining area. Wyo. Agr. Exp. Sta. Mimeo. 8 p.

7A3. Antelope

7A3 680501 Gibbens, R.P., H.G. Fisser, and M. May. 1968. Influence of grazing management systems on vegetation and wildlife habitat. Annual report of progress, July, 1967 - July, 1968. Univ. of Wyo. Coop. Res. Rep. to the BLM. Wyo. Agr. Exp. Sta. Sci. Rep. 102. 9 p.

7A3 690502 Gibbens, R.P., H.G. Fisser and M. May. 1969. Influence of grazing management systems on vegetation and wildlife habitat. Annual report of progress, July, 1968 - July, 1969. Univ. of Wyo. Coop. Res. Rep. to the BLM. Wyo. Agr. Exp. Sta. Sci. Rep. 179. 132 p.

7A3 700502 Gibbens, R.P. and H.G. Fisser. 1970. Influence of grazing management systems on vegetation and wildlife habitat. Annual report of progress, July, 1969 - July, 1970. Univ. of Wyo. Coop. Res. Rep. to the BLM. Wyo. Agr. Exp. Sta. Sci. Rep. 241. 32 p.

7A3 700505 Gibbens, R.P., H.G. Fisser and M. May. 1970. Influence of grazing management systems on vegetation and wildlife habitat. Semiann. Rep. to the BLM. Wyo. Agr. Exp. Sta. Mimeo. 2 p.

7A3 710501 Fisser, H.G. and R.P. Gibbens. 1971. Influence of grazing management systems on vegetation and wildlife habitat. Annual report of progress, July, 1970 - July, 1971. Univ. of Wyo. Coop. Res. Rep. to the BLM. Wyo. Agr. Exp. Sta. Sci. Rep. 336. 92 p.

7A3 720501 Gibbens, R.P., and H.G. Fisser. 1972. Influences of grazing management systems on vegetation and wildlife habitat. Annual Progress Report. Wyo. Agr. Exp. Sta. Sci. Rep. 414. 69 p.

7A3 750501 Gibbens, R.P., and H.G. Fisser. 1975. Influence of grazing management systems on vegetation and wildlife habitat in the Red Desert Region of Wyoming. Wyo. Agr. Exp. Sta. Sci. Mono. 29. 23 p.

7A3 790503 Fisser, H.G. 1979. Rangeland and wildlife resources of the South Pass historic mining area. Wyo. Agr. Exp. Sta. Mimeo. 8 p.

7A3 790504 Fisser, H.G. 1979. Grazing management system studies of northern desert shrub vegetation of southern Wyoming. Wyo. Agr. Exp. Sta. Mimeo. 3 p.

7A3 800505 Fisser, H.G., D.E. Rodgers and W.E. Gerhart. 1980. Influences of grazing management systems on vegetation and wildlife habitat. 1975 and 1979 Results and Long-Term Summary. Univ. of Wyo. Coop. Res. Rep. to the BLM. Wyo. Agr. Exp. Sta. Sci. Rep. 1047. 188 p.

7A3 820503 Fisser, H.G. 1982. Vegetation response to grazing management strategies on semiarid shrublands of Wyoming. Wyo. Agr. Exp. Sta. Mimeo. prepared for Grazing Systems Conference, Texas A & M University. November 1982. 4 p.

7A5. Moose

7A5 790503 Fisser, H.G. 1979. Rangeland and wildlife resources of the South Pass historic mining area. Wyo. Agr. Exp. Sta. Mimeo. 8 p.

8. Range Watersheds

8A. WATER RESOURCES

8A 790503 Fisser, H.G. 1979. Rangeland and wildlife resources of the South Pass historic mining area. Wyo. Agr. Exp. Sta. Mimeo. 8 p.

8B. WATERSHED SOIL EROSION AND SEDIMENTATION

8B 640501 5 Fisser, H.G., J.T. Nichols, R.W. Hamner, and J.R. Wight. 1964. Halogeton Research. 1963 Results. Univ. of Wyo. Coop. Res. Rep. to the BLM. Sections I - VIII. Wyo. Agr. Exp. Sta. Mimeo. p. 1-61. Section V - H.G. Fisser, J.T. Nichols, and J.R. Wight. Erosion transect study. 2 p.

8B 660503 Fisser, H.G. and R.E. Steger. 1966. Microclimate studies in desert shrub types. Wyo. Agr. Exp. Sta. Mimeo. 14 p.

8B 800503 Fisser, H.G., D.C. Trueblood and C.H. Butterfield. 1980. Soil vegetation relationships on rangeland exclosures in the Overland Planning Unit of southcentral Wyoming. Univ. of Wyo. Coop. Res. Rep. to the BLM. Wyo. Agr. Exp. Sta. Sci. Rep. 1042. 75 p.

8B 800504 Fisser, H.G., and D.C. Trueblood. 1980. Soil vegetation relationships on rangeland exclosures on the Salt Wells Planning Unit of southwestern Wyoming. Univ. of Wyo. Coop. Res. Rep. to the BLM. Wyo. Agr. Exp. Sta. Sci. Rep. 1041. 68 p.

8D. WATERSHED RESTORATION

8D 750502 Fisser, H.G., F.K. Taha, G.A. Lymbery and J.R. Cox. 1975. Herbage structure, productivity and phenology on Black Thunder. Annual Progress Report to Atlantic-Richfield Company. 1974. results. Wyo. Agr. Exp. Sta. Sci. Rep. 662. 118 p.

8D 760502 Fisser, H.G., J.R. Cox and F.K. Taha. 1976. Herbage structure, productivity and phenology on Black Thunder. Annual Progress Report to Atlantic-Richfield Company. 1975 results. Wyo. Agr. Exp. Sta. Sci. Rep. 779. 161 p.

8D 760503 Ries, R.E., H.G. Fisser, and A.T. Harrison. 1976. Vegetation, Section IX. p. 1-85. In Bergman, H.L. and M.D. Marcus (eds.). Final Report Environmental Assessment of the Black Thunder Mine Site, Campbell, Colorado. University of Wyoming, Laramie, Wyoming.

8D 770501 Fisser, H.G., J.R. Cox and F.K. Taha. 1977. Herbage structure, production and phenology on Black Thunder. Annual Progress Report to Atlantic-Richfield Company. 1976 results. Wyo. Agr. Exp. Sta. Sci. Rep. 832. 208 p.

8D 780501 Fisser, H.G., J.R. Cox and M. Mecke. 1978. Herbage structure, production and phenology on Black Thunder. Annual Progress Report to Atlantic-Richfield Company. 1977 results. Wyo. Agr. Exp. Sta. Sci. Rep. 856. 157 p.

8D 780502 Fisser, H.G., and J.R. Cox. 1978. Herbage structure, production and phenology on Black Thunder. Final Summary Report to Atlantic-Richfield Company. Wyo. Agr. Exp. Sta. Sci. Rep. 968. 243 p.

9. Range Economics

9A. RANCH PLANNING

9A 790503 Fisser, H.G. 1979. Rangeland and wildlife resources of the South Pass historic mining area. Wyo. Agr. Exp. Sta. Mimeo. 8 p.

10. Range Education and Research

10D. RANGE RESEARCH MODELS

10D1. Plant-Soil Emphasis

10D1 610501 Fisser, H.G. 1961. The range condition survey of the Big Horn and Wind River Basins. University of Wyoming. Plant Sci. Div., Range Manage. Notes 4:1.

10D1 620501 Fisser, H.G., C. Williams, A.A. Beetle, and R.L. Lang. 1962. A study of range conditions in the Big Horn and Wind River Basins. Prel. Report III. Wyo. Agr. Exp. Sta. Mimeo. 7 p.

10D1 620505 Fisser, H.G. 1962. Information to be obtained from the range management vegetation measurement transects. Wyo. Agr. Exp. Sta. Mimeo. 2 p.

10D1 630501 Fisser, H.G. 1963. A study of range conditions in the Big Horn and Wind River Basins. Prel. Report IV. Wyo. Agr. Exp. Sta. Mimeo Report. 13 p.

10D1 630503 3 Fisser, H.G., J.T. Nichols, and R.W. Hamner. 1963. Halogeton Research. 1962 Results. Univ. of Wyo. Coop. Res. Rep. to the BLM. Sections I - III. Wyo. Agr. Exp. Sta. Mimeo. p. 1-38. Section III - H.G. Fisser and R.W. Hamner. Exclosure studies (production, cover, height, and precipitation). p. 12-38.

10D1 640501 3 Fisser, H.G., J.T. Nichols, R.W. Hamner, and J.R. Wight. 1964. Halogeton Research. 1963 Results. Univ. of Wyo. Coop. Res. Rep. to the BLM. Sections I - VIII. Wyo. Agr. Exp. Sta. Mimeo. p. 1-61. Section III - H.G. Fisser, J.T. Nichols, and J.R. Wight. Exclosure studies (production, cover, height and precipitation). 33 p.

10D1 640501 4 Fisser, H.G., J.T. Nichols, R.W. Hamner, and J.R. Wight. 1964. Halogeton Research. 1963 Results. Univ. of Wyo. Coop. Res. Rep. to the BLM. Sections I - VIII. Wyo. Agr. Exp. Sta. Mimeo. p. 1-61. Section IV - H.G. Fisser, J.T. Nichols, and J.R. Wight. Soil moisture and temperature study. 3 p.

10D1 640501 5 Fisser, H.G., J.T. Nichols, R.W. Hamner, and J.R. Wight. 1964. Halogeton Research. 1963 Results. Univ. of Wyo. Coop. Res. Rep. to the BLM. Sections I - VIII. Wyo. Agr. Exp. Sta. Mimeo. p. 1-61. Section V - H.G. Fisser, J.T. Nichols, and J.R. Wight. Erosion transect study. 2 p.

10D1 680502 Fisser, H.G. 1968. Exclosure studies with transects of permanent plots. 1967 results. Univ. of Wyo. Coop. Res. Rep. to the BLM. Wyo. Agr. Exp. Sta. Sci. Rep. 118. 52 p.

10D1 680503 Fisser, H.G. 1968. Long-term exclosure studies with 2 ft x 20 ft and 4 ft x 4 ft permanent plots. 1959-1964 results. Univ. Wyo. Coop. Res. Rep. to the BLM. Wyo. Agr. Exp. Sta. Sci. Rep. 131. 78 p.

10D1 680506 Fisser, H.G. 1968. Synthetic study area nomination and related analytic research proposals to the Desert Biome Ecosystem Analysis Program of the International Biological Program. Wyo. Agr. Exp. Sta. Univ. of Wyo. Mimeo. 82 p.

10D1 690503 Fisser, H.G., J.E. Lester, G.C. Mitchell, and L.J. Bledsoe. 1969. Preliminary report of methodology and results for analyses of plant pattern project research on the Pawnee Site. U.S. I.B.P. Grassland Biome Technical Report 9. 68 p.

10D1 700503 Whysong, G.L. and H.G. Fisser. 1970. Frequency, density and basal area measures on sagebrush-grass ranges. 1969 results. Univ. of Wyo. Coop. Res. Rep. to the BLM. Wyo. Agr. Exp. Sta. Sci. Rep. 265. 35 p.

10D1 710504 Fisser, H.G. 1971. Evaluation of Pawnee Site frequency data. Wyo. Agr. Exp. Sta. Mimeo. 6 p.

10D1 720504 Stroud, D.W. and H.G. Fisser. 1972. Plant pattern micro-relief and *Bouteloua gracilis* culm density studies on the Pawnee Site: 1970-1971. U.S. I.B.P. Grassland Biome Technical Report 186. 26 p.

10D1 800502 4 Fisser, H.G., M.K. Owens, J.W. Uhlich, W.J. Waugh, D.E. Rodgers, and N.E. Hargis. 1980. Phenology and production studies on semiarid shrub-types. 1979 Results. Univ. of Wyo. Coop. Res. Rep. to the BLM. Vol. II: Precipitation, Productivity and Production Section. Sections III, IV, and V. Wyo. Agr. Exp. Sta. Sci. Rep. 1106. 219 p. Section IV - Productivity-phenology relationships.

10D1 800503 Fisser, H.G., D.C. Trueblood and C.H. Butterfield. 1980. Soil vegetation relationships on rangeland exclosures in the Overland Planning Unit of south central Wyoming. Univ. of Wyo. Coop. Res. Rep. to the BLM. Wyo. Agr. Exp. Sta. Sci. Rep. 1042. 75 p.

10D1 800504 Fisser, H.G., and D.C. Trueblood. 1980. Soil vegetation relationships on rangeland exclosures on the Salt Wells Planning Unit of southwestern Wyoming. Univ. of Wyo. Coop. Res. Rep. to the BLM. Wyo. Agr. Exp. Sta. Sci. Rep. 1041. 68 p.

10D1 820502 4 Fisser, H.G., S.K. Dean, T.L. Deboodt, R.J. Kinucan. 1982. Phenology and production studies on semiarid shrub-types. 1982 results. Univ. of Wyo. Coop. Res. Rep. to the BLM. Vol. II: Precipitation, productivity, and production sections III, IV, and V. Wyo. Agr. Exp. Sta. Sci. Rep. 1231. 256 p. Section IV - Productivity-phenology interrelation study.

10D1 830502 4 Fisser, H.G., S.K. Dean, T.L. Deboodt, R.S. Kinucan and J.L. Mulholland. 1983. Phenology and production studies on semiarid shrub-types. 1982 results. Univ. of Wyo. Coop. Res. Rep. to the BLM. Vol. II: Precipitation, Productivity and Production, Sections III, IV, and V. Wyo. Agr. Exp. Sta. Sci. Rep. 1231. 256 p. Section IV - Productivity-phenology interrelation study.

11. Computer Storage, Programming and Modeling

11 620505 Fisser, H.G. 1962. Information to be obtained from the range management vegetation measurement transects. Wyo. Agr. Exp. Sta. Mimeo. 2 p.

11A. CLIMATIC

11A 840501 Fisser, H.G. and L.P. Germain. 1984. Phenology and production studies on semiarid shrub-types. Statistical tests and computer programs. Wyo. Agr. Exp. Sta. Sci. Rep. 15 p.

11A 840502 Fisser, 1984. Statistics and interpretation. Wyo. Agr. Exp. Sta. Sci. Rep. 5 p.

11C. EDAPHIC

11C 840501 Fisser, H.G. and L.P. Germain. 1984. Phenology and production studies on semiarid shrub-types. Statistical tests and computer programs. Wyo. Agr. Exp. Sta. Sci. Rep. 15 p.

11C 840502 Fisser, H.G. 1984. Statistics and interpretation. Wyo. Agr. Exp. Sta. Sci. Rep. 5 p.

11E. RANGELAND INVENTORY

11E 830503 Fisser, H.G. 1983. Aridland shrub ecology and phenology research program. Wyo. Agr. Exp. Sta. Mimeo. 3 p.

11E1. Sites and Mapping

11E1 840501 Fisser, H.G. and L.P. Germain. 1984. Phenology and production studies on semiarid shrub-types. Statistical tests and computer programs. Wyo. Agr. Exp. Sta. Sci. Rep. 15 p.

11E1 840502 Fisser, H.G. 1984. Statistics and interpretation. Wyo. Agr. Exp. Sta. Sci. Rep. 5 p.

11E2. Condition and Trend

11E2 840501 Fisser, H.G. and L.P. Germain. 1984. Phenology and production studies on semiarid shrub-types. Statistical tests and computer programs. Wyo. Agr. Exp. Sta. Sci. Rep. 15 p.

11E2 840502 Fisser, H.G. 1984. Statistics and interpretation. Wyo. Agr. Exp. Sta. Sci. Rep. 5 p.

11E3. Production and Utilization

11E3 840501 Fisser, H.G. and L.P. Germain. 1984. Phenology and production studies on semiarid shrub-types. Statistical tests and computer programs. Wyo. Agr. Exp. Sta. Sci. Rep. 15 p.

11E3 840502 Fisser, H.G. 1984. Statistics and interpretation. Wyo. Agr. Exp. Sta. Sci. Rep. 5 p.

11F. PREDICTIVE MODELING

11F1. Herbage Yield

11F1 840501 Fisser, H.G. and L.P. Germain. 1984. Phenology and production studies on semiarid shrub-types. Statistical tests and computer programs. Wyo. Agr. Exp. Sta. Sci. Rep. 15 p.

11F1 840502 Fisser, H.G. 1984. Statistics and interpretation. Wyo. Agr. Exp. Sta. Sci. Rep. 5 p.

11F2. Phenology

11F2 770502 2 Fisser, H.G., L.W. Young, J.E. Orpet and N.E. Hargis. 1977. Phenology and production studies on semiarid shrub-types. 1976 results. Univ. of Wyo. Coop. Res. Rep. to the BLM. Sections I, II, III and IV. Wyo. Agr. Exp. Sta. Sci. Rep. 930. 438 p. Section II - Phenology step-wise regression for predictive modeling.

11F2 780503 2 Fisser, H.G., L.W. Young, J.E. Orpet and N.E. Hargis. 1978. Phenology and production studies on semiarid shrub-types. 1977 results. Univ. of Wyo. Coop. Res. Rep. to the BLM. Sections I, II, III and IV. Wyo. Agr. Exp. Sta. Sci. Rep. 970. 439 p. Section II - Phenology step-wise regression for predictive modeling.

11F2 790502 2 Fisser, H.G., L.W. Young, N.E. Hargis and K.E. Spaeth. 1979. Phenology and production studies on semiarid shrub-types. 1978 results. Univ. of Wyo. Coop. Res. Rep. to the BLM. Sections I, II, III and IV. Wyo. Agr. Exp. Sta. Sci. Rep. 985. 489 p. Section II - Phenology step-wise regression for predictive modeling.

11F2 800501 2 Fisser, H.G., M.K. Owens, W.J. Waugh, N.E. Hargis and J. Uhlich. 1980. Phenology and production studies on semiarid shrub-types. 1979 Results. Univ. of Wyo. Coop. Res. Rep. to the BLM. Vol. I: Phenology Study. Sections I and II. Wyo. Agr. Exp. Sta. Sci. Rep. 1044. 387 p. Section II - Phenology step-wise regression for predictive modeling.

11F2 820501 2 Fisser, H.G., S.K. Dean, T.L. Deboodt, R.J. Kinucan. 1982. Phenology and production studies on semiarid shrub-types. 1982 Results. Univ. of Wyo. Coop. Res. Rep. to the BLM - Vol. I: Phenology study sections I and II. Wyo. Agr. Exp. Sta. Sci. Rep. 1230. 415 p. Section II - Phenology multiple regression for predictive modeling.

11F2 830501 2 Fisser, H.G., S.K. Dean, T.L. Deboodt, R.S. Kinucan and J.L. Mulholland, 1983. Phenology and production studies on semiarid shrub-types. 1982 results. Univ. of Wyo. Coop. Res. Rep. to the BLM. Vol. I: Phenology Study, Sections I and II. Wyo. Agr. Exp. Sta. Sci. Rep. 1230. 430 p. Section II - Phenology multiple regression for predictive modeling.

S6. GRADUATE ADVISEE COMPLETION

1. Range Plants

1D. PLANT INSECTS

1D 670601 Kirkham, D.R. 1967. Harvester ant (*Pogonomyrmex owyheei* Cole) abundance in relation to grazing, vegetation and soil in north central Wyoming. M.S. Thesis. University of Wyoming. 83 p.

1E. PLANT PHYSIOLOGY

1E 620601 Vosler, L.C. 1962. An ecological study of *Atriplex nuttallii* in the Big Horn Basin of Wyoming. M.S. Thesis. University of Wyoming. 43 p.

1E1. Reproduction and Germination

1E1 760602 Kleinman, L.H. 1976. Phenodynamics and ecology of sagebrush-grass rangelands. Ph.D. Dissertation. University of Wyoming. 143 p.

1E1 770601 Taha, F.K. 1977. Ecological interrelationships of vegetation and soils in a shortgrass prairie ecosystem. Ph.D. Dissertation. University of Wyoming. 207 p.

1E1 770603 Hughes, H.G. 1977. Factors influencing vegetative and reproductive phenology of *Artemisia nova* Nels. M.S. Thesis. University of Wyoming. 59 p.

1E1 840601 Mulholland, J.L. 1984. Big sagebrush response and re-establishment following chemical control in western Wyoming. M.S. Thesis. University of Wyoming. 72 p.

1E2. Growth and Phenology

1E2 670603 Russey, G.R. 1967. The effect of grazing intensities on the roots of Nuttall's Saltbush (*Atriplex nuttallii* S. Wats.) M.S. Thesis. University of Wyoming. 49 p.

1E2 680601 Noller, G.L. 1968. The relationship of forage production to precipitation, cover and soils in north central Wyoming. Ph.D. Dissertation. University of Wyoming. 153 p.

1E2 680603 Wambolt, C.L. 1968. Size, density, growth rate and age of Rocky Mountain juniper and interrelationships with environmental factors. M.S. Thesis. University of Wyoming. 62 p.

1E2 710601 Wambolt, C.L. 1971. Moisture stress in woody plants of Wyoming as influenced by environmental factors. Ph.D. Dissertation. University of Wyoming. 158 p.

1E2 710603 Mitchell, G.C. 1971. Spatial distribution and successional state of grassland vegetation related to grazing intensity treatments. M.S. Thesis. University of Wyoming. 64 p.

1E2 720602 Garland, C.B., III. 1972. Ecology of aridland vegetation in western Wyoming. M.S. Thesis. University of Wyoming. 106 p.

1E2 730602 Ries, R.E. 1973. Quantitative evaluation of herbage production and site factors on ecologically diverse sagebrush-grass locations in Wyoming. Ph.D. Dissertation. University of Wyoming. 101 p.

1E2 750601 Lymbery, G.A. 1975. Vegetation re-establishment potential of strip-mined soils, Black Thunder Basin, Wyoming. M.S. Thesis. University of Wyoming. 62 p.

1E2 760602 Kleinman, L.H. 1976. Phenodynamics and ecology of sagebrush-grass rangelands. Ph.D. Dissertation. University of Wyoming. 143 p.

1E2 770601 Taha, F.K. 1977. Ecological interrelationships of vegetation and soils in a shortgrass prairie ecosystem. Ph.D. Dissertation. University of Wyoming. 207 p.

1E2 770603 Hughes, H.G. 1977. Factors influencing vegetative and reproductive phenology of *Artemisia nova* Nels. M.S. Thesis. University of Wyoming. 59 p.

1E2 770605 Cox, J.R. 1977. Phenology, carbohydrate storage, and annual biomass of five shortgrass prairie species in relation to environmental factors. Ph.D. Dissertation. University of Wyoming. 203 p.

1E2 800603 Hargis, N.E. 1980. Phenological comparison of two sagebrush species. M.S. Thesis. University of Wyoming. 110 p.

1E2 820601 Elwakeel, A.S. 1982. Effects of environmental factors on perennial grass production in five vegetation types in north central Wyoming. M.S. Thesis. University of Wyoming. 81 p.

1E2 830602 Kinucan, R.J. 1983. Phenological relationships and patterns of semiarid plant communities in central and western Wyoming. M.S. Thesis. University of Wyoming. 110 p.

1E3. Photosynthesis and Plant Nutrients

1E3 770605 Cox, J.R. 1977. Phenology, carbohydrate storage, and annual biomass of five shortgrass prairie species in relation to environmental factors. Ph.D. Dissertation. University of Wyoming. 203 p.

1E4. Herbage Removal

1E4 680602 Probasco, G.E. 1968. Diet preferences and utilization patterns of elk in the northern Big Horn Mountains, Wyoming. M.S. Thesis. University of Wyoming. 36 p.

1E4 690601 Todd, Lynn D. 1969. Preference and utilization trends by cattle on grass-forb vegetation in the northern Big Horn Mountains, Wyoming. M.S. Thesis. University of Wyoming. 73 p.

1E4 770605 Cox, J.R. 1977. Phenology, carbohydrate storage, and annual biomass of five shortgrass prairie species in relation to environmental factors. Ph.D. Dissertation. University of Wyoming. 203 p.

2. Range Ecology

2A. CLIMATIC FACTORS

2A 620601 Vosler, L.C. 1962. An ecological study of *Atriplex nuttallii* in the Big Horn Basin of Wyoming. M.S. Thesis. University of Wyoming. 43 p.

2A 680601 Noller, G.L. 1968. The relationship of forage production to precipitation, cover and soils in north central Wyoming. Ph.D. Dissertation. University of Wyoming. 153 p.

2A 680603 Wambolt, C.L. 1968. Size, density, growth rate and age of Rocky Mountain juniper and interrelationships with environmental factors. M.S. Thesis. University of Wyoming. 62 p.

2A 700601 Steger, R.E. 1970. Soil moisture and temperature relationships of six salt desert shrub communities in north central Wyoming. Ph.D. Dissertation. University of Wyoming. 142 p.

2A 710601 Wambolt, C.L. 1971. Moisture stress in woody plants of Wyoming as influenced by environmental factors. Ph.D. Dissertation. University of Wyoming. 158 p.

2A 710602 Jacoby, P.W. 1971. Interrelationships of vegetation and environmental factors on a mountain watershed in southeastern Wyoming. Ph.D. Dissertation. University of Wyoming. 127 p.

2A 730601 Whysong, G.L. 1973. Development and evaluation of statistical models for predicting perennial grass production in north central Wyoming. Ph.D. Dissertation. University of Wyoming. 150 p.

2A 760602 Kleinman, L.H. 1976. Phenodynamics and ecology of sagebrush-grass rangelands. Ph.D. Dissertation. University of Wyoming. 143 p.

2A 770603 Hughes, H.G. 1977. Factors influencing vegetative and reproductive phenology of *Artemisia nova* Nels. M.S. Thesis. University of Wyoming. 59 p.

2A 770604 Gdara, A.O. 1977. Interpretation of long-term vegetation data from permanent quadrats on semiarid shrublands of western Wyoming. M.S. Thesis. University of Wyoming. 114 p.

2A 770605 Cox, J.R. 1977. Phenology, carbohydrate storage, and annual biomass of five shortgrass prairie species in relation to environmental factors. Ph.D. Dissertation. University of Wyoming. 203 p.

2A 780601 Elmekki, A.O. 1978. Evaluation of brush control methods on rangelands. A study of rangeland ecology and use in Sudan. M.S. Professional Papers. University of Wyoming. 39 & 35 p., respectively.

2A 780602 Orpet, J.E. 1978. Long-term herbage production response following the chemical control of big sagebrush. M.S. Thesis. University of Wyoming. 118 p.

2A 800601 Mesdaghi, M. 1980. Vegetation of saline and alkali habitats of the Red Desert, Wyoming. Ph.D. Dissertation. University of Wyoming. 140 p.

2A 800602 Trueblood, D.C. 1980. Soil-vegetation relations at rangeland exclosures in the Gardner saltbush vegetation types of the Big Horn Basin, Wyoming. M.S. Thesis. University of Wyoming. 106 p.

2A 800603 Hargis, N.E. 1980. Phenological comparison of two sagebrush species. M.S. Thesis. University of Wyoming. 110 p.

2A 820601 Elwakeel, A.S. 1982. Effects of environmental factors on perennial grass production in five vegetation types in north central Wyoming. M.S. Thesis. University of Wyoming. 81 p.

2A 820602 Uhlich, J.W. 1982. Vegetation dynamics of non-grazed and grazed sagebrush grassland communities in the Big Horn Basin of Wyoming. M.S. Thesis. University of Wyoming. 120 p.

2A 830601 Dean, S.K. 1983. Plant response to livestock grazing and big sagebrush chemical control in western Wyoming 1962 - 1980. M.S. Thesis. University of Wyoming. 84 p.

2A 830602 Kinucan, R.J. 1983. Phenological relationships and patterns of semiarid plant communities in central and western Wyoming. M.S. Thesis. University of Wyoming. 110 p.

2A 840601 Mulholland, J.L. 1984. Big sagebrush response and re-establishment following chemical control in western Wyoming. M.S. Thesis. University of Wyoming. 72 p.

2B. BIOTIC FACTORS

2B 670601 Kirkham, D.R. 1967. Harvester ant (*Pogonomyrmex owyheei* Cole) abundance in relation to grazing, vegetation and soil in north central Wyoming. M.S. Thesis. University of Wyoming. 83 p.

2B 770605 Cox, J.R. 1977. Phenology, carbohydrate storage, and annual biomass of five shortgrass prairie species in relation to environmental factors. Ph.D. Dissertation. University of Wyoming. 203 p.

2B 790601 Shawesh, O.M. 1979. Vegetation analysis and ecology of rangelands in northwest Libya, Africa. M.S. Thesis. University of Wyoming. 76 p.

2B 820602 Uhlich, J.W. 1982. Vegetation dynamics of non-grazed and grazed sagebrush grassland communities in the Big Horn Basin of Wyoming. M.S. Thesis. University of Wyoming. 120 p.

2B 840601 Mulholland, J.L. 1984. Big sagebrush response and re-establishment following chemical control in western Wyoming. M.S. Thesis. University of Wyoming. 72 p.

2C. EDAPHIC FACTORS

2C 620601 Vosler, L.C. 1962. An ecological study of *Atriplex nuttallii* in the Big Horn Basin of Wyoming. M.S. Thesis. University of Wyoming. 43 p.

2C 640602 Hamner, R.W. 1964. An ecological study of *Sarcobatus vermiculatus* communities of the Big Horn Basin, Wyoming. M.S. Thesis. University of Wyoming. 75 p.

2C 640603 Miller, W.B. 1964. An ecological study of the mountain mahogany community and related biotic associations of the Big Horn Mountains. M.S. Thesis. University of Wyoming. 152 p.

2C 640604 Nichols, J.T. 1964. Soil-vegetation relationships of the 15-mile drainage, Washakie County, Wyoming. Ph.D. Dissertation. University of Wyoming. 120 p.

2C 660601 Wight, J.R. 1966. Site and vegetation characteristics and relationships of juniper communities in northwestern Wyoming. Ph.D. Dissertation. University of Wyoming. 130 p.

2C 660602 Robinson, L.D. 1966. The vegetation and small mammals of the juniper zone in north central Wyoming. M.S. Thesis. University of Wyoming. 115 p.

2C 670601 Kirkham, D.R. 1967. Harvester ant (*Pogonomyrmex owyheei* Cole) abundance in relation to grazing, vegetation and soil in north central Wyoming. M.S. Thesis. University of Wyoming. 83 p.

2C 670602 King, T.W. 1967. A study of vegetation, soils and small mammals of limber pine stands in north central Wyoming. M.S. Thesis. University of Wyoming. 58 p.

2C 680603 Wambolt, C.L. 1968. Size, density, growth rate and age of Rocky Mountain juniper and interrelationships with environmental factors. M.S. Thesis. University of Wyoming. 62 p.

2C 690602 Forbis, L.A. 1969. Ecological relationships of vegetation, soils, and vertebrates on paired limber pine and spruce-fir sites, Big Horn Mountains, Wyoming. M.S. Thesis. University of Wyoming. 105 p.

2C 700601 Steger, R.E. 1970. Soil moisture and temperature relationships of six salt desert shrub communities in north central Wyoming. Ph.D. Dissertation. University of Wyoming. 142 p.

2C 710601 Wambolt, C.L. 1971. Moisture stress in woody plants of Wyoming as influenced by environmental factors. Ph.D. Dissertation. University of Wyoming. 158 p.

2C 730601 Whysong, G.L. 1973. Development and evaluation of statistical models for predicting perennial grass production in north central Wyoming. Ph.D. Dissertation. University of Wyoming. 150 p.

2C 750601 Lymbery, G.A. 1975. Vegetation re-establishment potential of strip-mined soils, Black Thunder Basin, Wyoming. M.S. Thesis. University of Wyoming. 62 p.

2C 770601 Taha, F.K. 1977. Ecological interrelationships of vegetation and soils in a shortgrass prairie ecosystem. Ph.D. Dissertation. University of Wyoming. 207 p.

2C 770605 Cox, J.R. 1977. Phenology, carbohydrate storage, and annual biomass of five shortgrass prairie species in relation to environmental factors. Ph.D. Dissertation. University of Wyoming. 203 p.

2C 780601 Elmekki, A.O. 1978. Evaluation of brush control methods on rangelands. A study of rangeland ecology and use in Sudan. M.S. Professional Papers. University of Wyoming. 39 & 35 p., respectively.

2C 780602 Orpet, J.E. 1978. Long-term herbage production response following the chemical control of big sagebrush. M.S. Thesis. University of Wyoming. 118 p.

2C 790601 Shawesh, O.M. 1979. Vegetation analysis and ecology of rangelands in northwest Libya, Africa. M.S. Thesis. University of Wyoming. 76 p.

2C 800601 Mesdaghi, M. 1980. Vegetation of saline and alkali habitats of the Red Desert, Wyoming. Ph.D. Dissertation. University of Wyoming. 140 p.

2C 800602 Trueblood, D.C. 1980. Soil-vegetation relations at rangeland exclosures in the Gardner saltbush vegetation types of the Big Horn Basin, Wyoming. M.S. Thesis. University of Wyoming. 106 p.

2C 800603 Hargis, N.E. 1980. Phenological comparison of two sagebrush species. M.S. Thesis. University of Wyoming. 110 p.

2C 820601 Elwakeel, A.S. 1982. Effects of environmental factors on perennial grass production in five vegetation types in north central Wyoming. M.S. Thesis. University of Wyoming. 81 p.

2C 820602 Uhlich, J.W. 1982. Vegetation dynamics of non-grazed and grazed sagebrush grassland communities in the Big Horn Basin of Wyoming. M.S. Thesis. University of Wyoming. 120 p.

2C 830601 Dean, S.K. 1983. Plant response to livestock grazing and big sagebrush chemical control in western Wyoming 1962 - 1980. M.S. Thesis. University of Wyoming. 84 p.

2C 830602 Kinucan, R.J. 1983. Phenological relationships and patterns of semiarid plant communities in central and western Wyoming. M.S. Thesis. University of Wyoming. 110 p.

2C 840601 Mulholland, J.L. 1984. Big sagebrush response and re-establishment following chemical control in western Wyoming. M.S. Thesis. University of Wyoming. 72 p.

2D. PYRIC FACTORS

2D 780601 Elmekki, A.O. 1978. Evaluation of brush control methods on rangelands. A study of rangeland ecology and use in Sudan. M.S. Professional Papers. University of Wyoming. 39 & 35 p., respectively.

2E. AUTECOLOGY

2E 610601 Williams, C.S. 1961. Distribution of vegetation in the Wind River Canyon, Wyoming. M.S. Thesis. University of Wyoming. (directed A.A. Beetle) 90 p.

2E 620601 Vosler, L.C. 1962. An ecological study of *Atriplex nuttallii* in the Big Horn Basin of Wyoming. M.S. Thesis. University of Wyoming. 43 p.

2E 640602 Hamner, R.W. 1964. An ecological study of *Sarcobatus vermiculatus* communities of the Big Horn Basin, Wyoming. M.S. Thesis. University of Wyoming. 75 p.

2E 640603 Miller, W.B. 1964. An ecological study of the mountain mahogany community and related biotic associations of the Big Horn Mountains. M.S. Thesis. University of Wyoming. 152 p.

2E 660601 Wight, J.R. 1966. Site and vegetation characteristics and relationships of juniper communities in northwestern Wyoming. Ph.D. Dissertation. University of Wyoming. 130 p.

2E 660602 Robinson, L.D. 1966. The vegetation and small mammals of the juniper zone in north central Wyoming. M.S. Thesis. University of Wyo. 115 p.

2E 670602 King, T.W. 1967. A study of vegetation, soils and small mammals of limber pine stands in north central Wyoming. M.S. Thesis. University of Wyoming. 58 p.

2E 670603 Russey, G.R. 1967. The effect of grazing intensities on the roots of Nuttall's Saltbush (*Atriplex nuttallii* S. Wats.) M.S. Thesis. University of Wyoming. 49 p.

2E 680603 Wambolt, C.L. 1968. Size, density, growth rate and age of Rocky Mountain juniper and interrelationships with environmental factors. M.S. Thesis. University of Wyoming. 62 p.

2E 690602 Forbis, L.A. 1969. Ecological relationships of vegetation, soils, and vertebrates on paired limber pine and spruce-fir sites, Big Horn Mountains, Wyoming. M.S. Thesis. University of Wyoming. 105 p.

2E 700601 Steger, R.E. 1970. Soil moisture and temperature relationships of six salt desert shrub communities in north central Wyoming. Ph.D. Dissertation. University of Wyoming. 142 p.

2E 710601 Wambolt, C.L. 1971. Moisture stress in woody plants of Wyoming as influenced by environmental factors. Ph.D. Dissertation. University of Wyoming. 158 p.

2E 710603 Mitchell, G.C. 1971. Spatial distribution and successional state of grassland vegetation related to grazing intensity treatments. M.S. Thesis. University of Wyoming. 64 p.

2E 720601 Gibbens, Robert P. 1972. Vegetation pattern within northern desert shrub communities. Ph.D. Dissertation. University of Wyoming. 272 p.

2E 720603 Burrough, L.N. 1972. Artificial revegetation trials on juniper game ranges in north central Wyoming. M.S. Thesis. University of Wyoming. 135 p.

2E 750601 Lymbery, G.A. 1975. Vegetation re-establishment potential of strip-mined soils, Black Thunder Basin, Wyoming. M.S. Thesis. University of Wyoming. 62 p.

2E 760602 Kleinman, L.H. 1976. Phenodynamics and ecology of sagebrush-grass rangelands. Ph.D. Dissertation. University of Wyoming. 143 p.

2E 770603 Hughes, H.G. 1977. Factors influencing vegetative and reproductive phenology of *Artemisia nova* Nels. M.S. Thesis. University of Wyoming. 59 p.

2E 770604 Gdara, A.O. 1977. Interpretation of long-term vegetation data from permanent quadrats on semiarid shrublands of western Wyoming. M.S. Thesis. University of Wyoming. 114 p.

2E 770605 Cox, J.R. 1977. Phenology, carbohydrate storage, and annual biomass of five shortgrass prairie species in relation to environmental factors. Ph.D. Dissertation. University of Wyoming. 203 p.

2E 780601 Elmekki, A.O. 1978. Evaluation of brush control methods on rangelands. A study of rangeland ecology and use in Sudan. M.S. Professional Papers. University of Wyoming. 39 & 35 p., respectively.

2E 780602 Orpet, J.E. 1978. Long-term herbage production response following the chemical control of big sagebrush. M.S. Thesis. University of Wyoming. 118 p.

2E 800601 Mesdaghi, M. 1980. Vegetation of saline and alkali habitats of the Red Desert, Wyoming. Ph.D. Dissertation. University of Wyoming. 140 p.

2E 800602 Trueblood, D.C. 1980. Soil-vegetation relations at rangeland exclosures in the Gardner saltbush vegetation types of the Big Horn Basin, Wyoming. M.S. Thesis. University of Wyoming. 106 p.

2E 800603 Hargis, N.E. 1980. Phenological comparison of two sagebrush species. M.S. Thesis. University of Wyoming. 110 p.

2E 820601 Elwakeel, A.S. 1982. Effects of environmental factors on perennial grass production in five vegetation types in north central Wyoming. M.S. Thesis. University of Wyoming. 81 p.

2E 820602 Uhlich, J.W. 1982. Vegetation dynamics of non-grazed and grazed sagebrush grassland communities in the Big Horn Basin of Wyoming. M.S. Thesis. University of Wyoming. 120 p.

2E 830601 Dean, S.K. 1983. Plant response to livestock grazing and big sagebrush chemical control in western Wyoming 1962 - 1980. M.S. Thesis. University of Wyoming. 84 p.

2E 840601 Mulholland, J.L. 1984. Big sagebrush response and re-establishment following chemical control in western Wyoming. M.S. Thesis. University of Wyoming. 72 p.

2F. SYNECOLOGY

2F 790601 Shawesh, O.M. 1979. Vegetation analysis and ecology of rangelands in northwest Libya, Africa. M.S. Thesis. University of Wyoming. 76 p.

2F1. Great Plains

2F1 710603 Mitchell, G.C. 1971. Spatial distribution and successional state of grassland vegetation related to grazing intensity treatments. M.S. Thesis. University of Wyoming. 64 p.

2F1 720604 Shaver, J.C. 1972. Evaluation of vegetation types differentiated by aerial photo interpretation of panchromatic, color, and infrared imagery. M.S. Thesis. University of Wyoming. 86 p.

2F1 740603 Gordon, R.C. 1974. Multispectral imagery for range vegetation type mapping and estimation of above-ground green biomass. M.S. Thesis. University of Wyoming. 45 p.

2F1 750601 Lymbery, G.A. 1975. Vegetation re-establishment potential of strip-mined soils, Black Thunder Basin, Wyoming. M.S. Thesis. University of Wyoming. 62 p.

2F1 770601 Taha, F.K. 1977. Ecological interrelationships of vegetation and soils in a shortgrass prairie ecosystem. Ph.D. Dissertation. University of Wyoming. 207 p.

2F2. West

2F2 610601 Williams, C.S. 1961. Distribution of vegetation in the Wind River Canyon, Wyoming. M.S. Thesis. University of Wyoming. (directed A.A. Beetle) 90 p.

2F2 640601 Blankenship, J.O. 1964. Indirect estimation of standing crop for selected alpine species. M.S. Thesis. University of Wyoming. 52 p.

2F2 640602 Hamner, R.W. 1964. An ecological study of *Sarcobatus vermiculatus* communities of the Big Horn Basin, Wyoming. M.S. Thesis. University of Wyoming. 75 p.

2F2 640603 Miller, W.B. 1964. An ecological study of the mountain mahogany community and related biotic associations of the Big Horn Mountains. M.S. Thesis. University of Wyoming. 152 p.

2F2 640604 Nichols, J.T. 1964. Soil-vegetation relationships of the 15-mile drainage, Washakie County, Wyoming. Ph.D. Dissertation. University of Wyoming. 120 p.

2F2 660601 Wight, J.R. 1966. Site and vegetation characteristics and relationships of juniper communities in northwestern Wyoming. Ph.D. Dissertation. University of Wyoming. 130 p.

2F2 660602 Robinson, L.D. 1966. The vegetation and small mammals of the juniper zone in north central Wyoming. M.S. Thesis. University of Wyoming. 115 p.

2F2 670602 King, T.W. 1967. A study of vegetation, soils and small mammals of limber pine stands in north central Wyoming. M.S. Thesis. University of Wyoming. 58 p.

2F2 690602 Forbis, L.A. 1969. Ecological relationships of vegetation, soils, and vertebrates on paired limber pine and spruce-fir sites, Big Horn Mountains, Wyoming. M.S. Thesis. University of Wyoming. 105 p.

2F2 710601 Wambolt, C.L. 1971. Moisture stress in woody plants of Wyoming as influenced by environmental factors. Ph.D. Dissertation. University of Wyoming. 158 p.

2F2 710602 Jacoby, P.W. 1971. Interrelationships of vegetation and environmental factors on a mountain watershed in southeastern Wyoming. Ph.D. Dissertation. University of Wyoming. 127 p.

2F2 720601 Gibbens, R.P. 1972. Vegetation pattern within northern desert shrub communities. Ph.D. Dissertation. University of Wyoming. 272 p.

2F2 720602 Garland, C.B., III. 1972. Ecology of aridland vegetation in western Wyoming. M.S. Thesis. University of Wyoming. 106 p.

2F2 760602 Kleinman, L.H. 1976. Phenodynamics and ecology of sagebrush-grass rangelands. Ph.D. Dissertation. University of Wyoming. 143 p.

2F2 770601 Taha, F.K. 1977. Ecological interrelationships of vegetation and soils in a shortgrass prairie ecosystem. Ph.D. Dissertation. University of Wyoming. 207 p.

2F2 770603 Hughes, H.G. 1977. Factors influencing vegetative and reproductive phenology of *Artemisia nova* Nels. M.S. Thesis. University of Wyoming. 59 p.

2F2 770604 Gdara, A.O. 1977. Interpretation of long-term vegetation data from permanent quadrats on semiarid shrublands of western Wyoming. M.S. Thesis. University of Wyoming. 114 p.

2F2 780602 Orpet, J.E. 1978. Long-term herbage production response following the chemical control of big sagebrush. M.S. Thesis. University of Wyoming. 118 p.

2F2 800601 Mesdaghi, M. 1980. Vegetation of saline and alkali habitats of the Red Desert, Wyoming. Ph.D. Dissertation. University of Wyoming. 140 p.

2F2 800602 Trueblood, D.C. 1980. Soil-vegetation relations at rangeland exclosures in the Gardner saltbush vegetation types of the Big Horn Basin, Wyoming. M.S. Thesis. University of Wyoming. 106 p.

2F2 800603 Hargis, N.E. 1980. Phenological comparison of two sagebrush species. M.S. Thesis. University of Wyoming. 110 p.

2F2 820601 Elwakeel, A.S. 1982. Effects of environmental factors on perennial grass production in five vegetation types in north central Wyoming. M.S. Thesis. University of Wyoming. 81 p.

2F2 820602 Uhlich, J.W. 1982. Vegetation dynamics of non-grazed and grazed sagebrush grassland communities in the Big Horn Basin of Wyoming. M.S. Thesis. University of Wyoming. 120 p.

2F2 830601 Dean, S.K. 1983. Plant response to livestock grazing and big sagebrush chemical control in western Wyoming 1962 - 1980. M.S. Thesis. University of Wyoming. 84 p.

2F2 840601 Mulholland, J.L. 1984. Big sagebrush response and re-establishment following chemical control in western Wyoming. M.S. Thesis. University of Wyoming. 72 p.

3. Range Resources

3A. RANGE INVENTORY AND SURVEY

3A1. Site, Mapping, Remote Sensing

3A1 610601 Williams, C.S. 1961. Distribution of vegetation in the Wind River Canyon, Wyoming. M.S. Thesis. University of Wyoming. (directed A.A. Beetle) 90 p.

3A1 640602 Hamner, R.W. 1964. An ecological study of *Sarcobatus vermiculatus* communities of the Big Horn Basin, Wyoming. M.S. Thesis. University of Wyoming. 75 p.

3A1 640603 Miller, W.B. 1964. An ecological study of the mountain mahogany community and related biotic associations of the Big Horn Mountains. M.S. Thesis, University of Wyoming. 152 p.

3A1 640604 Nichols, J.T. 1964. Soil-vegetation relationships of the 15-mile drainage, Washakie County, Wyoming. Ph.D. Dissertation. University of Wyoming. 120 p.

3A1 670602 King, T.W. 1967. A study of vegetation, soils and small mammals of limber pine stands in north central Wyoming. M.S. Thesis. University of Wyoming. 58 p.

3A1 710602 Jacoby, P.W. 1971. Interrelationships of vegetation and environmental factors on a mountain watershed in southeastern Wyoming. Ph.D. Dissertation. University of Wyoming. 127 p.

3A1 720604 Shaver, J.C. 1972. Evaluation of vegetation types differentiated by aerial photo interpretation of panchromatic, color, and infrared imagery. M.S. Thesis. University of Wyoming. 86 p.

3A1 740601 Spraker, T.H. 1974. An ecological study of the mule deer *(Odocoileus hemionus* ssp. *hemionus)* on Sheep Mountain, Albany County, Wyoming. M.S. Thesis. University of Wyoming. 81 p.

3A1 740603 Gordon, R.C. 1974. Multispectral imagery for range vegetation type mapping and estimation of above-ground green biomass. M.S. Thesis. University of Wyoming. 45 p.

3A1 750601 Lymbery, G.A. 1975. Vegetation re-establishment potential of strip-mined soils, Black Thunder Basin, Wyoming. M.S. Thesis. University of Wyoming. 62 p.

3A1 790601 Shawesh, O.M. 1979. Vegetation analysis and ecology of rangelands in northwest Libya, Africa. M.S. Thesis. University of Wyoming. 76 p.

3A1 800601 Mesdaghi, M. 1980. Vegetation of saline and alkali habitats of the Red Desert, Wyoming. Ph.D. Dissertation. University of Wyoming. 140 p.

3A2. Condition and Trend

3A2 660601 Wight, J.R. 1966. Site and vegetation characteristics and relationships of juniper communities in northwestern Wyoming. Ph.D. Dissertation. University of Wyoming. 130 p.

3A2 660602 Robinson, L.D. 1966. The vegetation and small mammals of the juniper zone in north central Wyoming. M.S. Thesis. University of Wyoming. 115 p.

3A2 700601 Steger, R.E. 1970. Soil moisture and temperature relationships of six salt desert shrub communities in north central Wyoming. Ph.D. Dissertation. University of Wyoming. 142 p.

3A2 710602 Jacoby, P.W. 1971. Interrelationships of vegetation and environmental factors on a mountain watershed in southeastern Wyoming. Ph.D. Dissertation. University of Wyoming. 127 p.

3A2 710603 Mitchell, G.C. 1971. Spatial distribution and successional state of grassland vegetation related to grazing intensity treatments. M.S. Thesis. University of Wyoming. 64 p.

3A2 720601 Gibbens, R.P. 1972. Vegetation pattern within northern desert shrub communities. Ph.D. Dissertation. University of Wyoming. 272 p.

3A2 720604 Shaver, J.C. 1972. Evaluation of vegetation types differentiated by aerial photo interpretation of panchromatic, color, and infrared imagery. M.S. Thesis. University of Wyoming. 86 p.

3A2 720605 Butler, G.B. 1972. Ecological inventory and analysis of the Laramie Peak big game winter range. M.S. Thesis, Range Management, University of Wyoming. 170 p.

3A2 740602 Hanson, C.E. 1974. An evaluation of the sagebrush and juniper types as mule deer winter range at three locations in Wyoming. M.S. Thesis. University of Wyoming. 122 p.

3A2 740603 Gordon, R.C. 1974. Multispectral imagery for range vegetation type mapping and estimation of above-ground green biomass. M.S. Thesis. University of Wyoming. 45 p.

3A2 750601 Lymbery, G.A. 1975. Vegetation re-establishment potential of strip-mined soils, Black Thunder Basin, Wyoming. M.S. Thesis. University of Wyoming. 62 p.

3A2 760601 Gerhart, W.A. 1976. Mule deer food habitat and forage quality studies on juniper ranges in Wyoming. M.S. Thesis. University of Wyoming. 220 p.

3A2 760602 Kleinman, L.H. 1976. Phenodynamics and ecology of sagebrush-grass rangelands. Ph.D. Dissertation. University of Wyoming. 143 p.

3A2 770601 Taha, F.K. 1977. Ecological interrelationships of vegetation and soils in a shortgrass prairie ecosystem. Ph.D. Dissertation. University of Wyoming. 207 p.

3A2 770602 Hohl, R.J. 1977. Summer habitat preference and vegetation utilization of elk in the Snowy Range, Medicine Bow National Forest, Wyoming. M.S. Thesis. University of Wyoming. 69 p.

3A2 770603 Hughes, H.G. 1977. Factors influencing vegetative and reproductive phenology of *Artemisia nova* Nels. M.S. Thesis. University of Wyoming. 59 p.

3A2 770604 Gdara, A.O. 1977. Interpretation of long-term vegetation data from permanent quadrats on semiarid shrublands of western Wyoming. M.S. Thesis. University of Wyoming. 114 p.

3A2 780602 Orpet, J.E. 1978. Long-term herbage production response following the chemical control of big sagebrush. M.S. Thesis. University of Wyoming. 118 p.

3A2 800601 Mesdaghi, M. 1980. Vegetation of saline and alkali habitats of the Red Desert, Wyoming. Ph.D. Dissertation. University of Wyoming. 140 p.

3A2 800602 Trueblood, D.C. 1980. Soil-vegetation relations at rangeland exclosures in the Gardner saltbush vegetation types of the Big Horn Basin, Wyoming. M.S. Thesis. University of Wyoming. 106 p.

3A2 820601 Elwakeel, A.S. 1982. Effects of environmental factors on perennial grass production in five vegetation types in north central Wyoming. M.S. Thesis. University of Wyoming. 81 p.

3A2 820602 Uhlich, J.W. 1982. Vegetation dynamics of non-grazed and grazed sagebrush grassland communities in the Big Horn Basin of Wyoming. M.S. Thesis. University of Wyoming. 120 p.

3A2 830601 Dean, S.K. 1983. Plant response to livestock grazing and big sagebrush chemical control in western Wyoming 1962 - 1980. M.S. Thesis. University of Wyoming. 84 p.

3A2 840602 Deboodt, T.L. 1984. Vegetation composition and production dynamics in grazed versus non-grazed areas of central Wyoming. M.S. Thesis. University of Wyoming. 87 p.

3A3. Production and Utilization

3A3 640601 Blankenship, J.O. 1964. Indirect estimation of standing crop for selected alpine species. M.S. Thesis. University of Wyoming. 52 p.

3A3 680601 Noller, G.L. 1968. The relationship of forage production to precipitation, cover and soils in north central Wyoming. Ph.D. Dissertation. University of Wyoming. 153 p.

3A3 730601 Whysong, G.L. 1973. Development and evaluation of statistical models for predicting perennial grass production in north central Wyoming. Ph.D. Dissertation. University of Wyoming. 150 p.

3A3 730602 Ries, R.E. 1973. Quantitative evaluation of herbage production and site factors on ecologically diverse sagebrush-grass locations in Wyoming. Ph.D. Dissertation. University of Wyoming. 101 p.

3A3 780602 Orpet, J.E. 1978. Long-term herbage production response following the chemical control of big sagebrush. M.S. Thesis. University of Wyoming. 118 p.

3A3 800602 Trueblood, D.C. 1980. Soil-vegetation relations at rangeland exclosures in the Gardner saltbush vegetation types of the Big Horn Basin, Wyoming. M.S. Thesis. University of Wyoming. 106 p.

3A3 800603 Hargis, N.E. 1980. Phenological comparison of two sagebrush species. M.S. Thesis. University of Wyoming. 110 p.

3A3 820601 Elwakeel, A.S. 1982. Effects of environmental factors on perennial grass production in five vegetation types in north central Wyoming. M.S. Thesis. University of Wyoming. 81 p.

3A3 830601 Dean, S.K. 1983. Plant response to livestock grazing and big sagebrush chemical control in western Wyoming 1962 - 1980. M.S. Thesis. University of Wyoming. 84 p.

3A3 840602 Deboodt, T.L. 1984. Vegetation composition and production dynamics in grazed versus non-grazed areas of central Wyoming. M.S. Thesis. University of Wyoming. 87 p.

4. Grazing Management

4A. STOCKING RATES

4A 670603 Russey, G.R. 1967. The effect of grazing intensities on the roots of Nuttall's Saltbush (*Atriplex nuttallii* S. Wats.) M.S. Thesis. University of Wyoming. 49 p.

4A 720602 Garland, C.B., III. 1972. Ecology of aridland vegetation in western Wyoming. M.S. Thesis. University of Wyoming. 106 p.

4B. GRAZING SEASONS AND SYSTEMS

4B 840602 Deboodt, T.L. 1984. Vegetation composition and production dynamics in grazed versus non-grazed areas of central Wyoming. M.S. Thesis. University of Wyoming. 87 p.

4C. GRAZING HABITS AND DISTRIBUTION

4C 690601 Todd, L.D. 1969. Preference and utilization trends by cattle on grass-forb vegetation in the northern Big Horn Mountains, Wyoming. M.S. Thesis. University of Wyoming. 73 p.

4C 700602 Gunderson, J.J. 1970. A discussion of some important aspects of rangeland carrying capacity with emphasis toward deer of the genus *Odocoileus*. A discussion of some nutritional aspects of several browse species with special emphasis on the value of these plants as food for deer of the genus *Odocoileus*. M.S. Professional Papers. University of Wyoming. 53 & 55 p., respectively.

5. Range and Ranch Development

5A. PLANT CONTROL

5A 780601 Elmekki, A.O. 1978. Evaluation of brush control methods on rangelands. A study of rangeland ecology and use in Sudan. M.S. Professional Papers. University of Wyoming. 39 & 35 p., respectively.

5A1. Methods of Control

5A1B. BIOLOGICAL

5A1B670601 Kirkham, D.R. 1967. Harvester ant (*Pogonomyrmex owyheei* Cole) abundance in relation to grazing, vegetation and soil in north central Wyoming. M.S. Thesis. University of Wyoming. 83 p.

5A1D. CHEMICAL

5A1D720603 Burrough, L.N. 1972. Artificial revegetation trials on juniper game ranges in north central Wyoming. M.S. Thesis. University of Wyoming. 135 p.

5A1D760602 Kleinman, L.H. 1976. Phenodynamics and ecology of sagebrush-grass rangelands. Ph.D. Dissertation. University of Wyoming. 143 p.

5A1D770603 Hughes, H.G. 1977. Factors influencing vegetative and reproductive phenology of *Artemisia nova* Nels. M.S. Thesis. University of Wyoming. 59 p.

5A1D780602 Orpet, J.E. 1978. Long-term herbage production response following the chemical control of big sagebrush. M.S. Thesis. University of Wyoming. 118 p.

5A1D800602 Trueblood, D.C. 1980. Soil-vegetation relations at rangeland exclosures in the Gardner saltbush vegetation types of the Big Horn Basin, Wyoming. M.S. Thesis. University of Wyoming. 106 p.

5A1D830601 Dean, S.K. 1983. Plant response to livestock grazing and big sagebrush chemical control in western Wyoming 1962 - 1980. M.S. Thesis. University of Wyoming. 84 p.

5A1D840601 Mulholland, J.L. 1984. Big sagebrush response and re-establishment following chemical control in western Wyoming. M.S. Thesis. University of Wyoming. 72 p.

5A1D840602 Deboodt, T.L. 1984. Vegetation composition and production dynamics in grazed versus non-grazed areas of central Wyoming. M.S. Thesis. University of Wyoming. 87 p.

5B. RANGE REVEGETATION

5B 790601 Shawesh, O.M. 1979. Vegetation analysis and ecology of rangelands in northwest Libya, Africa. M.S. Thesis. University of Wyoming. 76 p.

5B4. Intermountain

5B4 720603 Burrough, L.N. 1972. Artificial revegetation trials on juniper game ranges in north central Wyoming. M.S. Thesis. University of Wyoming. 135 p.

5B4 780602 Orpet, J.E. 1978. Long-term herbage production response following the chemical control of big sagebrush. M.S. Thesis. University of Wyoming. 118 p.

5B4 800602 Trueblood, D.C. 1980. Soil-vegetation relations at rangeland exclosures in the Gardner saltbush vegetation types of the Big Horn Basin, Wyoming. M.S. Thesis. University of Wyoming. 106 p.

5B4 830601 Dean, S.K. 1983. Plant response to livestock grazing and big sagebrush chemical control in western Wyoming 1962 - 1980. M.S. Thesis. University of Wyoming. 84 p.

5C. WATER SPREADING, FURROWING, PITTING

5C 720602 Garland, C.B., III. 1972. Ecology of aridland vegetation in western Wyoming. M.S. Thesis. University of Wyoming. 106 p.

5C 770605 Cox, J.R. 1977. Phenology, carbohydrate storage, and annual biomass of five shortgrass prairie species in relation to environmental factors. Ph.D. Dissertation. University of Wyoming. 203 p.

5C 800601 Mesdaghi, M. 1980. Vegetation of saline and alkali habitats of the Red Desert, Wyoming. Ph.D. Dissertation. University of Wyoming. 140 p.

6. Range Livestock

6A. GENERAL MANAGEMENT

6A 840602 Deboodt, T.L. 1984. Vegetation composition and production dynamics in grazed versus non-grazed areas of central Wyoming. M.S. Thesis. University of Wyoming. 87 p.

6A1. Cattle

6A1 690601 Todd, L.D. 1969. Preference and utilization trends by cattle on grass-forb vegetation in the northern Big Horn Mountains, Wyoming. M.S. Thesis. University of Wyoming. 73 p.

7. Range Wildlife

7A. BIG GAME

7A 720603 Burrough, L.N. 1972. Artificial revegetation trials on juniper game ranges in north central Wyoming. M.S. Thesis. University of Wyoming. 135 p.

7A1. Deer

7A1 700602 Gunderson, J.J. 1970. A discussion of some important aspects of rangeland carrying capacity with emphasis toward deer of the genus *Odocoileus.* A discussion of some nutritional aspects of several browse species with special emphasis on the value of these plants as food for deer of the genus *Odocoileus.* M.S. Professional Papers. University of Wyoming. 53 & 55 p., respectively.

7A1 720605 Butler, G.B. 1972. Ecological inventory and analysis of the Laramie Peak big game winter range. M.S. Thesis, Range Management, University of Wyoming. 170 p.

7A1 740601 Spraker, T.H. 1974. An ecological study of the mule deer *(Odocoileus hemionus* ssp. *hemionus)* on Sheep Mountain, Albany County, Wyoming. M.S. Thesis. University of Wyoming. 81 p.

7A1 740602 Hanson, C.E. 1974. An evaluation of the sagebrush and juniper types as mule deer winter range at three locations in Wyoming. M.S. Thesis. University of Wyoming. 122 p.

7A1 760601 Gerhart, W.A. 1976. Mule deer food habitat and forage quality studies on juniper ranges in Wyoming. M.S. Thesis. University of Wyoming. 220 p.

7A2. Elk

7A2 680602 Probasco, G.E. 1968. Diet preferences and utilization patterns of elk in the northern Big Horn Mountains, Wyoming. M.S. Thesis. University of Wyoming. 36 p.

7A2 720605 Butler, G.B. 1972. Ecological inventory and analysis of the Laramie Peak big game winter range. M.S. Thesis, Range Management, University of Wyoming. 170 p.

7A2 770602 Hohl, R.J. 1977. Summer habitat preference and vegetation utilization of elk in the Snowy Range, Medicine Bow National Forest, Wyoming. M.S. Thesis. University of Wyoming. 69 p.

7D. NON-GAME

7D1. Rodents and Rabbits

7D1 660602 Robinson, L.D. 1966. The vegetation and small mammals of the juniper zone in north central Wyoming. M.S. Thesis. University of Wyoming. 115 p.

7D1 690602 Forbis, L.A. 1969. Ecological relationships of vegetation, soils, and vertebrates on paired limber pine and spruce-fir sites, Big Horn Mountains, Wyoming. M.S. Thesis. University of Wyoming. 105 p.

8. Range Watersheds

8A. WATER RESOURCES

8A 800601 Mesdaghi, M. 1980. Vegetation of saline and alkali habitats of the Red Desert, Wyoming. Ph.D. Dissertation. University of Wyoming. 140 p.

8C. HYDROLOGY AND WATER YIELD

8C 790601 Shawesh, O.M. 1979. Vegetation analysis and ecology of rangelands in northwest Libya, Africa. M.S. Thesis. University of Wyoming. 76 p.

8C 800601 Mesdaghi, M. 1980. Vegetation of saline and alkali habitats of the Red Desert, Wyoming. Ph.D. Dissertation. University of Wyoming. 140 p.

8D. WATERSHED RESTORATION

8D 750601 Lymbery, G.A. 1975. Vegetation re-establishment potential of strip-mined soils, Black Thunder Basin, Wyoming. M.S. Thesis. University of Wyoming. 62 p.

8D 770601 Taha, F.K. 1977. Ecological interrelationships of vegetation and soils in a shortgrass prairie ecosystem. Ph.D. Dissertation. University of Wyoming. 207 p.

9. Range Economics

9A. RANCH PLANNING

9A 840602 Deboodt, T.L. 1984. Vegetation composition and production dynamics in grazed versus non-grazed areas of central Wyoming. M.S. Thesis. University of Wyoming. 87 p.

10. Range Education and Research

10D. RANGE RESEARCH METHODS

10D 780602 Orpet, J.E. 1978. Long-term herbage production response following the chemical control of big sagebrush. M.S. Thesis. University of Wyoming. 118 p.

10D 820602 Uhlich, J.W. 1982. Vegetation dynamics of non-grazed and grazed sagebrush grassland communities in the Big Horn Basin of Wyoming. M.S. Thesis. University of Wyoming. 120 p.

10D 830601 Dean, S.K. 1983. Plant response to livestock grazing and big sagebrush chemical control in western Wyoming 1962 - 1980. M.S. Thesis. University of Wyoming. 84 p.

10D1. Plant-Soil Emphasis

10D1 620601 Vosler, L.C. 1962. An ecological study of *Atriplex nuttallii* in the Big Horn Basin of Wyoming. M.S. Thesis. University of Wyoming. 43 p.

10D1 640601 Blankenship, J.O. 1964. Indirect estimation of standing crop for selected alpine species. M.S. Thesis. University of Wyoming. 52 p.

10D1 640602 Hamner, R.W. 1964. An ecological study of *Sarcobatus vermiculatus* communities of the Big Horn Basin, Wyoming. M.S. Thesis. University of Wyoming. 75 p.

10D1 640603 Miller, W.B. 1964. An ecological study of the mountain mahogany community and related biotic associations of the Big Horn Mountains. M.S. Thesis. University of Wyoming. 152 p.

10D1 660601 Wight, J.R. 1966. Site and vegetation characteristics and relationships of juniper communities in northwestern Wyoming. Ph.D. Dissertation. University of Wyoming. 130 p.

10D1 660602 Robinson, L.D. 1966. The vegetation and small mammals of the juniper zone in north central Wyoming. M.S. Thesis. University of Wyoming. 115 p.

10D1 670602 King, T.W. 1967. A study of vegetation, soils and small mammals of limber pine stands in north central Wyoming. M.S. Thesis. University of Wyoming. 58 p.

10D1 670603 Russey, G.R. 1967. The effect of grazing intensities on the roots of Nuttall's Saltbush (*Atriplex nuttallii* S. Wats.) M.S. Thesis. University of Wyoming. 49 p.

10D1 680601 Noller, G.L. 1968. The relationship of forage production to precipitation, cover and soils in north central Wyoming. Ph.D. Dissertation. University of Wyoming. 153 p.

10D1 680603 Wambolt, C.L. 1968. Size, density, growth rate and age of Rocky Mountain juniper and interrelationships with environmental factors. M.S. Thesis. University of Wyoming. 62 p.

10D1 710601 Wambolt, C.L. 1971. Moisture stress in woody plants of Wyoming as influenced by environmental factors. Ph.D. Dissertation. University of Wyoming. 158 p.

10D1 710603 Mitchell, G.C. 1971. Spatial distribution and successional state of grassland vegetation related to grazing intensity treatments. M.S. Thesis. University of Wyoming. 64 p.

10D1 730601 Whysong, G.L. 1973. Development and evaluation of statistical models for predicting perennial grass production in north central Wyoming. Ph.D. Dissertation. University of Wyoming. 150 p.

10D1 730602 Ries, R.E. 1973. Quantitative evaluation of herbage production and site factors on ecologically diverse sagebrush-grass locations in Wyoming. Ph.D. Dissertation. University of Wyoming. 101 p.

10D1 740601 Spraker, T.H. 1974. An ecological study of the mule deer (*Odocoileus hemionus* ssp. *hemionus*) on Sheep Mountain, Albany County, Wyoming. M.S. Thesis. University of Wyoming. 81 p.

10D1 740602 Hanson, C.E. 1974. An evaluation of the sagebrush and juniper types as mule deer winter range at three locations in Wyoming. M.S. Thesis. University of Wyoming. 122 p.

10D1 760601 Gerhart, W.A. 1976. Mule deer food habitat and forage quality studies on juniper ranges in Wyoming. M.S. Thesis, University of Wyoming. 220 p.

10D1 770602 Hohl, R.J. 1977. Summer habitat preference and vegetation utilization of elk in the Snowy Range, Medicine Bow National Forest, Wyoming. M.S. Thesis, University of Wyoming. 69 p.

10D1 770603 Hughes, H.G. 1977. Factors influencing vegetative and reproductive phenology of *Artemisia nova* Nels. M.S. Thesis. University of Wyoming. 59 p.

10D1 770604 Gdara, A.O. 1977. Interpretation of long-term vegetation data from permanent quadrats on semiarid shrublands of western Wyoming. M.S. Thesis. University of Wyoming. 114 p.

10D1 770605 Cox, J.R. 1977. Phenology, carbohydrate storage, and annual biomass of five shortgrass prairie species in relation to environmental factors. Ph.D. Dissertation. University of Wyoming. 203 p.

10D1 800601 Mesdaghi, M. 1980. Vegetation of saline and alkali habitats of the Red Desert, Wyoming. Ph.D. Dissertation. University of Wyoming. 140 p.

10D1 800602 Trueblood, D.C. 1980. Soil-vegetation relations at rangeland exclosures in the Gardner saltbush vegetation types of the Big Horn Basin, Wyoming. M.S. Thesis. University of Wyoming. 106 p.

10D2. Animal Emphasis

10D2 710602 Jacoby, P.W. 1971. Interrelationships of vegetation and environmental factors on a mountain watershed in southeastern Wyoming. Ph.D. Dissertation. University of Wyoming. 127 p.

11. Computer Storage, Programming and Modeling

11A. CLIMATIC

11A 800603 Hargis, N.E. 1980. Phenological comparison of two sagebrush species. M.S. Thesis. University of Wyoming. 110 p.

11C. EDAPHIC

11C 800603 Hargis, N.E. 1980. Phenological comparison of two sagebrush species. M.S. Thesis. University of Wyoming. 110 p.

11E. RANGELAND INVENTORY

11E 770604 Gdara, A.O. 1977. Interpretation of long-term vegetation data from permanent quadrats on semiarid shrublands of western Wyoming. M.S. Thesis. University of Wyoming. 114 p.

11E2. Condition and Trend

11E2 800603 Hargis, N.E. 1980. Phenological comparison of two sagebrush species. M.S. Thesis. University of Wyoming. 110 p.

11E2 830602 Kinucan, R.J. 1983. Phenological relationships and patterns of semiarid plant communities in central and western Wyoming. M.S. Thesis. University of Wyoming. 110 p.

11E2 840601 Mulholland, J.L. 1984. Big sagebrush response and re-establishment following chemical control in western Wyoming. M.S. Thesis. University of Wyoming. 72 p.

11E2 840602 Deboodt, T.L. 1984. Vegetation composition and production dynamics in grazed versus non-grazed areas of central Wyoming. M.S. Thesis. University of Wyoming. 87 p.

11E3. Production and Utilization

11E3 730601 Whysong, G.L. 1973. Development and evaluation of statistical models for predicting perennial grass production in north central Wyoming. Ph.D. Dissertation, University of Wyoming. 150 p.

11E3 730602 Ries, R.E. 1973. Quantitative evaluation of herbage production and site factors on ecologically diverse sagebrush-grass locations in Wyoming. Ph.D. Dissertation. University of Wyoming. 101 p.

11E3 800603 Hargis, N.E. 1980. Phenological comparison of two sagebrush species. M.S. Thesis. University of Wyoming. 110 p.

11E3 830601 Dean, S.K. 1983. Plant response to livestock grazing and big sagebrush chemical control in western Wyoming 1962 - 1980. M.S. Thesis. University of Wyoming. 84 p.

11E3 830602 Kinucan, R.J. 1983. Phenological relationships and patterns of semiarid plant communities in central and western Wyoming. M.S. Thesis. University of Wyoming. 110 p.

11E3 840601 Mulholland, J.L. 1984. Big sagebrush response and re-establishment following chemical control in western Wyoming. M.S. Thesis. University of Wyoming. 72 p.

KEY WORD INDEX

CHAPTER 3: KEY WORD INDEX

ALPHABETIC KEY WORD INDEX TABLE OF CONTENTS

The keyword index is based on the list of over 800 key word topics. A table of contents for pagination by alphabetic letter is given below. The total list of keywords is given immediately following to provide readers a quick means of scanning for presence or absence of specific terms.

KEY WORD LIST

KEY WORD INDEX OF CITATION TITLES

The keyword subjects are organized alphabetically. Multiple titles under a keyword are organized alphabetically and in some cases modified slightly to bring a primary word to title beginning. The six digit identifying code number is given for each title.

670105. Harvester ant ecology, Big Horn Basin and Wind River Basin.

670601. Harvester ant (*Pogonomyrmex owyheei* Cole) abundance in relation to grazing, vegetation and soil in north central Wyoming.

840309. Long-term Wyoming rangeland monitoring and research cooperative studies of the USDI-BLM & University of Wyoming, Department of Range Management.

700201. Reliability of harvester ant mound counts from aerial photographs.

700302. Reliability of harvester ant mound counts from aerial photographs.

840602. Vegetation composition and production dynamics in grazed versus non-grazed areas of central Wyoming.

670402. Western harvester ant abundance in relation to grazing, vegetation and soil in north central Wyoming.

PRONGHORN, ANTELOPE

700107. Baggs Grazing systems study.

750105. Baggs Grazing systems study, impacts of grazing systems, BLM trend plots.

720402. Ecological description of biotic communities of the Wagon Wheel environs.

770302. Evaluation of three grazing management systems on northern desert shrub vegetation of southern Wyoming.

700505. Influence of grazing management systems on vegetation and wildlife habitat - semiannual report.

760202. Influences of grazing systems upon vegetation and wildlife.

800505. Influences of grazing management systems on vegetation and wildlife habitat.

780311. Range condition and trend review in the Lander Resource Area.

770406. Rangeland and wildlife resources of the South Pass Historic Area.

720601. Vegetation pattern within northern desert shrub communities.

820401. Vegetation response to grazing management strategies on semiarid shrublands of Wyoming.

ANTELOPE BITTERBRUSH

720605. Ecological inventory and analysis of the Laramie Peak Big Game Winter Range.

AQUATIC

750407. Ecological considerations for Project Wagon Wheel and hydrolic fracturing activities.

ARIDISOL

840314. Herbage yield dynamics and seasonal precipitation relationships.

800504. Soil-vegetation relationships on rangeland exclosures on the Salt Wells Planning Unit of southwestern Wyoming.

800503. Soil-vegetation relationships on rangeland exclosures in the Overland Planning Unit of south central Wyoming.

800602. Soil-vegetation relations at rangeland exclosures in the Gardner saltbush vegetation types of the Big Horn Basin, Wyoming.

ARIDLAND ECOLOGY

610104. Aridland ecology studies of vegetation, cover, composition, production, biochemical composition, precipitation, and site relationships under grazing, shrub control, seeding, and pitting treatments.

680504. Aridland ecology, 1967 results.

690501. Aridland ecology research, 1968 results.

700501. Aridland ecology research, 1969 results.

700506. Aridland ecology semiannual progress report.

710502. Aridland ecology research, 1970 results.

720502. Aridland ecology research, 1971 results.

730501. Aridland ecology research, 1972 results.

760301. Aridland ecology and phenology research in western Wyoming.

760303. Aridland ecological research activities in Wyoming and relationships to Libyan rangelands.

ARTEMISIA TRIDENTATA

790202. Competition as a control for *Halogeton glomeratus.*

840301. Detailed numerical phenology of sagebrush with visual description statements.

840203. Dynamics of vegetative plant part phenodynamics and related stress adaptations.

620201. An ecological study of the *Artemisia tripartita* subsp. *rupicola* and related shrub communities in Wyoming.

840405. Grazing management, longevity, and benefits of sagebrush control on semiarid rangelands.

710601. Moisture stress in woody plants of Wyoming as influenced by environmental factors.

760602. Phenodynamics and ecology of sagebrush-grass rangelands.

760501. Phenology and production studies on semiarid shrub-types.

770502. Phenology and production studies on semiarid shrub-types, 1976 results.

830501. Phenology and production studies on semiarid shrub-types, 1982 results.

810307. Plant composition response to picloram application in the Wyoming juniper type.

840204. Productivity dynamics of sagebrush-grass ecosystems as determined from biweekly sampling of a single grass and shrub species.

640502. Range survey in Wyoming's Big Horn Basin.

790305. Soils-phenology relations on some arid rangelands of western Wyoming.

780406. Structural definition of semiarid *Artemisia* and *Atriplex* rangeland ecosystems.

810604. Successional trends on revegetated *Juniperus osteosperma* communities in north central Wyoming.

770305. The effect of environmental factors on carbohydrate storage and herbage production of five shortgrass prairie plant species.

650401. Vegetation changes from cold desert to subalpine communities.

720601. Vegetation pattern within northern desert shrub communities.

ARTEMISIA TRIPARTITA

620201. Ecological study of the *Artemisia tripartita* subsp. *rupicola* and related shrub communities in Wyoming.

620204. Ecological study of the *Artemisia tripartita* subsp. *rupicola* and related shrub communities in Wyoming - Abst.

620503. Ecological study of the *Artemisia tripartita* subsp. *rupicola* and related shrub communities in Wyoming.

630502. Ecological studies of the *Artemisia tripartita* subsp. *rupicola* and related shrub communities in Wyoming - See 620201.

610202. *Trirhabda attenuata* (Say) (Coleoptera: Chrysomelidae), a leaf feeding beetle found on threetip sagebrush (*Artemisia tripartita* Rydb. subsp. *rupicola* Beetle).

650401. Vegetation changes from cold desert to subalpine communities.

ASPECT

640502. Range survey in Wyoming's Big Horn Basin.

ASPEN

710602. Interrelationships of vegetation and environmental factors on a mountain watershed in southeastern Wyoming.

770602. Summer habitat preference and vegetation utilization of elk in the Snowy Range, Medicine Bow National Forest, Wyoming.

ASTER SPECIES

770602. Summer habitat preference and vegetation utilization of elk in the Snowy Range, Medicine Bow National Forest, Wyoming.

840309. Long-term Wyoming rangeland monitoring and research cooperative studies of the USDI-BLM and University of Wyoming, Department of Range Management.

660503. Microclimate studies in desert shrub-types.

830303. Perennial grass production dynamics among five vegetation types in north central Wyoming.

790502. Phenology and production studies on semiarid shrub-types, 1978 results.

820502. Phenology and production studies on semiarid shrub-types, 1981 results.

830502. Phenology and production studies on semiarid shrub-types, 1982 results, Vol. II.

830601. Plant response to livestock grazing and big sagebrush chemical control in western Wyoming, 1962-1980.

640202. Plants of the 15-Mile Drainage, Washakie County, Wyoming.

840404. Population structure anomalies of *Juniperus osteosperma* invasion in north central Wyoming, U.S.A.

670201. Proper winter sheep grazing on saltbush rangeland.

600101. Range condition class survey - Big Horn and Wind River Basins.

610101. Range condition class survey - Big Horn and Wind River Basins.

610301. Range condition class survey in the Big Horn Basin of Wyoming.

590502. Range survey.

640502. Range survey in Wyoming's Big Horn Basin.

840312. Regional variation of phenology.

840311. Reinvasion following sagebrush control.

680205. Relationship of forage production to precipitation, cover and soils in north central Wyoming.

680601. Relationship of forage production to precipitation, cover and soils in north central Wyoming.

600104. Sagebrush control and seeding, Two Mile, Bud Kimball, Smilo, Upper and Lower Government.

840407. Saltbush community changes with and without grazing over a 20-year period in north central Wyoming.

610404. Seeding salt desert shrublands in Wyoming.

660302. Sheep grazing intensity studies and vegetation changes on *Atriplex nuttallii.*

680204. Size, density, growth rate, and age of Rocky Mountain juniper and interrelationships with environmental factors.

680603. Size, density, growth rate, and age of Rocky Mountain juniper and interrelationships with environmental factors.

690305. Size, density, growth rate, and age of Rocky Mountain juniper and interrelationships with environmental factors.

660116. Small mammal abundance and distribution in juniper.

650104. Soil moisture and temperature study.

680105. Soil moisture and temperature study.

790501. Soil vegetation relationships on rangeland exclosures on the Grass Creek Planning Unit of north central Wyoming.

640304. Soil-vegetation relations of the 15-Mile Drainage area in the Big Horn Basin of Wyoming.

640402. Soil-vegetation relations of the 15-Mile Drainage area in the Big Horn Basin of Wyoming - See 640304.

780106. Soil-vegetation relation study.

800602. Soil-vegetation relations at rangeland exclosures in the Gardner saltbush vegetation types of the Big Horn Basin, Wyoming.

590501. Study of range conditions in the Big Horn and Wind River Basins.

600201. Study of range condition classes and methods of range improvement in Wyoming.

600502. Study of range conditions in the Big Horn and Wind River Basins.

620501. Study of range conditions in the Big Horn and Wind River Basins.

630501. Study of range conditions in the Big Horn and Wind River Basins.

670602. Study of vegetation, soils and small mammals in limber pine stands in north central Wyoming.

620204. Ecological study of the *Artemisia tripartita* subsp. *rupicola* and related shrub communities in Wyoming - Abstract.

720602. Ecology of aridland vegetation in western Wyoming.

770305. Effect of environmental factors on carbohydrate storage and herbage production of five short-grass prairie plant species.

740102. Exclosure study, phenology and growth.

680101. Game range improvement program by control of juniper and seedling establishment studies of preferred browse species.

840405. Grazing management, longevity, and benefits of sagebrush control on semiarid rangelands.

700505. Influence of grazing management systems on vegetation and wildlife habitat, semiannual report.

730103. Intensive investigation of phenology and production relationships of vegetation.

780602. Long-term herbage production response following the chemical control of big sagebrush.

790403. Long-term herbage production response following the chemical control of big sagebrush.

760201. Methods of measuring environmental parameters on a pre-disturbed coal site.

660503. Microclimate studies on desert shrub-types.

650113. Microclimatic studies on desert shrub-types.

710601. Moisture stress in woody plants of Wyoming as influenced by environmental factors.

760601. Mule deer food habitat and forage quality studies on juniper ranges in Wyoming.

820201. Numerical phenology of big and black sagebrush from visual description statements.

820301. Numerical phenology of big and black sagebrush from visual interpretation statements.

820305. Numerical phenology scoring and use with quantitative variables.

760602. Phenodynamics and ecology of sagebrush-grass rangelands.

770306. Phenodynamics and ecology of sagebrush-grass rangelands.

800603. Phenological comparison of two sagebrush species in Wyoming.

810302. Phenological comparison of two sagebrush species in Wyoming.

750503. Phenology and production on semiarid shrub-types, 1974 results.

760501. Phenology and production on semiarid shrub-types, 1975 results.

770502. Phenology and production on semiarid shrub-types, 1976 results.

790502. Phenology and production on semiarid shrub-types, 1978 results.

810502. Phenology and production on semiarid shrub-types, 1980 results, Vol. II.

830501. Phenology and production on semiarid shrub-types, 1982 results.

760305. Physical and chemical analyses of range sites and interrelationships with herbage production and forage value potentials.

810307. Plant composition response to picloram application in the Wyoming juniper type.

830601. Plant response to livestock grazing and big sagebrush chemical control in western Wyoming, 1962-1980.

840204. Productivity dynamics of sagebrush grass ecosystems as determined from biweekly sampling of a single grass and shrub species.

840302. Productivity index procedures for sagebrush-grass rangeland.

640502. Range survey in Wyoming's Big Horn Basin.

800301. Shrub establishment, dominance and ecology on the juniper and sagebrush-grass types in Wyoming.

660601. Site and vegetation characteristics and relationships of juniper communities in northwestern Wyoming.

640104. Soil moisture and temperature study.

660301. Soil moisture and temperature characteristics under grazed and non-grazed conditions.

840408. Big sagebrush response and re-establishment following chemical control in Wyoming, U.S.A.

840601. Big sagebrush response and re-establishment following chemical control in western Wyoming.

640108. Ecology and control of juniper.

650105. Erosion transect study.

780601. Evaluation of brush control methods on rangelands.

680109. Forb control with chemical on subalpine range - Big Horn Mountains.

840405. Grazing management, longevity, and benefits of sagebrush control on semiarid rangelands.

640106. Harvester ant control and related study of vegetation response and economic feasibility of the procedure.

650110. Harvester ant control.

650112. Juniper control study on LU.

680102. Juniper control chemical, burn, mechanical on GRI exclosures.

780602. Long-term herbage production response following the chemical control of big sagebrush.

790403. Long-term herbage production response following the chemical control of big sagebrush.

830601. Plant response to livestock grazing and big sagebrush chemical control in western Wyoming 1962 - 1980.

760204. Production of a shortgrass prairie ecosystem as related to soil characteristics in eastern Wyoming.

640502. Range survey in Wyoming's Big Horn Basin.

830307. Re-establishment rates of sagebrush following chemical control on grazed and ungrazed ranges in western Wyoming.

830411. Re-establishment rates of sagebrush following chemical control on grazed and ungrazed ranges in western Wyoming.

840311. Reinvasion following sagebrush control.

600104. Sagebrush control and seeding, Two mile, Bud Kimball, Smilo, Upper and Lower Govt.

610405. Sagebrush spraying tour.

650108. Seeding in juniper and juniper control.

800301. Shrub establishment, dominance and ecology on the juniper and sagebrush-grass types in Wyoming.

780301. Soil erosion and movement studies in western Wyoming.

780406. Structural definition of semiarid *Artemisia* and *Atriplex* rangeland ecosystems.

610105. Study of biotic communities - juniper, saltbush, greasewood, blue grama, sagebrush and limber pine.

670304. Use of picloram for juniper control.

690301. Vegetation response to chemical control of low sagebrush *(Artemisia arbuscula)* in western Wyoming.

820303. Vegetation responses following strip chemical control of sagebrush at the Lower Government Draw site.

820304. Vegetation response to strip chemical control in the Government Draw area.

820402. Vegetation response to strip chemical control in the Government Draw area.

CHEMICAL TREATMENT LONGEVITY

830307. Re-establishment rates of sagebrush following chemical control on grazed and ungrazed ranges in western Wyoming.

CHRYSOTHAMNUS

840414. Biology and ecology of sagebrush in Wyoming: I. Herbage production and community dynamics.

840415. Biology and ecology of sagebrush in Wyoming: II. Reinvasion following sagebrush control.

840416. Biology and ecology of sagebrush in Wyoming: III. Regional variation of phenology.

840417. Biology and ecology of sagebrush in Wyoming: IV. Validation of a rangeland production model (ERHYM) for sagebrush sites.

840418. Biology and ecology of sagebrush in Wyoming: V. Herbage yield dynamics and seasonal precipitation relationships.

760601. Mule deer food habitat and forage quality studies on juniper ranges in Wyoming.

COMPUTER

COMPUTER PROGRAM

COMPUTER TAPE

730501. Aridland ecology research, 1972 results.

790303. Arid rangeland sampling.

830503. Aridland shrub ecology and phenology research program.

790202. Competition as a control for *Halogeton glomeratus*.

830104. Compilation of herbage production and cover data from Wind River Basin exclosures.

840104. Compilation of herbage production and cover data from Wind River Basin exclosures.

640201. Cover, composition, and production of contour-furrowed and seeded range as compared to native saltsage ranges.

630108. Cumberland district exclosures, production, cover.

750407. Ecological considerations for project Wagon Wheel and hydrolic fracturing activities.

760404. Ecological considerations for Project Wagon Wheel and hydrolic fracturing activities.

810603. Ecological effects of grazing systems management on sagebrush-grassland vegetation of Wyoming.

780308. Ecology of rangeland communities in northwestern Libya, Africa.

730302. Evaluation of furrowing and seeding in north central Wyoming Nuttall saltbush area.

620103. Exclosure studies, production, cover, height, and moisture.

630103. Exclosure studies, production, cover, height, precipitation.

630302. Exclosure studies, production, cover, height, and moisture phase.

630402. Exclosure studies production, cover, height, and moisture phases - See 630302.

640103. Exclosure studies - production, cover, height, precipitation.

650103. Exclosure studies - production, cover, height, precipitation.

660118. Exclosure studies - production, cover, precipitation.

670110. Exclosure studies - production, cover, precipitation.

680106. Exclosure studies - production, cover, precipitation.

690101. Exclosure studies - production, cover and precipitation.

700101. Exclosure studies - production, cover and precipitation.

720104. Exclosure studies - production, cover and precipitation.

730102. Exclosure studies - production, cover and precipitation.

740101. Exclosure studies - production, cover and precipitation.

750101. Exclosure studies - production, cover, precipitation.

760101. Exclosure studies - production, cover, precipitation.

770101. Exclosure studies - production, cover and precipitation.

780101. Exclosure studies - production, cover and precipitation.

790101. Exclosure studies - production, cover and precipitation.

800101. Exclosure studies - production, cover and precipitation.

810101. Exclosure studies - production, cover and precipitation.

820101. Exclosure studies - production, cover and precipitation.

650304. Exclosure studies - responses to sagebrush spraying.

810306. Grazing system impacts on semiarid vegetation of Wyoming.

630503. Halogeton research, 1962 results.

640501. Halogeton research, 1963 results, See 630103, 630104, 630105, 630106, 630108.

650501. Halogeton research, 1964 results.

660501. Halogeton research, 1965 results, See 650101, 650102, 650103, 650107.

640106. Harvester ant control and related study of vegetation response and economic feasibility of the procedure.

840310. Herbage production and community dynamics.

750502. Herbage structure, productivity and phenology on Black Thunder.

760502. Herbage structure, productivity and phenology on Black Thunder.

780501. Herbage structure, production and phenology on Black Thunder.

780502. Herbage structure, production and phenology on Black Thunder.

770604. Interpretation of long-term vegetation data from permanent quadrats on semiarid shrublands of western Wyoming.

810503. Long-term vegetation data from permanent quadrats on semiarid shrublands of Western Wyoming, 2 ft x 20 ft and 4 ft x 4 ft series.

830301. Long-term vegetation responses to sagebrush control in western Wyoming.

840309. Long-term Wyoming rangeland monitoring and research cooperative studies of the USDI-BLM & University of Wyoming, Department of Range Management.

770304. Long-term vegetation studies with permanent quadrats.

760601. Mule deer food habitat and forage quality studies on juniper ranges in Wyoming.

740502. Phenology and production on semiarid shrub-types, 1973 results.

750503. Phenology and production on semiarid shrub-types, 1974 results.

760501. Phenology and production studies on semiarid shrub-types.

770502. Phenology and production studies on semiarid shrub-types, 1976 results.

780503. Phenology and production studies on semiarid shrub-types, 1977 results.

790502. Phenology and production studies on semiarid shrub-types, 1978 results.

800502. Phenology and production studies on semiarid shrub-types, 1979 results.

810502. Phenology and production studies on semiarid shrub-types, 1980 results, Vol. II.

820502. Phenology and production studies on semiarid shrub-types, 1981 results.

830502. Phenology and production studies on semiarid shrub-types, 1982 results, Vol. II.

840501. Phenology and production studies on semiarid shrub-types.

750304. Pre-disturbance ecological studies improve and define potential for surface mine reclamation.

780311. Range condition and trend review in the Lander Resource Area.

720201. Rangeland relations and harvester ants in north central Wyoming.

830307. Re-establishment rates of sagebrush following chemical control on grazed and ungrazed ranges in western Wyoming.

840311. Reinvasion following sagebrush control.

680205. Relationship of forage production to precipitation, cover, and soils in north central Wyoming.

680601. Relationship of forage production to precipitation, cover, and soils in north central Wyoming.

840407. Saltbush community changes with and without grazing over a 20-year period in north central Wyoming.

780307. Semiarid rangeland and grazing evaluation of shrubland from permanent quadrats.

790501. Soil-vegetation relationships on rangeland exclosures on the Grass Creek Planning Unit of north central Wyoming.

790302. Soil-vegetation relationships on rangeland exclosures in the Grass Creek Planning Unit of north central Wyoming.

800602. Soil-vegetation relations at rangeland exclosures in the Gardner saltbush vegetation types of the Big Horn Basin, Wyoming.

810308. Soil-vegetation relationships in the Gardner saltbush vegetation types of Wyoming.

810314. Stagnation and decadency of *Cercocarpus montanus* in a southeastern Wyoming big game exclosure.

740501. Summary of data, 1962-1970. Aridland ecology research.

770602. Summer habitat preference and vegetation utilization of elk in the Snowy Range, Medicine Bow National Forest, Wyoming.

600102. Summer steer grazing on saltbush rangeland with reseeding and water spreader treatments.

640102. Summer steer grazing on saltbush range.

840601. Big sagebrush response and re-establishment following chemical control in western Wyoming.

670305. Cooperative research with the University of Wyoming.

630303. Does weather influence rangeland potential?

740407. Ecological considerations for Project Wagon Wheel.

750407. Ecological considerations for Project Wagon Wheel and hydrolic fracturing activities.

760404. Ecological considerations for Project Wagon Wheel and hydrolic fracturing activities.

720402. Ecological description of biotic communities of the Wagon Wheel environs.

810603. Ecological effects of grazing systems management on sagebrush-grassland vegetation of Wyoming.

720605. Ecological inventory and analysis of the Laramie Peak Big Game Winter Range.

770601. Ecological interrelationships of vegetation and soils in a shortgrass prairie ecosystem, Snowy Range, Medicine Bow National Forest, Wyoming.

690602. Ecological relationships of vegetation, soils, and vertebrates on paired limber pine and spruce-fir site, Big Horn Mountains, Wyoming.

620201. Ecological study of the *Artemisia tripartita* subsp. *rupicola* and related shrub communities in Wyoming.

620204. Ecological study of the *Artemisia tripartita* subsp. *rupicola* and related shrub communities in Wyoming.

620503. Ecological study of the *Artemisia tripartita* subsp. *rupicola* and related shrub communities in Wyoming.

630502. Ecological study of the *Artemisia tripartita* subsp. *rupicola* and related shrub communities in Wyoming - See 620201.

620206. Ecological study of *Atriplex nuttallii* in the Big Horn Basin of Wyoming - abstract.

620601. Ecological study of *Atriplex nuttallii* in the Big Horn Basin of Wyoming - abstract.

640602. Ecological study of *Sarcobatus vermiculatus* communities of the Big Horn Basin, Wyoming.

640603. Ecological study of the mountain mahogany community and related biotic associations of the Big Horn Mountains.

740601. Ecological study of the mule deer *(Odocoileus hemionus)* on Sheep Mountain, Albany County, Wyoming.

640108. Ecology and control of juniper.

780202. Ecology of rangeland communities in northwestern Libya, Africa.

780308. Ecology of rangeland communities in northwestern Libya, Africa.

830302. Economic feasibility of resting rangeland to improve forage production in central Wyoming.

780601. Evaluation of brush control methods on rangelands.

660107. Harvester ant ecology, Big Horn Basin.

670105. Harvester ant ecology, Big Horn Basin and Wind River Basin.

790304. Important rangeland plants of Wyoming and the West.

710602. Interrelationships of vegetation and environmental factors on a mountain watershed in southeastern Wyoming.

660108. Juniper ecology - permanent plot transects.

670106. Juniper ecology - permanent plot transects and seeding trials.

720302. Juniper ecology and response to protection from browsing.

690302. Juniper game range improvement studies in western Wyoming.

660111. Limber pine ecology.

670109. Limber pine ecology.

760601. Mule deer food habitat and forage quality studies on juniper ranges in Wyoming.

730101. On an area near Reno Junction, Wyoming, on which a strip-mine operation is proposed, the ecology, vegetation, and soils inventory studies were initiated in 1973.

760602. Phenodynamics and ecology of sagebrush-grass rangelands.

770306. Phenodynamics and ecology of sagebrush-grass rangelands.

840315. Productivity dynamics of sagebrush ecosystems as determined from biweekly sampling of a single grass and shrub species.

700305. Significance of plant patterns in ecosystem functioning.

710503. Significance of plant patterns in ecosystem functioning.

830309. Utah juniper invasion in north central Wyoming and historical climatic and management relationships.

650401. Vegetation changes from cold desert to subalpine communities.

820602. Vegetation dynamics of non-grazed and grazed sagebrush grassland communities in the Big Horn Basin of Wyoming.

ECOSYSTEM MODELING

840301. Detailed numerical phenology of sagebrush with visual description statements.

730601. Development and evaluation of statistical models for predicting perennial grass production in north central Wyoming.

680601. Relationship of forage production to precipitation, cover and soils in north central Wyoming.

720601. Vegetation pattern within northern desert shrub communities.

820602. Vegetation dynamics of non-grazed and grazed sagebrush grassland communities in the Big Horn Basin of Wyoming.

EDAPHIC FACTORS

700301. Distribution of plant biomass to plant organs.

640602. Ecological study of *Sarcobatus vermiculatus* communities of the Big Horn Basin, Wyoming.

720602. Ecology of aridland vegetation in western Wyoming.

810312. Habitat and vegetation of semiarid rangeland, Libya, Africa.

670601. Harvester ant (*Pogonomyrmex owyheei* Cole) abundance in relation to grazing, vegetation and soil in north central Wyoming.

710302. Moisture stress in woody plants as influenced by environmental factors.

710601. Moisture stress in woody plants of Wyoming as influenced by environmental factors.

700202. Predictive value of factors influencing perennial grass production on western Wyoming sagebrush-grass ranges.

720201. Rangeland relations and harvester ants in north central Wyoming.

700601. Soil moisture and temperature relationships of six salt desert shrub communities in north central Wyoming.

640604. Soil-vegetation relationships of the 15-mile drainage, Washakie County, Wyoming.

ELEVATION

840314. Herbage yield dynamics and seasonal precipitation relationships.

710602. Interrelationships of vegetation and environmental factors on a mountain watershed in southeastern Wyoming.

840501. Phenology and production studies on semiarid shrub-types.

690601. Preference and utilization trends by cattle on grass-forb vegetation in the northern Big Horn Mountains, Wyoming.

790503. Rangeland and wildlife resources of the South Pass Historic Mining Area.

670602. Study of vegetation, soils and small mammals in limber pine stands in north central Wyoming.

650401. Vegetation changes from cold desert to subalpine communities.

820602. Vegetation dynamics of non-grazed and grazed sagebrush grassland communities in the Big Horn Basin of Wyoming.

720601. Vegetation pattern within northern desert shrub communities.

820304. Vegetation response to strip chemical control in the Government Draw area.

820402. Vegetation response to strip chemical control in the Government Draw area.

ELEVATIONAL GRADIENT

610101. Range condition class survey - Big Horn & Wind River Basin.

770605. Phenology, carbohydrate storage and annual biomass of five short-grass prairie species in relation to environmental factors.

830601. Plant response to livestock grazing and big sagebrush chemical control in western Wyoming, 1962 - 1980.

700202. Predictive value of factors influencing perennial grass production on western Wyoming sagebrush-grass ranges.

830305. Productivity differences of grasslands in semidesert western Wyoming — A re-examination.

840302. Productivity index procedures for sagebrush-grass rangeland.

730602. Quantitative evaluation of herbage production and site factors on ecologically diverse sagebrush-grass locations in Wyoming.

740303. Relationship with soils and other environmental factors to rangeland production and rehabilitation.

740304. Relationships of range production and rehabilitation to soils and other environmental factors.

830308. Size and density parameters of western wheatgrass on diverse range sites in western Wyoming.

690403. Size, density, growth rate and age of Rocky Mountain juniper and interrelationships with environmental factors.

790501. Soil-vegetation relationships on rangeland exclosures on the Grass Creek Planning Unit of north central Wyoming.

800302. Soil-vegetation relationships on a semiarid rangeland exclosure in north central Wyoming - long-term studies.

720101. Study of biological and physical environmental characteristics as represented by remote imagery obtained from the N.A.S.A. Earth Resources Technology Satellite (ERTS).

770305. The effect of environmental factors on carbohydrate storage and herbage production of five shortgrass prairie plant species.

810309. Vegetation dynamics of non-grazed semiarid sagebrush grassland communities in the Big Horn Basin of Wyoming.

820302. Vegetation productivity and phenology in the Wind River Basin.

820404. Vegetation productivity and phenology in the Wind River Basin.

760503. Vegetation, Section IX. Final Report Environmental Assessment of the Black Thunder Mine Site.

ERHYM MODEL

830108. Yield prediction with ERHYM model.

840108. Yield prediction with ERHYM model.

EROSION

730601. Development and evaluation of statistical models for predicting perennial grass production in north central Wyoming.

770303. Development of snowfence systems on strip-mine spoils to increase seedling establishment.

650304. Exclosure studies - Responses to sagebrush spraying.

710503. Significance of plant pattern in ecosystem functioning.

EROSION TRANSECTS

830503. Aridland shrub ecology and phenology research program.

750407. Ecological considerations for project Wagon Wheel and hydrolic fracturing activities.

630105. Erosion transect study.

650105. Erosion transect study.

660113. Erosion transects for study of soil movement and relationship to climatic variation and grazing management.

700108. Erosion transects.

830106. Erosion transect data summarization and input to computers.

840106. Erosion transect data summarization and input to computers.

640501. Halogeton Research, 1963 results, See 630103, 630104, 630105, 630106, 630108.

830104. Compilation of herbage production and cover data from Wind River Basin exclosures.

840109. Compilation of U.S. Weather Service precipitation records - Wyoming.

720602. Ecology of aridland vegetation in western Wyoming.

810303. Effect of community changes on precipitation/production relationships in exclosures of saltbush vegetation.

810403. Effect of community changes on precipitation/production relationships in exclosures of saltbush vegetation.

640203. Effect of protection from grazing on two vegetation types in the Big Horn Basin of Wyoming.

820601. Effects of environmental factors on perennial grass production in five vegetation types in north central Wyoming.

700105. Exclosure soil profile studies with SCS.

710108. Exclosure soil profile studies with SCS.

720107. Exclosure soil profile studies with SCS.

730107. Exclosure soil profile studies with SCS.

770104. Exclosure soil profile studies with SCS.

790105. Exclosure soil profile studies with SCS.

620103. Exclosure studies, production, cover, height, and moisture.

630103. Exclosure studies, production, cover, height, precipitation.

640103. Exclosure studies, production, cover, height, precipitation.

650103. Exclosure studies: Production, cover, height, precipitation.

650304. Exclosure studies - Responses to sagebrush spraying.

660118. Exclosure studies - production, cover, precipitation.

670110. Exclosure studies - production, cover, precipitation.

670503. Exclosure study with transects of permanent plots - See 660106.

680106. Exclosure studies - production, cover, precipitation.

680502. Exclosure studies with transects of permanent plots.

690101. Exclosure studies - production, cover, precipitation.

700101. Exclosure studies, production, cover and precipitation.

700504. Exclosure studies with transects of permanent plots.

730102. Exclosure studies, production, cover and precipitation.

740101. Exclosure studies, production, cover and precipitation.

740102. Exclosure study, phenology and growth.

750101. Exclosure studies, production, cover and precipitation.

750102. Exclosure study, phenology and growth.

760101. Exclosure studies, production, cover and precipitation.

760102. Exclosure study, phenology and growth.

770101. Exclosure studies, production, cover and precipitation.

770102. Exclosure study, phenology and growth.

780101. Exclosure studies, production, cover and precipitation.

780102. Exclosure study, phenology and growth.

790101. Exclosure studies, production, cover and precipitation.

790102. Exclosure study, phenology and growth.

790103. Exclosure studies, productivity/standing crop sampling.

800101. Exclosure studies, production, cover and precipitation.

800102. Exclosure study, phenology and growth.

800103. Exclosure studies, productivity/standing crop sampling.

810101. Exclosure studies, production, cover and precipitation.

810102. Exclosure study, phenology and growth.

810103. Exclosure studies, productivity/standing crop sampling.

820101. Exclosure studies, production, cover and precipitation.

820102. Exclosure study, phenology and growth.

820103. Exclosure studies, productivity/standing crop sampling.

820104. Exclosure study, woody plant density.

790504. Grazing management system studies of northern desert shrub vegetation of southern Wyoming.

650501. Halogeton research, 1964 results.

750502. Herbage structure, productivity and phenology on Black Thunder.

770501. Herbage structure, productivity and phenology on Black Thunder.

780501. Herbage structure, productivity and phenology on Black Thunder.

730103. Intensive investigation of phenology and production relationships of vegetation.

770604. Interpretation of long-term vegetation data from permanent quadrats on semiarid shrublands of western Wyoming.

680102. Juniper control chemical, burn, mechanical on GRI exclosures.

670106. Juniper ecology - permanent plot transects and seeding trials.

720302. Juniper ecology and response to protection from browsing.

680503. Long-term exclosure studies with 2 ft x 20 ft and 4 ft x 4 ft permanent plots, 1959-1964 results.

780602. Long-term herbage production response following the chemical control of big sagebrush.

790301. Long-term herbage production response following chemical sagebrush control in western Wyoming.

770304. Long-term vegetation studies with permanent quadrats.

770306. Phenodynamics and ecology of sagebrush-grass rangelands.

740502. Phenology and production on semiarid shrub-types, 1973 results.

750503. Phenology and production on semiarid shrub-types, 1974 results.

760501. Phenology and production on semiarid shrub-types, 1975 results.

770502. Phenology and production on semiarid shrub-types, 1976 results.

780503. Phenology and production on semiarid shrub-types, 1977 results.

790502. Phenology and production on semiarid shrub-types, 1978 results.

800502. Phenology and production on semiarid shrub-types, 1979 results.

810502. Phenology and production on semiarid shrub-types, 1980 results.

820502. Phenology and production on semiarid shrub-types, 1981 results.

830501. Phenology and production on semiarid shrub-types, 1982 results.

830502. Phenology and production on semiarid shrub-types, 1982 results, Vol. II.

840501. Phenology and production on semiarid shrub-types, 1983 results.

760305. Physical and chemical analyses of range sites and interrelationships with herbage production and forage value potentials.

830601. Plant response to livestock grazing and big sagebrush chemical control in western Wyoming, 1962-1980.

700202. Predictive value of factors influencing perennial grass production on western Wyoming sagebrush-grass ranges.

760204. Production of a shortgrass prairie ecosystem as related to soil characteristics in eastern Wyoming.

740301. Productivity differences of semidesert grasslands in western Wyoming.

830305. Productivity differences of grasslands in semidesert western Wyoming - A re-examination.

780311. Range condition and trend review in the Lander Resource Area.

590502. Range survey.

830307. Re-establishment rates of sagebrush following chemical control on grazed and ungrazed ranges in western Wyoming.

740303. Relationship with soils and other environmental factors to rangeland production and rehabilitation.

840407. Saltbush community changes with and without grazing over a 20-year period in north central Wyoming.

650108. Seeding in juniper and juniper control.

780106. Soil-vegetation relation study.

800302. Soil-vegetation relationships at semiarid rangeland exclosures in north central Wyoming — long-term studies.

800401. Soil-vegetation relationships at semiarid rangeland exclosures in north central Wyoming — long-term studies.

FRINGED SAGEWORT

770601. Ecological interrelationships of vegetation and soils in a shortgrass prairie ecosystem, Snowy range, Medicine Bow National Forest.

760305. Physical and chemical analyses of range sites and interrelationships with herbage production and forage value potentials.

740503. Seed production ecology in the shortgrass prairie of the central United States.

750303. Seed production ecology in the shortgrass prairie of the central United States.

FURROW DEPTH

730302. Evaluation of furrowing and seeding in north central Wyoming Nuttall saltbush area.

FURROW SEEDING

730301. Methods of revegetation and restoration following a juniper wildfire.

FURROWING

740201. Contour-furrowing and seeding on Nuttall saltbush rangelands of Wyoming.

720602. Ecology of aridland vegetation in western Wyoming.

730302. Evaluation of furrowing and seeding in north central Wyoming Nuttall saltbush area.

730401. Evaluation of furrowing and seeding in north central Wyoming Nuttall saltbush area - See 730302.

GAMBEL OAK

710602. Interrelationships of vegetation and environmental factors on a mountain watershed in southeastern Wyoming.

GAME ANIMALS

750303. Seed production ecology in the shortgrass prairie of the central United States.

720101. Study of biological and physical environmental characteristics as represented by remote imagery obtained from the N.A.S.A. Earth Resources Technology Satellite (ERTS).

GAME RANGE IMPROVEMENT

720603. Artificial revegetation trials on juniper game ranges in north central Wyoming.

680101. Game range improvement program by control of juniper and seedling establishment studies of preferred browse species.

790108. Juniper and limber pine population dynamics and life history.

800106. Juniper and limber pine population dynamics and life history.

680102. Juniper control chemical, burn, mechanical on GRI exclosures.

690302. Juniper game range improvement studies in western Wyoming.

810307. Plant composition response to picloram application in the Wyoming juniper type.

810604. Successional trends on revegetated *Juniperus osteosperma* communities in north central Wyoming.

GAME RANGE INVENTORY

720605. Ecological inventory and analysis of the Laramie Peak Big Game Winter Range.

710110. Laramie Peak Game Range Inventory.

720113. Laramie Peak Game Range Inventory.

GAME WINTER RANGE

720605. Ecological inventory and analysis of the Laramie Peak Big Game Winter Range.

800301. Shrub establishment, dominance and ecology on the juniper and sagebrush-grass types in Wyoming.

GAME-PROOF EXCLOSURES

720302. Juniper ecology and response to protection from browsing.

GENETIC

770603. Factors influencing vegetative and reproductive phenology of *Artemisia nova* Nels.

760108. Grass seed sources at A.C. Hull 1940's plantings.

750304. Pre-disturbance ecological studies improve and define potential for surface mine reclamation.

830311. Grazing, vegetation and soils interactions since 1967 in the Powder Rim grazing system.

670601. Harvester ant (*Pogonomyrmex owyheei* Cole) abundance in relation to grazing, vegetation and soil in north central Wyoming.

610403. Intensity of grazing study.

710602. Interrelationships of vegetation and environmental factors on a mountain watershed in southeastern Wyoming.

810503. Long-term vegetation data from permanent quadrats on semiarid shrublands of Western Wyoming, 2 ft x 20 ft and 4 ft x 4 ft series.

830301. Long-term vegetation responses to sagebrush control in western Wyoming.

830601. Plant response to livestock grazing and big sagebrush chemical control in western Wyoming, 1962 - 1980.

640502. Range survey in Wyoming's Big Horn Basin.

830307. Re-establishment rates of sagebrush following chemical control on grazed and ungrazed ranges in western Wyoming.

840311. Reinvasion following sagebrush control.

740302. Remote sensing applied to the grassland biome.

670202. Rotation grazing on Wyoming's Big Horn Mountains.

840407. Saltbush community changes with and without grazing over a 20-year period in north central Wyoming.

660117. Saltbush root development and relation to grazing intensity.

780307. Semiarid rangeland and grazing evaluation of shrubland from permanent quadrats.

780404. Semiarid rangeland and grazing evaluation of shrubland from permanent quadrats.

660302. Sheep grazing intensity studies and vegetative changes on *Atriplex nuttallii.*

780301. Soil erosion and movement studies in western Wyoming.

720101. Study of biological and physical environmental characteristics as represented by remote imagery obtained from the N.A.S.A. Earth Resources Technology Satellite (ERTS).

820602. Vegetation dynamics of non-grazed and grazed sagebrush grassland communities in the Big Horn Basin of Wyoming.

690301. Vegetation response to chemical control of low sagebrush *(Artemisia arbuscula)* in western Wyoming.

820401. Vegetation response to grazing management strategies on semiarid shrublands of Wyoming.

760304. Vegetative response of *Bouteloua gracilis* to four levels of utilization.

670306. Western harvester ant abundance in relation to grazing, vegetation and soils in north central Wyoming.

670402. Western harvester ant abundance in relation to grazing, vegetation and soil in north central Wyoming.

GRAZING INTENSITY

840601. Big sagebrush response and re-establishment following chemical control in western Wyoming.

670305. Cooperative research with the University of Wyoming.

670603. Effect of grazing intensities on the roots of Nuttall's Saltbush (*Atriplex nuttallii* S. Wats.).

660304. Effect of sheep grazing intensities on vegetative change in Nuttall saltbush.

680109. Forb control with chemical on subalpine range - Big Horn Mountains.

660102. Grazing intensity influence on dietary intake by use of fistulated sheep.

650303. Halogeton abundance and distribution as influenced by degree of saltbush utilization.

670601. Harvester ant (*Pogonomyrmex owyheei* Cole) abundance in relation to grazing, vegetation and soil in north central Wyoming.

740403. Influence of livestock grazing on growth patterns, plant density and soil microrelief of a *Bouteloua gracilis* prairie.

750501. Influence of grazing management systems on vegetation and wildlife habitat in the Red Desert region of Wyoming.

610403. Intensity of grazing study.

670303. Plant control on salt desert shrub rangeland by sheep.

640502. Range survey in Wyoming's Big Horn Basin.

720201. Rangeland relations and harvester ants in north central Wyoming.

840311. Reinvasion following sagebrush control.

670202. Rotation grazing on Wyoming's Big Horn Mountains.

740503. Seed production ecology of the shortgrass prairie of the central United States.

750303. Seed production ecology in the shortgrass prairie of the central United States.

660302. Sheep grazing intensity studies and vegetative changes on *Atriplex nuttallii.*

660401. Sheep grazing intensity studies and vegetative changes on *Atriplex nuttallii* - See 660302.

680201. Soil moisture and temperature changes following sagebrush control.

710603. Spatial distribution and successional state of grassland vegetation related to grazing intensity treatments.

760401. Vegetative response of *Bouteloua gracilis* to four levels of utilization.

GRAZING MANAGEMENT

650304. Exclosure studies - responses to sagebrush spraying.

840308. Grazing management longevity and benefits of sagebrush control in western Wyoming.

820503. Vegetation response to grazing management strategies on semiarid shrublands of Wyoming.

GRAZING PRESSURE

640301. Winter sheep grazing studies on saltbush rangeland of Wyoming.

GRAZING STRESS

770304. Long-term vegetation studies with permanent quadrats.

830305. Productivity differences of grasslands in semidesert western Wyoming — A re-examination.

750303. Seed production ecology in the shortgrass prairie of the central United States.

GRAZING SYSTEMS

830503. Aridland shrub ecology and phenology research program.

660112. Baggs grazing systems study.

750105. Baggs grazing systems study, impacts of grazing systems, BLM trend plots.

670305. Cooperative research with the University of Wyoming.

810603. Ecological effects of grazing systems management on sagebrush-grassland vegetation of Wyoming.

770302. Evaluation of three grazing management systems on northern desert shrub vegetation of southern Wyoming.

770401. Evaluation of three grazing management systems on northern desert shrub vegetation of southern Wyoming.

790504. Grazing management system studies of northern desert shrub vegetation of southern Wyoming.

760302. Grazing systems research on northern desert shrub rangelands of Wyoming.

810306. Grazing system impacts on semiarid vegetation of Wyoming.

810406. Grazing system impacts on semiarid vegetation of Wyoming.

670302. Influence of summer steer use under rotation and season-long grazing systems - See 670202.

680501. Influence of grazing management systems on vegetation and wildlife habitat, July, 1967 - July, 1968.

690502. Influence of grazing management systems on vegetation and wildlife habitat, July, 1968 - July, 1969.

700502. Influence of grazing management systems on vegetation and wildlife habitat, July, 1969 - July, 1970.

HALOGETON PASTURES

HARVESTER ANT

HARVESTER ANT CONTROL

HEIGHT

HEIGHT/WIDTH RATIO

HERBAGE PRODUCTION

HERBAGE PRODUCTIVITY

760305. Physical and chemical analyses of range sites and interrelationships with herbage production and forage value potentials.

820304. Vegetation response to strip chemical control in the Government Draw area.

820402. Vegetation response to strip chemical control in the Government Draw area.

JUNIPER

710201. A greenhouse analysis of variance of picloram retention in soil and duff from juniper sites.

710301. A greenhouse study of variance of picloram retention in soil and duff from juniper sites.

830503. Aridland shrub ecology and phenology research program.

720603. Artificial revegetation trials on juniper game ranges in north central Wyoming.

680117. Baggs grazing system study.

730201. Conifer water potential as influenced by stand density and environmental factors.

740601. Ecological study of the mule deer *(Odocoileus hemionus)* on Sheep Mountain, Albany County, Wyoming.

640108. Ecology and control of juniper.

740602. Evaluation of the sagebrush and juniper types as mule deer winter range at three locations in Wyoming.

680101. Game range improvement program by control of juniper and seedling establishment studies of preferred browse species.

690303. Improvement studies on juniper sites in western Wyoming - See 690302.

790108. Juniper and limber pine population dynamics and life history.

800106. Juniper and limber pine population dynamics and life history.

660204. Juniper communities and range management in northwestern Wyoming.

650112. Juniper control study on L.U.

670102. Juniper control study of L.U.

690102. Juniper control study of L.U.

670106. Juniper ecology - permanent plot transects and seeding trials.

680111. Juniper ecology 20 ft x 50 ft permanent plots on GRI exclosures.

690302. Juniper game range improvement studies in western Wyoming.

680114. Juniper GRI exclosure seedings.

690109. Juniper GRI exclosure seedings.

690112. Juniper GRI Exclosure Tordon retention studies.

700103. Juniper GRI seedings.

710102. Juniper GRI seedings.

720105. Juniper GRI seedings.

720302. Juniper ecology and response to protection from browsing.

760601. Mule deer food habitat and forage quality studies on juniper ranges in Wyoming.

730108. Mule deer winter range evaluation.

810307. Plant composition response to picloram application in the Wyoming juniper type.

810407. Plant composition response to picloram application in the Wyoming juniper type.

840404. Population structure anomalies of *Juniperus osteosperma* invasion in north central Wyoming, U.S.A.

650108. Seeding in juniper and juniper control.

800301. Shrub establishment, dominance and ecology on the juniper and sagebrush-grass types in Wyoming.

810413. Shrub establishment, dominance and ecology on the juniper and sagebrush-grass types in Wyoming.

660601. Site and vegetation characteristics and relationships of juniper communities in northwestern Wyoming.

660116. Small mammal abundance and distribution in Juniper.

800503. Soil-vegetation relationships on rangeland exclosures in the Overland Planning Unit of south central Wyoming.

610105. Study of biotic communities - juniper, saltbush, greasewood, blue grama, sagebrush and limber pine.

680104. Study of elk-livestock competition in terms of utilization by class of animal and influence on vegetation.

680103. Study of moisture stress in woody species using pressure bomb.

810604. Successional trends on revegetated *Juniperus osteosperma* communities in north central Wyoming.
670404. Use of picloram for juniper control - See 670304.
830309. Utah juniper invasion in north central Wyoming and historical climatic and management relationships.
660205. Vegetation and small mammals of the juniper zone in north central Wyoming-abst.
660602. Vegetation and small mammals of the juniper zone in north central Wyoming.

JUNIPER AGE
690305. Size, density, growth rate and age of Rocky Mountain juniper and interrelationships with environmental parameters.

JUNIPER CONTROL
720603. Artificial revegetation trials on juniper game ranges in north central Wyoming.
650112. Juniper control study on L.U.
660104. Juniper control study on L.U.
680102. Juniper control chemical, burn, mechanical on GRI exclosures.
690102. Juniper control study on L.U.
700102. Juniper control study on L.U.
650108. Seeding in juniper and juniper control.
800301. Shrub establishment, dominance and ecology on the juniper and sagebrush-grass types in Wyoming.
610105. Study of biotic communities - juniper, saltbush, greasewood, blue grama, sagebrush and limber pine.
810604. Successional trends on revegetated *Juniperus osteosperma* communities in north central Wyoming.
670304. Use of picloram for juniper control.
670404. Use of picloram for juniper control - See 670304.

JUNIPER ECOLOGY
660108. Juniper ecology - permanent plot transects.
670106. Juniper ecology - permanent plot transects and seeding trials.
720302. Juniper ecology and response to protection from browsing.

JUNIPER EXCLOSURES
780104. Juniper GRI exclosure 20′ x 50′ permanent quadrats.

JUNIPER INVASION
660204. Juniper communities and range management in northwestern Wyoming.
810310. Population structure of Utah juniper in north central Wyoming.
660601. Site and vegetation characteristics and relationships of juniper communities in northwestern Wyoming.
610105. Study of biotic communities - juniper, saltbush, greasewood, blue grama, sagebrush and limber pine.
830309. Utah juniper invasion in north central Wyoming and historical climatic and management relationships.
830413. Utah juniper invasion in north central Wyoming and historical climatic and management relationships.

JUNIPER WILDFIRE
730301. Methods of revegetation and restoration following a juniper wildfire.
730402. Methods of revegetation and restoration following a juniper wildfire.
790106. Juniper wildfire in Wind River Canyon 13 miles south of Thermopolis.

JUNIPERUS OSTEOSPERMA
680505. *Juniperus osteosperma* in northwestern Wyoming - their distribution and ecology.
710601. Moisture stress in woody plants of Wyoming as influenced by environmental factors.
810307. Plant composition response to picloram application in the Wyoming juniper type.
840404. Population structure anomalies of *Juniperus osteosperma* invasion in north central Wyoming, U.S.A.
810604. Successional trends on revegetated *Juniperus osteosperma* communities in north central Wyoming.
660205. Vegetation and small mammals of the juniper zone in north central Wyoming - abst.
660602. Vegetation and small mammals of the juniper zone in north central Wyoming.

NONRANDOMNESS

700203. *Opuntia* clump size variation on the shortgrass plains, Pawnee Site, Nunn, Colorado.

680301. Patterns of live and dead vegetation in grassland ecosystems and their relationship to function.

690401. Plant pattern and distribution in ecosystems and relationships to function.

690503. Preliminary report of methodology and results for analyses of plant pattern project research on the Pawnee Site.

710603. Spatial distribution and successional state of grassland vegetation related to grazing intensity treatments.

NORTH FORK PASTURES

670201. Proper winter sheep grazing on saltbush rangeland.

610402. Studies of mechanically treated watersheds.

610102. Summer steer grazing on saltbush rangeland with reseeding and water spreader treatments.

650102. Summer steer grazing on saltbush range.

NORTHERN DESERT SHRUB

720601. Vegetation pattern within northern desert shrub communities.

NUCLEAR DETONATION

750302. Vegetation response to seismic motion from underground nuclear detonation.

750404. Vegetation response to seismic motion from underground nuclear detonation.

NUCLEAR STIMULATION

740407. Ecological considerations for Project Wagon Wheel.

NUMERICAL INDEX

830602. Phenological relationships and patterns of semiarid plant communities in central and western Wyoming.

NUMERICAL PHENOLOGY

840301. Detailed numerical phenology of sagebrush with visual description statements.

840304. Detailed numerical phenology of sagebrush with visual description statements.

840307. Detailed numerical phenology of sagebrush with visual description statements.

840402. Detailed numerical phenology of sagebrush with visual description statements.

840411. Detailed numerical phenology of sagebrush with visual description statements.

840203. Dynamics of vegetative plant part phenodynamics and related stress adaptations.

820201. Numerical phenology of big and black sagebrush from visual description statements.

820301. Numerical phenology of big and black sagebrush from visual interpretation statements.

820305. Numerical phenology scoring and use with quantitative variables.

820403. Numerical phenology scoring and use with quantitative variables.

830304. Phenology description and numerical scoring of sagebrush.

NUMERICAL SCORING

830408. Phenological description and numerical scoring of sagebrush.

NUTRIENT CONTENT

760601. Mule deer food habitat and forage quality studies on juniper ranges in Wyoming.

770601. Ecological interrelationships of vegetation and soils in a shortgrass prairie ecosystem, Snowy Range, Medicine Bow National Forest, Wyoming.

NUTRIENT INTAKE

760601. Mule deer food habitat and forage quality studies on juniper ranges in Wyoming.

NUTRIENT LEVELS

810304. Esophageal-fistulated steers for determination of dietary nutrient compositions on large pasture native rangeland.

720106. 2 ft x 20 ft and 4 ft x 4 ft permanent plot study.

760104. 2 ft x 20 ft and 4 ft x 4 ft permanent plot study.

660114. 2 ft x 20 ft and 4 ft x 4 ft permanent plots.

690106. 2 ft x 20 ft and 4 ft x 4 ft permanent plots.

700506. Aridland ecology semiannual progress report.

680117. Baggs grazing system study.

690105. Baggs grazing system study.

750105. Baggs grazing system study, impacts of grazing systems, BLM trend plots.

740104. Black Thunder pre-mining inventory.

760105. Black Thunder pre-mining inventory.

690110. Block juniper site, 1 ft x 1 ft permanent plot studies.

670503. Exclosure study with transects of permanent plots - See 660106.

680502. Exclosure studies with transects of permanent plots.

700504. Exclosure studies with transects of permanent plots.

810306. Grazing system impacts on semiarid vegetation of Wyoming.

670105. Harvester ant ecology, Big Horn Basin and Wind River Basin.

750502. Herbage structure, productivity and phenology on Black Thunder.

700502. Influence of grazing management systems on vegetation and wildlife habitat, July, 1969 - July, 1970.

700102. Juniper control study.

660108. Juniper ecology - permanent plot transects.

670106. Juniper ecology - permanent plot transects and seeding trials.

680111. Juniper ecology 20 ft x 50 ft permanent plots on GRI exclosures.

680503. Long-term exclosure studies with 2 ft x 20 ft and 4 ft x 4 ft permanent plots, 1959-1964 results.

650111. Sagebrush cover estimation with angle gauge.

740503. Seed production ecology of the short-grass prairie of the central United States.

650108. Seeding in juniper and juniper control.

830309. Utah juniper invasion in north central Wyoming and historical climatic and management relationships.

840602. Vegetation composition and production dynamics in grazed versus non-grazed areas of central Wyoming.

820602. Vegetation dynamics of non-grazed and grazed sagebrush grassland communities in the Big Horn Basin of Wyoming.

750302. Vegetation response to seismic motion from underground nuclear detonation.

710107. Wagon Wheel range survey.

PERMANENT PLOT TRANSECTS

660106. 1 ft x 1 ft permanent plot transects.

730109. 1 ft x 1 ft permanent plot transects.

750407. Ecological considerations for Project Wagon Wheel and hydrolic fracturing activities.

720605. Ecological inventory and analysis of the Laramie Peak Big Game Winter Range.

780308. Ecology of rangeland communities in northwestern Libya, Africa.

780501. Herbage structure, production and phenology on Black Thunder.

760202. Influences of grazing systems upon vegetation and wildlife.

730103. Intensive investigation of phenology and production relationships of vegetation.

660108. Juniper ecology - permanent plot transects.

770202. Vegetation analysis and ecology of rangelands in northwest Libya, Africa.

820602. Vegetation dynamics of non-grazed and grazed sagebrush grassland communities in the Big Horn Basin of Wyoming.

810316. Vegetative response to grazing systems management in Wyoming's Red Desert.

PERMANENT POINTS

750407. Ecological considerations for Project Wagon Wheel and hydrolic fracturing activities.

PERMANENT QUADRATS

780107. 1 ft x 1 ft permanent plot study.

690104. 1 ft x 1 ft permanent plot transects.

610104. Aridland ecology studies of vegetation, cover, composition, production, biochemical composition, precipitation, and site relationships under grazing, shrub control, seeding, and pitting treatments.

830503. Aridland shrub ecology and phenology research program.

750407. Ecological considerations for Project Wagon Wheel and hydrolic fracturing activities.

640108. Ecology and control of juniper.

630302. Exclosure studies, production, cover, height, and moisture phase.

790504. Grazing management system studies of northern desert shrub.

640106. Harvester ant control and related study of vegetation response and economic feasibility of the procedure.

650110. Harvester ant control.

760502. Herbage structure, production and phenology on Black Thunder.

770501. Herbage structure, production and phenology on Black Thunder.

780502. Herbage structure, production and phenology on Black Thunder.

770604. Interpretation of long-term vegetation data from permanent quadrats on semiarid shrublands of western Wyoming.

660104. Juniper control study on L.U.

780104. Juniper GRI exclosure 20 ft x 50 ft permanent quadrats.

810503. Long-term vegetation data from permanent quadrats on semiarid shrublands of western Wyoming, 2 ft x 20 ft and 4 ft x 4 ft series.

840309. Long-term Wyoming rangeland monitoring and research cooperative studies of the USDI-BLM and University of Wyoming, Department of Range Management.

770304. Long-term vegetation studies with permanent quadrats.

770405. Long-term vegetation studies with permanent quadrats.

780310. Rangeland vegetation, improvement and management in Libya, Africa.

780307. Semiarid rangeland and grazing evaluation of shrubland from permanent quadrats.

780404. Semiarid rangeland and grazing evaluation of shrubland from permanent quadrats.

790501. Soil-vegetation relationships on rangeland exclosures on the Grass Creek Planning Unit of north central Wyoming.

780406. Structural definition of semiarid *Artemisia* and *Atriplex* rangeland ecosystems.

PERMANENT RECORD PHOTOGRAPHS

780311. Range condition and trend review in the Lander Resource Area.

PERMEABILITY

640604. Soil-vegetation relationships of the 15-Mile Drainage, Washakie County, Wyoming.

PEROMYSCUS MANICULATUS

690602. Ecological relationships of vegetation, soils, and vertebrates on paired limber pine and spruce-fir site, Big Horn Mountains, Wyoming.

PHENODYNAMICS

840201. Culm weight dynamics as an index of sagebrush-grass community productivity.

750306. Phenodynamics and environmental characteristics of sagebrush-grass rangelands.

750403. Phenodynamics and environmental characteristics of sagebrush-grass rangelands.

760602. Phenodynamics and ecology of sagebrush-grass rangelands.

770306. Phenodynamics and ecology of sagebrush-grass rangelands.

770402. Phenodynamics and ecology of sagebrush-grass rangelands.

760203. Phenology and environmental relationships of sagebrush-grass rangelands.

840302. Productivity index procedures for sagebrush-grass rangeland.

780406. Structural definition of semiarid *Artemisia* and *Atriplex* rangeland ecosystems.

PHENOLOGIC PROGRESSION

820305. Numerical phenology scoring and use with quantitative variables.

820403. Numerical phenology scoring and use with quantitative variables.

830602. Phenological relationships and patterns of semiarid plant communities in central and western Wyoming.

PHENOLOGY

790303. Arid rangeland sampling.

760301. Aridland ecology and phenology research in western Wyoming.

830503. Aridland shrub ecology and phenology research program.

810605. Correlation of soil factors and environmental conditions with phenophase development of western wheatgrass.

830306. Dynamics of plant community development for diverse sagebrush sites in western Wyoming.

840203. Dynamics of vegetative plant-part phenodynamics and related stress adaptations.

780203. Effects of water stress on phenology and carbohydrate storage in the shortgrass prairie.

790401. Effects of water stress on phenology and carbohydrate storage in the shortgrass prairie.

790103. Exclosure studies, productivity/standing crop sampling.

740102. Exclosure study, phenology and growth.

750102. Exclosure study, phenology and growth.

760102. Exclosure study, phenology and growth.

770102. Exclosure study, phenology and growth.

780102. Exclosure study, phenology and growth.

790102. Exclosure study, phenology and growth.

800102. Exclosure study, phenology and growth.

810102. Exclosure study, phenology and growth.

820102. Exclosure study, phenology and growth.

770603. Factors influencing vegetative and reproductive phenology of *Artemisia nova* Nels.

750502. Herbage structure, productivity and phenology on Black Thunder.

760502. Herbage structure, productivity and phenology on Black Thunder.

770501. Herbage structure, productivity and phenology on Black Thunder.

780501. Herbage structure, productivity and phenology on Black Thunder.

780502. Herbage structure, productivity and phenology on Black Thunder.

720103. In 1972 the Aridland ecology studies were amplified to include detailed phenological studies.

730103. Intensive investigation of phenology and production relationships of vegetation.

840309. Long-term Wyoming rangeland monitoring and research cooperative studies of the USDI-BLM and University of Wyoming, Department of Range Management.

760201. Methods of measuring environmental parameters on a pre-disturbed coal site.

810602. Niche differentiation and species response comparisons between bluebunch wheatgrass and western wheatgrass.

800603. Phenological comparison of two sagebrush species.

810302. Phenological comparison of two sagebrush species in Wyoming.

810402. Phenological comparison of two sagebrush species in Wyoming.

830304. Phenological description and numerical scoring of sagebrush.

830408. Phenological description and numerical scoring of sagebrush.

760305. Physical and chemical analyses of range sites and interrelationships with herbage production and forage value potentials.

820304. Vegetation response to strip chemical control in the Government Draw area.

820402. Vegetation response to strip chemical control in the Government Draw area.

PRECIPITATION

610104. Aridland ecology studies of vegetation, cover, composition, production, biochemical composition, precipitation, and site relationships under grazing, shrub control, seeding, and pitting treatments.

670502. Aridland ecology research, 1966 results.

680504. Aridland ecology research, 1967 results.

690501. Aridland ecology research, 1968 results.

700501. Aridland ecology research, 1969 results.

710502. Aridland ecology research, 1970 results.

720502. Aridland ecology research, 1971 results.

730501. Aridland ecology research, 1972 results.

700506. Aridland ecology semi-annual progress report.

830503. Aridland shrub ecology and phenology research program.

830310. *Atriplex*/grass and forb relationships under no-grazing and shifting precipitation patterns in north central Wyoming.

840413. *Atriplex*/grass and forb relationships under no-grazing and shifting precipitation patterns in north central Wyoming.

840408. Big sagebrush response and re-establishment following chemical control in Wyoming, U.S.A.

840109. Compilation of U.S. Weather Service precipitation records - Wyoming.

840102. Computer program development for long-term weather record compilation.

840103. Computer program development for estimation of missing climatic data.

840301. Detailed numerical phenology of sagebrush with visual description statements.

710101. Determination of shrub production interrelated with herbaceous production and precipitation.

840203. Dynamics of vegetative plant-part phenodynamics and related stress adaptations.

750407. Ecological considerations for Project Wagon Wheel and hydrolic fracturing activities.

620201. Ecological study of the *Artemisia tripartita* subsp. *rupicola* and related shrub communities in Wyoming.

620204. Ecological study of the *Artemisia tripartita* subsp. *rupicola* and related shrub communities in Wyoming.

720602. Ecology of aridland vegetation in western Wyoming.

770305. Effect of environmental factors on carbohydrate storage and herbage production of five short-grass prairie plant species.

820601. Effects of environmental factors on perennial grass production in five vegetation types in north central Wyoming.

840303. Estimation of missing precipitation data in Wyoming using traditional, kriging, and best guess estimates.

730302. Evaluation of furrowing and seeding in north central Wyoming Nuttall saltbush area.

710102. Exclosure studies, phenology and growth.

620103. Exclosure studies, production, cover, height, and moisture.

630103. Exclosure studies, production, cover, height, precipitation.

640103. Exclosure studies, production, cover, height, precipitation.

650103. Exclosure studies, production, cover, height, precipitation.

660118. Exclosure studies, production, cover, height, precipitation.

670110. Exclosure studies, production, cover, height, precipitation.

680106. Exclosure studies, production, cover, height, precipitation.

690101. Exclosure studies, production, cover, height, precipitation.

700101. Exclosure studies, production, cover, height, precipitation.

710101. Exclosure studies, production, cover, height, precipitation.

720104. Exclosure studies, production, cover, height, precipitation.

730102. Exclosure studies, production, cover, height, precipitation.

740101. Exclosure studies, production, cover, height, precipitation.

750101. Exclosure studies, production, cover, height, precipitation.

760101. Exclosure studies, production, cover, height, precipitation.

770101. Exclosure studies, production, cover, height, precipitation.

780101. Exclosure studies, production, cover, height, precipitation.

790101. Exclosure studies, production, cover, height, precipitation.

800101. Exclosure studies, production, cover, height, precipitation.

810101. Exclosure studies, production, cover, height, precipitation.

820101. Exclosure studies, production, cover, height, precipitation.

840412. Grazing management longevity and benefits of sagebrush control in western Wyoming, U.S.A.

630503. Halogeton research, 1962 results.

640501. Halogeton research, 1963 results, See 630103, 630104, 630105, 630106, 630108.

650501. Halogeton research, 1964 results.

840310. Herbage production and community dynamics.

780501. Herbage structure, production and phenology on Black Thunder.

780502. Herbage structure, production and phenology on Black Thunder.

840314. Herbage yield dynamics and seasonal precipitation relationships.

780602. Long-term herbage production response following the chemical control of big sagebrush.

840309. Long-term Wyoming rangeland monitoring and research cooperative studies of the USDI-BLM and University of Wyoming, Department of Range Management.

830303. Perennial grass production dynamics among five vegetation types in north central Wyoming.

750306. Phenodynamics and environmental characteristics of sagebrush-grass rangelands.

740502. Phenology and production on semiarid shrub-types, 1973 results.

750503. Phenology and production on semiarid shrub-types, 1974 results.

760501. Phenology and production on semiarid shrub-types, 1975 results.

770502. Phenology and production on semiarid shrub-types, 1976 results.

780503. Phenology and production on semiarid shrub-types, 1977 results.

790502. Phenology and production on semiarid shrub-types, 1978 results.

800502. Phenology and production on semiarid shrub-types, 1979 results.

810305. Phenology and productivity comparisons of western wheatgrass *(Agropyron smithii)* and bluebunch wheatgrass *(A. spicatum)*.

810315. Phenology and culm weight variation between two wheatgrass species: *Agropyron smithii* and *Agropyron spicatum*.

810502. Phenology and production studies on semiarid shrub-types, 1980 results, Vol. II.

820502. Phenology and production studies on semiarid shrub-types, 1981 results.

830501. Phenology and production studies on semiarid shrub-types, 1982 results.

830502. Phenology and production studies on semiarid shrub-types, 1982 results, Vol. II.

760305. Physical and chemical analyses of range sites and interrelationships with herbage production and forage value potentials.

650107. Precipitation pattern study.

830305. Productivity differences of grasslands in semidesert western Wyoming — A re-examination.

PRECIPITATION PATTERN

PRE-MINING INVENTORY

PRIMARY SUCCESSION

PRIME SPECIES

PRODUCTION

820502. Phenology and production studies on semiarid and shrub-types, 1981 results, Vol. II.

830501. Phenology and production studies on semiarid shrub-types, 1982 results, Vol. I.

830502. Phenology and production studies on semiarid shrub-types, 1982 results, Vol. II.

840501. Phenology and production studies on semiarid shrub-types.

840503. Phenology and production studies on semiarid shrub-types.

760305. Physical and chemical analyses of range sites and interrelationships.

760402. Physical and chemical analyses of range sites and interrelationships.

750301. Pre-disturbance vegetation inventory and ecological monitoring procedures for rangeland strip-mining.

760204. Production of a shortgrass prairie ecosystem as related to soil characteristics in eastern Wyoming.

760310. Production of a shortgrass prairie ecosystem as related to soil characteristics in eastern Wyoming.

740301. Productivity differences of semidesert grasslands in western Wyoming.

830305. Productivity differences of grasslands in semidesert western Wyoming — A re-examination.

630301. Progress of cattle and sheep grazing trials on halogeton infested ranges.

670201. Proper winter sheep grazing on saltbush rangeland.

780311. Range condition and trend review in the Lander Resource Area.

740303. Relationship with soils and other environmental factors to rangeland production and rehabilitation.

740304. Relationships of range production and rehabilitation to soils and other environmental factors.

660301. Soil moisture and temperature characteristics under grazed and non-grazed conditions.

660402. Soil moisture and temperature characteristics as related to grass production following shrub control under grazed and ungrazed conditions.

670301. Soil moisture and temperature characteristics as related to grass production following shrub control under grazed and non-grazed conditions.

790501. Soil-vegetation relationships on rangeland exclosures on the Grass Creek Planning Unit of north central Wyoming.

800602. Soil-vegetation relations at rangeland exclosures in the Gardner saltbush vegetation types of the Big Horn Basin, Wyoming.

810308. Soil-vegetation relationships in the Gardner saltbush vegetation types of Wyoming.

810314. Stagnation and decadency of *Cercocarpus montanus* in a southeastern Wyoming big game exclosure.

740405. Statistical models for description and prediction of perennial grass production.

740501. Summary of data, 1962-1970, Aridland ecology research.

600102. Summer steer grazing on saltbush rangeland with reseeding and water spreader treatments.

640102. Summer steer grazing on saltbush range.

770305. The effect of environmental factors on carbohydrate storage and herbage production of five shortgrass prairie plant species.

810309. Vegetation dynamics of non-grazed semiarid sagebrush grassland communities in the Big Horn Basin of Wyoming.

820302. Vegetation productivity and phenology in the Wind River Basin.

820404. Vegetation productivity and phenology in the Wind River Basin.

820304. Vegetation response to strip chemical control in the Government Draw area.

820402. Vegetation response to strip chemical control in the Government Draw area.

610103. Winter sheep grazing studies on saltbush rangeland.

630101. Winter sheep grazing on saltbush range.

REPRODUCTIVE PHENOLOGY

840301. Detailed numerical phenology of sagebrush with visual description statements.

820305. Numerical phenology scoring and use with quantitative variables.

820403. Numerical phenology scoring and use with quantitative variables.

800603. Phenological comparison of two sagebrush species.

810501. Phenology and production studies on semiarid shrub-types, 1980 results, Vol I.

770605. Phenology, carbohydrate storage and annual biomass of five shortgrass prairie species in relation to environmental factors.

RESEARCH

650502. University of Wyoming and Bureau of Land Management Cooperative research in the Wind River Basin.

RESEEDING

730106. CIG pipeline seed plot study.

670305. Cooperative research with the University of Wyoming.

730301. Methods of revegetation and restoration following a juniper wildfire.

640502. Range survey in Wyoming's Big Horn Basin.

590102. Summer steer grazing on saltbush rangeland with reseeding and water spreader treatments.

600102. Summer steer grazing on saltbush rangeland with reseeding and water spreader treatments.

610102. Summer steer grazing on saltbush rangeland with reseeding and water spreader treatments.

RESOURCE MANAGEMENT

700304. The I.B.P. and its implications for natural resource management.

REST ROTATION

700505. Influence of grazing management systems on vegetation and wildlife habitat, semiannual report.

REST ROTATION UNIT

810306. Grazing system impacts on semiarid vegetation of Wyoming.

810316. Vegetation response to grazing systems management in Wyoming's Red Desert.

RESTING RANGELAND

840602. Vegetation composition and production dynamics in grazed versus non-grazed areas of central Wyoming.

REVEGETATION

720603. Artificial revegetation trials on juniper game ranges in north central Wyoming.

660303. Effect of control of harvester ants on revegetation.

730301. Methods of revegetation and restoration following a juniper wildfire.

730402. Methods of revegetation and restoration following a juniper wildfire.

730202. Revegetation of pipeline areas.

730303. Revegetation of pipeline areas.

770107. Snow catchment and revegetation on Jim Bridger mine spoil.

810604. Successional trends on revegetated *Juniperus osteosperma* communities in north central Wyoming.

660502. Summary of research during 1964 and 1965 for the control of the western harvester ants.

RHIZOMATOUS

840408. Big sagebrush response and re-establishment following chemical control in Wyoming, U.S.A.

730601. Development and evaluation of statistical models for predicting perennial grass production in north central Wyoming.

700503. Frequency, density and basal area measures on sagebrush-grass ranges.

RIO BLANCO

750302. Vegetation response to seismic motion from underground nuclear detonation.

RIPARIAN

830311. Grazing, vegetation and soils interactions since 1967 in the Powder Rim grazing system.

830304. Phenological description and numerical scoring of sagebrush.

830408. Phenological description and numerical scoring of sagebrush.

760501. Phenology and production studies on semiarid shrub types.

840315. Productivity dynamics of sagebrush ecosystems as determined from biweekly sampling of a single grass and shrub species.

640502. Range survey in Wyoming's Big Horn Basin.

840312. Regional variation of phenology.

820306. Sagebrush and perennial grass dynamics at the Granite Mountain shrub control site.

820405. Sagebrush and perennial grass dynamics at the Granite Mountain shrub control site.

610405. Sagebrush spraying tour.

660115. Sagebrush-wind-snow-pattern relation study.

650104. Soil moisture and temperature study.

610105. Study of biotic communities - juniper, saltbush, greasewood, blue grama, sagebrush and limber pine.

680103. Study of moisture stress in woody species using pressure bomb.

650401. Vegetation changes from cold desert to subalpine communities.

800601. Vegetation of saline and alkali habitats of the Red Desert, Wyoming.

SAGEBRUSH-BLUE GRAMA COMMUNITY

640604. Soil vegetation relationships of the 15-Mile Drainage, Washakie County, Wyoming.

830302. Economic feasibility of resting rangeland to improve forage production in central Wyoming.

SAGEBRUSH CONTROL

840415. Biology and ecology of sagebrush in Wyoming: II. Reinvasion following sagebrush control.

650103. Exclosure studies: Production, cover, height, precipitation.

650304. Exclosure studies - Responses to sagebrush spraying.

810301. Forage production and longevity of sagebrush control on semiarid rangelands in western Wyoming.

810401. Forage production and longevity of sagebrush control on semiarid rangelands in western Wyoming.

840308. Grazing management longevity and benefits of sagebrush control in western Wyoming.

840405. Grazing management, longevity, and benefits of sagebrush control on semiarid rangelands.

840412. Grazing management longevity and benefits of sagebrush control in western Wyoming, U.S.A.

750501. Influence of grazing management systems on vegetation and wildlife habitat in the Red Desert region of Wyoming.

830301. Long-term vegetation responses to sagebrush control in western Wyoming.

830405. Long-term vegetation responses to sagebrush control in western Wyoming.

790301. Long-term herbage production response following chemical sagebrush control in western Wyoming.

830601. Plant response to livestock grazing and big sagebrush chemical control in western Wyoming, 1962 - 1980.

840311. Reinvasion following sagebrush control.

600104. Sagebrush control and seeding, Two mile, Bud Kimball, Smilo, Upper and Lower Govt.

660301. Soil moisture and temperature characteristics under grazed and non-grazed conditions.

670301. Soil moisture and temperature characteristics as related to grass production following shrub control under grazed and non-grazed conditions.

680201. Soil moisture and temperature changes following sagebrush control.

820303. Vegetation responses following strip chemical control of sagebrush at the Lower Government Draw site.

SAGEBRUSH CONTROL BENEFITS

SAGEBRUSH CONTROL LONGEVITY

SAGEBRUSH COVER

SAGEBRUSH ECOLOGY

SAGEBRUSH-GRASS

SALT WELLS PLANNING UNIT

800504. Soil-vegetation relationships on rangeland exclosures on the Salt Wells Planning Unit of southwestern Wyoming.

SALTBUSH

640107. 2 ft x 20 ft and 4 ft x 4 ft permanent plot sampling.

830503. Aridland shrub ecology and phenology research program.

670305. Cooperative research with the University of Wyoming.

630108. Cumberland district exclosures, production, cover.

620206. Ecological study of *Atriplex nuttallii* in the Big Horn Basin of Wyoming - Abst.

620601. Ecological study of *Atriplex nuttallii* in the Big Horn Basin of Wyoming.

810403. Effect of community changes on precipitation/production relationships in exclosures of saltbush vegetation.

640203. Effect of protection from grazing on two vegetation types in the Big Horn Basin of Wyoming.

820601. Effects of environmental factors on perennial grass production in five vegetation types in north central Wyoming.

650305. Grass seeded spreader systems can increase carrying capacity.

660102. Grazing intensity influence on dietary intake by use of fistulated sheep.

650303. Halogeton abundance and distribution as influenced by degree of saltbush utilization.

650402. Halogeton abundance and distribution as influenced by degree of saltbush utilization.

630503. Halogeton research, 1962 results.

660501. Halogeton research, 1965 results, See 650101, 650102, 650103, 650107.

670601. Harvester ant (*Pogonomyrmex owyheei* Cole) abundance in relation to grazing, vegetation and soil in north central Wyoming.

790304. Important rangeland plants of Wyoming and the West.

660503. Microclimate studies on desert shrub types.

650113. Microclimatic studies on desert shrub types.

630301. Progress of cattle and sheep grazing trials on halogeton infested ranges.

640502. Range survey in Wyoming's Big Horn Basin.

680601. Relationship of forage production to precipitation, cover, and soils in north central Wyoming.

660302. Sheep grazing intensity studies and vegetative changes on *Atriplex nuttallii.*

640104. Soil moisture and temperature study.

650104. Soil moisture and temperature study.

640304. Soil-vegetation relations of the 15-Mile Drainage area in the Big Horn Basin of Wyoming.

610402. Studies of mechanically treated watersheds.

610105. Study of biotic communities - juniper, saltbush, greasewood, blue grama, sagebrush and limber pine.

810303. The effect of community changes on precipitation/production relationships in exclosures of saltbush vegetation.

670306. Western harvester ant abundance in relation to grazing, vegetation and soils in north central Wyoming.

SALTBUSH COMMUNITY

840407. Saltbush community changes with and without grazing over a 20-year period in north central Wyoming.

640604. Soil-vegetation relationships of the 15-Mile Drainage, Washakie County, Wyoming.

SALTBUSH ROOTS

670603. Effect of grazing intensities on the roots of Nuttall's Saltbush (*Atriplex nuttallii* S. Wats.).

660117. Saltbush root development and relation to grazing intensity.

SALTGRASS

640502. Range survey in Wyoming's Big Horn Basin.

800601. Vegetation of saline and alkali habitats of the Red Desert, Wyoming.

SEEDING

610104. Aridland ecology studies of vegetation, cover, composition, production, biochemical composition, precipitation, and site relationships under grazing, shrub control, seeding, and pitting treatments.

720603. Artificial revegetation trials on juniper game ranges in north central Wyoming.

740201. Contour-furrowing and seeding on Nuttall saltbush rangelands of Wyoming.

720602. Ecology of aridland vegetation in western Wyoming.

730302. Evaluation of furrowing and seeding in north central Wyoming Nuttall saltbush area.

730401. Evaluation of furrowing and seeding in a north central Wyoming Nuttall saltbush area - See 730302.

650501. Halogeton research, 1964 results.

700106. Inspect SCS seeding along highway west of Moneta.

670106. Juniper ecology - permanent plot transects and seeding trials.

680114. Juniper GRI exclosure seedings.

710102. Juniper GRI seedings.

680112. Kane Deer exclosure seeding analysis.

730301. Methods of revegetation and restoration following a juniper wildfire.

720102. Reseeding trials of grasses, forbs and shrubs on the bladed and disturbed area of a 120-mile long pipeline.

600104. Sagebrush control and seeding, Two mile, Bud Kimball, Smilo, Upper and Lower Govt.

640109. Seeding and fertilizer studies.

660103. Seeding and fertilizer studies.

650108. Seeding in juniper and juniper control.

800301. Shrub establishment, dominance and ecology on the juniper and sagebrush-grass types in Wyoming.

610402. Studies of mechanically treated watersheds.

SEEDING SALT DESERT

610404. Seeding salt desert shrublands in Wyoming.

SEEDLING ESTABLISHMENT

770303. Development of snowfence systems on strip-mine spoils to increase seedling establishment.

770404. Development of snowfence systems on strip-mine spoils to increase seedling establishment.

680101. Game range improvement program by control of juniper and seedling establishment studies of preferred browse species.

SEISMIC MOTION AND TESTING

750407. Ecological considerations for Project Wagon Wheel and hydrolic fracturing activities.

750302. Vegetation response to seismic motion from underground nuclear detonation.

750404. Vegetation response to seismic motion from underground nuclear detonation.

SELENIUM

720402. Ecological description of biotic communities of the Wagon Wheel environs.

SEMIARID DESERT

720402. Ecological description of biotic communities of the Wagon Wheel environs.

SEMIARID RANGELAND

610104. Aridland ecology studies of vegetation, cover, composition, production, biochemical composition, precipitation, and site relationships under grazing, shrub control, seeding, and pitting treatments.

830310. *Atriplex*/grass and forb relationships under no grazing and shifting precipitation patterns in north central Wyoming.

830302. Economic feasibility of resting rangeland to improve forage production in central Wyoming.

810301. Forage production and longevity of sagebrush control on semiarid rangelands in western Wyoming.

810401. Forage production and longevity of sagebrush control on semiarid rangelands in western Wyoming.

810306. Grazing system impacts on semiarid vegetation of Wyoming.

810406. Grazing system impacts on semiarid vegetation of Wyoming.

810312. Habitat and vegetation of semiarid rangelands, Libya, Africa.

810412. Habitat and vegetation of semiarid rangelands, Libya, Africa.

840314. Herbage yield dynamics and seasonal precipitation relationships.

840309. Long-term Wyoming rangeland monitoring and research cooperative studies of the USDI-BLM & University of Wyoming, Department of Range Management.

770304. Long-term vegetation studies with permanent quadrats.

830602. Phenological relationships and patterns of semiarid plant communities in central and western Wyoming.

750503. Phenology and production on semiarid shrub-types, 1974 results.

760501. Phenology and production studies on semiarid shrub-types.

800501. Phenology and production studies on semiarid shrub-types, 1979 results.

800502. Phenology and production studies on semiarid shrub-types, 1979 results.

810305. Phenology and productivity comparisons of western Wheatgrass *(Agropyron smithii)* and bluebunch wheatgrass *(A. spicatum)*.

810501. Phenology and production studies on semiarid shrub-types, 1980 results, Vol I.

810502. Phenology and production studies on semiarid shrub-types, 1980 results, Vol. II.

820501. Phenology and production studies on semiarid shrub-types, 1981 results, Vol. I.

820502. Phenology and production studies on semiarid shrub-types, 1981 results, Vol. II.

830501. Phenology and production studies on semiarid shrub-types, 1982 results, Vol. I.

830502. Phenology and production studies on semiarid shrub-types, 1982 results, Vol. II.

840501. Phenology and production studies on semiarid shrub-types.

840503. Phenology and production studies on semiarid shrub-types.

840204. Productivity dynamics of sagebrush-grass ecosystems as determined from biweekly sampling of a single grass and shrub species.

840312. Regional variation of phenology.

780307. Semiarid rangeland and grazing evaluation of shrubland from permanent quadrats.

780404. Semiarid rangeland and grazing evaluation of shrubland from permanent quadrats.

800401. Soil-vegetation relationships at semiarid rangeland exclosures in north central Wyoming - long-term studies.

800302. Soil-vegetation relationships on a semiarid rangeland exclosure in north central Wyoming - long-term studies.

780406. Structural definition of semiarid *Artemisia* and *Atriplex* rangeland ecosystems.

770202. Vegetation analysis and ecology of rangelands in northwest Libya, Africa.

810309. Vegetation dynamics of non-grazed semiarid sagebrush grassland communities in the Big Horn Basin of Wyoming.

810409. Vegetation dynamics of non-grazed semiarid sagebrush grassland communities in the Big Horn Basin of Wyoming.

840406. Vegetative phenology patterns of semiarid plant communities in Wyoming, U.S.A.

SEMIARID SHRUB-TYPES

680101. Game range improvement program by control of juniper and seedling establishment studies of preferred browse species.

740502. Phenology and production on semiarid shrub-types, 1973 results.

770502. Phenology and production studies on semiarid shrub-types, 1976 results.

780503. Phenology and production studies on semiarid shrub-types, 1977 results.

790502. Phenology and production studies on semiarid shrub-types, 1978 results.

750303. Seed production ecology in the shortgrass prairie of the Central United States.

750406. Seed production ecology in the shortgrass prairie.

710104. Shortgrass prairie blue grama culm density study.

710105. Shortgrass prairie seed production study.

720108. Shortgrass prairie blue grama culm density study.

720109. Shortgrass prairie seed production study.

730110. Shortgrass prairie blue grama culm density study.

730111. Shortgrass prairie seed production study.

740107. Shortgrass prairie blue grama culm density study.

740108. Shortgrass prairie seed production study.

750110. Shortgrass prairie blue grama culm density study.

750111. Shortgrass prairie seed production study.

770403. The effect of environmental factors on carbohydrate storage and herbage production of five shortgrass prairie plant species.

760304. Vegetative response of *Bouteloua gracilis* to four levels of utilization.

SHORTGRASS-CACTUS

680110. Vegetation pattern on shortgrass-cactus range.

690108. Vegetation pattern on shortgrass-cactus range.

700110. Vegetation pattern on shortgrass-cactus range.

SHOSHONI ANT

650501. Halogeton Research, 1964 results.

670105. Harvester ant ecology Big Horn Basin and Wind River Basin.

SHRUB COMMUNITIES

670502. Aridland ecology research, 1966 results.

690501. Aridland ecology research, 1968 results.

700506. Aridland ecology, semiannual progress report.

720502. Aridland ecology research, 1971 results.

830503. Aridland shrub ecology and phenology research program.

840301. Detailed numerical phenology of sagebrush with visual description statements.

620201. Ecological study of the *Artemisia tripartita* subsp. *rupicola* and related shrub communities in Wyoming.

620204. Ecological study of the *Artemisia tripartita* subsp. *rupicola* and related shrub communities in Wyoming.

620503. Ecological study of the *Artemisia tripartita* subsp. *rupicola* and related shrub communities in Wyoming.

630502. Ecological study of the *Artemisia tripartita* subsp. *rupicola* and related shrub communities in Wyoming.

780308. Ecology of rangeland communities in northwestern Libya, Africa.

770101. Factors affecting growth and productivity of vegetation stratum in eastern Wyoming.

810301. Forage production and longevity of sagebrush control on semiarid rangelands in western Wyoming.

670102. Juniper control study of L.U.

650113. Microclimatic studies on desert shrub-types.

710302. Moisture stress in woody plants as influenced by environmental factors.

760601. Mule deer food habitat and forage quality studies on juniper ranges in Wyoming.

750503. Phenology and production on semiarid shrub-types, 1974 results.

670303. Plant control on salt desert shrub rangeland by sheep grazing.

670403. Plant control on salt desert shrub rangeland by sheep grazing - See 670303.

780311. Range condition and trend review in the Lander Resource Area.

840407. Saltbush community changes with and without grazing over a 20-year period in north central Wyoming.

650108. Seeding in juniper and juniper control.

SNOWY RANGE

770602. Summer habitat preference and vegetation utilization of elk in the Snowy Range, Medicine Bow National Forest, Wyoming.

SOCIOLOGICAL PATTERN

720601. Vegetation pattern within northern desert shrub communities.

SODIUM

640602. Ecological study of *Sarcobatus vermiculatus* communities of the Big Horn Basin, Wyoming.

640604. Soil-vegetation relationships of the 15-Mile Drainage, Washakie County, Wyoming.

SOIL

710301. A greenhouse study of variance of picloram retention in soil and duff from juniper sites.

830503. Aridland shrub ecology and phenology research program.

810311. Correlation of soil factors and environmental conditions with phenophase development of western wheatgrass *(Agropyron smithii).*

810411. Correlation of soil factors and environmental conditions with phenophase development of western wheatgrass *(Agropyron smithii).*

810605. Correlation of soil factors and environmental conditions with phenophase development of western wheatgrass *(Agropyron smithii).*

720503. Detailed vegetation mapping of the I.B.P. Pawnee Site.

770601. Ecological interrelationships of vegetation and soils in a shortgrass prairie ecosystem, Snowy Range, Medicine Bow National Forest, Wyoming.

690602. Ecological relationships of vegetation, soils, and vertebrates on paired limber pine and spruce-fir sites, Big Horn Mountains, Wyoming.

620206. Ecological study of *Atriplex nuttallii* in the Big Horn Basin of Wyoming - Abst.

620601. Ecological study of *Atriplex nuttallii* in the Big Horn Basin of Wyoming.

640602. Ecological study of *Sarcobatus vermiculatus* communities of the Big Horn Basin, Wyoming.

800402. Evaluation of water and soil samples for forage and other agricultural development in the Tabuk region.

830311. Grazing, vegetation and soils interactions since 1967 in the Powder Rim grazing system.

650402. Halogeton abundance and distribution as influenced by degree of saltbush utilization.

670601. Harvester ant (*Pogonomyrmex owyheei* Cole) abundance in relation to grazing, vegetation and soil in north central Wyoming.

790201. Influence of environmental factors upon sagebrush and grass production in Wyoming.

710602. Interrelationships of vegetation and environmental factors on a mountain watershed in southeastern Wyoming.

720302. Juniper ecology and response to protection from browsing.

690112. Juniper GRI exclosure Tordon retention studies.

840309. Long-term Wyoming rangeland monitoring and research cooperative studies of the USDI-BLM and University of Wyoming, Department of Range Management.

650113. Microclimatic studies on desert shrub-types.

730101. On an area near Reno Junction, Wyoming, on which a strip-mine operation is proposed, the ecology, vegetation and soils inventory were initiated in 1973.

720301. Panchromatic, color and infrared photography for detailed vegetation mapping.

770502. Phenology and production studies on semiarid shrub-types, 1976 results.

760305. Physical and chemical analyses of range sites and interrelationships with herbage production and forage value potentials.

780305. Plant succession on plains pocket gopher mounds.

700202. Predictive value of factors influencing perennial grass production on western Wyoming sagebrush-grass ranges.

760204. Production of a shortgrass prairie ecosystem as related to soil characteristics in eastern Wyoming.

760310. Production of a shortgrass prairie ecosystem as related to soil characteristics in eastern Wyoming.

600101. Range condition class survey - Big Horn and Wind River Basins.

640502. Range survey in Wyoming's Big Horn Basin.

670504. Range condition class survey - concepts and application.

680205. Relationship of forage production to precipitation, cover, and soils in north central Wyoming.

680601. Relationship of forage production to precipitation, cover, and soils in north central Wyoming.

740304. Relationships of range production and rehabilitation to soils and other environmental factors.

670108. Shrub production estimation study.

680204. Size, density, growth rate, and age of Rocky Mountain juniper and interrelationships with environmental factors.

680603. Size, density, growth rate, and age of Rocky Mountain juniper and interrelationships with environmental factors.

640304. Soil-vegetation relations of the 15-Mile Drainage area in the Big Horn Basin of Wyoming.

800302. Soil-vegetation relationships on a semiarid rangeland exclosure in north central Wyoming - long-term studies.

810408. Soil-vegetation relationships in the Gardner saltbush vegetation type of Wyoming.

740405. Statistical models for description and prediction of perennial grass production.

670602. Study of vegetation, soils and small mammals in limber pine stands in north central Wyoming.

660205. Vegetation of small mammals of the juniper zone in north central Wyoming - Abst.

660602. Vegetation of small mammals of the juniper zone in north central Wyoming.

750601. Vegetation re-establishment potential of strip-mined soils, Black Thunder Basin, Wyoming.

760503. Vegetation, Section IX. Final Report Environmental Assessment of the Black Thunder Mine Site.

670306. Western harvester ant abundance in relation to grazing, vegetation and soils in north central Wyoming.

670402. Western harvester ant abundance in relation to grazing, vegetation and soils in north central Wyoming.

SOIL DEPTH

660601. Site and vegetation characteristics and relationships of juniper communities in northwestern Wyoming.

SOIL DESCRIPTION

790302. Soil-vegetation relationships on rangeland exclosures in the Grass Creek Planning Unit of north central Wyoming.

810309. Vegetation dynamics of non-grazed semiarid sagebrush grassland communities in the Big Horn Basin of Wyoming.

SOIL DISTURBANCE

800301. Shrub establishment, dominance and ecology on the juniper and sagebrush-grass types in Wyoming.

SOIL EROSION

750407. Ecological considerations for Project Wagon Wheel and hydrolic fracturing activities.

780301. Soil erosion and movement studies in western Wyoming.

780401. Soil erosion and movement studies in western Wyoming.

SOIL HORIZON

730601. Development and evaluation of statistical models for predicting perennial grass production in north central Wyoming.

740405. Statistical models for description and prediction of perennial grass production.

SOIL MOVEMENT

SOIL PROFILE

SOIL SURFACE

SOIL TEMPERATURE

770603. Factors influencing vegetative and reproductive phenology of *Artemisia nova* Nels.

640501. Halogeton research, 1963 results, See 630103, 630104, 630105, 630106, 630108.

650501. Halogeton research, 1964 results.

660501. Halogeton research, 1965 results, See 650101, 650102, 650103, 650107.

660503. Microclimate studies in desert shrub-types.

750306. Phenodynamics and environmental characteristics of sagebrush-grass rangelands.

760602. Phenodynamics and ecology of sagebrush-grass rangelands.

770306. Phenodynamics and ecology of sagebrush-grass rangelands.

800603. Phenological comparison of two sagebrush species.

810302. Phenological comparison of two sagebrush species in Wyoming.

790502. Phenology and production studies on semiarid shrub-types, 1978 results.

810305. Phenology and productivity comparisons of western wheatgrass *(Agropyron smithii)* and bluebunch wheatgrass *(A. spicatum).*

810315. Phenology and culm weight variation between two wheatgrass species: *Agropyron smithii* and *Agropyron spicatum.*

830502. Phenology and production studies on semiarid shrub-types, 1982 results, Vol. II.

840312. Regional variation of phenology.

710505. Semiannual progress report to the Bureau of Land Management, January 1971.

630104. Soil moisture and temperature study.

640104. Soil moisture and temperature study.

650104. Soil moisture and temperature study.

660105. Soil moisture and temperature.

660301. Soil moisture and temperature characteristics under grazed and non-grazed conditions.

660402. Soil moisture and temperature characteristics as related to grass production following shrub control under grazed and ungrazed conditions - See 660301.

670103. Soil moisture and temperature.

670301. Soil moisture and temperature characteristics as related to grass production following shrub control under grazed and non-grazed conditions.

670401. Soil moisture and temperature characteristics as related to grass production following shrub control under grazed and non-grazed conditions.

680105. Soil moisture and temperature.

680201. Soil moisture and temperature changes following sagebrush control.

690103. Soil moisture and temperature.

700104. Soil moisture and temperature.

700601. Soil moisture and temperature relationships of six salt desert shrub communities in north central Wyoming.

710103. Soil moisture and temperature.

790305. Soils-phenology relations on some arid rangelands of western Wyoming.

830105. Summary of soil moisture and temperature data from Granite Mountain exclosure.

840105. Summary of soil moisture and temperature data from Granite Mountain Exclosure.

SOIL TEXTURE

670601. Harvester ant (*Pogonomyrmex owyheei* Cole) abundance in relation to grazing, vegetation and soil in north central Wyoming.

760201. Methods of measuring environmental parameters on a pre-disturbed coal site.

720201. Rangeland relations and harvester ants in north central Wyoming.

800601. Vegetation of saline and alkali habitats of the Red Desert, Wyoming.

720601. Vegetation pattern within northern desert shrub communities.

SOIL TYPE

740302. Remote sensing applied to the grassland biome.

740404. Productivity potential differences of grasslands of semidesert western Wyoming.

STEM CROSS SECTIONS

STIPA COMATA

STOCHASTIC

STOCKING RATES

STREAMBANK WHEATGRASS

STRESS ADAPTATIONS

STRIP CHEMICAL CONTROL

STRIP-MINE

STRIP-MINE SPOILS

STRIP TRANSECTS

SUBALPINE COMMUNITIES

TOPOGRAPHY

770603. Factors influencing vegetative and reproductive phenology of *Artemisia nova* Nels.

650402. Halogeton abundance and distrubution as influenced by degree of saltbush utilization.

710302. Moisture stress in woody plants as influenced by environmental factors.

710601. Moisture stress in woody plants of Wyoming as influenced by environmental factors.

750304. Pre-disturbance ecological studies improve and define potential for surface mine reclamation.

740404. Productivity potential differences of grasslands of semidesert western Wyoming.

740302. Remote sensing applied to the grassland biome.

TOPSOILS

750601. Vegetation re-establishment potential of strip-mined soils, Black Thunder Basin, Wyoming.

760306. Vegetation re-establishment potential of strip-mined soils, Black Thunder Basin, Wyoming.

TORDON

640108. Ecology and control of juniper.

660204. Juniper communities and range management in northwestern Wyoming.

650112. Juniper control study on L.U.

660104. Juniper control study on L.U.

690102. Juniper control study on L.U.

680102. Juniper control chemical, burn, mechanical on GRI exclosures.

690302. Juniper game range improvement studies in western Wyoming.

680112. Kane deer exclosure seeding analysis.

710601. Moisture stress in woody plants of Wyoming as influenced by environmental factors.

800301. Shrub establishment, dominance and ecology on the juniper and sagebrush-grass types in Wyoming.

610105. Study of biotic communities - juniper, saltbush, greasewood, blue grama, sagebrush and limber pine.

670304. Use of picloram for juniper control.

TORDON RETENTION

710201. Greenhouse analysis of variance of picloram retention in soil and duff from juniper sites.

690112. Juniper GRI exclosure Tordon retention studies.

TRADITIONAL ESTIMATION

840403. Estimation of missing precipitation data using traditional, kriging, and subjective techniques.

TRANSECTS

650109. 1 ft x 1 ft permanent plot transects.

670104. 1 ft x 1 ft permanent plot transects.

680113. 1 ft x 1 ft permanent plot transects.

690104. 1 ft x 1 ft permanent plot transects.

700109. 1 ft x 1 ft permanent plot transects.

710106. 1 ft x 1 ft permanent plot transects.

770105. 1 ft x 1 ft permanent plot transects.

780107. 1 ft x 1 ft permanent plot transects.

700506. Aridland ecology semiannual progress report.

680117. Baggs grazing system study.

690105. Baggs grazing system study.

740104. Black Thunder pre-mining inventory.

760105. Black Thunder pre-mining inventory.

720110. Contour-furrow and seedings study.

750407. Ecological considerations for Project Wagon Wheel and hydrolic fracturing activities.

720602. Ecology of aridland vegetation in western Wyoming.

630105. Erosion transect study.

620103. Exclosure studies, production, cover, height, and moisture.

630302. Exclosure studies, production, cover, height, and moisture phase.

640103. Exclosure studies, production, cover, height, precipitation.

670110. Exclosure studies - production, cover, precipitation.

670503. Exclosure study with transects of permanent plots - See 660106.

680106. Exclosure studies - production, cover, precipitation.

UNGRAZED

830411. Re-establishment rates of sagebrush following chemical control on grazed and ungrazed ranges in western Wyoming.

830310. *Atriplex*/grass and forb relationships under no grazing and shifting precipitation patterns in north central Wyoming.

840413. *Atriplex*/grass and forb relationships under no grazing and shifting precipitation patterns in north central Wyoming.

830307. Re-establishment rates of sagebrush following chemical control on grazed and ungrazed ranges in western Wyoming.

660301. Soil moisture and temperature characteristics under grazed and non-grazed conditions.

660402. Soil moisture and temperature characteristics as related to grass production following shrub control under grazed and ungrazed conditions - See 660301.

670301. Soil moisture and temperature characteristics as related to grass production following shrub control under grazed and non-grazed conditions.

670401. Soil moisture and temperature characteristics as related to grass production following shrub control under grazed and non-grazed conditions.

750305. Study of rangeland soil movement characteristics.

UPLAND COMMUNITIES

640604. Soil-vegetation relationships of the 15-Mile Drainage, Washakie County, Wyoming.

UTAH JUNIPER

830312. Deer browse impacts on Utah juniper and true mountain mahogany in the Poison Butte unit west of Baggs, Wyoming.

610501. Distribution of vegetation in the Wind River Canyon, Wyoming.

660204. Juniper communities and range Management in northwestern Wyoming.

680505. *Juniperus osteosperma* in northwestern Wyoming - their distribution and ecology.

760601. Mule deer food habitat and forage quality studies on juniper ranges in Wyoming.

810310. Population structure of Utah juniper in north central Wyoming.

810410. Population structure of Utah juniper in north central Wyoming.

840404. Population structure anomalies of *Juniperus osteosperma* invasion in north central Wyoming, U.S.A.

650108. Seeding in juniper and juniper control.

800301. Shrub establishment, dominance and ecology on the juniper and sagebrush-grass types in Wyoming.

820401. Vegetation response to grazing management strategies on semiarid shrublands of Wyoming.

UTAH JUNIPER-BLACK SAGEBRUSH

830309. Utah juniper invasion in north central Wyoming and historical climatic and management relationships.

UTILIZATION

830310. *Atriplex*/grass and forb relationships under no grazing and shifting precipitation patterns in north central Wyoming.

680602. Diet preferences and utilization patterns of elk in the northern Big Horn Mountains, Wyoming.

680203. Dietary preference and utilization patterns of elk in the Big Horn Mountains, Wyoming.

680302. Dietary preferences and utilization of elk on spring range in the northern Big Horn Mountains of Wyoming.

680401. Dietary preferences and utilization of elk on spring range in the northern Big Horn Mountains of Wyoming.

670603. Effect of grazing intensities on the roots of Nuttall's saltbush (*Atriplex nuttallii* S. Wats.).

650304. Exclosure studies - Responses to sagebrush spraying.

810306. Grazing system impacts on semiarid vegetation of Wyoming.

VEGETATION PRODUCTION

VEGETATION SOIL RELATIONS

VEGETATION STAGNATION

VEGETATION STRUCTURE

VEGETATION TYPE

WHEATGRASS

700508. Frequency, density and basal area measures on sagebrush-grass ranges.

840405. Grazing management, longevity, and benefits of sagebrush control on semiarid rangelands.

650501. Halogeton research, 1964 results.

670105. Harvester ant ecology Big Horn Basin and Wind River Basin.

840310. Herbage production and community dynamics.

840314. Herbage yield dynamics and seasonal precipitation relationships.

840309. Long-term Wyoming rangeland monitoring and research cooperative studies of the USDI-BLM and University of Wyoming, Department of Range Management.

830502. Phenology and production studies on semiarid shrub-types, 1982 results, Vol. II.

840501. Phenology and production studies on semiarid shrub-types.

830601. Plant response to livestock grazing and big sagebrush chemical control in western Wyoming 1962 - 1980.

600101. Range condition class survey - Big Horn and Wind River Basins.

610101. Range condition class survey - Big Horn and Wind River Basins.

590502. Range survey.

840312. Regional variaton of phenology.

840311. Reinvasion following sagebrush control.

680205. Relationship of forage production to precipitation, cover, and soils in north central Wyoming.

680601. Relationship of forage production to precipitation, cover, and soils in north central Wyoming.

600104. Sagebrush control and seeding, Two mile, Bud Kimball, Smilo, Upper and Lower Government.

610404. Seeding salt desert shrublands in Wyoming.

590501. Study of range conditions in the Big Horn and Wind River Basins.

600201. Study of range condition classes and methods of range improvement in Wyoming.

600502. Study of range conditions in the Big Horn and Wind River Basins.

620501. Study of range conditions in the Big Horn and Wind River Basins.

630501. Study of range conditions in the Big Horn and Wind River Basins.

590102. Summer steer grazing on saltbush rangeland with reseeding and water spreader treatments.

840313. Validation of a rangeland production model (ERHYM) for sagebrush sites.

820302. Vegetation productivity and phenology in the Wind River Basin.

820404. Vegetation productivity and phenology in the Wind River Basin.

WIND RIVER MOUNTAINS

620201. Ecological study of the *Artemisia tripartita* subsp. *rupicola* and related shrub communities in Wyoming.

620204. Ecological study of the *Artemisia tripartita* subsp. *rupicola* and related shrub communities in Wyoming - Abst.

620503. Ecological study of the *Artemisia tripartita* subsp. *rupicola* and related shrub communities in Wyoming.

620202. List of plants observed on the Owl Creek and Wind River Ranges in Wyoming.

610101. Range condition class survey - Big Horn and Wind River Basins.

650401. Vegetation changes from cold desert to subalpine communities.

WINTER

780406. Structural definition of semiarid *Artemisia* and *Atriplex* rangeland ecosystems.

830310. *Atriplex*/grass and forb relationships under no grazing and shifting precipitation patterns in north central Wyoming.

WINTER KILL

820401. Vegetation response to grazing management strategies on semiarid shrublands of Wyoming.

WINTER RANGE

720605. Ecological inventory and analysis of the Laramie Peak Big Game Winter Range.

ANNOTATED BIBLIOGRAPHY

CHAPTER 4: ANNOTATED BIBLIOGRAPHY

Some 700 bibliographic items are given in this unit. They represent the total of all research activities by year and the citations for all forms of publications and presentations by Dr. Fisser and all associated authors. The annotations in most instances are brief. When the research and findings were new and informative, considerable descriptive and specific data were incorporated and added to expand beyond the normal context of an annotation to a citation. The items are in chronological order overall, with citations in order by source category within each year. The alphanumeric code items for the Topic Outline and six digit identifying code number are given for each citation. A Table of Contents by year is given below to enhance ready access to units within.

BIBLIOGRAPHIC TABLE OF CONTENTS

5901 Research Activities

3A1 590101 Range condition class survey - Big Horn & Wind River Basin, Wyoming.
3A2
2F2

The range condition class survey was initiated during 1959. The study was conducted in cooperation with the Wyoming Natural Resource Board, the Federal Land Range Users Association of Worland, the Lower Little Popo Agie Stockgrowers Association of Lander and the Bureau of Land Management. The primary objectives of the studies were to determine range conditions classes for major vegetation types throughout the area.

6A1 590102 Summer steer grazing on saltbush rangeland with reseeding and water spreader treatments.
4B
5B4
5A1C
5C
1E4

Summer steer grazing pastures were located along 15-Mile creek some 12 or 13 miles northwest of Worland. The summer steer grazing trials began during 1960.

6A2 590103 Winter sheep grazing studies on saltbush rangeland.
4A
4B
1E4

The winter sheep grazing trials were conducted on the Halogeton research pastures located northwest of Greybull. The pastures were initiated in 1956 and during 1959 continued to receive differential grazing intensities. Moderate grazing produced the most desirable weight gains in mature ewes. Halogeton *(Halogeton glomeratus)* had invaded new areas within the pastures.

5905 Wyoming Agricultural Experiment Station Publications

3A2 590501 Fisser, H.G., R.L. Lang and A.A. Beetle. 1959. A study of range conditions in the Big Horn and Wind River Basins. Pre. Report I. Wyo. Agr. Exp. Sta. Mimeo Circ. 119. 5 p.
2F2
3A1

This paper described the activities of the range survey conducted in the Big Horn and Wind River basins and identified basic concepts of condition class determination on an ecological basis.

3A2 590502 Fisser, H.G. 1959. Range survey. In Crop Care in Wyoming. p. 22, December.
2F2

The paper identified the activities of the range survey conducted in the Big Horn and Wind River Basins. The condition class concept was identified. Relic areas and fenced exclosures were identified as being a necessary part of the ecological research of the area.

6001 Research Activities

3A1 600101 Range condition class survey - Big Horn & Wind River Basin, Wyoming.

3A2

2F2 This was another statement of the condition class survey being conducted in the Big Horn and Wind River Basins during the second year of the study. Transects were established in all of the major areas. Soils information was collected at a number of the transect locations.

6A1 600102 Summer steer grazing on saltbush rangeland with reseeding and water spreader treatments.

4B

5A1C

5B4 Yearling steers grazed the experimental pastures during the summer of 1960. This

5C was a study to identify the relationships of the value of water spreading devices on

1E4 saltbush *(Atriplex gardneri)* rangelands as compared to grazing on native rangelands. Animal gain, vegetation cover, production and utilization data were collected.

6A2 600103 Winter sheep grazing studies on saltbush rangeland.

4A

4B The differential grazing intensities were continued and the data show that the cover

1E4 response to grazing at this point was rather insignificant. Although grazing intensity response of the saltbush *(Atriplex gardneri)* was not noted, it was noted that there was a greater amount of saltbush within the exclosures than in the grazed pastures. A very droughty situation occurred during the year which tended to confound the data.

5B4 600104 Sagebrush control and seeding, 2 mile, Bud Kimball, Smilo, Upper and Lower Govt.

5A1D

5A1C During late October, plantings of a variety of grasses and coordinated sagebrush *(Artemisia tridentata)* control, both mechanical and chemical, were conducted at exclosures in the Big Horn Basin and Wind River Basin.

5A1D 60010401 Chemical

Chemical was applied at several exclosures in the Big Horn and Wind River Basins. The chemical was 2,4-D applied by aerial application.

5C 60010402 Mechanical

5A1C

5B4 Mechanical control of sagebrush *(Artemisia tridentata)* was accomplished with a rangeland seeder and a pitting device. Seedings were conducted with the range seeder, and crested wheatgrass *(Agropyron cristatum)* seed was broadcast on the pitted site.

5B4 60010403 Seeding

Seeding activities included a mixture of 16 different grasses that were judged to perhaps have some potential in planting in the sagebrush-saltbush *(Artemisia-Atriplex)* rangelands of western Wyoming.

6002 Journal Publications

3A1 600201 Fisser, H.G., A.A. Beetle, R.L. Lang and D.R. Smith. 1960. A study of range condition classes and methods of range improvement in Wyoming. In Range Research

3A2

2F2 Progress in 17 Western States. State Exp. Sta. Div., Agr. Res. Serv., USDA, p. 133-134.

This publication identified the rangeland survey activities being carried on by the Range Management Department of the University of Wyoming. Primary interest was related to the survey work conducted in the Big Horn and Wind River Basins.

6005 Wyoming Agricultural Experiment Station Publications

2A 600501 Fisser, H.G. 1960. The writings of Lincoln Ellison. Wyo. Range Manage. Issue 136.

Lincoln Ellison was one of the very early persons strongly oriented to range management within the Forest Service. His publications dealt with rangeland ecology on forest lands. His unfortunate death at a young age stopped a very promising career in rangeland research and development.

3A2 2F2 2A 600502 Fisser, H.G., A.A. Beetle and R.L. Lang. 1960. A study of range conditions in the Big Horn and Wind River Basins. Prel. Report II. Wyo. Agr. Exp. Sta. Mimeo. 10 p.

This publication primarily was a compilation of the data collected during the summer of 1960 including also some analysis of precipitation data and isohyatal line development for differentiation of trends in precipitation from one location to another in the Big Horn and Wind River Basin areas.

3A2 3A1 600503 Fisser, H.G., A.A. Beetle, R.L. Lang and D.R. Smith. 1960. Range management survey studies. In Crop Care in Wyoming. p. 20-22, December.

This item simply identified the fact that rangeland survey activities had been conducted by the Range Management Department. This topic stresses the importance of ecological relationships in range survey analyses.

6101 Research Activities

3A1 610101 Range condition class survey - Big Horn and Wind River Basins, Wyoming.

3A2

2F2 During the three years of the range condition class survey of the Big Horn and Wind River Basins, over 400 vegetation cover composition transects were established in the various areas of study. The basic range survey was initially established at the Tensleep grazing allotment area, at the Buffalo Creek area just northeast of Thermopolis and at the Lander Grazing Association area located to the east of Lander. In addition, a considerable number of transects were established along the slopes of the Owl Creek and Wind River mountains relating to the vegetation classes that occur along the elevational gradient of the mountains.

6A1 610102 Summer steer grazing on saltbush rangeland with reseeding and water spreader treatments.

4B

5A1C As in 1960, the steers were grazed on the North Fork Experimental pastures for 43

5B4 days; but during 1961 only 13 head were used while during 1960, 18 head were utilized.

5C The daily gains during 1961 showed that the animals gained an average of 0.70 pounds

1E4 per animal on the untreated saltbush *(Atriplex nuttallii)* rangeland pasture, but on the treated pasture with the water spreaders and seeded grass areas gain was 1.60 pounds per animal. Pounds of gain per acre on the check pasture was only 0.87 during 1960 while on the treated pasture this value was 4.64.

6A2 610103 Winter sheep grazing studies on saltbush rangeland.

4A

4B The Dry Creek Halogeton research pastures near Greybull continued to receive

1E4 differential grazing intensities. Even under complete protection within the exclosures, halogeton *(Halogeton glomeratus)* had become established with densities comparable to most of the pastures. Cover of saltbush *(Atriplex gardneri)* showed a nonsignificant, inverse relationship with intensity of grazing. Cover on all pastures was considerably below that of the exclosures. Production of saltbush was inversely proportional to the degree of utilization. This relationship was very pronounced in the drouth year of 1960-61.

3A1 610104 Aridland ecology studies of vegetation, cover, composition, production, biochemical

3A2 composition, precipitation, and site relationships under grazing, shrub control and

3A3 seeding and pitting treatments.

4

5 During the three years of 1959, 1960 and 1961, in conjunction with the range

6B condition class survey and the halogeton *(Halogeton glomeratus)* research studies, some 30 exclosures were constructed in the semiarid regions of western Wyoming. On these exclosure locations, vegetation cover, composition and production were obtained with quadrats of permanent location and with other sampling techniques.

3A 610105 Study of biotic communities-saltbush, greasewood, blue grama, sagebrush, juniper and

2E limber pine.

2F2

2D Juniper *(Juniperous osteosperma)* community studies include soil-plant relationship,

2E site-plant-mammal relationship, long term juniper invasion, juniper control by

2A mechanical and chemical (Tordon) procedures with incorporated seeding studies, and forage production-precipitation relationship study under grazing and juniper control treatments.

During the three years of 1959, 1960 and 1961, several intensive studies of autecological studies of several important species in the semiarid rangelands of Wyoming were developed. The activity of this work was primarily related initially to the saltbush *(Atriplex gardneri)*, greasewood *(Sarcobatus vermiculatus)*, blue grama *(Bouteloua gracilis)* and sagebrush *(Artemisia tridentata)* sites. Juniper *(Juniperus osteosperma)* studies were to be developed in the future. Juniper control by mechanical and chemical (Tordon) procedures was conducted at several locations.

6102 Journal Publications

3A3 10D 610201 Fisser, H.G. 1961. Variable plot, square foot plot and visual estimate for shrub crown cover measurements. J. Range Manage. 14(3):202-207.

Three methods of shrub cover estimation were evaluated. These included visual estimate, square foot plot and variable plot. Three individuals obtained data. The most accurate data appeared to have been obtained with the use of the variable plot procedure, but errors associated with height of observer and comparable height of shrubs were noted.

1D 2E 5A1B 610202 Fisser, H.G. and R.J. Lavigne. 1961. *Trirhabda attenuata* (Say) (Coleoptera: Chrysomelidae), a leaf-feeding beetle found on threetip sagebrush (*Artemisia tripartita* Rydb. subsp. *rupicola* Beetle). J. Range Manage. 14(4):278-279.

A leaf-feeding beetle *(Trirhabda attenuata)* was found on threetip sagebrush *(Artemisia tripartita* spp. *rupicola)* on the northern slopes of the Owl Creek mountains some 10 miles south of Thermopolis. The beetle in some locations had almost completely defoliated the sagebrush plants.

6103 Presentations

3A1 3A2 2F2 610301 "Range condition class survey in the Big Horn Basin of Wyoming." Sept. 14, 1961. Range Weed Control Conf. Worland, Wyoming. by H.G. Fisser.

This presentation was given to identify the range condition class survey program in the Big Horn Basin and identified some interesting and unique vegetation-soil ecological interrelationships of importance.

3A 10D1 610302 "Theories and methods of range surveys." December 4, 1961. Wyo. Agr. Exp. Sta. Superintendents Conference. Lander, Wyoming. by H.G. Fisser.

The audience of non-rangeland and non-ecologically oriented individuals was given a very basic presentation relating the basic ecological concepts and theories of range management and range survey.

6104 Published Proceedings/Abstracts

1E1 2E 5A2D 1B4 610401 Vosler, L.C. 1961. Halogeton seed burial study. p. 1. In Proc. Range Weed Research Tour. Worland, Wyoming. September 14-15, 1961.

The speaker here identified the halogeton *(Halogeton glomeratus)* seed burial study which had been initiated as a 10 year cooperative study between the ARS, USDA and the experiment station personnel from universities from Idaho, Nevada, Utah, Washington and Wyoming.

5B4 5A1C 5C 610402 Vosler, L.C. 1961. Studies of mechanically treated watersheds. p. 1-3. In Proc. Range Weed Research Tour. Worland, Wyoming. September 14-15, 1961.

In his presentation Mr. Vosler identified the various dike and contouring activities which had been carried on primarily on saltbush *(Atriplex gardneri)* rangelands and the associated seeding of crested wheatgrass *(Agropyron cristatum)* and tall wheatgrass *(Agropyron elongatum)* on those areas. His emphasis related to the North Fork experimental area northwest of Worland upon which the summer steer grazing trials have been conducted.

4A 610403 Smith, D.R. 1961. Intensity of grazing study. p. 4-6. In Proc. Range Weed Research Tour. Worland, Wyoming. September 14-15, 1961.
4B
6

Dr. Smith identified the data of vegetation response to sheep grazing intensity trials on the Dry Creek halogeton pastures north northwest of Greybull. Vegetation response to grazing intensities was not significantly different among grazing intensity treatments.

5B4 610404 Hull, A.C. Jr. 1961. Seeding salt desert shrub lands in Wyoming. p. 7-11. In Proc. Range Weed Research Tour. Worland, Wyoming. September 14-15, 1961.
2A
5C

This presentation by Mr. Hull of the Agriculture Research Service (Utah State University at Logan) related to the activities that he conducted during 1948-1951 on rangelands of Wyoming. In this study he planted a variety of species of grasses on a number of locations in the Big Horn and Wind River Basins. He identified 15 sites in the western Wyoming area upon which he had planted a variety of species and conducted several kinds of land treatment activities. In general, excellent seedling establishment was noted; but during the following four to five years extreme drouth conditions occurred. In most situations, the seeded species were obliterated.

5A1D610405 Colton, R.D. 1961. Sagebrush spraying tour. p. 12-14. In Proc. Range Weed Research Tour. Worland, Wyoming. September 14-15, 1961.

This presentation was conducted on an area northwest of Hyattville on the slope of the Big Horn Mountains. Sagebrush *(Artemisia tridentata)* had been chemically controlled with 2,4-D and the information presented showed the excellent steer weight increases associated with the sagebrush control and increased vegetation. An important point made was that proper management of the sprayed lands was absolutely essential in order to establish grass increases needed to maintain the herbaceous species and in order to withhold the invasion of sagebrush.

6105 Wyoming Agricultural Experiment Station Publications

3A2 610501 Fisser, H.G. 1961. The range condition survey of the Big Horn and Wind River Basins. University of Wyoming. Plant Sci. Div., Range Manage. Notes 4:1.
3A1
2F2
10D1

This was a short note identifying the range conditions survey areas and the field activities. Over 550 transect lines had been established during the past three years.

6106 Graduate Advisee Completions

2F2 610601 Williams, C.S. 1961. Distribution of vegetation in the Wind River Canyon, Wyoming. M.S. Thesis. University of Wyoming. (directed A.A. Beetle) 90 p.
2E
3A1

This thesis addressed the topic of vegetation distribution as effected by geological formations. The formations addressed were: Phosphoria, Tensleep, Amsden, Madison, Dolomite, Gallatin, Gros Ventre and Flathead formations. Although geological formations and their impact upon the vegetation was of special importance, it was also noted that old burns had significant effect upon distribution of vegetation, especially the woody species including the junipers *(Juniperus osteosperma* and *J. scopulorum).* Bluebunch wheatgrass *(Agropyron spicatum)* and Utah juniper *(J. osteosperma)* were the most widely distributed species and each varied in density and amount as influenced by some of the geological factors.

The following conclusions were drawn: 1) On mezic, undisturbed areas of the Wind River Canyon, the dominant species were juniper and bluebunch wheatgrass and were commonly associated with one another. These two plants represented climax conditions in the canyon. 2) Geologic formations affected distribution of vegetation with a western exposure (as shown by differences in means and variance between formations) but did not affect distribution of vegetation with eastern exposure (work was done primarily with reference to bluebunch wheatgrass). 3) Exposure was important in affecting distribution as is indicated by the presence of Idaho fescue *(Festuca idahoensis)* on the west side of the canyon and the lack of this grass on the east side. 4) Portions of the canyon were relic areas grazed only by wildlife, the rugged terrain imposed by the geologic formations being the key factor in preventing domestic use. 5) Bluebunch wheatgrass differed under the effects of disturbance and tended to decrease in abundance while cheatgrass brome *(Bromus tectorum)* tended to increase. 6) Although canyons are not important as grazing regions, they may often indicate the type of vegetation which could be present in the surrounding areas. Vegetation zonation in this canyon provided a vertical stratification not apparent on adjacent level areas. Much future work needs to be done on areas like the Wind River Canyon so that distribution patterns of vegetation may be predicted.

6201 Research Activities

6A2 620101 Winter sheep grazing on saltbush range.

4A

4B A total of 650 sheep were grazed for 18 days on the pastures northwest of Greybull.
1E4 The response to the light, moderate and heavy use was significantly different at the 5 percent level with a Duncan's multiple range test. The visual appearance was readily obtained in terms of indicating the difference as a result of heavy use in which the saltbush *(Atriplex gardneri)* clumps were breaking down and halogeton *(Haolgeton glomeratus)* had become more abundant than in other pastures.

6A1 620102 Steer grazing study on saltbush range.

4B

5A1C The summer steer trials grazing on saltbush rangeland began in 1960. One pasture
5B4 was a native pasture of 489 acres. On the other, of 521 acres, about 90 acres had been
5C treated with water spreaders and with plantings of crested wheatgrass *(Agropyron*
1E4 *cristatum)*, tall wheatgrass *(A. elongatum)* and fireweed *(Kochia scoparia)* which was planted on the dikes. Animals were grazed for 73 days during 1962. Average daily gain of steers on native pasture was 0.98 pounds and on the treated pasture 1.30 pounds. Pounds of gain per acre on the native pasture was 1.75 and on the treated 5.64. Forage utilization on the treated pasture was 38 percent of crested wheatgrass and 7.5 percent of the tall wheatgrass. Other species were only occasionally grazed. Saltbush utilization on the non-treated check pasture was only 6.9 percent. Most utilization occurred on vegetation in the drainage ways and on the perimeter of the watering reservoir. The cottonwood *(Populus angustifolia)* trees were noted to have been highlined by the intensive use by the steers.

3A3 620103 Exclosure study: production, cover, height and moisture.

5A1D

5B Work was initiated during the 1962 field season in which a new phase of vegetation
5C study was begun on the halogeton *(Halogeton glomeratus)* exclosures most of which had
10D1 been constructed during 1959-1960. Production studies were initiated on 20 exclosures
2A during the 1962 field season. Only one transect was recorded within exclosures which
2C had not been subjected to range improvement practices. Areas within some of the exclosures had been sprayed to control sagebrush or seeded and pitted. Transects were established within each of the different treatment areas when they were present. Area cover, herbage production and height data studies on the vegetation were conducted with a 12 in x 12 in quadrat located systematically along a randomly located 100 ft steel tape. The term transect was applied to evaluation of 10 quadrat placements on a line. On sites dominated by saltbush, data were obtained in a like manner except that the plot size was 1 ft x 10 ft. The plot was placed at right angles to the steel tape and vegetation data were subdivided into 10 subplots. Species were individually clipped, sacked and oven dried prior to weighing. Simple aluminum rain gauges were installed at each exclosure. Precipitation data were recorded four times a year - April 15, July 1, September 1 and October 15. Future studies were to include the evaluation of the relationship of precipitation differences, both in time and amount, to variations in forage production.

6202 Journal Publications

2E 620201 Fisser, H.G. 1962. An ecological study of the *Artemisia tripartita* subsp. *rupicola* and
2A related shrub communities in Wyoming. Ph.D. Diss., Univ. of Wyo. 166 p.

2C

2F2 The distribution of four sagebrush species - big, *(Artemisia tridentata)* black *(A.*
3A *nova)*, three tip *(A. tripartita)* and silver sagebrush *(A. cana)* - on the slopes of the Owl
10D1 Creek and Wind River mountains was found to be controlled by and associated with interrelationships of total moisture components relative to variation in summer rainfall, variations in winter precipitation in the form of snowpack and the variations due to the

influence and intensity of air movement both in summer and winter. Time and amount of precipitation appeared to be the major controlling factor.

1B1 620202 Fisser, H.G. 1962. List of plants observed on the Owl Creek and Wind River Ranges in Wyoming. Wyo. Range Manage. Issue 159. 10 p.
3A
1A2

Plants which had been collected from the sagebrush controlled areas of the Owl Creek and Wind River ranges of Wyoming were identified and listed by family giving scientific and common names for almost 400 plants. This was the earliest comprehensive list of range plants obtained from these areas. One new state record and many county records were established. Voucher specimens were filed in the University of Wyoming Rocky Mountain and Range Management herbaria.

3A2 620203 Fisser, H.G. 1962. A comparison of line intercept, inferred point contact and weight measurement methods on foothill grassland vegetation. M.S. Thesis Abst. J. Range Manage. 15:124.
10D1

A comparison of the line interception and basal point contact methods was conducted by extracting point data from line intercept records. Line interceptions were conducted on 350 transects, each five feet long, located in 70 clusters in foothill grassland on the Red Bluff Research Station during the summer of 1958. Transects were read with an accurate mechanism constructed for repeatable location.

Analysis of intercept data consisted of summarizing the basal area estimations of each transect into 12 vegetation groupings. Data from each of these groupings were subjected to inference of systematic and random selection of 100, 50 and 25 points per transect. Clip plots were located at random in each cluster, as were the transects and analyzed in like manner.

Standard statistical measures were determined. In addition the following tests were conducted: Bartlett's test of homogeneity of variance, "t" test of means, analysis of variance on transects of "All Live Vegetation," number of transects required to sample within 10 percent of the mean, and regression and correlation analyses with standard tests of significance.

Little variability was indicated in the groupings which occupied the major portion of the total basal area estimate. Whether the point units were selected systematically or at random appeared to have little influence upon the data. Numbers of point units per transect appeared to have little effect on such values as percentage occurrence per transect and mean values per transect but measures of dispersion increased greatly in value with a decrease of numbers of point units per transect.

Comparison of data on a cluster and transect basis indicated more variability between clusters than within clusters. The number of transects required to sample within 10 percent of the means of the "All Live Vegetation" and "All Grass and Grasslike" groups was less than the number of transects actually recorded. Required cluster number was more than the number of field sampled clusters. The relatively low variation among transects indicates that the short five foot lines have sufficient length for accurate basal estimation in the vegetation type studied.

Regression and correlation analyses indicated significant relationships between weight and basal area for the "All Grass and Grasslike," "Bunchgrass and Grasslike," and "Sodgrass and Grasslike" groups. Expressed on a percentage basis, the correlation coefficient indicated variability in weight explained by regression of area to range from 10 to 20 percent. This low value may be due to the different transect and plot locations in the one acre cluster areas. The relationships of spike selaginella *(Selaginella densa)* to basal area and weight sampling procedures requires further study. Pattern and size effects apparently were responsible for differences between systematically and randomly selected measurements of bluebunch wheatgrass *(Agropyron spicatum)* but require additional investigation for complete analysis.

2E 620204 Fisser, H.G. 1962. Abst. of Ph.D. Thesis. Wyo. Range Manage. Issue 160. 2 p.
2A
2C See 620201.
2F2
3A

2E 620205 Vosler, L.C. 1962. Bibliography of *Halogeton glomeratus.* Wyo. Range Manage. Issue
1B4 158, 6 p.
5A

This bibliography contained 104 citations of comments and publications concerning halogeton *(Halogeton glomeratus)* which have been developed in the less than 30 years since this poisonous plant has been noted present in this country. The purpose of the bibliography was to bring together the vast number of publications into a complete listing. There were, in addition, numerous government news letters, mimeographed papers and progress reports which were not obtained and, of course, were not listed in this bibliography.

2E 620206 Vosler, L.C. 1962. Abstract of MS Thesis. Wyo. Range Manage. Issue 162. 1 p.
1E
2C See 620601.
2A

6203 Presentations

3A1 620301 "Concepts of range condition class survey" Nov. 30, 1962. Wyoming Wool Growers
3A2 Association Convention. Worland, Wyoming. by H.G. Fisser.
2F2
10D1 This was a presentation to range users identifying basic ecological concepts and field procedures related to the range condition class survey that was being conducted in the Big Horn and Wind River Basins.

6205 Wyoming Agricultural Experiment Station Publications

3A1 620501 Fisser, H.G., C. Williams, A.A. Beetle and R.L. Lang. 1962. A study of range conditions
3A2 in the Big Horn and Wind River Basins. Prel. Report III. Wyo. Agr. Exp. Sta.
2F2 Mimeo. 7 p.
10D1

The information presented in the report simply was a continuation of the data as reported previously for the range condition class survey in the Basin area of Wyoming.

6A2 620502 Nichols, J.T., L.C. Vosler, D.R. Smith and H.G. Fisser. 1962. Halogeton studies. In
3A3 Coop. Res. Rep. to the BLM. 1961 results. Wyo. Agr. Exp. Sta. Mimeo. p. 1-17.
2F2

The various aspects of the cooperative research program with the Bureau of Land Management primarily oriented toward the halogeton studies, but beginning with some of the work related to the exclosure production studies, were presented in the report of 1961 results.

2E 620503 Fisser, H.G. 1962. An ecological study of the *Artemisia tripartita* subsp. *rupicola* and
2A related shrub communities in Wyoming. In Crop Care in Wyoming. p. 1-3. June.
2C
2F2 This was a condensation of some of the work related to the Fisser dissertation and
3A identified some of the ecological relationships of the shrub communities on the Owl Creek and Wind River Mountains.

4B 620504 Fisser, H.G., M. May, W.M. Johnson, P. Stratton and D.R. Smith. 1962. Comparison of
4A rotation and season-long summer grazing on subalpine range in Wyoming. Wyo. Agr. Exp. Sta. Mimeo Circ. 177. 3 p.

Summer long grazing appeared to be more detrimental, especially to the perennial forbs, than was the rotation grazing procedure. This brief mimeo summarized, in general, the activities of the grazing studies conducted on the Big Horn Mountains for some six or seven years.

3A2 620505 Fisser, H.G. 1962. Information to be obtained from the range management vegetation measurement transects. Wyo. Agr. Exp. Sta. Mimeo. 2 p.
11
10D1

This was an essay which described the procedures for location and placement of quadrats for determination of cover and for the clipping of herbaceous species for determination of forage production. The intent of the clipping program in 1962 did not involve determination of total herbage production. The paper also strongly advocated the development of computer programs and transfer of data to computer accessible cards and analysis by computer programs. This was one of the earliest statements strongly advocating the computer analysis of rangeland research data.

6206 Graduate Advisee Completion

2E 620601 Vosler, L.C. 1962. An ecological study of *Atriplex nuttallii* in the Big Horn Basin of Wyoming. M.S. Thesis. Univ. of Wyo. 43 p.
1E
2C
2A
10D1

This early study of rangeland ecology of the Gardner saltbush *(Atriplex nuttallii)* type addressed a variety of topics. Seed germination studies showed that scarification was necessary because of the very poor germination of the seeds. Sampling was conducted on 40 different locations in the Big Horn Basin in which vegetation cover, soil and grazing stress were evaluated. Soil sampling was conducted and texture pH and total soluable salts (EC x 10^3). With regression analysis, it was shown that there was no specific limiting soil characteristic with regard to distribution of the saltbush communities in the Big Horn Basin. The saltbush plants even in this northern latitude arid region, often began growth as early as mid-February if temperatures were relatively mild for a few days. Intensive grazing was noted to have reduced saltbush cover by a significant amount.

6301 Research Activities

6A2 630101 Winter sheep grazing on saltbush range.
4A
4B The 1963 winter sheep grazing studies were continued with 920 head of ewes on the
1E4 pastures for 21 days from November 2-23, 1963. The 160-acre light use pastures had been divided during the summer of 1963 to make two 80-acre pastures, one of which retained the 20 percent rate of saltbush *(Atriplex gardneri)* utilization and the other one was increased to a level of 30 percent utilization. This was done because the data up through 1962 appeared to indicate a minimum effect of grazing at the 20 percent level, but in the 40 percent pastures the response appeared to be relatively strongly detrimental and it then seemed advantageous to divide the pastures and develop a 30 percent use grazing set. The heavy rate of use was obviously detrimental to the saltbush vegetation. Herbage production was determined within the pastures and exclosures in early July.

6A1 630102 Summer steer grazing on saltbush range.
4B
5A1C The 1963 summer steer grazing trials were continued with 49 head of yearling steers
5B4 grazing the pastures for 83 days. Average daily gain per animal was 0.66 pounds on the
5C non-treated check pasture and 1.03 pounds on the pasture treated with water spreaders
1E4 and seeded with grass. Pounds of gain per acre were four times greater on the treated pasture with 5.92 pounds, than the check pasture with 1.46 pounds. The erosion transects were installed in the steer pastures during the summer of 1963.

3A3 630103 Exclosure study: production, cover, height and precipitation.
5A1D
5B Exclosure studies on 29 sites were continued in 1963 as in 1962, and some additional
5C exclosures, specifically in the Kemmerer area, were established and clipping studies
10D1 initiated on them. Further information may be found in the annual report of 1963 data.
2A
2C

2A 630104 Soil moisture and temperature study.
2C
2F2 Soil moisture and temperature data during 1963 were accumulated at the Granite
10D1 Mountain and the Smilo exclosures. At each site two neutron probe access tubes were located in each of four treatment sites, the in-spray, out-spray, in-native and out-native. Soil moisture was read at seven depths (6 in, 12 in, 18 in, 24 in, 36 in, 48 in, and 60 in) to a maximum of 60 inches and soil temperature was read at four depths (1 in, 8 in, 15 in, and 22 in) to a maximum of 22 inches.

8B 630105 Erosion transect study.
2C
10D1 Erosion transects were installed during 1963 on the steer pastures, on the halogeton sheep pastures and the Cumberland exclosures. At the halogeton sheep pastures, 15 transects were located within each pasture, five transects each on a ridge, level and drainage location. On the steer pastures, a similar design was utilized except that one series was located on the bare ground just below the large spreader dikes. At the Cumberland exclosures, series of five transects were established in each treatment site, for instance, in-spray, out-spray, in-native, out-native. These transects consisted of permanently located four foot bars which were driven deeply into the soil to minimize or avoid any frost heaving problem. Measurements were taken from an angle iron placed atop and the other at ground surface downward to the soil surface. Each angle iron was marked, that is the upper and the lower, and a pin was moved downward at each of the 10 marks thus allowing replacement of points with very excellent precision.

1B4 630106 Germination and longevity of halogeton seeds.

1E

2E

5A2D

Germination and longevity of halogeton *(Halogeton glomeratus)* seeds in 1963 were tested by excavation of some of the seeds which had been planted in 1956 and noted in an earlier trial reported by Vosler in 610401. Seeds still retained some viability and the brown form of seed was still able to germinate in 1963.

2A 630107 Precipitation pattern study.

3A

The precipitation pattern study in 1963 was continued with readings of the rain gauges at the four seasonal dates throughout the year. Additional gauges were located at new exclosures with a total of 94 UW/BLM gauges. Data, by season, for 34 U.S. Weather Service stations were given also.

3A3 630108 Cumberland district exclosures, production, cover.

5A1D

3A1

The new Cumberland exclosures, in 1963, named Cumberland 1, 2, 3 and 4, were established in cooperation with the Bureau of Land Management. Cumberland 1 was located on a bottomland of mixed greasewood *(Sarcobatus vermiculatus)*, big sagebrush *(Artemisia tridentata)*, saltbush *(Atriplex gardneri)*, horsebrush *(Tetradymia canescens)* and rabbitbrush *(Chrysothamnus viscidiflorus)* vegetation about 20 miles south of Kemmerer. Cumberland 2 was located on the slopes of Elk Mountain to the southwest of Kemmerer on a type consisting primarily of low sagebrush *(Artemisia arbuscula)* with a scattering of big sagebrush. Cumberland 3 was located about five miles south of Sage, Wyoming, in a big sagebrush complex. Cumberland 4 was located 25 miles or so south southwest of Kemmerer in a saltbush-winterfat *(Atriplex gardneri - Ceratoides lanata)* type.

6302 Journal Publications

3A3 630201 Van Dyne, G.M., W. Vogel and H.G. Fisser. 1963. Influence of plot size and shape on range herbage production estimates. Ecology 44(3):746-759.

10D1

This publication was a summarization of the thesis work during 1958 and 1959 by Vogel and Fisser while at Bozeman. These data showed that both size and shape had an effective influence upon vegetation sampling accuracy. These data also showed, however, that with reasonable care, any researcher should be able to accomplish acceptable levels of sampling accuracy and incorporate this with acceptable levels of manpower requirements.

6303 Presentations

6A1 630301 "Progress of cattle and sheep grazing trials on halogeton infested ranges" March 19, 1963. Range Weed Control Conf. Portland, Oregon. by H.G. Fisser.

6A2

5A2D

4A

4B

This presentation was primarily a descriptive characterization of the grazing trial activities through 1962. Some specific notes of interest were 1) the fact that the two steer pastures began in 1960 with their grazing activities and 2) the fact that precipitation during some of the years since initiation had significant effect upon the vegetation. During good years, herbage production was very good. During dry years, herbage production was very low.

The sheep grazing trials were noted to have statistically significant differences of saltbush *(Atriplex gardneri)* production on the light use pasture at almost 300 pounds as compared to only 200 pounds on the heavy use pasture. The detrimental effect of the heavy use upon the saltbush was noted as being very evident in terms of the breakdown and disintegration of the large saltbush clumps.

3A3 630302 "Exclosure study: production, cover, height and moisture phase" March 19, 1963. Range
5A1D Weed Control Conf. Portland, Oregon. by H.G. Fisser.
5B
5C This presentation of 1959-1962 research itemized primarily that the exclosure studies
10D1 with herbage production and precipitation recording at exclosures was a new phase of
2A earlier work begun with the introduction and construction of the new exclosures in
2C 1959-1960. The herbage production transects of quadrats along a 100 foot tape was described in some detail and was shown as an add-on to the permanent plot quadrats which were to be sampled only occasionally or at best every four to five years.

2A 630303 "Does weather influence rangeland potential?" September 10, 1963. Rangeland User
3A Group. Worland, Wyoming. by H.G. Fisser.

This presentation was intended to bring out the basic ecological relationships of perennial native vegetation and its relationship and ability to grow and live within the semiarid regions of the Big Horn Basin. It was stated that the vegetation of the area was adapted to droughty conditions whether short-term or long-term. An important consideration was discussed in some detail regarding the limitation to increasing use during a period of better than average moisture because the plants very often required that excess moisture and extra growth to be able to increase root nutrient storages and to establish new plants. On the other hand, during a short-term period of drought, it probably would not be necessary to change grazing strategies either in time or numbers of animals simply because of short-term, one year or more droughty effect, because the perennial vegetation can usually overcome the drought effect if the grazing was at a moderate level and was not excessively detrimental to the individual plants.

6304 Published Proceedings/Abstracts

6A1 630401 Fisser, H.G. 1963. Progress of cattle and sheep grazing trials on halogeton infested
6A2 ranges. p. 7-8. In Proc. Range Weed Control Conf. Portland, Oregon. March 19,
5A2D 1963.
4A
4B See 630301.

3A3 630402 Fisser, H.G. 1963. Exclosure study: production, cover, height and moisture phases. p.
3A2 23-24. In Proc. Range Weed Control Conf. Portland, Oregon. March 19, 1963.
2F2
10D1 See 630302.

6305 Wyoming Agricultural Experiment Station Publications

3A1 630501 Fisser, H.G. 1963. A study of range conditions in the Big Horn and Wind River Basins.
3A2 Prel. Report IV. Wyo. Agr. Exp. Sta. Mimeo Report. 13 p.
2F2
10D1 This summarization of research from 1959-1962 of the Big Horn and Wind River Basin project showed that the vegetation in the Tensleep grazing allotment, by and large, was about an average condition although there were some areas with poor condition and one or two very small areas of excellent condition. In the Kirby Creek area, condition classification seldom was better than fair on any of the areas except on the Roundtop Mountain Relic Area. In the Lander area, the condition classification there was rated from fair to good and it was noted that the increased moisture levels of that area, compared to the Big Horn Basin areas, resulted in greater herbage productive potential than in the two northern units.

2E 630502 Fisser, H.G. 1963. Ecological studies of the *Artemisia tripartita* subsp. *rupicola* and related shrub communities in Wyoming. In Twelve Year Report to the Wyo. Natural Res. Board. Wyo. Agr. Exp. Sta. Mimeo Part VI. 2 p.
2A
2C
2F2
3A

See 620201.

6A2 630503 Fisser, H.G., J.T. Nichols and R.W. Hamner. 1963. Halogeton research. 1962 results. Univ. of Wyo. Coop. Res. Rep. to the BLM. Sections I - III. Wyo. Agr. Exp. Sta. Mimeo. p. 1-38.
3A3
2F2

6A2 63050301 Section I - H.G. Fisser and J.T. Nichols. Winter sheep grazing on saltbush range (Greybull) p. 1-6.
4A
4B
1E4

This was simply a presentation of data from the 1962 grazing season and showed the changes for the years 1960, 1961 and 1962. Spring of 1962 was moist and saltbush production on the heavy, moderate and light use pastures was 200, 237 and 292 lbs./A, respectively.

6A1 63050302 Section II - H.G. Fisser and J.T. Nichols. Summer steer grazing on saltbush range and seeded water spreaders (Worland). p. 7-11.
4B
5A1C
5B4
5C
1E4

This related to the 1962 steer grazing pastures and again, presented information of the grazing effects and relationships of animals and vegetation up through the 1962 period.

3A3 63050303 Section III - H.G. Fisser and R.W. Hamner. Exclosure study: production, cover, height and precipitation phase. p. 12-38.
5A1D
5B
5C
10D1
2A
2C

This described the 1962 activities of the exclosure study and presented some of the data from the 20 exclosures with clipping and production information as well as some of the precipitation data. A list of plant names found at the exclosures was included.

6401 Research Activities

6A2 640101 Winter sheep grazing on saltbush range.
4A

4B Sheep grazing studies were continued in 1964 on the rate of use pastures northwest
1E4 of Greybull. Approximately 900 head of sheep were grazed from October 27 to November 18, 1964. Studies were continued to evaluate production and cover on the pastures and exclosures within the pastures.

6A1 640102 Summer steer grazing on saltbush range.
4B

5A1C A new phase of study in 1964 was initiated on the spreader system steer pastures
5B4 west of Worland. In place of the summer steer use study, investigations were changed to
5C evaluate spring use by steers. Steers were on the pastures from May 12, 1964, to July
1E4 21, 1964. Weights of each of the 104 steers were obtained as well as measures of herbage production, cover, and utilization.

3A3 640103 Exclosure study: production, cover, height and precipitation.
5A1D

5B The 1964 exclosure studies were continued and expanded with multiple transects (In,
5C Out, Spray, Nospray) at each exclosure. Additional rainfall gauges and exclosures
10D1 included six Shoshoni Ant locations during 1964. Two maps of exclosure locations were
2A included as well as the plant list and production cover data.
2C

2A 640104 Soil moisture and temperature study.
2C

2F2 The 1964 soil moisture and temperature studies were continued at the Granite
10D1 Mountain and Smilo exclosures of the Wind River and Big Horn Basins, respectively. During the 1964 season, some additional work related to soil moisture and temperature was begun with the establishment of various sites complimentary to those of Gardner saltbush *(Atriplex gardneri)*. Vegetation communities of bud sagewort *(Artemisia spimescens)*, big sagebrush *(A. tridentata)*, spiny hopsage *(Grayia spinosa)*, birdfoot sagewort *(A. pedatifida)* and winterfat *(Ceratoides lanata)* were located in areas west and northeast of Worland. Within these sites, soil moisture, temperature and vegetation studies were established to determine relationships of water infiltration rates and temperature to the various vegetation types. In addition, at the Cumberland -1 and -3 exclosures, three access tubes were located in the spray and nospray areas in the exclosures.

2A 640105 Precipitation pattern study.
3A

The 1964 precipitation pattern study was increased to 99 UW-BLM rain gauges and 39 U.S. Weather Service stations, and showed a very distinct patterning of the precipitation and changes that occurred associated with elevation and topography effects.

1D 640106 Harvester ant control and related study of vegetation response and economic feasibility
5A1B of the procedure.
10D1

The harvester ant *(Pogonomyrmex occidentalis)* control study in 1964 was initiated on a unit about 10 miles south of Shoshoni. Within this area, 12 treatment sites were located. Three rates of chemical control to the harvester ant were applied. An intensive study site (600 ft x 600 ft) of 8 1/4 acres was established in the center of each 100 acre study area. Six were enclosed with livestock-proof fence and six were unfenced. Permanent quadrats were located at each of the study sites to measure, first of all, the general vegetation cover and composition and also to measure the differentiation of vegetation with distance from harvester ant colony mounds.

3A2 640107 2 ft x 20 ft and 4 ft x 4 ft permanent plot sampling.
2F2
10D1

In 1964 all of the quadrats at the 29 exclosures at which they had been established during 1959 and 1960 were reread during the summer of 1964. At these quadrats, the shrub cover, of sagebrush *(Artemisia tridentata)* and saltbush *(Atriplex gardneri)*, was outlined and mapped within the 2 ft x 20 ft quadrats. During the 1964 season, mat form species were recorded also and were shown on those data. In the 4 ft x 4 ft plots, the 100 points were initally set to record only cover of herbaceous species, not the shrub species. In 1964, the program was modified to include first hits on all species on every pin.

5A1C 640108 Ecology and control of juniper.
2E
10D1

During 1964 at the Kane Deer exclosure east of Lovell on the slopes of the Big Horn Mountains in the Utah juniper *(Juniperus osteosperma)* type, chemical control of juniper with Tordon was accomplished during late July. The exclosure was fenced. It was 400 ft x 400 ft. Half of it was sprayed. Within each of the two inside treatments and both in the sprayed outside and non-sprayed outside, permanent quadrats 20 ft x 50 ft in size were located, 10 quadrats in each treatment. Within each of the 20 ft x 50 ft quadrats, two permanently located 1 ft x 10 ft quadrats were located for the inventorying and continuing monitoring of understory herbaceous species changes. Herbage production clippings of understory species were initiated also. A juniper clearing study was established just north of the exclosure with permanent sampling slots.

5B4 640109 Seeding and fertilizer studies.
5C
10D1

Seedings of several species were attempted on October 29, 1964, on an area of the holding pasture at the Dry Creek halogeton *(Halogeton glomeratus)* sheep grazing pastures. A bulldozer was on hand and was utilized to excavate some pits 8 ft x 30 ft long x 2 ft deep. Seeds were planted within these pits. Excessive moisture during the March and April 1965, period caused continuous standing water in these pits for some period over a month in length. All of the seeds were lost because of this effect.

6402 Journal Publications

5A1C 640201 Nichols, J.T. 1964. Cover, composition and production of contour-furrowed and seeded range as compared to native saltsage ranges. Wyo. Range Manage. 187:27-38.
5C
5B4
3A2
10D1

Herbage production and cover of vegetation on a native saltbush *(Atriplex gardneri)* rangeland site and an adjacent site that had been treated with contour furrowing and with seeding of crested wheatgrass *(Agropyron cristatum)* were studied to determine the effect of the treatment upon herbage production. The furrowing and seeding had been conducted in 1955 and by the summer of 1963, the pits were about half filled with silt, but the herbage production was still greatly in excess of that of the native pasture indicating continued long-term effect of at least seven or eight years on pitted treatments and these soils of silty texture.

1B1 640202 Nichols, J.T. 1964. Plants of the 15-Mile Drainage Washakie County, Wyoming. Wyo. Range Manage. 190:44-50.
1A2

This 1964 list of 156 species of plants from 28 families which were found on the drainages of the 15-Mile area north northwest of Worland was one of the very early comprehen sive lists of species for the region. Many of the plants on the list were new collections for the Big Horn Basin and there was one plant which was a first collection for the state.

4A 640203 Nichols, J.T. 1964. Effect of protection from grazing on two vegetation types in the Big Horn Basin of Wyoming. Wyo. Range Manage. 192:50-69.
3A3

This study was initiated in 1963 to evaluate changes in species composition and forage production due to protection from grazing in three exclosures on two vegetation types in the western Big Horn Basin. The three exclosures are Dutch Nick Flats, Burnt Wagon and West Pasture. Considerable data are given in seven different tables. The data are analyzed to determine the effect of seven years of protection from grazing on vegetation composition and production. Points of conclusions are as follows:

1) Percentage cover, composition and production of the desirable forage species increased within all three exclosures.
2) Annual forbs were more prevalent outside the exclosures than inside, which may indicate that the perennial vegetation outside the exclosure failed to utilize the available moisture as efficiently as that inside.
3) Vigor of plants was better within the exclosures.
4) A small reduction of grazing pressure over the area dominated by the two vegetative types would result in increased productivity.
5) Vegetative change was very slow, thus important changes and trends may not appear for several more years. Future analysis would be valuable to determine if important forage species such as Indian ricegrass *(Oryzopsis hymenoides)* may become sufficiently established to become important.

6403 Presentations

6A2 640301 "Winter sheep grazing studies on saltbush rangeland of Wyoming" February 14, 1964. 17th Ann. Meeting Amer. Soc. of Range Manage. Wichita, Kansas. by H.G. Fisser.
4A
4B
1E4

Winter sheep grazing studies of the halogeton *(Halogeton glomeratus)* pastures in the Big Horn Basin were described defining the three original rates of use, 80, 40 and 20 percent utilization of the saltbush *(Atriplex gardneri)* vegetation and the general procedures of those trials. Little significant variation due to grazing treatment was noted during the first five years of the study although a general trend was established. Following the sixth season of grazing in 1963, deteriorating in condition of the heavy use pastures was readily visible. Production of saltbush during 1962 was significantly affected by all treatments as well (Duncan's Multiple Range Test). Distribution and abundance of halogeton appeared to be directly related to grazing pressure.

6A2 640302 "The Dry Creek Halogeton Pastures and response to winter sheep use." July 7, 1964. BLM Interagency Tour, Greybull, Wyoming. by H.G. Fisser.
4B
3A3
1E4

See 640301.

6A1 640303 "Summer steer grazing on saltbush rangeland" November 20, 1964. Wyoming Section American Society of Range Management Annual Winter Meeting. Casper, Wyoming. by H.G. Fisser.
4B
5A1C
5B4
5C
1E4

This was merely continuation of information which had been described earlier in 640102.

2C 640304 "Soil-vegetation relations of the 15-Mile Drainage area in the Big Horn Basin of Wyoming." February 14, 1964. 17th Ann. Meeting. Amer. Soc. Range Manage. Wichita, Kansas. by J.T. Nichols.
2F2
2E
8B

Seven vegetation communities were identified by the author in this presentation. Findings showed that basin big sagebrush (*Artemisia tridentata* ssp *tridentata)* was specifically located in bottomlands and in areas in which the taproots contacted the lower permanent aquifer. Greasewood *(Sarcobatus vermiculatus)* occurring in similar situations was found to have a similar taproot and water aquisition system.

On the uplands, blue grama *(Bouteloua gracilis)*, saltbush *(Atriplex gardneri)* and the Wyoming big sagebrush (*Artemisia tridentata* ssp *Wyomingensis*) also were studied, and relationships included those related to soil and water holding capacity. Differences in soil physical and chemical characteristics often modified potential for vegetation establishment and of a special importance were the availability of water to plants during drouthy periods. The soils with sodic chemistry and silty to clayey soil texture allowed little water availability or infiltration while the soils with little sodium problem or with sandy and perhaps only silty soil often allowed significant water availability which resulted in significant modification to the vegetation complex.

6404 Published Proceedings/Abstracts

6A2 4A 4B 1E4 — 640401 Fisser, H.G. 1964. Winter sheep grazing studies on saltbush rangeland of Wyoming. p. 53. In Proc. 17th Ann. Meeting, Am. Soc. Range Manage. Wichita, Kansas. February 14, 1964.

See 640301.

2C 2F2 2E 8B — 640402 Nichols, J.R. 1964. Soil-vegetation relations of the 15-Mile Drainage area in the Big Horn Basin of Wyoming. p. 25. In Proc. 17th Ann. Meeting Am. Soc. Range Manage. Wichita, Kansas. February 14, 1964.

See 640304.

6A1 4B 5A1C 5B4 5C 1E4 — 640403 Fisser, H.G. 1964. Summer steer grazing on saltbush rangeland. p. 7. In Proc. Wyo. Sec. Am. Soc. Range Manage. Casper, Wyoming. November 19-20, 1964.

See 640102.

6405 Wyoming Agricultural Experiment Station Publications

640501 Fisser, H.G., J.T. Nichols, R.W. Hamner and J.R. Wight. 1964. Halogeton Research. 1963 Results. Univ. of Wyo. Coop. Res. Rep. to the BLM. Sections I - VIII. Wyo. Agr. Exp. Sta. Mimeo. p. 1-61.

6A2 4A 4B 1E4 — 64050101 Section I - H.G. Fisser and R.W. Hamner. Winter sheep grazing on saltbush range (Greybull). 8 p.

The 1963 sheep grazing studies were continued with 920 head of ewes on the pastures for 21 days from November 2-23, 1963. The 160 acre light use pastures were divided to allow application of a 30 percent utilization rate in addition to the original 20, 40 and 80 percent use rates. The heavy rate of utilization was obviously detrimental to range vegetation.

6A1 4B 5A1C 5B4 5C 1E4 — 64050102 Section II - H.G. Fisser and R.W. Hamner. Summer steer grazing on saltbush range and seeded water spreaders (Worland). 6 p.

The 1963 summer steer grazing trials were continued with 49 head of yearling steers grazing the pastures for 83 days. Average daily gain per animal was 0.66 pounds on the non-treated check pastures and 1.03 pounds on the pasture treated with water spreaders and seeded with grasses. Pounds of gain per acre were four times greater on the treated pasture with 5.92 pounds than the check pasture with 1.46 pounds.

3A3 64050103 Section III - H.G. Fisser, J.T. Nichols and J.R. Wight. Exclosure study: production,
5A1D cover, height and precipitation phase. 33 p.
5B
5C See 630103.
10D1
2A
2C

2A 64050104 Section IV - H.G. Fisser, J.T. Nichols and J.R. Wight. Soil moisture and temperature
2C study. 3 p.
2F2
10D1 See 630104.

8B 64050105 Section V - H.G. Fisser, J.T. Nichols and J.R. Wight. Erosion transect study. 2 p.
2C
10D1 See 630105.

1B4 64050106 Section VI - H.G. Fisser and R.W. Hamner. Germination and longevity of halogeton
1E1 seeds. 1 p.
2E
5A2D See 630106.

2A 64050107 Section VII - H.G. Fisser. Precipitation pattern study. 5 p.
3A
See 630107.

3A3 64050108 Section VIII - H.G. Fisser, R.W. Hamner and J.T. Nichols. Cumberland district
5A1D exclosures. 3 p.
5A1
See 630108.

3A1 640502 Fisser, H.G. 1964. Range survey in Wyoming's Big Horn Basin. Wyo. Agr. Exp. Sta.
3A2 Bull. 424. 44 p.
2F2

This was a publication oriented specifically to an evaluation of the condition class survey program conducted in the Big Horn Basin from 1959-1962. The publication gives descriptive information of the location, geology and soils, climate, grazing history and other information of general interest relating to the Big Horn Basin and its place in the grazing aspects of north central Wyoming.

The methods of study in this program were essentially oriented toward the ecological basis of range survey. Methods of vegetation inventory were primarily oriented toward transects of 10 1 ft. x 1 ft. quadrats located in areas of rangeland vegetation selected because of definitive site characteristics including soil, slope, aspect, moisture relationship and kinds of vegetation types. Geological structures sometimes were significant. Soil samples were taken at each of the transect locations and for the permanent record files, photographs were taken whenever the transects were read. A study of precipitation pattern began in 1960 by installing rain gauges throughout the southeastern portion of the Big Horn Basin.

The condition class analysis was based upon vegetative potential of rangeland in terms of the species of vegetation, composition and the cover as they would have been under natural or ungrazed conditions. Three major associations of vegetation were identified for the Big Horn Basin. These were identified on the basis of the dominant shrub species which were 1) big sagebrush *(Artemisia tridentata)*, 2) Gardner saltbush *(Atriplex nuttallii)* and 3) greasewood *(Sarcobatus vermiculatus)*.

A number of relic areas were located, mostly near Thermopolis, on North Butte and on Roundtop Mountain as well as the Little Butte Thermopolis semi-relic and the

Buffalo pasture semi-relic. Farther to the east at the Red Hole area, a semi-relic area was located and two areas in the Bud Kimball portion of the Basin to the southwest of Tensleep were the Windy Point Valley semi-relic and the Castle Garden semi-relic. These semi-relic areas were utilized as bases from which to compare to adjacent grazed rangelands.

In addition, eight exclosures in the southeastern quadrant of the Big Horn Basin were utilized for comparison of vegetation response to protection from grazing within the fenced exclosure sites. At three exclosures, chemical herbicide 2,4-D were applied for control of big sagebrush and reseeding with pitting was conducted also. This work was done during late 1959. The water spreader and steer grazing study northwest of Worland and the grazing intensity study on the Dry Creek drainage northwest of Greybull were discussed in this report also.

For the sagebrush/grass type, excellent condition range was identified as that with at least 5 percent basal area of decreaser grasses and grass-like species. In addition, these plants would make up 80 percent of the total understory composition. Those species most abundant throughout the Big Horn Basin were wheatgrasses, (*Agropyron* species) needleandthread *(Stipa comata)* and threadleaf sedge *(Carex filifolia)*. Increaser species occupied less than 1 percent of the soil surface and invaders generally are not present in the excellent condition range. Crown cover of big sagebrush would not exceed 15 percent on an excellent condition range. As condition deteriorates, many changes may occur. As the decreaser species are reduced, the increasers and invaders may become dominant, or sagebrush may become abundant.

In the saltbush type, excellent condition range was identified with having 10 percent cover of the Gardner saltbush halfshrub, at least on sites of heavy and impermeable soils. On lighter textured soils, saltbush cover will normally be greater, perhaps up to as much as 20 percent cover. Decreaser species, such as western wheatgrass *(Agropyron smithii)*, Indian ricegrass *(Oryzopsis hymenoides)* and winterfat *(Ceratoides lanata)* will be present but usually in small quantities. As condition decreases, the saltbush clumps become smaller and broken down. Increaser species obtain some prominance, but only annual weeds become abundant.

In the greasewood type, soils were often a very restraining factor to the vegetation because of alkaline and impermeable characteristics. Areas have been observed with a dense understory of such grasses as western wheatgrass, alkali sakaton *(Sporobolus airoides)*, Nevada bluegrass *(Poa nevadensis)* and inland saltgrass *(Distichlis spicata)*. These have been observed in bottom lands in which the stream channel was near the level of the greasewood site. Depth of soil and water table were important factors determining understory cover and composition.

An Appendix to the publication contains information relating to 260 transects which were located in the Kirby Creek and Tensleep rangelands areas. For each transect, the information given was county location, legal description, percent slope, exposure, elevation, soil texture percentages, pH, total soluable salt, and percent area cover of decreaser grasses, increaser grasses, invader grasses, forbs, shrubs and the condition class evaluation for that transect area. A list of 100 plant species identified in the southeastern quadrant of the Big Horn Basin were presented in the Appendix as well. These plants occurred in 26 families. Almost half of the transects were located in the poor to very poor condition classification areas. Almost 20 percent were located on excellent or primarily good condition sites, and about 35 percent were located on fair condition sites.

6406 Graduate Advisee Completion

3A3 2F2 10D1 640601 Blankenship, J.O. 1964. Indirect estimation of standing crop for selected alpine species. M.S. Thesis. Univ. of Wyo. 52 p.

This work addressed the idea of determining standing crop by measurement of characteristics of the plants including the height, width, length and number of leaves of grasses. It was analyzed with regression analyses. The findings were not exceptionally strong and have not been utilized because of the large sampling error and the excessive numbers of individual plant units that needed to be sampled for determination of standing crop by this procedure. In addition, the standing crop indirect estimation procedure would have to be developed each year because of different growth characteristics of the plants.

2E 3A1 2C 2F2 10D1 — 640602 Hamner, R.W. 1964. An ecological study of *Sarcobatus vermiculatus* communities of the Big Horn Basin, Wyoming. M.S. Thesis. Univ. of Wyo. 75 p.

Four major communities of black greasewood *(Sarcobatus vermiculatus)* of Wyoming's Big Horn Basin were studied to determine their ecological relationships. Three characteristics of each community were studied: the overstory (canopy cover and composition), the understory (composition and numerical representation) and edaphic factors (pH, soluble salts and soil texture).

From this study the similarities and dissimilarities, vegetatively and edaphically, were shown. The heaviest soils supported the sparsest cover of overstory shrubs. But as the percentage of sand in the soil increased, the percentage of canopy cover also increased in an almost linear relationship. In some areas pH and total soluble salts through restriction of plant growth of certain species undoubtedly played an important role in delimiting the boundaries of various communities. In places of lesser amounts of salts and lower pH, a greater compliment of plants were found. This increased the floral complexity of the community. Thus it can be said that the various edaphic factors had a very marked control on the complexity of the different greasewood communities.

2E 3A1 2C 2F2 10D1 7B1 — 640603 Miller, W.B. 1964. An ecological study of the mountain mahogany community and related biotic associations of the Big Horn Mountains. M.S. Thesis, Univ. of Wyo. 152 p.

Curlleaf mountain mahogany *(Cercocarpus ledifolius)* was studied in its extreme northeastern geographical extension on the Big Horn Mountains of Wyoming. The species characterizes a shrubby belt of vegetation surrounding the base of the mountain range.

The community was studied in order to describe the physical habitat, plant association and small mammal and avian inhabitants. Three sites based on degree of parent material, disintegration and decomposition were distinguished within the curlleaf mahogany community. The location of these sites was dependent upon physiographic characteristics and type of geological deposition.

Most of the community was found to be present on colluvial sites which were composed of a mixture of poorly disintegrated debris and finer materials. The second site, exposed bedrock which exhibited varying stages of fracture, supported a large portion of a community. The last site occurred on areas which supported aeolian deposits of sandy material.

2C 2F2 3A1 — 640604 Nichols, J.T. 1964. Soil-vegetation relationships of the 15-Mile drainage, Washakie County, Wyoming. Ph.D. Thesis. Univ. of Wyo. 120 p.

Vegetation and soils of the 15-Mile drainage, Washakie County, Wyoming, were investigated to determine major plant communities and the relationships to certain edaphic factors. Five plant communities were recognized on the basis of physiognomy.

Sampling was conducted along 50′ transects within each community to determine percentage species cover and composition. Low growing vegetation was sampled by the point-frame method and shrubby vegetation by the line-intercept method for overstory vegetation. Individual points were placed at 1′ intervals along the same 50′ tape for understory vegetation.

Two broad categories of communities were recognized, based on different soil moisture relationships: 1) Lowland communities which were situated near the main drainage channels and had permanent subsoil moisture at lower depths from 8 to 15 feet below the surface and 2) Upland communities located on sites which were dependent upon soil moisture near the surface.

The lowland communities consisted of two kinds which were positioned in relation to one another in a commonly associated pattern: 1) the sagebrush *(Artemisia)*, rabbitbrush *(Chrysothamnus viscidiflorus)* community which was located adjacent to the stream channel on the first terrace and 2) the greasewood *(Sarcobatus vermiculatus)* community located on the second terrace from the stream channel. The stream terraces were colluvial deposits of past stream depositions. However, no significance was attributed to the differences of terrace level in terms of distribution of the two communities. Dominant species of both communities were rooted in the soil substrata of permanent moisture, and varied as to depth of root penetration depending on the deepness of channel cut. Permanent soil moisture in which the dominants were rooted was the principle factor responsible for existence of the lowland communities.

The principal differences in soils which could account for the delineation of the two lowland communities was in relation to the chemical soil properties and their effect on moisture relationship. The greasewood community had higher soil reaction, greater amounts of soluble salts and of sodium than that found in the sagebrush rabbitbrush communities.

The upland communities were of three types: 1) blue grama *(Bouteloua gracilis)* community with an abundance of forbs and a small amount of midgrasses, 2) sagebrush/ blue grama community with an understory vegetation similar to the dominant species of the blue grama community but with an overstory of mountain big sagebrush *(Artemisia tridentata* ssp. *vaseyana)* and 3) saltbush community dominated by Nuttall saltbush *(Atriplex nuttallii)* and very few associated species. Variable parent materials of the Willwood Formation resulted in soils with different properties. Breakdown of sandstone layers resulted in sandy soils with low pH values, and small amounts of soluble salts and sodium. These soils were characteristic of the gramagrass and sagebrush-gramagrass communities. In contrast, soils of the saltbush community developed primarily from clays and shales which resulted in finer texture, higher salt content and higher pH.

The effect of texture on water permeability was pronounced. Highest rates of water movement into the soil were recorded for the sagebrush-gramagrass community with somewhat lower rates for soils of the gramagrass community. The more permeable soils of the former were considered to be the principle factor which was responsible for mountain sagebrush as an overstory dominant. Factors of pH and soluble salts were approximately the same and were low for both communities. High soil permeability allowed maximum use of precipitation with little loss from runoff and evaporation. In contrast, soils of the saltbush community had low water permeability resulting in runoff and evaporation loss.

Another factor which reduced favorable water relations of the saltbush community was sodium dispersed soils which further reduced permeability. Also, soluble salts reduced water intake by plants by lowering the osmotic gradient. This was much the same situation described for the greasewood community but not as extreme. Differences in water relationships were considered the primary factor responsible for differentiation into the different upland communities. Root systems of dominant plants in the upland communities were distributed in the soil profile in relation to depth of moisture penetration and differences in root morphology.

Roots of Nuttall saltbush were shallow with lateral spreading, reflecting the concentration of moisture nearer the soil surface. Roots of most grasses occupied greater depths in the sandy permeable soils, and were concentrated beneath the plants with little lateral spreading. Blue grama had the most extensive root system of the grasses and Sandberg bluegrass *(Poa secunda)*, the most limited. Winterfat *(Ceratoides lanata)* and mountain big sagebrush, associated with the grasses, had root systems extending to depths of 4 feet with much lateral spreading. The most extensive region of absorption for the grasses was near the soil surface and at lower depths of the soil profile for winterfat and sagebrush.

6501 Research Activities

6A2 650101 Winter sheep grazing on saltbush range.

4A

4B The 1965 sheep grazing had 1,717 head on the pastures beginning November 23,
1E4 1965. Production on the pastures was 429, 434, 264, 182 lbs/acre on the slight, light, moderate and heavy use pastures respectively. On the slight and light use pastures, the saltbush *(Atriplex gardneri)* comprised from 85 to 95 percent of the total vegetation. On the moderate use pastures this value was 68 percent. Precipitation during 1965 was only 2.2 inches and production decreased somewhat on the slight and light use pastures but not as much as that on the heavy use pastures. Halogeton *(Halogeton glomeratus)* plants were counted within quadrats and were found to have the following numbers of plants per square foot for the slight, light, moderate and heavy use pastures respectively: 1.7, 2.4, 3.1 and 13.3. Other annual plants were abundant on the heavy use pastures. During July 1965, vegetational analyses were conducted by the point-frame method on 10 permanent 50 ft line transects that were located in each pasture. Vegetation was evaluated then on the basis of 300 point recordings on each transect by a systematic placement of the point frame which contained 10 pins.

6A1 650102 Summer steer grazing on saltbush range.

4B

5A1C During 1965, grazing on the North Fork experimental pastures consisted of a mixture
5B4 of cows, calves, yearlings, long yearlings and bulls instead of the steers which had been
5C used during previous years. Vegetation analyses of cover and composition were conducted
1E4 by the point-frame method. Permanent line transects 50 ft long were established in each pasture and read on June 10, 1965. In the treated pasture, analyses were conducted on three kinds of sites: 1) seeded areas adjacent to the dikes where a stand of grass was established, 2) unaffected areas of native saltbush range remote from the immediate dike area and 3) inter-dike areas where the vegetation was destroyed during dike construction but where seeded grasses failed to establish. In the native pasture, the transects were established on typical saltbush range. Production and utilization estimates were not conducted during the 1965 season. The erosion transects established in the pasture in 1963 were re-read on June 10, 1965. The animals which grazed the pastures during 1965 consisted of 17 cows, 13 animals older than yearlings and 36 yearlings (33 heifers and 3 bulls). The livestock were put into the pastures beginning May 15 and removed on July 13. A total of 66 cattle and six calves were in the pastures. On June 18, the gates to the upper pastures (check or untreated) were opened and the cattle made intermittent use of the area until removal on July 13. Six calves were born while grazing the pastures.

3A3 650103 Exclosure study: Production, Cover, Height and Precipitation.

5A1D

5B The 1965 exclosure studies of production and cover and precipitation research
5C activities were conducted on 41 different locations including each of the Dry Creek
2A Halogeton pastures and exclosures and the six Shoshoni Ant exclosure areas. Production
2C clipping in 1965 was conducted both inside and outside the exclosures and study areas
10D1 to determine the influence of grazing upon the vegetation during the previous years. In some instances, significant variations in production were noted outside the exclosures as compared to inside. Some areas, however, showed very little difference. At the exclosures where sagebrush *(Artemisia tridentata)* had been controlled, response of understory vegetation to the procedures was easily visible both inside and outside. Production values were lower in the grazed areas, but was greater on the sprayed sites than on the non-sprayed. The influence of the cool, moist spring of 1965, which retarded rate of plant growth, was evident in the greater production values of many areas. Sandberg bluegrass *(Poa secunda)* production values, and other early maturing species also, were above normal since they had not reached the stage of deterioration normally expected at the clipping dates. Later maturing plants, like wheatgrasses (*Agropyron* spp), were often in a comparably early stage of growth.

2A 650104 Soil moisture and temperature study.
2C
2F2
10D1

During 1965 soil moisture data were collected on eight dates at the Smilo and Granite Mountain exclosures, and on four dates on the Cumberland-1 and Cumberland -3 exclosures as well as on six dates at the sagebrush *(Artemisia tridentata)*, saltbush *(Atriplex gardneri)*, bud sagewort *(Artemisia spinescens)*, spiny hopsage *(Grayia spinosa)* and birdfoot sagewort *(A. pedatifida)* communities located in the Big Horn Basin as further described in 650113.

8B 650105 Erosion transect study.
2C
10D1

Erosion transects at the Dry Creek rate of use pastures north of Greybull and at the summer steer grazing pastures west of Worland and at the Cumberland -1, 2, 3 and 4 exclosures were reread. Much soil movement was noted on the shrub chemical control sites.

1B4 650106 Germination and longevity of halogeton seeds.
1E1
2E
5A2D

See 630109.

2A 650107 Precipitation pattern study.
3A

During 1965 several additional gauges were established at some bird guzzlers which had been located and constructed by Wyoming Game and Fish personnel in the Rock Springs/Farson area and a gauge was established at a pit reservoir constructed by the Bureau of Land Management west of Kemmerer. Precipitation during winter of 63-64 was minimal, an open winter, and vegetation was particularly harmed by the lack of snow. During spring, however, several heavy rain storms occurred throughout the Basin regions so many of the areas recorded precipitation amounts greater than had ever before been recorded. Until the middle of October, however, precipitation was very low and vegetation was not really able to take full advantage of the abundant moisture during the spring period. As in previous reports seasonal precipitation from 39 U.S. Weather Service Stations was included in the annual report.

5B4 650108 Seeding in juniper and juniper control.
5A1C
5A1D

During the early spring of 1964 several species of shrubs were planted in the areas, both within the Kane Deer exclosure east of Lovell and on an area outside the exclosure from which the juniper *(Juniperus osteosperma)* was removed by sawing all stems and trunks off at or near ground level. Permanent plots for cover and composition were established in this area, which is located north of the exclosure. The Utah juniper in treated areas of the Kane Deer exclosure have exhibited significant response to the chemical spray application with leaves turning brown.

3A2 650109 1 ft x 1 ft permanent plot transects.
2F2
5C
10D1

A new series of transects with permanent quadrats during summer 1965 was established at 18 exclosures. The basic process involved establishment of permanent plots with spikes at the corners of each of the quadrats located systematically at five-foot intervals along a 100 ft tape. This provided significant information that in the future would be of value because it would allow avoidance of a major problem of sampling - the random sampling error associated with relocating quadrats in non-permanent quadrat sampling. These quadrats were not to be destructively sampled and would only be sampled for cover and density.

1D 650110 Harvester ant control study.
5A1B
10D1

The harvester ant *(Pogonomyrmex occidentalis)* control study was in full operation with 400 transects of 10 quadrats located in each of the twelve 600 ft x 600 ft study areas, six of them exclosures, six of them not. Four of the exclosures were treated with the chemical Myrex and four of the open non-fenced areas were treated with the chemical Myrex. There had also been an equivalent number of permanent quadrats at harvester ant mounds both in treated and non-treated areas to measure the vegetation changes with distance from the mound center.

3A2 650111 Sagebrush cover estimation with angle gauge.
10D1

An angle gauge was constructed out of a lathe and a couple of crosspieces and tested on sagebrush cover. The angle gauge worked relatively well, but it had a high random sampling error value. Although the data were useful, it required an additional operation during the determination of cover, especially along with the transects of 1 ft x 1 ft permanent plots. As such, it was probably not worth the extra time and effort.

5A1C 650112 Juniper control study on L.U.
2E
10D1

The L.U. ranch was located some 50 or 60 miles west of Worland. The chemical Tordon, in combination with 2,4-D and by itself, was applied to several areas of juniper *(Juniperus osteosperma)* and limber pine *(Pinus flexilis)* during the summer. Harold Alley was in charge of the chemical application. We established several 20 ft x 50 ft quadrats with interior 1 ft x 10 ft quadrats to study vegetation response.

2A 650113 Microclimatic studies on desert shrub types.
3A1
2C
8B
10D1

The microclimatic studies of desert shrub types involved the comparison of vegetation types adjacent to saltbush *(Atriplex gardneri)* types to 1) determine the restraints that soils may have upon the location and distribution of saltbush and 2) to determine the relationships of additional vegetation types in their relationships to soils also. Those vegetation types, indicated earlier, were big sagebrush *(Artemisia tridentata)*, bud sagewort *(Artemisia spinescens)*, birdfoot sagewort *(A. pedatifida)* and spiny hopsage *(Grayia spinosa)*, as well as winterfat *(Ceratoides lanata)*. These sites were initiated in the 15-Mile area, in the Torchlight area out of Manderson, northeast of Worland and on a site southwest of Shell.

6503 Presentations

2F2 650301 "Vegetation changes from cold desert to subalpine communities" January 18, 1965. IX International Grassland Congress. Sao Paulo, Brazil. Presented in absentia by A. Ried.
2A
2C
3A1

See 650401.

4B 650302 "Comparison of rotation and season-long grazing on the Big Horn Mountains" July 22, 1965. Soil Conservation Society Annual Meeting, Wyoming Chapter. Big Horn Mountains. by H.G. Fisser.
4A

This presentation summarized the five year grazing management study which was carried on at the Big Horn Mountains from 1959-1964. A Wyoming Agricultural Experiment Station Research Journal was produced as a result of these studies and is found in item 670202. The grazing management study compared moderate and heavy rotation use with moderate season long use. These three treatments were evaluated in terms of livestock production and maintenance of the forage resource.

In general, the cover of the grass Idaho fescue *(Festuca idahoensis)* was maintained best in the season long unit as well was the production of that grass. The differences of steer daily gain between the moderately stocked units was not significant. Lower daily gain in the heavy rotation system indicates that stocking rates had more affect on animal unit production than systems of grazing.

4A 650303 "Halogeton abundance and distribution as influenced by degree of saltbush utilization" September 14, 1965. Range Weed Control Conf. Elko, Nevada. by H.G. Fisser.
3A3
1E1

See 650402.

5A1D650304 "Exclosure studies - Responses to sagebrush spraying" November 20, 1965. Wyoming Section American Society of Range Management Annual Winter Meeting. Worland, Wyoming. by H.G. Fisser.
2E
1E1

This was a detailed study of the vegetation response to sagebrush control at three exclosures in the Big Horn Basin. The basic questions asked with regard to their relationship to the potential for sagebrush control were:

1) Will sagebrush control increase forage production or will erosion occur and weeds become abundant?

2) How much increase can be expected and how soon can it be expected to occur?

3) What actually happens to the existing vegetation to give us an expected forage production increase?

4) Will the sagebrush control be permanent or will the area have to be resprayed after a few years?

5) In view of the possible sagebrush increase and the lack of permanency of the procedure, will the sagebrush control be economically worthwhile?

6) Will there have to be necessary changes in grazing management to be able to efficiently utilize the increased forage?

The three Bighorn Basin exclosures considered in this presentation were the Bud Kimball, the Smilo and the Cochran Exclosures, and the Granite Mountain Exclosure in the Wind River Basin. With slides taken at the exclosures and slides showing the data responses of herbage production and herbage cover changes from 1962-65, it was generally shown that the forage grasses increased considerably following sagebrush control. The answers to the questions were as follow:

1) (Will sagebrush *(Artemisia tridentata)* control increase production?) - YES
   a) In some cases cheatgrass *(Bromus tectorum)* will be abundant but should be replaced by western wheatgrass *(Agropyron smithii)* in time,
   b) Erosion has not been a problem.

2) (How much increase and how soon?)
   a) Increase should be at least double the original and be up four times the original production levels,
   b) Some initial increase will occur but this takes several years to stabilize depending on the condition of the rangeland.

3) (What happens to vegetation to give the increase?)
   a) In the spray areas there was an increase in soil moisture, a decrease in soil temperature, increased moisture in the plants and increased rate of growth in the plants.

b) Under the above conditions, both Sandberg bluegrass *(Poa secunda)* and western wheatgrass increased tremendously, but western wheatgrass replaced the bluegrass in time as condition classification improved. Other grasses, such as Indian ricegrass *(Oryzopsis hymenoides)*, Junegrass *(Koeleria cristata)*, bottlebrush squirreltail *(Sitanion hystrix)* and needleandthread *(Stipa comata)* also become more abundant as condition classification increased.

4) (Will sagebrush control be permanent?)
   a) All indications in the Big Horn Basin indicate that under moderate use sagebrush reinvasion should not be a serious problem.

5) (Is control economical?)
   a) The implication given was that a doubled carrying capacity on a permanent basis was likely and should be able to have been obtained at that time. If this were able to be done at a cost of only $3.25 to $3.50 per acre, the implication was that obviously it should be economical.

6) (Will management need to be changed?) No, except for:
   a) During the first few years there should be some management modification and control because livestock prefer the grass in the sprayed areas, probably because of moisture content. For this reason they should be restricted in their ability to overuse those areas.
   b) There may be some need for management change because of government regulations. The manager should allow for maturity during the first year in order to achieve the greatest grass seed production possible.

5B4 650305 "Grass seeded spreader systems can increase carrying capacity." May 26, 1965.
5C Washakie County Agr. Ext. Service Range Improvement Conf. Worland, Wyoming.
4A by H.G. Fisser.
10D1

A 90-acre spreader system constructed by the BLM on the northfork of the 15-Mile drainage west of Worland was seeded to crested wheatgrass *(Agropyron cristatum)* and tall wheatgrass *(A. elongatum)* in 1960. This and the surrounding saltbush *(Atriplex gardneri)* range was fenced to form a 521-acre pasture. A similar area of 489 acres was fenced along the same drainage with no spreader system or seeded grass. Detention dams were constructed and animals utilized the rangeland during the summer.

These two pastures were grazed by yearling steers during the summers of 1960-63. Crested wheatgrass during summer was much more palatable than saltbush. Cottonwoods *(Populus angustifolia)* and willows (*Salix* spp) also were browsed very strongly during the summer. Four times as many pounds of gain/acre were produced on the treated pasture as on the native range and carrying capacity was increased 2 1/2 times.

6504 Published Proceedings/Abstracts

2F2 650401 Fisser, H.G. 1965. Vegetation changes from cold desert to subalpine communities. p.
2A 425-427. In Proc. IX International Grassland Congress. Sao Paulo, Brazil. January
2C 18, 1965.
3A1

This was a report of the study of sagebrush (*Artemisia* spp) vegetation types on the slopes of the Owl Creek and Wind River ranges of western Wyoming. The basic data established in the dissertation showed that the big sagebrush *(A. tridentata)* type began at the lower elevations and with an increase in elevation, vegetation type changes occurred with the development, with increased elevation, of black sagebrush *(A. nova)*, silver sagebrush *(A. cana)*, extensive ecotone of silver and three-tip sagebrush *(A. tripartita)*, and finally a three-tip sagebrush ecosystem at the highest elevations.

This paper was specifically associated with the general concept of the change of vegetation composition and type from lower elevations at about 4000 ft to the highest elevations of about 10,000 ft. Changes were shown to be associated with temperature, decreasing with increased elevation, amount and time of precipitation, more summer precipitation and greater amounts with increased elevation, and air movement, which became more and more effective with increased elevation.

An important concept was that illucidated with regard to boundaries of vegetation types or ecosystems. In effect, it can be assumed that any ecosystem will exhibit a great deal of vegetation non-uniformity and insufficient habitat variation to dispel a given species or group of species. The term "community" may well be applied for practical identification of a given vegetational composite. But care must be taken to attach a connotation of gradational variation in habitat characteristics within and beyond the assumed boundaries.

4A 1B4 1E4 650402 Fisser, H.G. 1965. Halogeton abundance and distribution as influenced by degree of saltbush utilization. p. 11-13. In Proc. Range Weed Control Conf. Elko, Nevada. Aug. 2, 1965.

This paper was presented to show the differential influence of grazing intensity upon distribution and abundance of halogeton. This study was conducted at the Dry Creek pastures northwest of Greybull, Wyoming.

Several factors affected abundance and distribution of halogeton *(Halogeton glomeratus)*: 1) annual weather variation, 2) rangeland topography and associated variation in soil materials, 3) drainage patterns and water movement and 4) intensity of utilization. In 1964 on the halogeton pastures, the density by level of use from 20 to 30 to 40 to 80 percent was 1.0, 1.8, 2.0 and 7.5 plants per square foot. In 1965, the numbers of halogeton plants in the same order had increased to 1.8, 2.3, 3.1 and 13.3 plants per square foot in the 80 percent use pastures.

6505 Wyoming Agricultural Experiment Station Publications

650501 Fisser, H.G., J.R. Wight, J.R. Flesland, L.D. Robinson. 1965. Halogeton Research. 1964 Results. Univ. of Wyo. Coop. Res. Rep. to the BLM. Section I - VI. Wyo. Agr. Exp. Sta. Mimeo. p. 1-82.

6A2 4A 4B 1E4 65050101 Section I - H.G. Fisser and J.R. Wight. Winter sheep grazing on saltbush range (Greybull). 13 p.

Sheep grazing studies during 1964 were continued on the rate of use pastures northwest of Greybull. Approximately 800 head of sheep were grazed during November. Studies were continued to the next year.

6A1 4B 5A1C 5B4 5C 1E4 65050102 Section II - H.G. Fisser and J.R. Wight. Spring steer grazing on saltbush range and seeded water spreaders (Worland). 7 p.

A new phase of study in 1964 was initiated on the spreader system pastures west of Worland. In place of the summer steer use study, investigations were changed to evaluate spring use by steers. Steers were on the pastures from May 12, 1964, to July 21, 1964. Weights of steers were obtained as well as measures of herbage production, cover and utilization. Exclosure studies were continued to evaluate production, basal area and rainfall relationships. The 35 sites were essentially the same as those of previous years, but they listed the individual exclosures and pastures of the halogeton *(Halogeton glomeratus)* unit as separate individuals.

3A3 65050103 Section III - H.G. Fisser, J.R. Wight, J.R. Flesland and L.D. Robinson. Exclosure study:
5A1D production, cover, height and precipitation phase. 49 p.
5B
5C During 1964 herbage production of forage species, that is perennial grasses, annual
2A grasses, perennial forbs and annual forbs, was determined with 20 plot transects on 46
2C transects located at exclosures primarily in the Big Horn Basin, Wind River Basin and a few locations in the western Wyoming area. These included a transect series at each of the six fenced exclosures of the Shoshoni Ant study. Two maps showing locations of exclosures at which these data were obtained are included in the report.

2A 65050104 Section IV - H.G. Fisser, J.R. Wight and J.R. Flesland. Soil moisture and temperature
2C study. 6 p.
2F2

During 1964 soil moisture and temperature data were collected on seven dates at the Smilo Exclosure and on five dates at the Granite Mountain exclosure. Soil moisture was recorded on three dates at the Cumberland -1 and -3 exclosures.

2A 65050105 Section V - H.G. Fisser. Precipitation pattern study. 5 p.
3A

Precipitation pattern study in 1964 was conducted on some 96 UW-BLM rain gauges with seasonal recording. Seasonal precipitation was compiled for 39 US Weather Service stations.

5B4 65050106 Section VI - H.G. Fisser. Seeding and fertilizer trials in saltbush range. 2 p.
5C

The seeding and fertilizer trials on saltbush *(Atriplex gardneri)* range at the Dry Fork halogeton rate of use pastures was investigated and discontinued because of excessive moisture.

3A1 650502 Fisser, H.G. 1965. University of Wyoming and Bureau of Land Management
2F2 Cooperative research in the Wind River Basin. Wyo. Agr. Exp. Sta. Mimeo. 3 p.

This three page mimeo described the generalized characteristics of the rangeland research program titled at the present time "Arid Land Shrub Ecology Research" which involved the various activities which were listed in each of the sections of the annual report given to the Bureau of Land Management.

6601 Research Activities

6A2 660101 Winter sheep grazing on saltbush rangeland.
4A
4B The grazing intensity influence upon vegetation, cover and composition of saltbush
1E4 *(Atriplex gardneri)* vegetation was conducted during November 1966. This was the last year of the winter grazing studies on the Dry Creek halogeton pastures northwest of Greybull, Wyoming.

6B2B 660102 Grazing intensity influence on dietary intake by use of fistulated sheep.
6A2
4A Several of the sheep utilized by Dr. R.E. Rice of the University of Wyoming Animal Science Division, who had had the animals on his Carter mountain high elevation research study in the Cody area were brought to our pastures in November 1966. After a suitable period to adjust them to the saltbush *(Atriplex gardneri)* vegetation while being held on the holding pasture, they were fitted with the esophaegeal fistulas for several days to try to get some expression of the dietary intake of sheep on these saltbush pastures. The information was disappointing. The variation between animals was great and little definitive relationship of forage intake could be interpreted from differences in the kinds of pastures that the fistulated sheep had been placed into. A report was prepared.

5B4 660103 Seeding and fertilizer studies.
5C
Because of the continued snow cover during March and April 1966, and excessive moisture in the pit within which the seeds had been located, the entire seeding was wasted.

5A1C 660104 Juniper control study on L.U.
2E
10D1 Vegetation studies during 1966 continued with study of the permanent quadrats and with some clip quadrats to determine herbage production at some of the Tordon and 2,4-D treated areas.

2A 660105 Soil moisture and temperature.
2C
2F2 Soil mositure and temperature were read a number of times at the various exclosures during 1966.

3A2 660106 1 ft x 1 ft permanent plot transects.
2F2
5C Several series of permanent plot 1 ft x 1 ft transects were installed at exclosures in the Big Horn Basin and in the Casper area.

1D 660107 Harvester ant ecology, Big Horn Basin.
5A1B
See 670601.

2E 660108 Juniper ecology - permanent plot transects.
2F2
10D1 Permanent plot transects were located in the juniper *(Juniperus osteosperma)* ecology study at the Kane location east of Lovell.

7A2 660109 Elk utilization and dietary preference study - Bull Elk Park.
1E4
10D1 At Bull Elk Park, George Pobasco conducted a study with sampling to identify characteristics of elk utilization related to distance from the forest margin and dietary preferences. These data were presented in a later thesis.

3A3 660110 Shrub production estimation study.
10D1
1E2 Fisser began studies of the shrub production estimation process with Jim Bell of the Soil Conservation Service of Worland. This was a process utilizing a weight unit estimation procedure with a large quadrat, 4 ft x 5 ft, within which the number of weight units were estimated. Only three actual weight estimates were obtained.

2E 660111 Limber pine ecology.
3A1
2C Limber pine *(Pinus flexilis)* ecology was studied by Tom King at several locations
2F2 primarily in the western Big Horn Basin area.
10D1

4B 660112 Baggs grazing systems study.
4A
3A2 Developmental discussion was conducted regarding the 68,000 acre unit located west
1E4 of Baggs, Wyoming, with selection and possible locations for game and livestock exclosures and identification of grazing systems.

8B 660113 Erosion transects for study of soil movement and relationship to climatic variation and
2C grazing management.

Erosion transects located at several of the study sites were reread - Halogeton Pastures, Steer Pastures, Cumberland Exclosures.

3A2 660114 2 ft x 20 ft and 4 ft x 4 ft permanent plots.
2F2

During 1966, considerable time was spent visiting exclosures with the 2 ft x 20 ft and 4 ft x 4 ft permanent quadrats on them to verify shrub identification, many had been misidentified during the earlier readings, and to clarify the relationship of the data from the 1959-1960 recordings to those of the 1964 recordings.

2A 660115 Sagebrush-wind-snow pattern relation study.
10D1

Efforts were aimed at determining a relationship of snow depth to treatment within the Bud Kimball, Smilo and Granite Mountain exclosures by location of rods spaced diagonally across each of the treatment sites. These data were read by measuring snow depth in the immediate vicinity of each of the rods. In addition, some work was done at the Upper Government Draw exclosure by simply recording the location, direction, and depth of snow drifts from the sagebrush bushes. This was a matter of trying to interpret the impact of the shrubs upon the deposition of snow.

7D1 660116 Small mammal abundance and distribution in Juniper.
2F2
10D1 At a number of locations in the Big Horn Basin, but primarily at the Block study site and several others including the Kane juniper *(Juniperus osteosperma)* exclosure, Larry Robinson conducted an intensive study of small mammals with snap traps.

1E2 660117 Saltbush root development and relation to grazing intensity.
4A
10D1 Reagan Russey's thesis field work addressed development of saltbush *(Atriplex gardneri)* roots in the rate of use pastures northwest of Greybull. He used a 4 ft diameter cylinder within which water was placed and held for two or three days until it had soaked into the profile. He then excavated the very soft soil root mass and separated the root systems.

3A3 660118 Exclosure study: production, cover and precipitation.
5A1D
5B The production, cover and precipitation studies were continued at the 27 research
5C locations. This also included the three exclosures and eight pastures of the halogeton
2A research unit.
2C

6602 Journal Articles

3A2 660201 Fisser, H.G. and G.M. Van Dyne. 1966. Influence of number and spacing of points on
10D1 accuracy and precision of basal cover estimates. J. Range Manage. 19(3):205-211.

This paper reported a comparison of point techniques for analyzing basal cover of grasslands. A population of 350 line intercept transect records each recorded to 0.01 foot intervals for 5 feet, was sampled both randomly and systematically with 25, 50, and 100 points (500-0.01 feet units, per line). Statistical analyses were made to compare point densities, point placements, and their interactions with each other and with 12 plant categories. Comparisons were also made of the various point-sample estimates to the population parameters.

Some practical implications from these studies for sampling foothill grasslands similar to those described herein were: 1) To sample sod forming species, such as needleleaf sedge, random placement of points would be better than systematic placement. 2) To sample the dominant bunchgrasses, systematic placement would be better, especially if the distance between points is greater than the average basal diameter of the plants. 3) To sample plant categories with high basal cover, either of the point placement procedures would suffice. 4) Generally, greater efficiency in sampling would occur with 25 than with more points per line. 5) Basal cover of all species was estimated accurately by both point placements in all point densities.

1D 660202 Lavigne, R.J. and H.G. Fisser. 1966. Controlling western harvester ants. Mtn. States
5A1B Regional Public. 3. 4 p.

This brief publication described major characteristics of western harvester ants *(Pogonomyrmex occidentalis)* with sections describing the distribution, life stages, colony activities, swarming, mound construction, economic impact and control measures. The primary control mechanism indicated by this publication was the use of Mirex, a newly developed insecticide impregnated bait which proved successful for the harvester ant as well as the fire ant in the southern states.

1D 660203 Wight, J.R. and J.T. Nichols. 1966. Effects of harvester ants on production of a saltbush
1E4 community. J. Range Manage. 19(1):68-71.
5A1D
3A2 The experimental area was located in the University of Wyoming's experimental sheep pasture in the Big Horn Basin. This study compared saltbush *(Atriplex gardneri)* production around the perimeters of the denuded areas of anthills with the production of areas not influenced by the border effect. By considering forage lost due to the denuded areas, it sought to evaluate the overall effect of harvester ants *(Pogonomyrmex occidentalis)* on the total saltbush *(Atriplex gardneri)* production of the area.

Increased vigor and foliage cover of saltbush adjacent to the denuded areas around anthills resulted in an average increase of 166 percent in production over the zones farther from the ant mound. Average diameter of denuded areas was 30 feet.

The results of this study in the Big Horn Basin of Wyoming indicated the border effect can significantly increase production of vegetation adjacent to the denuded area around anthills in the saltbush community. This increased production may compensate for lack of production in the denuded areas under certain conditions. Therefore, production losses due to the harvester ant may not be as great as expected by only considering the reduction caused by denuded areas, without taking into account the compensitory effect of the perimeter area.

2F2 660204 Wight, J.R. 1966. Juniper communities and range management in northwestern Wyoming. Wyo. Range Manage. 218:117-119.
2E
2C
3A

This paper was concerned with the juniper communities of the Big Horn Basin which were comprised predominantly of *Juniperus osteosperma*, the Utah juniper and *J. scopulorum*, the Rocky Mountain juniper. The former occurred at lower elevations in more xeric situations and the latter at higher elevations in more mesic situations.

The paper gave generalized descriptions of site characteristics, vegetation characteristics of the juniper trees and of interrelated vegetation, and addressed the problems of management and use of the juniper rangelands. Juniper invasion of grass and sagebrush grass communities continued to be an important management problem. While the cause of juniper invasion was not thoroughly understood, overgrazing was probably one of the principle causes. In this study past juniper invasion was widespread, but currently did not appear to be a serious problem. However, once a young juniper seedling becomes established, it is likely to live a normal lifespan unless removed by mechanical or chemical means. As the young junipers develop into mature trees, they crowd out the plant species that are eaten by game and livestock. An effective herbicide in the use of Tordon has been developed.

2F2 660205 Robinson, L.D. 1966. Abstract of M.S. Thesis. Wyo. Range Manage. 229:168.
7D1
3A2

See 660602.

2C
2E

6603 Presentations

2A 660301 "Soil moisture and temperature characteristics under grazed and non-grazed conditions" Accepted for presentation at X th. International Grassland Congress. Helsinki, Finland. July, 1966.
2C
2F2

Soil temperature, soil moisture, precipitation and herbage production were measured under big sagebrush control and non-grazing treatments on two sites, one a mesic upland and the other an arid, cold desert community. Soil temperatures were generally warmer under sagebrush. Soil moisture would decrease on the mesic upland sagebrush control in non-grazed areas but tended to be slightly higher on the arid, cold desert site. A rapid herbage increase on the mesic site resulted in a lowered soil moisture under sagebrush *(Artemisia tridentata)* control. On the xeric site herbage increase, limited by moisture and temperature, was not able to increase sufficiently to utilize the moisture released by sagebrush removal.

6A2 660302 "Sheep grazing intensity studies and vegetative changes on *Atriplex nuttallii*" August 4, 1966. Symposium on Management of Salt Desert Shrub Vegetation. Cedar City, Utah. by H.G. Fisser.
4A
4B
1E4

This was a broad based presentation of the long-term sheep grazing intensity studies on the Dry Creek pastures northwest of Greybull. The grazing study with winter sheep use was initiated to determine the effect of different grazing intensities upon the spread of halogeton *(Halogeton glomeratus)* and influence upon saltbush *(Atriplex gardneri)* production.

The data indicated distinct limitation to the allowable use of saltbush although 30 percent use was apparently not detrimental, 40 percent resulted in significant forage reduction. Use over 40 percent for an extended period of favorable years would probably not deplete the vegetation too severely.

Greatest indication of overuse occurred during dry years as was indicated by the 1966 data. Optimum use in the Big Horn Basin of Wyoming of saltbush should be set at no more than 35 percent of the total annual herbage production available at the time of grazing.

1D 660303 "Effect of control of harvester ants on revegetation" November 11, 1966. Wyoming Woolgrowers Association Annual Meeting, Casper, Wyoming. by H.G. Fisser.
5A1B
1E4

Three questions concerning harvester ants *(Pogonomyrmex occidentalis)* must be answered by range managers when aiming toward maximum sustained production for domestic animals. 1) Does the presence of the harvester ants reduce the available forage supply for the grazing animal? 2) What is the relationship of harvester abundance to livestock grazing intensity? 3) Is it economically feasible to control harvester ants - assuming that a need for the control exists? The study was located 12 miles south of Shoshoni, Wyoming, in a uniform sagebrush grass rangeland. Twelve study plots of 100 acres each were established. Eight of these were sprayed with Mirex and four were left unsprayed. Within each of the 100-acre plots, an eight-acre intensive study unit was established. Six were fenced and six were left open.

In answer to question number 1, it seemed logical to assume that ants do reduce the available forage, especially in those many areas where ant colonies exist in numbers of 50 or more per acre. The answer to question number 2 seemed to fit into the area in which the harvester ant locations increased greatly on ridgetops which often had been used as bedgrounds. Otherwise, in many areas, their abundance is not excessive. The answer to the third question regarding economic feasibility can be addressed similarly to that of sagebrush control by spraying, since ant control by aerial application of Mirex costs about an equivalent amount. Whether this cost would be paid for by increased forage production is not known. It could be hypothesized, though, that increased forage production would result following control on areas with dense colonies of ant mounds due to the following factors: 1) revegetation of the bare mounds, 2) elimination of seed loss to foraging ants and perhaps small mammals and 3) reduction of small herbivorous mammals concomitant with the control of the harvester ant as a result of the ant control.

6A2 660304 "Effect of sheep grazing intensities on vegetative change in Nuttall saltbush" December 3, 1966. Wyoming Section, American Society of Range Management Annual Winter Meeting. Riverton, Wyoming. by H.G. Fisser.
4A
4B
1E4

See 660302.

6604 Published Proceedings/Abstracts

6A2 660401 Fisser, H.G. 1966. Sheep grazing intensity studies and vegetative changes on *Atriplex nuttallii*. p. 226-237. In Proc. Salt Desert Shrub Symposium. Cedar City, Utah. Aug. 1-3, 1966.
4A
4B
1E4

See 660302.

2A 660402 Fisser, H.G. 1966. Soil moisture and temperature characteristics as related to grass production following shrub control under grazed and ungrazed conditions. In Proc. Xth. Intern. Grassl. Congress. Helsinki, Finland. July 1966.
2C
2F2

See 660301.

6605 Wyoming Agricultural Experiment Station Publications

660501 Fisser, H.G., L.D. Robinson, D.R. Kirkam, R. Steger and G. Noller. 1966. Halogeton Research. 1965 Results. Univ. of Wyo. Coop. Res. Rep. to the BLM. Sections I - V. Wyo. Agr. Exp. Sta. Mimeo. 134 p.

6A2 66050101 Section I - Winter sheep grazing on saltbush range (Greybull). 11 p.
4A
4B
1E4

See 650101.

6A1 66050102 Section II - Steer grazing studies on saltbush range and seeded water spreaders
4B (Worland). 7 p.
5A1C
5B4 See 650102.
5C

3A3 66050103 Section III - Exclosure study: production, cover, height and precipitation phase. 105 p.
5A1D
5B See 650103.
5C
2A

2A 66050104 Section IV - Soil moisture and temperature study. 7 p.
2C
2F2 See 650104.

2A 66050105 Section V - Precipitation pattern study. 4 p.
3A
See 650107.

1D 660502 Lavigne, R.J. 1966. Summary of research during 1964 and 1965 for the control of the
5A1B western harvester ants, *Pogonomyrmex occidentalis* (Cresson). Wyo. Agr. Exp. Sta. Mimeo. 12 p.

This was a summary of a five-year study on the effects of control of harvester ants *(Pogonomyrmex occidentalis)* on subsequent revegetation of open rangeland on a study site established about 12 miles south of Shoshoni, Wyoming. This summary, prepared after two seasons, indicated significant response in terms of harvester ant control with minimal vegetation response.

2A 660503 Fisser, H.G. and R.E. Steger. 1966. Microclimate studies in desert shrub types. Wyo.
3A1 Agr. Exp. Sta. Mimeo. 14 p.
2C
8B

This two-year summary of the microclimate studies in the desert shrub types of the Big Horn Basin related to the soil moisture and temperature characteristics measured as indicated in earlier publications on arid sites with Nuttall saltbush *(Atriplex nuttallii)* sites. These arid sites included big sagebrush *(Artemisia tridentata)*, bud sagewort *(A. spinescens)*, birdfoot sagewort *(A. pedatifida)*, winterfat *(Ceratodes lanata)*and spiny hopsage *(Grayia spinosa)*.

6606 Graduate Advisee Completions

2C 660601 Wight, J.R. 1966. Site and vegetation characteristics and relationships of juniper
2E communities in northwestern Wyoming. Ph.D. Dissertation. University of Wyo. 130 p.
2F2
3A2
10D1

Site and vegetation characteristics and relationships were studied in the juniper *(Juniperus osteosperma)* communities along the western foothills of the Big Horn mountains and on the north and south sides of the Owl Creek range. The juniper zone occupied a belt of a mile or more in width and it covered approximately 345,000 acres. The zone occurred primarily between elevations of 4,500-6,300 feet on slopes and escarpments of the tilted strata along the flanks of the mountain ranges. Juniper distribution appeared to be related to specific geological formations only as the geological formations were related to other site characteristics, principally soil depth. They were found growing on all geological formations within the study area where other environmental factors were conducive for their growth.

Several chemical properties of the soil were evaluated. The soluable salts (saturated extract conductivity) and paste pH showed very little variation and were well within the desirable limits for normal plant growth. The saturated extract conductivity averaged 0.66 millimhos/cm and 7.67 was the average pH value. The percentage soil organic matter was relatively uniform throughout the study area, averaging 2.52 percent. Available phosphorus was quite low and averaged about 20 lbs. of P_2O_5 per acre with a range from 4 to 53 lbs per acre.

Soils of the juniper zone were characteristically shallow with an average profile depth of 14.3 inches on study sites with mature juniper. These same sites had an average soil depth (profile depth - volume of rock) of only 8.2 inches. The texture of the soils throughout the study area were predominantly coarse. Most of the soils analyzed were loams, sandy loams, or sandy clay loams. The coarse textures combined with the shallow depths of soil resulted in a low available moisture holding capacity for the juniper communities, an average of 1.24 inches.

The understory cover within the juniper zone exibited great variability and was inversely related to the density of the junipers. Crown cover of shrubs ranged from 0 to 26.39 percent with an average of 4.42 percent. Grass and forbs averaged 2.76 and 0.92 percent cover respectively, with ranges from 0 to 21.19 for grasses and from 0.01 to 3.30 for forbs. Big sagebrush *(Artemisia tridentata)* and snakeweed *(Gutierrezia sarothrae)* were the principle woody species. The major grasses were blue grama *(Bouteloua gracilis)* and sandberg bluegrass *(Poa secunda)*. Less abundant species were bluebunch wheatgrass *(Agropyron spicatum)* and others. Although many forbs were found, they contributed little to the total forage production. Hood phlox *(Phlox hoodii)* and Plains pricklypear *(Opuntia polyacantha)* were the dominant species of forbs.

Overstory of the zone studied was comprised almost entirely of *Juniperus osteosperma*. The junipers varied from single stem tree-like form to the multiple stem shrubby types. The majority were of the latter type. The mean height was 5.9 feet. Height/width ratio varied with age, increasing as the trees matured. The average height/width ratio was 0.90 for all sites.

Juniper roots appeared well adapted to the sites on which they grew, producing extensive lateral patterns as well as verticle extension where the soil moisture and substrata permitted. Thus, they were able to utilize the moisture trapped in the cracks and fissures of the rocks and that which falls upon the surrounding surface.

Stands with tree densities up to 776 trees per acre were recorded. The highest canopy cover measured was 54.1 percent. The majority of the juniper stands were composed primarily of mature trees and usually appeared stable. There were a small number of stands that appeared to be dying. These occurred on sites which had developed a relatively thick soil mantle. A few new stands that were developing were either on areas where junipers were invading the sagebrush/grass communities or on sites of exposed unweathered rock.

The average age of the juniper stands studied was about 150 years. Some trees encountered were more than 500 years old. Juniper invasion was widespread through the study area but low in intensity.

Sagebrush (*Artemisia* spp)/grass communities were closely associated with the juniper communities. Some curlleaf mountain mahogany *(Cercocarpus ledifolius)* and occasionally ponderosa pine *(Pinus ponderosa)* occurred at the upper boundaries.

The juniper zone in the area studied was a low value, but nevertheless important, component of the range complex. It provided forage and shelter for both domestic livestock and wildlife. It required very careful management because of its steep topography and wide variability. Overgrazing can result in a complete loss of soil as well as vegetation. The relationships between vegetation within the juniper zone and several site characteristics were studied. Results indicated that slope and soil depth were the major controlling factors in the distribution of the juniper. Slope appeared to be related to vegetation through both its relationship to soil depth and its effect on water runoff and erosion. The presence of juniper was definitely related to soil depth. As soil depth increased, the density of junipers decreased. Boundaries between juniper and sagebrush/grass communities were nearly always associated with a change in soil depth. Understory varied inversely with the overstory cover and thus was proportional to soil depth.

The functional relationship between soil depth and vegetation appeared to be one of moisture availability. The amount of moisture held in the soil profile is directly related to the depth of soil and this factor, available moisture, accounted for nearly as much variation in the density of mature junipers as slope, profile depth, percentage rock and texture combined. Soil texture had some relationship with vegetation, but it appeared that this was a result of a mutual association with soil depth.

Geological formations, exposure, soluable salts, pH, available phosphorus and soil organic matter had no measurable relationship to the distribution of junipers and other vegetation within the study area.

Although vegetation distribution is a product of the total environment and is dependent upon many environmental factors and their interactions, there are usually a few factors that have a major influence. In the case of the distribution of junipers within the juniper zone, soil depth appeared to be the main controlling factor. Soil depth exerted an influence on juniper distribution primarily through its affect on the growth of competing plants. It controlled available moisture and provided support for the roots of grasses and forbs. An adequate amount of soil was necessary for the grasses and forbs to become established whereas junipers were adapted to sites with little or no soil. Once grasses and forbs were established, they appeared to restrict the establishment of juniper seedlings, and in this manner, competed successfully with the junipers.

2F2 660602 Robinson, L.D. 1966. The vegetation and small mammals of the juniper zone in north central Wyoming. M.S. Thesis. Univ. of Wyo. 115 p.
7D1
3A2
2C
2E
10D1

An investigation of the vegetation and small mammals in the juniper zone of north-central Wyoming was conducted during the summer of 1965. All sites contained either Utah Juniper *(Juniperus osteosperma)* or Rocky Mountain Juniper *(J. scopulorum)* as overstory dominants. The vegetation and small mammals were described within each site. Nine stands were investigated.

The juniper zone ranged from 4,800 ft to 6,960 ft with *J. scopulorum* limited to the upper end of the zone where environmental conditions were more mesic. Tree densities ranged from 190 to 632 per acre.

Percent of soil surface covered by the tree crown cover ranged from 19 percent through 43 percent. The majority of the soils were shallow, but the presence of the junipers on deep soils demonstrated that neither species was limited to shallow soils. All soils were derived from sandstone or sandstone interbedded with some other materials such as limestone or siltstone. Soil textures were highly variable, ranging from silty clay loam to a sandy loam. Total soluble salts and pH values both appeared to be within ranges suitable for normal plant growth. Levels of organic matter were highly variable and did not appear to be of significance. Available phosphorus was low for all sites with the exception of one.

Percent cover of herbaceous vegetation was low throughout the juniper zone with means of 2.60 and 1.78 for the *J. scopulorum* and *J. osteosperma* communities respectively. An age classification was developed to be able to characterize different sizes and shapes of the Utah and Rocky Mountain junipers. Class I was immature and this consists of trees from 0 to 12 inches in height. Class II, immature also, trees 6 inches or more, crown diameter more than 12 inches in diameter, tree trunk generally not more than 2 inches in diameter at ground level. Crowns somewhat symmetrical and either rounded or pointed. Class III, mature, crown profile somewhat irregular in outline ranging from obtusely pointed to rounded or flattened. Branches heavier, stouter and tending to be somewhat gnarled. Crown more open than in the immature classes due to a dying out of smaller interior twigs. Class IV, overmature, crown rounded or flattened appearing very irregular in profile. Branches heavy and often gnarled and twisted. Crown open with most of the foliage occurring at ends of branches. Considerable latitude was allowed in the description of the classes. Unfortunately, trees located on good quality sites were found to differ from the descriptions given in the classification.

6701 Research Activities

4B 670101 Study of grazing systems and influence on vegetation and wildlife habitat (field work
1E4 initiated in 1967).
2F2
3A2 Several personnel from the Rawlins BLM, the Denver BLM, Mort May and Fisser
7A1 went out on the rangeland area west of Baggs to inspect this area in terms of possibility
7A3 for a grazing system study. This was subsequently implemented with the Baggs grazing
4C system study. Fisser went out in the field with Chris Vosler later during the summer
6A3 and observed the BLM procedures on their trend plot study activities.

5A1D 670102 Juniper control study of L.U.
2E

During 1967 some clipping of 1 ft x 1 ft plots on some of the treatment sites and completion of the data recordings, primarily overstory counts and measurements of shrubs and trees in the 20 ft x 50 ft plots, was completed.

2A 670103 Soil moisture and temperatures.
2C
2F2 During 1967 soil moisture and temperature data were obtained at the Granite
10D1 Mountain and Smilo exclosures on nine dates. The earliest was March 23 and the latest, at the Smilo, was December 18 (but the December 18 was in 1966). The latest in 1967 was October 14. Soil moisture and temperature were also obtained at the Cumberland -1 and -3 exclosures on seven dates. The earliest was April 20 and the latest October 14. Soil moisture values reflected the abundant rainfall during spring and early summer of 1967. But moisture levels during latter summer and into the fall were low, apparently a combined result of limited precipitation during August and September and increased capability for moisture utilization following the very favorable growth conditions which existed during spring and early summer. The low soil temperatures during May, June and July 1967 reflected the influence of high soil moisture content during that period. At the 15-inch and 22-inch depths, temperatures remained low during the latter part of the year, although air temperatures were high and precipitation minimal. With increasing depth, soil moisture variation was less subject to short term weather changes. To attain the greatest temperatures at the 15-inch and 22-inch depths, weather conditions must be warm and dry during the entire growing season. A short period of extremely warm and dry weather during midsummer would not cause an adequate temperature change at the lower depths.

3A2 670104 1 ft x 1 ft permanent plot transects.
2F2
5C During 1967 the cover data on 1 ft x 1 ft permanent plot transects were obtained on 24 exclosure locations primarily in the Natrona, Hot Springs, Big Horn, Washakie and Park County areas. These data were reproduced in a subsequent University of Wyoming Agricultural Experiment Station scientific report.

1D 670105 Harvester ant ecology, Big Horn and Wind River Basins.
5A1D

During 1967 work was conducted relating to the permanent plot transects at each of the 12 study sites of the Shoshoni ant program and the study of harvester ants *(Pogonomyrmex occidentalis)* and reinvasion potential.

2E 670106 Juniper ecology - permanent plot transects and seeding trials.
2F2
3A2 During 1967 data were obtained on the Kane Deer exclosure permanent plot transect
10D1 sites both relating to the native vegetation especially on the outside cleared area in which it was apparent that snakeweed *(Gutierrezia sarothrae)* was beginning to reinvade and also on the seeded area within the juniper *(Juniperus osteosperma)* site. Some

further developmental study was taken on location of other sites to address other problems and ways of controlling juniper.

7A2 670107 Elk utilization and dietary preference study - Bull Elk Park.
1E4

During 1967 the field data on Bull Elk Park was completed by Probasco. This data subsequently appeared in his thesis.

3A3 670108 Shrub production estimation study.
10D1
1E2

During late August of 1967 Fisser and Jim Bell of the SCS in Worland did some more trial work in the measurement of shrub production at the Bud Kimball exclosure. The work was done in the non-spray area and addressed the needs of sample size for weight unit estimates, both in terms of the quadrat size and the number of quadrats. It was finally determined that a 4 ft x 5 ft foot quadrat located 20 times would give adequate data. In addition, during the same period, we went out with soils people and went to quite a number of exclosures both in the Wind River Basin and the Big Horn Basin and determined soil characteristics at the exclosures. This included all six of the Shoshoni ant exclosures.

2E 670109 Limber pine ecology.
3A1
2C
2F2
10D1

During 1967 work was conducted by Larry Forbis on the Big Horn Mountains on the high elevation limber pine *(Pinus flexilis)* types and in this study he compared the ecology of abiotic and biotic components of adjacent paired sites of limber pine sites compared to spruce-fir *(Picea-Abies)* sites.

3A3 670110 Exclosure study: production, cover, precipitation.
5A1D
5B
5C
2A
2C

During 1967 herbage production was obtained at 37 different transect locations. Production clipping in 1967 was conducted both inside and outside the exclosures to give us a better idea of grazing impact upon the vegetation. Production was lower in the grazed areas but was greater on the sprayed sites than on the non-sprayed. On the Upper Government and Lower Government Draw study areas, moderate and heavy use outside the exclosure was evident. The influence of the wet spring and summer of 1967, which increased plant growth, was evident in the production values of the study areas. However, production of the areas was probably adversely affected by the early dry spring and summer of the previous year, 1966.

A separate activity during latter July involved travel to several exclosures during which time Fisser tape recorded comments concerning apparent relationship of vegetation changes and appearances as a result of site and grazing treatments. These exclosures were the Cumberlands, Farson, Ant Hill Lander, Boysen, Grass Creek Divide, Upper Grass Creek and Dutch Nick Flat.

6702 Journal Publications

6A2 670201 Fisser, H.G. 1967. Proper winter sheep grazing on saltbush rangeland. Univ. of Wyo. Range Manage. 244:201-206.
4A
4B
2A
3A2

This paper was a compilation of the information derived from the North Fork experimental sheep intensity study area northwest of Greybull, Wyoming. The summarized overall program, the activities, the kinds of research data that have been conducted on the site and the general summary and conclusions indicated that a very distinct limitation to the use of saltbush *(Atriplex gardneri)* was necessary.

Although 30 percent use was apparently not greatly detrimental to saltbush, 40 percent use resulted in a significant forage reduction of the shrub. Use at 40 percent or more for an extended period of favorable years would probably not severely deplete the vegetation. Greatest indication of overuse occurred during dry years as indicated by the 1966 data. Very little precipitation had fallen (0.74 inches from April 15 to July 1), in fact annual weeds even failed to grow during that period. Production was reduced on all pastures but most significantly on the 80 and 40 percent units. The grazing study showed finally that optimum winter use of saltbush in the Big Horn Basin should be set at no more than 35 percent of the total annual herbage production available at grazing time.

4B 670202 Smith D.R., H.G. Fisser, N. Jeffries and P.O. Stratton. 1967. Rotation grazing on Wyoming's Big Horn Mountains. Univ. of Wyo. Agr. Exp. Sta. Res. J. 13. 26 p.
4A
6A

The study had been grazed from 1951-58 to compare three intensities of grazing - light, moderate and heavy. In the spring of 1955 the pastures were modified to incorporate three kinds of grazing use - a moderate rotation, a heavy intensity rotation and a moderate season long.

Rotation grazing was predicated on the basis of 1) more uniform grazing and 2) beneficial effects to the plants of a rest from grazing, even though the plants are grazed more intensively during a shorter length of time. Small pasture studies such as that recorded here are poorly adapted to testing uniformity of grazing, but they may be useful for testing plant response. Rotation grazing failed to benefit Idaho fescue *(Festuca idahoensis)* on these ranges but this was because the season long grazing was in reality a misnomer because the grazing was limited to the vegetation on the alluvial soils until mid-August. Thus, Idaho fescue was deferred for a good share of the summer season each year. The actual deferrment, which occurred in the season long units, equaled or overshadowed any beneficial effect of periodic rest in the rotation systems.

In the study period, range deteriorated in the heavy rotation unit even though utilization was generally below 43 percent, the level at which earlier researchers had found the grass able to maintain itself under season long use. The difference of daily gain between the moderately stocked units, whether rotation or season long, was not significant. Lower daily gain of the steers in the heavy rotation system indicated that stocking rates have more affect on animal unit production than did systems of grazing.

1D 670203 Kirkham, D.R. 1967. Biological control as a tool in range management. Wyo. Range Manage. 246:254-262.
2B
5A1B

The objective of this publication was to describe biological control as a tool for regulating populations of undesirable range plants. Organisms used as biological agents are generally considered to be of three categories: 1) parasites and predators, 2) pathogens and 3) individuals forming an undesired population which possess man-induced sterility or lack of sterility.

The primary impact of undesirable plants on range areas is through competition with desired species. As competitors for nutrients, water and space, they not only reduce the total amount, but also the quality of useable forage. Many undesirable range plants are noxious which may cause physiological disturbance or death to range livestock.

The concept of biological control is based on natural phenomenon that many species prey on and live at the expense of other organisms. Populations of numerous species are regulated and sometimes irradicated by attacking organisms. Biological control, like all other methods, contains certain disadvantages. The location, testing and dispersal agents require a relatively long time, usually several years. Some pests may require immediate suppression. Since most campaigns only reduce pest populations at lower levels, the biological approach may be useless wherein all products must be completely free of the pest material.

Considerable difficulty is involved in finding and establishing natural enemies. Successful examples in the United States include the klameth weed *(Hypericum perforatum)*, pricklypear *(Opuntia polyacantha)* and rabbitbrush *(Chrysothamnus viscidiflorus)*. The biological approach to plant control is essentially a discipline of applied ecology. Its success depends upon an appreciation of both plant and animal ecology and the community relationships where the pest occurs. Application of this method involves several steps. The critical phase, in the case of plant control, is the selection of an agent that will not harm other plants. Biological control could be a useful tool for applied ecological management of rangelands.

3A3 10D1 670204 Wight, J.R. 1967. The sample unit and its effect on saltbush yield estimates. J. Range Manage. 20(6):323-325.

A 60 ft x 60 ft plot of Nuttall saltbush *(Atriplex nuttallii)* was completely harvested using 1 ft x 1 ft sample units. Yield data for sampling units of various sizes and shapes were obtained by combining contiguous 1 ft x 1 ft plots.

Sampling units size and shape both had an important effect on sampling efficiency. The coefficient of variation decreased rapidly as sampling unit size increased, until a sampling unit of about 60 feet square was reached. For the smaller sampling units the rectangular units generally had a lower coefficient of variation than those nearly square. Data are presented showing the relationship between sampling units of various sizes and shapes and 1) their coefficient of variation, 2) the number of samples needed to obtain a yield estimate that is within 20 percent of the population mean 90 percent of the time and 3) the total area needed to be sampled to obtain this yield estimate.

6703 Presentations

2A 2C 2F2 670301 "Soil moisture and temperature characteristics as related to grass production following shrub control under grazed and non-grazed conditions" February 14, 1967. 20th Ann. Meeting, Amer. Soc. Range Manage. Seattle, Washington. by H.G. Fisser.

Soil temperature, soil moisture, precipitation and herbage production were measured under big sagebrush *(Artemisia tridentata)* control and non-grazing treatments on two sites, one a mesic upland and the other an arid, cold desert community. Big sagebrush and livestock grazing caused suppression of understory herbage. Following sagebrush control grass production increased coincidentally with moisture release and was restricted by environmental limitations which were exhibited primarily by the amount and time of growing season precipitation, soil temperature and air temperature near the ground surface.

Following sagebrush control on a mesic site, production of herbaceous species can be increased so much that they are able to utilize all of the soil moisture released from use by the shrub, as well as moisture that was not utilized at all prior to the control treatment. Under these conditions an apparent decrease in soil moisture is noted. On xeric sites, although understory species increased also following sagebrush control, limitations caused by environmental characteristics such as high temperatures and low precipitation, restricted both the rate and total increase of herbaceous production. Subsequently, some of the moisture released from use by the shrubs was not utilized and an apparent increase in soil moisture was noted.

Non-grazing and sagebrush control treatments result in similar vegetative responses. A rehabilitation program must consider the influences of environmental limitations caused by differences in soil moisture, soil temperature and precipitation. These factors strongly affect the rate of understory re-establishment and production increase.

6A1 4B 4A 670302 "Influence of summer steer use under rotation and season-long grazing systems" July 29, 1967. Wyoming Section, American Society of Range Management Annual Summer Meeting. Big Horn Mountains. by H.G. Fisser.

See 670202.

5A1B670303 "Plant control on salt desert shrub rangeland by sheep grazing" August 21, 1967. Range Weed Control Conf. Las Cruces, New Mexico. by H.G. Fisser.
3A2
2B

This presentation was a modification and compilation of the work done at the grazing rate of intensity of use pastures northwest of Greybull which have been reported in some detail in 660302 and are also published in 670501.

5A1D670304 "Use of picloram for juniper control" August 21, 1967. Range Weed Control Conf. Las Cruces, New Mexico. by H.G. Fisser.
2E

On rangelands, control measures have been developed because of an apparent increase of undesirable woody species. A number of causal factors responsible for the plant increase have been postulated. Whatever the cause, we are concerned at present, with the efficient use of rangeland under the multiple use concept. Any program of woody plant control must, therefore, be developed with a recognition of the subsequent multiple use needs of the land and the basic ecology of the shrubland areas.

Control of woody species in the western United States has been conducted by a variety of procedures - the more common are chaining, burning and plowing. Major disadvantages of all these practices is that they require manpower, machinery and time. Aerial application of chemical herbicides, although often more costly, is usually much more rapid and efficient, especially in terms of manpower.

The increase of juniper (*Juniperus* spp.) has been widely studied. In many areas of the west, extensive control programs have been initiated both on areas of stable, mature stands and on areas which are apparently being invaded by the trees.

In the Big Horn Basin of Wyoming, studies have been conducted to evaluate control of juniper by use of Tordon. Tests with Tordon applied from backpack sprayers have been successful. In these tests, the trees were completely wetted with the chemical applied at a rate of 2 lb/acre in 40 gallons of water.

In 1964 a study was initiated by Harold Alley to test helicopter spraying of Tordon when applied at one-half, one and two lb/acre in a water solution at 5 gallons per acre. The chemical was applied on July 2. Results were discouraging. Most trees showed response to the chemical, but only on one side. Apparently the low solution volume of 5 gallons per acre was insufficient to obtain the dispersion necessary for complete control of the trees. Observation of the trees for three years following the application showed that most had regained vigor and are growing. Some leaf and stem distortion was still present.

Since aircraft are usually unable to apply more than 5 gallons of solution per acre, a new study was initiated in 1965. Tordon was applied by fixed wing plane, and to overcome the low dispersion value of the previous test, 2 lb of chemical was added to sufficient water to obtain 10 gallons of solution. With the spraying equipment of the plane set to apply 5 gallons of solution per acre, the desired chemical application of 2 lb/acre was obtained by having the plane fly twice over each path applying solution from two directions. Spraying was conducted on July 28, 1965. No response was visible in late October. In April 1966, however, most trees were at least partially brown with dead leaves. By July 1967, approximately 90 percent of the juniper trees were completely dead and the remainder were partially brown.

10D1 670305 "Cooperative research with the University of Wyoming." March 10, 1967. Bureau of Land Management Wyoming State Advisory Board Meeting. Riverton, Wyoming. by H.G. Fisser.
6A1
6A2
3A3
2A
2E

This presentation addressed seven research programs being carried on by Fisser and his researchers in conjunction with the BLM. These are listed in sequence from one to seven. There are additionally many categorizations of different kinds of information on a number of the topics that can be obtained from the listing of the outline of this presentation: 1) Summer steer grazing on saltbush rangeland with reseeding and water spreader treatments. 2) Winter sheep grazing intensity of use trials on saltbush rangeland. 3) Aridland ecology studies. 4) Soil moisture and temperature studies. 5) Biotic communities. 6) The Shoshoni-ant control study. 7) Analysis of grazing systems in the Baggs area of southern Wyoming.

1D 670306 "Western harvester ant abundance in relation to grazing, vegetation and soils in north central Wyoming." February 14, 1967. 20th Ann. Meeting, Amer. Soc. Range Manage. Seattle, Washington. February 14, 1967. by D.R. Kirkham.
2B
2C
4A

The study was conducted northwest of Greybull, Wyoming, to quantify correlation of western harvester ant *(Pogonomyrmex occidentalis)* abundance, determined by density and relative size of colonies, with grazing, vegetation and edaphic factors. The study area was located in Nuttall saltbush *(Atriplex nuttallii)* rangeland, an area which contained eight pastures of variable size and grazing treatment and included distinct soil differences.

Data were collected on the number and size of ant colonies, cover and composition of vegetation, and soil properties. Changes in the number of ant colonies per pasture were determined from aerial photographs taken before and after 10 years of grazing treatment.

Numbers of western harvester ant colonies were found to be only slightly different under the various intensities of grazing. Soil characteristics, as opposed to vegetation and grazing intensity, were assessed to be more influential as factors in harvester ant density. Ant colonies were more dense on sandy ridges than on the heavier soils adjacent to the drainages. Although the distributional range of the western harvester ant has been reported to be broad, locally existing environmental factors were found to be important in determining the abundance of the species within a given area.

6704 Published Proceedings/Abstracts

2A 670401 Fisser, H.G. 1967. Soil moisture and temperature characteristics as related to grass production following shrub control under grazed and non-grazed conditions. p. 23-25. In 20th Ann. Meeting Proc. Amer. Soc. Range Manage. Seattle, Washington. February 14, 1967.
2C
2F2

See 670301.

1D 670402 Kirkham, D.R. 1967. Western harvester ant abundance in relation to grazing, vegetation and soil in north central Wyoming. p. 28. In 20th Ann. Meeting Proc. Am. Soc. Range Manage. Seattle, Washington. February 14, 1967.
2B
2C

See 670306.

5A1B670403 Fisser, H.G. 1967. Plant control on salt desert shrub rangeland by sheep grazing. p. 48-54. In Proc. Range Weed Control Conf. Las Cruces, New Mexico. August 21, 1967. p. 48-54.
3A2
2B

See 670303.

5A1D670404 Fisser, H.G. 1967. Use of picloram for juniper control. p. 37-39. In Proc. Range Weed Control Conf. Las Cruces, New Mexico. August 21, 1967.
2E

See 670304.

6705 Wyoming Agricultural Experiment Station Publications

6A2 670501 Fisser, H.G. and R.E. Steger. 1967. Winter sheep grazing study on saltbush rangeland. 1966 Results. Univ. of Wyo. Coop. Res. Rep. to the BLM. Univ. of Wyo. Agr. Exp. Sta. Sci. Rep. 36. 9 p.
4A
4B
1E4

See 660101.

670502 Fisser, H.G., G. Noller, R.E. Steger, D.R. Kirkham and T. King. 1967. Aridland Ecology Research. 1966 Results. Univ. of Wyo. Coop. Res. Rep. to the BLM. Section I - IV. Univ. of Wyo. Agr. Exp. Sta. Sci. Rep. 41. 137 p.

2A 67050201 Section I - Soil moisture and temperature study.
2C
2F2

See 660105.

2A 67050202 Section II - Microclimate studies in desert shrub communities.
2C
3A2
2E

See 660105.

2A 67050203 Section III - Precipitation pattern study.
3A

See 660118.

3A3 67050204 Section IV - Exclosure study: production, cover, precipitation phase.
5A1D
5B
5C
2C

See 660118.

3A2 670503 Fisser, H.G. 1967. Exclosure study with transects of permanent plots. Univ. of Wyo. Coop. Res. Rep. to the BLM. Univ. of Wyo. Agr. Exp. Sta. Sci. Rep. 69. 94 p.
2F2
5C

See 660106.

3A1 670504 Fisser, H.G. 1967. Range condition class survey - concepts and application. In Crop Care in Wyoming. p. 2-5. March.
3A2
2F2

Determination of range health by a condition class survey is accomplished under the ecological concept that, given time, a plant community will attain stability. This degree depends upon uniformity of climate, edaphic and biotic factors. Variation in these factors often causes changes that are measurable in terms of differing forage potential.

Introduction of game and stock animals to harvest and forage may result in significant vegetational changes. These changes occur because animals trample, trail, eat and, therefore, alter habitat factors. Different species of plants have different tolerance levels both to animal utilization and to the altered habitat.

As condition decreases, the palatable species, and those least tolerant to disturbance, are reduced in number, vigor and abundance, allowing establishment and increase of inferior plants. Herbage production is not a determining factor in evaluating condition, although it is certainly related. Climate and soil are significant in determining excellent or climax condition, whatever the vegetation type.

6706 Graduate Advisee Completion

1D 670601 Kirkham, D.R. 1967. Harvester ant (*Pogonomyrmex occidentalis* Cole) abundance in relation to grazing, vegetation and soil in north central Wyoming. M.S. Thesis. Univ. of Wyo. 83 p.
2B
5A1B
2C

The study was conducted northwest of Greybull to correlate harvester ant (*Pogonomyrmex occidentalis*) abundance with grazing, vegetation and edaphic factors. Colony density, percent of area denuded, average mound volume and average area denuded were used as indices of ant abundance. The study area was located on a Nuttall saltbush (*Atriplex nuttallii*) rangeland type which was in essence the eight-pasture rate of use grazing system northwest of Greybull. Field data collected on the number and size of ant colonies, cover and composition of vegetation, and soil properties were examined statistically by a factorial regression and correlation analysis. Changes in number of ant colonies per pasture were determined from aerial photographs taken before and after 10 years of grazing treatments.

Locally existing environmental factors were found to be of importance in determining the abundance of the species within a given area. Different grazing intensities applied to the 10-year period did not affect harvester ant abundance. Soil texture, especially percent sand, was the most important factor influencing harvester ant abundance. Ant colonies were more dense and had larger mounds on sandy ridges than on heavier soils adjacent to drainages. Cover of bud sagewort *(Artemisia spinescens)*, Indian ricegrass *(Oryzopsis hymenoides)*, and perennial forbs were found to be possible factors determining ant abundance because of their importance as sources of food for the harvester ants.

2E 3A1 2C 2F2 10D1 670602 King, T.W. 1967. A study of vegetation, soils and small mammals of limber pine stands in north central Wyoming. M.S. Thesis. Univ. of Wyo. 58 p.

The study was conducted to characterize the sites, vegetation and small mammals of limber pine stands in north central Wyoming. Study areas included the foothills and slopes of the mountains which encompass the Big Horn Basin. Criterion for selection was that each site have a predominance of limber pine *(Pinus flexilis)*. Vegetation, site characteristics and small mammals were described.

At seven sites, vegetation was sampled with 10 20 ft x 50 ft quadrats located systematically so that overstory, understory and site data were collected. Small mammals were trapped. A square foot plot was used to sample the understory along two 50 ft transects at each quadrat. Trees were classified as mature, saplings and seedlings. Data were taken to determine density, frequency and basal area. The age of each stand was determined with the aid of an increment borer.

Soils were analyzed to determine texture, organic matter, pH, soluble salts, phosphorus and lime concentrations. Slope, exposure and elevation were also recorded. Soils were generally coarse, shallow and relatively low in organic matter. Soil pH was slightly acidic but well within the range considered normal for plant growth.

Moisture appeared to be the controlling factor in the distribution of limber pine. Available moisture was affected by soil type, slope and exposure. The areas which had either Douglas fir *(Pseudotsuga menziesii)* or Engeleman spruce *(Picea engelmannii)* in the immediate vicinity appeared more mesic than the sites with only limber pine. Limber pine acts as a pioneer tree species and on most areas would probably be replaced by more mesic conifers sometime in the future. On the one site where limber pine trees were unusually old, the site was so arid and the soil so undeveloped that the presence of limber pine had not created enough effect on environmental conditions to make this type of site favorable for the growth of any other tree.

1E2 4A 10D1 2E 670603 Russey, G. R. 1967. The effect of grazing intensities on the roots of Nuttall's saltbush (*Atriplex nuttallii* S. Wats.) M.S. Thesis. Univ. of Wyo. 49 p.

The study was conducted to determine the influence of different grazing intensities on the root systems of Nuttall saltbush *(Atriplex nuttallii)*. Saltbush root weights, root number and top weights of aerial standing crop were the major factors used in comparing the different treatments.

Rates of utilization greater than 20 percent caused a reduction in the size and weight of saltbush roots. The 0 percent rate of use caused a detrimental effect on saltbush plants that was equal to the detrimental effects of the 80 percent rate of use.

Conclusions applicable to this study were 1) Percent cover of saltbush was maintained at the highest possible level by harvesting the shrub at the rate of 20 percent annual growth yearly. Rates of harvest that exceed 20 percent cause a loss in vegetative cover. 2) Lateral saltbush roots extended over a wide area exceeding 14 ft in diameter, but the maximum range could not be determined. 3) The major form of reproduction for the species appeared to be suckers growing from shallow, lateral roots. 4) The 80 percent rate of use of saltbush prevented reproduction of the species as shown by fewer number of plants that existed in the heavily used pastures and the absence of young plants. 5) Variations in saltbush size and structure in the studied pastures was due chiefly to the past rate of harvest, rather than variations in soil characteristics. 6) Harvesting of saltbush at 20 percent rate of use produced the maximum root and top growth. As percent utilization increased, saltbush size and weight decreased. 7) Saltbush must be lightly grazed to promote healthy, productive growth and an extended period of non-utilization was as detrimental to the plant as excessive heavy utilization.

6801 Research Activities

7A 680101 Game range improvement program by control of juniper and seedling establishment studies of preferred browse species (initiated in 1968).

5A1C
5A1D
5A1E During the spring of 1968 a number of primarily browse species were planted at the three Game Range Improvement (GRI) exclosures. One forb, Palmer penstemon *(Penstemon palmeri)*, was planted; and the others included big *(Artemisia tridentata)* and black *(A. nova)* sagebrush, silver *(A. cana)* sagebrush, winterfat *(Ceratoides lanata)*, bitterbrush *(Purshia tridentata)* and serviceberry *(Amelanchier almifolia)*. Response of these plantings were to be reported in later citations.

5A1D680102 Juniper control chemical, burn, mechanical on GRI exclosures.

5A1E
5A1C At the Game Range Improvement exclosures at Kane, Medicine Lodge and Big
7A Trails, three treatments were applied to the juniper within the exclosures. The chemical control with Tordon was applied during late 1968 with helicoptor and encompassed one half of the exclosure. Second, one-quarter of the exclosure was subjected to juniper (*Juniperus* spp) control by mechanical means with a bulldozer which moved the trees out and piled them in a long string of dead materials through the center of the exclosure. This was similar to a chaining treatment. In the third treatment, on one quarter of each of the exclosures, the juniper was burned with torches.

1E2 680103 Study of moisture stress in woody species using pressure bomb.

2E A pressure bomb was constructed to determine moisture stress of plants. This work was carried on primarily by Wambolt. His primary effort was related to the moisture potential of juniper (*Juniperus* spp), but he also worked with sagebrush (*Artemisia* spp) and some of the pines (*Pinus* spp) in the Big Horn Basin area.

7A2 680104 Study of elk-livestock competition in terms of utilization by class of animal and influence
4C on vegetation.

The work here related primarily to the efforts by several graduate students including Probasco, Todd, Gerhart and Hanson. They studied activities and the relationships of vegetation in the juniper *(Juniperus)* type. This research also included study of the relationships to wildlife use with primary implication toward the elk.

2A 680105 Soil moisture and temperature.

2C
2F2 Soil moisture, temperature and precipitation data were recorded on four dates at the Granite Mountain and Smilo exclosures during 1968. Moisture levels on September 3 were greater in 1968 than in 1967 because of the high frequency and amount of precipitation which occurred during late summer. Greatest increase in soil moisture was noted in the inside spray treatment area at Granite Mountain, especially at the 6-inch and 12-inch depths on both sites. At Cumberland -1 and Cumberland -3 exclosures, soil moisture values did not vary significantly from the 1967 data. Those data did show a slight complimentary increase of soil moisture levels at the 6 and 12-inch depth related to the slight increase in precipitation. Soil moisture and temperature data at Granite Mountain and Smilo were low in April, but had increased by July. Readings in early September were lower than previous years and reflected the higher, late summer precipitation. Interestingly, the temperatures in October at the 1-inch and 8-inch depths were higher than in 1967 and 1966. This could probably be correlated with the cool temperatures and precipitation during September and October and the low soil moisture values at those depths.

The microclimate studies in the Big Horn Basin with regard to soil moisture and temperature also were carried on during 1967-68. These were located on sites with saltbush *(Atriplex gardneri)* and paired locations with big sagebrush *(Artemisia tridentata)*, birdfoot sagewort *(A. pedatifida)*, bud sagewort *(A. spinescens)*, winterfat *(Ceratoides lanata)* and spiny hopsage *(Grayia spinosa)*. Subsequent to the frequent summer rain showers during 1968, soil moisture values at the surface levels were higher than during the preceding years. In some cases, total moisture through the five ft depth of measure, was somewhat greater than that previously recorded, but these differences appeared minimal. Lower precipitation values during the 1968 winter and 1969 spring periods as compared to 1967 may well have been more influential to the overall soil moisture regime than the summer rainfall, much of which did not effectively enter the soil because of runoff and evaporation. Late summer temperatures during 1968 at the 15-inch depth were 4 - 5 °C cooler than in 1966. This probably occurred as a result of the frequent rain showers which would tend to cool the soil, and a slight increase in the soil moisture which would have the same effect (690501).

3A3 680106 Exclosure study: production, cover and precipitation.
5A1D
5B
5C
2A
2C

During 1968 herbage production by clipping of 1 ft x 1 ft quadrats was again carried on at the 40 exclosures. The dry spring of 1968 retarded the production of grass species while the wet summer increased production of annual and perennial forbs. These results indicate that annual and perennial forbs may utilize summer moisture more than grass species.

To obtain some production measure of shrub species which had not been clipped in the normal field procedure, a modified double sampling technique was devised and conducted on a number of the exclosures. This sampling was done during September. The technique involved a rapid, accurate weight unit estimation. The estimates were made on transects consisting of 20 quadrats, 4 ft x 5 ft, spaced from 6 to 12 steps apart. This enabled the entire exclosure to be sampled. The number of weight units at each plot was estimated individually by two men and then checked for accuracy. Weight units were determined prior to the estimation procedure. These weight units were subsequently clipped, taken to the laboratory, oven dried and weighed and then applied to the numbers of weight unit estimates for determination of an estimate of total standing crop. The list of plant names on the production study areas has been continued and has become lengthy.

3A2 680107 2 ft x 20 ft and 4 ft x 4 ft permanent plot study.
2F2

During 1968 all of the 2 ft x 20 ft and 4 ft x 4 ft quadrats at all locations were reread.

5A1D680108 Atrazine application to control annual weeds on saltbush ranges.
4A

During spring of 1968 Atrazine was applied to a portion of the heavy use pasture -2 and in pasture -3 (holding pasture) at the Dry Creek halogeton *(Halogeton glomeratus)* unit. This chemical is presumably useful for control of annual weeds which have become quite abundant in the holding pasture. This includes both halogeton and the annual aster *(Machaeranthera tanacetifolia)*.

5A1D680109 Forb control with chemical on subalpine range Big Horn Mountains.
4A

During spring of 1968 2,4-D was applied on one of the pastures of the grazing intensity study on the Big Horn Mountains.

3A2 680110 Vegetation pattern on shortgrass-cactus range.
2E
2F2

To gain more information relating to pattern of vegetation, especially in the prairie region, an instrument 1 m long by 1/2 m wide was constructed to provide information which would locate any given point within the 1/2 m^2 unit. The x,y and z axis data were being carried on by Lester originally and then taken over by Stroud who developed some analyses of those data.

2E 680111 Juniper ecology 20 ft x 50 ft permanent plots on GRI exclosures.

5A1D
5A1E
7A

The juniper *(Juniperus osteosperma)* ecology 20 ft x 50 ft quadrats were established and read on the three GRI exclosures. It was very interesting to note that in some locations, extreme numbers of big sagebrush *(Artemisia tridentata)* seedlings were recorded at the time of placement; and because of such high numbers, their number and general location on the quadrat map being made of each quadrat, were simply designated by a circle and an estimate of the total number of seedlings within the circle.

5B4 680112 Kane Deer Exclosure seeding analysis.

7A
5A1D

The seeding analysis in this item related to the plantings on the Kane deer exclosure. These were planted in 1965 and, unfortunately, were placed in an area which had received a large amount of Tordon in the application of the chemical control of the juniper. As a result, very few seedlings established, or even emerged, and these died.

3A2 680113 1 ft x 1 ft permanent plot transects.

2F2
5C

All of the transects of 1 ft x 1 ft permanent quadrats were reread during 1968.

5B4 680114 Juniper GRI exclosure seedings.

7A
5A1D
2E

During 1968, seedings of grass were established late in the fall at the three GRI exclosures. These seedings were a variety of grasses which had been obtained from several different locations and the work was continued by Burrough.

3A3 680115 Shrub production estimation trials.

1E2

During fall of 1968, Forbis and Fisser conducted shrub estimation trials with the use of the weight unit estimate procedure developed with Bell during earlier years. Data were obtained at all of the exclosures for which herbage production was determined by clipping. This shrub unit estimation process was not carried out, however, on the juniper *(Juniperus osteosperma)* exclosures. See 680106 also.

2A 680116 Precipitation pattern study 1968.

3A

Precipitation during the winter period of October 15, 1967, to April 15, 1968, was greater than in the previous year. Spring precipitation from April 15 to June 30 was less than the previous year also, with most areas receiving from 3 to 4 inches of precipitation. Generally cool temperatures associated with the lesser amounts of precipitation proved inhibitory to annual forbs and grasses as well as the perennials. Summer precipitation during July and August was greater than normal and also occurred with greater frequency than usual, resulting in late summer germination and growth of annuals. Fall precipitation from September 1 to October 15 was less than usual, with most areas receiving less than 1 inch. Some regrowth occurred, but this was minimal.

An unusual hail storm in early September south of Shoshoni resulted in almost complete degradation of the herbaceous standing grass crop. Observations at the site a few days later indicated a very marked and rapid regrowth of sandberg bluegrass *(Poa secunda)* and needleandthread *(Stipa comata)*. Western wheatgrass *(Agropyron smithii)* did not respond as much as these two.

A total of 128 UW-BLM gauges had been installed, a number of them at Guzzlers in the Sweetwater County area and some additional located in the grazing systems area west of Baggs. Moisture from 37 of the US Weather Service stations was compiled seasonally and produced in the annual report.

4B 680117 Baggs grazing system study.

The study of three grazing systems was implemented during 1968 with field work conducted primarily by R.P. Gibbens. At the eight exclosures permanent plot (1 ft x 1 ft) transects were established inside and outside of each, in each vegetation type and treatment area. Additional transects were distributed throughout the pasture units to identify key sites related to vegetation changes and grazing. They were also located near each of the BLM trend plots. Permanent strip browse transects were installed in juniper *(Juniperus osteosperma)*, true mountain mahogany *(Cercocarpus montanus)* and antelope bitterbrush *(Purshia tridentata)* areas.

6802 Journal Publications

5A2D680201 Fisser, H.G. 1968. Soil moisture and temperature changes following sagebrush control. J. Range Manage. 21 (4): 283-287.

2A
2C
2F2

Soil moisture and temperature were measured for a five-year period on a mesic foothill grassland and on an arid cold desert shrub type in western Wyoming. Treatments imposed were sagebrush *(Artemisia tridentata)* and grazing control. Herbage production increased on both the arid and mesic sites following treatments. The greatest increase occurred on the mesic site.

Average annual soil temperature combined for all depths was 11°C at the mesic site and 15°C at the arid site. Soil temperatures increased rapidly in the spring at the arid site and remained high throughout the summer. At the mesic site, they increased more slowly in the spring and decreased earlier and more rapidly in the fall than at the arid site.

Average annual soil temperatures at both sites combined for all depths, were almost 1°C warmer in the shrub dominanted areas than in the non-shrub, grassland sites. On a seasonal basis this relationship was reversed on the mesic site during the winter and summer periods, but on the arid site the only reversal occurred during the early summer period.

The greatest difference in subsurface temperatures occurred at the 15-inch depth where temperatures were warmer in the native shrub dominated areas than those subjected to sagebrush control. At the 8-inch depth, however, temperatures were greater under the shrub control treatment at the mesic site and almost equal at the arid site.

Soil moisture recharge during the spring period at the mesic site was much greater under the non-use treatment than in the grazed area. At the arid site, an insignificant variation occurred as a result of the grazing treatment. At the mesic site, little variation in soil moisture recharge resulted from the shrub control treatment. At the arid site, however, soil moisture recharge was much less in the shrub controlled grassland area than in the non-controlled shrub area.

Soil moisture withdrawal was similar at both the arid and mesic sites. Less moisture was taken from the soil under the grazed treatment and the non-controlled shrub treatment than in the areas which were non-grazed and shrub controlled.

Variation of seasonal moisture levels combined for all depths and years showed a much greater variation at the mesic site than at the arid site. During the spring period from mid-March to mid-June, soil moisture levels by depth were greatest from 24 inches to 60 inches below the soil surface at the mesic site. Greatest values occurred in the shrub controlled grassland area. Low moisture levels occurred above 24 inches in the grass root zone. At the arid site, high moisture levels during the spring period occurred only down to the 12 inches depth. At greater depths, moisture was low, apparently resulting from utilization of most of the available moisture at the upper levels by grasses.

At the mesic site, grasses utilized almost all available moisture in the upper 2 ft of soil, but that which was not utilized moved downward to accumulate in the 36 inches to 60 inches zone from which sagebrush depends for its major water supply. At the arid site, the same pattern was exhibited through the 6 inches level as that exhibited through the 24 inches depth at the mesic site, and a similar pattern at the arid site from the 12 inches through 36 inches depth as that at the 24 inches through 60 inches depth at the mesic site.

1E1 2E 680202 Julian, T.D. 1968. *Atriplex nuttallii.* Wyo. Range Manage. 260:94-101.

Over 6,000 seeds from nine collections at different locations in Wyoming, ranging from the Big Horn Basin to the Red Desert Area south of Rawlins, were studied to determine the viability and germination characteristics of these seeds. The basic objectives of the study were: 1) to determine if there were inhibitors in the seed coat, 2) to determine if a dormancy period was necessary for germination, 3) to find ways to increase germination and 4) to determine what would be the best depth of planting for best establishment.

Germules were carefully dissected from the seeds by hand under a microscope. With this work it was determined that less than 30 percent of the seeds contained germules although there did seem to be some inherent inhibition to seed germination caused by the chemicals characteristic of the seed coat.

The greatest importance of findings here were that there was a great number of seeds without viable germules. This relatively low seed fill seems to be characteristic of Nuttall saltbush *(Atriplex nuttallii).* Only slightly more than 26 percent of over 6,000 seeds representing nine collections contained germules. The low percentage should be considered when determining seeding rates. Cutting tests of individual seeds are recommended on samples of seeds so that seeding rate can be adjusted for what may be a high percentage of empty seeds. There did not seem to be a dormancy necessary for germination although there were definitely inhibitors in the seed coat which prevented germination.

7A2 1E4 680203 Probasco, G.E. 1968. Dietary preference and utilization patterns of elk in the Big Horn Mountains, Wyoming. Wyo. Range Manage. 258:90

The data were collected during the summer of 1967, on dietary preference and grazing patterns of elk in the northern Big Horn Mountains of Wyoming. Forest openings where only elk grazing occurred were studied to determine preferred plant species for both the spring and summer seasons.

One forest opening of approximately 300 acres was stratified to determine whether there was a correlation between elk grazing patterns and distance from forest margin. It was evident from the data that elk grazing patterns were not correlated with distance from forest margin. However, there was a definite correlation between elk grazing patterns and percent total basal cover of herbaceous vegetation.

Data on diet preferences indicated that elk utilized grasses during the spring period but shifted their preference to forbs during the summer. Preferred species for the spring period were mountain brome *(Bromus marginatus)*, other bromes (*Bromus* spp), Idaho Fescue *(Festuca idahoensis)* and bluegrass *(Poa)* spp. Preferred for the summer period were pale agoseris *(Agoseris glauca)*, cutleaf balsamroot *(Balsamorhiza incana)*, cutleaf cinquefoil *(Potentilla diversifolia)* and weedy milkvetch *(Astragalus miser)*.

2E 3A1 2A 680204 Wambolt, C.L. 1968. Size, density, growth rate and age of Rocky Mountain juniper and interrelationships with environmental factors. Wyo. Range Manage. 259:93.

An investigation was made of three Rocky Mountain juniper *(Juniperus scopulorum)* communities in the Big Horn Basin of north central Wyoming. Each of the study sites was chosen to exhibit an age gradation of the juniper trees so that the sites could be further subdivided into age zones. Juniper size and form were used as criteria on which to base age differences between zones. This was done so that the various age zones could be compared to one another on the basis of any differences or similarities found in this investigation.

Data were collected on vegetation, soils and other site characteristics. The only site with a northerly exposure had a slower juniper crown diameter growth rate in all three age zones than did the other two sites; but the slower growth rate occurred in only the mature age zone.

A strong correlation was evident between juniper age and soil depth with the older trees found on shallower soils. Soil pH was higher in samples taken from directly beneath juniper trees than from samples taken from openings between trees.

Shrub and grass cover increased and forb cover decreased from the mature zones to the younger zones downslope. There was evidence at two of the study sites that the growth rate of junipers increased with the deeper soil, higher organic matter and better soil moisture conditions in the younger zones found downslope.

A fair correlation was evident between juniper size and age for particular sites; but due to site differences, this does not seem consistent from site to site. It was also noted that as tree density increased, understory cover decreased.

3A3 680205 Noller, G.L. 1968. The relationship of forage production to precipitation, cover and soils in north central Wyoming. Wyo. Range Manage. 263:116-136.
2A
1E2
10D1

The study was conducted in the Big Horn and Wind River Basins of northwestern Wyoming, for a six-year period beginning in 1962. Precipitation, cover and certain soil factors were studied and related to vegetational production.

The study sites were divided into four vegetation types, sagebrush *(Artemisia)*/grass, saltbush *(Atriplex gardneri)*, blue grama *(Bouteloua gracilis)* and greasewood *(Sarcobatus vermiculatus)*, on the basis of general aspect. Soils of each type were analyzed to determine texture, ph, soluble salts, organic matter of the top soil and 1 ft samples.

The vegetation cover, composition and yield were determined during each year of the study and were separated into categories of shrub, grasses and forbs. Precipitation data were collected four times a year.

Considering all vegetation types, 50 percent of the total precipitation came in the spring and almost 75 percent of the total came in the combined winter/spring period. The summer period received a smaller amount of precipitation than any other period. In the blue grama and greasewood areas precipitation was greater in the fall than in the winter. Loam textured soils produced a large volume of forage, but these soils were also higher in organic matter and received more precipitation than other soils. Certain other soil factors, ph, amount of soluble salts and clay content were associated with the presence of specific types of vegetation, but were not closely related to total forage production.

Considering all study areas, total cover was not closely related to total forage production. The blue grama vegetation areas had the highest total cover values but the lowest forage yield. The greasewood areas had the lowest total cover values but the highest yield.

Total precipitation was more closely related to total production than to total cover. Total cover increased when big sagebrush *(Artemisia tridentata)* was chemically controlled and decreased with grazing. Total forage production on all vegetation types decreased with grazing and on the sagebrush grass areas increased with the chemical control of sagebrush.

Based on all areas, total production was best estimated with the regression equation using the amount of spring precipitation. Grass production was best indicated with the amount of winter/spring precipitation considering all areas. Forb yield was best estimated by the amount of spring precipitation based on all areas. Shrub production on the saltbush area was best indicated by spring precipitation.

6803 Presentations

2F 3A2 2E 680301 "Patterns of live and dead vegetation in grassland ecosystems and their relationship to function." September 24, 1968. I.B.P. Grasslands Biome Synthesis Meeting. Halsey, Nebraska. by H.G. Fisser.

Historically, distributional aspects of vegetation on a multitude of levels have been evaluated on a subjective basis, with little emphasis given to factors responsible for observed characteristics of location or functions inducing variations in dispersal. The contention for the analysis of pattern lies with the implication that the distribution of the organisms may be utilized to evaluate the forces responsible for non-random dispersion. Pattern thus becomes a legitimate object of ecological investigation if used to determine causative factors in plant distribution, or as an aid in the establishment of tolerance ranges.

Quantitative analyses of intracommunal distributional characteristics emphasized that random distribution was the exception rather than the rule. The form of vegetation pattern can be related to two general classes of non-randomness: contagious and regular. In most cases, analyses of distribution properties have shown that contagious distributions are the most common.

The term pattern may imply repeatability, but it does not imply uniform repeatability, since the factors causing non-random plant dispersions need not repeat themselves uniformly over geographic space. The evaluation of vegetation distributional form must be conducted in light of the kinds of possible causative factors and their form of pattern as related to the configuration of plant dispersal.

For the detection of non-randomness, by application of mathematical expectations, early workers utilized the Poisson and binomial distributions with a great number of modifications. Analysis of association has been conducted to determine the existence of similar biological groups and the degree to which they shared similar ecological amplitude to environmental forcing functions.

Common analytical procedures for association testing are the contingency and correlation analyses. A new and promising development in the description and analysis of organism dispersion criteria has been the use of multivariate analysis. By integration of many forms of quantitative data, the derived ecologically meaningful groups lead to a technique for a determination of pattern causality, more sensitive than earlier procedures. Studies to evaluate patterns and their causal relation seem doomed to failure if not incorporated into an integrated systems analysis approach.

7A2 1E4 680302 "Dietary preferences and utilization of elk on spring range in the Big Horn Mountains of Wyoming." February 14, 1968. 21st Ann. Meeting Soc. Range Manage. Albuquerque, New Mexico. by G.E. Probasco.

See 680203 and 680602.

6804 Published Proceedings/Abstracts

7A2 1E4 680401 Probasco, G.E. 1968. Dietary preferences and utilization of elk on spring range in the northern Big Horn Mountains of Wyoming. p. 22. In Proc. 21st Ann. Meeting Soc. Range Manage. Albuquerque, New Mexico. February 14, 1968.

See 680203 and 680602.

6805 University of Wyoming Agricultural Experiment Station Publications

4B 4C 1E4 2F2 3A2 7A1 7A3 6A3 680501 Gibbens, R.P., H.G. Fisser and M. May. 1968. Influence of grazing management systems on vegetation and wildlife habitat. Annual report of progress, July, 1967 - July, 1968. Univ. of Wyo. Coop. Res. Rep. to the BLM. Univ. of Wyo. Agr. Exp. Sta. Sci. Rep. 102. 9 p.

The grazing system study which was initiated during the summer of 1967 was applied to a system of three grazing management systems. A season long unit, a two pasture rotation unit and a four pasture deferred rotation system.

3A2 680502 Fisser, H.G. 1968. Exclosure studies with transects of permanent plots. 1967 results. Univ. of Wyo. Coop. Res. Rep. to the BLM. Wyo. Agr. Exp. Sta. Sci. Rep. 118. 52 p.
2F2
5C
10D1

See 670104. This report gave the data by transect and species from each of the 1 ft x 1 ft permanent plot transects on 24 exclosures in the Natrona County and Big Horn Basin areas. A plant list of some 170 different plants which occurred on the exclosures with the permanent plot transects was included in the report. Also included were small sketch maps showing general locations of transects in the exclosure areas.

3A2 680503 Fisser, H.G. 1968. Long-term exclosure studies with 2 ft x 20 ft and 4 ft x 4 ft permanent plots. 1959-1964 results. Univ. Wyo. Coop. Res. Rep. to the BLM. Wyo. Agr. Exp. Sta. Sci. Rep. 131. 78 p.
2F2
10D1

This report presented data of the 2 ft x 20 ft and 4 ft x 4 ft quadrats which were established during 1959 and 1960 and reread again in 1964. There were 30 locations and in addition to the three exclosures at the Halogeton Dry Creek research units, each of the individual pastures had quadrats within it also. Maps of Wyoming showing general locations of the exclosures as well as the tabular material which shows the counties, exclosure names, legal descriptions, the exclosure code names, the year established for the exclosure and the kinds of data obtained at each were included. A list of the plants encountered at the exclosures was included and comprised over 100 species. A map of each exclosure was drawn showing general location of each of the 2 ft x 20 ft and 4 ft x 4 ft quadrat locations. General directions of travel to each exclosure were given also.

680504 Fisser, H.G. and G.L. Noller. 1968. Aridland Ecology. 1967 results. Univ. of Wyo. Coop. Res. Rep. to the BLM. Sections I, II and III, Univ. of Wyo. Agr. Exp. Sta. Sci. Rep. 132. 93 p.

2A 68050401 Section I - Soil moisture and temperature study.
2C
2F2

This report presented data from 1967 for the nine dates during which soil moisture and temperature were taken at the Granite Mountain and Smilo exclosures and for the eight dates on which the data were obtained at the Cumberland -1 and -3 exclosures. Some soil temperature data for 1966 was included in this report as well. For additional information see 67050201 and 660105.

2A 68050402 Section II - Precipitation pattern study.
3A

The report of 1967 precipitation pattern data was described in greater detail in 660118. Growing conditions during 1967 in the Big Horn Basin were favorable. The report included precipitation data from some 120 UW-BLM gauges; 20 of those were established on the Baggs grazing system study during 1967. Thirty-four U.S. Weather Service Station data sets have been compiled to seasonal values as with those of the UW-BLM gauges and have been given in the report also.

3A3 68050403 Section III - Exclosure study: production, cover and precipitation phase.
5A1D
5B
5C
2C

The exclosure study activity for 1967 included also a list of all exclosures at which herbage production clippings were conducted and for each exclosure a list of the transect type and location was given. This list showed that 85 transects of data were accumulated on the 38 exclosure locations. The plant list was included also. The wet spring and summer of 1967 resulted in significant plant growth because of the very favorable growing conditions during the spring period.

2E2 680505 Wight, J.R. and H.G. Fisser. 1968. *Juniperus osteosperma* in northwestern Wyoming - Their distribution and ecology. Univ. of Wyo. Agr. Exp. Sta. Sci. Mono. 7. 31 p.
2C
3A2

This experiment station publication was essentially a portion of the Wight dissertation in which he described the habitat and vegetation characteristics of the Utah juniper *(Juniperus osteosperma)* type in the Big Horn Basin and on the Owl Creek Mountains on the south side of the Big Horn Basin. For more information see 660601.

2F2 680506 Fisser, H.G. 1968. Synthetic study area nomination and related analytic research proposals to the Desert Biome Ecosystem Analysis Program of the International Biological Program. Univ. of Wyo. Agr. Exp. Sta. Mimeo. 82 p.
10D1

This publication included a description of the Alkali Flat study area located 25 miles east of Rawlins. The descriptive characterization of the site included location and general description, geology and soils, climate, vegetation, native fauna, grazing history and use, ownership, and communication and facilities.

The synthetic study areas specifically addressed in this project proposal were the sagebrush synthetic area on the upland type and the saltbush synthetic type at the lower semi-permanent lakebed area.

Interest in and capability for cooperative activities in this program was exhibited by personnel in the following units of the University of Wyoming: Division of Research Development, Botany Department, Zoology Department, Chemistry Department, Civil Engineering Department, Electrical Engineering Department, Natural Resources Research Institute, Veterinary Science Division, Microbiology Section, Plant Science Division, Range Management Section, Weed Control Section, Agricultural Engineering Section, Entomology Section, Crops Section, Soils Section, Biochemistry Division, and Animal Science Division. In addition, the Wyoming Game and Fish Commission has expressed a vital interest in the Alkali Flat desert biome study. Because of the dynamics of the antelope populations in that area, the Bureau of Land Management, which has administrative authority on the area, has also exhibited significant interest and assistance in development of the publication. Research proposals were submitted by 30 scientists from the University of Wyoming addressing a variety of initial state and internal parameter studies.

6806 Graduate Advisee Completions

3A3 680601 Noller, G.L. 1968. The relationship of forage production to precipitation, cover and soils in north central Wyoming. Ph.D. Dissertation. Univ. of Wyo. 153 p.
2A
1E2
10D1

The study was conducted in the Big Horn and Wind River Basins of northwestern Wyoming utilizing six years of data which began in 1962. Precipitation, vegetation cover and certain soil factors were studied and related to the vegetation production. The study sites were divided in four vegetation types: sagebrush *(Artemisia)* grass, saltbush *(Atriplex gardneri)*, blue grama *(Bouteloua gracilis)* and greasewood *(Sarcobatus vermiculatus)*. Soils were analyzed for texture, pH, soluble salts and organic matter. Considering all study areas, total cover was not related to the total forage production.

The blue grama areas had the highest total cover value but the lowest forage yield, while the greasewood area had the lowest cover but the highest yield. Total precipitation was more closely related to total production than to total cover. Total cover generally increased when sagebrush was chemically controlled and decreased with grazing. Total forage production on all vegetation types decreased with grazing and on the sagebrush grass areas increased with the chemical control of sagebrush.

Based on all areas, total production was best estimated with a regression equation using the amount of spring precipitation. Grass production was best indicated with the amount of winter/spring precipitation considering all areas. Forb yield was best estimated by the amount of spring precipitation based on all areas.

Shrub production on the saltbush area was best indicated by a spring precipitation. Differences in the amount of spring precipitation explained 17 percent of the differences in total forage production. Each inch of spring precipitation accounted for 53 lb./acre of total forage production. The amount of winter/spring precipitation considering all areas together explained 22 percent of the variation in grass production and each inch of the combination accounted for 52 and 30 lb./acre for the winter and spring periods respectively. Winter precipitation accounted for a greater increase in grass production probably because it was the most limited amount for the two periods.

7A2 680602 Probasco, G.E. 1968. Diet preferences and utilization patterns of elk in the northern Big
1E4 Horn Mountains, Wyoming. M.S. Thesis. Univ. of Wyo. 36 p.

Data were collected to determine dietary preferences and grazing patterns of elk in the northern Big Horn Mountains of Wyoming. Forest openings where only elk grazing occurred were studied to determine preferred plant species for both the spring and summer season.

One forest opening of approximately 300 acres was stratified to determine if there was a correlation between elk grazing patterns and distance from forest margin. It was evident from the data that elk grazing patterns were not correlated with distance from forest margin. However, there was a definite correlation between elk grazing patterns and percent total basal cover of herbaceous vegetation. Data on diet preferences indicated that elk utilized grasses during the spring period, but shifted their preference to forbs during the summer season.

1E2 680603 Wambolt, C.L. 1968. Size, density, growth rate and age of Rocky Mountain juniper and
2A interrelationships with environmental factors. M.S. Thesis. Univ. of Wyo. 62 p.
2C

2E Investigation was made of three communities of Rocky Mountain juniper *(Juniperus*
10D1 *scopulorum)* in the Big Horn Basin of Wyoming. Size and form were used as criteria to base age differences between zones. This was done so that the various age zones could be compared to one another on the basis of any differences or similarities found in this investigation. Vegetation, soils and other site characteristics were studied.

There was evidence at two of the study sites that growth rate of the junipers increased with deeper soil, higher organic matter and better soil moisture condition in the younger zones found downslope. A fair correlation was evident between Utah juniper size and age for particular sites, but due to site differences this does not seem consistent from site to site. It was also noted that as tree density increased, understory cover decreased.

6901 Research Activities

3A3 690101 Exclosure study: production, cover and precipitation.
5A1D
5B During 1969 the early spring period during April and May was quite dry and caused
5C early cessation and restricted growth of cool season plants. Because of this dryness,
2A many plants obtained maturity by the end of June when normally they would have been
2C growing well into August. The presence of annuals and forbs was less than during the two previous years. All annuals demonstrated low vigor and early maturity. The perennial warm season grasses were benefited by the late spring/early summer moisture but this effect was dampened by high later summer temperatures in combination with, again, a dry period and little moisture. Including all of the individual pastures and the three exclosures of the Halogeton Drycreek experimental unit north of Greybull, production data were obtained on 41 locations.

5A1D690102 Juniper control study of L.U.
2E

During 1969 the activities on the L.U. Ranch areas, which had been treated with Tordon and 2,4-D to control juniper *(Juniperus osteosperma)* and limber pine *(Pinus flexilis)*, consisted primarily of rereadings on the 20 ft x 50 ft macro-quadrats located in the treatment areas.

2A 690103 Soil moisture and temperature.
2C
2F2 During 1969 soil moisture and temperature data were recorded on four dates at the Granite Mountain, Smilo, Cumberland -1, and Cumberland -3 exclosure locations. Despite the major portion of the spring precipitation occurring in late June, soil moisture values of the upper soil region showed a definite increase in July as compared to April on all study areas except Smilo. Soil moisture data collected at Smilo, at this period, indicated no appreciable difference between April and July. Data for the rest of 1969 appeared to follow the same pattern as that from previous years. Soil temperature data collected in July at the Granite Mountain and Smilo exclosures appeared lower than normally expected, which may be attributed to the late spring precipitation in June.

3A2 690104 1 ft x 1 ft permanent plot transects.
2F2
5C During 1969 the transects with 1 ft x 1 ft permanent quadrats were reread on all 26 exclosures on which they had been established. Permanent quadrats have also been established on some of the exclosures in which herbage production is collected. There have also been some permanent plot transects established in the Baggs grazing unit area.

The primary use of exclosures is that related to the evaluation of vegetation response to grazing influence by livestock and game and in measuring changes associated with improvement practices such as shrub control, pitting, contouring and seeding. Vegetation changes with time at these exclosures reflect response to short-term weather variations and intensity of animal use pressure.

It is often incorrectly assumed that an exclosure is of no value unless inside and outside differences exist, with a concommitent implication that grazing, whether on herbaceous or woody species, results in drastic change to the vigor, composition, and productivity of the native species. True, intensive use results in range condition deterioration, but under moderate use impact, very little, if any damage occurs. Thus, exclosures can be utilized as indicators of moderate, nondamaging use as well as overuse.

In opposition to the above concept, the commonly slow response of vegetation especially in the arid regions of Wyoming, to grazing pressure changes and improvement practices, often require several to many years for the manifestation of a change to be visable to the observer. For this reason, permanent plot transects were installed, from which are obtained cover, composition, and frequency data that aid in early determination of vegetation changes. In addition, by use of the permanent plots, a major source of random sampling error is removed, allowing for greater precision of estimate for evaluation of any vegetation changes.

4B 690105 Baggs grazing systems study.

1E4
2F2
3A2
7A1
7A3
4C
6A3

During 1969 all of the permanent plot transects and trend plots were reread at the eight exclosures established in the pastures. Within each of the pastures some 100 permanent plot transects have been established in the grazing system area. There was no apparent vegetation response to the different grazing treatments.

3A2 690106 2 ft x 20 ft and 4 ft x 4 ft permanent plot.

2F2

During 1969 considerable time was taken to recheck data, especially in the 2 ft x 20 ft quadrats, at the exclosures where these permanent plots had been established. A matter of species misidentification has now been essentially cleared up and the data finalized for publication. Photographs were retaken of the photo point quadrats also.

5A1D690107 Atrazine application to control annual weeds on saltbush range.

4A

In early June 1969 the atrazine spray application was reviewed on the Halogeton pastures. In the pasture 2 spray application area, very few annuals were present and those present were only halogeton *(Halogeton glomeratus)* and tansyleaf aster *(Machaeranthera tanacetifolia)*. The saltbush *(Atriplex gardneri)* had vigorous growth in the spray areas. In pasture three the response was similar; but there were dead strips of saltbush which occurred from the heavy rate of use which was noted during the spraying process because a nozzle had been lost from one of the spray units allowing an excessively great amount of chemical to be applied.

This chemical has a hormone effect on the saltbush and excessive amounts kills all vegetation. The sprayed saltbush in pasture three generally had a yellowish color and the non-sprayed had a greenish color, but the sprayed vegetation had obviously greater amounts of growth than the non-sprayed. The grass was still alive in the gullies, from the water effect on the chemical perhaps, otherwise the grasses have primarily been killed off by the spray and this includes bottlebrush squirreltail *(Sitanion hystrix)*, Indian ricegrass *(Oryzopsis hymenoides)* and sandberg bluegrass *(Poa secunda)*.

3A2 690108 Vegetation pattern on shortgrass-cactus ranges.

2E
2F1
10D1

On the Pawnee site, south of Cheyenne, during 1969 a study site was established to determine the specific location of plains pricklypear *(Opuntia polyacantha)* in a 30 m x 30 m sample area. Subplots were at 1 decimeter squares and each decimeter within the 900 square meters was observed as to its location and whether cactus was present or absent. The question of vegetation pattern on shortgrass-cactus ranges was also addressed by Mitchell with strip transects 1 dm wide and 30 m long. His data was presented in a subsequent thesis (710603).

5B4 690109 Juniper GRI exclosure seedings.

7A
5A1D
2E

The seedings of the Utah juniper *(Juniperus osteosperma)* game range improvement exclosures was reviewed. These data were subsequently in a thesis by Burrough. See 720603.

3A2 690110 Block juniper site, 1 ft x 1 ft permanent plot studies.
2F2
5C During 1969 the 1 ft x 1 ft permanent plot transects established in the block juniper
10D1 (*Juniperus* spp) site were reread. Some additional work was established by Robinson in 1965 with long strip transects in the Block Juniper site as well as some other sites relating to his thesis activity.

2C 690111 Boars Tusk Sand Dune Area Tour.
2F2
8B Fisser had a tour of the Boars Tusk Sand Dune area with Charles Dahlin on August
10D1 13, 1969. There was potential for vegetation stabilization practices in the Sand dune area and the location of a couple of large, that is 20-30 acre, exclosures was suggested. In addition to big sagebrush and Douglas rabbitbrush in the area, a considerable amount of bitterbrush and buffaloberry was present as well. This area supported a sizable elk herd which was observed while traveling over the Dune areas.

5A1D690112 Juniper GRI Exclosure Tordon retention studies.
7A
2E During 1969, Burroughs and Fisser collected samples of duff and soil from beneath
1E2 and between Utah juniper *(Juniperus osteosperma)* trees treated with tordon. These samples were to be analyzed to determine the tordon retention levels still evident in the materials by planting beans *(Phaseolus vulgaris)* on the collected soil and duff samples.

6903 Presentations

5A1D690301 "Vegetation response to chemical control of low sagebrush (*Artemisia arbuscula*) in
2F2 western Wyoming." August 26, 1969. Range Weed Control Conference. Reno,
2E Nevada. by H.G. Fisser.
3A2

Following chemical control of low sagebrush *(Artemisia arbuscula)* at the Cumberland -2 exclosure a first flush of increased basal cover and foraged production was noted. Basal cover appeared to average about twice as much in the sprayed area as in the native non-sprayed. Grazing had a strong influence on the vegetation and overshadowed much of the response to the chemical treatment. Western wheatgrass (*Agropyron smithii*) was greatly influenced by the summer grazing, probably more so than the bluegrass (*Poa* spp). This can readily be visualized, since the *Poa* species generally tend to mature and shatter by midsummer, while the western wheatgrass requires another month to six weeks before it attains its full growth complement. Thus the grazing would certainly tend to damage the western wheatgrass more than the bluegrass species.

An interesting addendum may be pointed out from the data obtained in 1968 from shrub production values. It will be noted that big sagebrush produced about 160 lbs/a on the native site protected from grazing, but only 145 lbs/a in the native site subject to grazing. Total difference in production was about 45 lbs/a; 245 lbs/a on the site protected from grazing and only 200 lbs/a on the grazed area. The productivity of the shrubs is an important constituent of the total predictive potential of the site. Certainly the grazing treatment has had a strong influence on the production, herbaceous basal area, and shrub ground cover.

The value of the chemical treatment cannot be looked at solely in terms of the herbaceous species response to the chemical and subsequent removal of sagebrush, because the productive potential of the site includes as much shrub production as about half of the herbaceous production, or approximately one half of what can probably be called the total productive potential of the site.

Whatever the findings, in years to come it will be important to evaluate the net results of a shrub control program in terms of the increase of herbaceous species production as compared to the replaced herbaceous production of the shrubs and woody species. If the increased herbaceous production of the forbs and grasses does not at least equal the production of the woody species, it would seem that chemical treatment may be questioned as to its validity, unless we are absolutely sure that the shrub production is of no inherent importance toward utilization and stabilization of these rangelands. However, this probably cannot be the case because few rangelands are utilized solely by animals which would use only herbaceous materials.

Various species of game animals are present on practically all ranges. Their forage needs must be met as well as those for domestic animals. The relationship of the kind of herbaceous production that we seek to obtain from rangelands must be dependent on the variable aspects of the requirements of both domestic and wildlife animals.

5A1D690302 "Juniper game range improvement studies in western Wyoming." November 18, 1969.
7A Wyoming Game and Fish Commission Biologists Meeting, Powell, Wyoming. by
2E H.G. Fisser.

This presentation was complemented by 72 slides and addressed six major topics. These topics are as follows.

1) Concepts of Game Range Improvement Studies in Wyoming Juniper (*Juniperus* spp.) Types.
   a) Kinds of Juniper Types.
   b) General Ecology of Juniper and Relationship to Game Range Improvement Possibilities.
      i) Arid Site - 7 to 14 inches of precipitation, little snow.
      ii) Soils - poorly developed, thin, rocky.
      iii) Vegetation - very few perennials in juniper stands because of strong competition for water by the junipers.
      iv) Temperature - very hot in the summer.
      v) Winter Game Range - immediately above the sagebrush type and irrigated hay and meadow fields.
      vi) Little Livestock Use - at present.

   The concept of rancher and public response to the general idea of increasing numbers of game animals and subsequent competition with livestock creates some problem. By publicizing the concept of game range improvement, we may be able to minimize the conflict of interests.

2) The Initial UW Studies of the Kane Deer Exclosure.
   a) The fencing occurred in 1964 and spraying was done on July 28, 1965. Production studies and permanent plots have been established on that site.
   b) The clearing work was done in 1964 and seeding was done on May 14, 1965.
   c) Seeding inside the spray in October 1965. No seedling establishment because of the impact of Tordon apparently.

3) Game Range Improvement Studies on the new game-proof exclosures at Kane, Medicine Lodge, and Big Trails.
   a) Fencing 1968
   b) Spray July 15, 1968
   c) Burn June 1968
   d) Doze June 1968
   e) Fall 1968 Seeding on November 17
   f) Spring 1969 Seeding on April 15
   g) Fall 1969 Seeding on October 15

4) Tordon Retention in Soil and Duff.

5) Carrying Capacity - Game and Animals.

6) Underground Nuclear Detonation and Possible Effects to Wildlife.

5A1D690303 "Improvement studies on juniper sites in western Wyoming." November 21, 1969. Amer. Soc. of Range Manage. Wyoming Section, Annual Winter Meeting. Sheridan, Wyoming. by H.G. Fisser.
7A
2E
1E2

This presentation gave essentially the same materials as that in 690302.

2E 690304 "Soil moisture relationships of certain salt desert shrubs." Feb. 12, 1969. 22nd Ann. Meeting Soc. Range Manage. Calgary, Alberta. by R.E. Steger.
2A
2C
3A2

This study was conducted by Steger from 1965-67 to determine soil moisture relationships of six salt desert shrub species while determining phenological and other soil and plant characteristics as well. The soil moisture regimes of the plant communities were expressed as a mean of soil moisture per foot of soil of three holes as determined by a neutron scattering instrument. The recordings were taken at 6, 12, 18, 24, 36, 48 and 60 in depths monthly during the winter and twice monthly during the plant growing season. During 1966, a dry year, and 1967, one of the wettest on record, data extremes were noted. The similarities and differences of soil moisture relationships of each of the six plant communities studied is given below.

1) Nuttall saltbush *(Artiplex nuttallii)* growth occurred in a rather narrow time period. Soil moisture penetration was only to the six inch level. The plant growth period was essentially the same for 1966 and 1967. Soil water recharge occurred only slightly before the plant growing season in the spring.

2) Birdfoot sagebrush *(Artemisia pedatifida)* soil moisture content was higher in 1965 and 1966 than in 1967. Plant growth duration and time of growth varied little regardless of the amount of precipitation received. Moisture changes were much more dramatic in these soils than in the saltbush. Moisture changes occurred at both the six and 12 inch depths. Recharge of soil water occurred only slightly before plant growth initiation.

3) Bud sagebrush *(Artemisia spinescens)* soil moisture level was clearly a reflection of the amount of precipitation received that year. The length of the plant growing season more strongly reflected the amount of moisture available. This was apparently due to leaf shed during stress periods and that the plant was able to green up very early and quickly matured. Soil moisture recharge started at mid-winter in this community. Most of the changes in moisture content were at the 6 inch depth but occurred as deep as the 24 inch depth.

4) Winterfat *(Ceratoides lanata)* soil moisture content changes were noted to be equal at about the 6 and 12 inch depths and to change slightly also at the 18 inch depth. Recharge occurred during mid-winter.

5) Big sagebrush *(Artemisia tridentata)* changes were approximately equal at the 6 and 12 inch depth, less pronounced at the 18 inch depth though variable even to the 24 inch depth. Soil water recharge started at mid-winter. Plant duration and season of growth was approximately the same each year.

6) Spiny hopsage *(Grayia spinosa)* exhibited the least spectacular changes in moisture content observed in this plant community than in others. Comparable changes in soil moisture content were noted at the 6, 12, and 18 inch depths and to a lesser extent at the 24 inch depth. Growth duration was more sensitive to soil moisture content as this plant, as did bud sagebrush, shed its leaves when moisture levels were low. Soil moisture recharge again started in mid-winter.

The soil moisture relationships of the six plant communities were thus shown to be influenced by season of recharge, the rate of use, the depth of moisture penetration, the reaction of the plants to moisture stress, and the phenological changes of plant growth during the growing season.

2E 690305 "Size, density, growth rate and age of Rocky Mountain juniper and interrelationships
1E2 with environmental parameters." February 12, 1969. 22nd Ann. Meeting Soc. Range
2A Manage. Calgary, Alberta. by C.L. Wambolt.
2C

An investigation of three Rocky Mountain juniper *(Juniperus scopulorum)* communities in the Big Horn Basin of north central Wyoming was conducted in which each site was chosen to exhibit an age gradation of the juniper trees so sites could be divided into age zones. Juniper size and form were used as criteria on which to base age differences between zones. This was done so that the various zones could be compared to one another on the basis of any differences or similarities found in this investigation.

Data were collected on vegetation, soil, and other site characteristcs. The only study site with a northerly exposure had a slower juniper crown diameter growth rate in all three age zones than did the other two sites, but it only had a slower growth rate in height in the mature age zone. A strong correllation between juniper age and soil depth was noted, with older trees found on shallower soils. Soil pH was higher in samples taken from directly beneath juniper trees than it was from samples taken from openings between the trees. Shrub and grass cover increased and forb cover decreased from the mature zones to the younger zones found downslope.

There was evidence at two of the study sites that the growth rate of junipers increased with deeper soil, higher organic matter and better soil moisture conditions in the younger zones found down slope. A fair correlation was found between juniper size and age for particular sites, but due to site differences, this does not appear to be consistent from site to site. It was also noted that as tree density increased, understory cover decreased. See also 690403 and 680603.

6904 Published Proceedings/Abstracts

10D1 690401 Fisser, H.G. 1969. Plant pattern and distribution in ecosystems and relationships to
3A2 function. p. 183-196. In R.L. Dix and R.G. Beidleman (eds.) Proc. Information
2E Synthesis Project Grassland Biome. U.S. I.B.P. Range Science Series No. 2. Colo.
2F2 State Univ. Ft. Collins, Colorado.

This paper considered the role of pattern in grassland ecosystems by a review of literature pertinent to the evaluation of pattern presence. Most critical was the evaluation of pattern causality as functional to the existence of dynamic ecosystems. Historically, distributional aspects of vegetation were evaluated at many levels on a subjective basis, with little emphasis given to factors responsible for observed characteristics of location or functions inducing variations in dispersal. The contention for the analysis of pattern lies with the implication that the distribution of the organisms may be utilized to evaluate the forces responsible for non-random dispersion. Pattern becomes a legitimate object of ecological investigation if used to determine causative factors in plant distribution, or as an aid in establishing tolerance ranges.

Quantitative analyses of intracommunal distributional characteristics has emphasized that random distribution was the exception rather than the rule. The form of vegetation pattern can be related to two general classes of non-randomness: contageous and regular. In most cases analysis of distribution properties have shown that contageous distributions are the most common.

The term "pattern" may imply repeatability. It does not necessarily imply uniform repeatability, since the factors causing non-random plant dispersions need not repeat themselves uniformly over geographic space. The evaluation of vegetation distributional forces must be conducted in light of the kinds of possible causative factors and their form of pattern as related to the configuration of plant dispersal.

For detecting non-randomness by application of mathematical expectation, early workers utilized the Poisson and binomial distributions with a great number of modifications. Analysis for association has been conducted to determine the existence of similar biological groups and the degree to which they shared similar ecological amplitude to environmental forcing functions. Common analytical procedures for association testing are the contigency and correlation analysis. A new and promising development in the description and analysis of organism dispersion criteria has been the use of multivariate analysis. By integration of many forms of quantitive data the derived ecologically meaningful groups lead to a technique to determine pattern causality, more sensitive than earlier procedures. Studies to evaluate patterns and their causal relations seem doomed to failure if not incorporated into an integrated systems analysis approach.

2E 2A 2C 3A2 690402 Steger, R.E. Soil moisture relationships of certain salt desert shrubs. p. 50. In Proc. 22nd Ann. Meeting Soc. Range Manage. Calgary, Alberta. February 10-13, 1969.

See 690304.

2E 1E2 2A 2C 690403 Wambolt, C.L. 1969. Size, density, growth rate and age of Rocky Mountain juniper and interrelationships with environmental factors. p. 57. In Proc. 22nd Ann. Meeting Soc. Range Manage. Calgary, Alberta. February 10-13, 1969.

See 690305.

6905 Wyoming Agricultural Experiment Station Publications

690501 Fisser, H.G. and G.L. Whysong. 1969. Aridland Ecology Research. 1968 results. Univ. of Wyo. Coop. Res. Rep. to the BLM. Sections I - IV. Wyo. Agr. Exp. Sta. Sci. Rep. 178. 114 p.

2A 2C 2F2 69050101 Section I - Soil moisture and temperature.

This report of soil moisture and temperature data for 1968 showed that data were collected on four dates at the Granite Mountain, Smilo, and Cumberland -1 and -3 exclosures. Soil moisture levels in September 1968 were greater than in 1967 because of the high frequency and amount of precipitation during summer. Greatest increase in soil moisture was noted in the inside spray treatment area at Granite Mountain. See also 680105.

2A 2E 2C 3A2 69050102 Section II - Microclimate studies in desert shrub communities.

This report of 1967 and 1968 data included soil moisture and temperature values collected at the six vegetation community sites in the Big Horn Basin. Data were collected on 11 dates during 1967 and on four dates during 1968. See also 690402.

2A 3A 69050103 Section III - Precipitation pattern study.

This report of 1968 data presents all of the seasonal data for precipitation at the UW/BLM gauges and also at some 35 US Weather Service stations as well.

3A3 5A1D 5B 5C 2C 69050104 Section IV - Exclosure study: production, cover and precipitation phase.

This report of 1968 data gave herbage production as in previous years, but in addition, at the sagebrush *(Artemisia tridentata)* grass and saltbush *(Atriplex gardneri)* sites, estimates of total community herbage production were obtained with the use of the weight unit estimate procedures conducted during September. The report included the plant list, the list of exclosures, and individual sheets showing production by species and groups of species throughout. The dry spring of 1968 retarded production of grass species while the relatively wet summer showed an increase in production of annual and perennial forbs. These results indicated that the latter species may be more capable of utilizing summer moisture more efficiently than the grass species. See also 680106.

4B 4C 1E4 2F2 3A2 7A1 7A3 6A3 690502 Gibbens, R.P., H.G. Fisser and M. May. 1969. Influence of grazing management systems on vegetation and wildlife habitat. Annual report of progress, July 1968 - July 1969. Univ. of Wyo. Coop. Res. Rep. to the BLM. Wyo. Agr. Exp. Sta. Sci. Rep. 179. 132 p.

This report of the 1968-69 field research activities of the Baggs grazing systems study included data, vegetation cover and composition data from the transects located inside and outside each of the eight exclosures and also transects located throughout the pasture units in areas similar and comparable to the location of the BLM trend plots.

2F1 10D1 2E 3A0 690503 Fisser, H.G., J.E. Lester, G.C. Mitchell, and L.J. Bledsoe. 1969. Preliminary report of methodology and results for analyses of plant pattern project research on the Pawnee Site. U.S. I.B.P. Grassland Biome Technical Report 9. 68 p.

Distributional forms and dispersion characteristics of plains pricklypear *Opuntia polyacantha* were examined in areas subjected to three grazing rates (light, medium, and heavy). Determinations of non-randomness were made from comparisons of observed numbers of individuals per quadrat to the expected per quadrat derived from a Poisson series. Frequency data were obtained from transects of 256 contiguous dm^2 quadrats in 30 m x 30 m study plots within each grazing treatment. The analysis of variance technique applied to the frequency data gave an estimate of the clump sizes within each treatment. Mean area of clumps for the light, medium and heavy grazing treatments were: 4, 16, and 128; 4 and 128; and 8, 32, and 128 dm, respectively.

As a step toward developing adequate conceptual interactions with other research segments of the grassland biome program, four scales of plant measurements and dispersion, based upon common vegetational characteristics, were developed. The four scales of measurement were developed upon subjective characterization of common shortgrass plains vegetation.

The "microscale" level of measurement applied to the distribution of individual culms of herbaceous mat forming species such as blue grama *(Bouteloua gracilis)* and buffalograss *(Buchloe dactyloides)* and perhaps annuals which grow in a dense or mat form. The "mesoscale" measurement applied to the distribution of grass mat and bare ground interspaces. The "macroscale" measures would evaluate dispersion characteristics of species which form distinct clumps within the matrix of grass and bare ground interspaces. Distribution of shrubs and plains pricklypear cactus would fit this level of measurement. "Megascale" measurements would be concerned with the evaluation of vegetation types within the broad concept of grasslands. Criteria of measurement, characterization, and suggested relationships to ecosystem components of the four measurement levels are presented.

To provide precise and detailed data of plant distribution characteristics, a field microplotter was designed to record positions based upon distance measures along two perpendicular axes. In addition, vertical displacement or microrelief can be measured at each point as a distance value above or below a horizontal plane extending from the point of origin. Data thus are given in terms of X, Y, and Z coordinates in units of .001 m. A description of this microplotter mechanism is given in the report.

6906 Graduate Advisee Completions

4C 1E4 6A1 690601 Todd, Lynn D. 1969. Preference and utilization trends by cattle on grass-forb vegetation in the northern Big Horn Mountains, Wyoming. M.S. Thesis. Univ. of Wyo. 73 p.

Vegetative preference and trend of utilization were shown by cattle on the northern Big Horn mountains of Wyoming. Two study areas were selected. The first was at an elevation of 8,600 ft while the second was at an elevation of 9,300 ft.

The livestock prefer grasses and sedges over forbs. There was some difference in preference for forbs between the two elevational sites. Some degree of utilization trend was shown for most species. The most definite and most explainable trends were found to be in forbs. Utilization trends in grasses and sages were not as explicit. Transects were taken immediately after cattle had grazed an area. Only freshly grazed plants were recorded as grazed on any particular date. In this way, only utilization by cattle was recorded and that of all other animals was ignored.

An index, including frequencies of utilization and recurrence, was constructed to determine preference and utilization trends within each species. The results of this single grazing period study provided a simple representation of utilization trend and preference for certain species. There are many variables which may influence livestock preference and trend of utilization. The general patterns of use recorded in this study represent animal preference on specific days during the summer. Utilization preference and trend may vary with changes of weather, grazing duration, and species composition, but can be expected to remain relatively uniform under normal conditions.

2E 690602 Forbis, L.A. 1969. Ecological relationships of vegetation, soils and vertebrates on paired limber pine and spruce-fir sites, Big Horn Mountains, Wyoming. M.S. Thesis. Univ. of Wyo. 105 p.
2F2
7D1
2C

On the northern Big Horn Mountains of Wyoming, four paired sites of limber pine *(Pinus flexilis)* were compared with adjacent similar sites of spruce/fir *(Picea/Pseudotsuga)* overstory. Data of abiotic and biotic characteristics were measured. The objectives were 1) to examine the ecological relationships of the differences of the limber pine type from adjacent spruce/fir stands and 2) to determine the probable position of the limber pine community in relation to ecological succession in the Big Horn mountains.

Soils ranged from loam to clay loam in texture. pH values were generally slightly higher on the limber pine site than on the spruce/fir. pH values ranged from 6.6 to 7.2 on the limber pine site. The amount of duff was significantly less on the limber pine site than on the spruce/fir, with 203 to 718 grams per square foot on the limber pine site.

Overstory data indicated that the number of trees per acre increased as the percent composition of limber pine decreased. The average age of trees on the paired site showed that limber pine was generally older than the adjacent spruce and fir. The limber pine site averaged from 65 to 235 years of age. The growth index showed that limber pine trees increased their circumference much slower than spruce and fir trees on adjacent sites. The growth index, circumference divided by age, varied from 0.16 to 0.296 on limber pine sites.

Limber pine sites had more understory plant species associated with them than did the spruce/fir sites. Based on similarity indexes, study areas consisting of both limber pine and spruce/fir overstory were found to be more similar in understory species composition than the understory species composition on all limber pine sites or on all spruce/fir sites.

Snap trap lines were set simultaneously on paired sites of each area. Nine species of mammals were collected. The deer mice *(Peromyscus maniculatus)* and the red-backed voles *(Clethrionomys gapperi)* comprised 88 percent of the total taken. The former was a dominant in the limber pine site and the latter comprised 50 percent of the total. They occurred on both sites, however. The least chipmunks *(Eutamias minimus)*, which comprised only 8 percent of all small mammals, was the only species which was consistently present only in limber pine sites.

Pellet group counts were used to show that deer and elk demonstrated no selectivity between limber pine and spruce/fir areas during spring/summer or fall/winter periods. The data showed there were more deer than elk.

There was a positive relationship between a growth index value derived by dividing average tree circumference by average tree age and soil pH on limber pine and spruce/fir sites. A positive relationship existed between the amount of duff and a duff accumulation index derived by relative site overstory age x relative site overstory density/percent slope, on limber pine and spruce/fir sites. A negative relationship existed between soil pH and an index derived by site elevation/site percent slope, on limber pine and spruce/fir sites.

The limber pine community was an early transitional successional stage. The trees themselves can be pioneers and exist on bare, rocky sites along with Crustose lichens. In time, the trees may alter the site to allow the rest of the associated community to establish. However, the advance of succession depends on site characteristics. There are sites that could be called limber pine climax areas.

Limber pine trees also have the ability to invade grasslands and brushy areas, to pave the way for higher successional stages. In this role, limber pine may be considered a grassland climax breaker. In summary, limber pine can be a pioneer, intermediate sere, or climax type in relation to ecological succession.

7001 Research Activities

3A3 5A1D 5B 5C 2A 2C 700101 Exclosure study: production, cover and precipitation.

The 1970 research activities of exclosure studies to determine production and cover were conducted as in past years except that during 1970 the field activities did not include the estimate of shrub and mat form production as had been done during 1968.

It was noted in presenting these data that a new approach had been developed in that all of the cover data and weight data were coded on computer data cards and entered into the University of Wyoming computer system. The programs have been written for analysis of these data and the data is to be presented as taken from the computer printouts.

Most spring precipitation occurred during late June, which caused the cool season perennial grasses to exhibit a restricted growth and early maturation due to the dry conditions from April to June. Perennial warm season grasses benefitted from the late spring moisture, however. It was noted at the halogeton pastures that on June 8, 1970, the crested wheatgrass *(Agropyron cristatum)* and Russian wildrye *(Elymus junceus)* were already headed out and the heads were over two feet tall and that in the number two halogeton *(Halogeton glomeratus)* pasture, the saltbush *(Atriplex gardneri)* plants were over 9 inches tall and in some areas in the Big Horn Basin, cheatgrass *(Bromus tectorum)* was particularly dense.

5A1D 2E 700102 Juniper control study.

At the Kane and Big Trails Utah juniper *(Juniperus osteosperma)* areas, a new chemical application using an arsenic base compound was used. This application was called the Big Trails and Kane herbicide trials and a number of permanent plots were located on these.

5B4 7A 5A1D 2E 700103 Juniper GRI seedings.

During 1970, seeding was done by broadcast methods over large areas and in large plots of all three of the Utah juniper *(Juniperus osteosperma)* game range improvement exclosures.

2A 2C 2F2 700104 Soil moisture and temperature.

During 1970, soil moisture and temperature data were collected on four dates during the year. On April 15, 1970, soil moisture was lower than usual, especially in the upper 18 inches of the soil. Concomitant with the reduction of soil moisture, temperatures in the upper 15 inches of soil were 2 to 4 degrees warmer than expected. Frequent showers during spring and summer recharged the soil to about average levels. At Granite Mountain during 1970 precipitation was about three quarters of an inch above the seven year average. The heavy snow storm in early April closed the road and soil moisture and temperature were not obtained until July. Soil moisture tended to be a little lower than usual, especially in the sagebrush spray sites. Soil temperatures were slightly less than the long-term means. At the Cumberland -1 and -3 exclosures annual precipitation was about 1.5 inches less than the long-term mean. Soil moisture in the upper 18 inches of soil was considerably below average, especially during spring. This difference was not as evident at the greater depth in the spray shrub control sites as on the native sites. This is apparently a reflection of greater moisture use by shrubs as compared to grasses, at depths from two to five feet.

2C 8B 700105 Exclosure soil profile studies with SCS.

Soil profile studies at exclosures were conducted with personnel of the SCS and the BLM.

5B4 700106 Inspect SCS seeding along highway west of Moneta.
1E2

During the summer of 1970 the SCS seedings along the highway about eight miles west of Moneta were inspected. A few grasses had done exceptionally well. One of them was mammoth wildrye *(Elymus giganteus)*.

4B 700107 Baggs grazing systems study.
1E4
2F2 During 1970 comprehensive data were obtained at all of the transects in the Baggs
3A2 grazing systems study. During the tour of the pasture system on July 28 with personnel
7A1 of the BLM and the University, several topics were discussed. Important considerations
7A3 were suggestions for increased use on the three grazing systems. There are inadequate
4C livestock on the pastures to put a stress on the vegetation. A moderate 15 percent
6A3 increase has been put into effect on the Poison Buttes as of this spring, Thane Johnson suggested holding use at the present level on the Powder Rim until the end of this cycle and that would require another year before an increase in the use. Increase of use on the Oppenheimers is complicated by lack of water. Any increases must consider the sheep winter use override on the Powder Rim. Questions considered were: what about adding sheep use on the Oppenheimer and Poison Buttes? what about juniper manipulation and seeding studies? and what about considering expansion of the program? Discussions addressed sage hens, deer, antelope, their impacts and whether there should be some new exclosures to further study the impact of wildlife. Any stock increase must be related with the antelope dynamics as was indicated by Mr. Townsend of the Game and Fish department. It was suggested that the BLM have an extra person on the site for management of wildlife and livestock as well because of the override from one period to another, and the lack of concern by the ranchers for movement of the animals from one pasture to another, especially at the times when they are supposed to be out of a pasture. The transect work on the pastures including the browse transects installed by Gibbons were observed in late July also.

8B 700108 Erosion transects.
2C

During 1970 all of the erosion transects on the six general locations on which they had been established were reread on the Dry Creek halogeton *(Halogeton glomeratus)* pastures, on the 15-Mile summer steer grazing pastures and on each of the four Cumberland exclosures in the Kemmerer area.

3A2 700109 1 ft x 1 ft permanent plot transects.
2F2
5C Permanent plot transects were established on the Hyatt ranch northwest of Hyattville and on a number of the exclosures at which herbage production is annually collected including Horse Creek, Smilo, Demer, McGraw Flat, Upper Government Draw, Lower Government Draw, Granite Mountain and others.

3A2 700110 Vegetation pattern on shortgrass-cactus range.
2E
2F1 The vegetation pattern activities during 1970 included work with the pattern meter and also with strip transects of 256 contiguous decimeter square quadrats along a 30 meter line. Some work was also conducted on the 30 m x 30 m plot location in which all of the plains pricklypear *(Opuntia polyacantha)* locations had been identified by decimeter location.

7002 Journal Publications

1D 700201 Fisser, H.G. and D.R. Kirkham. 1970. Reliability of harvester ant mound counts from
10D1 aerial photographs. J. Colo.-Wyo. Acad. Sci. 7(1):42-43.

These data were primarily an analysis and description of those presented by Kirkham in his thesis which is given in 670601.

3A3 2A 11E2 700202 Whysong, G.L. 1970. Predictive value of factors influencing perennial grass production on western Wyoming sagebrush-grass ranges. J. Colo.-Wyo. Acad. Sci. 7:(1) 36-37.

This presentation was essentially a portion of the work conducted by Whysong in his Ph.D. dissertation work which is described in more detail in 730601. Long-term exclosure studies on western Wyoming sagebrush *(Artemisia)* grass rangelands had yielded information concerning the relationships between environmental, edaphic, vegetative factors and annual production of perennial grasses. Time and amount of precipitation were most highly correlated with perennial grass production, while soil depth and depth of soil horizons were lowest. Phenological characteristics and basal cover of perennial grasses were intermediate in this respect. The use of environmental, edaphic and vegetative factors in regression and correlation analysis techniques indicated that a large proportion of the variation in perennial grass production could be explained. Predictions of annual production could be made by July 1 and could prove beneficial to livestock producers and land management agencies in administering their rangeland resources.

3A2 2E 2F1 10D1 1E2 700203 Lester, J.E. and H.G. Fisser. 1970. *Opuntia* clump size variation on the shortgrass plains, Pawnee Site, Nunn, Colorado. J. Colo.-Wyo. Acad. Sci. 7(1):37-38.

Plains pricklypear *(Opuntia polyacantha)* was examined in upland sites, subjected to three grazing treatments (light, medium and heavy) for variations in mean area of clump size. A quantitative determination of non-randomness was established on each site by comparing the observed number of individuals per quadrat to the expected number per quadrat derived from a Poisson series.

Frequency data were obtained from transects of 256 contiguous decimeter square quadrats, in 30 m x 30 m study plots within each grazing treatment. The analysis of variance technique applied to the frequency data gave an estimate of clump sizes for each transect. Mean area of Plains pricklypear clump size in each grazing treatment was determined from the consistent appearance of peaks in a set of replicate transects. Clump sizes for the light, medium and heavy grazed treatments were found at scales of 4, 16 and 128; 4 and 128; and 8, 32 and 128 dm, respectively.

7003 Presentations

1E2 2E 1E1 700301 "Distribution of plant biomass to plant organs." January 3, 1970. I.B.P. Desert Biome Plant Specialists Meeting. Las Vegas, Nevada. by H.G. Fisser.

This presentation was, in effect, a discussion of 80 references evaluated in terms of plant distribution characteristics with special emphasis toward desert ecosystem vegetation. In a discussion concerned with distribution of biomass to plant organs, complexities of the entire biological cycle in the ecosystem must be considered. A broad concept of this cycle may be described as constituting the uptake of elements from the soil and the atmosphere by living plants, biosynthesis involving the formation of new complex compounds, and the return of elements to the soil and atmosphere with the litterfall, including both above and below ground constituents, of part of the organic matter or with the death of the organism.

The biological cycle must be viewed, in addition, as an intricate, polycyclic process comprising cycles differing in duration and effect on the environment. Periods of cycles may be designated as intraseasonal, seasonal, annual, long-term and secular. When addressing energy transfer and distribution of biomass, definitions such as the following have been given for the various components: 1) biomass, 2) dead organic matter, 3) the total reserves of organic matter, 4) the annual increment (net primary production), 5) the litterfall and 6) the true increment.

Biomass accumulation, or the amount of organic matter present on a given site over a long-term period has been shown to range from 20,000 kg/ha on desert and tundra biome systems while forest systems may range as high as over 500,000 kg/ha. A parallelism is exhibited by the desert and tundra biomes with respect to similarity of the proportion of biomass found in the perennial aerial plant parts as related to that found in the below ground root systems. The desert and tundra biomes have a greater proportion of biomass in the roots than do the forests. The root fraction is thus a very important constituent in nutrient cycling and often constitutes more than 80 percent of the total biomass. Some authors suggest that ratios of various plant parts within a community are relatively constant. On that basis, suggestions are made to the effect that initial determination of such ratios may then be extrapolated to additional data comprised of the above ground biomass.

Many publications are available concerning biomass distribution to various aboveground portions of the plant. The proportions are dependent upon growth form and functionally related environmental criteria and are arrived at by the plant as a result of adaptation. As aerial growth form is a characteristic of adaptation to environmental factors, so is root form, depending upon various factors of soil environment. Several categories of root form and structure and morphology have been described by a great number of authors. The distribution of biomass to the various portions of the plant are obviously caused by environmental forcing factors. Morphological and physiological adaptations are exhibited by a variety of different species and can be noted with regard to different primary stress characteristics of specific environmental situations. It is suggested that there were linear relationships of age and dry weight of big sagebrush *(Artemisia tridentata)*. This indicates that more work should be done to determine whether this is also true with regard to the root system of this shrub.

A list of ecological phenomenon which have been noted to influence productivity and biomass distribution are: 1) habitat factors a) radiation, b) humidity, c) evaporation, d) dew, e) rainfall and f) edaphic factors and 2) biological and ecological phenomena a) succession and development, b) community and climax, c) physiological and autecological characteristics and d) the role of fire. Definitive ecosystem analysis seems to require information of both the above ground and below ground biomass components of the ecosystem. This work then should evaluate the accumulation dynamics of the biotic components as influenced by environmental factors, stature and structure and from these data then an evaluation of the biomass characteristics of various successional stages can be determined.

1D 700302 "Reliability of harvester ant mound counts from aerial photographs." May 1, 1970. Colo.-Wyo. Acad. Sci. Greeley, Colorado. by H.G. Fisser.

See 700201 and 670601.

3A2 700303 "*Opuntia* clump size, variation on the shortgrass plains - Pawnee Site, Nunn, Colorado." May 1, 1970. Colo.-Wyo. Acad. Sci., Greeley, Colorado. by H.G. Fisser.
2E
2F1

See 700203.

2F1 700304 "The I.B.P. and its implications for natural resource management." June 24, 1970. Wyoming Game and Fish Commission. Biologists Meeting. Lander, Wyoming. by H.G. Fisser.
3A2

This presentation was given in response to questions from Wyoming Game and Fish personnel regarding the I.B.P. program. The topics and general discussions were given as follows:

1) The International I.B.P. Program.

2) The U.S. National Program, including the National Committee and the Grassland Biome Program, with which we are most closely associated. The grassland biome program was shown to be only a part of the analysis of ecosystems studied which included grasslands biomes and other biomes which were tundra, deciduous forest, coniferous forest, tropical and desert. Other activities of the U.S. National Committee on the I.B.P. program were human physiology, human ecology in the Artic and human genetics.

3) The basic intent of the I.B.P. program was aimed at the general description and characterization of physical and biological properties of biomes and application of the properties to mathematical formulas. From these, generalization of data can be applied to the management of natural resources.

4) The initial activities of I.B.P. Grassland Biome Program were developed at the Pawnee site south of Cheyenne, and at comparable grassland sites of the grassland biome in the western United States.

5) The various research projects and research concepts were integrated to emphasize four major categories of ecosystem function: 1) abiotic, 2) producers, 3) consumers and 4) decomposers.

6) The desert biome study portion of the I.B.P. program was initiated primarily at Utah State University. The Range Management Department at the University of Wyoming has submitted for consideration a proposal to study the Alkali Flat Desert Biome Site, which is located a few miles east of Rawlins, Wyoming, and to add into that the antelope dynamics as a major functional program.

2F1 700305 "Significance of plant patterns in ecosystem functioning." November 14, 1970. I.B.P. Grassland Biome Annual Review Meeting. Fort Collins, Colorado. by H.G. Fisser.
2E
3A2

See 710503.

3A3 700306 "Predictive value of factors influencing perennial grass production on western Wyoming sagebrush-grass ranges." May 1, 1970. Colo.-Wyo. Acad. Sci. Greeley, Colorado. by G.L. Whysong.
1E2
11E3

See 700202.

7004 Published Proceedings/Abstracts

1E2 700401 Fisser, H.G. 1970. Distribution of plant biomass to plant organs. p. 34-42 and 110-115. In Proc. Intern. Biol. Prog. Desert Biome Plant Specialists Meeting. Las Vegas, Nevada. January 3-4, 1970.
2E
1E1

See 700301.

7005 Wyoming Agricultural Experiment Station Publications

700501 Fisser, H.G. and G.L. Whysong. 1970. Aridland Ecology Research. 1969 results. Univ. of Wyo. Coop. Res. Rep. to the BLM. Sections I - IV. Wyo. Agr. Exp. Sta. Sci. Rep. 240. 128 p.

2A 70050101 Section I - Soil moisture and temperature.
2C
2F2

This is the report of the 1969 data collection of the soil moisture and temperature study, given in more detail in 69050101.

2A 70050102 Section II - Microclimate studies.
2C
2E These are the 1969 data report given in more detail in 69050102.
3A2

2A 70050103 Section III - Precipitation pattern study.
3A

This record of the 1969 precipitation data is given in more detail in 69050103.

3A3 70050104 Section IV - Exclosure study: production, cover and precipitation phase.
5A1D
5B This is given in greater detail in 69050104.
5C
2C

4B 700502 Gibbens, R.P. and H.G. Fisser. 1970. Influence of grazing management systems on vegetation and wildlife habitat. Annual report of progress, July 1969 - July 1970. Univ. of Wyo. Coop. Res. Rep. to the BLM. Wyo. Agr. Exp. Sta. Sci. Rep. 241. 32 p.
4C
1E4
2F2
3A2
7A1 This annual report of 1969 and 1970 data included basic transect data of permanent
7A3 plot transects and also of the browse transects. A comprehensive plant list was included.
6A3

2F2 700503 Whysong, G.L. and H.G. Fisser. 1970. Frequency, density and basal area measures on sagebrush *(Artemisia)* grass ranges. 1969 results. Univ. of Wyo. Coop. Res. Rep. to the BLM. Wyo. Agr. Exp. Sta. Sci. Rep. 265. 35 p.
3A2
10D1

This report was a special analysis addressed to the phyto-sociological relationship of frequency, density and basal area of species to one another. The intent of this report was to analyze the data and to determine Importance Values. This work was conducted at 15 exclosures in the sagebrush-grass sites in the Big Horn Basin, the Wind River Basin and in the Kemmerer vicinity. Transects were established to obtain frequency, density and basal area data. With these phytosociological relationships mathematical models for prediction of perennial grass production may be constructed and may well prove useful in terms of analysis of vegetation composition changes through time.

Sampling consisted of measuring frequency within 1 ft x 1 ft quadrats located along the transect line of 20 quadrats. Rooted presence was required to constitute frequency. Basal area, which in this case represented the portion of the quadrat occupied by a plant at the intersection of the stem and root with the soil surface, was also obtained from the 1 ft x 1 ft quadrats. These values were determined for each species and each plot.

Density refered to numbers of plants per unit area. Density of annuals was easily determined by counting individuals, although in the case of cheatgrass some care was needed since individual culms would often arise by tillering from a single seed. These culms were counted as a single plant. A similar condition existed with perennial bunchgrasses. Clumps of culms, which appeared to live as a single plant unit, were counted as individuals. Because of the difficulty of determining physiological individuality of rhizomatous species, each culm or small group of culms was counted as an individual plant although they were often connected by rhizomes. Clumps of plains pricklypear *(Opuntia polyacantha)* pads, which appeared to arise from a common root source, were counted as single plant units.

The size of quadrat, within which density counts were obtained, were dependent upon plant growth form and abundance. Density of herbaceous species was obtained from 6 in x 6 in quadrats located within the 1 ft x 1 ft quadrats such that the upper left hand quarter of the quadrats coincided when viewed from the right hand side and perpendicular to the transect line. Shrub density was obtained from two sizes of plots:

2 ft x 2 ft and 2 ft x 4 ft. These were located such that the upper right hand corner, as described above, coincided with the same corner of the 1 ft x 1 ft quadrats. The 2 ft x 4 ft frame was located perpendicularly to the transect line. Subjective evaluation was utilized to determine which size was most appropriate on each transect. The smaller was used when shrub density was high and larger on sites with low density.

The proportion an individual species contributed to all species in each of the three measures was termed relative frequency, basal area and density. The sum of all relative values described earlier gives the Importance Value which designates the importance of that plant species occurrance within the community. The larger the Importance Value the more important is that plant species.

The range of Importance Values extended from 176 for western wheatgrass *(Agropyron cristatum)* at the Horse Creek exclosure to 0.01 for green rabbitbrush *(Chrysothamnus viscidiflorus)* on the Granite Mountain exclosure. These data can be useful, especially with regard to trends in vegetation changes following different kinds of treatment. The most striking observations as related to vegetation response to the influences of grazing use and protection from grazing were noted at the Buffalo Creek exclosure where casual observation indicated a plant community charactized by low shrub density and an abundance of desirable perennial grasses. However, data derived from sampling the area indicated high big sagebrush *(Artemisia tridentata)*density both inside and outside the exclosure due to numerous sagebrush seedlings which had become established. As many as 20-30 individual plants were not uncommon in a 2 ft x 4 ft plot. Although the density of big sagebrush was somewhat higher on the outside portion of the exclosure, the Importance Value indicated no real difference between the inside and outside. It appeared obvious that sagebrush was tending to increase on the area. However, it will involve continuing the sampling procedure described herein or similar methods to determine the influence and extent of increased sagebrush density upon the plant community.

Of separate interest was the fact that bluebunch wheatgrass *(Agropyron spicatum)* was the most important plant species on the inside portion of exclosures, while sandberg bluegrass *(Poa secunda)* and western wheatgrass *(Agropyron smithii)* were the most important on the outside. These results indicated vegetation response was not only due to the influence of protection from grazing on the inside, but also the effects of grazing on the outside.

3A2 700504 Fisser, H.G. 1970. Exclosure studies with transects of permanent plots. 1969 results.
2F2 Univ. of Wyo. Coop. Res. Rep. to the BLM. Wyo. Agr. Exp. Sta. Sci. Rep. 275.
5C 84 p.

This was a compilation of the 1969 sampling of the 1 ft x 1 ft permanent plot transects. This report gave the compilation of data by exclosure and transect.

4B 700505 Gibbens, R.P., H.G. Fisser and M. May. 1970. Influence of grazing management systems
4C on vegetation and wildlife habitat. Semiann. Rep. to the BLM. Wyo. Agr. Exp. Sta.
1E4 Mimeo. 2 p.
2F2
3A2
7A1
7A3
6A3

This was a short report describing the activities of the Baggs grazing system study program through 1969. During the 1969 season vegetative cover determinations were made on 21 permanent transects paired with BLM trend plots in the 4 pasture rest-rotation unit, the 2 pasture rotation unit and the season long unit. Based on cover determinations made for the past three seasons, western wheatgrass *(Agropyron smithii)* showed an increase in cover at 15 of the 18 sampling sites where it was present. No correlation of increased cover with grazing treatment was evident. Mean precipitation for the period October 1968 through October 1969 was determined from the 17 stations within the study area and was 10.27 inches. This was the second consecutive year of relatively good precipitation and accounts, in part, for the general increase of western wheatgrass cover.

Utilization checks made in spring 1969 showed that sheep utilization within the four pastures was spotty. Average utilization within pastures ranged from 16-17 percent for big sagebrush to 44-67 percent of Nuttall saltbush *(Atriplex nuttallii)*. Utilization of other shrub species was no greater than 12 percent in any pasture. Sheep utilization of dry grass forage was light to moderate and use of new grass growth in spring did not exceed 21 percent for any species. In pasture D, which received the most uniform spring grazing by sheep, average utilization of new grass growth was as follows: western wheatgrass 11 percent, Indian ricegrass *(Oryzopsis hymenoides)* 11 percent, Sandberg bluegrass *(Poa secunda)*11 percent, bottlebrush squirreltail *(Sitanion hystrix)* 14 percent, prairie junegrass *(Koeleria cristata)* 19 percent. The highest value, 21 percent, was recorded for muttongrass *(Poa fendleriana)* in pasture A.

Determination of cattle use of western wheatgrass following the summer 1969 grazing season showed average utilization values (percent of stems grazed) as follows: pasture A, 71 percent; pasture B, 31 percent; Oppenheimer west pasture, 43 percent; Oppenheimer east pasture, 47 percent; Poison Buttes pasture, 65 percent.

As in previous years, deer utilization of true mountain mahogany *(Cercocarpus montanus)* was heavy in areas of winter concentration. Use of winter twig growth of mountain mahogany ranged up to 90 percent. Observations in pastures not grazed by cattle during the summer indicated that the small resident deer population grazed the sparse stand of Indian ricegrass within the juniper type to a considerable extent.

The maximum number of antelope observed in the study area in a given day was 80 animals. This was in comparison to a maximum of 61 counted in the previous summer. Summer-long observation led to the conclusion that there had been an increase in the antelope population.

3A3 700506 Fisser, H.G. 1970. Aridland ecology, semiannual progress report to the Bureau of Land Management, March 1970. Wyo. Agr. Exp. Sta. Mimeo. 1 p.
3A2
2F2

Vegetation production was sampled in 38 study areas. Precipitation gauges were read on four dates except for a few read later than April 15 due to adverse weather and road conditions.

At 26 study areas vegetation cover and composition were determined from 20 permanently located 1 ft x 1 ft plots. In addition, permanent transects were established at 15 sagebrush *(Artemisia)* grass study areas located in the Kemmerer, Bighorn and Wind River Basin vicinities. Vegetation sampling consisted of measuring frequency, density and basal area of all species present including shrubs. The study sites at which transects were established were; Bud Kimball, Buffalo Creek, Cochran, Cumberland -1, Cumberland -3, Demer, Farson, Granite Mountain, Horse Creek, Lander Ant, Lower Government Draw, McGraw Flat, North Butte, Sweetwater and Upper Government Draw.

Soil moisture and temperature data were collected at the Smilo, Granite Mountain, Cumberland -1 and Cumberland -3 exclosures on four dates during 1969. Data were obtained from five shrub communities for study of vegetation type relationships to soils, soil moisture, soil temperature and precipitation on the dates during 1969.

7006 Graduate Advisee Completions

2A 700601 Steger, R.E. 1970. Soil moisture and temperature relationships of six salt desert shrub communities in north central Wyoming. Ph.D. Dissertation. Univ. of Wyo. 142 p.
2E
2C
3A2

This soil moisture and temperature study was conducted on six salt desert shrub plant communities in north central Wyoming from 1965-67. The objectives were to characterize the soil moisture and soil temperature relationships and to study certain edaphic and floristic characteristics of these communities.

The depth of moisture penetration and the amount of available soil moisture, by plant community, were of great importance in this study. The ranking by depth of soil moisture penetration was Nuttall saltbush *(Atriplex nuttallii)* - 6 inches, birdfoot sagewort *(Artemisia pedatifida)* - 12 inches, winterfat *(Ceratoides lanata)* - 18 inches, bud sagewort *(A. spinescens)* - 18 inches in the dry year and 24 inches in the wet year, big sagebrush *(Artemisia tridentata)* - 24 inches and spiny hopsage *(Grayia spinosa)* - 36 inches in the dry year and 48 inches in the wet year.

The amount of available water followed a similar ranking except that big sagebrush was the second driest study area in 1966. Soil moisture relationships were found to be affected by soil characteristics of texture and salt content, which in turn influenced moisture-holding capacity, infiltration and percolation. The soils occupied by Nuttall saltbush, birdfoot sagewort and the 6-12 inch sample of the bud sagewort study areas, were characterized by fine, textured soils, high salt contents and high moisture tension relationships. Soils in the big sagebrush and spiny hopsage study areas were coarse and held water at low tensions. The winterfat study had soils that were intermediate.

Plant characteristics of active plant growth, stage of growth, root depth and type, appeared to be important soil moisture regulating factors. Soil moisture recharge and discharge appeared to be regulated by plant activity. In general, soil moisture discharge occurred during the period of active plant growth from early April to August. The rate of discharge increased during the period of plant flowering and fruiting in May and June. Bud sagewort started growth earlier than other species and stopped growth during dry periods. Spiny hopsage started growth at the same time of most species studied but ended growth during dry periods. Soil moisture recharge was most evident when plants were not actively growing.

Spiny hopsage root depth was limited to bedrock while roots of other species appeared to stop due to lack of moisture penetration. Moisture penetration was limited by a sandstone layer on the big sagebrush study area but not on the spiny hopsage study area. The spiny hopsage roots were able to use the moisture contained in this sandstone layer.

The Nuttall saltbush, birdfoot sagewort, winterfat and bud sagewort strong lateral root habits made best use of shallow moisture. The more specialized root habits of big sagebrush and spiny hopsage were adapted to deeper moisture.

Plant cover and number of species increased directly with an increase in soil moisture. Within species competition was greater than that among species, as evidenced by a greater distance between individuals. Grass plants and plains pricklypear *(Opuntia polyacantha)* were found to occupy positions closest to the shrubs and half shrubs.

Soil temperature was important under three major situations. First, plant growth of all species, except bud sagewort, was initiated at 10°C at the 8-in depth. Second, soil moisture recharge was not possible on frozen soils of fine texture found on the Nuttall saltbush and birdfoot sagewort study areas. Third, 0.20 in of soil moisture per 12 inches of soil was lost from the Nuttall saltbush and birdfoot sagewort study area soils during the period of freezing temperatures.

7A1 700602 4C Gunderson, J.J. 1970. A discussion of some important aspects of rangeland carrying capacity emphasizing deer of the genus *Odocoileus*. A discussion of some nutritional aspects of several browse species with special emphasis on the value of these plants as food for deer of the genus *Odocoileus*. M.S. Professional Papers. Univ. of Wyo. 53 & 55 p., respectively.

The first paper produced by Gunderson in the Plan B thesis addressed the topic of nutritional aspect of browse species for deer. His conclusions derived from the literature view were: 1) The number of plant species utilized by deer depends upon the geographic area, 2) The browse species utilized will depend upon the geographical area of the deer species considered, 3) The number of browse species and the type of browse species utilized depends upon seasonal changes and use intensities of the different seasons, 4) Browse species vary in nutritional aspects, 5) The browse species utilized not only depend upon their availability but upon their nutritional qualities, 6) Nutritional aspects within a species of browse will vary according to the season and 7) Deer seemingly select and utilize those browse species that are the highest in crude protein and change species as the nutritional aspects of the browse changes.

The second paper addressed the concepts and problems associated with determining of rangeland carrying capacity in the concept of deer numbers and use. The exact determination of deer carrying capacity depends on the specific area within a specific range. Therefore, no generalizations can really be made concerning deer populations over extremely large geographical areas.

The most important limiting factor affecting the number of deer within any area is the quantity and quality of browse. It is therefore important that accurate and meaningful data are collected on the forage. Among the different vegetation types browse is one of the most difficult to measure. More accurate techniques in the area of browse sampling should be developed to get better indications of production and utilization, which in turn will be beneficial and essential in determining correct carrying capacity.

Another factor to consider is that with regard to the successional stage of vegetation. It is important that management practices coincide with management objectives in plants. A grazing range that may be in excellent condition for one class of herbivore, may conversely be nonsuitable for another class. In this respect just as acceleration of succession, a disclimax, or even a reversal in succession could cause carrying capacity to change.

The preservation of existing deer range must also be considered in terms of long range management objectives. The range and its flora must be receiving protection not only from man but from the animals, in this instance, deer. Therefore it is important that numbers of deer are accurately determined, manipulated and controlled. It is important that sampling and statistical methods be appropriate in the censusing of deer numbers.

7101 Research Activities

3A3 710101 Determination of shrub production interrelated with herbaceous production and
1E2 precipitation on some 50 sites in western Wyoming encompassing wide geographical
5B and environmental variation (Additional to Aridland Ecology program). (Initiated in 1971).

This topic and heading number was written in detail because of the initiation of shrub production as a regular part of the herbage production and cover determination. This activity was detailed in 1968 when it was used in an experimental mode and was described at that time under topic code number 680115. The herbage production activities at the exclosures during 1971 were essentially the same as those during previous years. Although this heading implies that precipitation pattern is included, a new heading will be introduced below (710111) as the heading for precipitation pattern and discussion related to that.

5B4 710102 Juniper GRI seedings.
7A
5A1D Activities at the game range improvement exclosures involved seedings on
2E large plots 16 ft x 68 ft in size both during late March and November. During late November, there were seedings carried on in the Big Trails burn area, aerial seeding from a fixed-wing airplane. This work was done with Burrough. In late March the Kane Herbicide plots from the road appeared to be very little, if any different than the non-sprayed. In essence, the arsenical obviously had very little impact on anything, probably because of the arid conditions and was dismissed as an unuseable form of plant control in this area. On the Big Trails herbicide plot, there was a slight degree of yellowing and browning of juniper needles, but minimal and of no significance as a treatment effect.

2A 710103 Soil moisture and temperature.
2C
2F2 These activities with the neutron probe and measurement of soil temperatures were discontinued during 1971.

1E2 710104 Shortgrass prairie blue grama culm density study.
10D1
2F1 This work was conducted on the I.B.P. Pawnee site south of Cheyenne with the
2E mechanism constructed by the engineering firm in Denver with which to be able to identify x, y and z axes of individual culms of blue grama *(Bouteloua gracilis)* or any other individual species within a 1/2 m by 1 m quadrat. Lester began the work and Stroud took over.

1E2 710105 Shortgrass prairie seed production study.
10D1
2F1 The seed production study was initiated with Shrader. The work utilized a strip
2E transect. Width of the clipped plot was identical to width of the electric clipper. That
3A3 was utilized to measure the entirety of all herbage production along the transect length. This work was reported in a subsequent meeting at the SRM meetings in Mexico City in February 1975. The code number for this presentation is 750303.

3A2 710106 1 ft x 1 ft permanent plot transects.
2F2
5C During 1971 permanent plot 1 ft x 1 ft transects were established on eight new exclosures in the Big Horn Basin and all transects at other exclosures were reread.

3A1 710107 Wagon Wheel range survey.
3A2
5C This work was carried on independent of University participation and involved
10D1 permanent plot location of a downwind and upwind site, each of these consisting of 90
8D permanently located plots on which both 1 ft x 1 ft and 4 ft x 4 ft permanent locations were established. These may be of future interest for University research in subsequent years.

2C 710108 Exclosure soil profile studies with SCS.
8D

Exclosure soil profiles were conducted during 1971 and a considerable number of exclosures were visited with the SCS and BLM people. Clarence Fowkes did the primary soil profile identification.

3A1 710109 Sheep Mountain deer habitat study.
7A1

Spaker conducted the Sheep Mountain deer habitat study. His thesis was a result of this. It is found in 740601.

3A2 710110 Laramie Peak Game Range inventory.
7A1
7A2 The inventory of rangeland vegetation on the Laramie Peak Game Range area was conducted by Butler during the summer of 1971. This resulted in a thesis, reported in 720605.

2A 710111 Precipitation Pattern Study.
3A

The precipitation pattern study continued with precipitation recorded seasonally at 110 UW/BLM gauges. Seasonal data for some 38 US Weather Service Stations were recorded also.

7102 Journal Publications

5A1D710201 Burrough, L.N. and H.G. Fisser. 1971. A greenhouse analysis of variance of picloram
7A retention in soil and duff from juniper sites. J. Colo.-Wyo. Acad. Sci. Vol. 7(2):103.
2E
1E2 Picloram treatments were applied July 1968 in three gameproof exclosures in the juniper *(Juniperus)* zone on the western foothills of the Big Horn Mountains in north central Wyoming. Precipitation in the zone ranges from about 13-17 in annually. Half of each of the 400 ft x 400 ft exclosures was sprayed with picloram at the rate of 2 lb/acre applied with 10 gallons of water on two consecutive opposite passes by a helicopter. An almost complete juniper kill resulted from the application by the following spring.

On September 29, 1969 and on December 5, 1970, samples of soil and juniper duff were taken randomly from the sprayed plots of each exclosure. In the greenhouse these samples were mixed in pots with uncontaminated potting soil in concentrations of 12.5 percent, 25 percent, 50 percent, 75 percent and 100 percent. Beans *(Phaseolus vulgaris)* were planted in the pots and grown for 3 weeks under optimum moisture, temperature and light conditions.

Picloram retention in soils and duff was apparent from both sample sets taken 13 and 26 months after application. Responses were detected among sites reflecting differences in degree of soil development and acidity. Picloram retention decreased with soil development and with an increased mesic site character. No significant differences in bean growth were noted from plants growing in potting soil or the uncontaminated site soil and duff samples.

7103 Presentations

5A1D710301 "A greenhouse study of variance of picloram retention in soil and duff from juniper
7A sites". April 30, 1971. Colo.-Wyo. Acad. Sci. Pueblo, Colorado. by H.G. Fisser.
2E
1E2 See 710201.

1E2 710302 "Moisture stress in woody plants as influenced by environmental factors." February 17, 1971. 24th Ann. Meeting Soc. Range Manage. Reno, Nevada. by C.L. Wambolt.
2E
2A
2C
2F2

Several species of trees and shrubs were used to investigate influence of atmospheric, edaphic, topographic and vegetation variables of the environment upon plant moisture stress. Diurnal variation in moisture stress was highly correlated to changes in temperature, relative humidity and a vapor pressure deficit. The correlation coefficients for this relationship approached 1.0 for most analyses. The diurnal stress fluctuated up to values of four times the minimum stress values on days of observation. As the summer progressed, both maxima and minima stresses increased for the 24 hour periods. Stress values for the same plants sampled at intervals throughout the summer increased significantly as the season progressed and soil moisture decreased.

Correlation coefficients were near 1.0 for the relationship of plant density to moisture stress when other environmental influences were held constant. This was interpreted as quantitative proof of the intraspecific competition for available moisture by plants.

Moisture stress values of 21 stands of coniferous trees, three each of seven species, were quite highly correlated (R^2=0.84) to elevation, percent sand, temperature, relative humidity and vapor pressure deficit. These factors were well correlated even though the stands were found in widely divergent habitats. High cation content in the xylem sap was associated with high moisture stress for the stands from which the sap was collected.

7104 Published Proceedings/Abstracts

1E 710401 Wambolt, C.L. 1971. Moisture stress in woody plants as influenced by environmental factors. p. 25-26. In Proc. 24th Ann. Meeting. Soc. Range Manage. Reno, Nevada. February 14-18, 1971.
2E
2A
2C
2F2 See 710302.

7105 Wyoming Agricultural Experiment Station Publications

4B 710501 Fisser, H.G. and R.P. Gibbens. 1971. Influence of grazing management systems on vegetation and wildlife habitat. Annual report of progress, July, 1970 - July, 1971. Univ. of Wyo. Coop. Res. Rep. to the BLM. Wyo. Agr. Exp. Sta. Sci. Rep. 336. 92 p.
4C
1E4
2F2
3A2
7A1
7A3
6A3

This report presented data from 1970 to early 1971 for the Baggs grazing system study. More information is available under the code number 700107.

710502 Fisser, H.G. and R.E. Ries. 1971. Aridland ecology research. 1970 results. Univ. of Wyo. Coop. Res. Rep. to the BLM. Sections I - IV. Wyo. Agr. Exp. Sta. Sci. Rep. 350. 133 p.

2A 71050201 Section I - Soil moisture and temperature studies.
2C
2F2

This report of the 1970 soil moisture and temperature studies is described in more detail under code number 700104.

2A 71050202 Section II - Soil moisture and temperature studies on microclimate study sites in the Big Horn Basin.
2C
2E
3A2

The 1970 activities of soil moisture and temperature on the microclimate study sites is given greater description in item 700104. Included in this annual report are portions of the abstract of the Steger thesis and the summary of information taken from his thesis.

2A 71050203 Section III - Precipitation pattern study.
3A

The 1970 activities of precipitation pattern study are given some additional dicussion under item number 700101. The annual report includes seasonal data from some 110 UW/BLM gauges and from some 38 US Weather Service Stations which data is compiled into a seasonal form similar to the dates of the recordings for the UW/BLM gauges.

3A3 71050204 Section IV - Exclosure study: production, cover and precipitation phase.
5A1D
5B
5C
2C

Additional description of the 1970 activities of exclosure studies are given under 700101. This annual report, as in previous years, gave the herbage production data and shrub production values as well for each of the some 40 exclosures at which herbage production is being obtained.

1E2 710503 Fisser, H.G. 1971. Significance of plant pattern in ecosystem functioning. Wyo. Agr. Exp. Sta. Mimeo. 2 p.
2E
2F1
3A2

Patterns (or spatial arrangements) are legitimate objects of ecological investigation, especially those related to ecosystem function and process studies, if utilized to define functional relationships of environmental factors and biological tolerance limits. Procedures and publications related to simple and often largely mechanical characterization of pattern studies are numerous. Expression of spatial arrangement data by functional utility as an integration of both physical and biological phenomenon through time is largely non-existent.

This is due to several factors, such as those associated with the common research limitations of time, funding and limited research capabilities. Of greatest significance, development of a simple conceptual hypothesis relating position data to the function processes of an ecosystem is difficult. This difficulty was recognized by discussions among a variety of Pawnee site researchers, administrators, modelers and programmers. Although a general concensus agreeing to the importance of pattern study exists, little if any headway has been made toward conceptualizing the role of plant patterns as an integral function of the biome model development.

By defining the dimensional form of vegetation distribution in various treatment areas, we should be able to identify the functional integration of grazing stress expressed by specific vegetation response of the multitude of interrelated biological and physical environmental modifications to such processes and functions as 1) herbage productivity and production, 2) plant competition, 3) transpiration, translocation, photosynthesis, carbon dixoide synthesis and oxygen production, 4) plant structure, 5) native and domestic animal distribution and food intake characteristics, 6) microsoil variation and erosion, 7) soil surface characteristics including temperatures and net radiation and 8) soil moisture, precipitation, infiltration, and runoff. We are now able to quantify many of the above criteria, but specification of the role each play in the dynamic processes and functions of the biological system is not well understood.

3A2 710504 Fisser, H.G. 1971. Evaluation of Pawnee Site frequency data. Wyo. Agr. Exp. Sta. Mimeo. 6 p.
2E
2F1
1E2
10D1

Blue grama *(Bouteloua gracilis)* was the dominant species on all five treatment areas. It was least abundant on the old exclosures and most abundant on the heavy use pastures. Although it decreased with utilization rate, its abundance on the 10-year-old exclosure was almost as great as on the heavy use pasture, indicating an interacting response phenomenon resulting from grazing pressure release. Most important, it showed a lack of interspecific competition from other plants not yet as abundant as on the old exclosures. Threadleaf sedge *(Carex filifolia)* abundance differences among treatments indicated a similar response, but also decreased in the 10-year exclosure. This apparently indicated an inability to compete successfully with blue grama in a non-use situation imposed in a low successional state. Conversely, the greater abundance of threadleaf sedge with grazing pressure indicates 1) that it is not utilized heavily, or if it is, that it is able to escape the detrimental effect of herbivore utilization and 2) that it can successfully compete with blue grama in situations of greater blue grama abundance. Where the sedge is utilized significantly, it apparently becomes less competitive.

Red threeawn *(Aristide longiseta)* and sand dropseed *(Sporobolus cryptandrus)* attained greatest abundance in the old exclosure, and tended to decrease with a shorter period of protection and increased utilization. An increase of the threeawn in the medium use pasture as compared to the light use however, indicates an interacting causality due to differential grazing use, at variance with the administered levels of intensity, and differential response to competitive capability changes in blue grama due to the influence of utilization.

The annual sixweeks grass *(Vulpia octoflora)* was least abundant in the old exclosure and exhibited an increase under grazing. We may assume an interacting causality function relating to competition for available moisture, time of seedling establishment, or period of maximum growth rate, vigor reduction due to previous grazing as well as the occurrence of an occupied perennial grass niche to which the annual grass can more readily invade than a perennial.

Annual forbs were least abundant in the 30-year-old exclosure and increased by a factor of 2 1/2 to 3 times under the grazing treatments, but were most abundant on the 10-year-old exclosures. This response was not manifested in terms of drastic changes of total herbage productivity, abundance, or vigor. Although abundance of annuals is largely dependent from year to year upon seasonal weather conditions, the commonly accepted concept of minimal disturbance resulting from light grazing utilization must be reconsidered. Whether the plants replaced by annuals contributed little to the total biomass productivity is insignificant. Changes in the present context must also be considered in the light of: 1) ecosystem species diversity, 2) capability of an ecosystem to adapt to environmental modification, 3) factors influencing competition capability among and within species, 4) differential capability of species to re-establish and/or invade unoccupied niches, 5) influential seasonal micro-environmental conditions which may modify response capability and 6) modifications caused by physical disturbance to vegetation, soils and moisture relationships resulting from the mere presence and movement of large herbivores.

The perennial forb scarlet globemallow *(Sphaeralcea coccinea)* was most abundant in the 30-year exclosure. It retained its abundance on the light use pasture indicating that, although probably stressed by grazing at most levels of utilization intensity, it was capable of quickly establishing new plants to replace those removed and severely damaged by the assumed minimal disturbance and stress associated with the least intensive level of grazing. In the medium use pasture, the globemallow frequency was reduced to 1/4 that in the light use, indicating insufficient competitive capability to overcome the influence of grazing, although herbivore stress at this level of use would obviously result in suitable niche openings. A similar situation exists in the heavy use pastures. Abundance of globemallow in the 10-year exclosure was intermediate to the values occurring in the 30-year and the light use pasture as compared to the medium and heavy use units. This further indicated the extensive time requirements necessary for developing of successional improvement in climax status.

Another perennial forb, scarlet gaura *(Gaura coccinea)*, sparse but occurring in all study units, was most abundant in the light use pasture by a factor of 4 to 5 as compared to the other pastures and exclosures. This would appear to indicate a competitive capability to withstand stresses caused by light rates of livestock utilization. It may also indicate a lack of stress under light use, in which case we could assume non-use of this plant. Its sparseness in the exclosures certainly seems to indicate a low level of competitive capability with respect to associated species.

3A3 710505 Fisser, H.G. and R.E. Ries. 1971. Semi-annual progress report to the Bureau of Land
2C Management, January 1971. Wyo. Agr. Exp. Sta. Mimeo. 1 p.
2F2
2E During 1970 vegetation production was sampled on 38 sites and precipitation was
3A2 read four times at each location. Soil moisture and temperature were measured at Smilo,
5A1D Granite Mountain, Cumberland -1 and Cumberland -3 on four dates during 1969 and
5B also from the six shrub communities of the microclimate research program in the Big
5C Horn Basin.

7106 Graduate Advisee Completions

1E2 710601 Wambolt, Carl L. 1971. Moisture stress in woody plants of Wyoming as influenced by
2E environmental factors. Ph.D. Dissertation. Univ. of Wyo. 158 p.
2C
2F An investigation of moisture stress in several woody plants was conducted to
2A2 evaluate the influence of certain environmental factors on stress values. These
10D1 environmental factors included atmospheric, edaphic, topographic and vegetation
variables. Data were obtained in the following plant communities: Utah juniper *(Juniperus osteosperma)*, Rocky Mountain juniper *(J. scopulorum)*, Nuttall saltbush *(Atriplex nuttallii)*, big sagebrush *(Artemisia tridentata)*, ponderosa pine *(Pinus ponderosa)*, Douglas fir *(Pseudotsuga menziesii)*, lodgepole pine *(Pinus contorta)*, limber pine *(Pinus flexilis)* and Engelmans spruce *(Picea engelmannii)*. A number of preliminary trials were conducted to develop techniques necessary for standardized field procedures.

It was found that sample twigs should have a cluster of twigs at least 2 1/2 inches in diameter and several years of secondary growth for minimum variance of stress values. Stress determinations had to be taken within five minutes after the sample was cut from the plant to avoid abnormally high values. Increased effective solar radiation resulted in higher stresses on the sunny side of plants than on the shaded side. Except for relation to sunlight, position of the sample twig on the plant had no influence on stress values as produced by the pressure chamber obtaining field measurements of stress.

Pressure chamber, allowed for accurate comparisons of moisture stress among plants growing in natural use environments and made this study possible. A more thorough knowledge of plant/water physiological process relations would lead to a better understanding of plant system functions. Studies such as this may eventually result in a number of applied resource management techniques concerned with the production and control of plants.

The objectives of this study were: 1) to develop techniques needed to conduct standardized field procedures for the study of plant moisture stress, 2) to study the influence of certain atmospheric, edaphic, topographic and vegetation variables of environment on moisture stress and 3) to demonstrate potential uses of moisture stress data for resource management.

The problems involved in this study were examined in a survey of the literature. A description of the study area search described the nature of the country from which these data were collected while the materials and methods employed were cited to clarify field procedure. The results of this study were presented and discussed with their highlights emphasized in the summary. Additional details are given under code numbers 690403 and 690305. Utah juniper trees sprayed with the herbicide Tordon were found to have a much lower moisture stress than unsprayed trees nearby. It is likely that reduction in living tissue was responsible for the lower stress in sprayed junipers. Measurements of the effects of two resource management practices were taken in order to illustrate the potential of moisture stress knowledge for management purposes. Moisture stress values of young lodgepole pine growing in an area previously clear-cut were compared to stress of young pine in an adjacent old growth stand of lodgepole. The stress values in the clear-cut area were lower despite a higher tree density and full exposure to sunlight and wind in the clear-cut.

2A 710602 Jacoby, P.W. 1971. Interrelationships of vegetation and environmental factors on a mountain
2F2 watershed in southeastern Wyoming. Ph.D. Dissertation. Univ. of Wyo. 127 p.
3A1

3A2 An ecological study of a forested mountain watershed in southeastern Wyoming
10D2 investigated certain interrelationships between environment and vegetation. Data were collected along an altitudinal gradient ranging from 8,400 to 10,600 feet in elevation. Soils were found to be coarser in the upper part of the area, indicating an influence from erosion and differences in parent material. Soils also showed a lower pH in the upper watershed, suggesting the effects of leaching. While organic matter content was not found to correlate closely with altitude, soil moisture holding capacity showed a trend of higher values with increased elevation, although individual sites varied from this trend.

Elevational zonation, accompanied by differences in temperature and precipitation, was evident for both tree and understory vegetation. Forests generally showed a compositional change from lodgepole pine *(Pinus contorta)* to subalpine fir *(Abies lasiocarpa)* and finally, Englemann spruce *(Picea engelmannii)*, with increasing elevation.

Lodgepole pine forest was the most prevalent vegetation type, lying between 8,400 and 9,800 feet elevation. This type included species such as quaking aspen *(Populus tremuloides)*, limber pine *(Pinus flexilis)*, subalpine fir, Gambel oak *(Quercus gambelii)*, and Engleman spruce, the first three species of which dominated occasional sites within the lodgepole pine type. Aspen was found to prevail on sites at lower elevations which accumulated snow drifts and was associated with the abundance in distribution of Colorado columbine *(Clematis columbiana)*, Fremont geranium *(Geranium fremontii)* and heartleaf arnica *(Arnica cordifolia)*. Limber pine sites were associated with areas having a high amount of surface rock and loam textured soils. Gambel oak *(Quercus gambelii)* and Douglas fir *(Pseudotsuga menziesii)* were sparsely distributed near the lower range of the lodgepole forest.

Subalpine fir was given an arbitrary zone of dominance above the elevational range of the lodgepole pine type (9,000 ft-10,000 ft) within which it mixed with lodgepole and limber pine in the middle to lower portions, and with Engleman spruce in the upper ranges. As a species, subalpine fir was the most widespread tree across the elevational gradients and was second only to lodgepole pine in importance.

Engleman spruce forests occupied a well defined zone from 10,000 ft to 10,600 ft, in which it appeared to be the long-term dominant, being generally more abundant than subalpine fir, despite the disturbance to this zone by fire and timber cutting. Engleman spruce was found in a mature, pure stand on one site which contained trees averaging nearly 300 years of age, as compared with much younger stands in disturbed areas.

Information derived from a study of tree growth rings, supported observations of wide spread occurrence of fire within the watershed which have apparently reoccurred several times in the past, the most recent being approximately 50 years ago, followed by a period of little to no fire up to the present time.

Understory was not generally as abundant within stands having abundant trees as in the more sparse stands of timber. Stands of trees having a high total basal area contained less total plant cover in the understory, and thus, the grazing value of this type of stand was negligible. Grazing use by domestic stock and wild game was felt to be concentrated on meadow, rather than forest types.

The coniferous forest of this watershed was considered to be in secondary plant succession following the effects of fire and logging, with subalpine fir reclaiming areas temporarily dominated by lodgepole pine.

2E 710603 Mitchell, G.C. 1971. Spatial distribution and successional state of grassland vegetation
2F1 related to grazing intensity treatments. M.S. Thesis. Univ. of Wyo. 64 p.
3A2

2E The study was conducted on the IBP Pawnee site to measure the pattern of several
2F1 plant species in relation to grazing intensity. Five study sites were selected: a light
1E2 grazed, a medium grazed, a heavy grazed, a 10-year exclosure and a 30-year exclosure. A
10D1 variance procedure analysis determined the pattern, scale and intensity of blue grama

(Bouteloua gracilis), needleleaf sedge *(Carex eleocharis)*, plains pricklypear *(Opuntia polyacantha)* and scarlet globemallow *(Sphaeralcea coccinea)*. The sample areas were in differing stages of secondary succession due to grazing pressure or lack of it. The four species for the pattern analysis were determined to be non-randomly distributed. Small scale patterns which could be contributed to morphology and seed dispersal characteristics were exibited for the pricklypear and globemallow. At medium scales, the reciprocal pattern forced upon neighboring species by pricklypear seemed to be dominant. Large scale pattern was found, but could not be attributed to grazing influences. The pattern intensity of all rhizomatous species decreased as the site approached a climax condition.

7201 Research Activities

3A1 720101 Study of biological and physical environmental characteristics as represented by remote
2F1 imagery obtained from the NASA Earth Resources Technology Satellite (ERTS) with emphasis on grazing impact of livestock and associated impact of browsing game animals on rangelands of southern Wyoming. (Initiated in 1972.)

In conjunction with the Geology Department, we have a research program with Rob Gordon doing the field work and addressing the relationship of remote sense information of high level aerial photography and low level aerial photography compared to the ERTS imagery from the satellites and the relationship of the ability to determine the vegetation types and characteristics via the three forms of imagery are to be addressed. The question is whether there is some difference in the vegetation as influenced by the differences of grazing impact under the grazing systems of the Baggs area where this work is to be carried out.

8D 720102 Reseeding trials of grasses, forbs and shrubs on the bladed and disturbed area of a 120
5B4 mile long pipeline, extending across the Big Horn Basin. Funding by Colorado
5C Interstate Pipeline Corp. Primary emphasis of trials for aesthetic improvement as
1E2 well as erosion control and forage and browse production. (Initiated in 1972).

Work established seeding plots along the 120 mile CIG pipeline which travels from the north northwestern part of the Big Horn Basin, south southeastward past Worland and out of the Big Horn Basin in the south southeast portion of the Big Horn Basin. Sixteen species of grasses and a few shrubs were planted on a total of 17 different sites with the planter given for our use by the Plant Science Division. Plantings were done in September and November 1972.

1E2 720103 In 1972 the aridland ecology studies were amplified to include detailed phenological
2E studies of native vegetation over a broad geographical and ecological site
2F2 characterization. This intensive study is specifically aimed at evaluating factors
3A2 affecting phenodynamics of major native vegetation species. Initial discussions
1E1 addressed the possibility of developing significant phenological studies on a number
5A1D of the exclosures throughout the state. This work will be distributed over all of the
2A districts in the state and will probably also include some exclosures which at this
3A2 time have not had herbage production studies on them.

3A3 720104 Exclosure study: production, cover and precipitation.
5A1D
5B The exclosure studies during 1972 were carried on in the same manner as in previous
5C years.
2C
2A

5B4 720105 Juniper GRI seedings.
7A
5A1D During mid-April 1972, seedings were conducted on the GRI exclosures and also on
2E the new Rice Burn exclosure.

3A2 720106 2 ft x 20 ft and 4 ft x 4 ft permanent plot study.
2F2

The permanent 2 ft x 20 ft and 4 ft x 4 ft quadrats were reread at all of the exclosures during the summer of 1972.

2C 720107 Exclosure soil profile studies with SCS.
8B

Some exclosure soil profiles were conducted with SCS personnel during the summer of 1972.

1E2 720108 Shortgrass prairie blue grama culm density study.

2F1

2E The culm density study on the Pawnee site south of Cheyenne was carried on with some field work. It primarily addressed the idea of cleaning up the tape data not handled previously.

5B4 720109 Shortgrass prairie seed production study.

2F1

2E The shortgrass prairie seed production study by Shrader was carried on during this summer.

3A3

1E2

3A2 720110 Contour furrow and seedings study.

5A1D

5C A follow-up of the work done by Nichols several years ago was conducted on the Burnt Wagon contour furrow and seeding plot. In these activities, we reread all of the permanent point transects that he had established and also did clippings of total production as he had done also.

5B

3A2 720111 Exclosure condition analyses.

2F2

1E2 Condition classification at several exclosures was observed and interpreted by Fisser by traveling to the exclosures with camera and tape recorder, taking pictures and recording comments while viewing various parts of a number of exclosures.

3A1 720112 Wagon Wheel range survey.

3A2

5C During late October 1972 the standing crop of vegetation on the Wagon wheel sites, both the downwind and upwind sites, was conducted and also weight unit estimates of shrubs was established.

8D

3A2 720113 Laramie Peak Game Range Inventory.

7A1

7A2 The rangeland inventory of the Laramie Peak Game Range was conducted by Gary Butler during 1972. His M.S. thesis involved evaluation of the inventory data.

7202 Journal Publications

1D 720201 Kirkham, D.R. and H.G. Fisser. 1972. Rangeland relations and harvester ants in northcentral Wyoming. J. Range Manage. 25(1):55-60.

2B

5A1B

2C This study evaluated the relationships of harvester ant *(Pogonomyrmex occidentalis)* abundance and activity with grazing intensity, vegetation cover and composition and edaphic factors. Ant colony density, percent to soil surface denuded by ants and average area denuded per ant colony were used as indices of ant abundance and activity. The study area, located in Nuttall saltbush *(Atriplex nuttallii)* range type, was a pasture system of 600 acres involving four grazing intensities with distinct variations in vegetation cover and composition values.

After 10 years of grazing at four intensities of use, harvester ant abundance was positively correlated with some plants, the correlations appeared incidental to the ability of a species to establish and grow in specific soil conditions, and were not related to any ant preference for available food source.

Soil texture was the most important factor influencing harvester ant abundance. If harvester ant abundance was affected by grazing and range condition, then such responses would be slow, requiring more than the 10 years in the study.

7203 Presentations

3A1 3A2 2F1 10D1 720301 "Panchromatic, color and infrared photography for detailed vegetation mapping." February 8, 1972. 25th Ann. Meeting Soc. Range Manage. Washington, D.C. by H.G. Fisser.

The Pawnee site rangelands, although relatively uniform and dominated by blue grama, exhibit phytosociological variation resulting from soil, moisture, relief and grazing treatment influences upon vegetation. The spatial modeling effort of the grassland biome program requires a detailed vegetation map with definitive type units no larger than 0.1 ha (10 m x 10 m). To this end, three forms of remote sensing imagery were utilized: panchromatic photographs at a scale of 1:5800, as well as color and infrared prints at scales of about 1 inch to the mile. (1: 63,360).

Utilizing mirror and lens stereoscopes, first order delineations were applied to the panchromatic photographs with further detail added following comparison to the color and infrared prints. Type lines were drawn on the black and white prints based on the photographic tone, texture, pattern, topographic location and color variations on the three imagery forms.

Through secondary photographic procedures, the typelines were transferred to a topographic base map with a scale of 1:4800. Ground truth data were based on leaf area indices (LAI) derived from point hits utilizing a frame with a motorized and moveable pin. Some type lines were modified on the basis of leaf area index values following statistical evaluation.

2F2 2E 3A3 720302 "Juniper ecology and response to protection from browsing." July 6, 1972. 1st Wyo. Shrub Ecology Workshop, Casper, Wyo. by H.G. Fisser.

Juniper *(Juniperus)* ecology and response to protection from grazing was addressed in this topic at the first Wyoming shrub ecology workshop. The discussion was carried out at the Owl Draw exclosure where a game-proof exclosure and a livestock-proof exclosure were constructed. Sixteen transects have been located at the exclosure over a variety of sites and conditions caused by differences in slope, exposure and soil characteristics.

After several years protection from grazing, it is apparent that a variety of additional species will become abundant under the juniper in time. This increase in shrubs such as bitterbrush *(Purshia tridentata)* shows that utilization, whether by wildlife alone or with combinations of wild horses, livestock and deer, have caused significant deterioration to the ecological condition of the juniper type in that area south of Casper.

7204 Published Proceedings/Abstracts

3A1 3A2 2F1 720401 Shaver, J.C. and H.G. Fisser. 1972. Panchromatic, color and infrared photography for detailed vegetation mapping. p. 33. In Proc. 25th Ann. Meeting Soc. Range Manage. Washington, D.C. Feb. 8, 1972.

See 720301.

3A1 3A2 5C 8D 720402 Fisser, H.G. 1972. An ecological description of biotic communities of the Wagon Wheel environs. p. I-1 thru I-48. In Rogers L.A. (ed.). Project Wagon Wheel Study Report Number Two. El Paso Natural Gas Co., PME-WW-13.

A base line study primarily concerning vegetation, but with some reference to animals and man, was made in the Wagon Wheel environs. Detailed inspection at some 43 sites was made along with general observation of overall ecosystem criteria. The area is a temperate semi-arid desert subject to harsh natural environmental fluctuation. The natural biotic communities are quite stable and capable of withstanding much stress.

The native vegetation is typical of other areas with similar conditions. Of the 18 vegetation types described by the U.S. Interagency Range Survey Committee, at least 11 of them are found within 25 miles of Wagon Wheel. Several varieties of sagebrush and grass are the principle vegetation in the immediate area of Wagon Wheel. Some 11 plant types generally considered poisonous were observed. No significant selenium concentration was found, however.

A number of native animals, including the pronghorn antelope and sage grouse, inhabit the area. Alterations to the environment exist from farming, ranching, construction and other activities in the area.

The report describes ecological attributes of biological systems and related physical components of the Wagon Wheel environs as deduced from observations combined with cognitive principles which characterize how organisms exist in their environment. These principles are predicated upon known energy- dependent interrelationships of living organisms with abiotic environmental substances and gradients. The development of a descriptive ecological statement thus involves, in addition to definition and inventory of biological components, an inherent, two-fold objective: first are derived the generic level relationships of physical environmental components with the organisms; the second objective involves the development of conceptual implications defining those environmental criteria most influential to control of ecosystem functions.

Biological ecosystems, considered in the broadest context, are regulated by three factors - geological materials (including parent materials, relief and ground water), macroclimate, and organisms. These regulating and controlling factors are at least partially independent of one another. Each is a composite of many components, all variable in time and space. Ecosystems are dynamic, changing with fluctuations of the controlling factors and internal modulations. Because they vary temporally and spatially, disruptions by man causing modifications to the controlling factors can induce temporary or permanent innovations which may be deleterious or beneficial. Energy flow and nutrient cycling rate, as modified by floral and faunal structural attributes, are basic to ecosystem regulation. The resulting products can be observed and portrayed in terms of definitive structural characteristics of an ecosystem. The structural characteristics can be categorized by life form, species composition and infraspecific variation, community morphology and species diversity.

During the field investigation period, basic ecological criteria were noted by inspection and observation at 43 Observation Sites located within a 25-mile radius of the Wagon Wheel installation. Sites were selected partially on the basis of geographical distribution in order to sample the area most subject to influence by the proposed underground detonation (most sites were within a 10-mile radius). Specific observation sites were selected primarily on the basis of any apparently unique vegetational, edaphic or relief characteristics. Those observed most often were simple composition or productivity differences of the almost ever-present sagebrush type. Observation and evaluation of these diverse indicators of biophysical structure and function was utilized to hypothesize the causative factors responsible for the existant biological conditions.

The program intent has been to document by location, the physical and biological criteria common to the observation sites, and to detail and note unique characteristics and their hypothesized causative functions. Supporting this information is a permanent photographic record of some 120 color slides verifying the observed site conditions. Notes obtained at each of the sites are attached in an appendix.

7205 Wyoming Agricultural Experiment Station Publications

4B 720501 Gibbens, R.P. and H.G. Fisser. 1972. Influences of grazing management systems on
4C vegetation and wildlife habitat. Annual Progress Report. Wyo. Agr. Exp. Sta. Sci.
1E4 Rep. 414. 69 p.
2F2
3A2 This report of the 1971 data collection procedures and compilation of data of the
7A1 Baggs grazing system study provides information which will be used for the analysis of
7A3 influences of different grazing systems upon the vegetation in the southern Wyoming
6A3 area.

720502 Fisser, H.G. and R.E. Ries. 1972. Aridland ecology research. 1971 results. Univ. of Wyo. Coop. Res. Rep. to the BLM. Sections I and II. Wyo. Agr. Exp. Sta. Sci. Rep. 428. 130 p.

2A 72050201 Section I - Precipitation pattern study.
3A

The precipitation pattern study in 1971 was carried on as in previous years.

3A3 72050202 Section II - Exclosure study: production, cover and precipitation phases.
5A1D
5B The 1971 activities of exclosure studies as noted previously included the
5C determination of total community herbage production with the addition of a weight unit
2C estimate procedure for determination of growth of shrubs and mat-form species.

3A1 720503 Shaver, J.C. and H.G. Fisser. 1972. Detailed vegetation mapping of the I.B.P. Pawnee
3A2 Site. U.S. I.B.P. Grassland Biome Technical Report 150. 44 p.
2F1

The Pawnee Site rangelands, although relatively uniform and dominated by blue grama *(Bouteloua gracilis)*, exhibit phytosociological variation resulting from soil, moisture, relief and grazing treatment influences upon vegetation. The spatial modelling effort of the grassland biome program requires a detailed vegetation map with definitive type units no larger than 0.1 ha (10 m x 10 m). To this end, three forms of remote sensing imagery were utilized: panchromatic photographs at a scale of 1:5800, and color and infrared prints at a slightly smaller scale.

Utilizing mirror and lens stereoscopes, first order deliniations were applied to the panchromatic photographs with further detail added following comparison to the color and infrared prints. Type lines were drawn on the black and white prints, based on photographic tone, texture, pattern, topographic location and color variations observable on the three imagery forms. Through a secondary photographic procedure, the type lines were transferred to a topographic base map with a scale of 1:4800. Ground truth data were based on leaf area indices (LAI) derived from point hits utilizing a motorized point frame with a movable pin.

1E2 720504 Stroud, D.W. and H.G. Fisser. 1972. Plant pattern microrelief and blue grama
10D1 *(Bouteloua gracilis)* culm density studies on the Pawnee Site: 1970-71. U.S. I.B.P.
2F1 Grassland Biome Technical Report 186. 26 p.
2E

Plant distribution, microrelief and blue grama *(Bouteloua gracilis)* culm density were intensively measured in a 30-year exclosure and in the pastures with light, medium and heavy rates of utilization at the I.B.P. Pawnee Site on the Pawnee National Grasslands. Measurements were made using a .5 m x 1 m microplotter with horizontal X and Y axes and a vertical Z axis.

Blue grama was the dominant species with composition of 55.6 percent on medium, 44.4 percent on light and 32.5 percent on the heavy grazed pasture. The degree of roughness was found to be similar for the 30 year exclosure as well as the light and medium grazed pastures, but showed a definite decrease on the heavy grazed site. Culm density within clumps of blue grama showed a definite increase from 3.5 to 5 per cm^2 surface area on the light grazed to a net 7.3 culms per cm^2 on the heavy grazed site.

7206 Graduate Advisee Completions

2F2 720601 Gibbens, R.P. 1972. Vegetation pattern within northern desert shrub communities. Ph.D.
2E Dissertation. Univ. of Wyo. 272 p.
3A2

This study of intercommunity pattern was conducted on several plant communities common in the Red Desert of Wyoming and also occurring throughout the larger area of the Northern Desert Shrub Formation. The study sites were located west of Baggs,

Wyoming at elevations ranging from 6,400 ft to 7,200 ft. The study determined whether patterns existed within plant communities and to determine whether causative factors of such patterns could be assigned to morphological, sociological or environmental factors.

Three stands, homogeneous in appearance, were studied in each of the following communities: Gardner saltbush *(Atriplex gardneri)*, birdfoot sagewort *(Artemisia pedatifida)*, woody aster *(Machaeranthera glabriuscula)* and big sagebrush *(A. tridentata)*. Two stands of bud sagewort *(Artemisia spinescens)* and bluebunch wheatgrass *(Agropyron spicatum)* communities were studied plus a sprayed and unsprayed *Artemisia tridentata/Agropyron* species site and a single site dominated by *Agropyron* species. All of the study sites were subjected to grazing by domestic livestock, antelope, deer, and wild horses.

Pattern analyses by the mean square-block size method were made on data from replicated cross-slope and with-slope transects 25.6 m in length at each site. The basic sampling unit was a decimeter squared quadrat. Canopy frequency of shrubs, basal area frequency of bunchgrasses and density of single stem forbs and grasses was recorded. A limited number of transects using a 50 cm square quadrat as the basic unit were sampled in four communities and grid type sampling was tried in the big sagebrush and woody aster communities. Pattern intensity statistics and covariance values for species pairs were calculated for each direction on each 30 m x 30 m study plot. Soils at each study site were analyzed for pH, total soluable salts, and texture. Root systems of the principle species were excavated to determine horizontal spread. Point centered quarter samples were taken on study sites on the Gardner saltbush, birdfoot sagewort, woody aster, and bud sagewort communities to obtain a measure of plant densities.

The woody aster community occurred on shallow, highly alkaline soils. The birdfoot sagewort community occurred on strongly alkaline, deep soils. The Gardner saltbush and bud sagewort communities occurred on moderately alkaline, deep soils. Clay soil texture prevailed in each of the above communities. The big sagebrush community occurred on soils with a low soluable salt content and clay loam and sandy clay loam textures in the surface 6 inches. The bluebunch wheatgrass community occurred on a rocky, relatively mesic site where soils were low in soluable salts. Soil pedestals, formed under shrub canopies partly by erosion and partly by deposition, were prominant microtopographic features in the Gardner saltbush, bud sagewort and big sagebrush communities. The community found on a given area seemed to be dictated by the soil type and salt content. However, many species - Gardner saltbush, Hoods phlox *(Phlox hoodii)*, Indian ricegrass *(Oryzopsis hymenoides)*, bottlebrush squirreltail *(Sitanion hystrix)*, wheatgrass species, sandburg bluegrass *(Poa secunda)* - were well represented in several communities indicating that they possessed a wide range of tolerances for soil conditions.

Morphological pattern due to plant size was not strongly expressed in any community, even by shrubs such as Gardner saltbush and big sagebrush which developed relatively large canopies. This is believed due to the fact that the stands were all even aged and a wide range of plant sizes was present. The clearest example of sociological caused pattern occurred in the birdfoot sagewort and big sagebrush communities. Clusters of birdfoot sagewort plants with a dimension of about 40 cm imposed a mirror image pattern on associated species. In the big sagebrush community, wheatgrass species exhibited pattern scales indicating it grew more densely in open areas between shrub canopies while Sandburg bluegrass showed the opposite condition, growing more densely under the shrub canopies. In most comparisons, the covariance between species pairs indicated a nill or indeterminate association even at block sizes equivalent to the horizontal root spread of dominant species, and few significant associations occurred. In fact, no patterns could be definitely attributed to extent of root systems. Plant densities in all communities were such that root systems were highly overlapped in horizontal space.

A total of 20 species were abundant enough in one or several communities to permit pattern analyses. All exhibited at least one scale, and often more, of pattern, which seemed to be induced by environmental factors. These pattern scales ranged from 160 to 1280 cm on the cross-slope and with-slope analyses respectively, and indicated that some were isodiametric and others non-isodiametric. No one scale of pattern was characteristic of a given community or of a given species in different communities. The fact that pattern scales varied from site, both within and among communities, was convincing evidence that the patterns were determined by environmental factors peculiar to each site. This hypothesis was supported by the measure of pattern intensity.

Pattern intensity differed among sites in each community studied, yet the pattern intensity of different species on a given site were nearly the same. This indicated that all species were controlled by some environmental factor or factors with operationally different intensities at different sites. The pattern intensity measure was not correlated with density of the dominant species except on the Gardner saltbush sites. Thus, the differences in pattern intensity could not be attributed to increased influence of dominant plants resulting from higher densities.

Scales of pattern exhibited by wheatgrass species, Sandberg bluegrass and Engelman fleabane *(Erigeron engelmanni)* were almost identical on an area sprayed seven years previously to kill big sagebrush and on a contiguous unsprayed area. However, pattern intensity for each of the three species was about twice as high on the sprayed as on the unsprayed area. This example lended support to the hypothesis that pattern intensity decreased as climax condition was approached, if a sprayed area was considered as one setback in successional stage. The other communities studied were all climax, or at least stable "grazing climax," so no further evidence on scale or intensity of pattern in seral stages was available. It was established that environmentally induced patterns existed in the communities studied, but only guesses could be made as to the specific causes of pattern. The communities existed in a climate where moisture was limiting to plant growth. Hence, the soil moisture realm was a logical place in which to look for patterns and controlling factors. Yet, to sample even one site for variation in infiltration rate, moisture retention and availability, soil salinity and texture, etc. at an intensity which would allow the testing of correlation between pattern scales of vegetation and environmental factors would be a tremendous task. Further complication was added by the fact that the dominant species were long-lived and their current patterns could still reflect conditions no longer operative. That plant patterns contain ecological information is undeniable, but the information is so confounded through time and space that little can be said with certainty. An experimental proof of hypothesis would be a long, arduous task. The demonstration of pattern within the communities has useful implication with regard to sampling techniques. Since all of the species exhibit phases or patterns of low and high densities, representative samples must include both phases. Transects which are at least twice as long as a pattern scale or an adequate number of truly random quadrats would be best. A selected or "haphazard" quadrat might sample only a high or low density patch, yielding a biased picture of vegetative parameters. Regularly spaced quadrats should be avoided as they can all too easily be "in phase" with high or low density areas.

Pattern analysis using the mean square-block size method would be of little value in studying range condition and trend. It was a safe assumption that pattern intensities would vary under different intensities of grazing and perhaps pattern scale as well. However, the required sampling procedure was slow and tedious and the computational load heavy. No unique predictive qualities to either scale or intensity of pattern were found to justify the labor involved.

2F2 720602 Garland, C.B. III. 1972. Ecology of aridland vegetation in western Wyoming. M.S. Thesis. Univ. of Wyo. 106 p.
4A
1E2
5C

Nineteen livestock exclosures were studied within the Big Horn Basin. They were selected because of the location and the similarity of grazing use on adjacent ranges. The herbaceous and shrub layers were measured on protected and unprotected ranges by the use of permanent 1 ft x 1 ft quadrat transects. These helped to discover cover and composition changes related to livestock grazing and associated vegetation treatments. The treatments included non-grazing, seeding and furrowing, seeding and blading, and big sagebrush *(Artemisia tridentata)* spraying. The major species of each exclosure which were common to more than one protected area have been analyzed by the changes in average percent cover per plot along transects. This was done to help detect a significant or insignificant vegetation trend for a given exclosure.

The general effects upon the vegetation from a treatment or treatments are explained in the following.

1) Non-grazing usually showed an increase of sagebrush and grass inside the exclosure. Generally there was not a significant change. However, some species increased more than others. The vegetative cover of saltbush ranges usually remained similar in cover inside and outside the exclosures.

2) Seeding and furrowing was used primarily on the saltbush vegetation type. Crested wheatgrass *(Agropyron cristatum)* was used in the seeding. This type of combination treatment usually resulted in a slight increase of grass cover. The overall effect, however, did not seem to increase the forage production significantly.

3) Seeding and blading was done on Gardner saltbush *(Atriplex gardneri)* vegetation types. This treatment showed a similar response in grass cover as the above treatment, but greatly reduced the saltbush cover. This can be debated regarding its beneficial effect, because saltbush provides a majority of feed for livestock on this type of range. However, grass reseeding does give livestock a better chance for a well-balanced diet.

4) Chemical spraying was conducted upon big sagebrush. This treatment had the most significant effect upon increases in grass cover, both inside and outside the exclosures. In many cases, sagebrush was completely eliminated, thus allowing for abundant increases in grass cover.

The two most important climatic factors affecting vegetation growth in the Big Horn Basin were temperature and precipitation. Most of the exclosures studied were located in areas of high summer heat which causes plants to transpire rapidly. Annual precipitation ranged from a high of 15″ at the Upper Enos Creek exclosure to a low of 6″ at the Big Flat exclosure. Both factors discourage the establishment of many grass species. The native species of the semi-desert shrub vegetation types were well able to support themselves in these situations. Gardner saltbush and bottlebrush squirreltail *(Sitanion hystrix)* were two species with outstanding abilities to adapt to the vagaries of adverse climatic conditions. Usually a sagebrush grass association was found on more mesic locations as compared to the saltbush areas. Two of the more important species of the sagebrush area were big sagebrush and western wheatgrass *(Agropyron smithii)*.

The basin's edaphic factors were of two general types. Saltbush vegetation was usually found on poorly developed soils with different textural groups. These soils were basic in nature with the pH in the range of 7.2 and 8.5. There were also a fairly large number of soluble salts present in the soil. Often there was a hard vescicular crust on the surface, which, when broken, became powdery. Thus water penetration in these soils was poor. Also, the overall water holding capacity was poor because of soil, precipitation, and heat conditions.

Sagebrush type vegetation occurred on fairly well developed soils of usually a loam or sandy loam texture with moderate pH and low total soluable salt values. The water penetration and water holding capacity of these soils was somewhat better than those of the saltbush areas. As a result, the sagebrush vegetation types usually exhibited greater vegetation productivity, or potential for greater productivity, than the saltbush sites.

Because of the slow growth and response of vegetation, especially in arid regions of Wyoming, to grazing pressure changes and improvement practices, several years are often required for the manifestation of change to be readily visible to the observer. As a result, some of the exclosures studied may not yet show a true picture of a vegetational trend. Some exclosures may never show inside and outside differences because the grazing intensity may be so light and moderate that subsequent effect on vegetation is unmeasureable.

5B4 720603 Burrough, L.N. 1972. Artificial revegetation trials on juniper game ranges in north central Wyoming. M.S. Thesis. Univ. of Wyo. 135 p.
7A
5A1D
2E

This study was conducted on the game range improvement exclosures located on the perimeter of the Big Horn Basin of northern Wyoming. Areas dominated by thick stands of scrubby Utah juniper *(Juniperus osteosperma)* and *(J. scopulorum)* trees in Wyoming offer potential for improved wildlife habitat following removal of these species and revegetation with more desirable plant species. Improvement of mule deer winter range would be a benefit in supporting higher numbers needed to satisfy increased future hunting demands.

A study of seeding trials in the juniper zone was conducted to determine preferred seasons for seeding, preferred juniper control treatments, preferred methods of planting, and preferred species for reseeding. The exclosures were constructed in juniper stands treated with chemical control by spraying, bulldozing, and burning. Spring versus fall seeding, broadcasting versus planting, and differences due to spray, burn and bulldozed treatments were noted on 29 species of shrubs and forbs in the first study evaluation in August 1970. The 1971 evaluation considered spring versus fall seeding dates of 10 shrubs, two forbs, and one grass species on the burned and scraped portions in each exclosure and on the Mills burn study site.

The 1970 evaluation showed that planted seed outperformed the seed that was broadcast and left uncovered. The data show that fall seeding applications were superior to spring applications for most species. The burned and bulldozed treatment areas were superior to the sprayed treatment areas in seedling production. The continued toxicity of the Picloram chemical greatly reduced the establishment of most of the species. Only 30 seedlings were observed in the sprayed plot collectively while 149 seedlings were observed on the burn treatment areas, 143 seedlings were observed on the bulldozed treatment areas. Seedling numbers were extremely low due to severe competition from weedy annuals, low rainfall and utilization of the seedlings by small mammals. The species which were observed to be best adapted to the study sites were cicer milkvetch *(Astragalus cicer)*, silver sagebrush *(Artemisia cana)*, black sagebrush *(A. nova)*, big sagebrush *(A. tridentata)*, serviceberry *(Amelanchier alnifolia)*, Mexican cliffrose *(Cowania mexicana)*, alfalfa *(Medicago sativa)* and Palmer penstemon *(Penstemon palmeri)*.

By the time of the August 1971 evaluation, collectively there were more seedlings surviving on the spray treatments within the exclosures than on the burn treatment areas. High competition from numerous fast growing, weedy plants coupled with low moisture and high summer temperatures tended to make the burn treatment areas poorer sites for seedling establishment than the sprayed treatment areas. On the Mills burn study site, however, many more seedlings occurred than on any or all of the exclosures combined. The burn on the Mills ranch was a more recent burn than those caused in the exclosures. Competition was less than that on other exclosures. It was a natural and apparently a very hot burn, because many of the perennial grasses and forbs were completely eliminated from the burn area. As in the previous evaluation, fall seeding appeared to be preferred for most species. The planted species which seemed to be best adapted to the broadcast situation were mammoth wildrye *(Elymus giganteus)*, cicer milkvetch, fourwing saltbrush *(Atriplex canescens)*, bitterbrush *(Purshia tridentata)*, winterfat *(Eurotia lanata)*, rough menodora *(Menodora scabra)* and Palmer penstemon.

In a subsequent work, a thesis was completed by Spaeth in 1981 and these data are reported in code number 810604. His thesis evaluated essentially the same sites and locations and seedings over a longer 10-year period.

3A1 720604 Shaver, J.C. 1972. Evaluation of vegetation types differentiated by aerial photo interpretation of panchromatic, color, and infrared imagery. M.S. Thesis. Univ. of Wyo. 86 p.
3A2
2F1

See 720301.

3A2 720605 Butler, G.B. 1972. Ecological inventory and analysis of the Laramie Peak Big Game Winter Range. M.S. Thesis, Range Management, Univ. of Wyo. 170 p.
7A1
7A2
7B1

This study was conducted during the summer of 1972 on the Laramie Peak Big Game Winter Range in southeastern Wyoming. This area provided habitat for deer, antelope and elk, especially during winter months. The objectives of the study were to describe vegetation types, determine condition and trend, establish trend transects, pellet group transects, and determine key wintering areas.

Vegetation was mapped with an aerial photograph mosaic of 1:20,000 scale. Soil series were correlated with range sites and soil samples taken within each vegetation type. Vegetative ground cover contributed by plant species was determined for each vegetation type. Each plant species encountered within the vegetation type was identified and collected. The range condition was determined on the basis of the SCS procedures. Permanent trend transects were established in all of the vegetation types to detect future changes in range condition. Key areas were determined on the basis of abundance of use as indicated by 1/10 acre permanent pellet group transects which were 8 feet wide and 544 feet long. These pellet group transects were placed where observations by Wyoming Game and Fish personnel had shown wildlife to concentrate in winter. The transects which indicated considerable wildlife use, as indicated by density of pellet groups were considered key areas.

Six vegetative types were mapped and described: conifer-savanna-sagebrush-grass type, grassland type, sagebrush *(Artemisia)*-grass type, coniferous forest type, wet meadow type and wasteland. The first type contributed the second largest acreage of any type and the most important type of habitat for wildlife during the winter months. The grassland type provided important antelope habitat throughout most of the year and comprised the largest type acreage. Range sites were mapped and described and they are associated with their associated soil series. The range sites were shallow loamy, shallow igeous, igneous, loamy and wetland.

Generally the shallow loamy, loamy and wetland range sites were in the good condition class while the shallow igneous and igneous range sites, which supported the conifer-savannah-sagebrush-grass type were in the fair condition class. Over-grazing of these thin soiled sites quite possibly caused this low condition. Future management should strive to better the condition of the conifer-savannah-sagebrush-grass type which provides the best wildlife habitat on the winter game range. This goal should prove attainable by using the results of the last three objectives of this study to monitor the wildlife habitat and game numbers.

Eight key wildlife wintering areas were picked from pellet group density determinations. By evaluating vegetation use on these winter concentration areas of wildlife, proper use of the game range as a whole should be maintained.

Two key species, antelope bitterbrush *(Purshia tridentata)* and Idaho fescue *(Festuca idahoensis)*, were selected for monitoring wildlife utilization of the vegetation on the key areas. Assessing utilization on these two species prior to spring growth should provide an estimate of the pressure exerted by wildlife on the key species and key areas the previous winter.

Continued use of utilization checks on key species in key areas, the established permanent trend transects, the permanent pellet group counts, and winter observation censusing now being carried on by the Wyoming Game and Fish Department should provide important tools for monitoring and managing the Laramie Peak Big Game Winter Range.

Several recommendations were developed by the author and these have been briefed to the following.

1) Land trades should be made with adjacent ranchers to close in the state owned land. This would help hunting management of the area a great deal.
2) No livestock grazing should be allowed on the conifer-savannah-sagebrush-grass type until condition has increased considerably.

3) Cattle grazing should be allowed on the grassland areas unless elk populations increase substantially.

4) Hay contracting to adjacent ranchers should remain a policy practice.

5) Annual clearing and counting of the permanently established pellet group transects should be conducted.

6) Annual spring utilization checks should be conducted on the key areas and species to monitor wildlife use and pressure on the winter range and to verify the key species selected.

7) Permanent trend transects should be included on all of the verified key areas to provide a double check along with the utilization estimations. Fence repair and construction should be initiated to prevent trespass by cattle. Water development could easily be improved in the area. Although water distribution is good, shallow springs and low water tables throughout the unit lend themselves to easy development.

7301 Research Activities

8D 730101 On an area near Reno Junction, Wyoming, on which a strip-mine operation was
3A1 proposed, the ecology, vegetation and soils inventory studies were initiated in 1973.
3A2 Continuing work was involved with the establishment of permanent monitoring
2F1 locations for development and evaluation of best post-mining rehabilitation procedures.

3A3 730102 Exclosure study: production, cover and precipitation.
5A1D

5B The 1973 activities of herbage production data - at the exclosures in which also was
5C incorporated the shrub production for estimates of total community production - was
2A carried on as in past years.
2C

1E2 730103 Intensive investigation of phenology and production relationships of vegetation as
2F2 influenced by site characteristics of aridland sites representing diverse habitats of
3A2 western Wyoming with specific attention given to big sagebrush *(Artemisia*
2A *tridentata)*, black sagebrush *(A. nova)*, western wheatgrass *(Agropyron smithii)* and
2E bluebunch wheatgrass *(A. spicatum)* (initiated in 1973).
1E1

5A1D Twelve exclosure locations were identified for use in the intensive phenological studies program. They included the Horse Creek, Bud Kimball, Smilo and Demer exclosures in the Big Horn Basin, the Shoshoni Ant -7, Upper Government Draw and Sweetwater exclosures in the Wind River Basin, the Owl Draw exclosure in the area south of Casper, the Farson exclosure, Mesa Antelope exclosure in the Pinedale region, the Cumberland -3 exclosure located west of Kemmerer, the Black Mountain exclosure south of Green River and the Red Wash -2 exclosure southwest of Wamsutter.

Permanent transects were established for the study of 20 permanently identified plants of each species along each transect with identification simply by measurement coordinates. Phenology of the four prime species was given primary emphasis but, in addition, phenological interpretation and evaluation of each of the individual species on an exclosure site was also recorded. The primary, important developmental feature of this program, related to the fact that phenology dynamics were soon found to require both a vegetative and a reproductive characterization. Data recordings with numerical scores and data sheets were developed to utilize this kind of phenology identification.

2A 730104 Precipitation pattern study.
3A

The precipitation pattern activities during 1973 included continuation of precipitation recordings at the UW-BLM gauges four times during the year and also in the reports the 38 U.S. Weather Service Station records of precipitation were compiled on a seasonal basis.

3A1 730105 Range inventory with remote sense analyses on Baggs grazing system.
2F1

Early in the spring, by study with Rob Gordon, it was determined that the 1 to 40,000 aerial photographs as well as the imagery from the ERTS satellite could be easily coordinated because the definitive nature of the aerial photographs allowed identification of location of individual exclosures. The satellite-based imagery was easily recognizable for identification of major vegetation types with reasonably good resolution.

8D 730106 CIG pipeline seed plot study.
5B4

5C On May 22 and 23 a number of the CIG pipeline plot locations were investigated.
1E2 The winter of 1972-73 was very dry throughout the middle area of the Big Horn Basin,
especially west and north of Worland, and on many areas very little seedling response

was noted. On some areas, where moisture had occurred, a number of species exhibited good to excellent response.

2C 730107 Exclosure soil profile studies with SCS.
8B

The soil exclosure study during 1973 were concentrated primarily in the Rock Springs area and the exclosures of that southern Wyoming region.

3A2 730108 Mule deer winter range evaluation.
2F2
5C

Initial studies were developed on analysis of mule deer-vegetation relationships on winter ranges of Twin Creek east of Lander and an area east of Worland and on the west face of the Bighorn Mountains in the juniper type adjacent to the Medicine Lodge Game Range Improvement exclosure.

3A2 730109 1 ft x 1 ft plot permanent transects.
2F2
5C

1 ft x 1 ft permanent plot transects were reviewed in a few situations where some data corrections were necessary.

1E2 730110 Shortgrass prairie blue grama culm density study.
2F1
2E

The blue grama *(Bouteloua gracilis)* culm density study with Stroud's fieldwork and tape cleanup was continued.

1E2 730111 Shortgrass prairie seed production study.
5B4
2F1
2E
3A3

Shrader's work on seed production study was completed.

7302 Journal Publications

1E2 730201 Wambolt, C.L. 1973. Conifer water potential as influenced by stand density and environmental factors. Canad. J. Bot. 51(12):2333-2337.
2E
2A
2C
2F2
10D1

This report was prepared as a supplement to the dissertation by Wambolt given in 710601. The data were obtained in the Big Horn Basin on sites with juniper *(Juniperus)* and associated coniferous species as well as sagebrush *(Artemisia)* and addressed the topic of water stress relationships to stand density and environmental factors.

The investigation of several conifer species was conducted to evaluate the influence of atmospheric, edaphic, topographic and vegetation variables of the environment on water potential. Correlation coefficients averaged close to 1.0 for the relationship of plant density to water potential when other environmental factors were static. This was interpreted as quantitative evidence of the intraspecific competition for available moisture by plants. Water potential values of 21 stands of coniferous trees, three each of seven species, were highly correlated ($R = 0.84$) to elevation, percentage sand, temperature, relative humidity and vapor pressure deficit, even though the stands were found in widely divergent habitats.

8D 730202 Bedell, T.E. 1973. Revegetation of pipeline areas. J. Colo.-Wyo. Acad. Sci. 7(4):29-30.
5B4
5C
1E2

Industrial uses of range landscapes are vastly different from most other uses. These uses may range from minimum disturbances such as splicing in an underground telephone cable or seismograph trails for various kinds of mineral exploration, through larger pipeline, transmission line, road/highway disturbances, to such large landscape changes as occur from strip-mining. Revegetation of such disturbed areas is often necessary in order to minimize ecological changes, protect the environment and be aesthetically pleasing. Because of the kind of disturbance, e.g., pipelines which cut across the landscape regardless of topography, soils, geology or vegetation conditions, prevention of soil erosion constitutes the major potential problem. Since most revegetation research has been concerned with forage with livestock, little information exists regarding recommendations for the many and varied site conditions resulting from industrial uses.

An example of the kind of effort needed is a project at the University of Wyoming in cooperation with the Colorado Interstate Pipeline Company and the Bureau of Land Management. Various grass, forb and shrub species are being tested on 17 different sites along a 130 mile pipeline which extends from the northwestern portion of the Big Horn Basin south-southeastward diagonally across the entirety of the Big Horn Basin. Many types of soil/site conditions will exist on most industrial use projects and research leading to the most appropriate revegetation and management procedures must be conducted.

7303 Presentations

5A1E 730301 "Methods of revegetation and restoration following a juniper wildfire." February. 6, 1973. 26th Ann. Meeting Soc. Range Manage. Boise, Idaho. by D.L. Layton.

5B4
5C
1E1
1E2

The information and research of this report were conducted on the Rice Burn exclosure, an area southeast of Hyattville, on the slope of the Big Horn Mountains. A 5-acre juniper burn was used as the base for the location of this exclosure and the revegetation study.

In April 1972, eight months after the fire, a 240 ft x 77 ft area was established and fenced for the revegetation treatments. Methods of reseeding included broadcast seeding, broadcast rake seeding and furrow seeding. Twelve species were planted including one grass (Mammoth wildrye *(Elymus giganteus)*), two forbs (Cicer milkvetch *(Astragalus cicer)* and Palmer penstemon *(Penstemon palmeri)*) and nine species of shrubs commonly utilized for browse by wildlife and livestock in the intermountain region. Seed of all species was applied in a mixture at a rate estimated at 10 lb/acre. Establishment of at least a few seedlings of each species was expected.

During the summer of 1972, the study area was divided into 24 plots of 10 ft x 77 ft. Seedlings were mapped by location within each plot. Seedlings of seven species were observed. These were subjected to measurements of height, diameter and the numbers of leaves. These species counts were compiled into success ratio charts and correlated into each type of seeding.

A total of 1,036 plant seedlings were observed, 756 in the furrow treatment and only 280 in the broadcast area. These represented seven species, including mammoth wildrye, cicer milkvetch, Palmer penstemon, winterfat *(Ceratoides lanata)*, fourwing saltbush *(Atriplex canescens)*, antelope bitterbrush *(Purshia tridentata)* and rough menodora *(Mendora scabra)*.

Over the entire study area of 19,000 square feet the seed application resulted in an average seedling establishment rate of one plant per 9 square feet.

5B4 730302 "Evaluation of furrowing and seeding in a north central Wyoming Nuttall saltbush area." February 6, 1973. 26th Ann. Meeting Soc. Range Manage. Boise, Idaho. by M.H. Mackey.

5A1D
3A2
5C

The research and activities of this report were taken from data obtained at the Burnt Wagon contour furrowing and seeding study area which has been earlier reported under code 720110. The Gardner saltbush *(Atriplex gardneri)* area of the Burnt Wagon site was furrowed and seeded with crested wheatgrass *(Agropyron cristatum)* in 1957. In 1962 and again in 1972, data on foliage cover, species composition, frequency and production were obtained from plots inside and outside the furrow seeded area. Yearly precipitation averaged 6.5 inches with about 50 percent occurring during May and June in short duration high intensity showers. The area had been subject to spring and fall cattle use, winter sheep use and yearly horse use. Since 1957, furrow depth decreased approximately 65 percent.

The total aboveground herbage was as follows: Treated, 590 lb/acre; Native, 350 lb/acre. Livestock strongly preferred the furrow seeded area to the native. This was the probable cause for the herbage cover and production decline over the past 10 years. One could assume that furrows were less effective in trapping moisture currently than after initial establishment. Because of the current excessive, overuse this study suggested a potential need for fencing or other management practices. These are necessary in order to limit and define the most useful periods for livestock grazing which would be least detrimental to the vegetation.

8D 730303 "Revegetation of pipeline areas". April 30, 1973. Colo.-Wyo. Acad. Sci. Ft. Collins, Colorado. by T.E. Bedell.
5B4
5C
1E2 See 730202.

7304 Published Proceedings/Abstracts

5B4 730401 Mackey, M.H. 1973. Evaluation of furrowing and seeding in a north central Wyoming Nuttall saltbush area. p. 13. In Proc. 26th Ann. Meeting. Soc. Range Manage. Boise, Idaho. February 4-9, 1973.
5A1D
3A2
5C

See 730302.

5A1E 730402 Layton, D.L. 1973. Methods of revegetation and restoration following a juniper wildfire. p. 12. In Proc. 26th Ann. Meeting Soc. Range Manage. Boise, Idaho. February 4-9, 1973.
5B4
5C
1E1
1E2 See 730301.

7305 Wyoming Agricultural Experiment Station Publications

730501 Fisser, H.G. and R.E. Ries. 1973. Aridland ecology research. 1972 results. Univ. of Wyo. Coop. Res. Rep. to the BLM. Sections I and II. Wyo. Coop. Exp. Sta. Sci. Rep. 521. 128 p.

2A 73050101 Section I - Precipitation pattern study.
3A

During 1972 precipitation was recorded seasonally as in past years on a total of 147 UW-BLM precipitation gauges. In addition, the data from 37 U.S. Weather Service Stations was compiled on a similar seasonal basis.

3A3 73050102 Section II - Exclosure study: production, cover and precipitation phases.
3A1
5A1D
5B
5C
2C

During 1972 production studies were conducted on 47 exclosure locations throughout the western Wyoming area, primarily in the Big Horn Basin, Wind River Basin and with some locations in areas to the south and in western Wyoming. With the clipping transects located both inside and outside of exclosures and with multiple sites at some exclosures, there were a total of 106 transects of vegetation which were clipped and sampled at these various exclosures during the summer period.

During the fall, shrub, cactus and mat-form species production was estimated with a 4 ft x 5 ft quadrat with weight unit sampling at all exclosures except those in the juniper type. A list of plants recorded at the exclosures is provided in the report as well as a section relating to the computation of utilization and adjustment of production for that.

3A1 730502 Shaver, J.C. and H.G. Fisser. 1973. Evaluation of vegetation types differentiated by aerial photo interpretation of panchromatic, color and infrared imagery. U.S. I.B.P. Biome Technical Report 238. 86 p.

5A1D
5B
5C
2C

The study was conducted on the I.B.P. Pawnee Site in northeastern Colorado to evaluate vegetation types differentiated by aerial photo inspection. The imagery used was found to be quite adequate for minute delineation work. A total of 13 vegetation types were recognized through use of photo inspection, leaf area index data and personal field observations. A total of 66 subtypes were designated on the aerial photographs by inspection through stereoscopic imagery studies. Each of the 66 subtypes represented a real situation of homogeneity which was different from adjacent areas. However, these differences were slight in many instances, allowing subtypes to be lumped together with other similar subtypes to arrive at the final 13 types. In most instances it could be readily seen through visual inspection of the photographs and in the field which of the subtypes were similar. The H-Group analysis was implemented to augment the visual inspection.

7306 Graduate Advisee Completions

3A3 730601 Whysong, G.L. 1973. Development and evaluation of statistical models for predicting perennial grass production in north central Wyoming. Ph.D. Dissertation. Univ. of Wyo. 150 p.

11E3
2A
2C
10D1

This study was conducted on sagebrush-grass ranges in north central Wyoming to determine the interrelationships between vegetation, soils, precipitation and perennial grass production, and to estimate production by means of statistical models. Data of vegetation cover and production were obtained from the inside locations of 12 exclosures over an average of 5 years. Two years of additional data were used to assist in evaluating the models constructed.

Thickness of the B soil horizon, depth of solum and depth to caliche layers were found to increase with precipitation. Conversely, thickness of the A horizon decreased with increasing moisture on the mesic areas (average annual precipitation between 9 and 12 inches). This was attibuted to accelerated erosion resulting from heavy grazing in previous years. Deeper soils on the xeric areas (average annual precipitation between 6 and 9 inches) had greater amounts of shortgrass cover and less of midgrass cover.

On mesic areas tall, mid and bunchgrass cover were found to increase with deeper soil development. Season and amount of precipitation were correlated with perennial grass production. All periods of precipitation considered correlated with production on the mesic sites. Precipitation occurring during winter and spring and winter-plus-spring from the previous year were important on xeric sites. Winter-plus-spring precipitation from the previous year was associated with maximum root depth and played a role in determining perennial grass production. Greater grass cover was found as wetter locations and shrub cover was higher on the more xeric sites. Tallgrass cover was found to increase with precipitation and shortgrass cover to decrease.

Depths to the concentrated root zone were greater when maximum root penetration was high. Deeper concentrated root zones were associated with increased amounts of shortgrasses and forbs. Perennial grasses located on the more xeric areas were shorter, tended to have a rhizomatous growth form and produced less herbage than those on the mesic areas.

Regression analysis was used to construct prediction and linear approximations of functional models to describe perennial grass production on xeric, mesic and all areas combined. Soil depths, root depths and precipitation were the major factors that explained variation in perennial grass production on the xeric areas. Total annual and winter-plus-spring precipitation from the previous year were equally important in influencing production on the mesic areas. Time and amount of precipitation was found to be more important on the xeric areas. Vegetation factors were least important in determining perennial grass production.

Functional models explained 86, 73 and 77 percent of the variation in perennial grass production on xeric, mesic and all areas combined, respectively. The prediction models accounted for 96, 91 and 83 percent of the variation, respectively. An evaluation of the models indicated that no one model was consistently superior in providing the best predictions of perennial grass production on an individual site basis.

There appeared to be a limit on the number of sites to include in any model in order to maintain practicality. Standard error of these models ranged was 39, 104 and 91 lb/acre respectively. The prediction models were composed of empirical equations which did not bear a resemblance in structure or function to the actual system but were abstract mathematical expressions whose purpose was to predict perennial grass production.

3A3 730602 Ries, R.E. 1973. Quantitative evaluation of herbage production and site factors on
11E3 ecologically diverse sagebrush-grass locations in Wyoming. Ph.D. Dissertation. Univ.
1E2 of Wyo. 101 p.
10D1

Seventeen sagebrush grass locations in north central, central and southwestern Wyoming were studied to evaluate the relationship between environmental factors and annual herbage production. The objectives of the study were: 1) to determine significant simple correlations between environmental factors and annual herbage production, 2) to isolate and quantify the affects of significant environmental factors on annual production, 3) to use these factors in a predictive equation to describe their relationship between environmental factors and annual production and 4) to predict a productive herbage yield potential for each of the 17 sagebrush grass study areas.

Forty-two independent variables (environmental factors) and one dependent variable (annual herbage production) were analyzed by regression analysis procedures. Significant simple correlations were found between 19 of the environmental factors tested and annual herbage production.

The 42 environmental factors were further analyzed by stepwise multiple regression analysis to isolate and quantify that group of factors best correlated with annual production. Those independent variables which explained the greatest amount of variation in annual production as a group were: 1) fall precipitation, 2) spring precipitation, 3) summer potential evapotransporation, 4) percentage very fine sands, 5) percentage course fragments and 6) percentage grass and forb composition.

All the important independent variables, except for summer potential evapotransporation, were positively correlated with annual production. The six respective multiple regression coefficients were 45, 64, -185, 17, 6, 4. These six independent variables were statistically significant at greater than the 90 percent probability level. With these variables in the multiple regression equation a multiple correlation coefficient of 0.80 and a coefficient of determination of 0.64 were observed and were statistically significant at more than the 95 percent probability level.

The statistically important environmental factors appeared biologically sound. Fall precipitation had a beneficial effect on the fall soil moisture. Some of the fall soil moisture may be carried through the winter in a frozen state and can become available for plant growth during the following spring. Spring precipitation, as defined in this study, occurred during active growth and could be readily used by the plant community for production. Moisture losses reflected by high summer evapotranspiration resulted in earlier depletion of soil moisture and termination of growth, thus reducing production. Percentage coarse fragments speeded intake of water into the soil profile, while very fine sands had caused a distribution of macropores and micropores which resulted in good regulation of water movement. Grass and forb composition was important in part because of the way annual production was defined, but may also have been important as the only plant community index in the equation from which production can be predicted.

Because of their significance levels, the environmental factors isolated in this study should be considered as possible factors in future site evaluation work. They offer the opportunity for continued study on an experimental basis to further substantiate or disprove their importance.

The overall regression equation, calculated in this study, may be summarized in words as follows. More annual herbage production, as defined in this study, can be expected from sites with greatest fall precipitation, spring precipitation, percentage very fine sand, percentage coarse fragments, percentage grass and forb composition and with lowest summer potential evapotransporation. The relationship between annual production and the six significant environmental factors described by this equation was limited by the range of the independent and dependent variables in its construction.

The predictability of the derived equation was evaluated and it appeared satisfactory as a prediction tool. An average annual production, representing an average productive potential, was predicted by the derived equation for each study area. The 17 study areas were ranked in order of decreasing productive potential which ranged from a high of 860 lb/acre to a low of 370 lb/acre. This ranking provided a method of comparing the intrinsic potential of each study site based on environmental factors found important by stepwise multiple regression analysis.

7401 Research Activities

3A3 740101 Exclosure study: production, cover and precipitation.

5A1D
5B
5C
2A
2C

The production, cover and precipitation data for the clip plot transects were again conducted as in previous years and the data reported in the subsequent annual report. Precipitation totals for 1974 averaged much lower than 1973. In the Sweetwater area, the decrease from the 1973 total was as much as eight inches (less than half). The average decrease was usually about three inches. Even though total precipitation was less than in 1973, forage production was similar for most areas dominated by cool season species. The above average forage production for the cool season species could have been the result of spring (April-June) moisture. It was noted also that cool season grasses cured early in the summer because of hot temperatures and moisture stress.

Areas dominated by blue grama *(Bouteloua gracilis)* had much lower forage production than 1973, and even lower than the long-term average. The western half of Wyoming received very little moisture during the summer from July 1 to September 1. Blue grama, being a warm season grass, needs the summer moisture for adequate growth. The precipitation records indicated that spring precipitation was about normal and that summer precipitation was much below normal.

1E2 740102 Exclosure study - phenology and growth.

2E
2F2
3A2
1E1
2A
5A1D

The exclosure study with the phenology work was conducted on 12 locations. Several of them which had not had clipping procedures conducted on them previously have been incorporated into the production system. These were the Red Wash -2, Black Mountain, Mesa Antelope and Owl Draw exclosures. Phenological classes have been identified both for vegetative and reproductive characteristics of the plant development and we have a separate characterization for the sagebrushes (*Artemisia* spp) as compared to those for the grasses. Included in the activities and data analysis of the phenology program was the regression analysis of 17 independant variables including environmental factors and annual production values with the two dependent variables which were defined as reproductive and vegetative phenophases. The simple relationships between plant growth and the environmental and productive characteristics was found to be inadequate and the stepwise regression was utilized for isolating and quantifying factors that may have had some kind of cause and effect relationship with plant growth rate. The variables which did appear to have a great amount of influence upon the variation of plant growth rate were: soil temperature at 15 in depth, percentage soil moisture at 6-12 in depth, percent soil moisture the previous month at 0-6 in depth, current air temperature and minimum air temperature since the last sampling date. As was expected, soil temperature and soil moisture were the more important driving factors in the phenology of the big *(Artemisia tridentata)* and black *(A. nova)* sagebrush and western *(Agropyron smithii)* and bluebunch wheatgrass *(A. spicatum)* complexes.

During late 1973 and primarily during early 1974, Fisser established photo location points of individual culms of grasses and branches of sagebrushes at a number of locations. Photos were taken at about 2-week intervals during the year and this continued for several years. Rods were driven in the ground upon which the camera was rested from which the photograph was taken. Spikes were driven into the soil halfway between the camera rest rod and the branch or culm that was to be photographed under magnification with the close-up lens. The locations of these sites were in the order for which the photographs were taken for a one-day series which began in the Mesa Antelope exclosure, Farson exclosure, Sweetwater exclosure, Shoshoni Ant -7 exclosure, the North Butte Relic area just north of Thermopolis and the Smilo exclosure.

2A 740103 Precipitation pattern study.

3A

The precipitation pattern studied includes continuing seasonal data collection at the UW/BLM gauges and compilation of seasonal data for almost 40 of the U.S. Weather Service Stations, primarily in the western half of Wyoming.

3A1 740104 Black Thunder pre-mining inventory.
8D
3A2 During 1974 considerable effort was carried on at the Black Thunder site by Lymbery with studies relating to the potential for grass re-establishment and work by Taha related to the physiological and photosynthetic relationship of the plants on the Black Thunder area. The permanent plot transects which had been located at each of the 22 study sites within and adjacent to the exclosures during 1973 by Ries were reread by Lymbery and Taha.
2F1

3A2 740105 Mule deer winter range evaluation.
7A1

The deer winter range studies during this year were primarily addressed by Gerhart with his studies near Tensleep and in the Medicine Lodge GRI exclosure area.

8D 740106 CIG pipeline seed plot study.
5B4
5C On July 17 and 18, Fisser and Johnson did an evaluation of all of the 17 pipeline plots and found that in some areas excellent grass establishment had been accomplished, but in some the seedings had been a complete loss.
1E2

1E2 740107 Shortgrass prairie blue grama culm density study.
2F1
2E The blue grama *(Bouteloua gracilis)* culm density studies were conducted for some period of time by Stroud but were discontinued during this year.

1E2 740108 Shortgrass prairie seed production study.
5B4
2F1 The seed production study was discontinued.
2E
3A3

7402 Journal Publications

5B4 740201 Fisser, H.G., M.H. Mackey and J.T. Nichols. 1974. Contour-furrowing and seeding on Nuttall saltbush rangelands of Wyoming. J. Range Manage. 27(5):459-462.
5A1D
3A2
5C At the Burnt Wagon research area northwest of Worland, the effects of a 1957 contour furrowing and seeding with crested wheatgrass *(Agropyron cristatum)* trial was re-evaluated. Vegetation analyses in 1962 indicated that total forage production was increased greatly by furrowing and seeding. Subsequent analysis in 1972 has shown that treatment effects were still evident but had declined in magnitude. In 1962, five years after the treatment, total herbage production on the treated areas was 972 lb/acre compared to 412 lb/acre for the untreated range. Greater production was due both to the yield of crested wheatgrass and improved vigor of Nuttall saltbush *(Atriplex nuttallii)*. By 1972 total production of the treated area had declined to 590 lb/acre but still was 54 percent greater than the control site.

Coincident with decreased production, foliage cover of crested wheatgrass decreased by 74 percent and Nuttall saltbush by 50 percent, part of which can be attributed to reduced water holding capacity of the furrows by about 30 percent from their original capability. The untreated native range produced 384 lb/acre in 1972 which was not appreciably different from production 10 years previously. Likewise, foliage cover percentages on the untreated native range remained relatively stable.

3A1 740202 Fisser, H.G. and J.C. Shaver. 1974. Remote sensing applied to the grassland biome. p. 293-295. In Proc. First International Congress of Ecology, The Hague, The Netherlands. September 10, 1974.
3A2
2F1

See 740302.

7403 Presentations

3A3 1E2 2A 2E 2B 740301 "Productivity differences of semi-desert grasslands in western Wyoming." February 5, 1974. 27th Ann. Meeting Soc. Range Manage. Tucson, Arizona. by H.G. Fisser.

The productivity differences of two blue grama *(Bouteloua gracilis)* grasslands in western Wyoming were investigated. Although several hundred kilometers from the recognized western boundary of the shortgrass prairie of the central U.S., these areas in the semi-desert portion of western Wyoming exhibited many of the characteristics of the prairies, in addition to environmental criteria normally associated with the intermountain region to the west.

Fenced exclosures at two locations were constructed in 1958, and all livestock grazing has been excluded since then. The study was initiated to evaluate the impact of livestock grazing on native grassland ecosystems, especially in terms of species composition, cover and production. Data obtained for 14 years from the sites showed that average annual herbage production was 236 kg/ha and 219 kg/ha under the protected and grazed treatments, respectively, at the xeric site. At the more mesic location, 360 kg/ha was produced on the protected site and 336 kg/ha outside the fenced exclosure area under grazing stress. On both sites, herbage production was consistently less on the grazed location and greatest within the fenced and protected areas.

To a casual observer, there appeared to be little visual difference between the vegetation within the exclosure and that outside with the livestock grazing treatment, since no significant vegetation composition changes resulted from the grazing stess. Much of the quantitative data exhibited insignificant differences resulting with protection from grazing treatment. Observation and subjective interpretation, however, indicated that many biologically significant changes had occurred.

The mesic site exhibited significantly greater productivity potential than the xeric. Production on the xeric site was limited not only by physical site characteristics but also because so few species were associated with the blue grama. A greater variety of grasses and forbs on a site, such as at the mesic location, significantly aids the potential for increased productivity by maintaining the productive level during dry years. It also utilizes added moisture during wet years.

3A1 3A2 2F1 740302 "Remote sensing applied to the grassland biome." September 19, 1974. First International Congress of Ecology, The Hague, The Netherlands. by H.G. Fisser.

The work for this report was conducted on the Pawnee site of the grassland biome program located south of Cheyenne about 25 miles. From the study of panchromatic, color and infrared aerial photographs, 13 separate vegetation types were distinguished on the Pawnee site. Some were difficult to recognize visually on the ground because they were distinguished by quantitative measures to a greater extent than by qualitative, visual aspects.

The weighted average LAI of blue grama *(Bouteloua gracilis)* were found to be highest in the light use pasture and lowest in the heavy use pasture, while that of buffalograss *(Buchloe dactyloides)* was found to be the opposite. Grazing had a lowering effect on LAI as did seasonal advancement during which growing conditions, especially availability of moisture, became less favorable.

Upland sites offered more favorable conditions for growth of blue grama than did lowland sites. Consequently, LAI was greater for this species on the upland sites. The opposite was true for buffalograss except that ridges also supported an abundant amount of this grass.

Soil type was evidently the major deciding factor in vegetation type. The influences of soil and topography intermingled, however, to cause the large scale pattern of vegetation types found on the Pawnee site. Spectral variation was readily visible for delineation of vegetation types with significant LAI differences. Inadequate processing control limited the derivation of LAI values from the photographic prints.

2C 2A 3A3 5B4 740303 "Relationship with soils and other environmental factors to rangeland production and rehabilitation." August 22, 1974. Bureau of Land Management Workshop Conference of Soils and Geology Specialists. Lander, Wyoming. by H.G. Fisser.

This presentation utilized some of the data from a number of the rangeland herbage production exclosures and related to work especially derived from that by Steger in the microclimate and the soil moisture studies. The primary stress of the presentation was that relating to the extreme importance of soil and other environmental factors that must be considered in the interpretation of rangelands and the determination of production, especially with regard to rehabilitation of drastically disturbed soils. Data taken from some of the Black Thunder work and soil plant relationships there were included in this presentation as well.

3A2 3A1 8D 2F1 740304 "Relationships of range production and rehabilitation to soils and other environmental factors." August 22, 1974. Workshop of BLM Soils and Geology Specialists. Lander, Wyo. by H.G. Fisser.

The materials for this presentation were in essence a restatement of the basic findings of the initial rangeland survey of the Black Thunder site near Douglas, Wyoming, and related to the basic concepts relating to range site criteria and factors of importance with regard to production and rehabilitation.

2F2 11E2 2A 2C 1E2 740305 "Environmental factors affecting annual herbage production at 17 sagebrush-grass locations in Wyoming." February 5, 1974. 27th Ann. Meeting Soc. Range Manage. Tucson, Ariz. by R.E. Ries.

Seventeen sagebrush/grass locations in north central, central, and southwestern Wyoming were studied for three growing seasons to evaluate the relationship between environmental factors and annual herbage production. Forty-two independent variables (environmental factors) and one dependent variable (annual herbage production) were analyzed by a multiple stepwise regression procedure.

Fall precipitation, spring precipitation, summer potential evapotranspiration, percentage very fine sand, percentage coarse fragments and percentage grass and forb composition explained the greatest amount of variation in annual herbage production. The six respective multiple regression coefficients were 45, 64, -185, 17, 6 and 4. All the important independant variables except for summer potential evapotranspiration, were positively correlated with annual herbage production. Each of these six independent variables was statistically significant at more than the 90 percent probability level. With these variables in the multiple regression equation, a multiple correlation coefficient of .80 and a coefficient of determination of .64 were observed; both are statistically significant at more than the 95 percent probability level.

These statistically significant environmental factors also appear biologically sound. Fall precipitation increases fall soil moisture. Some of this moisture can be carried through the winter in a frozen state and then become available for plant growth during the following spring. Spring precipitation, as defined in this study, occurred during active growth and should be used readily by the plant community for production. Moisture loss as reflected by higher summer potential evapotranspiration results in faster depletion of soil moisture and termination of growth, thus reducing production. Coarse fragments feed the intake of water into the profile, while very fine sands may cause a distribution of macropores and micropores, which results in beneficial regulation of soil water. Grass and forb composition was important because annual herbage production, as defined in this study, was greater in communities with a higher grass and forb composition.

7404 Published Proceedings/Abstracts

2B 740401 Fisser, H.G. 1974. Productivity differences of semidesert grasslands in western
3A3 Wyoming. p. 22-23. In Proc. 27th Ann. Meeting, Soc. Range Manage. Tucson,
1E2 Arizona. February 5, 1974.
2A
2E See 740301.

3A1 740402 Shaver, J.C. and H.G. Fisser. 1974. Detailed grassland type delineation with
3A2 panchromatic, color and infrared aerial photographs. p. 369-373. In V.G. Iglovikov,
2F1 A.P. Movsissyants, F.F. Sarganov, A.A. Stepanenko (Compilers). Improvement of natural and production of seeded meadows and pastures. XII International Grassland Congress, Moscow, USSR. June 12-16, 1974.

This study was conducted on the IBP grassland biome location, the Pawnee site, in northeastern Colorado, U.S.A. The Pawnee site was a relatively uniform grassland with blue grama *(Bouteloua gracilis)* abundant and dominant over most of the area. Significant phytosociological variation existed on the site resulting from soil, moisture, relief and grazing treatment influences on the vegetation. The study was initiated to develop, for the spatial modeling effort of the grassland biome program, a detailed vegetation map with definitive type units no larger than 0.1 ha (10 m x 10 m). To define these types in terms of weighted average LAI values, three forms of remote sensing, stereoscopic imagery - panchromatic, color and infrared photographs at a scale of approximately 1:5800 - were utilized for first order delineations. Ground truth definition was based on leaf area indices (LAI) derived from point contact data. The study area of 405 ha was separated into 13 sampling units related to management and utilization criteria. Each sampling unit was delineated into vegetation "subtypes" by stereoscopic inspection of pancromatic color and infra-red aerial photographs. A total of 66 subtypes were delineated on the 13 sampling units. Within each subtype vegetation was sampled with a point frame. The point hit data obtained was used in calculation of estimated LAI on each subtype.

The procedures of stereoscopic subtype delineation, transfer of these from the photographs to the baseline map, and computation of area occupied by each type, involved a number of innovations derived from other disciplines. These are discussed, as well as the use of the computer H-Group Analysis, which was utilized in combination with the photo inspection and general knowledge of the area to form the 13 significant vegetation types from the original 66 subtypes.

The weighted average LAI of blue grama was highest in the light use pasture and lowest in the heavy use pasture, while that of buffalograss *(Buchloe dactyloides)* was found to be the opposite. The total average LAI was nearly equal in all three of the grazing treatment pasture units. Implications which can be drawn from these data are discussed. Large scale vegetation pattern on the study site is discussed in relation to soil type and topographic location. Also see 740302.

4A 740403 Stroud, D.W. and H.G. Fisser. 1974. Influence of livestock grazing on growth patterns,
2E plant density and soil microrelief of a Bouteloua gracilis prairie. p. 347-351. In V.G.
1E2 Iglovikov, A.P. Movsissyants, F.F. Sarganov and A.A. Stepanenko (Compilers).
2C Biological and physiological aspects of the intensification of grassland utilization.
2F1 XII International Grassland Congress, Moscow, USSR. June 12-16, 1974.

A 0.5 m x 1 m field microplotter, from which were recorded locations along three axes, was used to determine the abundance and distribution of plant culm density, distribution and soil microrelief. The data were obtained from a fenced area of approximately 1 ha with a history of 30 years protection from grazing and from pastures of 130 ha each, subjected to light, medium and heavy rates of utilization for a similar period of time. The ratings from the digital meters on the microplotter were in inches at 0.1 inches (2.54 mm) intervals. Determination of plant locations was conducted by recording, on a tape recorder, the x and y-axis readings. Single stem plants were located by the x and y coordinates taken at their base. Plants forming mats or clumps, including blue grama *(Bouteloua gracilis)*, were located by x and y coordinates taken at the periphery of their bases (similar to that done by pantographing). Microrelief readings were taken on 20 parallel lines spaced at 2.5 cm along the x axis and run the 1 m length of the y axis. The horizontal x and y readings and the elevation readings from the vertical z axis pin were recorded at about 600 locations within a 0.5 m x 1 m quadrat. Location and elevation data along the parallel lines was recorded whenever a microrelief change occurred. Each location point was identified as to whether the surface consisted of mineral (bare soil or rock) or organic (living or dead plant) material.

Culm density within clumps of blue grama were measured using the field microplotter with a 5 cm^2 frame (2.24 x 2.24 cm) attached to the z axis rod. Two methods were used to sample culm density; first the 0.5 m^2 area of the microplotter quadrat was divided into 50 subplots, each 1 dm^2. When blue grama occupied at least 1/3 of a 1 dm^2 subplot a 5 cm^2 frame was placed with the base line midpoint located nearest the center of the subplot where the entire frame would be within a clump. The x and y coordinates were then recorded for the baseline midpoint and the number of culms counted.

The second procedure of measuring culm density was conducted with the same equipment. The culm density variation due to location within the clump was determined. Distinct clumps were identified within the permanent quadrats, then the 5 cm^2 frame was positioned at one edge of a clump. The readings were taken in a continuous manner, either parallel or perpendicular to the x axis, whichever direction was the shortest distance across the clump passing through the center.

Dynamics of culm growth and development were investigated to evaluate influence of grazing treatment. At each of the permanent upland quadrat locations 25 seed bearing culms were clipped at ground level at each of four location points near the base quadrat. Average weight of culms, numbers of spikes per culm and weights of spikes were determined.

The average total live vegetation cover of 50 percent was relatively uniform over the entire study area with little difference attributable either to soil type or grazing treatment. Blue grama was the dominant plant and occurred with a reduced cover percentage in the heavy use pasture. Buffalograss *(Buchloe dactyloides)* evidently increased in abundance under the heavy grazing rate. Clump sizes of blue grama were smaller on the lowland sites than on the upland sites. Grazing intensity apparently had little influence (reproductive culm density of blue grama was greatest on the heavy use pasture and on the lowland sites). Harvested culms from each of the three pastures indicated that weight per culm, numbers of spikes per culm and weight per spike decreased with increased grazing intensity. Culms were most abundant near the edge of the blue grama clumps and most sparse near the center. Measurements of microrelief variation exhibited greater roughness on the upland sites than on the lowland. Roughness was relatively uniform in the protected exclosure area and in the light and medium use pastures. In the heavy use pastures, both on the upland and lowland sites, the surface microrelief was significantly modified by grazing livestock impact.

3A3 740404 Fisser, H.G. 1974. Productivity potential differences of grasslands of semi-desert western
1E2 Wyoming. p. 116-124. In V.G. Iglovikov, A.P. Movsissyants, F.F. Sarganov and
2A A.A. Stepanenko (Compilers). Improvement of natural and production of seeded
2E meadows and pastures. XII International Grassland Congress, Moscow, USSR. June
2B 12-16, 1974.

See 740301.

3A3 740405 Whysong, G.L. and H.G. Fisser. 1974. Statistical models for description and prediction
11E3 of perennial grass production. p. 408-413. In V.G. Iglovikov, A.P. Movsissyants, F.F. Sarganov and A.A. Stepanenko (Compilers). Biological and physiological aspects

of the intensification of grassland utilization. XII International Grassland Congress, Moscow, USSR. June 12-16, 1974.

This study was initiated to develop statistical models capable of predicting perennial grass production based on relationships among vegetation, soils and precipitation. Vegetation data were obtained for seven years from 12 locations with significant variability of environmental conditions. Information concerning precipitation, soils and roots were obtained at each location. Sites were stratified into two groups and termed mesic and xeric based upon whether average annual precipitation was greater or less than 225 mm.

Investigation of factors which could significantly influence perennial grass production included evaluation of: 1) amount and distribution of precipitation, 2) thickness of soil horizons, 3) root depth of predominant perennial grass species, 4) kind and amount of vegetation and 5) growth form of perennial grasses inhabiting the study areas. Descriptive and prediction regression models were constructed to determine the relative importance of these factors and to predict perennial grass production on the xeric, mesic and all sites combined models. Five years of data were used for construction of the model and the additional two years of data were used for evaluation. The descriptive model for the five xeric sites included the listing of the perennial grass production as the dependant variable. The independant variables were: spring precipitation, winter plus spring precipitation lag one year, thickness of soil A horizon, depth of concentrated root zone, shrub cover, midgrass cover.

The two most important factors affecting perennial grass production were thickness of soil A horizon and concentrated root zone depth. A 1 cm increase in thickness of the A horizon between areas resulted in a 46 kg/ha increase in grass production. Similarly, as the depth of the concentrated root zone increased between sites by one cm, grass production was increased by 15 kg/ha. Precipitation occurring during spring and previous winter and spring were almost of equal importance when considered together but were of less importance than the soil and root factors. The model explained 86 percent of the variation in perennial grass production on the xeric sites and was statistically significant at the 0.01 probability level. The standard error of the model was 44 kg/ha which was about 18 percent of the mean production. Regression and correlation coefficients were statistically significant.

On the seven mesic sites, total annual precipitation was the most important variable, followed closely by precipitation from the previous winter and spring. The total model included the independent variables of: total annual precipitation, winter plus spring precipitation lag one year, shrub cover, tallgrass cover.

The model explained 73 percent of the variation in perennial grass production which was statistically significant at the 0.01 probability level and standard error was 116 kg/ha which was within 25 percent of the mean production. Coefficients were all statistically significant at the 0.01 level. The all sites model was intermediate in terms of quantitative values. For additional materials see citation 730601.

2A 740406 Ries, R.E. 1974. Environmental factors affecting annual herbage production at 17
3A2 sagebrush-grass locations in Wyoming. p. 21-22. In Proc. 27th Ann. Meeting Soc.
2C Range Manage. Tucson, Arizona. February 3-8, 1974.

See 740305.

3A1 740407 Whicker, F.W., A.W. Alldredge, H.G. Fisser, W.C. Hanson, J.E. Johnson and G. Post.
8D 1974. Ecological considerations for Project Wagon Wheel. Phase I: Pre-event
3A2 biospheric inventory, Impact analysis and mitigation, and baseline documentation. A
5C study conducted for El Paso Natural Gas Co. PNE-WW-35. 280 p.

This publication was a comprehensive report by all cooperators of the Wagon Wheel Project. The publication was divided into six major topics within which each of the major authors had some portion of input. The topics are:

1) Introduction

The increasingly serious shortage of natural gas in the United States led to a search by industry and government within the last several years for new methods of obtaining marketable gas. One such method, yet to be fully tested, involves nuclear stimulation of gas-bearing formations which are not sufficiently permeable to permit adequate gas flow into a conventional type well bore. Nuclear stimulation of such gas-bearing formations involves detonation of a nuclear device within the formation, which creates an underground chimney of broken rock in the vicinity, and causes increased permeability outside the chimney. Natural gas, diffusing into the large area of increased permeability, flows into the chimney and ultimately through the reentry well drilled into the chimney sometime following the detonation.

The principal questions surrounding the use of nuclear explosives to stimulate natural gas involves the effectiveness of the method, human safety, environmental considerations and the economic feasibility. This investigation was concerned mainly with environmental questions and human welfare from the standpoint of environmental amenities. The description and analysis of ecological systems associated with the Wagon Wheel program involved i) careful documentation and assessment of the populations and ecosystems prevailing within the area and region expected to be affected, ii) careful description and the nature and magnitude of potential disturbances expected to accompany the overall project, iii) predictions of the types and magnitudes of the expected effects and iv) careful pre- and post-event documentation of previously selected environmental criteria.

2) Environs of the Wagon Wheel Site

This section provided a brief description of the environmental setting of Project Wagon Wheel. It was prepared to place the study area into broad geographical, climatic and biological perspective and to provide a foundation for the specific biospheric inventory and analysis in section III. Portions of this section were quoted or paraphased from an earlier report by Fisser (see 720112).

3) Biospheric Inventory and Analysis

This section was divided into the following components: A. Natural and Cultivated Plant Communities, B. Aquatic Communities, C. Avian Populations, D. Non-domestic Mammalian Populations and E. Livestock and Agriculture. Each of these sections addressed the general ecological characteristics related to the specific kind of faunal and floral aspects noted in the region adjacent to the Wagon Wheel site.

4) Anticipated Ecological Effects of Project Wagon Wheel

This section had headings of: A. Ground Motion, B. Conventional Physical Activities and C. Radiation Effects.

5) Documentation of Anticipated Ecological Impacts

A. Natural Terrestial Plant Communities, B. Aquatic Communities, C. Avian Populations and Habitat, D. Non-domestic Mammalian populations and E. Agriculture.

6) Mitigation of Ecological Impact

Had topics and discussions similar to those of the two previous major categories. In addition, a topic of Timing Strategy is included.

7) Summary

Gave an overview statement of the characteristics noted by the scientists in the field work on site.

8) Literature

9) Appendices

7405 Wyoming Agricultural Experiment Station Publications

3A3 740501 Fisser, H.G. and R.E. Ries. 1974. Summary of data, 1962-1970. Aridland ecology research. Univ. of Wyo. Coop. Res. Rep. to the BLM. Wyo. Agr. Exp. Sta. Sci. Rep. 556. 132 p.
2A
3A
5A1D
2C
58
5C

This was a summary of production and cover and precipitation data from 19 exclosures for the period 1962-1970. The report included tabular material of vegetation production, cover and frequency and included precipitation data recorded at each of the exclosures. Maps were shown with locations of the various study sites. A list of plants identified as occurring on the sites included almost 200 species.

740502 Fisser, H.G. and L.H. Kleinman. 1974. Phenology and production on semiarid shrub types. 1973 results. Univ. of Wyo. Coop. Res. Rep. to the BLM. Sections I, II and III. Wyo. Agr. Exp. Sta. Sci. Rep. 607. 256 p.

2A 74050201 Section I - Phenology study.
2E
3A3
5A1D
1E2
2F2
1E1

During 1973 phenological recordings were taken at the 12 intensive phenology study sites for several periods during the year. The greatest number was six recordings during 1973. The annual report included the phenological scoring of the four prime species of each individual plant of the 20-plant series on transects for each of the prime species and phenological scoring by vegetative and reproductive character for every species which was located on each one of the 12 study locations.

2A 74050202 Section II - Precipitation pattern study.
3A

The precipitation pattern study was continued as in past years with recording of precipitation seasonally of four times a year at the some 120 UW-BLM gauges and the summarization of U.S. Weather Service Station data on a seasonal basis also. Precipitation during 1973 was much above average on the Boyson, Lower Government Draw, McGraw Flats, Sweetwater and Upper Government Draw exclosures. Average herbage production was above normal also. Other sites had varying precipitation and production relationships which may have resulted from the 1972 season of precipitation.

3A3 74050203 Section III - Exclosure study: production, cover and precipitation phase.
5A1D
5B
5C
2A
2C

During 1973 the production clipping and weight unit estimate procedures were carried on as in previous years. In the annual report the 1973 long-term averages of precipitation and clipped herbage production of the various treatments was compiled into a table for 20 of the exclosures.

1E2 740503 Shrader, J.A. and H.G. Fisser. 1974. Seed production ecology of the shortgrass prairie of the central United States. Wyo. Agr. Exp. Sta. Mimeo. 2 p.
5B4
2F1
2E
3A3

The production of seed by native species on the IBP grassland biome Pawnee site in northeast Colorado was investigated and the influences of grazing intensity, range site, fertilization, irrigation, long-term protection from grazing and abiotic factors were evaluated to determine the role of seeds by species as interrelated with biocontent and primary aerial productivity.

In the analysis of primary production and cycling of terrestrial ecosystems, the production of seeds as a biological process has shown that biocontent of the seeds and their associated reproductive organs contribute a significant portion of the total caloric value of the grass herbaceous production for a growing season. The amounts of seeds thus became important when quantifying energy flow through the ecosystem.

Permanent plots were established in each of three grazing treatment areas, fertilized and irrigated areas, two 30-year and one 10-year areas protected from grazing. The number of inflorescences per unit were recorded throughout the growing season. Six species: fringed sagewort *(Artemisia frigida)*, blue grama *(Bouteloua gracilis)*, buffalograss *(Buchloe dactyloides)*, needleleaf sedge *(Carex eleocharis)*, Plains pricklypear *(Opuntia polyacantha)* and scarlet globemallow *(Sphaeralcea coccinea)* were selected for detailed study because of their economic importance and extensive distribution over the grassland biome.

Phenological data was collected on the species under investigation as a supplement to and in conjunction with inflorescence counts and seed collection.

7406 Graduate Advisee Completions

3A1 7A1 10D1 740601 Spraker, T.H. 1974. An ecological study of the mule deer (*Odocoileus hemionus* ssp. *hemionus*) on Sheep Mountain, Albany County, Wyoming. M.S. Thesis. Univ. of Wyo. 81 p.

Ecological aspects of the resident and migratory herds of mule deer on Sheep Mountain were investigated during the period of February 3, 1972 to May 15, 1973. An estimate of 835 mule deer, made by ground accounts, was determined for the entire wintering herd.

Using histological analysis of fecal material, coupled with browse utilization surveys, the food habits of mule deer were shown for a complete year cycle. Fecal flotation, centrifugation and necropsy procedures were used in the isolation and identification of nine species of internal parasites, including two species of lungworm, for management and pathogenicity purposes. Thirty seven mule deer carcasses, found on the east slope of Sheep Mountain, were examined for probable cause of death with major emphasis on predation relationships.

Food habit studies showed that the mule deer utilized up to 80 percent shrub species in their diet during the winter period. During spring, summer and early fall, a mixture of shrubs, grasses and forbs was found in their diet, with grasses comprising the largest percentage. The preferred shrub species were antelope bitterbrush *(Purshia tridentata)*, true mountain mahogony *(Cercocarpus montanus)* and sagebrush (*Artemisia* spp.). Juniper (*Juniperus* spp.) was rarely, if ever, browsed by deer or elk.

3A2 7A1 10D1 740602 Hanson, C.E. 1974. An evaluation of the sagebrush and juniper types as mule deer winter range at three locations in Wyoming. M.S. Thesis. Univ. of Wyo. 122 p.

Because there has been little work in Wyoming evaluating the sagebrush (*Artemisia* spp.) and juniper (*Juniperus* spp.) mule deer winter ranges, this study was undertaken to evaluate these types by: 1) describing the vegetation of each area, 2) observing deer movements, 3) classifying deer into age and sex ratios, 4) relating nesting and feeding areas to vegetation, 5) determining diets and 6) evaluating the importance of specific browse species. The vegetation of each type was described by type and the canopy cover of juniper.

Deer moved onto the areas in late November and December. The daily movements peaked early and late in the day, except during periods of extreme cold or heavy snow cover. The majority of deer observed on all study areas were does and fawns.

The diet on all areas was primarily shrubs in the winter, but forb and grass use increased in early spring. Sagebrush, juniper and curlleaf mountain mahogany *(Cercocarpus ledifolius)*, which occurred on only one area, were the most important shrub species in the diet. Juniper was important as an emergency food during a harsh winter, but was also a major food item during a mild winter. The Twin Creek area, east of Lander, was of limited value as a winter range because of the poor condition of some sagebrush species. The Hall Creek Divide area, farther east of the Twin Creek area, was rated a fair winter range due to the lack of cover and a limited number of browse species. The Medicine Lodge area, above Hyatteville in the Big Horn Basin, was classified as fair to good winter range since it provided more of the food, water and cover requirements essential to mule deer.

Diet of deer varied a great deal from sample period to sample period, from year to year and from location to location. The variation in the diet between sample periods and years was linked to two possible factors. The first study year the winter was more severe at all locations, altering the feeding behavior of the deer. The movement of the deer on the area, and the collection of fecal samples in different locations on the study area, also could have altered the results of the diet analysis. Sagebrush was an important constituent of the diet on all study areas. The principle value of sagebrush was as a food source that was available when snow covered forbs and grasses. Mountain mahogany was only present on the Medicine Lodge area. The shrub was a substantial portion of the diet, over 30 percent, both study years. Mountain mahogany was available to the deer under most snow conditions. The value of juniper was as a food source and as cover. Juniper provided a substantial portion of the diets both study years. The increased use of juniper in the severe winter pointed out its importance as an emergency food. The juniper provided cover and exclusion for the deer.

3A1 740603 Gordon, R.C. 1974. Multispectral imagery for range vegetation type mapping and estimation of above-ground green biomass. M.S. Thesis. Univ. of Wyo. 45 p.
3A2
2F1

The work of this thesis was conducted on the area of the Baggs grazing system study west of Baggs, Wyoming, in the southwestern part of the state. The objectives of the study were: 1) to determine the usefulness of ERTS-1 multispectral imagery for mapping the rangeland resource and 2) to estimate the aboveground green biomass by relating it to plant reflectance. These projects were established with the idea of assisting the rangeland resource manager in his daily endeavors, not to eliminate him.

The first objective was successful, although hampered by color bias, cloud cover, loss of detail through enlargement and resultant poor location association caused by the latter. Both ERTS and the Skylab images provided useful maps for the range vegetation types. The ERTS-1 map contained representatives of all the vegetation types except one and it was the smallest type. The type not included was a seeded patch of crested wheatgrass *(Agropyron cristatum).*

The map produced from this image would serve as a useful survey base for the resource manager. Techniques for estimation of aboveground green biomass have not been confirmed due to the mechanical failure of the photometer. However the graphs of log transmittance versus above-ground green biomass indicated two things. First, it appears possible, for the Poison Buttes site, to estimate aboveground green biomass by relating it to log transmittance. Secondly, for pasture C, it may be possible to estimate the green biomass produced on this location by relating it directly to an area where aboveground green biomass can be estimated from the ERTS imagery. These two conclusions may be tested by relating the numbered 1974 summer imagery with the 1974 production data found in tabular materials.

For quick, accurate and detailed information on forage production, range condition and range vegetation types and subtypes, rangeland resource managers need a mechanical interpreter which will produce an unbiased hardcopy of the required dynamic information. To date remote sensing has been a useful but limited tool. The interpreter has only been able to view the images. The information is there; and until a mechanical method of interpretation is available to write down the results, this new tool cannot be utilized to its fullest potential.

7501 Research Activities

3A3 750101 Exclosure study: production, cover and precipitation.
5A1D
5B
5C
2A
2C

During 1975 the activity of determining production and cover at all of the exclosures was carried on as in previous years.

1E2 750102 Exclosure study: phenology and growth.
2E
2F2
3A2
1E1
5B

During 1975 the phenology activities were carried out at the 12 intensive sites as in the two years previous to this. Although phenology scorings were only taken three times at the Cumberland -3, Cedar Mountain and Red Wash -2 exclosures, data were recorded four times at Bud Kimball, Mesa Antelope, Owl Draw, and Upper Government Draw exclosures, but the recordings at Demer were taken 11 times beginning on January 31, 1975, and ending on December 29. At Farson, phenology was recorded 10 times with the first recording on February 6 and the last on November 26, while at Horse Creek the data also were taken 10 times with the first recording on January 24 and the last on December 30. At the Sweetwater exclosure, phenology was recorded eight times beginning with the first reading on January 25 and the last on October 25. There were only a few years during which recordings were taken during the months of November through March and 1974 for that reason is an important year with regard to the phenology recordings.

2A 750103 Precipitation pattern study.
3A

Precipitation records were maintained on all of the UW/BLM gauges for 1975 which was a much more moist year than usual with considerably greater amounts of precipitation occurring than normal. Most of the above normal precipitation occurred in March and April.

3A1 750104 Black Thunder pre-mining inventory.
8D
3A2
2F1

Cox and Taha did the primary field work on the Black Thunder pre-mining inventory during 1975. Most of the field activities centered on determining herbage production in the 22 exclosures and recording cover and composition data on the permanent plot transects associated in each of the 10 vegetation types determined by Ries during 1973 and 1974. Cox and Taha did additional field work specifically oriented toward the physiological interrelationships of different plants especially related to water stress. These data were utilized for their theses.

4B 75010501 Baggs grazing systems study - Impacts of grazing systems.
4C
1E4
2F2
3A2
7A1
7A3
6A3

During 1975 all of the permanent plot transects established in 1967 and 1968 by Gibbens were reread by Gerhart. He also reread most of the permanent browse transects established by Gibbens, but a number of them had not been adequately identified so that some new ones were re-established during 1975. Counts of antelope indicated an increase in number with about 120 on the pastures.

3A3 75010502 Baggs grazing systems study - BLM trend plots.
2F2
7A1
6A3
4B
4C

The BLM trend plots within the Baggs grazing system were reread by personnel of the BLM.

3A2 750106 1 ft x 1 ft permanent plot transects - Rawlins, Kemmerer, Pinedale.
2F2
5C

Permanent transects with plots of 1 ftx 1 ft were located in four exclosures of the Rawlins region in the Great Basin Divide area. These were at the Lamont, Lost Creek, Cyclone Ridge and Chain Lakes exclosures and also at three exclosures west of Kemmerer permanent plot transects were located. These were Boulder Ridge, Rock Creek Ridge and Mahogany exclosures.

8D 750107 CIG pipeline seedplot study.
5B4
5C
1E2

Fisser and Johnson did a re-evaluation of the 17 pipeline plots in the Big Horn Basin.

2C 750108 Exclosure soil profile studies with SCS Dutch Nick.
8B

Determination of soil profile characteristics was carried on at the Dutch Nick exclosure with the SCS on August 6.

3A1 750109 Wagon Wheel range survey.
3A2
5C
10D1
8D

The Wagon Wheel range survey work was carried on by Fisser during August, with determination of standing crop of the grass and other herbaceous species, and during October 11-12, with determination of weight unit estimates of shrubs and mat-form species.

1E2 750110 Shortgrass prairie blue grama *(Bouteloua gracilis)* culm density study.
2F1
2E

The blue grama *(Bouteloua gracilis)* culm density study was discontinued during 1975.

1E2 750111 Shortgrass prairie seed production study.
5B4
2F1
2E
3A3

The shortgrass prairie seed production study was discontinued during 1975.

7503 Presentations

8D 750301 "Pre-disturbance vegetation inventory and ecological monitoring procedures for rangeland strip-mining." February 12, 1975. 28th Ann. Meeting Soc. Range Manage. Mexico City, Mexico. by H.G. Fisser.
3A1
3A2
2F1

These data were collected as part of a multidisciplinary research team engaged by the Wyoming Environmental Institute for the Atlantic Richfield Company. ARCO has proposed to surface mine a coal lease block in the Powder River Basin of eastern Wyoming. We conducted vegetation inventory studies and established long-term monitoring procedures. The major objectives of the study were to: 1) develop descriptive information and define the vegetation types, 2) develop vegetation type map for the base area, 3) describe present and historical uses of the lease area and their impact on rangeland and 4) establish study locations to monitor the vegetation production.

The lease area rangelands survey was conducted in two conceptual categories. The first involved the inventory and classification of vegetation types with a history of grazing utilization but otherwise natural in aspects of soil condition and floristic character. The second category involved inspection and evaluation of areas subjected to massive physical disturbance. These had been identified as abandoned fields, pipelines, roadsides and oil well areas. The lease area was divided into 17 range sites as characterized by the Soil Conservation Service. Vegetation sampling was conducted within each of these range sites with sample lines. From these sample data, 10 vegetation types were delineated for the lease area by the described procedures. These were vegetation types which had been subjected only to grazing pressure and have not been influenced by other disturbances such as homesteading, oil development, or road construction. Six of these, or approximately 617 acres, were grasslands with the remaining four types, of about 5837 acres, exhibiting a shrub grassland aspect.

8D 750302 "Vegetation response to seismic motion from underground nuclear detonation." February 13, 1975. 28th Ann. Meeting Soc. Range Manage. Mexico City, Mexico. by H.G. Fisser.
3A1
3A2
2F1

This work involved documentation on predetonation permanently marked plots of vegetation productivity. The underground nuclear detonations of the Rio Blanco in northwestern Colorado failed to affect vegetation productivity. The permanently marked plots were established prior to the detonation for assessment of ground surface conditions and plant response to the underground nuclear detonations. Three nuclear devices were placed approximately 6000 feet below the surface and detonated on May 17, 1973. Prior to that time, Fisser established permanently located quadrats for study of vegetation responses to the detonation-induced ground movement. These quadrats were located from 150 yards to six miles from the surface point of the nuclear devices.

Prior to the detonation, vegetation cover was determined and recorded within each quadrat by use of photographic record. Subsequent photographs were taken of the exact quadrat locations on May 17, the day of the detonation, and then again on May 28, August 22 and finally, a year after the event, on May 27, 1974.

At the close-in location, seismic motion resulted in significant breaking up of the soil and some crumbling and falling of soil along steep creek banks. The soil movement failed, however, to significantly affect vegetation productivity. The vegetation loss from crumbling soil near the creek banks was minimal. The presence of plants throughout the detonation period and into the next year verified that seismic motion did not significantly modify or restrict plant growth.

1E2 750303 "Seed production ecology in the shortgrass prairie of the Central United States." February 12, 1975. 28th Ann. Meeting Soc. Range Manage. Mexico City, Mexico. by J.A. Shrader.
2E
2A
2C
2F2

Grazing stress tended to reduce seed production in three species of native shortgrass species as determined in the paper presented by Shrader and Fisser. In the study, production of seeds by native species in an area of northeast Colorado on the Pawnee grasslands research site was under investigation since 1967. Influences of grazing intensity, range site, fertilization, irrigation, long-term protection from grazing and abiotic factors such as weather were evaluated to determine the energy value and caloric content and above ground weight of seeds. The species examined included fringed sagewort *(Artemisia frigida)*, blue grama *(Bouteloua gracilis)* and Plains pricklypear *(Opuntia polyacantha)*. The importance of the study centers on the fact that the seeds of these species were utilized in grazing by livestock and game animals, and also supported small animals, insects and birds which live in the prairies. They also play a vital role in plant reproduction.

Throughout the growing season, the reproductive parts of the three species were collected at maturity. The flowers, seeds per flower head, weight of the seeds, caloric value and viability of the seeds were determined. Analysis was made of seed characteristics as influenced by grazing, range sites, season of use, irrigation and fertilization treatments. Results indicate the sagebrush was not affected after light and moderate grazing, but significantly affected when heavily grazed. The flowers, seeds per flower and total production were signficantly reduced when the grazing intensity of the blue grama increased. Finally, increased grazing also reduced the number of flowers of cactus which in turn reduced the total production of seeds.

8D 750304 "Pre-disturbance ecological studies improve and define potential for surface mine reclamation." October 22, 1975. Third NCA/BCR Coal Conference and Expo II. Third Symposium on Surface Mining and Reclamation. Louisville, Kentucky. by H.G. Fisser.

3A1
3A2
2F1

This report of the pre-mining inventory of the Black Thunder Site of the Atlantic Richfield Company in eastern Wyoming was given much more detail than in the presentation at the Society for Range Management meetings as noted under 750301. Following detailed description of the area, the climate, topographic characteristics, vegetation relationships of the range and inventory procedures which had been conducted on the site, the summary statement indicated that rangeland rehabilitation was possible and feasible following the mining operation and that native species were of significant importance. This was done with considerable discussion and presentation of quantitative data regarding ground cover, range condition class, carrying capacity and potential production of each range site as determined during the original survey.

The only logical procedure for establishing a stable biological system appeared to be one which would attain a native short mixed grass prairie vegetation. This alternative is considered the most desirable and most successful because of the following considerations: a) a mixed species grassland provides for a continuing grazing resource, b) native species are genetically and climatically adapted to normal environmental fluctuations such as seasonal and periodic drouth and seasonal temperature fluctuations, c) a mixture of native species provides a self-maintaining, ecologically stable growth cover resistant to grazing pressure, low water and nutrient availability and d) native species propagules are already present in the area in the form of seeds, root stalks and rhizomes which may be used to achieve an advanced successional stage which will be important in stabilizing soil against wind and water erosion, as well as allowing grazing productivity as soon as possible after disruption (presumably within two years).

2C 750305 "A study of rangeland soil movement characteristics." August 12, 1975. Amer. Soc. Civil Engineer. Watershed Manage. Symp. Logan, Utah. by H.G. Fisser.

8B
10D1

This study was established at a livestock-proof exclosure west of Kemmerer, Wyoming, in 1963 with an exclosure measuring 60 m x 120 m. The chemical 2,4-D was applied to one half of the fenced exclosure area and to several hectares outside the fenced unit. Treatment combinations thus included grazing, no grazing, shrub control and no shrub control. The procedures for measuring soil movement in the present study involved a simple mechanism termed an erosion transect, wherein the investigator measured the distance to the soil surface at relocatable points and from a constant height. Each transect consisted of two metal bars (1.2 m x 2 cm) which were driven vertically down into the soil until only about 2 dm remained visible above the soil surface. These were spaced 1.8 m apart. Data collection was obtained with the aid of two angle irons, attached horizontally to the vertical metal bars. One angle iron was notched to fit on to the permanent bars at the soil surface. The other angle iron had welded to it attachments to fit on to the top of the permanent bar such that no lateral movement could occur. Each of the angle irons had 10 marks based at 15 cm intervals. The vertical distance from the bar top to the soil surface was measured at each pair of marks. Five transects were installed in each of the four treatment combination areas in August 1963. Data were subsequently obtained in 1966 and 1970. Relocation position accuracy was adequate and measurements at each date were at essentially the same points. The permanent bars were apparently not influenced by frost heaving. Data interpretation involved study of treatment means and also of variance least squares analysis and appropriate F-tests of variance data.

Measurement of soil surface parameters indicated there was no significant soil movement response associated with the grazing treatment. A very strong statistical significance was associated with the chemical treatments, however, because of the definite soil movement which followed shrub control. The interaction of grazing impact and chemical control of shrubs resulted in significant soil movement which had a very important influence upon the ability of the native vegetation to express production changes.

1E2 750306 "Phenodynamics and environmental characteristics of sagebrush-grass rangelands." November 5, 1975. Wildland Shrubs Symp. Logan, Utah. by L.H. Kleinman.
2F
3A2
11F2 This paper described the early activities and developmental procedures in the phenodynamics and environmental relationship studies conducted in Wyoming for the Bureau of Land Management during 1973 and 1974. Intensive phenological studies have been located on 12 selected sagebrush/grass areas distributed throughout the western half of Wyoming. These areas express a broad ecological amplitude. Seventeen recognizable phenophases were devised for shrubs, forbs and grasses. All flowering species in each study were included in the study. Phenological data and environmental parameters (precipitation, maximum and minimum temperature, soil temperature and percent soil moisture) were measured year-round at two week intervals. Variation in times and amount of plant growth occurred within the same species, the same year and the same location. Soil temperature was one of the primary driving factors to phenology. Significant changes in plant growth occurred with only 0.5 degree centigrade change in soil temperature. The relationship between soil moisture and phenology was less obvious. Seed dispersal occurred about the same time through two complete growth cycles; whereas initiation of growth and bloom times were quite variable. Other environmental phenomena important in driving and regulating plant growth were precipitation from fall through spring, soil moisture at various depths and maximum-minimum air temperatures. Multiple regression and stepwise selection analysis procedures were used to substantiate or reject relationships observed from interpretation of field data, which included phenological information and 15 environmental variables.
1E1

7504 Published Proceedings/Abstracts

2C 750401 Fisser, H.G. 1975. A study of rangeland soil movement characteristics. p. 421-422. In Proc. Amer. Soc. Civil Engineers Watershed Manage. Symp. Logan, Utah. August 11-13, 1975.
8B

See 750305.

8D 750402 Fisser, H.G. and R.E. Ries. 1975. Pre-disturbance ecological studies improve and define potential for surface mine reclamation. p. 128-134. In Proc. Third NCA-BCR Coal Conference and Expo II. Vol. I. Louisville, Kentucky. October 21-23, 1975.
3A1
2A2
2F1

See 750304.

1E2 750403 Kleinman, L.H. and H.G. Fisser. 1975. Phenodynamics and environmental characteristics of sagebrush-grass rangelands. p. 161. In H.C. Stutz, (ed.) Proc. Wildland Shrubs Symp. U.S. Forest Service Shrub Sciences Lab. Provo, Utah. November 4-6, 1975.
2F
2F2
3A2
1E1

See 750306.

8D 750404 Fisser, H.G. 1975. Vegetation response to seismic motion from underground nuclear detonation. p. 37. In Proc. Soc. Range Manage. 28th Ann. Meeting. Mexico City, Mexico. February 10-14, 1975.
3A1
3A2
5C

See 750302.

8D 750405 Ries, R.E. and H.G. Fisser. 1975. Pre-disturbance vegetation inventory and ecological monitoring procedures for rangeland strip-mining. p. 40. In Proc. 28th Ann. Meeting Soc. Range Manage. Mexico City, Mexico. February 10-14, 1975.
3A1
3A2
2F1

See 750301.

1E2 750406 Shrader, J.A. and H.G. Fisser. 1975. Seed production ecology in the shortgrass prairie of the central United States. p. 29. In Proc. 28th Ann. Meeting Soc. Range Mange. Mexico City, Mexico. February 10-14, 1975.
5B4
2F1
2E
3A3

See 750303.

3A 750407 Whicker, F.W., A.W. Alldredge, H.G. Fisser and G. Post. 1975. Ecological Considerations for Project Wagon Wheel and Hydraulic Fracturing Activities. Phase II(a). Annual summary report for 1974. A study conducted for El Paso Natural Gas Co. PNE-WW-40. 33 p.
8D
3A1
3A2
5C

This report summarized those activities of the team during 1974 which were not completed in time to include in the Phase I report (PME-WW-35) which appeared in June 1974. This document contains data and observations from the vegetation, aquatic and mammalian studies. Most of the studies are for the purpose of pre-event documentation for Project Wagon Wheel. Some additional information was presented which was gathered to document ecological impacts of hydraulic fracturing experiments in the Pinedale Unit near the Wagon Wheel site. The main intent of this report is that of documentation rather than interpretation.

Ecological considerations relative to Project Wagon Wheel have been ongoing for more than three years. The vegetation studies were initiated on permanent quadrats in 1972. At that time, two macroplot locations were established. The experimental site (WW) was located immediately to the southeast of the Wagon Wheel site; and the control plot, identified as CWW, was located some eight miles to the west. These macroplot units consisted of nine transects, each with 10 permanently marked location points. In 1974 production and biomass of shrubs and half-shrubs and mat-forming woody plants was determined. In 1973 and 1974 production values of grasses and forbs were determined in addition to those of woody species. Whereas shrub production was determined in October, herbaceous production was measured with clipping plots during August near the close of the growing season.

Production decreases in 1974 reflected the low summer and fall precipitation. Such production fluctuations are expected conditions related principally to climate. Livestock grazing had comparatively little influence upon the vegetation. During August, cover data were determined within the quadrats prior to clipping and at the origin point quadrat of the permanently located sampling points. The cover data from these 180 quadrats were presented in tabular form. A minor decrease in woody plant cover was noted from 1973-74. Vegetation cover changes usually reflect only massive weather changes. Further evaluation of cover data on the 180 quadrats involved derivation of frequency percent data. Again little change occurred. If the true weather during the summer and fall of 1974 was significant to plant growth, these data might be significantly affected in 1975.

During the summer of 1974 additional studies on soil erosion were conducted. Erosion transects were established at selected locations by driving two concrete rebars (5/8 in x 4 ft) into the ground so that only 6 inches protruded above the soil surface. The bars were spaced 2 m apart. Readings were taken from 30 positions along the horizontal bar, resting on the tops of the driven bars. These readings were simply measurements in mm from the top of the bar to the ground surface. Changes in soil surface parameters can readily be expressed in terms of variance and appropriate statistical tests. Several of these transects were established at Bluff X and at Bluff 9.

An additional data set has been established at these two bluffs with the use of Bluff Edge transects. These consisted of 30 m transect lines permanently established with steel rods and were positioned about 2 m from the edge of the bluff. The distance of each meter position to the bluff's edge and any significant cracks in the soil surface was measured in cm.

Permanent record photographs were taken at all study locations. At the macroplot locations, a series of three photos was taken at the first sampling location of each of the nine transects. The photographs were taken at ground level at each of the erosion transects to present a soil surface profile. A general photo of each transect was taken also. At the Bluff Edge transects, a series of pictures was taken of each 5 m segment of the transect line.

In conjunction with the proposed hydro-frac operations, a series of transects, each with 20 permanently located (1 ft x 1 ft) quadrats, were established at the Pinedale -1 well and the Pinedale -7 well. There were two transects at the Pinedale -1 well and nine transects at the Pinedale -7 well. These transects were placed in several positions to represent existing vegetation, both on the well site area and on the undisturbed native vegetation in the immediate vicinity.

In conjunction with permanent transect lines, a series of slides were taken for a permanent photographic record. Photographs were taken along each line and oriented so that the well head of each location was visible. Additional photographic records were established at several seismic testing locations and along the road from the highway to the Pinedale -7 well. These data were a means of monitoring existing vegetation and conditions subject to the impact of vehicles for the hydro-frac operation. Except for the bladed area in the immediate vicinity of the wells, little disturbance was noted. During the spring and summer of 1974, a number of plants were collected for development of a series of herbarium specimens representing the plants, including trees, shrubs, forbs, grasses and grass-like species which occur in the Wagon Wheel area.

3A 750408 Whicker, F.W., A.W. Alldredge, H.G. Fisser, W.C. Hanson and G. Post. 1976. Ecological considerations for Project Wagon Wheel and hydraulic fracturing activities. Phase II(b): Annual summary for 1975 and project completion report. A study conducted for El Paso Natural Gas Company PNE-WW. 53 p.
8D
3A1
3A2
5C

The vegetation studies were initiated in permanent quadrats in 1972. At that time two macroplot locations were established. The experimental site (WW) was located immediately to the southeast of the Wagon Wheel site, and the control plot (CWW) was located eight miles to the west. Each of these macroplots consisted of nine transects, each with 10 permanently marked location points.

The macroplots were sampled using established range science techniques to provide estimates of annual production and cover of shrubs, grasses and forbs, and species frequencies based on 180 12 in x 12 in quadrats per macroplot. Production was measured by clipping and weight unit estimation, whereas cover was visually estimated. Frequencies were based on the presence of a given species within each quadrat.

The purpose of these measurements was that of pre-event baseline documentation. Since year-to-year fluctuations due to precipitation, grazing and other influences were expected, several years of pre-event and post-event data would be necessary to detect any subtle changes in these parameters which measure community structure and performance. The control macroplot allowed sorting of natural factors from the effects of the experiment. However, it was recognized, and the data bear out the fact, that the control plot was not a "perfect control," since precipitation, grazing and other features varied between the two plots.

The production data were summarized over four consecutive years for woody and mat-form species and over three consecutive years for herbaceous vegetation. While year-to-year fluctuations in production were evident, the variations were not major. Although the control plot seemed to consistently show higher grass production than the WW plot, the figures for total production were similar for the two plots. Annual variation in total production appeared rather minor for the years considered.

The cover percent data suggested higher values for woody and mat-form species in the WW plot but higher values for herbaceous cover in the CWW plot. Very little annual variation in shrub cover was evident, but a significant decrease in herbaceous cover was found in 1975.

Frequency data for individual species were summarized. As expected, the annual forbs tended to exhibit greater variation through time than the perennial grasses and shrubs.

In addition to the quantitative data, a library of photographs taken at fixed locations under specifically documented conditions was compiled. These photographs, which provided a separate aid to data interpretation, are available as needed.

7505 Wyoming Agricultural Experiment Station Publications

4B 750501 Gibbens, R.P. and H.G. Fisser. 1975. Influence of grazing management systems on
4C vegetation and wildlife habitat in the Red Desert Region of Wyoming. Wyo. Agr.
1E4 Exp. Sta. Sci. Mono. 29. 23 p.
2F2

7A1 This monograph report of the Baggs grazing systems study included the
7A3 comprehensive analysis of all data from 1967-1971. At the time of the completion of this
6A3 study in 1971, it was determined that additional field sampling was not necessary on an annual basis because of the very slow vegetation response to grazing treatments. In fact, it appeared that little response to the treatments had occurred during the five years of the study. The reasons were: 1) a substantial reduction of grazing use had been caused by the BLM adjudication near the beginning of the study, 2) a considerable amount of rangeland in all systems had been sprayed for sagebrush control and 3) the combination of reduced grazing intensity, increased forage production associated with the big sagebrush *(Artemisia tridentata)* control and the fact that during the few years immediately prior to and during the years of study, weather conditions were generally very favorable and resulted in good vegetation growth as well as significant amounts of grass establishment.

Since the vegetation were being modified primarily by climatic factors and very minimally by the grazing management, subsequent work would be continued on a four year basis; and fit with the cycles of the four pasture Powder Rim system. It was implied that the next data set would be obtained during the summer of 1975.

8D 750502 Fisser, H.G., F.K. Taha, G.A. Lymbery and J.R. Cox. 1975. Herbage structure,
3A1 productivity and phenology on Black Thunder. Annual Progress Report to Atlantic-
3A2 Richfield Company. 1974 results. Wyo. Agr. Exp. Sta. Sci. Rep. 662. 118 p.
2F1

This report of data through 1974 of the Black Thunder herbage structure, productivity and phenology study included a comprehensive compilation of the herbage production data from the 22 exclosures and the vegetation cover and composition data from the associated permanent plot transects. The permanent quadrat location in this study was a 30 cm x 60 cm frame upon which was mounted a point frame with 20 pins. The base frame was located along the transect line randomly and with permanent identification by long rods driven flush with the ground surface. The base of the point frame was located on the metal rods and leveled and then the point frame with 20 pins was set in five positions on the frame so that 100 points were recorded within each one of the quadrats. Relocation precision was generally less than 3 mm differences for each pin. This data was presented in greater detail in 820202.

750503 Fisser, H.G. and L.H. Kleinman. 1975. Phenology and production on semiarid shrub types. 1974 results. Univ. Wyo. Coop. Res. Rep. to the BLM. Sections I, II and III. Wyo. Agr. Exp. Sta. Sci. Rep. 729. 399 p.

1E1 75050301 Section I - Phenology study.
1E2

2E This 1974 field activities report of the phenological study in the western half of
2F2 Wyoming included a listing of the 12 exclosures and identified the prime species which
3A2 occurred in each exclosure. A list was given of some 100 plant names which occurred on the 12 exclosures. A figure was included which showed the generalized relationships of time of initiation of growth, of floral stalk appearance, of bloom initiation, of drying, seed dissemination starting and plant being dried. Tabular material included compilation, by exclosure, of phenological scores of all species on each of the 12 locations. Phenological

scores were obtained from each of the 20 plants of the prime species (western wheatgrass *(Agropyron smithii)*, bluebunch wheatgrass *(A. spicatum)*, black sagebrush *(Artemisia nova)* and big sagebrush *(A. tridentata)*). In addition, there was information relating to plant height dimensions of the clumps or bushes, the age class of the plant, the dead branch material, twig length, and seed stalk length. For additional comment, see 740102.

2A 75050302 Section II - Precipitation pattern study.
3A

The 1974 precipitation data were compiled in the report as in previous years with the seasonal precipitation for the 110 UW/BLM gauges and some 40 U.S. weather stations. Precipitation totals for 1974 average much lower than 1973. In the Sweetwater area the decrease from the 1973 total was as much as 8 inches (less than half). The average decrease was usually about 3 inches. For additional information, see 740103.

3A3 75050303 Section III - Exclosure study: production, cover and precipitation.
5A1D
5B
5C
2A
2C

The 1974 herbage production studies were presented in the continuing format by transect for each of the exclosures. Herbage production during 1974, even though precipitation was much less than the previous year, remained fairly similar for most of the areas dominated by the cool season species, which may have been a result of the early spring (April-June) moisture. It was noted also that the cool season grasses cured very early in the summer of 1974 because of high temperatures and associated moisture stress. Areas dominated by blue grama *(Bouteloua gracilis)* had much lower forage production than 1973 and even lower than the long-term average. The western half of Wyoming received little moisture during the summer from July 1 to September 1. A comparison of the 1974 precipitation and production with average precipitation and production for the years 1962-1972 was shown on 21 exclosures. For additional comment, see 740101.

7506 Graduate Advisee Completions

8D 750601 Lymbery, G.A. 1975. Vegetation re-establishment potential of strip-mined soils, Black Thunder Basin, Wyoming. M.S. Thesis. University of Wyoming. 62 p.
3A1
3A2
2F1
2E
2C
1E2

This work relating to the M.S. thesis developed by Lymbery was conducted on the Black Thunder site in the pre-mining inventory program. The purpose of this study for the thesis was to determine, through greenhouse observation and statistical analysis, the effect of shallow clayey, loamy, sandy topsoil and acid clay shale on 1) the species of native plants that will establish themselves from seeds and rhizomes, 2) the aerial production and below ground biomass of the naturally propagating species, 3) the aerial production and below ground biomass of nine species of grasses and forbs planted in these soils materials and 4) the effect of season of planting.

Topsoils were collected from transects 12 ft long by 1 ft wide and 2 in thick on each of the five soil types and standard soil analysis procedures were conducted on the soil materials. The potting arrangement was conducted in a randomized complete block with a two factor factorial design for determination of relationships of soils and species effect on root and aerial biomass.

To determine the effect of the date of soil removal on the type and quantity of vegetation produced, soil samples containing vegetative material were gathered at periodic intervals throughout April, May, June and July of 1974. Samples were collected from the same 12 ft x 1 ft x 2 in transect used originally in this experiment. These samples were also placed on the clay shale material and allowed to propagate naturally as were the original transplant control plots. Roots were weighed after being washed by the machine provided by the Rocky Mountain Forest and Range Experiment Station in Laramie.

Results of the species used and methods employed gave the following conclusions: 1) yellow sweetclover *(Melilotus officinalis)*, alfalfa *(Medicago sativa)* and transplant control pots dominated above ground and below ground production in all soil types, 2) pubescent wheatgrass *(Agropyron trichophorum)*, intermediate wheatgrass *(A. intermedium)* and blue grama *(Bouteloua gracilis)* produced satisfactory above ground and below ground biomass, but did not dominate production, 3) crested wheatgrass *(Agropyron cristatum)*, desert wheatgrass *(A. desertorum)*, streambank wheatgrass *(A. riparium)* and *(Onobrychis viciaefolia)* sainfoin produced very poor yield. The use of these species for acid spoil rehabilitation would be questionable, 4) clayey soil yielded the most forage, sandy soil the least which may be attributed to poor water holding capacity of the sandy soil and possible lack of nutrients due to leaching, 5) soil samples collected in April and May provided diverse species composition and satisfactory biomass production, 6) soil samples collected in June and July were marked by poor production and a preponderance of pepperweed *(Lepidium densiflorum)* and sixweek fescue *(Vulpia octoflora)*. Some pots produced nothing, 7) transplant control pot production was dominated by western wheatgrass *(Agropyron smithii)* and blue grama, any attempt at seeding should include these two species, 8) the acid nature of the parent material may have been responsible for poor growth of certain species, possibly due to a lack of nutrients, 9) favorable greenhouse conditions were responsible for the excellent production of certain species and 10) irrigation, fertilization, selected seeded species, use of topsoil as a plant source and adequate seedbed depth will enable strip-mined land to become fully productive within an acceptable time span.

7601 Research Activities

3A3 760101 Exclosure study: production, cover and precipitation.
5A1D

5B The herbage production research activities were continued during 1976 as in 1975
5C with data subsequently listed in the annual report of data for that year for each of the
2A 40 exclosure locations at which herbage production was obtained.
2C

1E2 760102 Exclosure study: phenology and growth.
2E

2F2 During 1976 the phenology studies were continued at 12 exclosures for intensive
3A2 analysis of study on the four prime species.
1E1
2A
5A1D

2A 760103 Precipitation pattern study.
3A

Precipitation recordings were continued on the 234 UW/BLM gauges and on the 44 U.S. Weather Service stations.

3A2 760104 2 ft x 20 ft and 4 ft x 4 ft permanent plot studies.
2F2

During 1976 all 2 ft x 20 ft and 4 ft x 4 ft permanent plots were reread at all of the exclosures upon which they are located, including the Cumberland exclosures.

3A1 760105 Black Thunder pre-mining inventory.
8D

3A2 Continued work on the Black Thunder mining inventory during 1976 included
2F1 the production and cover studies as previously established with the exclosures and permanent plot transects, as well as intensive moisture stress studies carried on by Cox.

3A2 760106 1 ft x 1 ft permanent plot transects - Kemmerer-Boulder Ridge exclosures.
2F2

5C At the Boulder Ridge exclosure west of Kemmerer, a series of 10 permanent
10D1 plot transects were located at the combination game and livestock exclosures of this site.

8B 760107 CIG pipeline seed plot study.
5B4

5C The pipeline seed plots were reread and re-evaluated by Fisser and Johnson again
1E2 during 1976.

1E1 760108 Grass seed sources at A.C. Hull 40s plantings.
1E2

5B4 In July Fisser met with Barker of North Dakota Agriculture Research Group out of
2F1 Mandan and collected seed from crested wheatgrass *(Agropyron cristatum)* and Russian
2E wildrye *(Elymus junceus)* at the Big Flat exclosure. The hope is that these seeds would
3A3 provide a new genetic base of materials derived from A.C. Hull's 1940 plantings, which currently would not be available. He was to take these seeds to his greenhouse in Mandan and establish new plantings and determine their relationship to existing genetic sources of these two species.

7602 Journal Publications

1E3 760201 Cox, J.R. and H.G. Fisser. 1976. Methods of measuring environmental parameters on a pre-disturbed coal site. J. Colo.-Wyo. Acad. Sci. 8(1):45-46.

8D
3A1
3A2
2F1

Studies were set up in 1973 by the University of Wyoming to gather baseline data for the Atlantic Richfield Company in the Little Black Thunder Creek area, 43 miles south of Gillette, Wyoming. Twenty-two study sites were selected from 10 vegetation types. Cattle-proof fences were constructed at each site and three permanent 100-foot transect lines were randomly selected. Along each line, five quadrat locations were randomly selected. Measurements of vegetation were obtained with a modified ocular point frame to determine species composition and frequency (see 820202). Total above ground biomass for each plant species were determined by destructive sampling. The 1975 results showed a 10 percent greater production over 1974, due primarily to precipitation increases in June 1975. Additional studies were initiated in 1974 to determine carbohydrate reserves in relation to available soil moisture, soil texture, plant phenology, soil surface temperature and precipitation. Five species were selected for this study: western wheatgrass *(Agropyron smithii)*, blue grama *(Bouteloua gracilis)*, Junegrass *(Koeleria cristata)*, needleandthread *(Stipa comata)* and big sagebrush *(Artemisia tridentata)*. Root samples were collected on the first and 15th of each month from May to September, and collected monthly for the remainder of the year. Soil moisture readings were taken at 6, 12 and 18 inch depths. Soil samples were taken for water potential, bulk density and soil texture. Preliminary results show carbohydrate reserves to be closely related to soil moisture and plant phenology.

4B 760202 Gerhart, W.A. and H.G. Fisser. 1976. Some influences of grazing systems upon vegetation and wildlife. J. Colo.-Wyo. Acad. Sci. 8(1):68.

4C
1E4
2F2
7A1
7A3
6A3

In 1967 a program was initiated to study the influences of various grazing systems upon the vegetation and wildlife on a 68,000 acre study site west of Baggs, Wyoming. The area was divided into three grazing systems including a rest-rotational unit, a deferred unit and a season-long unit. Prior to the initation of the study, several management projects were undertaken, including big sagebrush *(Artemisia tridentata)* spraying, a 25 percent reduction in stocking rates, fence and exclosure construction, and water developments. Vegetation cover was obtained on 95 permanent transects in the brush-grass type and on 1400 feet of permanent combination belt, line point and quadrat transects in the Utah juniper *(Juniperus osteosperma)* type. During the study period (1967-1975) only slight changes were noted in the Utah juniper type. The brush-grass type showed a steady increase in the cover of perennial grasses under all grazing systems, and no differences were readily observed between grazed and non-grazed (exclosures) areas. Grazing pressures apparently were not sufficiently heavy to stress the plant communities for detection of responses to management systems. With the possible exception of pronghorn antelope mortality, the grazing system had no readily apparent adverse effects upon wildlife species in the area, and may have actually improved the wildlife habitat.

1E2 760203 Kleinman, L.H. and H.G. Fisser. 1976. Phenology and environmental relationships of sagebrush-grass rangelands. J. Colo.-Wyo. Acad. Sci. 8(1):48-49.

1E1
2E
2F2
3A2
2A
5A1D

This was a paper relating to the phenology and environmental relationships study by the University of Wyoming. This was a report of the activity being carried on by the University addressed to the phenology and environmental relationships of sagebrush *(Artemisia)*-grass rangelands.

Time and rate of plant development is of primary importance when interrelated with vegetation production, environmental phenomena and management of rangeland. Little detailed work has been done relating phenology with forage productivity and environmental information of aridland species of the intermountain United States. The phenology of plants determines grazing pattern and actual planning for management. Because of the interrelationship of time and rate of plant development and the resource management decision making process, a cooperative research program was initiated between the BLM and the University of Wyoming.

The basic objective was determining interrelationships of vegetation production, environmental characteristics, grazing, management decision making and phenodynamics. The environmental data obtained at each sampling data included precipitation since the previous recording measured at gauges installed at the study area, maximum and minimum air temperature were recorded from thermometers installed at each location, three samples of soil temperature in each vegetation were taken at a common 15 in depth, three samples of percent soil moisture were determined by gravimetric sampling techniques at each vegetation type with a "king tube" at three different soil depths - 0-6 in , 6-12 in and 18-24 in ; close-up photographs for photographic record of the growth of the two sagebrush *(Artemisia tridentata* and *A. nova)* and two wheatgrass *(Agropyron smithii* and *A. spicatum)* species were taken, and there were at some exclosures a series of slides taken of identical branches for several years. The phenological scoring system has also been identified in the annual reports of the Wyoming Agricultural Experiment Station.

1E2 2F1 2C 3A2 8B 10D1 760204 Taha, F.K. and H.G. Fisser. 1976. Production of a short-grass prairie ecosystem as related to soil characteristics in eastern Wyoming. J. Colo.-Wyo. Acad. Sci. 8(1):50.

Physical and chemical soil properties as related to herbage production of major forage plant species were studied on the Black Thunder lease area study south of Gillette, Wyoming. Soils were divided into 13 management groups according to depth, texture, development and drainage. Further physical and chemical analyses were performed within each management group. At each group study site, an exclosure 30 ft x 30 ft was established. Ten clipping quadrats 30 cm x 90 cm, were randomly located within each exclosure for grass and forb production estimates. Shrub production was estimated by a modified double sampling technique. A stepwise multiple regression analysis was used to relate soil factors to herbage production.

Soil physical characters were found to significantly affect annual herbage production with air-water ratio, moisture equivalent, modulus of rupture showing positive correlations and texture (clay) and hydraulic conductivity showing negative correlations. Soil chemical characteristics appeared to exert little influence, if any, on herbage production. Organic matter was the only such parameter that showed a significant positive correlation.

1E3 2E 2F1 760205 Cox, J.R. 1976. A mixed method of carbohydrate analysis. J. Colo.-Wyo. Acad. Sci. 8(1):30.

The most widely used method for extracting and determining concentration of total available carbohydrates (TAC) are the enzyme, acid and anthrone methods. The enzyme method is time consuming and requires the mixture of several chemicals and a periodic standardization of chemical solutions. The acid method requires less time, but periodic standardization of solutions is also required. The anthrone method takes less time than the previous two methods, but the results are biased by the impurities in solution. Because of the shortcomings of the three methods used separately, an improved method for extracting TAC was developed, using a mixture of the enzyme and anthrone methods. The time required for TAC, per sample: 30 min. for the enzyme method, 24 min. for the acid method and 16 min. for the mixed method. The number of chemical reagents needed for each method: 10 for the enzyme, eight for the acid, three for the anthrone and five for the mixed. The cost to set up the acid method is approximately $3,000; the cost of the mixed method is approximately $1,200 (excluding the cost of the Beckman 25 spectrophotometer). The mixed method averaged 12 percent higher apparent carbohydrates than the enzyme method for crested wheatgrass *(Agropyron cristatum)* and 15 percent higher for big sagebrush *(Artemisia tridentata)*.

7603 Presentation

10D1 760301 "Aridland ecology and phenology research in western Wyoming." March 3, 1976.
1E2 Regional Rangeland Research Symposium. Bureau of Land Management. Denver,
2F2 Colorado. by H.G. Fisser.
3A2
2A This meeting was held with several individuals from a variety of disciplines from the
2E Bureau of Land Management and was concentrated on identifying aridland ecology
1E1 problems and possible related phenological research activities. In this presentation
5A1D information was given concerning field procedures of the University of Wyoming phenological research study, the characterization of the vegetative and reproductive phenology scoring systems and the information summarized through 1973, 1974 and 1975.

4B 760302 "Grazing systems research on northern desert shrub rangelands of Wyoming." March
4C 11, 1976. GP-9 Committee Annual Meeting. Denver, Colorado. by H.G. Fisser.
1E4
2F2 The information given here was a comprehensive presentation including the work
7A1 done from 1967-1971 by Gibbens and that done by Gerhart in 1975 on the Baggs
7A3 grazing systems unit west of Baggs, Wyoming. (See 760202.)
6A3
10D1
3A2 760303 "Aridland ecological research activities in Wyoming and relationships to Libyan
2F2 rangelands." May 15, 1976. College of Agriculture, University Al Fateh Special
5C seminar. Tripoli, Libya. by H.G. Fisser.
5C

This was a broad-based description of all the major aspects of the aridland ecological research activities conducted at the University of Wyoming and the various aspects of aridland vegetation structure, production and ecological relationships were related to the general kind of Libyan rangelands.

1E2 760304 "Vegetative response of *Bouteloua gracilis* to four levels of utilization." February 18,
2E 1976. 29th Ann. Meeting Soc. Range Manage. Omaha, Nebraska. by D.W. Stroud.
6A3
2F1 This study was conducted in conjunction with the US/IBP grassland biome program on the Pawnee site in northeastern Colorado. Blue grama *(Bouteloua gracilis)* was the dominant species on this shortgrass prairie region. The influence of grazing treatments on this grass was evaluated by study of shoot density, plant distribution and soil microrelief. Data were obtained from a fenced area with a history of 30 years protection from grazing, and from pastures subjected to light, medium and heavy rates of utilization for a like period of time.

Measurements were made using a .5 m x 1 m field microplotter with horizontal X and Y axes and a vertical Z axis. Blue grama was the dominant plant under all grazing treatments, with cover values of 38, 44 and 56 percent in the exclosure and light and medium use pastures respectively. In the heavy use pasture, this grass occurred with only 32 percent cover, an apparent response to grazing stress.

Measurement of blue grama clump sizes was conducted at all permanent quadrat locations. On the Ascalon upland soil sites, the mean clump size was 2.9 dc^2. Clump sizes on lowland sites of 47-AB Undifferentiated soils were smaller with a mean value of 2.3 dc^2. Clump size differences among pastures did not appear to be related to grazing treatment. In the protected exclosure area and light use pasture, the sizes were 3.1 and 2.0 dc^2 respectively, while in the heavy use pasture, the clump size was 2.3 dc^2. An apparent anomaly was noted with the very large clump size of 3.4 dc^2 in the medium use pasture. Density of blue grama shoots was determined on the basis of the numbers of live shoots per 5 cm² frame. The average density was slightly greater, with 5.7 shoots per frame on the lowland sites with the 47-AB Undifferentiated soils, than on the upland Ascalon soil sites, on which were found 5.0 shoots per frame. In the exclosure protected from grazing and in the light use pasture, only 3.6 and 3.8 shoots per frame, respectively, were recorded. The density was slightly greater in the medium use pasture with 4.6 shoots per frame. A significantly greater density was recorded in the heavy use pasture with 7.6 per 5 cm² frame. Shoot density was significantly related to location within clumps. Density was greater, with a value of 5.6 shoots per 5 cm² frame, in the first 2 cm from the plant unit margin. In the third and fourth cm from the edge, density was reduced to 3.0 shoots per frame, while on the fifth cm to the center of the clumps, density was only 2.6 shoots per frame.

The measurement of blue grama culm growth and development by clipping the seed-bearing culms, indicated a significant response to grazing treatment. The mean weight of the 25 clipped culms per sample was 2.4 g on the light use pasture, but only 2.0 and 1.3 g on the medium and heavy use units respectively. Mean numbers of reproductive spikes per sample of 25 clipped culms were 41, 31 and 35, on the light, medium and heavy use pastures. Although the numbers of spikes were not significantly different, weights of the spikes were greatly influenced by grazing treatment, with a mean value of 0.6 g per sample from the light use pasture, but much lower values of 0.5 and 0.3 g from the medium and heavy use areas.

1E2 760305 "Physical and chemical analyses of range sites and interrelationships with herbage production and forage value potentials." February 18, 1976. 29th Ann. Meeting Soc. Range Mange. Omaha, Nebraska. by F.K. Taha.

2C

2F1

3A2

8B Herbage production and forage value of plant species at Black Thunder, Wyoming, were investigated and influences of different site factors were evaluated. Major site factors studied were soil physical and chemical properties, precipitation and temperature. These site factors were shown to exert great influence on forage quantity and quality.

10D1

The investigation of the soils of the study area was conducted according to the latest accepted soil survey practices as used by the National Cooperative Soil Survey program. Soils were divided into 13 major management groups according to depth (very shallow, shallow, moderately deep and deep), texture (coarse, moderately coarse, fine and very fine), development (immature and developed) and drainage (well drained and imperfectly drained). Further, physical and chemical analyses were performed within each major soil group. This included hydraulic conductivity, modulus of rupture, air-water permeability ratio, moisture equivalent, 15-bar water, settling volume, total salts, soil pH, percent organic matter, cation exchange capacity and available nutrients (NO_3, NO_4, P, K, Ca, Mg, S, B, Zn, Fe, Mn and Cu). Other parameters of range sites investigated were precipitation and maximum and minimum temperatures. These two environmental factors were examined at the beginning of each month.

At each of the soil group types an exclosure 30 ft x 30 ft was established. Ten clipping quadrats, 30 cm x 90 cm, were randomly located within each exclosure for grass and forb estimates. Species were clipped at ground or crown level. Shrub production was estimated by a modified double sampling technique. Major forage species were collected from each soil managment group for chemical analysis and digestibility trials. These were blue grama *(Bouteloua gracilis)*, western wheatgrass *(Agropyron smithii)*, sandberg bluegrass *(Poa secunda)*, prairie junegrass *(Koeleria cristata)*, needleandthread *(Stipa comata)*, big sagebrush *(Artemisia tridentata)* and fringed sagewort *(A. frigida)*. Standard methods of chemical analysis were used for nutrient determination and a modified Tillian Turian two stage in-vitro technique was utilized for digestibility trials.

A stepwise multiple regression program was used to examine all site factors and their interrelationships to herbage production and forage value. The findings were being tested for development of a computer model to be used in similar areas.

1E2 760306 "Vegetation re-establishment potential of strip-mining soils, Black Thunder Basin,
8D Wyoming." February 19, 1976. 29th Ann. Meeting Soc. Range Manage. Omaha,
3A1 Nebraska. by G.A. Lymbery.
3A2
2F1
2E
2C

Current plans of Atlantic Richfield Company envisioned the strip-mining for coal of a 6,500 acre tract of land near Reno Junction, Campbell County, Wyoming. New mining laws will force the operators to eliminate permanent spoil piles at the mining site. As the seam is exhausted, the pit will be backfilled with the strata in the approximate order prior to removal. A layer of topsoil will be spread on the surface of the overburden and revegetated by natural propagation and seeded species. Some authors state that the practice of "top dressing" provides no particular advantage in establishment of plant cover, and that only slightly higher yields are obtained from top dressed sites.

This paper addressed the problem of soil value as a plant and seed source necessary for the re-establishment of the pre-mining community. The purpose of this study was to determine through greenhouse observation and statistical analysis, the effect of shallow clayey, loamy and sandy topsoil, and acid clay shale on: 1) the species of native plants that will establish themselves from seeds and rhizomes, 2) the aerial production and below ground biomass of the naturally propagating species, 3) the aerial production and below ground biomass of nine species of grasses and forbs planted in these soil materials and 4) the effect of season of planting. Further information concerning this study can be found in 750601.

1E3 760307 "Methods of measuring environmental parameters on a pre-disturbed coal site." April
8D 22, 1976. 47th Ann. Meeting Colo.-Wyo. Acad. Sci. Boulder, Colorado. by J.R. Cox.
3A1
3A2 See 760201.
2F1

1E3 760308 "A mixed method of carbohydrate analysis." April 22, 1976. 47th Ann. Meeting Colo.-
2E Wyo. Acad. Sci. Boulder, Colorado. by J.R. Cox.
2F1

See 760205.

4B 760309 "Some influences of grazing systems upon vegetation and wildlife." April 22, 1976. 47th
4C Ann. Meeting Colo.-Wyo. Acad. Sci. Boulder, Colorado. by W.A. Gerhart.
1E4
2F2 See 760202.
7A1
7A3
6A3

1E2 760310 "Production of a shortgrass prairie ecosystem as related to soil characteristics in eastern
2F1 Wyoming." April 22, 1976. 47th Ann. Meeting Colo.-Wyo. Acad. Sci. Boulder,
2C Colorado. by F.K. Taha.
3A2
8B See 760204.

1E2 760311 "Phenology and environmental relationships of sagebrush-grass rangelands." April 22,
1E1 1976. 47th Ann. Meeting Colo.-Wyo. Acad. Sci. Boulder, Colorado. by L.H.
2E Kleinman.
2F2
3A2 See 760203.
2A
5A1D

1A1 760312 "Rangeland plants and plant identification." July 13, 1976. Wyoming OWRC/SRM Youth Range Camp. Tensleep, Wyoming. by H.G. Fisser.
3A1

This was a combined oral presentation and field trip characterization for presentation to teenagers regarding rangeland plants and plant identification. The field trip traveled from the Tensleep camp to the Castle Garden State Park area southwest of Tensleep where the vegetation of the sagebrush *(Artemisia)* juniper *(Juniperus)* type were observed and inspected and then up Tensleep Canyon into the curlleaf mountain mahogany *(Cercocarpus ledifolius)* type in which the differences in understory species were stressed.

7604 Published Proceedings

1E2 760401 Stroud, D.W. and H.G. Fisser. 1976. Vegetative response of *Bouteloua gracilis* to four levels of utilization. p. 11. In Proc. Soc. Range Manage. 29th Ann. Meeting. Omaha, Nebraska. February 16-20, 1976.
2A
6A3
2F1

See 760304.

1E2 760402 Taha, F.K. and H.G. Fisser. 1976. Physical and chemical analyses of range sites and interrelationships with herbage production and forage value potentials. p. 13. In Proc. Soc. Range Manage. 29th Ann. Meeting. Omaha, Nebraska. February 16-20, 1976.
2C
2F1
3A2
8B

See 760305.

1E2 760403 Lymbery, G.A. and H.G. Fisser. 1976. Vegetation reestablishment potential of strip-mined soils, Black Thunder Basin, Wyoming. p. 49. In Proc. Soc. Range Manage. 29th Ann. Meeting. Omaha, Nebraska. February 16-20, 1976.
8D
3A1
3A2
2F1
2E
2C

See 760306.

3A 760404 Whicker, F.W., A.W. Alldredge, H.G. Fisser, W.C. Hanson and G. Post. 1976. Ecological considerations for Project Wagon Wheel and Hydraulic Fracturing Activities. Phase II(b): Annual summary for 1975 and project completion report. A study conducted for El Paso Natural Gas Co. PN-WW- . 53 p.

This document was an extension of previous reports (Whicker, et al. 1974 and Whicker, et al. 1975). This report extended the baseline documentary information relative to Project Wagon Wheel through 1975 and summarized observations on the massive hydraulic fracturing experiment at Pinedale Unit Well -5. The study described in this and previous reports has been suspended indefinitely because Project Wagon Wheel will likely not be carried out within the foreseeable future. Should the project begin to move forward again, ecological baseline studies could be resumed and information merged with that already in hand.

The data from the vegetation studies during the past four years are summarized in tabular form. Year to year fluctuation and production was minimal although the control plot seemed to be consistently higher with grass production than the WW plot. Figures for total production, however, are similar for both plots. Annual variation of total production appeared rather minor during the years considered.

The cover percent data suggested higher values for woody and mat-form species in the WW plot, but higher values for herbaceous cover in the CWW plot. Little annual variation in shrub cover was evident, but a significant decrease in herbaceous cover was noted in 1975.

Frequency data for individual species were summarized. As might be expected, the annual forbs tended to exhibit greater variation through time than the perennial grasses and shrubs.

In addition to the quantitative data, a library of photographs taken at fixed location points under specifically documented conditions was compiled. These photographs provide a separate aid to data interpretation.

7605 Wyoming Agricultural Experiment Station Publications

1E2 760501 Fisser, H.G. and L.H. Kleinman. 1976. Phenology and production studies on semiarid shrub types. 1975 results. Univ. of Wyo. Coop. Res. Rep. to the BLM. Sections I, II and III. Wyo. Agr. Exp. Sta. Sci. Rep. 831. 410 p.

1E1 76050101 Section I - Phenology study.
1E2
2E
2F2
3A2

The 1975 phenology study research data included the listing of each of the 12 study sites and the prime species located at each, a list of plant species' names which occurred on the study areas, the phenological survey of all species for each exclosure and date with an intratabular index, the data for prime species intensive phenological inventory by exclosure, plant species and data collection date for each permanently identified plant also with an intratabular index. The phenological development dates for all species for three major growth stages - growth initiation, full bloom, seed dissemination, by exclosure, for the years 1973-75 are given also with an intratabular index.

A separate section relating to the interrelationships of independent variables with the dependent phenological development variable was given in a separate set of tables, which gave the stepwise multiple regression data and the correlation matrix values for the four species. In addition, simple correlation coefficients between independent variables and plant species phenology, and the statistical summary for the derived regression equations was given.

Growth initiation for western wheatgrass *(Agropyron smithii)* varied from March 5 to April 26 with an average date of March 15. Full bloom dates varied from June 20 to July 20 with an average date of July 10. Start of seed dissemination varied from July 25 to November 10 with an average date of August 15.

Growth initiation of bluebunch wheatgrass *(A. spicatum)* varied from March 1 to April 15 with an average date of March 10. Full bloom varied from June 10 to July 25, with an average date of June 30. Start of seed dissemination varied from July 10 to August 15 with an average date of July 30. Seed dissemination always appeared to always be one month after full bloom.

Growth initiation of black sagebrush *(Artemisia nova)* varied from May 15 to May 30 with an average date of May 20. Full bloom varied from September 15 to September 20 with an average date of September 15. Start of seed dissemination varied from October 25 to November 5 with an average date of November 5.

Growth initiation of big sagebrush *(A. tridentata)* varied from April 22 to June 15 with an average date of May 15. Full bloom varied from August 5 to October 10 with an average date of September 15. Seed dissemination varied from October 14 to November 20 with an average date of November 10.

The sagebrush species produced new leaves in spring and summer. Many leaves persisted through the winter, but were shed before clipping started the next fall. The results of the stepwise selection process for western wheatgrass showed that 10 of the 17 independent variables accounted for 42 percent of the variation in phenological change. For bluebunch wheatgrass, four environmental variables were significant and explained 74 percent of the variation in its phenological change. Six environmental factors were significant, but explained only 32 percent of the variation in the phenological change of black sagebrush. Thirteen environmental and productivity factors were significant and explained 43 percent of the variation in the phenological change of big sagebrush.

Temperature extremes were important in influencing phenological change. In fact, both air and soil temperature extremes dictated the time of growth initiation for the four species studied. These data are also used in the dissertation compiled by Kleinman, code number 760602.

2A 76050102 Section II - Precipitation pattern study.
3A

The 1975 precipitation data were compiled for all of the UW-BLM gauges on a seasonal basis and also for some 40 U.S. Weather Service stations located in the western portion of Wyoming. Precipitation totals for 1975 averaged much higher than in 1974. The significant increase in moisture occurred in the early spring. The above normal spring moisture resulted in very high forage production, especially for shrub species. Areas dominated by Gardner saltbush *(Atriplex gardneri)*, such as the Burnt Wagon exclosure, increased in forage production by almost 300 percent. The average annual forage production at the Burnt Wagon exclosure for the eight year period, from 1962-1970, was 271 lb./acre. In 1975 the total forage production was 752 lbs, which is a 277 percent increase. Most of this increase was from the high increase in Gardner saltbush production, from an average of 252 to 680 lb./acre. On the Gardner saltbush areas there was no significant increase in perennial grass production. Areas dominated by big sagebrush *(Artemisia tridentata)* and western wheatgrass *(Agropyron smithii)* had an increase in grass and forb production of 76 percent. The shrub production increased by about 97 percent. The precipitation records show the precipitation of the October-April period and the April-July period were above normal. Closer study of U.S. Weather Service records showed that most of the above normal precipitation occurred in March and April.

3A3 76050103 Section III - Exclosure study: production, cover and precipitation.
5A1D
5B
5C
2A
2C

The 1975 activities of the herbage production and cover studies at the exclosures were continued as in previous years. As indicated in 76050102, herbage production on a number of the areas was greatly above average because of the very beneficial high moisture during spring.

3A1 760502 Fisser, H.G., J.R. Cox and F.K. Taha. 1976. Herbage structure, productivity and phenology on Black Thunder. Annual Progress Report to Atlantic-Richfield Company. 1975 results. Wyo. Agr. Exp. Sta. Sci. Rep. 779. 161 p.
8D
3A2
2F1

This report of the 1975 research on the Black Thunder site included all of the production data and permanent quadrat cover data collected during that year.

3A1 760503 Ries, R.E., H.G. Fisser and A.T. Harrison. 1976. Vegetation, Section IX. p. 1-85. In Bergman, H.L. and M.D. Marcus (eds.). Final Report Environmental Assessment of the Black Thunder Mine Site, Campbell, Co., University of Wyoming, Laramie, Wyoming.
8D
3A2
2F1

This publication was the final report of the initial environmental assessment of the vegetation types on the Black Thunder site which had been conducted as an overall environmental assessment of the vegetation, the range sites, soils and general environmental interrelationships.

7606 Graduate Advisee Completions

3A2 760601 Gerhart, W.A. 1976. Mule deer food habitat and forage quality studies on juniper ranges in Wyoming. M.S. Thesis, University of Wyoming. 220 p.
7A1
10D1

This study was initiated to evaluate the general nutrient content of diets of mule deer wintering on juniper woodland ranges in Wyoming. Forage quality, habitat description and use, and general mule deer ecology were also analyzed at three different study sites. A vegetation survey was conducted at the Ten Sleep area to describe the habitat available. Ground surface cover, plant cover and shrub and tree density were estimated in 1 sq ft plots and belt transects.

The vegetation of the Ten Sleep area was a mixture of Utah juniper *(Juniperus osteosperma)*, curlleaf mountain mahogany *(Cercocarpus ledifolius)* and big sagebrush *(Artemisia tridentata)*/grass communities. The density of juniper trees varied from 479 to 915 plants per acre and canopy cover averaged 22.5 percent. Forbs and half-shrubs contributed more understory cover than grasses. Mountain mahogany occupied shallow, rocky sites near ridgetops and was often codominant with juniper. Mountain mahogany plants varied in density from 218 to 1,699 individuals per acre and averaged 1,040. Grass and sedge cover was more important than forb and half-shrub cover. Big sagebrush/grass areas averaged about 7 percent big sagebrush canopy cover. Bare ground and litter provided most of the ground surface cover. Bluebunch wheatgrass *(Agropyron spicatum)* and needleandthread *(Stipa comata)* were the important grass species present.

Vegetation at the Medicine Lodge area was similar to that found at the Ten Sleep area. Utah juniper trees dominated the hillsides of the Twin Creek study area and the valley floor was a mixture of sagebrush, rabbitbrush (*Chrysothamnus* spp.), forbs and grasses. Big sagebrush, black sagebrush *(A. nova)*, silver sagebrush *(A. cana)*, Douglas rabbitbrush *(C. viscidiflorus)* and antelope bitterbrush *(Pushia tridentata)* were the important shrub species present (Hanson 1974).

Deer movement onto, above and within each study area was primarily a function of snow depth, snow condition and spring regrowth. Migratory herds moved onto the area immediately prior to or after heavy snowfalls, at higher elevations and from surrounding agricultural fields after livestock were put on the fields and utilized most of the available forage. Peak movement onto the study area occurred in November and December. Deer began moving off of the areas during late April and May as snow cover receded and new spring growth became available, either on surrounding fields or at higher elevations. Soft snow depth of 20 inches or less were tolerated but crusted snow over 12 to 15 inches deep was avoided. Whenever temperatures dropped below 15^{0} F, snow depths became excessive or snow crusted, deer moved to exposed south-facing slopes to feed and rest. During warm spells (temperature exceeding approximately 45^{0}) deer often fed and rested in areas not exposed to direct sunlight. Cattle utilized the open range at Twin Creek until late December, and a few elk were present on the Ten Sleep and Medicine Lodge areas during the winter.

The Ten Sleep study area was more important as an intermediate fall range rather than a winter range for deer. Approximately 83 deer per section utilized the Ten Sleep area during the fall, but only about half were present during the winter. Approximately 49 fawns and 19 bucks per 100 does occupied the Ten Sleep area.

On the Medicine Lodge area, there were about 122 deer per section. The wintering herd consisted of approximately 61 fawns and 12 bucks per 100 does.

Deer populations on the Twin Creek fluctuated a great deal but an average of about 83 deer per section were estimated. The ratio of fawns per 100 does was 76, and there were 19 bucks per 100 does.

Diets were based on fecal analysis and reported on a percent frequency basis. Browse provided the bulk of the foods consumed by the deer. Alfalfa *(Medicago sativa)* was eaten in small amounts, particularly during the fall and again during spring. In the spring, grasses and forbs became important food items.

Mule deer diets, during the winter, on the Ten Sleep area averaged 74.5 percent browse, 21 percent forbs, 3 percent succulents (little soapweed *(Yucca glauca)* and Plains pricklypear *(Opuntia polyacantha)*) and 1.5 percent grasses. On the Medicine Lodge area deer consumed 90 percent browse, 9 percent forbs and 1 percent graminoids and succulents. The shrub composition of the diet was 30 percent juniper, 26.5 percent mountain mahogany and 12 percent sagebrush. During the winter mule deer at Twin Creek ate about 88 percent browse, 9 percent forbs, 2.7 percent graminoids and about

0.3 percent succulents. Approximately 29 percent of the total diet was contributed by sagebrush, 18 percent by juniper and 13 percent by bitterbrush.

Some dietary overlap between cattle and deer was recorded for sagebrush and bitterbrush at Twin Creek. Present livestock numbers and lengths of time on the winter range were not great enough to harm either the range or the available deer forage.

Elk on the Medicine Lodge and Ten Sleep study areas competed with deer for mountain mahogany, sagebrush, rabbitbrush and little soapweed. Elk utilized mountain mahogany more efficiently than deer. Deer nipped current twig growth whereas elk utilized current, as well as previous year's, twig growth of mountain mahogany. Present populations appeared compatible, however, increased elk numbers may force deer to move to new wintering areas. Other areas may not provide sufficient cover, food quality, availability or variety and space to sustain a healthy, productive deer herd. Damage to crops may pose future problems also. A few elk on these winter ranges may benefit present and future deer herds. Elk often knock down tall mountain mahogany plants, thus making them more available for deer use and possibly creating sprouting, which would also then become available for future deer use.

Crude protein and detergent analysis of important browse species, forb mixtures and grass mixtures was conducted under early, mid- and late-winter to evaluate the approximate nutrient intakes received by mule deer on each study area. Digestible dry matter (DDM) percentages were based on equations proposed by Goering and Van Soest, 1970.

Juniper was a poor source of protein, a good source of acid detergent fiber (ADF) and was rated as excellent for digestibility. Big sagebrush provided a good source of protein, good to excellent sources of ADF, and DDM were excellent. Curlleaf mountain mahogany, an evergreen shrub, provided a fair to good source of protein for deer and the percentages of ADF and DDM were estimated at excellent. In comparison, antelope bitterbrush, a deciduous shrub, provided a poor source of protein in early winter and a fair source in mid and late winter. The amount of ADF was rated as fair for deer in early winter and good in mid and late winter. Percentage of DDM was rated as excellent throughout the winter. Rubber rabbitbrush *(Chrysothamnus viscidiflorus)* was a poor source of protein in early and mid-winter and a fair protein source in late-winter for deer. Percentage of ADF was considered good and DDM percentage was rated excellent throughout the winter. Black sagebrush provided a fair source of protein and was an excellent source of ADF and DDM for deer throughout the winter. Forbs provided a poor source of protein in early and mid-winter and a fair source in late-winter. Grasses were a poor source of protein in early winter and mid-winter and a good source in late winter, and were only a poor to fair source of ADF for deer. Date and site of collection, collection procedures and laboratory procedures accounted for most of the differences in the nutrient content reported for the plants analyzed.

The general, overall health of the deer observed on each study area was good to excellent throughout the winter and did not support the low protein intake estimated. Collection methods and laboratory procedures probably resulted in minimal protein and ADF calculations. The overall good condition of the deer observed and the generally high fawn:doe ratios suggested that deer were receiving adequate to good nutrient levels throughout the year. However, the low protein intakes calculated did warrant monitoring and possible future research if productivity and winter mortality became excessive. Severe winters or increased elk utilization on the study areas may cause nutrient intakes to fall below the minimum requirement necessary for deer maintenance and productivity. The Ten Sleep area was rated at fair winter range for mule deer and was not a critical wintering area, although the site provided a good fall intermediate range. The Medicine Lodge area was rated at good winter range and was a critical wintering area particularly at lower elevations. Heavy snow accumulations on the Twin Creek juniper areas reduced the accessibility and the forage available to deer, which in turn decreased the quality of nutrients available. Thus, Twin Creek was not considered a critical wintering area for deer and was only rated as a fair winter range. Lower lying winter sagebrush ranges near Twin Creek appeared to be more critical for the health and welfare of the deer herds present (Hanson 1974).

Critical wintering areas should be carefully monitored to insure that adequate quantity and quality of browse species are available for deer use. Livestock and elk populations, distribution and time of use on these sites should be controlled if present deer populations are to be maintained or increased and to minimize deer utilization of agricultural crops. Special hunting seasons to reduce deer numbers, if and when necessary, should be set when the deer have concentrated on the winter range and are more accessible. Earlier and longer hunting on the more inaccessible mountainous areas where heavy timber and canyons provide adequate cover and protection for deer would be good for more recreation for hunters and would not be detrimental to the present populations. Special early hunts arranged with local wildlife people and ranchers along the creek bottoms and surrounding fields could be initiated to control the resident lowland agricultural herds. Some juniper woodland ranges in Wyoming were critical mule deer wintering areas. They provided the cover and food necessary for maintaining the health and welfare of many herds and should be managed accordingly. For planning and evaluations of the effect of proposed juniper control, conversion measures would be beneficial. Juniper provide some food and good protective cover for deer during the winter and large scale control measures should be avoided on critical areas.

1E2 760602 Kleinman, L.H. 1976. Phenodynamics and ecology of sagebrush-grass rangelands. Ph.D. Dissertation. University of Wyoming. 143 p.

1E1
2E
2F2
3A2
2A
5A1D

Phenological studies were located on 12 selected livestock exclosures distributed throughout western Wyoming. Relationships between phenodynamics, vegetation production and environmental characteristics of *(Agropyron smithii)*, *(Agropyron spicatum)*, *(Artemesia nova)* and *(Artemesia tridentata)* were investigated. Specific objectives were: 1) to develop phenophase descriptions for native sagebrush species in western Wyoming, 2) to identify dates of a) initiation of growth, b) peak flowering and c) seed dissemination, 3) to determine relationships of plant species phenology with vegetation production and environmental characteristics and 4) to provide information for increased resource management decision making efficiency.

Phenophase descriptions with 17 numerical scores representing annual growth dynamics were developed. A phenological inventory of all flowering species in the 12 study areas was also established. Variations in phenological event dates were recorded from 1973-75. Local site variations in the phenodynamics of each prime species was observed and recorded during each year of study. Growth initiation dates ranged from 16 days for black sagebrush to 55 days for big sagebrush. Full bloom ranged from six days for black sagebrush to 67 days for big sagebrush. Start of seed dissemination ranged from 12 days for black sagebrush to 102 days for western wheatgrass. The other phenophases exhibited similar variation in time of occurrance.

The variation extremes in dates of phenological events characterized the magnitude of ecological tolerance for each species. Black sagebrush was restricted to side hill sites with shallow, rocky soil and exhibited very little variation in its phenological progression. The other three prime species were found on a variety of habitat sites from very fine textured soils in valley bottoms to coarse textured, sandy sites on hillsides, and exhibited large variations in their phenological progression. Seventeen independent variables (environmental and herbage production factors) and phenology were analyzed by stepwise multiple regression procedures. Simple correlation analysis resulted in 16 of the 17 independent variables being simply and significantly correlated with the phenodynamics of the two sagebrush and two wheatgrass species. On the basis of simple correlation analysis six environmental variables were common and significant to all four prime species. They were: soil temperature, soil moisture percent at 15 - 30 cm and 0 - 15 cm, previous month soil moisture percent at 15 - 30 cm, average annual precipitation and spring precipitation.

In the stepwise multiple regression analysis 16 of the 17 independent variables were also significant. Forty-two percent of the variation in phenodynamics of western wheatgrass, 74 percent of the variation in phenodynamics of bluebunch wheatgrass was explained by these stepwise multiple regression equations. The equations explained 32 percent of the variation of black sagebrush and 43 percent of the variation for big sagebrush. The most important variables common to all four species were: soil temperature, soil moisture percent at 0 - 15 cm and 15 - 30 cm, summer and fall precipitation, herbage production and air temperatures. Each of these environmental and productivity factors significantly influence the phenological progression rate of the two shrubs and two grasses. Air and soil temperature extremes dictated the time of growth initiation in the spring for all four species when the soil moisture supply was adequate. Shallow soil moisture and temperature most significantly influenced the phenodynamics of both bluebunch wheatgrass and black sagebrush. The phenodynamics of western wheatgrass and big sagebrush were most significantly influenced by soil moisture at medium depth (15-30 cm), summer precipitation and maximum air temperature. A very close relationship existed between the phenological progression and the herbage production of western wheatgrass and big sagebrush, but no statistically significant relationship was found between the phenological progression and herbage production of bluebunch wheatgrass and black sagebrush.

7701 Research Activities

3A3 770101 Exclosure study: production, cover and precipitation.

5A1D

5B Herbage production, cover and precipitation data were collected at the 40 exclosures
5C as during previous years and the reports prepared by exclosure and transect for the
2A production data by species and groups of species for herbaceous as well as for shrubby
2C species.

1E2 770102 Exclosure study: phenology and growth.

2E

2F2 Intensive phenology study was continued on the 12 exclosures distributed
3A2 throughout the western half of the state. These data were obtained on seven dates
1E1 during the growing season beginning on April 23 and ending on September 17. The
2A phenological response of the four prime species in 1977 were greatly affected by the
5A1D interrelationship of air temperature and amount and time of precipitation.

In general, spring moisture was deficient and high temperatures occurred throughout all the study areas. In particular, June was characterized by daily temperatures 5-7° F higher than normal and monthly precipitation was 1/2 to 2-inches below normal. The combination of these two environmental factors resulted in delayed initiation of growth of western wheatgrass *(Agropyron smithii)* and bluebunch wheatgrass *(A. spicatum)*. Temperatures in July remained above normal but fortunately there were 1-3 inches of precipitation. Due to the delayed initiation of the grass species, no reproduction was recorded, either for western or bluebunch wheatgrass at several of the exclosures. Also, as a result of the dry weather, growth and initiation of sagebrush (*Artemisia* spp) occurred at earlier dates than for the past four years.

The 1977 growing season was hot and dry. The combination of these two environmental factors resulted in delayed initiation of growth for the wheatgrass species and reproduction occurring at only half of the study sites. These same conditions resulted in an inverse relationship for the sagebrush species. Growth initiation and reproduction occurred at earlier dates than the previous four years.

2A 770103 Precipitation pattern study.

3A

The UW/BLM gauges were maintained through the year and recorded at the four seasonal dates, as during previous years, and in addition in the Big Horn Basin an additional three to four dates were recorded with precipitation then being taken on March 1, April 15, June 1, July 1, August 1, September 1, October 1 and November 15. The U.S. Weather Bureau Stations were recorded and compiled as previously also.

2C 770104 Exclosure soil profile studies with SCS.

8B

Soil profile descriptions and related range site characteristics were obtained on a considerable number of exclosure locations in the Big Horn Basin with Jim Stephens and Don Viktorin of the Soil Conservation Service.

3A2 770105 1 ft x 1 ft permanent plot study.

2F2

5C All of the permanent plot (1 ft x 1 ft) transects in the Big Horn Basin were reread
10D1 by Fisser, Trueblood and Samuelson. Fisser established permanent plot transects at the Carter Divide exclosure in the Wind River Basin and reread the permanent plot transects on the exclosures in the Great Basin northwest of Rawlins, at the Lamont, Cyclone Ridge, Chain Lakes and Lost Creek exclosures.

3A1 770106 Black Thunder pre-mining inventory.

8D

3A2 At the Black Thunder pre-mining area Fisser inventoried a small area just north of
2F1 the primary Black Thunder mine site.

8D 770107 Snow catchment and revegetation on Jim Bridger mine spoil.
2A
2C

At the Jim Bridger mine area, west of Rock Springs, Fisser and Dennis Cleaves (working on a thesis program on the relationship of snow catchment and revegetation on the mine spoil areas) established sampling procedures during late fall.

7702 Journal Publications

2E 770201 Taha, F.K. and H.G. Fisser. 1977. Factors affecting growth and productivity of vegetation stratum in eastern Wyoming. J. Colo.- Wyo. Acad. Sci. 9(1):36.
2C
2A
2B

Forty soil factors were investigated to determine their effect on growth and productivity of grasses, forbs and shrubs on the Black Thunder pre-mining inventory site in eastern Wyoming. Soil characteristics included 10 physical parameters of the "A" and "B" horizons, in addition to 20 chemical characteristics measured within the "A" horizon region. Vegetation production was determined at near peak biomass for grasses and forbs during early August and during late September and October for shrubs. Four soil-moisture related factors were shown to significantly affect grass production. These were air:water permeability ratio of the "A" and "B" horizons, organic matter and moisture equivalent of the "A" horizon. These factors were found to be responsible for 70 percent of the variation in grass production. 76 percent of the variation in forb production was affected by five soil physical factors. These were: percent silt of the "A" horizon, air:water permeability ratio of the "A" and "B" horizons, available moisture of the "A" horizon and percent clay of the "B" horizon.

Statistically significant soil factors responsible for shrub production, however, were not possible to isolate. It was concluded that the ability of shrub species to draw moisture and nutrients from depths not considered in this investigation might have contributed to the apparent non-existance effect of soil factors examined on shrub production. Soil moisture related factors were found, in general, to be the overiding influence on vegetation growth and production in eastern Wyoming. None of the nutrient factors seemed to be important in this region.

2C 770202 Shawesh, O.M. and H.G. Fisser. 1977. Vegetation analysis and ecology of rangelands in northwest Libya, Africa. J. Colo.-Wyo. Acad. Sci. 9(1):35-36.
2A
2B
8C

Quantitative description and ecology investigations of semiarid rangelands in northwestern Libya were initiated by both authors in May 1976. A 25,000 ha livestock exclosure near Assa, some 25 km from the Mediterranean sea coast, was utilized. Average annual precipitation was 120 ml, most occurring from October to March. Average annual temperature was about 35^{0} C. Soils were mostly azonal sands with pH values from 7.5 to 9.0. The water table, as shallow as 135 cm beneath the surface, was usually saline indicating possible seawater intrusion. Average elevation was about 5 m above sea level. Vegetation and soil relationships were investigated with permanent established transects for measures of plant cover and density, and by soil profile sampling for detailed analysis of physical and chemical characteristics. Stable rangeland sites were characterized by four communities. Annual grasses were common understory components of stable rangeland, but not on the disturbed soil sites.

7703 Presentations

2E 770301 "Rangeland site ecology and associated plant physiology." January 19, 1977. Central High Plains Research Review. U.S.D.A.-A.R.S. Fort Collins, Colo. by H.G. Fisser.
2A
2B
2C

The primary thrust of this presentation was related to the comparison of blue grama *(Bouteloua gracilis)* growth and potential at two exclosures in western Wyoming, one the Boysen exclosure in the Wind River Basin and the other the Dutch Nick exclosure in the Big Horn Basin. The former occurred on a site with less rainfall and warmer temperatures than the latter. This resulted in greater diversity in numbers of species at the more mesic site in the Big Horn Basin as compared to the more xeric site in the Wind River Basin.

4B 770302 "An evaluation of three grazing management systems on northern desert shrub
4C vegetation of southern Wyoming". February 17, 1977. 30th Ann. Meeting Soc.
1E4 Range Manage. Portland, Oregon. by H.G. Fisser.
2F2

7A1 This was a report concerning the Baggs grazing system study with an evaluation of
7A3 data obtained during 1975 compared to those obtained from the initiation of the study
6A3 in 1967-1971. W.A. Gerhart conducted the 1975 study. Data indicated an upward trend in range condition under the current management strategies. Vegetation responses were relatively slow and should be evaluated over a longer term study.

Antelope increased their range and population during the study period. The grazing systems had no apparent adverse affects on the wildlife observed, with a possible exception of indirectly contributing to pronghorn fawn mortality. Resident wildlife populations and the existing vegetative cover and composition would benefit most through the development of more reliable and permanent water sources.

8D 770303 "Development of snowfence systems on strip-mine spoils to increase seedling
2A establishment." February 16, 1977. 30th Ann. Meeting Soc. Range Manage.
2C Portland, Oregon. by D.T. Cleaves.

This study was concerned with the development of a site-specific snowfence system for dispersed entrapment of snow and subsequent distribution of water by development of a soil surface manipulation treatment. The project was a cooperative study with the Argon National Laboratories, the Jim Bridger Mine and the University of Wyoming Range Management Department. Short-term studies were to include seedling establishment measurements on slopes below the snowpack areas, soil moisture relationships, sediment runoff and erosion plots, different surface manipulations which may increase infiltration of the melt water, and weather monitoring with an on-site weather station to check effectiveness of the snowfence system. It was also to establish whether or not design improvements were necessary. These data were to be compared to similar sites at the mine which had not been so treated. The long-term monitoring of the site (proposed for at least five years) would also include total annual herbage production compared with untreated sites. Different heights, lengths and patterns of snowfencing were to be studied to determine the best distribution patterns for the meltwater.

3A2 770304 "Long-term vegetation studies with permanent quadrats." February 16, 1972. 30th Ann.
2F2 Meeting Soc. Range Manage. Portland, Oregon. by A.D. Gdara.
10D1

2E This report concerned the long-term studies of vegetation cover dynamics on grazed
2A and protected sites in western Wyoming, which were initiated in 1960 with a cooperative agreement between the BLM and the UW Range Management Department. Data were obtained from 29 livestock exclosures distributed throughout western Wyoming on semiarid, northern desert shrub rangelands and at an eight-pasture research grazing study area on a Gardner saltbush *(Atriplex gardneri)* vegetation type in north central Wyoming.

The vegetation cover data were obtained from permanently located quadrats inside and outside of exclosure areas. Within the 2 ft x 20 ft quadrats crown cover of shrubs, halfshrubs, Plains pricklypear *(Opuntia polyacantha)* and mat-form species were determined by position mapping procedures. Within the 4 ft x 4 ft quadrats herbaceous cover was determined utilizing a 10 point frame with pins spaced at 2-inch intervals. A point frame was positioned at 10 locations within each quadrat. First hits of all species as well as ground surface character were recorded. Data were obtained at four-year intervals during 1960, 1964, 1968, 1972 and 1976. Measures of annual herbage production and seasonal precipitation were recorded.

At the Ant Hill Lander location, an upland sagebrush *(Artemisia)*-grass type in west central Wyoming, significant changes in vegetation occurred. During the first eight years of the study little change occurred. During the latter years a significant decrease of Plains pricklypear occurred but big sagebrush *(A. tridentata)* was noted to have increased. Since grazing stress had not appeared to be significant, other factors which may have caused these changes were investigated. It appeared that moisture during the last years was much more abundant than normal, resulting in the sagebrush increase; while the cactus bug (*Chelinidea vittiger*) or the bluecactus borer (*Meliteria dentata*) appeared responsible for the decrease in cover of the pricklypear.

1E3 2C 1E2 5C 2A 2B 10D1 1E4 2E

770305 "The effect of environmental factors on carbohydrate storage and herbage production of five short-grass prairie plant species." February 16, 1977. 30th Ann. Meeting Soc. Range Manage. Portland, Oregon. by J.R. Cox.

This work was conducted on the Black Thunder pre-mining inventory investigation site in eastern Wyoming. During a two year period data were obtained on the shortgrass plains location to investigate plant responses to fluctuating environmental conditions.

Nine study sites were selected on the basis of soil and vegetation characteristics. Three areas had sandy soils, three loam soils and three clay soils. All areas had five plant species in common: big sagebrush *(Artemisia tridentata)*, western wheatgrass *(Agropyron smithii)*, blue grama *(Bouteloua gracilis)*, Junegrass *(Koeleria cristata)* and needleandthread *(Stipa comata)*.

Temperature and precipitation data were collected monthly at each site. Maximum and minimum thermometers were placed at ground level protected from direct radiation by containers perforated to allow air movement. Aluminum precipitation gauges were installed at each site. Soil samples were taken at depths of 6, 12 and 18 inches and analyzed for soil texture and water potential values.

Root samples were taken of the five species and soil moisture readings were collected biweekly throughout the growing season (May-September) and bimonthly throughout the remaining portion of the year. Root samples were quick-frozen with dry ice and later oven dried. Samples were ground and analyzed for total available carbohydrates. Soil moisture readings were taken at depths of 6, 12 and 18 inches.

A phenological classification system was utilized to monitor plant growth dynamics at each site. Grass production estimates were obtained from 10 randomly located clipping quadrats at each site. Shrub production estimates were obtained by a modified double sampling technique.

A stepwise multiple regression program was used to examine the interrelationships of environmental and biological components. Factors such as leaf water potential, available water, soil texture, soil moisture, precipitation and air temperature were compared to carbohydrate storage and herbage production.

In 1976 depletion of carbohydrates was associated with unseasonably warm temperatures in February and March and readily available soil moisture. Early spring weather apparently affected carbohydrate storage or depletion dynamics and total herbage production more than other parameters studied.

1E2 1E1 2E 2F2 3A2 2A 5A1D

770306 "Phenodynamics and ecology of sagebrush-grass rangelands." February 15, 1977. 30th Ann. Meeting Soc. Range Manage. Portland, Oregon. by L.H. Kleinman.

This paper related to the phenological studies located at 12 livestock exclosures in western Wyoming. The intent of this research was to determine and evaluate relationships between phenodynamics, vegetation production and environmental characteristics of two grasses and two shrubs. This report itemized a number of variations and characteristics of changes of phenodynamics of these species among the years during the previous study.

There were 17 independent variables which were applied through a regression program. Sixteen of these were found to be significantly correlated with phenodynamics of the two sagebrush (*Artemisia* spp) and two wheatgrass (*Agropyron* spp) species.

The most important variables common to all four species were soil temperature, soil moisture percent at 0-15 and 15-30 cm, summer and fall precipitation, herbage production and air temperatures. Each of these environmental and productivity factors significantly influenced the phenological progression of the two sagebrush and two wheatgrass species. Air and soil temperature extremes dictated the time of growth initiation in the spring for all four species when the soil moisture supply was adequate. Shallow soil moisture and temperature was significantly influenced by phenodynamics of bluebunch wheatgrass and black sagebrush.

The phenodynamics of western wheatgrass *(Agropyron smithii)* and big sagebrush *(Artemisia tridentata)* were most significantly influenced by soil moisture of medium depth (15-30 cm), summer precipitation and maximum air temperature. A close relationship existed between the phenological progression and the herbage production of western wheatgrass and big sagebrush. No statistically significant relationship was found between the phenological progression and herbage production of either bluebunch wheatgrass *(Agropyron smithii)* or black sagebrush *(Artemisia nova)*.

See 760602.

1E2 8D 3A2 2F1 1E1 1E3 770307 "Factors affecting growth and productivity of vegetation stratum in eastern Wyoming." April 23, 1977. 48th Ann. Meeting Colo.-Wyo. Acad. Sci. Colorado Springs, Colorado. by F.K. Taha.

See 770201.

2C 2A 2B 8C 770308 "Vegetation analysis and ecology of rangelands in northwest Libya, Africa." April 23, 1977. 48th Ann. Meeting Colo.-Wyo. Acad. Sci. Colorado Springs, Colorado. by O.M. Shawesh.

See 770202.

1A1 1A2 2A 2F2 770309 "Rangeland plants and plant identification." June 28, 1977. Wyoming OWRC/SRM Youth Range Camp. Ten Sleep, Wyoming. by H.G. Fisser.

This was a combined classroom and field presentation of range plants and plant identifying characteristics related to plants growing on the semi-desert area to the west of Ten Sleep and those found at two higher elevations along the highway eastward from Ten Sleep.

7704 Published Proceedings/Abstracts

4B 4C 1E4 2F2 7A1 7A3 6A3 770401 Gerhart, W.A. and H.G. Fisser. 1977. An evaluation of three grazing management systems on northern desert shrub vegetation of southern Wyoming. p. 52. In Proc. Soc. Range Manage. 30th Ann. Meeting. Portland, Oregon. February 17, 1977.

See 770302.

1E2 1E1 2E 2F2 3A2 2A 5A1D 770402 Kleinman, L.H. and H.G. Fisser. 1977. Phenodynamics and ecology of sagebrush-grass rangelands. p. 11. In Proc. Soc. Range Manage. 30th Ann. Meeting. Portland, Oregon. February 14-17, 1977.

See 770306 and 770602.

2E 770403 Cox, J.R. and H.G. Fisser. 1977. The effect of environmental factors on carbohydrate
1E3 storage and herbage production of five short-grass prairie plant species. p. 41. In
2C Proc. Soc. Range Manage. 30th Ann. Meeting. Portland, Oregon. February 14-17,
1E2 1977.

5C

2A See 770305.

2B

11D1

1E4

8D 770404 Cleaves, D.T. 1977. Development of snowfence systems on stripmine spoils to increase
2A seedling establishment. p. 32. In Proc. Soc. Range Manage. 30th Ann. Meeting.
2C Portland, Oregon. February 14-17, 1977.

See 770303.

3A2 770405 Gdara, A.O. 1977. Long-term vegetation studies with permanent quadrats. p. 8. In Proc.
2F2 Soc. Range Manage. 30th Ann. Meeting. Portland, Oregon. February 14-17, 1977.

10D1

2E See 770304.

2A

9A 770406 Fisser, H.G. 1977. Rangeland and wildlife resources of the South Pass Historic Area. p.
4B 77-88. In D.S. Harder. Management suggestions: South Pass Mining Area. Prepared
3B for the Bureau of Land Management contract YA-510-PH-110. March 31, 1977.
6A1 Parks and Recreation Department, University of Wyoming, Laramie.

7A1

7A2 The materials in this report were compiled to characterize the rangeland and wildlife
7A3 resources on the South Pass area because of its specific historic interest relating to the
7A5 mines and other activities which had been carried on in that area during the 1800s.
8A General vegetation types and classes and lists of plant species were included in the report. Description of livestock use, wildlife components, including large mammals, deer, antelope, moose, elk and other mammals, were described. The importance of a land with regard to present economic assessment of the livestock income from specific grazing use was derived and those of other related to more asthetic characteristics of the South Pass historic mining area were noted to be difficult to interpret.

It was noted that in terms of net profit to the livestock operator, the 3,500 AUMs of cattle and sheep grazing allotted to the area provided an economically derived return to land value of not much more than $5,000. In terms of importance to the economy of the region or state, an AUM is computed to contribute $25 - $30 of economic activity. On this basis, the 3,500 AUMs could reasonably be expected to promote some $100,000 in gross income and associated financial value to the local community and associated state economy.

Importance of wildlife components is often computed in terms of the value of the average numbers of hunter days per animals harvested. The value of a recreation day for such species of hunted animals is derived from the average number of hunter days expended for each animal taken. These values were expanded for the total number of harvested animals of each species from a given area. A value of $14,000 was estimated for pronghorn antelope, mule deer, elk and moose harvested from the South Pass historic mining area. Wildlife harvest was thus certainly not an insignificant use even of an area as small as this.

Recommendations were suggested that management must conceptually include requirements for maintenance and enhancement of the biological renewable resources with minimal conflict among various uses. Surface uses such as mine site development for visitor use should probably be excluded from large animal use. Those sites selected for development should be limited to areas of minimal impact upon livestock or wildlife. Site developments for recreation uses should be similarly evaluated.

7705 Wyoming Agricultural Experiment Station Publications

3A1 770501 Fisser, H.G., J.R. Cox and F.K. Taha. 1977. Herbage structure, production and phenology on Black Thunder. Annual Progress Report to Atlantic Richfield Company. 1976 results. Wyo. Agr. Exp. Sta. Sci. Rep. 832. 208 p.

8D
3A2
2F1

Vegetation analyses of the Black Thunder unit east of Reno Junction, Wyoming, were initiated during summer 1973 by cooperative agreement with the Atlantic-Richfield Company, the Wyoming Environmental Institute and the University of Wyoming. Those investigations resulted in description and mapped delineation of 10 biologically signficant vegetation types, as well as considerable information related to vegetation condition classification, livestock carrying capacity, plant identification and effects of disturbances such as pipeline and road construction, and historical farming activities.

Since the entire lease area of the Black Thunder unit was proposed to be surface mined, the Atlantic-Richfield Company had entered into agreement with the University of Wyoming, Plant Science Division, for continued ecologically oriented investigation of vegetation production and structure dynamics. Biological studies to monitor existing conditions and provide measures of any changes that might occur as a result of resource uses and impacts were recognized as needed adjuncts to environmental descriptions and statements. The specific objectives of this research program were: 1) to record vegetation structure and herbage production data within each of 10 vegetation type areas (22 cluster locations), 2) to establish and define baseline parameters of vegetation structure and herbage production for each of the vegetation type areas and 3) to develop statistical designs for data analyses.

Because of the long-term character of the studies required for the development of sound ecologically oriented baseline information of vegetation production, structure and phenology dynamics, permanent cluster locations were established in each of the 10 vegetation type areas described on the 7,000 acre lease area. On each of the two largest vegetation type areas, (4 ARTR BOGR/AGSM and 4 ARTR BOGR/STCO/OPPO) which occupied 37 percent and 40 percent of the total lease area, respectively, sampling cluster points were established at six locations. On a smaller vegetation type (4 ARTR BOGR), which encompassed 13 percent of the Black Thunder unit, three cluster locations were established. A single sampling site was established on each of the remaining seven vegetation types because they represented small acreages constituting in aggregate only 10 percent of the lease area.

A variety of methodological sampling activities are conducted at each of the 22 cluster locations, the center point of which was designated by a steel fence post. Three permanently established transect lines, located by restricted randomization procedures, were installed. The permanently located quadrats, for measurement of vegetation structure and determination of ground cover composition and frequency percentage values, were established at random distances along the transect lines. In order to measure herbage production and other soil and plant characteristics without the impact of periodic livestock utilization, a 30 ft x 30 ft wire fence exclosure was constructed at each cluster. Restricted randomization selection procedures were utilized for the location of each, specifically to avoid overlap onto the permanent transect lines.

The rain gauges and maximum-minimum thermometers were located within the exclosures as well as the clipping quadrats for measures of herbage production. Other forms of data such as determination of phenology values, soil samples and herbage materials (both above and below ground) were obtained from within the exclosures. Supplementary research, with selected objectives providing support information for the requirements of a general investigation program, was conducted in conjunction with the formal data gathering activities. Data were gathered in some instances at the exclosure locations and also at other sites in the lease area. A brief descriptive summary of these investigations was presented in the Fisser, et al, 1976 report.

Recently, the Ph.D. dissertation by Faisal Taha was completed and the doctoral degree awarded to him at the University of Wyoming spring graduation exercises. The complete dissertation is given in the publication list in the report. A preliminary citation for the Ph.D. dissertation by Jerry Cox is also given in the publication list. His degree requirements were expected to be completed for December 1977 graduation.

Information presented in this annual progress report was divided into seven sections which were as follows: 1) ground cover composition, 2) phenology study, 3) exclosure studies (production, precipitation and temperature), 4) carbohydrates, 5) soil moisture, 6) leaf water stress and 7) appendices. Included in the report was a publication list of all papers, presentations and manuscripts prepared related to the basic project effort by Dr. Fisser and others in the Range Management group.

770502 Fisser, H.G., L.W. Young, J.E. Orpet and N.E. Hargis. 1977. Phenology and production studies on semiarid shrub types. 1976 results. Univ. of Wyo. Coop. Res. Rep. to the BLM. Sections I, II, III and IV. Wyo. Agr. Exp. Sta. Sci. Rep. 930. 438 p.

1E2 2E 2F2 3A2 1E1 2A1D

77050201 Section I - Phenology study.

Phenology records were taken at the 12 intensive study site exclosure locations during 1976. Growth initiation, full bloom, and start of seeding dissemination for 1976 was compared with the three previous years. Comparison of western wheatgrass *(Agropyron smithii)* showed a later initiation of plant growth which progressed to an earlier seed maturity and plant drying. The floral bud stage for big sagebrush *(Artemisia tridentata)* appeared later, but full bloom to seed maturity occurred earlier.

In 1976, western wheatgrass in the southwestern part of the state at the Cedar Mountain, Cumberland -3 and Mesa Antelope exclosures exhibited full bloom earlier than the exclosures in the Big Horn Basin. At these exclosures, floral bud stage and full bloom was 15-20 days later than normal.

The average growth stage of bluebunch wheatgrass *(Agropyron spicatum)* was the same at all exclosures except Owl Draw, where the floral bud stage through full bloom was nearly a month late. Exclosures in the Big Horn Basin showed a 2-week earlier floral bud stage for big sagebrush. Full bloom for back sagebrush *(A. nova)* occurred 10-15 days earlier at Horse Creek and Sweetwater exclosures. At the Demer exclosure to the east of the Big Horn River on July 28, 1976, sagebrush leaves were still quite green and growing well, while at other locations, especially in southern and western Wyoming, most leaves of sagebrush were fading in color and often shedding much earlier than the previous year.

11F2 2E 2F2 3A2 1E1 2A 5A1D

77050202 Section II - Phenology step-wise regression for predictive modeling.

In this section are provided the responses of both vegetative and reproductive phenological development rates as influenced by 17 different environmental factors including soil, temperature, moisture, precipitation and forage production characteristics. For each of the four prime species, western wheatgrass *(Agropyron smithii)*, bluebunch wheatgrass *(A. spicatum)*, big sagebrush *(Artemisia tridentata)* and black sagebrush *(A. nova)*, a stepwise selection regression model was utilized to determine the linear relationship of dependant variables, that is phenology, upon the independent environmental factors. Significance of the multiple regression was tested by analysis of variance. Significance of the partial regressions of individual independent variables was tested by t-test. All independent variables significant at less than 75 percent confidence were rejected from the analysis.

2A 3A

77050203 Section III - Precipitation pattern study.

Winter and spring moisture values during 1976 were relatively uniform and in some instances greater than exhibited on long-term average records. In the Big Horn Basin area and the northern portion of the Wind River Basin, total annual precipitation at most gauges was considerably greater than long-term average values. It appeared that summer precipitation values, as well as fall precipitation, were considerably greater than normal. Southward through the Wind River Basin and into Carbon and Sweetwater Counties, precipitation data were considerably less than normal and seemed strongly affected because of low precipitation during summer and fall. To the east in Natrona County, precipitation appeared to be greater than average at most locations. This occurred apparently because of good moisture conditions during summer and fall as well as during winter and spring periods.

Several significant occurrences of snowfall were noted in the area southeast of Worland at the Demer exclosure. More than 8 inches of snow was present on January 10. At the Shoshoni Ant area there was approximately 5 inches of snow on the ground at that time. Most locations had significant amounts of snow during January, but in the second week of March significant variation was noted, in that the Sweetwater exclosure had none and the soil was relatively dry, while at the Demer exclosure some snow was present but very muddy soils were evident and also early spring growth of Sandberg bluegrass *(Poa secunda)*, western wheatgrass *(Agropyron smithii)* and big sagebrush *(Artemisia tridentata)* was noted. The snowfall occurrence and late winter warm temperature interaction obviously had significant effects upon the initiation of plant growth.

The summer of 1976, especially in the Big Horn and Wind River Basin areas could be called a year of yellow sweetclover *(Melilotus officinalis)* because of its great abundance along the highways in western Wyoming. Phenology progression along pipeline revegetation plots in the Big Horn Basin on July 15 was more than 2 weeks ahead of that which had occurred during 1975.

Obviously all areas in the Big Horn Basin had not sustained significant snowfall, especially to the west of the Big Horn River, and on those areas herbage production was very limited.

Tabular material are presented for all of the UW-BLM rain gauge data by seasonal units and for some 40 U.S. Weather Service stations for which the data were compiled on a seasonal basis similar to that of the UW-BLM gauges.

3A3 77050204 Section IV - Exclosure study: production, cover and precipitation phase.

5A1D

5B Herbage production estimates for 1976 were compared with the data from the previous
5C three years. The plant growth during 1976 was the same as 1973, 15 percent higher than
2A 1974 and 35 percent lower than 1975. Area cover percentage and production of annual grasses
2C and forbs were lower in 1976 than 1975. Perennial grass production for 1976 was higher than for previous years. Sandberg bluegrass *(Poa secunda)* and western wheatgrass *(Agropyron smithii)* showed the greatest increase in production, the latter the greatest increase in area cover. Shrub production for the year was considerably lower than for 1975. The greatest decrease was exhibited by big sagebrush *(Artemisia tridentata)*. Late June rains were beneficial to growth and subsequent total annual herbage production of perennial grasses. The accumulation of total precipitation did not rise above average until rains occurred in September and October. This was too late, however, to have any significant effect on shrub production.

The decrease in shrub and annual grasses production was responsible for most total herbage production differences between 1976 and 1975. The usual tabular material of the total list of 45 exclosure locations, the description of the computation for utilization and presentation of herbage production data for each of the exclosures by transect are given in the report.

7706 Graduate Advisee Completion

2F2 770601 Taha, F.K. 1977. Ecological interrelationships of vegetation and soils in a shortgrass
1E2 prairie ecosystem. Ph.D. Dissertation. University of Wyoming. 207 p.

8D

2C An ecological investigation of interrelationships between vegetation productivity and
3A2 nutrient content, and the soils' physical and chemical characteristics, was conducted in
2F1 northeastern Wyoming from early 1974 to mid 1976. Vegetation parameters studied
1E1 were grass production, forb production, shrub production, total vegetation production
1E2 and carbohydrates, and protein content of blue grama *(Bouteloua gracilis)*, western

wheatgrass *(Agropyron smithii)*, fringed sagewort *(Artemisia frigida)* and big sagebrush *(A. tridentata)*. Soil factors included a total of 40 physical and chemical soil characteristics. Specific objectives were:

1) to quantify the interrelationship of vegetation with physical and chemical soil characteristics by utilizing correlation coefficients and analysis of variance techniques.
2) to evaluate and separate the significant interrelationships by employing a step-wise multiple regression analysis.
3) to provide interpretation of biologically significant associations between vegetation and soil factors.
4) to use the established signficant relationships between vegetation and soil variables for the development of a predicted model for similar investigations in a shortgrass perennial system.

Grass production relationships were found to be affected mainly by soil physical characteristics. Simple correlation coefficients were significant for organic matter, air:water permeability ratio, organic nitrogen, settling volume A and percent silt of the A horizon. Further analysis by step-wise multiple regression showed that air:water permeability ratio of the A and B horizons, moisture equivalent of the A horizon and the organic matter explained 70 percent of the variation in grass production.

Eight factors of the 40 soil factors were simply and significantly correlated with forb production. These were (in order of importance): copper, percent silt of the A horizon, cation exchange capacity, sulphur, calcium, available moisture of the A horizon, available moisture of the B horizon and 15-bar moisture of the A horizon. Seventy-six percent of the variation in forb production was explained by five soil physical factors. These were percent silt of the A horizon, air:water permeability ratio of the A and B horizons, available moisture of the A horizon and percent clay of the B horizon.

Statistically significant soil factors responsible for shrub production, however, were not possible to isolate due to some compounding factors. It is suggested that the ability of shrub species to draw moisture and nutrients from depths not considered in this investigation might have contributed to the apparent non-existence of relationships between shrub production and soil factors.

Relationships between total vegetation production and soil factors were similar to that of grass production and soil variables. The vegetation making up most of the production was concluded to be the determining factor in the relationships between total production and soil parameters.

Of the four prime species studied, only western wheatgrass was shown to exhibit phenotypical responses to soil characteristics. Carbohydrate amount was influenced by four moisture related factors and two nutrient factors that contributed to 93 percent of the variation. Percent clay of the A and B horizons, 15-bar moisture of the A horizon and moisture equivalent of the B horizon were the prime moisture factors. Of the nutrient factors, only phosphorus and magnesium contributed significantly to carbohydrate content. Protein content of western wheatgrass was influenced by four moisture variables and seven nutrient factors. A high coefficient of determination of 99.8 percent was attributed to possible interaction of independent variables that were tested against protein content. Nevertheless, the variables separated by the multiple regression analysis to explain the variation in protein content were biologically sound. These were settling volume of the A horizon, moisture equivalent of the A and B horizons, and percent sand, pH, cation exchange capacity, sodium, nitrate, calcium, magnesium and mangenese of the B horizon.

Blue grama, fringed sagewort and big sagebrush carbohydrates and protein content did not show significant association with soil factors. This was attributed to possible genetically controlled mechanisms of mineral absorption, other environmental factors and variation in degree of maturity of vegetation material.

Statistical regression models were constructed for vegetation parameters that were significantly related to soil characteristics. An F-value four times or larger than the selected F-level was found to be most appropriate for the construction of models and thus was suggested for use in similar work.

3A2 770602 Hohl, R.J. 1977. Summer habitat preference and vegetation utilization of elk in the
7A2 Snowy Range, Medicine Bow National Forest, Wyoming. M.S. Thesis, University of
10D1 Wyoming. 69 p.

Research was conducted from June to September 1975 on a small resident elk herd within the Main Bow Division of the Medicine Bow National Forest. Major emphasis was placed upon habitat preference and forage utilization of the elk as influenced by the structure of the vegetation. Observations were also made on insect influence, recreational use, livestock grazing and elk calf age to help evaluate the utilization and preference patterns of the elk.

Data on forage utilization indicated graminoids were grazed early in the study period and comprised the bulk of the forage consumed well into July. Forb species provided the greatest portion of the diet from mid-July until mid-August. Grasses and sedges again became the desired vegetation from late summer into early fall. Increased use of browse, especially willow (*Salix* spp), occurred from mid-August until the end of the study period in mid-September. Preferred graminoids were: tufted hairgrass *(Deschampsia caespitosa)*, slender wheatgrass *(Agropyron trachycaulum)*, bluegrass (*Poa* spp), mountain timothy *(Phleum alpinum)*, sedges (*Carex* spp), bromes (*Bromus* spp), Idaho fescue *(Festuca idahoensis)*, reedgrasses (*Calamagrostis* spp) and blue wildrye *(Elymus glaucus)*. Forb preference included: hairy goldmaster *(Heterotheca villosa)*, pale agoseris *(Agoseris glauca)*, aster (*Aster* spp), dandelion *(Taraxacum officinale)*, heartleaf arnica *(Arnica cordifolia)* and lodgepole lupine *(Lupinus parviflorus)*.

The study type was divided into four major habitats: forest (lodgepole pine *(Pinus contorta)*, spruce-fir *(Picea-Abies)* and quaking aspen *(Populus tremuloides)* types), non-forest (parks or meadows and clearcuts), forest margins and riparians. Of the three forest elements, the lodgepole pine type was preferred, especially when calves were young. As the calves matured, utilization of parks and forest margins increased. Later in the summer, the elk preferred the riparian browse type and adjacent forest margin.

A definite relationship existed between habitat preference and structure of the vegetation. Stratification, spatial distribution and species abundance influenced the habitat preference. A vertical and horizontal placement of a closed vegetation inhibited the physical movement of the elk, but also supplied cover. The open vegetation patterns helped to furnish a diverse floral composition as a choice for forage utilization, while species abundance affected the availability of preferred forage. Insects, human presence and elk calf age had less influence on preference and utilization patterns than vegetation structure.

1E2 770603 Hughes, H.G. 1977. Factors influencing vegetative and reproductive phenology of
1E1 *(Artemisia nova)* Nels. M.S. Thesis. University of Wyoming. 59 p.
2E
2F2 A study was conducted to identify and quantitatively determine those
3A2 environmental, soil and topographical factors which influenced the phenology of black
2A sagebrush (*Artemisia nova*). Data were collected at three areas of western Wyoming
5A1D from April 1974 to December 1975.
10D1 A 17-stage numerical index was used for all phenological observations. The index described phenological progression from growth initiation to dormancy. Phenological observations and environmental data were collected at approximately 2-week intervals year round. Fifteen environmental, two productivity, 14 soil and two topographical factors were analyzed by a step-wise multiple regression program to identify those factors which influence vegetative growth and flowering.

Vegetative growth was found to be influenced by soil temperature at 38 cm, soil moisture percent at 15-30 cm on date, and soil moisture percent at 46-61 cm on date.

Soil temperature was important due to the earliness of growth and the effect which temperature has upon various processes (carbohydrate translocation, water uptake). Soil moisture percent at 15-30 cm and 46-61 cm was significant due to the importance of water in all physiological processes. The soil, being the primary reserve for the water, must contain adequate available soil moisture levels to insure plant growth. Low levels of winter recharge would easily affect vegetative growth until sufficient moisture levels are reached. These factors accounted for 54 percent of the variation of the vegetative growth of black sagebrush.

The flowering of black sagebrush was found to be influenced mainly by soil temperature, soil moisture percent at 15-30 cm on date, soil moisture percent at 15-30 cm one month previous and current air temperature. Soil temperature exhibited a strong positive correlation with flowering. The negative correlation of soil moisture percent at 15-30 cm on date and one month previous was interpreted as mechanisms enabling the plant to fully utilize the environment, yet still complete their life cycle. The significance of air temperature in influencing flowering phenology was possibly caused by genetic selection of those plants which did use air temperature to determine flowering dates. Some sort of hormonal interaction was also possible with air temperature. These factors accounted for 87 percent of the variation in the flowering phenology of black sagebrush.

Those factors which were not sampled as frequently as the environmental data, or which did not regress with time, appeared to have been underemphasized by the computer program. These factors consist of all precipitation, soil, herbage production and topographical factors. Potentially significant factors were determined to be not so, due to the fact of not regressing with each phenological observation.

3A2 770604 Gdara, A.O. 1977. Interpretation of long-term vegetation data from permanent quadrats on semiarid shrublands of western Wyoming. M.S. Thesis. University of Wyoming. 114 p.
2F2
10D1
2E
2A
11E

The long-term studies of vegetation cover dynamics on grazed and protected sites in western Wyoming were initiated in 1960 with a cooperative agreement between the BLM and the University of Wyoming. Data were obtained from 29 livestock exclosures distributed throughout western Wyoming on semiarid northern desert shrub rangelands, and at an eight pasture research grazing study area on a Gardner saltbush *(Artiplex gardneri)* vegetation type. Vegetation cover data were obtained from permanently located quadrats inside and outside of exclosure areas. Within the 2 ft x 20 ft quadrats, crown cover of shrubs, half shrubs, Plains pricklypear *(Opunta polyacantha)* and mat-form species was determined by position mapping procedures. Within the 4 ft x 4 ft quadrats, herbaceous cover was determined utilizing a 10-point frame with pins spaced at 2-inch intervals. The point frame was positioned at 10 locations within each quadrat. First hits of all species as well as ground surface character were recorded. Data were obtained at four intervals during the years 1960, 1964, 1968, 1972 and 1976.

Changes of vegetation cover on the study areas were due to a combination of several factors. Grazing influence was apparent in almost all the study areas. Vegetation type differences also played an important part in the dynamics of vegetation cover. Shrub and forage species generally increased, but grass species decreased in average cover on the grazed areas compared to the protected areas.

1E3 770605 Cox, J.R. 1977. Phenology, carbohydrate storage and annual biomass of five short-grass prairie species in relation to environmental factors. Ph.D. Dissertation. University of Wyoming. 203 p.
2C
1E2
5C
2A
2B
11E1
10D1
1E4
2E

Results of this Ph.D. dissertation program are partially shown previously under code number 770305. Early leaf growth of the cool season bunchgrasses was found on all soil types but was most evident on the sandy soils. Blue grama *(Bouteloua gracilis)* produced seed heads earlier on sandy soils than on the heavier textured soils, and all species showed a draw-down of carbohydrates with the initiation of growth and a corresponding replenishment in the summer or fall.

On an annual basis soil moisture (0-15 cm) and minimum temperature were found to influence vegetative and reproductive phenology for each of the species studied. The storage of carbohydrates in the roots of big sagebrush *(Artemisia tridentata)* was regulated by day length, temperature and soil moisture (0-15 cm). Carbohydrate storage in the grasses was not dominated by the environmental factors studied. Annual biomass accumulations of prairie Junegrass *(Koeleria cristata)* were related to increasing temperatures, 15-bar water holding capacity (0-15) and precipitation. Annual biomass of the remaining species was determined by environmental extremes during the summer months.

	7801	**Research Activities**
3A3 5A1D 5B 5C 2A 2C 2B 2E	780101	Exclosure study: production, cover and precipitation. During 1978 the production determination at the 44 exclosure research sites continued as in past years. These data are recorded in the 1978 results of the report to the Bureau of Land Management.
1E2 2E 2F2 3A2 1E1 2A 5A1D	780102	Exclosure study: phenology and growth. During 1978 phenology recordings were conducted seven times on each of the 12 intensive phenology study locations.
2A 3A	780103	Precipitation pattern study. During 1978 precipitation data were recorded on the UW/BLM gauges and on the seasonal basis as during previous years. The summary data from the U.S. Weather Bureau stations were also compiled on a seasonal basis. New data compilations shown in the 1978 data report include the monthly collection of data at the UW/BLM gauges in the Big Horn Basin and additional tabular computations showing distribution of precipitation by various areas in the western Wyoming region.
2E 5A1D 3A2	780104	Juniper GRI exclosure 20 ft x 50 ft permanent quadrats. During 1978 the number, location and sizes of Utah juniper *(Juniperus osteosperma)* trees and shrubs within the 20 ft x 50 ft permanent quadrats was rerecorded. The exclosure locations were Kane, Medicine Lodge and Big Trails.
5B4 8D 5B4 5C 1E2	780105	Juniper GRI seeding. During 1978 seeding trials were conducted at the GRI juniper exclosures.
2C 8B	780106	Soil-vegetation relation study. During 1978 a detailed study was begun to interrelate the soil and vegetation characteristics of the exclosures in the Big Horn Basin and at all exclosures. The soil components were then identified and samples taken when necessary.
3A2 2F2 5C	780107	1 ft x 1 ft permanent plot study. During 1978 the permanent 1 ft x 1 ft quadrat transects were read in the Big Horn Basin in conjunction with the soil vegetation study with the Bureau of Land Management.
2C 8B 10D1	780108	Exclosure soil profile studies with SCS. Soil profile studies were conducted during 1978 to complete all of the data necessary for the Grass Creek soil-vegetation relation study.

7802 Journal Publications

2C 780201 Fisser, H.G. 1978. Soil surface movement and relation to vegetation structure. p.
8B 706-711. In Mohan K. Wali (ed.) Ecology and Coal Resource Development. Vol. 2,
2F2 International Congress for Energy and the Ecosystem. University of North Dakota,
3A2 Grand Forks, N.D. June 12-16, 1978. Pergamon Press, Elmsford, NY.

See 780304.

2F2 780202 Shawesh, O.M. and H.G. Fisser. 1978. Ecology of rangeland communities in
2A northwestern Libya, Africa. p. 190-192. In D.N. Hyder (ed.) Proc. 1st Intern.
2B Rangeland Congress, Denver, Colorado. August 14-18, 1978.
8C

See 780308.

1E3 780203 Cox, J.R. and H.G. Fisser. 1978. The effects of water stress on phenology and
2C carbohydrate storage in the shortgrass prairie. p. 529-535. In J.R. Goodin and D.K.
1E2 Northington (eds.) 1979. Intern. Arid Lands Conference on Plant Resources. Intern.
5C Center Arid and Semiarid Land Studies. Lubbock, Texas. October 8-15, 1978.
2A
2B See 780309.
10D1
1E4
2E

7803 Presentations

2C 780301 "Soil erosion and movement studies in western Wyoming." February 8, 1978. 31st Ann.
8B Meeting Soc. Range Manage. San Antonio, Texas. by H.G. Fisser.
2F2
3A2

Soils of rangelands are subject to natural as well as man-induced erosive forces. Natural factors such as wind and rain, and man-related forces such as grazing livestock, vehicular traffic and even simply walking over an area can shift significantly the position of earth particles. A stable vegetation-soil complex will have essentially no sheet erosion and little soil movement. Such sites often exhibit soil surface roughness parameters in excess of those exhibited by communities subjected to sheet erosion. Sheet erosion, if it occurs, can be determined by a decrease or increase of the soil surface level. Soil movement within a community, however, is not as simply measured. A study area was established in a low sagebrush (*Artemisia arbuscula*) area in western Wyoming in 1963.

Measurement of soil surface parameters indicated there was no significant soil movement response associated with the grazing treatment. A strong statistical significance was associated with the chemical treatments because of the definite soil movement occurrences following shrub control. The interaction of grazing impact and chemical control of shrubs resulted in a highly significant soil movement which had an important influence upon the ability of the native vegetation to express production changes.

11F1 780302 "Models for prediction of natural vegetation potential in eastern Wyoming." February 8,
3A2 1978. 31st Ann. Meeting Soc. Range Manage. San Antonio, Texas. by H.G. Fisser.
2A
2C
2E
2F2

A functional regression model was developed using the stepwise multiple regression technique for prediction of natural vegetation production in eastern Wyoming. Vegetation production was segregated as grass production, forb production, shrub production and total production. A regression model was developed for each vegetation class, 40 soil factors investigated on 13 study site locations, were included in each of the four individual models. There was a grass production model, a forb production, a shrub production model and a total production model. The regression models helped suggest cause and effect relationships and provide useful predictive information on range ecosystems.

3A1 780303 "Status of remote sensing in Wyoming." March 9, 1978. Annual Meeting GPC-7 Committee. Brownsville, Texas. by H.G. Fisser.
3B
3C

At the annual meeting of the GPC-7 committee in Brownsville, Texas, Fisser presented a comprehensive report of existing remote sensing information, primarily aerial photography as well as some satellite imagery and photography in Wyoming. In terms of actual operation and application of remote sense information in Wyoming, little had been done either by state agencies, federal agencies, and especially individuals. In time there would likely be some further developments so that some state agency could monitor the remote sense information input and develop it for application and use by individual land users.

2C 780304 "Soil surface movement and relation to vegetation structure." June 15, 1978. Intern. Congress for Energy and the Ecosystem. University of North Dakota, Grand Forks, North Dakota. by H.G. Fisser.
8B
2F2
3A2

Vegetation and soil surface stability of reclaimed surface mined areas are important criteria for satisfactory rehabilitation. Soils of rangeland are subject to natural and man-induced erosive forces. Natural factors such as wind and rain, man-related forces such as grazing livestock, vehicular traffic and even simply walking over an area can significantly affect the position of earth particles. A stable vegetation-soil complex will have essentially no sheet erosion and little soil movement. Such sites often exhibit soil surface roughness parameters in excess of those exhibited by communities subjected to sheet erosion. The sheet erosion, if it occurs, then can be determined by a decrease or increase of the soil surface level. Soil movement within a community, however, is not as simply measured.

A study area was established in a low sagebrush (*Artemisia arbuscula*) area in western Wyoming in 1963. Measurement of soil surface parameters showed no significant soil movement response associated with the grazing treatment. A strong statistical significance was associated with the chemical treatments because of the definite soil movement occurrences following shrub control. The interaction of grazing impact and chemical control of shrubs resulted in a highly significant soil movement.

Surface mining and other energy industry activities will affect rangelands and the agricultural industry. Many reclamation and rehabilitation techniques have been devised, and have shown, in a few years, resulting vegetation production equal or superior to the original native vegetation. Long-term herbage production levels are inherently dependent upon ecological relationships, especially moisture and temperature dynamics. Restructured strip-mined lands are dependent on a soil-geology matrix often subject to erosive forces. Determination of an acquired soil stability can be expressed with study of soil surface dynamics such as those from the erosion transects described in this paper.

7D1 780305 "Plant succession on plains pocket gopher mounds." February 7, 1978. 31st Ann. Meeting Soc. Range Manage. San Antonio, Texas. by M.A. Foster.
2E
2F1
3A1

The plains pocket gopher *(Geomys bursarius)*, a common rodent in central and western United States, was found in association with a variety of vegetation and soil types. Gophers were a problem in crop and alfalfa *(Medicago sativa)* fields, and damaged rangelands as well. As a result of their burrowing activities, gophers deposit loose soil on the surface. Mounds commonly cover 5-15 percent of the surface of gopher infested rangeland. Plant succession on mounds progressed as it would on any denuded area.

Research was initiated in 1974 to determine the effects of plains pocket gophers on plant succession on two range sites (sands and silty). The age of 150 mounds was determined on each range site, and each mound was permanently marked. The mounds were separated into less than 1-year, 1-year and 2-year old categories. Fifty additional mounds were marked on each range site in 1975 and in 1976. Frequency of vegetation was determined on each age class of mound. Vegetation was sparse on less than 1-year-old mounds. All plants present were established at the original soil surface and were growing through the mound. Frequency of perennial grasses and forbs increased on mounds up to two years in age, then decreased. Vegetation was near stabilization on 4-year-old mounds. Plant succession progressed more slowly on the silty range site than on the sands range site.

10A 2F2 2C 2A2 780306 "Range management equals functional ecology." July 11, 1978. Wyoming OWRC/SRM Youth Range Camp. Ten Sleep, Wyoming. by H.G. Fisser.

The Youth Camp presentation, given to a group of high school students, was oriented toward persons with minimal backgrounds relating to organismic or biotic elements and to limited experience with concepts of environment and ecology. Therefore, the characterization of the presentation was a description of the major categories of kinds of vegetation classes, characteristics and anatomy situations that would be important to one interested in rangeland ecology.

1) Vegetation Types: a) Prairie, i) Tall grass, ii) Sandhills; b) Plains; c) Pacific bunchgrass; d) Southern desert grasslands; e) Northern desert shrub, i) Sagebrush grass, ii) Salt desert grass; f) Southern desert shrub, i) mequite grassland, ii) creosote; g) Mountain browse; h) Oak brush; i) Chapparal; j) Pondorosa Pine.
2) Plant Classification: a) Grass and grass-like; b) Herbs and forbs; c) Half-shrubs; d) Shrubs; e) Trees.
3) Plant Anatomy: a) Gymnosperms; b) Angiosperms; c) Monocots; d) Dicots.

3A2 2F2 10D1 2E 2A 780307 "Semiarid rangeland and grazing evaluation of shrubland from permanent quadrats." August 15, 1978. 1st Intern. Rangeland Congress. Denver, Colorado. by A.D. Gdara.

The long-term studies of vegetation cover dynamics on grazed and protected sites in western Wyoming were initiated in 1960 with a cooperative agreement between the Bureau of Land Management and the University of Wyoming Range Management Department. Data were obtained from 29 livestock exclosures distributed throughout western Wyoming on semiarid northern desert shrub rangelands, and at an eight pasture research grazing study area on a Gardner saltbush *(Atriplex gardneri)* vegetation type.

Vegetation cover data were obtained from permanently located quadrats inside and outside exclosure areas. Within the 2 ft x 20 ft quadrats, crown cover of shrubs, half-shrubs, Plains pricklypear *(Opuntia polyacantha)* and mat-form species such as Hood's plox *(Phlox hoodii)*, was determined by position mapping procedures; within 4 ft x 4 ft quadrats herbaceous cover was determined using a 10 point frame with pins spaced at 2 in intervals. The point frame was positioned at 10 locations within each quadrat. First hits of all species as well as ground surface characters were recorded.

Data were obtained at four year intervals during the years 1960, 1964, 1968, 1972 and 1976. Changes of vegetation cover on the study areas were due to a combination of several factors. Grazing influences in almost all of the study areas, and vegetation type difference, apparently played an important part in dynamics of vegetation cover. Shrub and forb species generally increased, but grass species decreased in average cover on the grazed areas as compared to the protected areas.

2F 2C 2B 8C 3A1 780308 "Ecology of rangeland communities in northwestern Libya, Africa." August 15, 1978. 1st Intern. Rangeland Congress. Denver, Colorado. by O.M. Shawesh.

Quantitative description and ecology investigations of semiarid rangelands in northwestern Libya were initiated in May 1976. A 25,000 ha livestock exclosure near Assa, some 25 km from the Mediterranean seacoast was utilized. The rangeland vegetation was predominantly characterized by a variety of low growing shrubs, with some small areas of perennial grass. Annuals were present but not abundant. Vegetation and soil relationships were investigated with permanently established transects for measures of plant cover and density and by soil profile sampling for detailed analysis of physical and chemical characteristics. The relatively stable rangeland sites were characterized by frequent presence of common sagewort *(Artemisia campestris)*, which comprised 33 percent of all vegetation composition. Rock rose *(Helianthemum lippii)*, saggy sparrowwort *(Thymelaea microphylla)* and hidma *(Echiochilon fruticosum)* were the most common additional shrubs. False esparto *(Lygeum spartum)* and hairgrass *(Koeleria pubescens)* were the most common perennial grasses. Height, diameter and density of all perennial vegetation was observed to have an average of 24.35 cm, 2.83 cm and 3.77 plants/m^2, respectively. The average percent cover for the entire life form was 7.46 percent. Initial investigation of soils at each transect revealed what is normally considered to be inconsequential differences of percent sand. Values ranged from a high 96.5 percent at Profile 1 to the lowest of 92.7 percent at Profile 6. These values, and those of interrelated silt and clay percentages, characterize the textural composition of azonal sands. Textural differences of silt ranged from a very low 1 percent in Profiles 1 and 4 compared to 3.4 percent and 3.7 percent on Profiles 2 and 6. Clay composition in the soils varied least with only 2.5 percent in Profile 1, and 3.6 and 4.0 percent in Profiles 5 and 6.

1E3 780309 "The effects of water stress on phenology and carbohydrate storage in the shortgrass
2C prairie." October 11, 1978. Intern. Arid Lands Conf. on Plant Resources. Texas Tech
1E2 University. Lubbock, Texas. by J.R. Cox.
5C

2A A study was conducted in 1976 in eastern Wyoming to determine seasonal variation
2B of leaf water potential (LWP), phenological development and total non-structural
10D1 carbohydrates; and how various soil textural groups affected plant water use,
1E4 morphology and storage of nutrients in western wheatgrass *(Agropyron smithii)* and blue
2E grama *(Bouteloua gracilis)*. Leaf water potential, phenological development and total non-
10D structural cover were similar on all soil textural types for both grasses. Root reserves were lowest on June 15 for western wheatgrass with the elongation of leaves (third leaf) whereas LWP was highest. Reserves were rapidly accumulated in early July and maximum levels occurred with the initiation of flowering. During the following sampling period, LWP increased (more negative) whereas the level of reserves increased. Plants of loamy soils matured earlier in late July, but individuals on sandy soils matured more rapidly in late August. Reserves in blue grama gradually increased as LWP increased in all soil types. Seed formation was generally constant on soil types, with maturation dates slightly earlier on clayey soils.

These two grasses were important forage species on rangelands of the shortgrass prairie region. Moisture is a crucial environmental factor on these semiarid grasslands. An insight into soil moisture availability differences among soil textural groups and related effects of plant phenology and carbohydrate reserves would help determine plant survival limits on soils to be surface mined.

5B 780310 "Rangeland vegetation, improvement and management in Libya, Africa." October 12,
2F 1978. Intern. Arid Lands Conf. on Plant Resources. Texas Tech University.
2C Lubbock, Texas. by O.M. Shawesh.
2B

8C Transects with permanent quadrats were located at 500 m intervals across the Assa
3A1 study area of northwestern Libya, Africa. Transects were 50 m long and utilized
11F1 permanently located quadrats of 30 cm x 30 cm, 1 m x 1 m, and 2 m x 5 m for determination of various vegetation parameters. Soils were primarily azonal sands with usually more than 95 percent sand. Soil pH averaged 8.5 and depth was approximately 135 cm. Most soils were reddish yellow or strong brown.

Average plant height ranged from 24 to 30 cm, the greatest occurred in areas of sparse density. Perennial plant basal diameter varied from 3 to 7 cm. Again, greatest size was noted in areas of least density.

Density values ranged from four to less than one perennial plant per sq. m. Common sagewort *(Artemisia campestris)* provided about 3 percent cover and total vegetation cover was about 8 percent.

Computer based data summarization and evaluation of biological and physical interrelationships were developed since little, if any, quantitative information of Libyan rangelands was available. These data, and subsequent studies on the permanently located transects, were of great importance. Continued investigation was essential for development of proper management and use criteria for animals such as game, goat, sheep and cattle which were common in north Africa.

3A1 780311 "Range condition and trend review in the Lander Resource Area." December 11, 1978.
3A2 Range survey report to Bureau Land Manage. Cheyenne, Wyoming. by H.G. Fisser.
2A
2B During fall 1978 Fisser was contacted by the Bureau of Land Management to
2C conduct an independent survey of condition classification of rangelands especially in the
2E Big Pasture allotment west of Lander in the area of Oregon Trail Crossing. The proposal
2F2 contained several topics, including:

1) Data base
   a) Personal records of range survey transects of vegetation cover and composition with specific map and site location information in the area west of Lander.
   b) At five exclosures, permanent vegetation inventory plots were established in 1959 and 1960.
   c) Herbage production and vegetation cover data have been obtained annually from exclosures in the area east of Lander.
   d) Rain gauges have been established at all of the exclosures.
   e) Permanent site record photographs were taken of some of the early survey transect locations in 1959-1961 and annually at the exclosures with herbage production. Sample quadrats of 2 ft x 20 ft and with 4 ft x 4 ft plots have been photographed at four year intervals.
2) Field procedures
   a) A reread of representative sampling of the 1959-1961 survey transects for vegetation cover and composition.
      i) Duplicate photographs at all locations where present.
      ii) Estimate vegetation composition by weight similar to SCS concepts.
      iii) Follow SCS technicians guides estimate for condition classification percentages.
      iv) Interpret vegetation and site character for determination of range trend.
   b) Inspect exclosures for apparent response to protection from grazing and sagebrush control.
   c) BLM personnel will accompany Dr. Fisser whenever possible and provide transportation when possible for him.
   d) Field work is expected to be completed by November 15, 1978.
   e) It is understood that snow cover may severely limit the time and location of field activities.
3) Data interpretation
   a) A final summary of condition classification and trend is to be presented on or about December 4, 1978, to the Bureau of Land Management in Cheyenne.
   b) Findings will be coordinated to BLM allotments by working with resource area personnel.
   c Data sheets, photographs, and other information will be available to the BLM.

Following 10 days of intensive field recording activity during October and November, and intensive interpretation of data and comparison of fall 1978 data to data accumulated during the previous 17 years, the report was prepared in a definitive format to provide BLM personnel with information and background which would assist them:

1) To describe and map the vegetation types,
2) To determine rangeland condition classification, and
3) To develop estimates of grazing capacity.

Data interpretation was based on itemization of the allotments, areas, the licensed use (AUMs), active stocking rate (A/AUM), and other information. The data of weight composition percentage of the individual species, groups of species by grasses, forbs and shrubs, and individual shrubs as well, were compiled as shown and as indicated by the SCS procedures so that a condition classification value was obtained for each transect data set, both early and late. These were computed then to an A/AUM for each of the individual areas. This interpretation for a compilation of a condition class value was utilized as item 1) to determine estimates of grazing capacity and generalized initial expression of trend.

Additional trend factors were identified as:

2) Grass distribution. This was rated from good to poor and was determined on the basis of the amount of bare areas present in the site. If the bare areas greater than one person's hand were common, the distribution was itemized as being poor and if distribution left areas no more than 2-3 inches in diameter at the most, then the distribution would be identified as good.
3) Shrub re-establishment. This was simply rated as minimum or maximum. If there was shrub re-establishment, it was rated as maximum. With no re-establishment, it was rated as minimum.
4) Vigor. Grasses were characterized in terms of the appearance of the grasses within the exclosures as compared to those on the outside. Shrubs were characterized on the basis of appearance of the non-grazed shrubs within the exclosures as well.
5) Water erosion was identified as minimum or maximum, based on appearance of actual water rivulet gulleys within the vegetation complex.
6) Wind erosion was identified as low, moderate or high based on whether any open areas showed active wind loss.
7) Litter was rated from poor to fair to good and was identified on the basis of natural litter existing on the ground.
8) Soil surface was a composite interpretative evaluation relating to slick spot presence or absence, trampling presence or absence and the existence of ped development on the soil surface and if all conditions appeared positive with regard to natural ecological surface relationship, it was characterized as good.

The actual trend then was characterized on the basis of all of the terms as downward, static or upward. Some additional rating modification was used for characterization in terms of limitation because of big sagebrush *(Artemisia tridentata)* control of the community whether upward, downward or slightly up with sagebrush control; or stable and upward with sagebrush control; or up static with sagebrush control indicating three condition levels of sagebrush importance on the ecological sites.

The following questions must be answered in the affirmative in order to justify animal stocking rate reduction actions: 1) Was there evidence that the recent (10-20 years) existing stocking rates have caused a deterioration of the soil-vegetation complex? 2) In terms of the existing soil-vegetation complex, could a stocking rate reduction be expected to result in significant improvement or change? 3) Would modification of forage allocation be imminent because of recent changes in populations of antelope, deer, or horses? 4) Could long-term shrub population dynamics through natural maturation and death be expected to allow natural increase of understory species, given continual stocking rates?

The data, reports, transparancies and almost 400 slides are filed at the U.W. Department of Range Management. These data comprise one of only a few efforts to comprehensively work at a large area with long-term data and interpret subjectively the response of the vegetation to the existing climatic and grazing treatments. To determine whether management modification could affect significantly the existing vegetation would require continuous monitoring of actual animal use of vegetation and the long-term responses to utilization.

7804 Published Proceedings

2C 8B 2F2 3A2 780401 Fisser, H.G. 1978. Soil erosion and movement studies in western Wyoming. p. 16. In Proc. Soc. Range Manage. 31st Ann. Meeting, San Antonio, Texas. February 7-9, 1978.

See 780301.

11F1 3A2 2A 2C 2E 2F2 780402 Taha, F.K. and H.G. Fisser. 1978. Models for prediction of natural vegetation potential in eastern Wyoming. p. 19. In Proc. Soc. Range Manage. 31st Ann. Meeting, San Antonio, Texas. February 7-9, 1978.

See 780302.

7D1 2E 2F1 3A1 780403 Foster, M.A., J. Stubbendieck and H.G. Fisser. 1978. Plant succession on plains pocket gopher mounds. p. 8. In Proc. Soc. Range Manage. 31st Ann. Meeting, San Antonio, Texas. February 7-9, 1978.

See 780305.

3A2 2F2 1021 2E 2A 780404 Gdara, A.O. and H.G. Fisser. 1978. Semiarid rangeland and grazing evaluation of shrubland from permanent quadrats. Manuscript submitted to 1st Intern. Rangeland Congress, Denver, Colorado. August 14-18, 1978.

See 780307.

5B 2F 2C 2B 8C 3A1 780405 Shawesh, O.M. and H.G. Fisser. 1978. Rangeland development and improvement in Libya, Africa. p. 65. In J.R. Goodin and D.K. Northington (eds.) abstracts Intern. Arid Lands Conf. Intern. Center for Arid and Semi-arid Land Studies. Texas Tech University, Lubbock, Texas. October 8-15, 1978.

See 780310.

2F2 3A3 2A 2C 2E 780406 Fisser, H.G. 1978. A structural definition of semiarid *Artemisia* and *Atriplex* rangeland ecosystems. Abstract for 2nd Intern. Congress of Ecology. Jerusalem, Israel. September 10-16, 1978.

Efficient management requires knowledge and descriptive information of rangeland grazing ecosystems. Studies to examine and define vegetation structural characteristics and interrelationships with environmental factors of semiarid ecosystems of western Wyoming were conducted from 1962-1977. Physiognomic expression was derived mainly from dominance of the shrubs big sagebrush *(Artemisia tridentata)* and Gardner sagebrush *(Atriplex gardneri)*. At 78 locations data was obtained within and outside livestock-proof exclosures ranging in area from 0.5 to 4.0 ha. At some locations, herbage production was determined annually from 1962. Other structural data including shrub mapping in permanent quadrats was initiated as early as 1959. Vegetation cover values were obtained at 2-4 year intervals at most locations with large permanent macroplots or with transects with permanent 0.3 m x 0.3 m quadrats. Phenodynamics of major shrubs and grasses were intensively observed and quantified at 12 locations from 1974-77. Amount and temporal distribution of precipitation was recorded. In addition, the study of the grazing treatment impacts to vegetation structure was investigated. Chemical control of sagebrush has been an important treatment at several locations.

Detailed, thorough analyses have often been instrumental for interpretation of vegetation responses. The geographical extent of the study area was approximately 500 km from north to south and 300 km from east to west, with altitudes ranging from 1350 m to 2000 m. Almost 50 percent of the precipitation occurred during April, May and June. Average annual precipitation ranged from as little as 125 mm to more than 400 mm.

Total annual herbage production as low as 90 kg/ha was common on some sites, while at locations with better soils and greater precipitation, average herbage production was as great as 400 kg/ha. On the sagebrush rangelands, understory species of abundant grasses such as western wheatgrass *(Agropyron smithii)* and needleandthread *(Stipa comata)* and others often provided more than 75 percent of the total annual herbage production. Gardner saltbush, however, often was almost a sole dominant and provided more than 90 percent of the annual herbage production. Annual species, including cheatgrass *(Bromus tectorum)*, were abundant only as a result of disturbance, or following an atypically long, moderately warm and moist spring period.

In the sagebrush ecosystems, shrub canopy cover sometimes exceeded 40 percent. Understory species were often sparse in such instances. When shrub crown cover was no more than 10-15 percent, understory species were usually, but not always, more abundant. The structurally simple saltbush ecosystems seldom provided more than 10-14 percent total vegetative cover. Shrubs on the sagebrush areas were more favored for grazing than were grasses and forbs, although a difference between vertical structural characteristics of grazed and nongrazed big sagebrush was often evident.

The predominantly winter utilization of saltbush sites was apparently equally detrimental to the shrubs and herbaceous species. Moisture and temperature dynamics from year to year were more often expressed by plant structural responses than were the stresses associated with moderate utilization by livestock and wildlife.

7805 Wyoming Agricultural Experiment Station Publications

3A1 780501 Fisser, H.G., J.R. Cox and M. Mecke. 1978. Herbage structure, production and phenology on Black Thunder. Annual Progress Report to Atlantic-Richfield Company. 1977 results. Wyo. Agr. Exp. Sta. Sci. Rep. 856. 157 p.

8D
3A2
2F1
1E2
2A

The table of contents for this report includes eight sections.

1) Introduction. Gives a preliminary statement of the program activity and lists the measures, methods, and procedures that have been conducted during the years of work on this project. It also identifies the kinds of data obtained during the 1977 growing season.

Vegetation analyses of the Black Thunder unit east of Reno Junction, Wyoming, were initiated during summer 1973 by cooperative agreement of the Atlantic-Richfield Company, the Wyoming Environmental Institute, and the University of Wyoming. Those investigations of the range management and botany departments resulted in description and mapped delineation for 10 biologically significant vegetation types as well as considerable information related to vegetation condition classification, livestock carrying capacity, plant identification. They also addressed the effects of disturbances such as pipeline and road construction and historical farming activities. Biological studies to monitor existing conditions and provide measures of any changes as a result of resource uses and impacts were recognized as needed adjuncts to environmental descriptions and statements.

2) Publication list from Black Thunder project - this list included 20 citations of written and oral reports relating to the Black Thunder Basin research program. These included the annual reports to the Atlantic-Richfield Company, oral presentations at Colorado-Wyoming Academy of Sciences, the Society for Range Management meetings, and dissertations and presentations at the International Arid Land Conference in Lubbock, Texas during 1978.

3) Ground cover composition. Percent cover composition was presented for each of the three permanent transect lines at each cluster location. The mean area cover composition for each exclosure was also presented. Exclosures were classified according to vegetation types, usually named after the most abundant species. A cumulative listing of species encountered with the point frame from 1974-77 was presented, with growth and life form groupings developed for summarization of data. Genus and species names for each plant were identified.

A summary of mean ground cover composition percentage was given for the entire lease area from data collected at 19 monitoring locations. Monitoring locations 69, 54, and 56 were eliminated due to mining activities during 1975 and 1977. Litter was 37.53 percent, bare soil 32.06 percent, erosion pavement 6.58 percent, lichens 3.23 percent, rock 0.41 percent, and vegetation 18.19 percent. The species listed below composed the majority of vegetation composition. These were, in order of importance, blue grama *(Bouteloua gracilis)* 3.38 percent, big sagebrush *(Artemisia tridentata)* 2.12 percent, western wheatgrass *(Agropyron smithii)* 2.17 percent, needleandthread *(Stipa comata)* 1.63 percent, threadleaf sedge *(Carex filifolia)* 1.58 percent, Kentucky bluegrass *(Poa pratensis)* 0.87 percent, Junegrass *(Koeleria cristata)* 0.63 percent, and Plains pricklypear *(Opuntia polyacantha)* 1.15 percent.

4) Phenology Study. A numerical designation of phenological development by month for each species and at each exclosure was presented. From this summary an annual chronological record of initial leaf growth, full leaf growth, flowering, seed development, and maturity was to be developed.

Leaf initiation in the cool season grasses was similar in 1976 and 1977. However, reduced quantities of moisture in April, May and June 1977 hastened plant maturity and reduced plant leaf area, resulting in a decreased percentage of ground cover composition. The initiation of forb growth and development was similar for 1976 and 1977, but leaf growth was reduced in 1977. With the early depletion of soil moisture there was little difference in the phenological progression of the shrubs and half-shrubs among the three years. Those plants had deep rooting systems to obtain water, as well as bark to reduce water transpirational losses.

5) Exclosure studies (production, precipitation and temperature). Ten clipping quadrats, 30 cm x 90 cm, were randomly located within each exclosure for grass and forb production estimates. Average herbage production for the entire lease area was 913.5 lb./acre. Grasses, forbs and shrubs produced 473.1, 92.9 and 347.5 lb./acre, respectively. Average annual precipitation was 11.10 inches for the 19 locations. The values ranged from a low of 9.53 inches, recorded at exclosure six in the north central area near the mine entrance site, to a high of 12.97 inches at exclosure 18, between sections 34 and 35 near the southern weather station.

6) Abstract of Taha dissertation. See 770601.

7) Abstract of Cox dissertation. See 770605.

8) Discussion. This section presented tabular material with summary listings of ground cover composition for the entire lease area for the four years from 1974-77 with characterization of cover by litter, bare ground, erosion pavement, rock, lichen (*Parmelia* spp.) and vegetation (grasses, forbs, shrubs, cactus). In addition, a summary of annual precipitation during the five years with specifics of April-June precipitation and a July-September precipitation for each of the four years was presented as well. The largest table included the annual herbage production estimates from all 22 exclosure sites from 1974-77 for groupings of total-grass, total-forbs, total-shrubs and total-all species. These listings by category of vegetation types indicated an overall average mean production among years of 657 lb./acre.

3A1 780502 Fisser, H.G. and J.R. Cox. 1978. Herbage structure, production and phenology on Black Thunder. Final Summary Report to Atlantic-Richfield Company. Wyo. Agr. Exp. Sta. Sci. Rep. 968. 243 p.

8D
3A2
2F1
1E2
2A

This report summarized the four years of field work which were initiated in late 1973 and carried on through 1974, 1975, 1976, and 1977. Data collection consisted primarily of monitoring vegetation in terms of cover, composition, structure and vegetation production. Significant interest was also attended toward soil-vegetation interrelationships because of the intimate relationship of vegetation with soils and other environmentally limiting factors.

Annual precipitation was approximately 10 inches with the lowest having occurred during the years 1974, 1975 and the greatest during 1977. The high amount of moisture received in 1977 was overridden to some extent because expected spring moisture was minimal and temperatures cool so that vegetation growth was significantly retarded until excessive summer precipitation and cool temperatures increased production.

Vegetation structure was investigated by determining cover composition with permanently located quadrats and sampled by a specially designed 100 point quadrat frame. Ground cover composition consisted of an average of about 34 percent non-vegetative, 33 percent litter and 33 percent live vegetation. These data varied among locations, and among years. The litter material varied less than the vegetational materials and the live vegetation varied more than the non-vegetative. A low in 1977 of 18.1 percent live vegetation plus lichens of 2.3 percent was recorded, while a high of 46.1 percent of live vegetation was found in 1975, including 2.6 percent of lichens (*Parmelia* spp.).

Tests of sampling adequacy utilizing an analysis of variance and associated F-tests indicated that for the most part, sampling was adequate for determination of amounts of ground cover materials based on standard F-test variables, usually at least at the 95 percent level. However, some were significant only at 90 percent, and in one instance, significance occurred at 75 percent. Because of the specific nature of the recently developed requirements, an analysis of these data indicated however, that individual transect exclosure data would not fit the Wyoming Department of Environmental Quality requirements, but when combined into exclosures for vegetation types, there was, for the most part, adequacy of sampling.

Grass production obtained from clipping quadrats during the summer growing seasons averaged 57 g/m^2 over all sampling locations of the Black Thunder unit. These data ranged from a low of 38 g/m^2 to a high of 88. Forb production was not exceptionally important in relation to the total herbage production. Total average shrub production on the Black Thunder unit was 42 g/m^2 with the majority occurring in the sagebrush-grass types. Plains pricklypear *(Opuntia polyacantha)* was fairly constant for the area and produced on the order of 2 g/m^2. Total annual herbage production of all species was 102 g/m^2. Analysis of variance of the production data indicated that production estimation was well within the limits of normal statistical variation.

Phenological investigations were conducted for the four year period. These data indicated, as expected, variations in initiation of growth, growth rate and seed maturity dates. Almost all forms of plant processes were variable among species, among sites and among years. Some initial detail analysis has shown that plant event occurrence dates can, and often do, vary from one year to another by at least one phenological stage. In a number of instances, they varied by at least two or three vegetation stages within any one month. Site characteristics have an overriding effect upon phenological progression of plants, and as such, will be of significant importance with regard to rehabilitation processes and the likelihood of acceptable rehabilitation procedures.

780503 Fisser, H.G., L.W. Young, J.E. Orpet and N.E. Hargis. 1978. Phenology and production studies on semiarid shrub types. 1977 results. Univ. of Wyo. Coop. Res. Rep. to the BLM. Sections I, II, III and IV. Wyo. Agr. Exp. Sta. Sci. Rep. 970. 439 p.

1E2 78050301 Section I - Phenology study.
2E
2F2 See 770102.
3A2
1E1
2A
5A1D

11F2 78050302 Section II - Phenology step-wise regression for predictive modeling.
2E
2F2 See 770102.
3A2
1E1
2A

2A 78050303 Section III - Precipitation pattern study.
3A
See 770103.

3A3 78050304 Section IV - Exclosure study: production, cover and precipitation phase.
5A1D
5B See 770101.
5C
2A
2C
2E
2F2

7806 Graduate Advisee Completions

5A 780601 Elmekki, A. 1978. Evaluation of brush control methods on rangelands. A study of
2A rangeland ecology and use in Sudan. M.S. Professional Papers. Univ. of Wyo. 39 &
2C 35 p., respectively.
2D
2E

Sound management principles are essential to the success of control methods on brush infested rangelands. Relatively little range improvement can be made without getting rid of the brush by burning, root plowing, chaining, spraying, or some other method. Each method has specific advantages and may have certain limitations. Once established, woody plants cannot usually be limited by good grazing management practices alone. Measures must be taken to convert brush dominated grazing lands to more productive types of vegetation. Brush control must be combined with other practices favorable to forage production that will alter the direction of successional trend.

Before starting a program of brush control, one should consider why control is needed on a particular range and what benefits may reasonably be expected. Control may be needed to increase the quantity of forage. Elimination of woody plants has often increased grazing capacity two and three fold. Better quality forage has usually accompanied these increases in production, resulting in a longer "green feed" period, when nutrients are above maintenance requirements of livestock. Control is needed in some areas to reduce fire hazards and prevent livestock losses.

The more important factors in selecting methods to control are: 1) density of the stands of shrubs, 2) stage and rate of infestation, 3) growth forms, whether trees are many stemmed or single trunk, 4) benefits that may be realized in view of soil moisture condition and potential productivity of land, 5) size of the area to be treated and 6) potential hazards of a control measure to livestock, grass cover, nearby crops, and human populations.

Controlled burning has been used widely in brush stands of non-sprouting species. Mechanical control is often used to break down brush so it will burn more rapidly and facilitate complete removal, and it is used where brush is all of the sprout species and where fire is too risky. Chemical control is used alone to kill sprouting species and may be used in combination with fire. Biological control includes insects and grazing animals.

The type of grazing following brush removal often governs the effectivness of brush control. Any system of grazing that weakens the herbaceous cover also lessens grass competition and favors the invasion of brush. A system of grazing that maintains herbaceous cover results in a strong competition and retards shrub invasion.

The second thesis paper presented by Elmekki described major ecological zones in Sudan and evaluated relationships of soil, climate and vegetation which cause differentiation. The objective was to describe the nomadic use patterns and their impact on plants and soils. Management suggestions were developed for a future improvement according to overall objectives of range management. The specialized use patterns in Sudan will require a comprehensive evaluation and study to optimize the interrelations of ecological limits with the nomad economy.

5A1D780602 Orpet, J. 1978. Long-term herbage production response following the chemical control of big sagebrush. M.S. Thesis. University of Wyoming. 118 p.
3A2
3A3
2A
2C
2E
2F2
5B4
10D

Grass and forb production was monitored for 15 consecutive years at five different study sites in western Wyoming where big sagebrush *(Artemisia tridentata)* had been chemically treated. Exclosures encompassed portions of both untreated and chemically treated rangeland for analysis of the effects that grazing may have had upon the sagebrush control projects. Sagebrush was successfully controlled and little reinvasion occurred on all but one of the study sites. Total grass and forb production at all sites averaged two times greater on the sprayed areas inside the exclosures compared to the untreated areas inside the exclosures. The sprayed areas outside the exclosures displayed only slightly greater production than the untreated sites. At the conclusion of the study period, three of the study sites produced twice as much perennial grass on the sprayed areas inside the exclosures as the untreated areas.

Many big sagebrush control projects have produced beneficial increases in livestock forage production over long periods of time. There are, however, many factors which contribute to the success or failure of spraying big sagebrush for range improvement. The vegetation and site selections should be analyzed before a program is initiated. For instance, Wyoming big sagebrush *(A. t. Wyominensis)* seems to be harder to control (Granite Mountain) and also seems to have better seedling re-establishment (Upper Government Draw) than mountain sagebrush *(A. t. vaseyama)*. Wyoming big sagebrush is also more palatable to livestock and wildlife than mountain big sagebrush; so in many cases, this shrub does not require control. Low precipitation in an area (or during a particular year) will decrease the effectiveness of big sagebrush control and will result in a lesser increase in production of desirable grasses and forbs following control. The ability of annual grasses and forbs to invade sprayed areas was another important consideration (Smilo and Bud Kimball). Also, the basics of chemical control should be known — what type of chemical should be used, how much should be applied, and what time of year it should be applied.

A range improvement practice such as chemical control of big sagebrush is only as good as the grazing management following it. Sprayed ranges should be managed separately from ranges not sprayed because livestock tend to congregate on those areas due to increased availibility and palatability of forage. This may cause overgrazing which increases the susceptibility to big sagebrush reinvasion. Overgrazing may also contribute to increases in undesirable annual grass and forb production. Proper management is, therefore, an important factor in the success and longevity of chemically treated big sagebrush grasslands.

If range improvement projects employing the chemical control of big sagebrush are properly planned and executed, livestock forage production can be greatly increased.

This study described desirable perennial grass increases averaging two-fold after a period of 15 years, with increases on the sprayed areas of up to nine times over the native areas recorded.

Longevity of the increased production for sprayed projects exceeded 18 years at certain locations. In most cases only limited big sagebrush reinvasion was evident.

7901 Research Activities

3A3 790101 Exclosure study: production, cover and precipitation.

5A1D 5B 5C 2A 2C 2B 2E

During 1979 forage production methodology was modified to include an early spring herbage production estimation of the 12 exclosures of the intensive phenology study locations. The herbaceous species were clipped within the 1 ft x 1 ft quadrats as commonly conducted during the summer production estimation and estimation of standing crop of shrub, mat-form and plains pricklypear *(Opuntia polyacantha)* was obtained with the use of weight-unit estimates and procedures utilizing the 4 ft x 5 ft quadrats for estimation of numbers of weight units per 20 sq ft quadrat.

The summer procedures for clipping herbaceous species and the fall procedures for determination of total standing crop by weight unit estimate were carried on as in previous years. The data, as in previous years, included a plant production listing of the study areas and a separation of the early spring data sets from the later sampling data sets. The spring sampling dates generally ranged from May 20 - June 5. More than 60 percent of the standing crop was already present by those spring dates as compared to the standing crop determined at the summer sampling dates.

1E2 790102 Exclosure study: phenology and growth.

2E 2F2 3A2 1E1 2A 5A1D

During 1979 intensive phenological measures at each of the 12 intensive study sites were conducted eight times during the growing season beginning on April 29, 1979, and the last on September 22, 1979. The phenological characterization of the plants was modified to separate the vegetative scoring values ranging from 1-8 and those for the reproductive values ranging from 8-18. Having had the vegetative values of 14 and 16 for vegetative parts introduced difficulty in interpretation, especially by computer evaluation of the phenological development.

1E2 790103 Exclosure study: productivity/standing crop sampling.

10D1 2E 2F2 3A2 1E1 2A 5A1D

In conjunction with the long-term intensive phenology-standing crop dynamics and relationships, range site productivity potential was addressed in 1979 by implementation of procedures addressing productivity of the four prime species. To complement the development of predictive modeling for interpretation and quantitative data for use in the range management decision processes, the productivity procedures could be utilized to estimate total standing crop throughout the season without having to conduct detailed and time consuming complete species clipping procedures many times during the growing year. Thus the complete set of procedures with regard to production and productivity involved four major activities: 1) early and late cover and standing crop determination, 2) recording phenology of prime species on existing permanent transect lines, 3) selection and clipping of prime species plant culm and bud units and 4) environmental variables.

The productivity units involved selection of 20 individual plant units of each prime species. These were selected to best represent the average phenology of a given prime species as indicated from the values of average phenology on the intensive transect lines. The specimens to be clipped were selected at locations near the corresponding phenology transect line. At the first date each year 40 specimens of each species were selected. Twenty were clipped and individually bagged with exclosure, species and date identification. The remaining 20 were identified with soft plastic-covered wire placed on the ground in the case of grasses and at the major twig base in the case of the sagebrushes. These were clipped at the next sampling date, when an additional set of 20 plants for subsequent sampling date clipping were identified with the soft, colored wire.

The selection of productivity units at the observation time prior to the clipping date was incorporated into the field procedures because of the intended extrapolation of the prime species productivity data to that of the plant community. Productivity measures of entire plant populations typically involve clipping of all species at frequent intervals throughout the growing season. These data estimate total standing crop at each date. They also require an extensive work force and time allocation requirements totally beyond the limits of the present study of 12 locations over an area representing the western two-thirds of Wyoming.

Productivity units of the wheatgrasses *(Agropyron smithii* and *A. spicatum)* were individual culms. Selection was based primarily on phenology average of the permanent transect lines. Height of culms, although often uniform for most plants, was specifically removed from the observer's selection process.

Productivity units of sagebrushes *(Artemisia tridentata* and *A. nova)* were individual twig tip initiation buds and included one node from the previous season's development. The inclusion of the added node was a minimal part of the productivity unit weight and was included to simplify the field procedures which required identification of a new bud twig as well as a uniform specified point at which it was to be removed from the shrub. The soft, colored wire was located at the base of the twig and did not affect the bud growth development. Phenologic development, representative of the adjacent permanent phenology transect, defined the observer's selection criteria. Length of current year twig development was used only if it related to identification of phenology which became definitive in some late season reproductive stages.

2A 790104 Precipitation pattern study.
3A

During 1979 precipitation was recorded at 190 UW/BLM rain gauges. The increase in numbers were mostly in the Big Horn and Wind River Basin areas. Data of 50 U.S. Weather Service stations was compiled in the annual report by the seasonal data periods as for the UW/BLM gauges. As in previous years, the precipitation at the gauges in the Big Horn Basin were recorded on the monthly dates during the growing season - April 15, June 1, July 1, August 1, September 1 and October 15. Several tables of various summations of the precipitation data from the UW/BLM and the Weather Service gauges are presented.

2C 790105 Exclosure soil profile studies with SCS.
3A

During June 1979 Jim Stevens, SCS Buffalo, was able to characterize soil profiles at the following exclosures: Kane GRI, Kane Seed, Big Trails ARTR, South Butte.

5A1E 790106 Juniper wildfire in Wind River Canyon 13 miles south of Thermopolis.
2D

During the summer of 1979 Fisser noticed a wildfire area in the Wind River Canyon 13 miles south of Thermopolis. This had occurred by someone pouring gasoline or diesel fuel along the ground for a considerable distance prior to a deposit of a great amount of fuel under a number of Utah juniper *(Juniperus osteosperma)* trees. This fire encompassed some 10 acres and was still smoldering on the day of his arrival. He did not take vegetation measures, but took a great number of photographs which can be used as reference points in the future, both of the burned area and of non-burned area and a considerable number of slides showed burned and non-burned situations.

3A2 790107 Baggs grazing systems study (BLM trend plots).
4B
4C The Bureau of Land Management personnel reread the trend plots in the Baggs
1E4 Grazing System study during the summer of 1979. These records and photographs are
2F2 available in the Range Management Department offices.
7A1
7A3
6A3

2E 790108 Juniper and limber pine population dynamics and life history.
3A2
2F2

Utah juniper *(Juniperus osteosperma)* and limber pine *(Pinus flexilis)* population dynamics and life history studies were conducted by Spaeth during 1979 in all of the juniper game range improvement exclosures and other locations.

7902 Journal Publications

2A 790201 Ries, R.E. and H.G. Fisser. 1979. Influence of environmental factors upon sagebrush and grass production in Wyoming. Agro-Ecosystems 5(1):41-55.

Of 42 environmental factors considered, the six most important for annual herbage production on 17 sagebrush *(Artemisia)*-grass range communities in Wyoming were identified and quantified by multiple stepwise regression. The factors were fall precipitation, spring precipitation, summer potential evapotranspiration, percentage coarse soil fragments (greater than 2 mm), percentage very fine sands (0.10 - 0.05 mm), and percentage grass and forb composition. These six factors were all related to water storage, availability and use. The results should aid in better understanding the interrelationship of environment and the vegetative communities of sagebrush-grass areas.

As with any derived equation, proof is difficult when all six factors were operating and interacting simultaneously. The factors isolated and quantified by multiple stepwise regression in this study were highly significant from a statistical standpoint. Furthermore, the relationships appeared biologically reasonable. All factors pointed to the importance of water to the vegetative community, and they supplemented and interacted well with each other. These factors appeared to be of prime importance in regulating water availability and production from the vegetative communities within the sagebrush-grass areas of Wyoming.

1E2 790202 Cleaves, D.T. and J. Taylor. 1979. Competition as a control for *Halogeton glomeratus*. J. Colo.-Wyo. Acad. Sci. 11(1):18.
1B4
2B
2E
2F2

During the 1978 growing season, preliminary sampling was initiated to determine the effectiveness of perennial vegetation and the control of halogeton *(Halogeton glomeratus)* through competition. The study was originated to determine possible alternative methods of control other than spraying with 2,4-D, which eliminates all broad leaf vegetation on the treated site.

Prior studies indicate that competitive control could be the most beneficial. At the Jim Bridger Mine in southwestern Wyoming, native stands of big sagebrush *(Artemisia tridentata)*, shadscale *(Atriplex confertifolia)*, Gardner saltbush *(A. gardneri)* and other shrubs adjacent to heavily infested mine spoils have prevented the establishment of halogeton. First year sampling of two mine spoil treatments, one with 7.5 percent cover of native perennials and the other with 1.2 percent cover of native perennials yielded 199.8 g/m^2 and 327.4 g/m^2 of halogeton (oven dried weight), respectively. Nearby stands of native vegetation had 22.8 percent cover of shrubs and yielded only trace amounts of halogeton (attributed to edge effect and the continued trampling of the area by investigators).

Cover data were obtained using 30 m steel tapes and a line-intercept format. Weight data were taken by clipping 0.5 m x 2.0 m quadrats. All transects and quadrats were placed using baselines and random number tables. Second and third year sampling will include two additional mine spoil treatments and more extensive sampling in order to reinforce the author's hypothesis.

7903 **Presentations**

5A1D790301 "Long-term herbage production response following chemical sagebrush control in western
3A2 Wyoming." February 13, 1979. 32nd Ann. Meeting Soc. Range Manage. Casper,
3A3 Wyoming. by H.G. Fisser.
2A
2C Grass and forb production was monitored for 15 consecutive years at five different
2E study sites where big sagebrush *(Artemisia tridentata)* had been chemically treated.
2F2 These sites, located in north central and central Wyoming, were aerially strip sprayed
5B4 with 2 lb. of 2,4-D butyl ester mixed in diesel oil. Exclosures were erected encompassing portions of both untreated and chemically treated rangeland for analysis of the effects grazing may have on these sagebrush control projects. Sagebrush was controlled and little invasion occurred on all but one of the study sites. The exception displayed an almost total reinvasion of sagebrush 12 years after treatment.

Total grass and forb production at all sites averaged two times greater on the sprayed areas inside the exclosures compared to the untreated areas inside the exclosures. Production differences outside the exclosures were not as great. The sprayed areas displayed slightly greater production than the untreated sites. Perennial grass production averaged twice as much on the sprayed areas inside the exclosures for all sites except one. At the exception, annual grass invasion inhibited perennial grass production. At the conclusion of the study period, three of the study sites were still producing twice as much perennial grass on the sprayed areas inside the exclosures as the untreated areas.

During the study period, perennial grass increases as much as nine-fold on the sprayed areas over the untreated areas were recorded. Major problems encountered dealt with the invasion of annual grass and the concentration of livestock on the sprayed sites outside the exclosures. The subspecies of sagebrush and different site characteristics were also important factors in determining the outcome of sagebrush control projects.

2C 790302 "Soil-vegetation relationships on rangeland exclosures in the Grass Creek Planning Unit
3A of north central Wyoming." September 5, 1979. Wyo. SCS Soil Survey Work
8B Planning Conference. Laramie, Wyoming. by H.G. Fisser.
2F2
3A2 To comprehensively define some of the soil-vegetation relationships in western
10D1 Wyoming, the University of Wyoming and Bureau of Land Management entered into an agreement which specifically related to three planning unit areas. Within these areas, several rangeland exclosures had been constructed in years past from which vegetation and site data have been obtained and recorded. Vegetation data included herbage production, cover, and composition values from as early as 1960; some obtained annually and some less often. Environmental information such as moisture, temperature and soil description were recorded at some exclosure sites.

The geographic distribution of exclosures provided great vegetation diversity and significant heterogeneity of environmental characteristics, including soils. Comprehensive study of soils and vegetation data from these exclosure sites presumably could provide important information not currently well defined, or even known. Definitive statements of objectives developed from this program of investigation were: 1) to inventory and characterize soils within rangeland exclosures and 2) to interrelate the soil characteristics with long-term vegetation data.

The soil inventory and soil vegetation relationship information from these exclosures provided a valuable data base for the Bureau of Land Management to increase effectiveness of decision processes applied to rangeland resource management and use. These data contributed to the better understanding of rangeland synecological relationships within the scientific community. They also enhanced information transfer to those concerned with use and management of natural resources.

3A 790303 "Arid rangeland sampling." September 27, 1979. University of Wyoming Zoo. Biol.
10D Sampling Seminar. Laramie, Wyoming. by H.G. Fisser.
2E

This seminar described sampling procedures utilized in rangeland research programs. The primary vegetation parameters of vegetation production, cover, density and phenology were estimated by a number of procedures which may be technically correct, but which may not provide data suitable for the project objectives. Researchers must evaluate the procedures they employ, and determine whether they meet the study's subjective and objective needs.

1A2 790304 "Important rangeland plants of Wyoming and the West." September 5 - November 21.
2E 1979. Special 10-week lecture session course for BLM employees. University of
2F2 Wyoming RgMgt 654M - Range Plant Identification and Ecology. Rock Springs,
3A2 Wyoming. by H.G. Fisser.

This lecture series was conducted in Rock Springs, Wyoming, for Bureau of Land Management employees. Course work involved field viewing and identification of vegetation types and classroom expression of rangeland ecology of the major sagebrush *(Artemisia)* and saltbush *(Atriplex)* communities in western Wyoming. Plant identification was important. The interpretative ability of the students to identify environmental factors responsible for minute environmental differences which often could result in very significant changes of plant ecosystems was stressed.

2A 790305 "Soil-phenology relations on some arid rangelands of western Wyoming." February 15,
10D1 1979. 32nd Ann. Meeting Soc. Range Manage. Casper, Wyoming. by L.W. Young,
1E2 III.
2C
2E Phenological development of plants is dependent upon seasonal weather conditions
2F2 and also several soil factors. The development of four species: big sagebrush *(Artemisia*
3A2 *tridentata)*, black sagebrush *(A. nova)*, western wheatgrass *(Agropyron smithii)* and bluebunch wheatgrass *(A. spicatum)* were measured at 12 arid rangeland sites in western Wyoming. The phenological stages were identified by vegetative stages, reproductive stages, and numerical scores. Data were obtained from May to October during 1973-78.

A number of soil characteristics have been identified as factors which often controlled growth rate of plants. At five of the 12 sites, detailed measurements were taken for soil moisture and temperature to a depth of 60 cm. At each of these sites, the soils had been described and analysis conducted for physical properties of structure and texture.

The soil-plant-phenology relations were being investigated with a stepwise multiple regression program. Soil moisture appeared to be the most important factor influencing initiation of growth and early vegetative stages. At the Farson exclosure in 1977, soil moisture values remained low throughout the summer, resulting in no vegetative growth until August. Plant reproduction was not recorded at any time through the 1977 growing season.

Soils which had a greater percentage of fine materials and a greater capability for water storage showed an earlier phenological development. Reproductive phenology initiation and seed development were dependent on adequacy of early summer soil moisture and a combination of summer temperatures and soil moisture depletion interrelationships. In 1977 precipitation at most sites was low and temperatures warm. Compared with 1976 and 1978, *Agropyron* species showed a delay in reproduction initiation and an earlier vegetative maturity. The *Artemisia* species showed a later vegetative initiation of growth in the spring followed by rapid maturation in the fall. Initiation and development of seedheads on the *Artemisia* species occurred later than usual.

1E2 790306 "Competition as a control for *Halogeton glomeratus*." April 30, 1979. 50th Ann. Meeting
1B4 Colo.-Wyo. Acad. Sci. Pueblo, Colorado. by D.T. Cleaves.
2B
2E See 790202.
2F2

7904 Published Proceedings

1E3 790401 Cox J.R. and H.G. Fisser. 1979. The effects of water stress on phenology and
2C carbohydrate storage in the shortgrass prairie p. 529-535. In J.R. Goodin and D.K.
1E2 Northington (eds.) Proc. of the International Arid Lands Conference. International
5C Center for Arid and Semi-Arid Land Studies. Texas Tech. University, Lubbock,
2A Texas. October 8-15, 1978.
2B
11D1 See 780203 and 780309.
1E4
2E

2A 790402 Young, L.W. and H.G. Fisser. 1979. Soil-phenology relations on some arid rangelands of
10D1 western Wyoming. p. 53-54. In Proc. 32nd Ann. Meeting, Soc. Range Manage.
1E2 Casper, Wyoming. February 12-15, 1979.
2C
2E
2F2 See 790305.
3A2

5A1D790403 Orpet, J.E. and H.G. Fisser. 1979. Long-term herbage production response following the chemical control of big sagebrush. p. 29. In Proc. 32nd Ann. Meeting, Soc. Range Manage. Casper, Wyoming. February 12-15, 1979.

See 790301.

7905 Wyoming Agricultural Experiment Station Publications

2C 790501 Fisser, H.G., D.C. Trueblood and D.D. Samuelson. 1979. Soil vegetation relationships on
5A1D rangeland exclosures on the Grass Creek Planning Unit of north central Wyoming.
3A2 Univ. of Wyo. Coop. Res. Rep. to the BLM. Wyo. Agr. Exp. Sta. Sci. Rep. 969.
3A3 279 p.
2A
2E Soil vegetation relationships have long been recognized as important to rangeland
2F2 management. To comprehensively define some of these relationships in western
5B4 Wyoming, specifically in the Grass Creek planning unit in the Big Horn Basin, the University of Wyoming and the Bureau of Land Management, Wyoming State Office, entered into an agreement relating to three planning unit areas. Within these areas, several rangeland exclosures were constructed, from which vegetation and site data had been obtained and recorded. The geographic distribution of exclosures provided great vegetation diversity and significant heterogeneity of environmental characteristics, including soils.

Definitive statements of this program and the objectives developed for the investigation were: 1) to inventory and characterize soils within rangeland exclosures, and 2) to interrelate the soil characteristics with long-term vegetation data. The soil inventory and the soil vegetation relationship information from these exclosures provided a valuable data base for the Bureau of Land Management to improve decision processes applied to rangeland resource management and use. In addition, the data aided understanding of rangeland synecological relationships within the scientific community and enhanced information transfer to those concerned with use and management of natural resources.

The study was conducted on 39 exclosures in the Grass Creek planning unit of the Big Horn Basin in north central Wyoming. For purposes of this study, the soils were given in three forms of discussion: 1) profile description, 2) lab analyses done at the University of Wyoming and 3) supplemental analyses done at the University of Idaho soils lab.

All vegetation data were taken from the annual progress report of an ongoing cooperative research project with the Bureau of Land Management and from research files of the University of Wyoming. For detailed discussions of the methods used, refer to these reports. The vegetation data were presented as long-term means for cover and production for shrubs, perennial grasses, annual grasses, perennial forbs, annual forbs and total. The report discussed soils of the 51 research sites and also the vegetation characteristics, the majority of which were cover data from permanent quadrat transects. Herbage production had been obtained at nine of these sites. The combination of interaction of soil and vegetation was evaluated and these data compiled to determine specific relationships of vegetation and soils. Soils data included complete descriptions by profile of all soil site identified areas.

790502 Fisser, H.G., L.W. Young, N.E. Hargis and K.E. Spaeth. 1979. Phenology and production studies on semiarid shrub types. 1978 results. Univ. of Wyo. Coop. Res. Rep. to the BLM. Sections I, II, III and IV. Wyo. Agr. Exp. Sta. Sci. Rep. 985. 489 p.

1E2 79050201 Section I - Phenology study.
2E
2F2
3A2
1E1
2A
5A1D

During 1978 intensive phenology recording was conducted as during previous years at the 12 phenology study exclosures. Data were obtained on eight dates during the year. The phenology development stages for 1978 were compared with the earliest and latest dates of the different stages from 1973-77. In general, the 1978 observations by date were nearer the mean average of the comparison years, with the exception of the Owl Draw exclosure site. At this location, the reproduction stages of full bloom and seed formation for all species occurred at later dates. Comparing this to 1977, which was very dry, reproduction and plant maturation drying occurred much earlier. In 1978, there was ample moisture in the fall, resulting in a longer reproduction period. The grass species remained greener for a longer period of time before leaf browning became apparent.

11F2 79050202 Section II - Phenology step-wise regression for prediction modeling.
2F
2F2
3A2
1E1
2A

The step-wise regression for predictive modeling used this statistical analysis procedure to interrelate independent variables and their significance upon the rate of phenological development of the four prime species identified in the intensive phenology study. The independent variables included a variety of soil temperature, soil moisture, precipitation, forage production and air temperature variables that were measured at the times of recording of phenological data. The step-wise multiple regression analysis was conducted both for the vegetative and the reproductive stage data for each of the four prime species (big and black sagebrush *(Artemisia tridentata* and *nova)* and western and bluebunch wheatgrass *(Agropyron smithii* and *spicatum)*).

Each of the multiple regression analyses identified limiting factors which enhanced and increased the rate of phenological progression, both vegetatively and reproductively. They also identified those factors which aided continued maintenance of plant growth which resulted in delayed progression of phenological development. Factors which tended to promote phenological development were high temperatures, both ambient air and soil, low soil moisture values and low precipitation values. Factors which slowed or prolonged phenological development were such characteristics as beneficial soil moisture and better than average annual precipitation as well as lower temperatures.

2A 79050203 Section III - Precipitation pattern study.
3A

Precipitation during 1978 was generally greater than long-term averages. The most significant increases in precipitation were recorded in the Big Horn Basin during spring. Some of those precipitation events have been classified as 100 year occurrences. In addition, the majority of all gauges exhibited had greater than normal precipitation values. In the Big Horn Basin, average precipitation was more than 2.50 inches above the long-term values, both for the UW gauges and the U.S. Weather Bureau state gauges. Data from the southwestern Wyoming rain gauges was less than average for the Weather Bureau recordings.

Preliminary evaluation of seasonal distribution of precipitation common to major geographic areas was shown in tabular form. The Big Horn Basin and Wind River Basin were recognized as areas with low winter precipitation and a strong concentration of precipitation during spring and early summer. Areas to the east and south and southwest across the continental divide did not often exhibit such extreme precipitation distribution characteristics.

3A3 79050204 Section IV - Exclosure study: production, cover and precipitation.
5A1D
5B
5C
2A
2C
2E
2F2

This report of the herbage production data obtained during 1978 was in the same format as past aridland ecology annual reports. Basal cover, frequency, production and precipitation data were presented in summary sheets, one summary for each study area treatment. Tabular listings of study areas, plant names, estimates of utilization and of the individual tabular production values were presented in this report.

Forage production in the Wind River Basin during 1978 was comparable to the long-term averages since 1962. Deviations above and below the average were due to differences in shrub production. In the Big Horn Basin, production of Gardner saltbush *(Atriplex gardneri)* was unusually high at the Halogeton and the Sand Gulch exclosures. This increase was due to precipitation in May, which was 2-4 inches above normal. The summer months of June and July were deficient in moisture; the exclosures where moisture was minimal showed low shrub production. Exclosures which received average moisture during the fall showed many seed heads present on the sagebrush *(Artemisia)* species and production totals were slightly higher than the long-term average. As was noted during previous years, total cover estimated did not correspond well with total production. However, proportionate changes between perennial grass cover and shrub cover showed positive correlation with appreciable changes in grass and shrub production.

At Lander Ant, grass production was low and shrub production high. Cover estimates reflected the same proportion of change. A similar correlation existed at Halogeton -3, a proportional increase in cover estimate to that of an increase of perennial grass and shrub production. Proportionate change in cover estimates and production estimates appeared to be more reliable for the perennial grasses than for shrub species.

3B 790503 Fisser, H.G. 1979. Rangeland and wildlife resources of the South Pass historic mining
9A area. Wyo. Agr. Exp. Sta. Mimeo. 8 p.
4B
6A1
7A1
7A2
7A3
7A5
8A

The South Pass historic mining area on the Wind River Mountains to the south of Lander, Wyoming, receives a great deal of multiple use. Wildlife, livestock and especially, recreational use have a strong impact on the area. The Bureau of Land Management contacted the University of Wyoming Department of Recreation and Parks and requested a review of the area to determine its importance as a resource. Fisser contributed an evaluative statement regarding the rangeland and wildlife resources of the South Pass historic mining area.

The occurrence of the South Pass historic mining area at elevations in excess of 8000 ft above sea level indicates the general nature of the native biological components and their uses. Rangelands with shrub, grass and forb vegetation occupy the greatest amount of area. They are utilized by livestock for a maximum of 4 1/2 months between the dates of May 15 and September 30. Wildlife use is largely restricted to the spring, summer, and fall periods. During a 3-4 month winter period, most large game animals leave the area for lower elevations of winter wildlife habitat.

Importance of a resource does not depend upon continuity of use. The limited periods of accessibility to livestock, wildlife and some forms of human use can enhance the significance of an area on a local basis, and in a regional and national context.

This report addressed the following characteristics for interpretation of rangeland use: 1) vegetation components, 2) livestock components, 3) wildlife components and 4) economic importance based on generalized livestock values and wildlife use characteristics.

The 3,500 AUMs of cattle and sheep grazing allotted to the area provided an economically derived return to land value of about $5,000. In terms of importance to the economy of the region or state, an AUM is computed to contribute $25-$30 of economic activity. On this basis, the 3,500 AUMs can reasonably be expected to promote some $100,000 in gross income and the associated financial value to the local community and state economy. Importance of wildlife components is often computed in terms of value of average numbers of hunter days per animals harvested. These data were not available at the time of report and have since been found to contribute a significant amount to the economy, especially to the Lander region.

The importance of the South Pass historic mining area rangelands and the many associated used applied to them is obviously very economically significant. This is true both in the context of the regionally related livestock industry as well as the wildlife hunter harvest aspect. Additional multiple use activities of an aesthetic, recreational, or industrial nature further enhance the use and importance of this natural resource.

4B 790504 Fisser, H.G. 1979. Grazing management system studies of northern desert shrub
4C vegetation of southern Wyoming. Wyo. Agr. Exp. Sta. Mimeo. 3 p.
1E4

2F2 This summary statement identifies the reports and activities of the Baggs Grazing
7A1 System study. The study was implemented under operational ranching conditions, rather
7A3 than as a strictly controlled research project. Permanent fenced exclosures for evaluation
6A3 of livestock and wildlife impact and comparison of native to shrub control vegetation
were installed. Transects of permanent quadrats were located within each pasture and exclosure. Vegetation cover and composition were measured at each location annually between 1967 and 1971 and again in 1975.

The base data records involved data sets from 1967, 1971 and 1975. In the latter, records showed a steady increase in the cover of perennial grasses under all grazing systems. No differences were readily observed between grazed and non-grazed (exclosure) areas. Grazing pressures were not sufficient to stress the plant communities to such a degree as to detect any differences during the short term, 9-year period of the study. Obviously, a longer study time was needed.

It was not implied that the grazing systems were ineffective. In the present situation, the apparent lack of response of vegetation to differential strategies of livestock use can be attributed to conditions resulting from significant areas of sagebrush control which were implemented shortly prior to the initiation of the grazing system, a 25 percent reduction of carrying capacity of all three areas, and also because precipitation and temperature combination during the years of study resulted in good growing conditions.

7906 Graduate Advisee Completions

2F 790601 Shawesh, O.M. 1979. Vegetation analysis and ecology of rangelands in northwest Libya,
5B Africa. M.S. Thesis. Univ. of Wyo. 76 p.
2C

2B Quantitative description and ecological investigations of semiarid rangeland in
8C northwestern Libya were initiated in May 1976. A 25,000 ha livestock exclosure near
3A1 Assa, some 25 km from the Mediterranean seacoast, was utilized. Average annual
precipitation was 138 mm, most occurring from October to March. Average annual temperature was about 30°C. Soils were mostly azonal sands with average pH value 8.51 and mean electrical conductivity of 1.1 mmhos/cm.

Vegetation and soil relationships were investigated with permanently established transects for measures of height, diameter and density of perennial plants and percent cover of all live vegetation; and by soil profile sampling for detailed analysis of physical and chemical characteristics.

The relatively stable rangeland sites were characterized by the frequent presence of common sagewort *(Artemisia campestris)*, which comprised 33 percent of all vegetation composition. Rock rose *(Helianthemum lippi)*, saggy sparrowort *(Thymelia microphylla)* and hidma *(Echiochilon fruticosum)* were the most common additional shrubs. The most common grasses were false esparto *(Lygeum sparteum)* and hairgrass *(Koeleria pubescens)*.

Height, diameter and density of all perennial vegetation was observed to have an average of 24.35 cm, 2.83 cm and 2.72 plants per m^2, respectively. The average percent cover for the the entire life form was 7.46 percent. Statistical analysis showed some variability among transects, percent cover for the entire vegetation in the analysis of variance indicated no significant differences among the transects. The relatively low variations among some transects may be attributed to uniform soil electrical conductivity, plant heights and diameter, a long grazing history of the study area and minimal areas of severe wind erosion effects.

The three most important climatic factors affecting growth of vegetation on the Assa rangelands were precipitation, wind, and temperature. Establishment of many grass species was inhibited by low precipitation, with an average of 138 mm annually, by the velocity of hot, southerly winds, and by the high average annual temperature of 30°C. Adapted native species of shrub vegetation and some perennial grasses were the only plants able to survive in this situation. The soils of the study area were azonal sands. Soil textural analysis indicated that soils were characterized by low clay, silt and high sand contents with averages of 3.39 percent, 2.38 percent and 94.23 percent, respectively. Levels of soluble salts were high and organic matter contents were low.

Response of vegetation to improvement practices was slow; several years were often required for the manifestation of a change to be readily visible to the observer.

8001 Research Activities

3A3 800101 Exclosure study: production, cover and precipitation.
5A1D
5B
5C
2A
2C

During 1980 the determination of production on the 12 phenology sites was conducted during the third week of May for the early production, then during July and August for the later production as during previous years.

1E2 800102 Exclosure study: phenology and growth.
2E
2F2
3A2
1E1
5B

Phenology data were obtained on the 12 study locations on 11 dates essentially at 2-week intervals during 1980.

1E2 800103 Exclosure study: Productivity/standing crop sampling.
10D1
2E
2F2
3A2
1E1
2A
5AD1

Productivity/standing crop estimation procedures sampling was conducted as during 1979 on all of the dates of phenology sampling during 1980, with culm clippings of western and bluebunch wheatgrass *(Agropyron smithii* and *spicatum)* and clippings of bud and leaders of big and black sagebrush *(Artemisia tridentata* and *nova).*

2A 800104 Precipitation pattern study.
3A

During 1980 seasonal precipitation was recorded at the more than 150 UW/BLM gauges. The Big Horn Basin stations were recorded monthly during the summer growing season as noted earlier.

4A 800105 Roadside intensive use site seven miles east of Powder River.
3A2
4B
4C
6A
1E4
2E
2F2

Fisser noticed an area seven miles east of Powder River, Wyoming, along the highway to the south in which intensive livestock use during the summer was harming the rangeland. Vegetation was being eliminated and even sagebrush was being taken down and broken off by the foraging cattle. At the northeast fence corner of the pasture, adjacent to the south side of the highway barrow pit, photographs were taken parallel and over the fence of the heavy use pasture unit, and of the fence line proceding southward from that corner which separated that intensive use pasture from another with moderate utilization to the east. These photographs will be maintained as a long-term record. In the future, someone may wish to evaluate the response of the livestock to this intensive summer use activity.

2E 800106 Juniper and limber pine population dynamics and life history.
3A2
2F2

The Utah juniper *(Juniperus osteosperma)* and limber pine *(Pinus flexilis)* population dynamics and life history studies continued with activity at the Block juniper site and at some of the other game range improvement exclosures in the Big Horn Basin.

8002 Journal Publications

2E 800201 Waugh, W.J. and H.G. Fisser. 1980. Vegetation stagnation and decadence in a southeastern
6A Wyoming big game exclosure: preliminary report. J. Colo.-Wyo. Acad. Sci. 12(1):29.
2F2

1E2 Permanent exclosures are often used by land resource managers to represent climax
1E3 vegetation and optimum forage production. In a true mountain mahogany (*Cercocarpus*
1E4 *montanus*), mountain big sagebrush (*Artemisia tridentata* ssp. *vaseyana*) stand on the
7A1 foothills of the Medicine Bow Mountains. Mountain mahogany plants protected from
7A2 large ungulate browsing for 30 years produced significantly less annual biomass per unit
10D of canopy cover than browsed plants. There was no significant difference in sagebrush production between samples inside and outside the exclosure. Subjective sampling of age class and percent dead material indicated a greater frequency of older, more decadent mountain mahogany plants inside the exclosure than outside.

4B 800202 Rodgers, D.E. and H.G. Fisser. 1980. Vegetational response to grazing systems in
4C Wyoming's Red Desert. J.Colo.-Wyo. Acad. Sci. 12(1):29-30.
1E4

2F2 A grazing systems study was initiated in 1967 to determine the effects on
7A1 vegetation cover and trend. The research area was divided into a four-pasture rest-
7A3 rotation allotment, a two-pasture deferred allotment and a year-round seasonal-use area.
6A3 Livestock exclosures were constructed in all pastures. The study was implemented under actual ranching conditions, with both sheep and cattle using the allotments.

Since 1967 grass cover had increased uniformly on all the pastures, with western wheatgrass (*Agropyron smithii*) and needleandthread (*Stipa comata*) showing the highest gains. The rate of increase did not differ significantly among pastures at the .05 level of confidence. Two factors might be responsible for nullifying any differential effects: 1) a 25 percent overall reduction in stocking rate since 1967 and 2) higher precipitation rates during recent years. Grass cover inside livestock and wildlife exclosures was also slow in responding, with only four of 10 comparisons showing significant increases after 12 years. Studies should be continued to determine the long range results.

8003 Presentations

2F2 800301 "Shrub establishment, dominance and ecology on the juniper and sagebrush-grass types
7A1 in Wyoming." December 3, 1980. Shrub Establishment Symposium. Wyoming Game
2E and Fish Department. Laramie, Wyo. by H.G. Fisser.
1E2

1E4 Essential components of big game winter range are food production and available
2A cover. These depend upon the relationships of overstory and understory plant species.
2C The juniper areas of Wyoming sagebrushes are characterized as understory species,
2D although they often do provide cover. Overstory species in the Wyoming juniper type
3A2 may include either Utah juniper (*Juniperus osteosperma*) or Rocky Mountain juniper (*J.*
3A3 *scopulorum*) associated with limber pine (*Pinus flexilis*). Understory production is greatly
4B suppressed by juniper abundance as low as 80 trees per acre - a spacing of 25 feet
4C between trees. Thus, the development of this program was initiated with the study of
5A1C juniper control techniques. Some objectives of the program were: 1) evaluate juniper
5A1D control techniques - chemical, burn, mechanical doze and sawing, 2) evaluate food value
5A1E importance of juniper for game, principally mule deer and elk, 3) evaluate plant seeding
5B4 techniques (broadcast and plant), season of year (spring and fall), and seedling
5C adaptability under juniper control techniques (spray, doze) and 4) subjectively plot successional trends on treated areas, of native and introduced species including shrubs, forbs, and grasses.

Results of juniper control techniques showed that two-way application of Tordon (2 lbs/A A.I.) during late August was lethal to Utah juniper, big sagebrush (*Artemisia tridentata*), and curlleaf mountain mahogany (*Cercocarpus ledifolius*). Black sagebrush (*Artemisia nova*), a root sprouter, exhibited significant positive growth during the next two years following treatment. Rocky Mountain juniper was not significantly affected.

Mechanical dozing of juniper was effective, but not acceptable because of soil disturbance and subsequent erosion. Sawing juniper near ground level allowed them to begin reinvasion. Understory establishment of the juniper sawing area of half-shrubs, grasses, forbs and shrubs was evident within 2 years. Burn treatment to control juniper was effective both in the tree by tree control in the exclosures and in the natural wildfire area. Controlled burn was the most satisfactory method of juniper removal. Tordon residual effects limited revegetation, dozing promoted erosion and sawing allowed juniper trees to re-establish.

Deer behavior monitoring and fecal analyses showed that juniper was a common diet component; but the most preferred shrubs during moderate winters were sagebrushes, mountain mahogany, skunkbush *(Rhus trilobata)*, serviceberry *(Symphoricarpos occidentalis)* and bitterbrush *(Purshia tridentata)*. During severe winters, deer actively selected specific juniper trees until all available green leaf material was removed. Differential diet quality and palatability of individual trees were probable, but had not yet been investigated.

Results of planting technique trials showed that planted seed usually resulted in better stands of seedlings. Broadcast plantings were most successful with small seeded species such as sagebrushes, Palmer penstemon *(Penstemon palmeri)*, cicer milkvetch *(Astragalus cicer)* and winterfat *(Cerotoides lanata)*. Fall season application of seed was more acceptable than that during spring. Seed application on chemically treated areas was successful with some grasses when planted three years after spraying. Forbs and shrubs were still suppressed, although some species exhibited ability to establish.

Skunkbush sumac, plains cottonwood *(Populus sargentii)*, and serviceberry appeared inside the exclosure by the 10th year following treatment. These species were present along nearby streams, but not on the upland juniper areas near the exclosures. Protection from game and livestock use thus indicated relatively intensive utilization of these shrubs, as well as the very palatable curlleaf mountain mahogany, an important vegetational component on the Big Horn Mountains.

Broom snakeweed *(Gutierrezia sarothrae)* responded strongly on the burn, doze and saw areas as compared to the spray and native sites. Soil disturbance provided an environment suitable for immediate expansion of the population of this non-palatable, short-lived half shrub.

The following conclusions became evident: 1) Most acceptable juniper control was selective burning in most cases. Chemical control can be effective and on sites with strong black sagebrush understory may be acceptable, 2) Juniper was of secondary preference to deer, thus increased game food production could be realized by control of this tree, 3) Fall season broadcasting of small seeded species including sagebrushes and others can be utilized in juniper or other stable shrubs if dominant competition was removed, 4) Juniper and sagebrush grass vegetation types were essentially closed communities into which addition of new and/or different species could be accomplished only by removal of the tree and shrub dominance, 5) Big, black, and silver sagebrush *(Artemisia cana)* are easily established with viable seed application to properly prepared sites, that is, those from which the controlling biological component such as juniper had been removed and 6) Long-term establishment of shrub species required protection from game as well as livestock use for several years. Either fencing or management of the animals was necessary to give the young plants adequate time to obtain a stature able to withstand foraging.

2C 800302 "Soil-vegetation relationships on a semiarid rangeland exclosure in north central
3A2 Wyoming - long-term studies." February 14, 1980. 33rd Ann. Meeting Soc. Range
8B Manage. San Diego, California. by D.C. Trueblood.
2F2

3A2 To comprehensively define some of the soil vegetation relationships in western
10D1 Wyoming, the University of Wyoming and the Bureau of Land Management entered into an agreement which specifically relates to the three planning unit areas. Within these areas, several rangeland exclosures were constructed in years past from which vegetation and site data had been obtained and recorded. Vegetation data included herbage

production, cover and composition values from as early as 1960. Some were obtained annually and some less often. Environmental information such as moisture, temperature, and soil description were recorded at some exclosure sites. A geographic distribution of exclosures provided great vegetation diversity and significant heterogeneity of the environmental characteristics, including soils. The comprehensive study of soils and vegetation data from these exclosure sites could provide important information not currently well defined, or in some instances, even known.

Objectives developed in this program of investigation were: 1) to inventory and characterize soils within rangeland exclosures and 2) to interrelate the soil charateristics with long-term vegetation data.

Soil inventory and soil vegetation relationship information from these exclosures provided a valuable data base to increase effectiveness of decision processes applied to rangeland resource management and use. These data contributed to the better understanding of rangeland synecological relationships within the scientific community, and they enhanced information transfer to those concerned with use and management of natural resources.

2E 800303 "Vegetation stagnation and decadence in a southeastern Wyoming big game exclosure: a
6A preliminary report." April 25, 1980. 51st Ann. Meeting Colo.-Wyo. Acad. Sci.
2F2 Denver, Colorado. by W.J. Waugh.
1E2
1E3 See 800201.
1E4
7A1
7A2
10D

4B 800304 "Vegetational response to grazing systems in Wyoming's Red Desert." April 25, 1980.
4C 51st Ann. Meeting Colo.-Wyo. Acad. Sci. Denver, Colorado. by D.E. Rodgers.
1E4
2F2 See 800202.
7A1
7A3
6A3

8004 Published Proceedings

2C 800401 Trueblood, D.C. and H.G. Fisser. 1980. Soil vegetation relationships at semiarid
3A2 rangeland exclosures in north central Wyoming — long-term studies. p. 31. In Proc.
8B 33rd Ann. Meeting Soc. Range Manage. San Diego, California. February 11-14, 1980.
2F2
3A2 See 800302.
10D1

2C 800402 Hajjar, S., G. Spitzer, E.G. Meyer and H.G. Fisser. 1980. An evaluation of water and
5C soil samples for forage and other agricultural development in the Tabuk region. A
5D report submitted to A.M. Aziz, Governor, Tabuk Province, Saudi Arabia. Amer. Res.
8A Design Scientists, Inc. 49 p.
8C
8D This report contained the results of laboratory analyses of water and soil samples
10D obtained on a trip to the Tabuk, Saudi Arabia region, during May 1980. Tabuk occurred in the northwestern portion of Saudi Arabia and this area of the province was adjacent to the Red Sea and to Jordon to the north. The climate of the area was extreme with average annual precipitation not much more than 2 inches per year and with high winds, high temperatures and poorly developed soils. Surprisingly, perennial forbs, grasses and shrubs palatable for livestock were able to exist in sheltered sites within this region. There were even sites of high-saline situations upon which palatable forage species

existed. The team of scientists were invited to Saudi Arabia to investigate the potential for development of forage plantings to enhance agricultural interests in that region.

This report has led the authors to conclude that although each site presented a unique situation of necessary treatment application, the addition of organic matter material, fertilizers and other amendments for improvement of soil quality would generally enhance the forage production potential to an acceptable level. Although the regeneration projects would be expensive on a per acre basis, the long-term benefits were thought to be economically acceptable.

The benefits of the suggested projects and improvement activities were: 1) The study sites would represent a variety of soil conditions. Based on the results of this initial work, conclusions could be drawn as to the feasability of transferring the techniques used to larger crop production areas in the Tabuk region. 2) During the two year period, the various plants, forage or crop oriented, would produce seed and plant materials acclimated to the specific large planting areas. The expense and time involved in transferring these materials to nearby forage and crop production situations will be greatly reduced. 3) The study would add a wealth of scientific knowledge to the libraries in the Tabuk region, benefitting everyone.

8005 Wyoming Agricultural Experiment Station Publications

800501 Fisser, H.G., M.K. Owens, W.J. Waugh, N.E. Hargis and J. Uhlich. 1980. Phenology and production studies on semiarid shrub types. 1979 Results. Univ. of Wyo. Coop. Res. Rep. to the BLM. Vol. I: Phenology Study. Sections I and II. Wyo. Agr. Exp. Sta. Sci. Rep. 1044. 387 p.

Vol. I

1E2 80050101 Section I - Phenology study.

2E
2F2
3A2
1E1
2A
5A1D

This report included compilation of all of the vegetative and reproductive data values obtained at the 12 intensive study sites on the eight dates of data collection during 1979. The 1979 phenology curves for the principle species in each exclosure were compared to curves developed from the earliest to the latest dates observed from 1973-78. The phenology scoring systems averages also were converted to the currently utilized scoring system which was a simple numerical shifting of recording so that all of the vegetative values ranged from 1-8 and all of the reproductive ranged from 8-18. Phenological development of grasses at study exclosures located in the Big Horn Basin progressed 1-3 weeks ahead of exclosures located in the Wind River and Green River Basins. This distinction was not as evident for regional comparisons of sagebrush phenology.

11F2 80050102 Section II - Phenology step-wise regression for prediction modeling.

2E
2F2
3A2
1E1
2A

The step-wise multiple regression analysis of environmental data as independent variables, and their effect upon the dependent variables of reproductive and vegetative phenophase, were analyzed for productive and for vegetative phenology of western wheatgrass *(Agropyron smithii)*, bluebunch wheatgrass *(A. spicatum)*, big sagebrush *(Artemisia tridentata)* and black sagebrush *(A. nova)*. Factors which promoted phenological progression rate and those which tended to delay progression while increasing forage production, were similar to those of previous years.

800502 Fisser, H.G., M.K. Owens, J.W. Uhlich, W.J. Waugh, D.E. Rodgers and N.E. Hargis. 1980. Phenology and production studies on semiarid shrub-types. 1979 Results. Univ. of Wyo. Coop. Res. Rep. to the BLM. Vol. II: Precipitation, Productivity and Production Section. Sections III, IV, and V. Wyo. Agr. Exp. Sta. Sci. Rep. 1106. 219 p.

Vol. II.

2A 80050203 Section III - Precipitation pattern study
3A

This report contained tabular material of the 150 UW/BLM gauges compiled from the seasonal data and similar data compiled for more than 60 U.S. Weather Service stations. During 1979, precipitation was commonly below average at most rain gauge stations. During winter and spring periods, moisture was light, but well-distributed. Because of the small amounts which occurred during summer and fall periods, the proportion of spring and winter was greater than usual. Northern Wyoming was relatively warm during the early growing season, but south of the Continental Divide, many locations exhibited fortunate cool temperatures and effective rain showers that enhanced growth of perennial forbs.

Precipitation data during 1979 in the Big Horn and Wind River Basins ranged from 70 to 85 percent of long-term average values.

1E2 80050204 Section IV - Productivity - phenology relationships.
10D1
2E
2F2
3A2
1E1
2A
5A1D

The productivity unit clipping procedures during 1979 were conducted at each of the 12 intensive phenology studies and on each of the eight dates during which phenology was obtained. The individual weight data was given in grams for each of the individual 20 weight units, for each of the species at each location on each date, and was summarized in tabular form.

3A3 80050205 Section V - Exclosure study: production, cover and precipitation.
5A1D
5B
5C
2A
2C
2E
2F2

The herbage production data for the 1979 sampling series were compiled in two series of individual transect data value compilations, the first for the 12 locations of the intensive phenology sites which were estimated for early spring standing crop for all species. The second series includes estimation for standing crop for all species at all locations. The latter work was conducted during June and July for the clip plot transects, and during September and October for the weight-unit estimate procedures.

The below-normal growing season precipitation on many of the study sites contributed to forage production values below the long-term averages. Exceptions to this were two sites in the Wind River Basin - Lower Government Draw and Upper Government Draw - where moisture from heavy winter snowfall increased forage production to above the long-term averages. Prairie junegrass *(Koeleria cristata)* was especially evident and made up the largest percent of this increase in production at Upper Government Draw. Most of the other sites in the Wind River Basin showed either average or slightly below average production.

Forage production in the Big Horn Basin varied by site with most showing only average production values. Low production values were most evident at the sites located in saltbush communities, while sites located in sagebrush communities tended to show forage values closer to the long-term average.

2C 800503 Fisser, H.G., D.C. Trueblood and C.H. Butterfield. 1980. Soil-vegetation relationships on rangeland exclosures in the Overland Planning Unit of south central Wyoming. Univ. of Wyo. Coop. Res. Rep. to the BLM. Wyo. Agr. Exp. Sta. Sci. Rep. 1042. 75 p.
3A2
8B
2F2
3A2
10D1

This report addressed soils and vegetation relationships of 11 exclosure locations in the Overland Planning Unit of south central Wyoming. This was an area of little precipitation, high summer temperatures, high evaporation and with soils that tended to be alkaline and calcareous. Much of the area was dominated by shrubs, particularly big sagebrush (*Artemisia tridentata*). Specific objectives were: 1) to inventory and characterize soils within rangeland exclosures and 2) to interrelate the soil characteristics with long-term vegetation data.

The soils of this area showed a low degree of development, probably due to the arid conditions. Because of this, the soils of all locations studied were classified into the orders Aridisol or Entisols. Of the 16 study sites however, the soils of only two were classified in the same soil series, which indicated great variation in both chemical and physical parameters.

At 44 percent of the study sites, the vegetation was dominated by sagebrush. At the remainder of the study sites, the vegetation was dominated by other species of drouth tolerant shrubs with the exception of one site at which the vegetation was dominated by perennial grasses.

The following were the relationships between soil and vegetation parameters that seemed to be important: 1) the effective rooting depth appeared to be related to vegetation type, 2) the effect rooting depth appeared to be directly related to the total percent cover in the juniper type, 3) the percentage of sand appeared to be directly related to total percent cover of the juniper type, 4) the percentage of very fine sand appeared to be inversely related to total percent cover in the juniper type, 5) the percentage of silt appeared to be inversely related to total percent cover in the juniper type, 6) the percentage of silt appeared to be related to vegetation type, 7) the pH of a 1:5 dilution appeared to be related to vegetation type, 8) the pH of a saturated soil paste appeared to be inversely related to total percent cover in the juniper type, 9) the percentage of calcium carbonate appeared to be inversely related to total percent cover in the juniper type and 10) the percent soil moisture in the profile at 15 bars appeared to be related to vegetation type.

Strong definitive expression was not developed for apparent trends or relations of soils and vegetation from the 16 study sites of this report. Substantive information can be derived from the data, of course, but because of the limited sample size, such expressions of relations over cause and effect responses, in a quantitative sense at least, cannot be finalized until the completion of additional analyses from the combined data sets of the three reports of this series.

2C 800504 Fisser, H.G., and D.C. Trueblood. 1980. Soil vegetation relationships on rangelands
3A2 exclosures on the Salt Wells Planning Unit of southwestern Wyoming. University of
8B Wyoming Coop. Res. Rep. to the BLM. Wyo. Agr. Exp. Sta. Sci. Rep. 1041. 68 p.
2F2

3A2 This study was conducted on nine exclosures in the Salt Wells planning unit in the
10D1 Red Desert of southwestern Wyoming. This is an area of little precipitation, high summer temperatures and high evaporation with soils that tended to be alkaline and calcareous. Much of the area was dominated by shrubs, principally big sagebrush *(Artemisia tridentata)*. Nine exclosure locations in the Salt Wells Planning Unit were studied to evaluate the relationships between soils and vegetation under rangeland conditions. Specific objectives were: 1) to inventory and to characterize soils within rangeland exclosures and 2) to interrelate the soil charateristics with long-term vegetation data.

Soils in this area showed a low degree of development due to arid conditions. Because of this, the soils at all but one of the 11 are classified into the order Aridisol and Entisol. Of the 11 study sites, however, the soils at only two were classified in the same soil series which provided great variation in both chemical and physical parameters. At 60 percent of the study sites, vegetation was dominated by sagebrush *(Artemisia tridentata)*. The remainder of the study sites were dominated by other species of drouth tolerant shrubs.

The following relationships between soil and vegetation parameters appeared to be important: 1) the percentage of sand appeared to be inversely related to total percent cover in the sagebrush type, 2) the percentage of silt appeared to be directly related to total percent cover in the sagebrush type, 3) the percentage of clay appeared to be directly related to total annual production in the sagebrush type, 4) the organic matter content of the soil appeared to be directly related to total annual production in the sagebrush type, 5) the electrical conductivity of a soil paste extract appeared to be related to vegetation type, 6) the bulk density appeared to be related to vegetation type and 7) the bulk density appeared to be inversely related to the total annual production in the sagebrush type.

4B 800505 Fisser, H.G., D.E. Rodgers and W.E. Gerhart. 1980. Influences of grazing management
4C systems on vegetation and wildlife habitat. 1975 and 1979 Results and Long-term
1E4 Summary. Univ. of Wyo. Coop. Res. Rep. to the BLM. Wyo. Agr. Exp. Sta. Sci.
2F2 Rep. 1047. 188 p.
7A1
7A3 The Baggs Grazing system study was initiated in 1967. The report compiled data
6A3 from all of the permanant plot transects locations in 1975 and 1979. These were the two series of data following the last data collection in 1971 at the end of the first 4-year section.

Perennial grass cover amounts of increase did not differ among pasture units. It was concluded that the management systems had no significant differential effects since the study began 12 years ago.

Western wheatgrass *(Agropyron smithii)* did not show increases in cover values, but needleandthread *(Stipa comata)*, Indian ricegrass *(Oryzopsis hymenoides)* and bluebunch wheatgrass *(Agropyron spicatum)* all made significant gains. Utilization on all perennials was light except for western wheatgrass, which received proper use or slightly above. Since utilization on most major forage species was slight, and actual use in AUMs was often less than allowable use, stocking rates could be considered as light to moderate.

The results then seemed to indicate that continuous grazing at lighter stocking rates could be as effective as rotational and deferred systems in restoring and maintaining a desirable range condition. At this particular stage of range recovery, continuous grazing has been effective. The results may be different at a more advanced stage of recovery. The study must continue to determine the long range results.

Effects of the grazing systems on wildlife habitat seemed minimal. Antelope had adapted fairly well to the fences, and had been able to extend their range since the study began. Water developments, though few in number, were beneficial to the antelope as well as to the other wildlife species.

8006 Graduate Advisee Completions

2C 800601 Mesdaghi, M. 1980. Vegetation of saline and alkali habitats of the Red Desert,
10D1 Wyoming. Ph.D. Dissertation. University of Wyoming. 140 p.
2A
2E During the summers of 1978 and 1979, studies were conducted on the saline-alkaline
2F2 habitats of the Red Desert in south central Wyoming to obtain information which would
3A1 contribute to a better understanding of why some plant species grow in saline
3A2 environments and what roles chemical and physical characteristics play in distribution.
5C Three saline habitats, named Chain of Lakes Flat, Circle Bar Lake and Battle Springs
8A Flat, were selected as representatives of saline-alkali areas in the Great Divide Basin.
8C The vegetation types of these areas were Baltic rush *(Juncus balticus)*, saltgrass *(Distichlis stricta)*, greasewood *(Sarcobatus vermiculatus)* and big sagebrush *(Artemisia tridentata)*. They often formed distinctive zones of vegetation around saline lakes. In each type, vegetation parameters of cover, soil characteristics of pH conductivity, soluble salts, texture and soil moisture were determined.

A significant difference between plant types was found for conductivity, exchangeable sodium and soil moisture. Greasewood and saltgrass were found on soils with highest salt content with averages of 12.5 and 12.2 mmhos/cm, respectively, and sagebrush occurred on soils with the lowest salt content with an average of 7 mmhos/cm. Soils under saltgrass had the highest mean value for exchangeable sodium percentage but had narrower ranges than the greasewood. Sagebrush had the lowest mean and range of exchangeable sodium percentages.

A gradient of water from more xeric to more mesic, and a gradation of soil texture from more coarse and sandy to finer textured clay loam, was found along the study area's outer perimeter inward to the margin of the saline lakes. Saltgrass and sagebrush showed relatively high correlations (r = 0.75 and 0.77) with soil water and no significant correlations were found between the pH, soil texture and vegetation parameters.

The high negative correlation between the cover of sagebrush and conductivity and exchangeable sodium (r = -0.89 and -0.97) indicated the adverse effects of salt to the survival of the species. The low negative correlations between the cover percentages of saltgrass, greasewood and these soil factors confirmed the better adaptation of these species in saline environments. Clustering of vegetation types showed high similarities of stands of the same type (50-90 percent). Results of ordination of stands along the watered gradient further confirmed the validity of initial vegetation mapping of the study area.

2C 800602 Trueblood, D.C. 1980. Soil-vegetation relations at rangeland exclosures in the Gardner saltbush vegetation types of the Big Horn Basin, Wyoming. M.S. Thesis. University of Wyoming. 106 p.

5A1D
10D1
3A2
3A3
2A
2E
2F2
5B4

The importance of interrelationships between soil and vegetation is obvious to the casual observer. In order to define some of these relationships, 15 rangeland exclosures located in the Big Horn Basin, dominated by Gardner saltbush *(Atriplex gardneri)*, were studied to determine relationships between soil physical and chemical properties and the vegetation parameters annual production, ground cover and occurrence of selected species. Relationships were analyzed using correlation coefficients and a one-way analysis of variance, completely randomized design. Eight soil parameters were correlated to one or more production parameters, six to one or more cover parameters and 16 to one or more species. Overall, the most important function of these factors was their ability to modify infiltration. Nearly all of the soil properties for which significant correlations were found affected infiltration in some way or were related to other factors that do.

The exclosures were located in the Big Horn Basin. The climate of this area with a mean annual precipitation of 7.3 inches, with large diurnal and seasonal temperature fluctuations were common. Soils were generally calcareous and alkaline. Some contained high levels of gypsum and salts.

Annual herbage production data were collected at seven of the 15 exclosures while percent cover by individual species was obtained at all 15. A soil pit was dug in each exclosure, the profile described, and samples taken from each horizon. These samples were analyzed in the laboratory for various physical and chemical properties.

The production data were quite variable, especially for individual species. Standard deviation equalled or exceeded the mean in many cases. Production on these sites was low, particularly of understory species. Cover data was less variable than production data, probably due to the increased sample size. Ground cover at these sites was also low, especially understory cover.

Soils were classified into two orders, Aridisol and Entisol, and 10 soil series. The soils were generally loamy in texture, saline, low in organic matter and high in calcium. Eight soil properties were found to be significantly correlated to one or more production parameters: calcium carbonate equivalent; soluable and exchangeable magnesium; cation exchange capacity; and percentage of sand, silt, and very fine sand and organic matter. Six soil properties were significantly correlated to one or more cover parameters: exchangeable calcium, gypsum, bulk density, the pH of a saturated base, exchangeable sodium and the percentage of clay.

It appeared that the most important function of the factors identified as significantly related to vegetation characteristics was in their ability to modify infiltration. Water holding capacity was probably not critical because these soils seldom receive adequate moisture at any time to exceed the water holding capacity. The amount of rainfall that actually entered the soil was critical; therefore, at these sites, the most important properties of the soil as related to vegetative growth were those affecting infiltration. Nearly all of the soil properties for which significant correlations were noted affected infiltration in some way or were related to other factors that did.

1E2 800603 Hargis, N.E. 1980. Phenological comparison of two sagebrush species. M.S. Thesis.
2A University of Wyoming. 110 p.
2C

2E On semiarid rangelands of northern deserts, moisture and temperature are often
2F2 factors that limit plant growth. While shortages of these factors will obviously limit
3A3 biomass production, whether or not shortages will constrain phenological development
11E2 was not as well established. Big sagebrush *(Artemisia tridentata)* and black sagebrush
11E3 *(A. nova)* are common and important shrubs in the arid western United States.
11C Occasionally, they cohabit the same range sites. But more commonly, they occupy
11A different sites.

The objectives of this study were: 1) determine whether or not sites dominated by big sagebrush differ significantly ($p \leq 0.10$), from adjacent sites dominated by black sagebrush with respect to soil moisture and temperature, 2) determine whether or not big and black sagebrush differed in their respective phenological patterns - a season long difference of 0.5 phenophase scoring units was considered a marked difference - determine statistically whether or not the same species differed in vegetative growth rates ($p \leq 0.10$) and 3) utilize multiple linear regression to select those environmental variables that best predict and describe phenological production and growth ($p \leq 0.05$).

The relationships between vegetative phenology, reproductive phenology and vegetative production with 13 environmental factors (primarily soil and moisture related) were investigated at two research areas in central and north central Wyoming. The following conclusions were drawn with regard to the relationship of big and black sagebrush phenology and growth with site environmental factors: 1) Based on percent soil moisture and soil temperature, the respective sites on which big and black sagebrush occurred did not significantly differ ($p \leq 0.10$). 2) Big and black sagebrush did not differ significantly in their respective vegetative or reproductive phenological patterns throughout the growing season. Twig growth on big sagebrush was greater than twig growth on black sagebrush ($p \leq 0.10$). 3) Multiple regression equations were calculated - a) multiple regression equations derived to predict vegetative and reproductive phenophase 3 weeks and 1 year in advance were not reliable in four out of six instances ($p \leq 0.05$). Coefficients of determination were low for these equations. b) equations were derived to describe vegetative twig elongation and weight gain. These equations were significant ($p \leq 0.05$) and minimum air temperature was the most important independent variable.

Differences in environmental parameters occurred between the two research areas. Soil moisture and rainfall were more favorable at Horse Creek than at Sweetwater; the results of which were greater twig elongation and weight gain at the Horse Creek research area. The moisture regime was important to vegetative twig elongation and weight gain; less important to vegetative and reproductive phenological progression.

8101 Research Activities

3A3 810101 Exclosure study: production, cover and precipitation.
5A1D

5B The herbage production activities for 1981 followed closely the activities of previous
5C years. Total standing crop was estimated at each of the 12 intensive phenology study
2A locations again at those locations, as well as the other 30 sites distributed primarily in
2C the Big Horn Basin, Wind River Basin and the far western area near Kemmerer, Wyoming.

1E2 810102 Exclosure study: phenology and growth.
2E

2F2 Identification of vegetative and reproductive phenology scores at the 12 intensive
3A2 study location exclosures for the four prime species, western and bluebunch wheatgrass
1E1 *(Agropyron smithii* and *spicatum)* and big and black sagebrush *(Artemisia tridentata* and
5B *nova)*, was conducted as in previous years at 2-week intervals. During 1981, sampling was conducted on 12 sample dates during the growing season.

1E2 810103 Exclosure study: productivity/standing crop sampling.
10D1

2E Productivity and standing crop were determined for each of the four prime species
2F2 during 1981 on 12 sample dates throughout the growing season. Productivity elements
3A2 of the two grasses were individual culms, leaf culms and the individual reproductive
1E1 culms as well, although reproduction was sometimes completely absent for one or both
2A of the grasses. The productivity elements for the sagebrushes were the terminal buds
5A1D and stem or reproductive stem leaders.

2A 810104 Precipitation pattern study.
3A

During 1981, seasonal precipitation was obtained on almost 200 UW/BLM precipitation gauges. Most of the increase occurred in the Big Horn and Wind River Basins. In the Big Horn Basin, as during the previous couple of years, precipitation was recorded monthly from June through September, giving six recording dates instead of four for the year.

2F2 810105 Woody plant density from 4 ft x 5 ft weight unit estimate plots.
10D

2E In conjunction with determining herbage production of the woody plant species
1E2 during 1981, the density of these plants including shrubs, phlox *(Phlox hoodii)*, cactus *(Opuntia polyacantha)* and other mat form species, was determined by counting plant units which occurred within the 4 ft x 5 ft weight unit estimate plots. The density units for shrubs were individual plants rooted within the perimeter of the 4 ft x 5 ft quadrat outline. The weight units of mat form species was less definitive, and essentially characterized as any discrete plant unit of the mat-form species which appeared to grow as a single plant unit. This same characterization applied to the cactus species.

8103 Presentations

5A1D 810301 "Forage production and longevity of sagebrush control on semi-arid rangelands in
3A3 western Wyoming." February 10, 1981. 34th Ann. Meeting Soc. Range Manage.
3A2 Tulsa, Oklahoma. by W.E. Bensel.
2E

2F2 Big sagebrush *(Artemisia tridentata)* control to enhance forage production is a
5A2D recognized means of range improvement. Documentation of long-term response to this
1E2 treatment was not abundant. The research in this program was undertaken to ascertain forage plant production and the extent of shrub re-establishment 17 years after control on an upland semiarid rangeland site containing grazed and ungrazed areas. Sagebrush removal by chemical control allowed nutrients, radiation and water sources formerly

utilized by the shrubs to be diverted for use by desirable forage species. The control of sagebrush increased production of forage species, especially grasses. This improved usability by grazing and browsing animals. The limited sagebrush re-establishment recurred as a function of understory species forming a closed community through which the shrubs could penetrate in disturbed situations. Results of this improvement program have shown a desirable outcome, as evidenced by increased forage production and limited sagebrush re-establishment. Shrub control programs on semiarid rangeland can consistently improve forage production over long durations of time.

1E2 810302 "Phenological comparison of two sagebrush species in Wyoming." February 11, 1981. 34th Ann. Meeting Soc. Range Manage. Tulsa, Oklahoma. by N.E. Hargis.

2E
2F2
11A
11B
11C
11E
2C
2A
11F2
3A2
1E1

Big sagebrush *(Artemisia tridentata)* and black sagebrush *(A. nova)* ecology has been well documented, but their comparative phenology has not. Therefore, research was undertaken to 1) compare reproductive and vegetative phenological patterns, twig elongation and weight gain, 2) compare soil moisture and temperature on separate sites of each species and 3) calculate multiple regression equations based on these factors to predict phenophase and describe twig growth.

Soil moisture was gravimetrically determined and soil and air temperatures recorded at two sagebrush/grass locations. Phenological stages were indexed by numerical scores. The patterns differed slightly between species, but more noticeably between climatic regimes. Twig elongation and weight gains were greater on big sagebrush than on black sagebrush.

Soil moisture and temperature differences between sites were not significant. By multiple regression, phenophase was predicted in advance with moderate success. Minimum air temperature was the most important limiting factor in twig growth. It was concluded that 1) the two species differed in amount of twig elongation but not in phenological pattern, 2) their respective sites did not differ with respect to soil moisture and temperature and 3) the environmental variables predicted phenophase moderately well and twig growth very well.

2A 810303 "The effect of community changes on precipitation/production relationships in exclosures of saltbush vegetation." February 11, 1981. 34th Ann. Meeting Soc. Range Manage. Tulsa, Oklahoma. by L.A. Joyce.

2B
2C
2E
2F2
3A2
3A3
1E2

Exclosures were constructed at seven previously grazed Gardner saltbush *(Atriplex gardneri)* sites in the Big Horn Basin in Wyoming during 1955-1960. Vegetation dynamics, monitored annually since 1962, were examined to develop a predictive equation for total production based on precipitation. Successional changes reduced the predictive ability of any equation at four sites. Total production had not increased within any exclosure since 1962. However, grass and forb production, as a percent of total production, increased with a corresponding decrease in saltbush production at four sites. That contrasts to no obvious changes at the other three. Total production was best predicted by October to July precipitation. Generally, the regressions were more significant to those sites not showing any community changes.

Saltbush production was highly correlated with precipitation received from April to July and often not significantly correlated with October to April precipitation, the period with which grass and forb production was most often correlated. As community structure changed in time, the relationships between production and precipitation also changed.

1B1 810304 "Esophageal fistulated steers for determination of dietary nutrient composition on large pasture native rangeland." February 11, 1981. 34th Ann. Meeting Soc. Range Manage. Tulsa, Oklahoma. by G.R. O'Connor.

10D
6A1
6B2

Most esophageal studies on western rangelands have been on small pastures seldom over 40 ha. In this study, six steers were placed with a commercial herd for most of the grazing season on an area typical of the intermountain region west of Cody, Wyoming, during the months of June through December 1977. Four pastures were utilized and esophageal samples were collected 15 times. In general, nutrient levels fell through the grazing season. Protein, however, peaked at 9.6 percent in late summer, then decreased to a low of 4.4 percent in early December. TDN peaked in mid-July at 37.0 percent, then decreased to a low of 25.0 percent in December. NDF and ADF remained fairly constant throughout the grazing season. Lignin values peaked in early September and atypically decreased to study termination. Moving cattle from one pasture to another, even during summer and early fall, had no effect on forage nutrient level. These data showed that necessary nutrients both for fetal calf growth and maintainance of the dam were inadequate during early winter when cows were maintained on native range.

1E2 1E1 2E 11F2 3A2 11E3 2C 2A

810305 "Phenology and productivity comparisons of western wheatgrass *(Agropyron smithii)* and bluebunch wheatgrass *(A. spicatum)*." February 11, 1981. 34th Ann. Meeting Soc. Range Manage. Tulsa, Oklahoma. by M.K. Owens.

The understanding of phenodynamics and productivity differences between major forage species is essential for proper management in semiarid rangelands. Twelve diverse sites in western Wyoming were selected and sampled intensively for study of phenological development and productivity comparisons of western wheatgrass *(Agropyron smithii)* and bluebunch wheatgrass *(A. spicatum)*.

Plants were ranked phenologically and marked for collection at the subsequent sampling date. Environmental parameters sampled included soil temperature, soil moisture, air temperature, precipitation and water potential. The sampling interval was maintained at 10-day to 2-week intervals during maximum growth periods.

Polar ordination was implemented to identify similar plots. These were blocked and analysis of variance was performed to determine significant differences between the species. The environmental differences were then analyzed to detect effects on phenodynamics and productivity.

Soil temperature in spring and soil moisture during the latter growing season strongly influenced phenology and productivity. Western wheatgrass responded to summer precipitation while bluebunch wheatgrass required deep soil moisture generally associated with the early spring period.

4B 4C 1E4 2F2 7A1 7A3 6A3 11E

810306 "Grazing system impacts on semiarid vegetation of Wyoming." February 12, 1981. 34th Ann. Meeting Soc. Range Manage. Tulsa, Oklahoma. by D.E. Rodgers.

Grazing systems have been studied to develop management techniques that will provide a sustained yield of forage for livestock. This particular study was unique in that it was conducted under actual ranching conditions. The research areas, with a four pasture rest rotation unit, a two pasture deferred unit, and a continuous seasonal use area, were designated in 1967.

Stocking rates were reduced to help restore a depleted range. Permanent plot transects were established on pasture and exclosure sites to obtain vegetation cover, composition and utilization data.

Total perennial grass cover increased uniformly on all three grazing systems during the 12-year study period, indicating an overall upward trend in range condition. Few exclosure comparisons showed higher grass cover values as a response to protection from livestock utilization. Perennial grass utilization was light to moderate, with western wheatgrass receiving the heaviest use (45-60 percent). Exclosure and utilization data are indicative of a light to moderate stocking rate. After 12 years, continuous grazing had been as effective in grazing management as rotational or deferred systems, at this particular stocking rate.

5A1D810307 "Plant composition response to picloram application in the Wyoming juniper type." February
1E2 12, 1981. 34th Ann. Meeting Soc. Range Manage. Tulsa, Oklahoma. by H.G. Fisser.
2A
2C Within the physiographic province of the Wyoming juniper (*Juniperus* spp.) type are
2E highly heterogeneous matrices of plant communities with different species and potentials.
2F2 Game range improvement must determinately affect the existing climax species. Three
3A2 deer-proof exclosures were constructed in 1968. Aerial spray application of picloram (2 lb. A.I./acre) was applied in early August. Half of each exclosure (about 1.85 acre) was treated. Control areas outside the exclosures were untreated. Cover of shrub and understory species was determined each year. Precipitation, soil and site data were recorded.

Basal area of Utah juniper (*Juniperus osteosperma*), big sagebrush (*Artemisia tridentata*), Hood's phlox (*Phlox hoodii*) and broom snakeweed (*Gutierrezia sarothrae*) decreased on the spray areas. Black sagebrush (*Artemisia nova*) increased on spray areas where the shrub was indigenous to the site. Grass and annual forbs and black sagebrush, a root sprouter, increased on spray areas due to reduced competition for light and water.

Picloram effectively controlled target shrubs and improved game range. Sites with high indigenous populations of cheatgrass (*Bromus tectorum*) are unsuited to this game range improvement practice.

2C 810308 "Soil-vegetation relationships in the Gardner saltbush vegetation types of Wyoming."
3A2 February 10, 1981. 34th Ann. Meeting Soc. Range Manage. Tulsa, Oklahoma. by
8B D.C. Trueblood.
2F2
3A2 The importance of soil-vegetation relationships is intuitively obvious. To define some
10D1 of these relationships, 15 rangeland exclosures in the Big Horn Basin of Wyoming
11A dominated by Gardner saltbush (*Atriplex gardneri*), were studied to compare soil
11C moisture and chemical properties and the vegetation parameters of annual production,
11E ground cover and occurrence of selected species.

Data were analyzed using correlation coefficients and a one-way analysis of variance, completely randomized design. Eight soil parameters (sand, silt, very fine sand, organic matter, calcium carbonate, cation exchange capacity, and soluable and exchangeable magnesium) were correlated to one or more production parameters. Six soil parameters (clay, bulk density, pH, gypsum, exchangeable sodium and exchangeable calcium) were correlated to one or more cover parameters. Sixteen soil parameters were correlated to the occurrence of one or more species.

The most important correlations were related to factors which modify infiltration. Vegetation parameters that were correlated to soil properties were: 1) total, understory, shrub, pricklypear (*Opuntia polyacantha*) and crested wheatgrass (*Agropyron cristatum*) production, 2) total, shrub and squirreltail (*Sitanion hystrix*) cover and 3) the occurrence of bud sagewort (*Artemisia spinescens*), winterfat (*Ceratoides lanata*) and sandberg bluegrass (*Poa secunda*).

2F2 810309 "Vegetation dynamics of non-grazed semiarid sagebrush grassland communities in the
2A Big Horn Basin of Wyoming." February 11, 1981. 34th Ann. Meeting Soc. Range
2C Manage. Tulsa, Oklahoma. by J.W. Uhlich.
2E
3A2 Knowledge of long-term rangeland plant response to climatic environmental
2B dynamics is limited because few extensive studies are available. This research was initiated in 1962 to annually document vegetation cover and forage production on relic areas and exclosures protected from livestock utilization. Information was obtained on plant responses to environmental changes and community interaction. Transects of 1 ft x 1 ft and 4 ft x 5 ft quadrats were utilized to obtain herbage data on different species and life forms. Environmental information included precipitation, temperature, soil description and site characteristics.

Amount and distribution of available moisture was the major factor which influenced plant fluctuations and response. Herbage production showed the greatest annual change, while cover values showed smaller responses. Successional trends were driven by large climatic changes such as moisture abundance or drouth.

Improvement was limited by the ecologically dominant presence of big sagebrush *(Artemisia tridentata)* in this ecosystem. The results indicated that management alternatives on these semiarid rangelands were limited due to slow natural changes within the vegetation community.

2E 3A2 2F2 2C 2D

810310 "Population structure of Utah juniper in north central Wyoming." February 12, 1981. 34th Ann. Meeting Soc. Range Manage. Tulsa, Oklahoma. by W.J. Waugh.

Juniper *(Juniperus* spp) invasion is an active process which modifies existing community structure on western rangelands. This research involved a long-term study, initiated in 1965, to monitor the rate of Utah juniper *(J. osteosperma)* invasion onto a black sagebrush *(Artemisia nova)*, bluebunch wheatgrass *(Agropyron spicatum)* community and to assess the related changes of understory vegetation of woody and herbaceous plants.

By 1979 no significant changes in juniper density or understory cover were observed. The invasion question was then approached by examining past juniper establishment dynamics. Population structure analysis, including sampling of annual growth rings, was conducted adjacent to the ongoing invasion study unit.

Soil development, fire frequency, and past variations in climate were analyzed. Root crown cross-sections were cut from trees located in plots, stratified along a gradient from an established juniper stand to an established sagebrush/grassland stand, and the distribution of the age classes was determined for each plot.

Many trees that appeared to be single, multistemmed individuals from the surface, were revealed upon grubbing, to be multiple individuals of various ages, twisted and gnarled together. Fire frequency was determined from cross-sections and climatic variation was inferred using methods of dendroclimatology.

2C 2A 2E 3A3 1E2 2F2

810311 "Correlation of soil factors and environmental conditions with phenophase development of western wheatgrass *(Agropyron smithii).*" February 10, 1981. 34th Ann. Meeting Soc. Range Manage. Tulsa, Oklahoma. by L.W. Young.

Phenological progression of western wheatgrass *(Agropyron smithii)* in western Wyoming, observed over a period of 4 years, was highly variable for numbers of days of annual growth cycle and in number of days of per stage. Four vegetative and four reproductive stages were selected for observation and related to five independent variables: cm available soil water for plant growth at 0-15 cm, soil temperature at 45 cm, average air temperature during the respective growth stage and cumulative sum of maximum air temperatures.

Discriminate analysis was used to explain variability in time between growth stages and to identify the limit of variables necessary for progression of phenology from one stage to the next and in particular, initiation of the reproductive stage. In the vegetative stage, initiation of growth and development to the fourth leaf stage was delayed by high availability of soil water. Vegetative browning was hastened by high maximum air temperatures.

Reproduction occurred only 40 percent of the time and was dependent upon available soil water during the period of growth initiation. The information from 1976-78 was used to formulate a predictive set of values for the independent variables which were then compared to actual phenological development during 1979 for each of the phenological stages.

2F 2E 2A 2C 11E

810312 "Habitat and vegetation of semiarid rangelands, Libya, Africa." February 11, 1981. 34th Ann. Meeting Soc. Range Manage. Tulsa, Oklahoma. by O.M. Shawesh.

Six plant communities representing major vegetation and edaphic variation on the Assa Range Management Research area of semiarid rangeland of northwestern Libya were recognized on the basis of changes in floristic composition, relative dominance of leading species and changes in edaphic features: 1) *Artemisia campestris*,

2) *Thymelaea-Helianthemum-Echiochilon*, 3) *Lygeum-Artemisia*, 4) *Suaeda-Halocnemum*, 5) dune sand area and 6) pavement eroded sites.

Stepwise multiple regression analysis showed that 76.5 percent of the annual forb cover variation was derived from relationships to only three independent soil variables: organic matter, phosphorus and manganese. Sand, total carbonate and manganese explained 77.6 percent of annual grass cover variation. Clay, phosphorus and manganese accounted for 76.0 percent of the variation in perennial grass cover. Total carbonate and calcium explained 74.7 percent of the shrub cover variation.

Percent sand, moisture equivalent and electrical conductivity were the only factors to exhibit no significant effects on the dominant species. Soil properties which influenced the moisture regime, soil structure and fertility showed a significant effect on the growth and distribution of the plants. Soil factors were also found to have interacting effects that could modify vegetation response functions.

2E 810313 "Some recent research findings in the true mountain mahogany types." May 28, 1981.
7A1 10th Wyoming Shrub Ecology Workshop. Sybille Canyon, Wyoming. by H.G. Fisser.
2A

2C True mountain mahogany *(Cercocarpus montanus)* has been the subject of diverse and
1E3 intensive research activities. It was a primary deer forage species of foothills in Wyoming,
2F2 Utah and Colorado. Most communities were developmental, although on some sites
10D communities were considered as climax. Distribution was within both the intermediate and higher elevation mesic brush and woodland vegetation types and the more mesic shrubland vegetation types at lower elevations. It often intermixed with big and black sagebrush *(Artemisia tridentata* and *nova)* and Utah juniper *(Juniperus osteosperma)* and occasionally with bitterbrush *(Purshia tridentata)* and more mesic complexes.

It was relatively short lived and exhibited definite age diameter structure relationships. Several minerals were concentrated, accounting for much of its importance for winter game use. Nutrient quality was superior to most associated woody species. Proper management and use criteria were not well defined, and further research has been strongly suggested.

2E 810314 "Stagnation and decadency of *Cercocarpus montanus* in a southeastern Wyoming big
6A game exclosure." May 28, 1981. 10th Wyoming Shrub Ecology Workshop. Sybille
2F2 Canyon, Wyoming. by W.J. Waugh.
1E2

1E3 The study was conducted to determine whether annual shrub production within a big
1E4 game exclosure, located within a mountain big sagebrush *(Artemisia tridentata* ssp. *vaseyana)*/true mountain mahogany *(Cercocarpus montanus)* association in southeastern Wyoming, have stagnated after 40 years non-use. Sampling for percent canopy cover, annual production, shrub density, age size class and percent dead material was done within a homogeneous stand and inside and outside the exclosure.

Differences in percent canopy cover and shrub density were insignificant for both species. Current year sagebrush production was no different inside versus outside the exclosure, but was significantly greater outside than inside for mountain mahogany.

The protected mountain mahogany produced much less biomass inside the exclosure than outside of it. It appeared that some utilization was necessary for optimum production of mountain mahogany, and that protected areas did not necessarily define the "climax" condition of mountain shrubs.

1E2 810315 "Phenology and culm weight variation between two wheatgrass species: *Agropyron*
1E *smithii* and *A. spicatum*." June 17, 1981. 14th Intern. Grassl. Congress. Lexington,
2E Kentucky. by M.K. Owens.
11F2

3A2 Knowledge of plant phenology and live weight dynamics are necessary to
11E3 manage semiarid rangelands. Thirteen sites in Wyoming were sampled for plant
2C phenologic development and the culm weights of western wheatgrass *(Agropyron smithii)*
2A and bluebunch wheatgrass *(A. spicatum)*. These species were synpatric and allopatric

within these rangelands. Environmental data were also collected for use in differentiating sites.

Data on all the variables were collected at 14 day intervals through the growing seasons of 1979 and 1980. Permanently marked plants were samples for phenology and plants of similar development were collected for culm weight estimates. Cumulative precipitation, degree days, soil water potential at three depths (0-15, 16-30 and 46-60 cm), soil temperature at 25 cm, and Julian dates were recorded.

A two-way analysis of variance performed on culm weight and phenology data identified seven groups of similar sites. For the culm weights, western wheatgrass formed four distinct groups, bluebunch wheatgrass formed two distinct groups, and both species together formed one group. The phenologic data revealed the same groupings of sites.

Discriminate analysis of physical site characteristics showed that bluebunch wheatgrass occurred on sites with low sand content, high soil pH, low soil moisture, and/or deep rooting habit depths. Western wheatgrass favored sites with moderate to high sand content, neutral pH, moderate soil moisture and/or shallow rooting depths.

Analysis of micrometerological variables showed western wheatgrass produced the greatest culm weights at moderate amounts of degree day accumulation and cumulative precipitation. Bluebunch wheatgrass culm weights were greatest at higher values of both variables. Phenologic development was species related, with bluebunch wheatgrass developing more rapidly. Phenology on sites with small culm weights developed more rapidly than sites with large culm weights.

4B 810316 "Vegetative response to grazing systems management in Wyoming's Red Desert." June
4C 18, 1981. 14th Intern. Grassl. Congress. Lexington, Kentucky. by D.E. Rodgers.
1E4
2F2 Over the past 60 years, livestock grazing systems have been studied to aid
7A1 development of range management techniques provide higher sustained forage yield of
7A3 native perennial grasses. On many rangelands, rest rotation and deferred systems have
6A3 proven superior to continuous grazing for management purposes. On others this has not been true. The objectives of this particular study were to determine the influences of different grazing systems on arid shrub-steppe vegetation of Wyoming.

The study was located on the Red Desert in south central Wyoming. In 1966 three grazing units were designated: a four pasture rest rotation unit, a two pasture deferred unit and a continuous seasonal use area. Cattle stocking rates were adjusted to reduce grazing pressure on a depleted range, and to equalize forage use over the three units. Vegetation cover and utilization were measured on 43 permanent cluster transects located on sagebrush/grass types on the various pastures. The transects, consisting of 20 30 cm x 30 cm quadrats each, were randomly placed within areas of moderate grazing intensity. Vegetation cover data was recorded in 1967, 1971, 1975 and 1979, and utilization data was collected annually from 1967-1971.

The changes in vegetation cover between 1967 and 1979 expressed an upward trend in range condition on all three grazing units. Perennial decreaser grasses such as Indian ricegrass (*Oryzopsis hymenoides*) and needleandthread (*Stipa comata*) made significant gains, while big sagebrush (*Artemisia tridentata*) and Plains pricklypear (*Opuntia polyacantha*), both invaders, either declined or remained unchanged.

Perennial increaser grasses such as western wheatgrass (*Agropyron smithii*) and sandberg bluegrass (*Poa secunda*) made gains on most pasture units. Changes in amounts of cover for decreasers, increasers and invaders were compared among the three units. All changes were equivalent, indicating that upward trend of range condition had been occurring at a uniform rate on all units. Utilization of desirable perennial grasses ranged from 27-57 percent, with approximately equal use among pastures on individual species.

Because decreaser plants exhibited cover gains and invaders declined, it was apparent that range condition was improving. Gains by increaser plants indicate range improvement had not yet progressed to the stage where their populations stabilized prior to the decline associated with trend movement toward the excellent condition class. Utilization data indicated equal grazing pressure on all sampling sites, and revealed that intensity was in the light to moderate range.

A rainfall-related bias could be eliminated because long-term precipitation over the area was relatively uniform. This study showed that, after 13 years, continuous grazing was as effective in range management as deferred and rest rotation grazing, under the following conditions: 1) the range type was northern arid shrub-steppe, 2) the condition class was in the lower range of condition classification and 3) the stocking rate was light to moderate.

8104 Published Proceedings

5A1D 810401 Bensel, W.E. and H.G. Fisser. 1981. Forage production and longevity of sagebrush
3A3 control on semiarid rangelands in western Wyoming. p. 9. In Absts. Soc. Range Manage.
3A2 34th Ann. Meeting. Tulsa, Oklahoma. February 8-13, 1981.
2E
1F2D See 810301.
5A2
1E2
3A1

1E2 810402 Hargis, N.E. and H.G. Fisser. 1981. Phenological comparison of two sagebrush species
2E in Wyoming. p. 21. In Absts. Soc. Range Manage. 34th Ann. Meeting. Tulsa,
2F2 Oklahoma. February 8-13, 1981.
11A
11B See 810302.
11C
11E
2C
2A
3F2
3A1
1E1

2A 810403 Joyce, L.A. and H.G. Fisser. 1981. The effect of community changes on
2B precipitation/production relationships in exclosures of saltbush vegetation. p. 22. In
2C Absts. Papers Soc. Range Manage. 34th Ann. Meeting. Tulsa, Feb. 8-13, 1981.
2E
2F2 See 810303.
3A2
3A3
1E2

1B1 810404 O'Connor, G.R., H.G. Fisser and J.W. Waggoner. 1981. Esophageal-fistulated steers for
10D determination of dietary nutrient composition on large pasture native rangeland. p.
6A1 26. In Absts. Soc. Range Manage. 34th Ann. Meeting. Tulsa, Feb. 8-13, 1981.
6B2
See 810304.

1E2 810405 Owens, M.K. and H.G. Fisser. 1981. Phenology and productivity comparisons of western
1E1 wheatgrass (*Agropyron smithii*) and bluebunch wheatgrass (*A. spicatum*). p. 21. In
2F2 Absts. Soc. Range Manage. 34th Ann. Meeting. Tulsa, Oklahoma February 8-13,
2C 1981.
2A
2E See 810305.
3A2
11E3

4B 810406 Rodgers, D.E. and H.G. Fisser. 1981. Grazing system impacts on semiarid vegetation of
4C Wyoming. p. 34. In Absts. Soc. Range Manage. 34th Ann. Meeting. Tulsa,
1E4 Oklahoma. February 8-13, 1981.
2F2
7A1 See 810306.
7A3
6A3

5A1D 810407 Spaeth, K.E. and H.G. Fisser. 1981. Plant composition response to picloram application
1E2 in the Wyoming juniper type. p. 35. In Absts. Soc. Range Manage. 34th Ann.
2A Meeting. Tulsa, Oklahoma. February 8-13, 1981.
2C
2E See 810307.
2F2
3A2

2C 810408 Trueblood, D.C. and H.G. Fisser. 1981. Soil-vegetation relationships in the Gardner
3A2 saltbush vegetation type of Wyoming. p. 12. In Absts. Soc. Range Manage. 34th
8B Ann. Meeting. Tulsa, Oklahoma. February 8-13, 1981.
2F2
3A2 See 810308.
10D1
11A
11C
11E

2F2 810409 Uhlich, J.W. and H.G. Fisser. 1981. Vegetation dynamics of non-grazed semiarid
2A sagebrush grassland communities in the Big Horn Basin of Wyoming. p.22. In
2C Absts. Soc. Range Manage. 34th Ann. Meeting. Tulsa, Oklahoma. February 8-13,
2E 1981.
3A2
2B See 810309.

2E 810410 Waugh, W.J. and H.G. Fisser. 1981. Population structure of Utah juniper in north
3A2 central Wyoming. p. 40. In Absts. Soc. Range Manage. 34th Ann. Meeting. Tulsa,
2F2 Oklahoma. February 8-13, 1981.
2C
2D See 810310.

2C 810411 Young, L.W. and H.G. Fisser. 1981. Correlation of soil factors and environmental
2A conditions with phenophase development of western wheatgrass (*Agropyron smithii*).
1E2 p. 12. In Absts. Soc. Range Manage. 34th Ann. Meeting. Tulsa, Oklahoma. February
2F2 8-13, 1981.
3A3
2E See 810311.

2F 810412 Shawesh, O.M. and H.G. Fisser. 1981. Habitat and vegetation of semiarid rangelands,
2E Libya, Africa. p. 23. In Absts. Soc. Range Manage. 34th Ann. Meeting. Tulsa,
2A Oklahoma. February 8-13, 1981.
2C
11E See 810312.

2F2 810413 Fisser, H.G. 1981. Shrub establishment, dominance and ecology on the juniper and
7A1 sagebrush-grass types in Wyoming. p. 23-27. In L.S. Stelter, E.J. DePuit, S.A. Mikol
2E (eds.). Proc. Symposium on shrub establishment on disturbed arid and semiarid
1E2 lands. December 3, 1980. Wyo. Game and Fish Dept. Publication.
1E4
2A See 800301.
2C
2D
2F2
3A2
3A3
4B
4C
5A1C
5A1D
5A1E
5B4
5C

4B 810414 Rodgers, D.E. and H.G. Fisser. 1981. Vegetative response to grazing systems
11E management in Wyoming's Red Desert. p. 451. In Summaries of Papers. J. Allan
4C Smith (ed.). XIV International Grassland Congress. Lexington, Kentucky. June
1E4 14-23, 1981.
2F2
7A1 See 810316.
7A3
6A3

1E2 810415 Owens, M.K. and H.G. Fisser. 1981. Phenology and culm weight variation between two
1E1 wheatgrass species: *Agropyron smithii* and *A. spicatum*. p. 205. In Summaries of
2F2 Papers. J. Allan Smith (ed.). XIV International Grassland Congress. Lexington,
11E3 Kentucky. June 14-23, 1981.
2C
2A See 810315.
2E
3A2

8105 Wyoming Agricultural Experiment Station Publications

810501 Fisser, H.G., M.K. Owens, D.E. Rodgers, W.J. Waugh, N.E. Hargis and J. Uhlich. 1981. Phenology and production studies on semiarid shrub types. 1980 Results. Univ. of Wyo. Coop. Res. Rep. to the BLM. Vol. I: Phenology Study. Sections I and II. Wyo. Agr. Exp. Sta. Sci. Rep. 1172. 496 p.

1E2 81050101 Section I - Phenology study.
2E
3F2 The annual report of the 1980 phenology study was given in the same form as
3A2 previous years listing the phenological stages of each of the categories that have been
1E1 defined. The listing of the 12 study areas and the prime species was presented on each total listing of all plant species noted on the 12 study sites and the various categories of phenological data which had been obtained during the 11 sampling dates mostly at 2-week intervals during 1980. The phenological sampling was begun on April 26 and the last for the year was done on September 20. The 1980 phenology ranking values had been modified from those used during 1973-79. The earlier phenology score averages were converted to those used in the currently modified and updated scoring system in order to be able to compare one year against all.

11F2 81050102 Section II - Phenology step-wise regression for predictive modeling.
2E
2F2 The phenology data for 1980, as in previous years, was subjected to the multiple
3A2 linear regression for predictive modeling for each of the prime species for vegetative and
1E1 for reproductive phenological development. The Hocking-LaMotte-Leslie method of
2A multiple regression analysis was implemented for the present study. This method was a
backward selection process as opposed to the previous forward selection process. This
method was deemed superior for this data set. As usual, warm temperature values
tended to promote rate of phenologic development, and good soil moisture values tended
to decrease the rate of growth.

810502 Fisser, H.G., M.K. Owens, J.W. Uhlich, W.J. Waugh, D.E. Rodgers and N.E. Hargis. 1981. Phenology and production studies on semiarid shrub types. 1980 Results. Univ. of Wyo. Coop. Res. Rept. to the BLM. Vol. II: Precipitation, Productivity and Production Sections III, IV, and V. Wyo. Agr. Exp. Sta. Sci. Rep. 1173. 220 p.

2A 81050203 Section III - Precipitation pattern study.
3A

The precipitation data for the 1980 period were again summarized for the 200 UW/BLM gauges and for some 60 U.S. Weather Service stations in western Wyoming. Precipitation for 1980 in the Big Horn Basin was slightly greater than the long-term average. Fremont, and especially Natrona, counties were below average with the former receiving 92 percent of normal and the latter only 73 percent. The latter was influenced by very low precipitation during the fall period. Weather Bureau stations exhibited similar relations. In Carbon County, precipitation of the UW/BLM gauges was 105 percent of normal. The counties of south central and western Wyoming received about 20 percent more moisture than normal.

1E2 81050204 Section IV - Productivity - phenology interrelation study.
10D1
2E The activity of culm clippings of the prime species at the intensive phenology study
2F2 exclosures was conducted at each of the sampling dates during 1980. These productivity
3A2 clippings were obtained at 2-week intervals during most of the growth season from late
1E1 April to latter September.
2A A disastrous break in the fence at the Horse Creek study site resulted in complete
5A1D loss of data from grasses and forbs. They were completely grazed by livestock that
penetrated the exclosure during the second week of June. Intensive livestock utilization
may have an effect on herbage production during later years; it was to be monitored to
determine whether response did occur.

Changes in culm weights were positive through mid-June. But at many locations, culm weight losses for grasses and for buds of sagebrushes *(Artemisia tridentata* and *nova)* were closely linked with warm summer temperatures and reduced effective precipitation. Data were shown by individual culm weights given in mg for each exclosure on each sample date for each of the prime species.

3A3 81050205 Section V - Exclosure study: production, cover and precipitation.
5A1D
5B The 1980 herbage production studies included the determination of early standing
5C crop at the 12 intensive phenology study sites, and determination of greatest standing
2A crop at all sites during the summer and fall. Herbage production values varied greatly
2C from one region of the state to another.

Precipitation in the southwestern section of the state was much above the long-term average. This increase in precipitation contributed to high forage production values of the Cumberland sites, where the highest ever recorded values were noted.

Precipitation in the Wind River Basin was mostly below average. Forage production on the sites in this area varied from very near average in the southern areas to below average in the central and northern parts of the Basin. The Lander Ant site continued to show a downward trend in herbage production. This degradation could be attributed to the decline in abundance of blue grama and the minimal response of the mid-grasses such as needleandthread *(Stipa comata)*, because of the dominance and control exhibited by Wyoming big sagebrush *(Artemisia tridentata* spp. wyomingensis).

In the Big Horn Basin, precipitation varied from above the long-term on sites dominated by Gardner saltbush *(Atriplex gardneri)*, to average on big sagebrush sites, to below average on sites dominated by blue grama *(Bouteloua gracilis)*. Sites dominated by big sagebrush had forage values mostly below average. Buffalo Creek exclosure again had herbage production much below its average. This degradation was found to be highly correlated with an increase in big sagebrush individuals. Gardner saltbush dominated sites had an abundance of precipitation this year. Many of these sites stayed green well into the summer. Despite the increase in precipitation and the extended period for growth, forage production on these sites did not show a large increase in production. The late spring and summer distribution of the precipitation probably came too late to effect the forage production of these early growing plants.

Horse Creek exclosure was not sampled during the summer sampling date this year. The exclosure was vandalized, and most of the vegetation inside the exclosure was grazed and heavily trampled. Shoshoni -8 outside treatment was also not reported since the data were misplaced.

3A2 810503 Fisser, H.G. and A.O. Gdara. 1981. Long-term vegetation data from permanent quadrats
2F2 on semiarid shrublands of Western Wyoming, 2 ft x 20 ft and 4 ft x 4 ft series.
10D1 Univ. of Wyo. Coop. Res. Rep. to the BLM. Wyo. Agr. Exp. Sta. Sci. Rep. 1114. 119 p.
2E
11E This was a summary of the long-term vegetation cover data which had been obtained
2A on 29 exclosures at 4-year intervals between 1960 and 1980. The data for 1980 were not compiled for this report.

Changes of vegetation cover on the study areas were due to a combination of several factors. Grazing influence was evident on almost all the study areas, and, depending on the vegetation type, apparently was often an important factor in dynamics of the vegetation cover. Additional study showed that climatic variables (especially precipitation) were important, as were physical and chemical soil characteristics.

Shrub and forb species generally increased on the grazed as compared to the protected areas, while grasses decreased. In the protected, non-grazed sites, grasses usually increased. Exchangeable sodium, carbon, texture, pH, and bulk density of soils were specifically related to vegetation characteristics, primarily because they affected, and often limited, water infiltration rates. Long-term records have also shown that relationships between precipitation and herbage production often changed with community structure modification through time.

These summarized data from the 29 exclosures and 8 pasture units of the halogeton grazing study can be interpreted in much greater detail than has been done in the past. This publication provided a permanent record to be utilized by others in the future. Of specific interest would be study of plant demography and survivorship curves from the 2 ft x 20 ft permanent quadrats.

8106 Graduate Advisee Completions

1B1 810601 O'Connor, G. R. 1981. Esophageal-fistulated steers for determination of dietary nutrient
10D composition on large pasture native rangeland. M.S. Thesis. University of Wyoming.
6A1 70 p.
6B2

Most esophageal studies on western rangelands have been on small pastures seldom over 40 ha. In this study six steers were placed with a commercial herd for most of the grazing season on an area typical of the intermountain region west of Cody, Wyoming, during the months of June through September 1977. Four pastures were utilized and esophageal samples were collected 15 times. In general, nutrient levels declined through the grazing season. Protein, however, peaked at 9.6 percent in late summer, then decreased to a low of 4.4 percent in early December. In Vitro Dry Matter Disappearance peaked in mid July at 37.0 percent, then decreased to a low of 25.0 percent in December. Neutral Detergent Fiber and Acid Detergent Fiber remained fairly constant throughout the grazing season. Lignin values peaked in early September and atypically decreased to study termination. Moving cattle from one pasture to another, even during summer and early fall, had no effect on forage nutrient levels. These data showed that necessary nutrients both for fetal calf growth and maintenance of the dam were inadequate during early winter, when cows were maintained on native range.

Environmental data collected by the nearby weather service station in 1977 indicated the moisture available for plant growth was 45 percent less than the 30-year average. The temperature data indicated an above normal temperature. The higher temperatures compounded the effect of low moisture by increasing transpiration rates. The lack of moisture and higher temperatures were reflected in low dietary protein and IVDMD values and high lignin and silica values early in the grazing season.

The condition class of the range seemed to have little effect on diet quality. The diet quality remained constant when cattle were moved from South Logan to the Horton pasture where the condition class decreased from fair to very poor. ADF, NDF, lignin, silica and the majority IVDMD values were not affected by this move. The move to a higher condition class pasture, the Horton to North Logan, did not introduce any increase in dietary quality, though an effect may have been nullified by the time of year. The final movement back into the Horton again produced no significant effect.

The management implications were basic. Cattlemen must take into consideration type of spring weather and its effect upon vegetative growth. Adverse growing conditions affect forage growth, necessitating a flexible grazing schedule. The nutrient values during early winter are marginal, and the producer must begin supplementing his cattle at that time.

1E2 810602 Owens, M. K. 1981. Niche differentiation and species response comparisons between
1E3 bluebunch wheatgrass and western wheatgrass. M.S. Thesis. University of Wyoming.
2A 84 p.
2C
2F2 Niches of bluebunch *(Agropyron spicatum,)* and western *(A. smithii)* wheatgrass were
3A2 differentiated using physical site characteristics and micrometeorological variables. Blue-
3A3 bunch wheatgrass occurred on sites with low sand content, high soil pH, low soil moisture
1E1 due to depletion, and deep rooting depths. Western wheatgrass occurred on sites with
2E moderate to high sand content, neutral pH, moderate soil moisture and shallow rooting depth.
11E3 Micrometeorological variables showed western wheatgrass produced the greatest culm weights at moderate amounts of degree day accumulation and cumulative precipitation. Bluebunch wheatgrass culm weights were greatest at higher values of both variables. Phenological development was species related with bluebunch wheatgrass developing more rapidly.

Both physical site characteristics and micrometeorological variables accounted for differences in species response of western and bluebunch wheatgrass. The use of the physical characteristics differentiated the sites which possessed bluebunch or western wheatgrass in the plant community. The micrometeorological variables divided the site to a finer degree to differentiate culm weight levels and phenological development rates. Further information may be found under citation code number 810315.

4B 810603 Rodgers, D.E. 1981. Ecological effects of grazing systems management on sagebrush-
4C grassland vegetation of Wyoming. M.S. Thesis. University of Wyoming. 90 p.
1E4
2F2 This thesis was an analysis of the data derived from the Baggs grazing system
7A1 study from 1967-1979. Cover comparison results showed gains in total grass and total
7A3 forb cover, and losses in total shrub cover on all three grazing units between 1967 and
6A3 1979. These responses were indicative of an upward trend in range condition.

Plant species were categorized according to their status as decreasers, increasers and invaders. Changes in cover for these classes were analyzed between years within each grazing system. Four-year comparisons revealed most decreasers and increasers gained substantially on all pastures during the 1971-75 period. Big sagebrush *(Artemisia tridentata)*, an increaser, initially declined over the area during the 1967-1971 period, and then again between 1975 and 1979. Invaders responded somewhat erratically, though the annuals increased significantly over much of the area during the 1971-75 interval. Few substantial changes occurred during the final four years of the study. The 1967-1975 and 1967-1979 comparisons produced similar results, as most significant changes had taken place by 1975.

Livestock reductions on the study site were partly responsible for the upward trend in range condition. Favorable moisture conditions prevailed during the early years of the study, and contributed further to vegetation recovery. The decline of big sagebrush could also have been a contributing factor.

Range condition classification on the study area was identified by fitting the behavior patterns of decreasers, increasers and invaders to a graph showing the response of these vegetation classes to increased grazing intensity. The results of this procedure, along with photographic evidence and SCS data, indicated range condition throughout the area had improved from low to high/fair during the study period.

Proportional changes in plant cover were compared among grazing units for the 1967-1975 and 1967-1979 intervals. Changes for total decreasers were equivalent, indicating range improvement was uniform over the three allotments. Though the Oppenheimer and Powder Rim units received less use than the Poison Buttes unit throughout the study area, they did not benefit in vegetation recovery.

After 12 years deferred rotation, deferred seasonal and continuous seasonal grazing systems were of equal value in range management under the conditions: 1) the stocking rate was light to moderate, 2) range condition was in the lower range of condition classification throughout the study and 3) the vegetation type was dominated by big sagebrush and perennial grasses. Under conditions similar to these, a continuous seasonal grazing plan would be recommendable because of its practicality.

5B4 810604 Spaeth, K. 1981. Successional trends on revegetated *Juniperus osteosperma* communities in north central Wyoming. M.S. Thesis. University of Wyoming. 261 p.
2F
2A
2C
2D
2E
3A2

Within the physiographic province of the Wyoming juniper (*Juniperus* spp.) type were highly heterogeneous matrices of plant communities with different species and potentials. Juniper control treatments (burn, spray, bulldozed), revegetation trials and successional transect data were initiated to evaluate the feasibility of the game range improvement concept.

Prescribed burn treatments which left indigenous understory shrubs and perennial grasses intact were the most effective in juniper shrub eradication, invasion retardation and seedling establishment. Aerial spray application of Picloram (2 lbs A.I./acre) and bulldoze treatments were also effective in initial juniper eradication but were detrimental to endemic non-target understory shrubs. The picloram treatment also inhibited seedling establishment due to the toxic residues of the chemical in the soil and the juniper litter.

Data comparison analysis on sprayed areas revealed significant increases of perennial and annual grasses, annual forbs and black sagebrush *(Artemisia nova)* - a root sprouting species, due to reduced competition for light and water.

Perennial forb composition remained static. Sites containing a Utah Juniper-black sagebrush-curlleaf moutain mahogany-bluebunch wheatgrass *(Juniperus osteosperma-Artemisia nova-Cercocarpus ledifolius-Agropyron spicatum)* community were the most adaptable in regard to vegetation response and seedling establishment. Big sagebrush *(Artemisia tridentata)*, silver sagebrush *(A. cana)*, black sagebrush, Mexican cliffrose *(Cowania mexicana)*, Stansbury cliffrose *(C. stansburiana)*, 4-wing saltbush *(Atriplex canescens)*, Palmer penstemon *(Penstemon palmeri)*, cicer milkvetch *(Astragalus cicer)*, skunkbush sumac *(Rhus trilobata)* and Arnold hawthorne *(Crataegus arnoldii)* were successfully established on this site.

Sites with indigenous populations of cheatgrass *(Bromus tectorum)* were not suited to picloram spray and bulldozed treatments because of the reversion of the successional trend to an annual grass-forb sere. Due to the vegetational diversity of the Wyoming juniper ecosystem, game range improvement practices must be planned on a site specific basis. Precise efforts in response to natural burns offered the best opportunity for improving wildlife habitat.

The following conclusions are evident: 1) Juniper-sagebrush-grass vegetation types were essentially closed communities, and new or different species could be added only by removing tree and shrub dominance. 2) Due to the diversity and heterogeneity of the Wyoming juniper type, game range improvement or restoration practices must be planned on a site specific basis. 3) Natural burn sites and prescribed burning offer the best opportunity for habitat restoration and revegetation practices. 4) Picloram control of juniper was acceptable with black sagebrush understory. 5) Perennial grasses increased significantly on burn and spray treatments. Doze areas promoted accelerated annual grass and forb establishment, and soil erosion. Dozing was also the most detrimental treatment to perennial forbs. 6) Areas where annual grasses and forbs were dominant constituents of the understory composition should be avoided. Any treatment will revert the successional trend further to an annual grass-forb sere which is not beneficial to deer or elk. 7) Prescribed burning of cheatgrass brome before seed drop may increase the success rate for seedling establishment. 8) In regard to juniper reinvasion, the burn treatment was most effective followed by the spray treatment. Juniper re-invasion and establishment was prominent on the bulldozed areas and were mature in many instances after a 10 year period. 9) Species for revegetation purposes in the Wyoming juniper type should be treated on an individual basis since their optimum season and method of seed application differ. Separate spring and fall application should be adopted instead of one major seed mixture applied at one time. 10) With competition removal including juniper and annuals, the following were readily established: big sagebrush, silver sagebrush, antelope bitterbrush, fourwing saltbush, winterfat, black sagebrush, Mexican cliff rose, cicer milkvetch, mammoth wildrye *(Elymus giganteus)*, Palmer penstemon. 11) Specific information dealing with size, length, and spacing of juniper removal areas on big game ranges and the pinyon-juniper type were applicable to the Wyoming juniper type. Large areas of juniper removal should be strictly avoided as this practice would ultimately be detrimental to wildlife, livestock, people, aesthetics, soils and watersheds.

2C 810605 Young II, L. W. 1981. Correlation of soil factors and environmental conditions with phenophase development of western wheatgrass. M.S. Thesis. University of Wyoming. 65 p.

2A
1E2
2E
2F2
3A3

The growth of western wheatgrass *(Agropyron smithii)* was observed in the north central and southern basins of Wyoming to relate phenostage development with influencing soil and climatic environmental factors. The environmental variables used in the study were: available soil water in the 0-15 cm and 15-30 cm profile, soil temperature at 38 cm, average and cumulative maximum air temperature. Measurements for these variables were recorded during the inclusive phenology stage phases described and identified as: 1) initiation of growth, 2) mature vegetative growth, 3) vegetative regreening, 4) vegetative browning, 5) initiation of reproduction, 6) seed formation and 7) seed maturity.

Discriminate analysis was used to identify the climate and soil variables which had the greatest significance upon progression of phenology and upon individual phenological stages. By identifying the significant variables, threshold values for available soil water and temperature can be related to the phenology stages, and parameters can be set for these environmental factors.

The conditions suitable for initiation of growth showed a mean value of 0.83 cm of available soil water in the 0-15 cm profile coupled with an average air temperature of 1.1 ^{0}C for five successive days. Available cm of soil water was the amount of water measured in cm for a given depth of a profile which was available for plant growth above a base value, which was the water potential of the soil at -15 bars. During the vegetative phase, soil moisture could be expected to decrease as air temperatures increased. The amount of available soil water in the 0-15 cm profile present at the end of the fourth leaf stage of development was critical for determining whether the plant remained in the vegetative stage or developed reproductive parts. The plants which developed reproductive parts were associated with mean value for available soil moisture of 1.03 cm, while those which remained in the vegetative state showed a mean value of 0.53 cm of available soil water.

The progression of phenology was most accurately described by the cumulative sum of daily temperatures above 1.1 °C. Variation and time periods between related phases of phenology from one year to the next were explained by the number of days required to reach the same temperature sum. During the reproductive phase only one of the values for either soil temperature or average air temperature were important for describing reproductive phenology. The analysis of the environmental variables and their influence upon describing separate stages of phenology and phenological progression showed that a moisture term coupled with a temperature term was highly effective.

Reproduction was predicted by classifying seven observed cases in which the plants remained in the vegetative state, and five cases in which the plant showed initiation of reproductive parts. Based upon the values of the environmental variables, 80 percent of the cases which showed reproduction were correctly classified and 57 percent of the cases which did not show initiation of reproductive parts were correctly classified. This led to an overall correct prediction for reproduction of 66 percent. This could be improved with greater emphasis on available soil water in the 0-15 cm profile.

The environmental variables of available soil water in the 0-15 cm and 15-30 cm profile, soil temperature and sums of maximum cumulative air temperature were effective for describing the phenological phases of growth in western wheatgrass. In terms of management making decisions, threshold values for these variables have been identified. The classification procedures of the discriminate analysis can be used for predicting reproduction probability. The results from this study have proven that measurements of a few environmental factors such as soil moisture, soil temperature and air temperature can be useful for making decisions for the use of western wheatgrass in reclamation and range management practices.

8201 Research Activities

3A3 820101 Exclosure study: production, cover and precipitation.

5A1D 5B 5C 2A 2C

Activities of the aridland shrub ecology production studies during 1982 were essentially the same as previous years with production, cover and precipitation being recorded at all of the study sites. The early standing crop was again obtained at the 12 intensive phenology study locations and added to the values obtained for the later total standing crop.

1E2 820102 Exclosure study: phenology and growth.

2E 2F2 3A2 1E1 5B

During 1982 the phenology and growth data were obtained on fewer sampling dates than during 1981. But activities and procedures were essentially the same as those for previous years.

1E2 820103 Exclosure study: productivity/standing crop sampling.

10D1 2E 2F2 3A2 1E1 2A 5A1D

The clipping of productivity weight standing crop units was again obtained whenever intensive phenology data recording was conducted at the 12 study locations in the project area.

2F2 820104 Exclosure study: woody plant density.

10D 2E 1E2

Woody plant density was incorporated into determining shrub herbage production by identifying the number of plant units rooted within the 4 ft x 5 ft quadrats utilized to determine weight unit elements for each of the quadrats.

2A 820105 Precipitation pattern study.

3A

Precipitation data were again recorded on the 200 UW/BLM rain gauges distributed primarily in the Big Horn and Wind River basins.

8202 Journal Publications

1E2 820201 Fisser, H.G. and N.E. Hargis. 1982. Numerical phenology of big and black sagebrush from visual description statements. J. Colo.-Wyo. Acad. Sci. 14(1):39.

2E 2F2 11A 11B 11C 11E 2C 2A 11F2 3A2 1E1

Detailed phenological scoring has been developed for six vegetative and nine reproductive stages of big and black sagebrush (*Artemisia tridentata* and *A. nova)*. The system utilized several scoring systems described in literature and was modified with additions developed from seven years of field study in western Wyoming.

Definitive expression of plant growth stages was enhanced with short descriptive statements describing visual appearance of both vegetative and reproductive structures through intermediate as well as major growth stages. Phenological development was a continuous process. But it could be characterized with a series of steps, sometimes artificial, for expressing biological relationship with environmental functions.

The six major vegetative stages were complemented with an additional 17 intermediate statements that assisted, by the use of digital units, with the quantification of the growth process. The nine major reproductive stages were complemented with 18 additional descriptive statements. No attempt was made to assign descriptive statements to every possible fraction. Statements were developed to describe easily recognizable characteristics of the plants which permitted untrained observers to use the scoring system. The phenological scoring system was a useful tool for management and use interpretation.

2F2 820202 Taha, F.K., H.G. Fisser and R.E. Ries. 1982. A modified 100-point frame for repetition inventory. J. Range Manage. 36(1):124-125.

3A1
3A2
2F1
1E2
10D

A modified point-frame was developed for vegetation sampling on a short-grass prairie ecosystem in eastern Wyoming. This sampling device recorded 100 point-hits within a 30 cm x 60 cm quadrat at a given sampling location. This technique provided quantitative data sensitive to the general ground cover for vegetative changes over time for each quadrat.

Field experience demonstrated that the point-frame was easily used with two crew members, one to read the pin hits and the other to record the information. Once the field crew were familiar with the technique, they were able to record 7500-8000 points per eight hour day. Vegetation inventory was conducted on a range ecosystem that covered 3100 ha and was comprised mainly of rough breaks, level upland sagebrush *(Artemisia)*-grass ephemeral clay lake bottoms and moist stream channel areas.

8203 Presentation

1E2 820301 "Numerical phenology of big and black sagebrush from visual interpretation statements." April 25, 1982. Colorado-Wyoming Academy of Science. Fort Collins, Colorado. by H.G. Fisser.

2E2
2F
11A
11B

See 820201.

11C
11E
2C
2A
11F2
3A2
1E1

3A2 820302 "Vegetation productivity and phenology in the Wind River Basin." May 27, 1982. 11th Wyoming Shrub Ecology Workshop. Lander, Wyoming. by H.G. Fisser.

1E2
10D
2E
2F2
1E1
2A
5A1D

Cumulative productivity on the basis of multiple clippings of a single species during a growing season was useful for extrapolation to community total standing crop dynamics. With two estimates of community standing crop at three sites in the Wind River Basin, the interpolated productivity elements of western wheatgrass *(Agropyron smithii)* showed logical cumulative biomass dynamics. These were representative of growing condition differences among the three sites during 1980. The peak standing crop was very early compared to unusual weather and plant phenodynamic progressions. This would not have been evident from the usual standing crop estimate in July or August.

At the time of early production estimates on May 22, 1980, mean standing crop (357 lb./acre) of the three sites was 78 percent of the total biomass attained on the summer clipping dates (427 lb./acre) during August (SH, August 15; SW, August 2; UG, August 1). However, by extrapolating to the date of greatest western wheatgrass culm weight attained in late June, it was apparent that maximum total community standing crop was much nearer 560 lb./acre, some 35 days earlier than usual. This latter extrapolation estimate was 24 percent and 32 percent greater than the summer and spring estimates, respectively.

With moderate effort related to environmental monitoring and multiple sampling procedures, it was apparent that multiple sampling during a growing season enhanced ability to estimate community biomass productivity estimates.

Precipitation, temperature, phenodynamics at approximately 14 date intervals, spring and summer estimates of total standing biomass and individual culm weights of western wheatgrass were examined from three research sites in the Wind River Basin. Productivity of total community biomass estimates were developed based on a time efficient procedure, simply clipping 20 culms of western wheatgrass. With the culm weight productivity data, it was shown that early warm weather caused an occurrence of maximum standing crop 35 days earlier than usual during 1980. These procedures provided a means for community productivity estimation with very modest time and manpower requirements. Temporal distribution of precipitation and temperature dynamics were examined and found to support vegetation and biomass dynamics.

5A1D820303 "Vegetation responses following strip chemical control of sagebrush at the Lower
3A2 Government Draw site." May 28, 1982. 11th Wyoming Shrub Ecology Field Tour,
3A3 Lander, Wyoming. by H.G. Fisser.
2A

2C The exclosure construction and chemical shrub control spray application at
2E the Lower Goverment Draw was applied during 1958. Chemical control was
2F2 excellent with almost 100 percent kill of the shrubs in the strip areas. Very little drift
5B4 occurred and the areas between the sprayed strips exhibited little harm to any of the
10D shrubs.

At the time of the field tour in May 1982, it was evident that the big sagebrush *(Artemisia tridentata)* control inside the exclosure was producing much greater grass production in the spray area than the non-sprayed area. The Lower Government Draw area was somewhat different from other sites because of the relatively large amount of sand in the soil profile. This soil characteristic, somewhat uncommon for a bottomland area but representative of run-on situations, was responsible for the abundant needleandthread *(Stipa comata)* occurrence on this site, as well as the cheatgrass *(Bromus tectorum)*.

During the 21 years of herbage production studies on the Lower Government Draw exclosure, it was apparent that the sagebrush spray area protected from grazing had responded significantly with grass production increase. The spray area outside the exclosure had increased somewhat and produced considerably greater forage production than the non-spray area inside the exclosure. The non-spray area outside the exclosure consistently produced the least amount of forage production. The area outside the exclosure had also been consistently utilized by summer cattle grazing which tended to use this area a short distance from an important water source.

Due to intensive livestock use, sagebrush invasion near the exclosure developed rapidly during the last eight years. During the first 10 years following chemical application, the perennial grass species of western wheatgrass and needleandthead among others were able to inhibit sagebrush establishment in the spray area. During the last eight years, intensive livestock use reduced the vigor of the grasses and that, in conjunction with good growing seasons for sagebrush establishment, had resulted in dense sagebrush establishment within 100-150 yards of the exclosure. As was noted from visual inspection of the area to the south and southeast, distances farther from the exclosure and areas which were not intensively utilized by the animals in their travels to and from the water source, had been able to exclude sagebrush establishment to a high degree.

5A1D820304 "Vegetation response to strip chemical control in the Government Draw area." May 27,
3A2 1982. 11th Wyoming Shrub Ecology Workshop. Lander, Wyoming. by S.K. Dean.
3A3

2A The Upper Government Draw exclosure was located in west central Wyoming,
2C approximately 10 miles northeast of Lander, at an elevation of 6,080 feet. The exclosure
2E was on a gentle south facing slope dominated by big sagebrush *(Artemisia tridentata)*,
2F2 western wheatgrass*(Agropyron smithii)*, Prairie junegrass*(Koeleria cristata)*, Sandberg
5B4 bluegrass*(Poa secunda)*, and needleandthread *(Stipa comata)*. Most of the predominant
10D

forbs are members of the Asteraceae family such as common salsify *(Tragopogon dubius)*, hawksbeard *(Crepis acuminata)* and low fleabane *(Erigeron pumilus)*. Annual forbs and grasses are also present. Soils were moderately fine textured and somewhat calcareous. The mean annual precipitation was approximately 9 inches with the majority falling as snow in the winter and spring.

The two chemical spray strip areas of the site were 273 feet wide and were applied in 1958. The exclosure was constructed in 1959 with the subsequent fencing unit constructed in 1961 to protect some of the seeding application trials. Inside the exclosure on the non-sprayed side, production had increased slowly over the past 20 year period (by 150 lb./acre). Prairie junegrass and western wheatgrass accounted for the majority of the increased production. The sprayed area out-produced the non-sprayed by a long-term average of 72 percent. Production increased during the last five year interval in both the sprayed and non-sprayed treatments coinciding with an increase in annual precipitation. Both treatments inside the exclosure were in an upward trend in terms of production. Grazing at the exclosure had been moderate. The sprayed area out-produced the non-sprayed by a long-term average of 20 percent. Trend in terms of production had been the same for both the sprayed and non-sprayed treatments. Current production trend was static outside the exclosure. The average herbaceous production at both treatments had not varied more than 10 lb./acre in the previous five years.

Based on a 20-year analysis of annual herbaceous production data, sprayed treatments still produced increased forage production compared to the non-sprayed treatment. Peak production in the sprayed areas occurred about 10 years following application of chemical in both the grazed and non-grazed cases.

Production increased inside the exclosure, but it was static outside the exclosure for both the sprayed and non-sprayed treatments. Production for all four treatments had increased over the past five years, probably because of an inch or more of precipitation above the long-term average of the previous 10 years.

1E2 820305 "Numerical phenology scoring and use with quantitative variables." May 27, 1982. 11th Wyoming Shrub Ecology Workshop. Lander, Wyoming. by R.J. Kinucan.
2E
2F2
11A
11B
11C
11E
2C
2A
11F2
3A2
1E1

Vegetative and reproductive phenological progression of big and black sagebrush *(Artemisia tridentata* and *A. nova)* were quantified with a numerical scoring system. Phases of vegetative development were identified with numerals 1.0-8.0. Decimal units were specified for a number of intermediate, recognizable subphases of herbage growth appearance. The numerals 8.0-18.0 identified phasic components of reproductive development, and also included subphase differentiation with decimal values. These definitive expressions of phenologic progression and sequential numerical identification units enhanced interpretation of plant development and environmental relationships.

5A1D820306 "Sagebrush and perennial grass dynamics at the Granite Mountain shrub control site." May 27, 1982. 11th Wyoming Shrub Ecology Workshop. Lander, Wyoming. by J.L. Mulholland.
3A2
3A3
2A
2C
2E
2F2
5B4
10D

The Granite Mountain exclosure, 30 miles east of Lander, was established in 1962 to monitor plant response to big sagebrush *(Artemisia tridentata)* control with 2,4-D and associated changes that might result from exclusion of livestock from within the exclosure. Total grass and forb production increased substantially for 10 years in response to the chemical treatment, and then began to decline. This decline was caused by the incomplete kill of sagebrush with an associated earlier than usual reinvasion of shrubs, combined with an intensive summer cattle grazing which severely restricted perennial grass and forb development, but which tended to enhance sagebrush establishment.

Without the annual monitoring of vegetation production, the interactive influences of poor sagebrush control and intensive livestock stress would have, at best, been more difficult to detect.

8204 Published Proceedings/Abstracts

4B 820401 Fisser, H.G. 1982. Vegetation response to grazing management strategies on semiarid shrublands of Wyoming. p. 162-165. In D.D. Briske and M.M. Kothman (eds.). Proc. Natl. Conf. on Grazing Techniques. Dept. Range Sci. Texas A & M University. College Station, Texas. November 10-13, 1982.

4C
1E4
2F2
7A1
7A3
6A3

This report concerned the long-term response of the grazing management research which examined actual ranching operations on semiarid rangeland dominated chiefly by big sagebrush *(Artemisia tridentata)* and Utah juniper *(Juniperus osteosperma)* in the area near Baggs, Wyoming. Total grass cover had shown a steady increase on all systems. Forb cover had shown no trend and shrub cover had declined from 1975-79, the result of sagebrush death during the severe winter of 1978-79. Cover changes of western wheatgrass *(Agropyron smithii)* were not statistically significant within individual pastures, grazing systems, or even when combined among all three management units. However, the combined values of western wheatgrass, needleandthread *(Stipa comata)*, Indian ricegrass *(Oryzopsis hymenoides)* and bluebunch wheatgrass *(Agropyron spicatum)* increased significantly from 1967-1979 on each grazing system.

Comparison among grazing systems of total perennial grass cover over all sites indicated no significant differences among management units. Further analysis of cover data stratified into vegetatively stable, shallow, juniper and sagebrush sites and the more productive and dynamic sagebrush/grass sites also indicated no differences among grazing regimes for either of these site types.

Percent frequency of the key species western wheatgrass remained unchanged or decreased slightly while needleandthread, Indian ricegrass, and even bluebunch wheatgrass increased on most transects. The total perennial grass cover gains thus were due mainly to increased needleandthread, bluebunch wheatgrass and Indian ricegrass. The 1979 data indicated significant increase of total perennial grasses inside the exclosure at only three of the 10 transect pairs. Western wheatgrass exhibited similar increases at only two. Exclosures showing least significances were located on the most xeric and unproductive sites. Juniper response was insignificant. Black sagebrush *(Artemisia nova)* and big sagebrush declined significantly due to environmentally induced winter kill.

Antelope, deer and sage grouse are economically important, primarily for revenue realized from sale of permits and costs of living expenditures by hunters while in the area. Rangeland treatments likely to deteriorate wildlife habitat were investigated prior to implementation. The pronghorn, and even the birds in this case, had consistently utilized the chemical shrub control areas. Water development had significantly modified wildlife distribution patterns. New fencing restricted antelope behavior for only a short time. Because of carry-over from the critical 1978-79, winter when more than 1,500 antelope, some six to eight times higher than usual, were forced to the area, counts in 1980 showed an almost tripling of summer antelope numbers. Deer losses (not antelope) were great during that time and they had not recovered.

Range recovery on all units had not reached more than high fair, and in some cases, low good, ecological condition classification. A combination of vegetation response to range improvements, favorable moisture during early years of the study and reduced stocking intensity of sheep and cattle brought uniform response to the influence of the three grazing systems. Since limited condition classification on most sites was caused by abundance of shrubs, future changes of both shrub cover and condition classification would be closely interrelated. Responses to specific grazing systems may then be expected to become apparent.

5A1D 820402 Dean, S.K. 1982. Vegetation response to strip chemical control in the Government Draw
3A2 Area. p. 35-36. In: Fisser, H.G. and K.L. Johnson, eds. 1984 Proceedings of the 11th
3A3 Wyoming Shrub Ecology Workshop. Lander, Wyoming. May 27-28, 1982. University
2A of Wyoming, Laramie.
2C
2E See 820304.
2F2
5B4
10D

1E2 820403 Kinucan, R.J. 1982. Numerical phenology scoring and use with quantitative
2E environmental variables. p. 36-39. In: Fisser, H.G. and K.L. Johnson eds. 1984.
2F2 Proceedings of the 11th Wyoming Shrub Ecology Workshop. Lander, Wyoming.
11A May 27-28, 1982. University of Wyoming, Laramie.
11B
11C See 820305.
11E
2C
2A
11F2
3A2
1E1

3A2 820404 Fisser, H.G. 1982. Vegetation productivity and phenology in the Wind River Basin. p.
1E2 42-45. In: Fisser, H.G. and K.L. Johnson, eds. 1984. Proceedings of the 11th
10D Wyoming Shrub Ecology Workshop. Lander, Wyoming. May 27-28, 1982. University
2E of Wyoming, Laramie.
2F2
1E1 See 820304.
2A
5A1D

5A1D 820405 Mulholland, J.L. 1982. Sagebrush and perennial grass dynamics at the Granite Mountain
3A2 shrub control site. p. 39-42. In: Fisser, H.G. and K.L. Johnson, eds. 1984.
3A3 Proceedings of the 11th Wyoming Shrub Ecology Workshop. Lander, Wyoming.
2A May 27-28, 1982. University of Wyoming, Laramie.
2C
2E See 820306.
2F2
5B4
10D

8205 Wyoming Agricultural Experiment Station Publications

820501 Fisser, H.G., S.K. Dean, T.L. Deboodt, R.J. Kinucan and J.L. Mulholland. 1982. Phenology and production studies on semiarid shrub types. 1981 results. Univ. of Wyo. Coop. Res. Rep. to the BLM. Vol. I: Phenology study sections I and II. Wyo. Agr. Exp. Sta. Sci. Rep. 1199. 489 p.

1E2 82050101 Section I - Phenology study.
2E
2F2 The phenology study report for the 1981 data, including 12 sampling dates at 12
3A2 intensive phenological study exclosures, was printed in the same format as previous
1E1 reports. Tabular presentation of data, primarily from computer output for each of the individual recordings of phenological scoring or vegetative and reproductive development for the four prime species and for all species, were given in the various tables. The list of exclosures and the prime species which occurs on each was given as well as a

comprehensive table of total plant species which were found on all of the combined 12 phenological study sites.

11F2 2E 2F2 3A2 1E1 2A 82050102 Section II - Phenology multiple regression for predictive modeling.

This was a multiple regression analysis report of phenology data compared to environmental factors for the 1981 data. This stepwise regression program was the same as that used with the 1980 data. For the most part, significant variables associated with phenological progression were those factors which tended to enhance the amount of moisture at the site, or those which tended to reduce the amount of moisture. Soil moisture values generally were beneficial during the year and tended to reduce the rate of phenological progression both for vegetative and reproductive characteristics of the prime species plant growth. Temperatures, especially during summer, were generally warm and tended to enhance progression of phenological growth.

820502 Fisser, H.G., S.K. Dean, T.L. Deboodt, R.J. Kinucan and J.L. Mulholland. 1982. Phenology and production studies on semiarid and shrub types. 1981 results. Univ. of Wyo. Coop. Res. Rep. to the BLM. Vol. II: Precipitation, productivity and production sections III, IV and V. Wyo. Agr. Exp. Sta. Sci. Rep. 1200. 292 p.

2A 3A 82050203 Section III - Precipitation pattern study.

The 1981 precipitation data were recorded in the tabular form similar to that of previous years, with the seasonal data from the 200 UW/BLM gauges compiled by county and region and the like data from U.S. Weather Service stations which was compiled in similar manner. The modification, as had been carried on for a few years, was with regard to monthly precipitation recording on the UW/BLM gauges in the Big Horn Basin. These data are also compiled on a monthly basis for the year as were the Weather Service records from that region.

In general, the precipitation during 1981 was about 10 percent greater than normal. Significant patterns of differences in amounts of precipitation were noted in the Big Horn Basin. The northern area in Park County received considerably less than normal precipitation, but other counties in the Big Horn Basin to the south received greater increases with distance southward. Fremont and Natrona counties were similar to Hot Springs and Washakie counties, while Carbon County exhibited a surprising 36 percent increase in precipitation for the year. Westward and in the southern portion of the state, precipitation was somewhat greater throughout the Sweetwater-Sublette county areas, but was significantly less than average in the Uinta and Lincoln county areas.

Precipitation patterns among seasons were presented but often were discussed very little. It may be noted that the 1981 precipitation exhibited, in the central and north central portions of the state, typically 45-50 percent of precipitation during the spring period from April 15 to July 1. In the southern and far western areas of the state, precipitation distribution usually exhibited a greater proportion of winter precipitation and large summer and fall values as well. As a result, spring period growing season precipitation in those areas was generally reduced to 30-35 percent of total annual moisture. These data have not been investigated in detail, but future analysis by computer processing should be able to exhibit significant differences that may occur. The relationship these precipitation differences may have with the general vegetation types may be closely associated with type differentiation and herbage potential differences.

A new series of sector maps showing locations of all of the gauges, both UW/BLM and Weather Service by sector areas in the state was completed and was included in the 1981 report. There were 21 sector maps which covered all of Wyoming, within which are all of the gauges reported in the tabular materials. Thus the maps extended westward from Casper and north and south from the Montana to the Colorado borders.

1E2 82050204 Section IV - Productivity-phenology interrelation study.
10D1
2E The productivity clippings, as during 1980, were obtained at approximately 2-week
2F2 intervals during the majority of the growing season, beginning in late April and ending
3A2 in late September or early October. Data were obtained at least 10 times on most
1E1 locations. The tabular material presented the complete listing of each individual culm or
2A bud weight unit in mg for each sample date for each of the four prime species at each of
5A1D the 12 exclosures. In some cases, the 20 vegetative culm weights are separated from the reproductive culm weights. Mean weight for each date and phenological development at that time were presented in the tabular report. Growth increments for each time period were calculated, as well.

3A3 82050205 Section V - Exclosure study: production, cover and precipitation.
5A1D
5B The herbage production data for 1981 were presented in a format similar to that of
5C previous years with data given for the 12 exclosure sites which were clipped early, from
2A late May to early June, as well as all of the clipping data for summer data collection and
2C fall field activities. Herbage production values varied greatly from one region of the state to another. Precipitation in southwestern Wyoming was much below the long-term average, but significant summer and fall precipitation was beneficial to grass and shrub production. In some instances, such as at Cumberland exclosure -1 and at the Farson exclosure, grass production was marginal while shrub production was much above normal.

The Wind River Basin inclement weather, with dry and cool early spring conditions, severely inhibited plant development. Although excellent moisture and warm temperatures developed later in June, most sites showed less than normal perennial grass production and only moderate shrub production.

Spring season moisture during 1981 in the Big Horn Basin was much above normal at almost all locations. This was followed with moderate summer temperatures and frequent rain showers which resulted in considerably greater amounts of perennial grass production than during the previous year.

4B 820503 Fisser, H.G. 1982. Vegetation response to grazing management strategies on semiarid
4C shrublands of Wyoming. Wyo. Agr. Exp. Sta. Mimeo. prepared for Grazing Systems
1E4 Conference, Texas A & M University. November 1982. 4 p.
2F2
7A1 See 820401.
7A3
6A3

8206 Graduate Advisee Completions

2A 820601 Elwakeel, A.S. 1982. Effects of environmental factors on perennial grass production in
1E2 five vegetation types in north central Wyoming. M.S. Thesis. University of
2C Wyoming. 81 p.
2E
2F2 This study was conducted on five sites in north central Wyoming, compared to
3A2 production of perennial grasses in five different vegetation types and production inside
3A3 and outside of exclosures. The primary objective was to examine the effects of temperature and precipitation on the annual production of the perennial grasses in each site.

The study sites were located in the Big Horn Basin. Wheatgrass (*Agropyron* spp), big sagebrush (*Artemisia tridentata*), blue grama (*Bouteloua gracilis*), greasewood (*Sarcobatus vermiculatus*) and saltbush (*Atriplex gardneri*) vegetation types were studied. Data included 19-year long-term production, temperature and precipitation sets for study of general relationships. Intensive analysis was applied to the 5-year data sets from 1976-1980.

Herbage production was studied in relation to grazing, temperature, precipitation and their interactions. Comparisons of the perennial grass production and related effects of climatic factors were studied with analysis of variance and regression analysis. Sites with greatest perennial grass production inside the exclosures of the 5-year period, were the blue grama and greasewood types with 365 and 343 lb./acre, respectively. The sagebrush type yielded the least perennial grass production with only 100 lb./acre. Wheatgrass and saltbush types produced equal amounts of perennial grass with 265 lb./acre.

The blue grama vegetation type was the highest yielding type as far as perennial grasses were concerned. There was light grazing around the site, having no water available at the vicinity. Production increased with temperature and precipitation while cool season species were negatively affected.

The sites with salt-affected soils, such as the greasewood and saltbush types, showed an increasing production pattern through time with a relative decrease in temperature and increase in precipitation. The site supporting cool season grasses responded positively to lower temperature and to increased precipitation, probably as a result of reduced salinity. Grazing was light around the greasewood site, but was heavy around the saltbush site.

For the wheatgrass type, production during the 5-year period declined comparatively to the 19-year long-term production average. The decrease in production on this site accompanied the drop in temperature below optimum growing conditions during the last 5-year period. Heavy grazing around the exclosure showed a contrast between the production inside the exclosure and outside. The effect of increased precipitation was masked by the negative effect of below optimum temperatures prevailing during the 5-year period.

The sagebrush type was the lower most perennial grass production among the five vegetation types for both 19-year and 5-year terms. That was attributed to sagebrush invasion and competition through time. Sagebrush contributed in rendering the amount of moisture less available to grasses through competition and interception. Higher temperatures during the last five years were not beneficial to the cool season grasses in the site. Production increased outside the exclosure because of the very light and even, at times, absence of grazing around the exclosure.

Higher temperatures, accompanied by light grazing and adequate amounts of precipitation, were the best conditions for highest perennial grass production in the blue grama vegetation types. Low temperatures with great amounts of precipitation were the conditions of obtaining high yields in the saline sites of greasewood and saltbush. Low temperatures with adequate amounts of precipitation could achieve the highest possible production in the wheatgrass sites. Low temperatures with control or removal of sagebrush could increase perennial grass production in the sagebrush site.

More accurate predictions about annual herbage production could be made if data on evaporation, evapotransporation and soil water holding capacity were available.

2B 820602 Uhlich, J.W. 1982. Vegetation dynamics of non-grazed and grazed sagebrush grassland communities in the Big Horn Basin of Wyoming. M.S. Thesis. University of Wyoming. 120 p.
2F2
2A
2C
2E
3A2
10D

The data from this thesis were obtained from permanently-located transects of long-term studies with initial data obtained in 1959. This thesis reports the vegetation development on six sites established during the study period. All six sites were located within the big sagebrush *(Artemisia tridentata)*/bluebunch wheatgrass *(Agropyron spicatum)* habitat site on homogeneous, topoedaphic sites. These sites were all found on level to easterly-sloping plateaus at an approximate elevation of 1,500 meters. Soils were Argid Aridisols, moderately deep, calcic and of a loamy texture. Average annual precipitation throughout the sites ranged from 21 cm, at the northwest section of the study area, to 28 cm at the southern and eastern portions. Five of the sites contained both a stand excluded from livestock utilization and an adjacent stand which was annually grazed by livestock. The sixth site, a relic area, contained only a non-grazed stand. The livestock exclosures had been free from grazing for approximately 20 years - 1960-1980. The relic stand had never been extensively grazed by livestock. Vegetation composition of big sagebrush and the understory herbaceous vegetation varied greatly among the sites and, in some cases, among the adjacent stands.

Several vegetation sampling designs were originally used to record changes in the vegetation. The annual sampling design measured cover in herbaceous standing crop weight on the non-grazed and relic sites during the period 1963-1980, and on adjacent grazed stands beginning in 1965 and lasting through 1980. Only cover data of big sagebrush, plains pricklypear *(Opuntia polyacantha)* and Hood's phlox *(Phlox hoodii)* weight data of the herbaceous understory were used in this study.

A long-term permanent plot sampling design measured cover data for big sagebrush, plains pricklypear, Hood's phlox and the perennial grasses for the Demer and Buffalo Creek sites. These data were collected in 1959, 1964, 1968, 1972, 1976 and 1980. The original sampling design placed definite limits on the analysis and conclusions that could be made.

In order to interpret vegetative change the vegetation stands needed to be characterized for their composition and abundance. Canopy cover of big sagebrush was found to vary from 4-16 percent throughout the study sites. Composition of the herbaceous understory differed between stands from one dominated by bluebunch wheatgrass to one dominated by western wheatgrass *(Agropyron smithii)*. Average standing crop biomass for the sites varied from 138 to 641 kg/ha throughout the sites.

A relationship of precipitation (P) and canopy cover of big sagebrush (CC) to average standing crop biomass (B) of the herbaceous understory component of $B = -499.0 - 24.7\ CC + 40.0\ P$ characterized the 11 stands. This relationship reiterated the concept that productivity on loamy sites was highly correlated to the amount of big sagebrush found in the stand and to the precipitation regime.

Vegetation changes were found to be highly variable among stands. Most notable was increased big sagebrush abundance. This phenomenon occurred both from an increase in density and from an increase in the size of the existing individuals. Understory changes were apparently controlled by the increase in density of big sagebrush but less so by the increases in the size of the plants. In stands where changes in big sagebrush did not occur, several patterns of variation occurred in the herbaceous understory. Some stands showed little absolute change in the average standing crop biomass, but relative weight changes and composition were found. Other stands, especially those that were grazed and where the canopy cover of the big sagebrush was relatively low, exhibited increases in the average standing crop biomass.

The inconsistent nature of the vegetation development of the stands indicated that trends of vegetation change were not always uniformly directional or predictable according to classical successional theory. The resulting patterns of change were best interpreted by the "initial floristic/relative stability" hypothesis of vegetation development. This model predicts that many different assemblages of the same species could form relatively stable communities on a given site. The relative abundance of the component species would depend largely on the nature of past disturbance and the vegetation composition at the time of disturbance.

The sagebrush ecosystem offers a tremendous resource for red meat production, wildlife, water, recreation and aesthetics. To efficiently develop this resource to its potential the ecosystem must be properly managed toward the goals of increased production of forage, livestock, wildlife, improved water quality, enhanced recreation opportunities and towards improvements in maintenance of aesthetic experiences. To accomplish these goals, the vegetation composition of the ecosystem must be managed. Unfortunately, present concepts of succession, as interpreted by the "range condition classification" method, do not provide a viable method to determine rates and directions of vegetation development.

The concept of "initial floristic/relative stability" may prove to be more useful, as an aid to managers, for interpreting vegetal patterns (Anderson & Holte, 1981). This model accepts that range improvements other than changes in grazing or rest will need to be found in order to manipulate vegetation development towards the climax state.

8301 Research Activities

2A 830101 Precipitation pattern study.
3A
11A

During 1983 the precipitation data for all gauges was recorded by the BLM as well as the University personnel at all of the locations during the April 15 precipitation pattern.. After that time, however, the University program for recording precipitation data was discontinued along with the other portions of the aridland shrub ecology field activities. Most of the Bureau of Land Management offices were sending people to record precipitation data at all of the gauges which had previously been recorded by the range management people in the shrub ecology program. The data will be compiled with the existing long-term records of precipitation for each of the exclosure and rain gauge stations as has been carried on in previous years.

11A 830102 Computer program development for long-term weather record compilation.
2A

During 1983 the Department of Range Management acquired computer tapes of long-term weather records from the National Weather Service in Ashville, North Carolina. The weather data are difficult to obtain from the NOOA tapes because of an inconsistency in which the data were entered on those tapes. A great deal of time was taken during the entire year to decipher some of the coding systems utilized on those tapes.

11A 830103 Computer program development for estimation of missing climate data.
2A

Because missing climatic data often occurred, a definitive analysis was conducted by the University program, especially by Larry Germain, to determine best ways for identifying objective estimations for missing data. He compiled a considerable record of information that showed that the kriging process for temporal and spatial estimations was the most accurate, and also allowed determination of statistical error associated with the estimation of a missing value. Unfortunately, this program required extensive computer time and accuracy was not significantly greater than that obtained with the reciprocal of the distance squared procedure. The latter was utilized for later estimates.

11E3 830104 Compilation of herbage production and cover data from Wind River Basin exclosures.
3A3
11F1

The herbage production and cover data from the Wind River Basin exclosures has been proposed for use by the Bureau of Land Management personnel in the Lander Resource Area office.

11C 830105 Summary of soil moisture and temperature data from Granite exclosure.
2A
2C

Soil moisture and temperature data were being obtained at several exclosures in the Wind River Basin by the Lander BLM personnel and in conjunction with activities associated with Roger Mertens of the Wyoming BLM State office.

11C 830106 Erosion transect data summarization and input to computer.
8B
2C

While at the Northwest Watershed Research Center in Boise, Idaho, the erosion transect data were summarized and compiled by Fisser with programs originally developed by Dr. Clifton Johnson of the Research Center. These data had been compiled, were on computer file and the graphic display program that he had developed for expression of those data was exceptionally valuable.

2F 830107 Wyoming rangeland reference sites.
3A
3B
3C
11E1
11E2
11E3

Very little work was completed on the Wyoming Rangeland reference work during 1983. Some activity in description of exclosures was conducted in the Big Horn Basin by Gary Blincow of the BLM.

11E2 830108 Yield prediction from ERHYM model.
11E3
2A
2C
3A2
11F1
11F2

Yield prediction from the ERHYM model was applied to data from the McGraw Flat exclosure with interpolation to climatic records at Lander. Although Lander was some distance from the McGraw Flat exclosure, it was the earliest data set which had reached total compilation of daily precipitation and temperature values for long-term record. The output of this program, with interpolation of the seasonal precipitation recorded at the McGraw Flat exclosure to a daily basis related to the occurrence of precipitation and proportional amounts of precipitation in the daily records at Lander, gave adequate results as a starting point for use of the ERHYM model. It appeared that this model should be very applicable to our range sites in Wyoming, and that it required only the completion of daily precipitation and temperature files for the climatic records to be able to continue with greater analysis during 1984.

11A 830109 Compilation of U.S. Weather Service precipitation records - Wyoming.
2A

The compilation of the U.S. Weather Service precipitation records for Wyoming proceeded fairly well except that there were so many large areas and many of the stations such as at the Oregon Trail Crossing station, Jeffrey City, Sand Draw, Muddy Gap, Gas Hills and others just in the Wind River Basin for which many missing days, often 2-3 days a week during entire years, and often entire blocks of 2, 3 or 4 years at a time were missing. The needs of the project were a completed compiled file of daily precipitation and daily maximum and minimum temperatures for each station. If data for precipitation or temperature were not obtained for a specific Weather Service station or exclosure location, then the values were interpolated from best values from the nearest recording stations which appeared to be most ecologically similar.

8303 Presentations

5A1D830301 "Long-term vegetation responses to sagebrush control in western Wyoming." February 15, 1983. 36th Ann. Meeting Soc. Range Manage. Albuquerque, New Mexico. by S.K. Dean.
3A2
3A3
2A
2C
2E
2F2
5B4
10D

Big sagebrush *(Artemisia tridentata)* in six diverse areas of western Wyoming was controlled with 2,4-D between 1958 and 1962. Exclosures were constructed to protect both shrub treated and non-treated areas from livestock grazing. Long-term records of herbage production, cover, and grazing management had been analyzed to compare the vegetation responses on sagebrush controlled sites to those not subjected to the shrub treatment.

Comparative responses of vegetation to different levels of grazing on shrub treated and non-treated sites had been completed. Combined livestock and wildlife grazing often modified responses of forage species, including sagebrush, which caused confounding to subsequent grazing management implications.

Sagebrush cover and density were still restricted on all treated sites even 20 years following control. Grass and forb production maintained a positive response to the shrub control treatment. Cheatgrass *(Bromus tectorum)* became a major part of the treated community and perennial grass production peaked from 5 to 10 years after 2,4-D application in both grazed and ungrazed sites. The magnitude of the production increase was dependent upon the initial sagebrush control, precipitation and the rain site potential.

4B 9A 3A3 3A2 6A 5A1D 11E2

830302 "Economic feasibility of resting rangeland to improve forage production in central Wyoming." February 16, 1983. 36th Ann. Meeting Soc. Range Manage. Albuquerque, New Mexico. by T.L. Deboodt.

Resting rangeland from forage utilization by livestock has been commonly practiced to improve ecological condition and herbage production. This study was developed to evaluate the potential forage increase and economic feasability relationships on two similar semiarid big sagebrush *(Artemisia tridentata)*-blue grama *(Bouteloua gracilis)* sites in western Wyoming. Exclosures were constructed on the study sites in 1964 to exclude livestock grazing. Baseline data of forage production was determined within the exclosures at that time. Continued studies from 1970-1982 provided data for analysis of grazing treatment effect upon forage production dynamics.

The combination of fall/winter precipitation was more significant in explaining production dynamics than was spring/summer precipitation. Production between sites was significantly different since one had been subjected to a chemical shrub control treatment in 1970. Production of forage species increased both in and out of the exclosures but at a decreasing rate over time.

Data on resting the site was statistically inconclusive as to whether rest was a factor in the production increase. Grazing during fall and winter for 18 years positively influenced production trends and provided a significant positive internal rate of return.

2A 1E2 2C 2E 2F2 3A2 3A3

830303 "Perennial grass production dynamics among five vegetation types in north central Wyoming." February 16, 1983. 36th Ann. Meeting Soc. Range Manage. Albuquerque, New Mexico. by H.G. Fisser.

Perennial grass production dynamics were investigated at five sites, each of a different vegetation type, for 5 years and compared to a 19-year data base in the Big Horn Basin of west central Wyoming. Grass production was positively related to increased precipitation and negatively related to increased summer temperatures, but positively related to warm spring temperatures.

At the big sagebrush *(Artemisia tridentata)* site, average annual temperatures were almost 2 °C lower than the long-term standard normal. Very warm summers resulted in a production decrease of 90 kg/ha although spring precipitation was above normal.

Grass production for all sites averaged 4 kg/ha/cm precipitation. Shrub invasion accompanied warm summer temperatures and cool season grass decreases. Warm season grasses were promoted by higher summer temperatures. Perennial grass production was negatively related to intensive grazing, but responded positively to light utilization.

1E2 2E 2F2 3A2 1E1 11A 11B 11C 11E 2C 2A 11F2

830304 "Phenological description and numerical scoring of sagebrush." February 16, 1983. 36th Ann. Meeting Soc. Range Manage. Albuquerque, New Mexico. by H.G. Fisser.

Ten years of work were consolidated and from this was developed over 50 specific statements with quantitative scoring describing readily observable states of reproductive and vegetative sagebrush *(Artemisia tridentata* and *A. nova)* development. Floral bud development, for example, with a numerical score of 9.0 was further distinguished by the following, 9.4 reproductive step development, no buds evident, 9.5 buds very small, 9.6 buds swelling, 9.7 buds enlarged, green color, 9.8 faint yellow at bud apex, and 9.9 bud closed but very yellow. The value 10.0 identified bloom initiation.

The vegetative development of sagebrushes was also identified by specific stages. For example, the stage twig elongation was scored 4.0. Additional further steps were 4.1 internode barely visible between new leaves, 4.3 internode clearly visible, 4.6 two or more internodes present, 4.7 lateral branchlets on new twigs, 4.8 lateral branchlets show internodes. For similarity through extremely detailed quantitative environment data, the definitive numerical phenology scoring derived by this work enabled a much closer application of plant development relationships with finite itemization of such measures as precipitation and temperature gradation, day length and a variety of soil characteristics.

3A3 830305 "Productivity differences of grasslands in semi-desert western Wyoming — A re-examination." February 16, 1983. 36th Ann. Meeting Soc. Range Manage. Albuquerque, New Mexico. by L.A. Joyce.

1E2
2A
2E
2B

The effect of environmental differences and grazing stress were reflected in species composition and production at two blue grama *(Bouteloua gracilis)* sites in Wyoming. Exclosures were constructed in 1959-1960; and vegetation dynamics were monitored annually on grazed and non-grazed pastures at each site. The average annual precipitation at the more arid site was 13.3 cm and total standing crop averaged 295 kg/ha in the exclosure. The less arid site received 17.7 cm mean annual precipitation and total standing crop averaged 387 kg/ha in the exclosure. Although the water use efficiency was the same in the exclosures at both sites, community composition differed and species diversity was greater at the less arid site. Annual forbs dominated the forb component of the exclosure at the less arid site whereas perennial forbs dominated the forb component at the more arid site. On the grazed pastures, annual forbs dominated the forb component. Shrubs were found only on the grazed site at the more arid site and on both grazed and ungrazed at the less arid site. Also see 740301 and 740404.

1E2 830306 "Dynamics of plant community development for diverse sagebrush sites in western Wyoming." February 15, 1983. 36th Ann. Meeting Soc. Range Manage. Albuquerque, New Mexico. by R.J. Kinucan.

2A
2C

Phenology is a common rangeland research procedure. Most reports, however, focus on floral development and not vegetative growth. Phenologic development was recorded for all species at biweekly intervals during the growing season since 1973 at 12 diverse sites in western Wyoming. Many aridland forage production species, such as western and bluebunch wheatgrass *(Agropyron smithii* and *A. spicatum)*, scarlet globemallow *(Sphaeralcea coccinea)* and winterfat *(Ceratoides lanata)*, failed to flower consistently. Identification of vegetation maturation stages was derived. Phenology and synecological productivity of all species was incorporated with indices of environmental variables. Community-wide development varied widely with differing environment conditions. Spring and summer 1981 growing conditions were cool and moist. Under these conditions community phenological development continued longer than normal.

The 1982 season was dry and cool, delaying phenologic progression, but temperatures increased to normal by mid-June, causing a rapid growth of most species. Big sagebrush *(Artemisia tridentata)* grew slowly due to its inability to extract moisture from either its taproots or shallow fibrous roots. Combined synecological indices for both environmental and biotic variables identified response behavior factors.

5A1D830307 "Re-establishment rates of big sagebrush *(Artemisia tridentata)* following chemical control on grazed and ungrazed ranges in western Wyoming." February 17, 1983. 36th Ann. Meeting Soc. Range Manage. Albuquerque, New Mexico. by J.L. Mulholland.

1E1
2A
2B
2C
2E
2F2
11E

Re-establishment rates of big sagebrush *(Artemisia tridentata)* onto shrub controlled areas was determined under grazed and ungrazed conditions. Data from five study areas in western Wyoming sprayed with 2,4-D 20-25 years previously was assessed. Exclosures were constructed at the time of the chemical control treatment to exclude livestock. Density, cover and age of sagebrush were measured in each treatment. Stem cross sections of sagebrush were taken to determine age through growth ring analysis.

Favorable spring and summer precipitation, as well as intensive livestock grazing, were shown to enhance sagebrush re-establishment. The longevity of chemical treatment was significantly affected by the initial control effectiveness, which strongly influenced forage production increase. Subsequent response of all species, including sagebrush re-establishment rate, was closely related to grazing use patterns. Intensive early spring grazing caused earlier and greater re-establishment by sagebrush than did summer and fall grazing.

2E 830308 "Size and density parameters of western wheatgrass on diverse range sites in western Wyoming." February 15, 1983. 36th Ann. Meeting Soc. Range Manage. Albuquerque, New Mexico. by H.G. Fisser.

10D
1E2
11E2
2A
2C
2E
2F2

Harsh environs are usually assumed to reduce numbers of plants of a species, but inherent biological response because of lesser density is also assumed to result in larger individuals. At 11 diverse rangeland sites of western Wyoming individual culms of western wheatgrass *(Agropyron smithii)* have been harvested during the growth season at biweekly intervals for biomass and related to the density of culms at each site.

Comparative environmental parameters including precipitation, temperature, soil physical and chemical characteristics, and soil moisture and temperature were expressed as an index to the vigorousness or inherent productive potential of each site. These indices were applied to the size-density characteristics of the grass culms.

Definitive site variables which modified biotic interactions then were utilized for extrapolation to assist with determination of forage production potential by site.

2F2 830309 "Utah juniper invasion in north central Wyoming and historical climatic and management relationships." February 16, 1983. 36th Ann. Meeting Soc. Range Manage. Albuquerque, New Mexico. by H.G. Fisser.

2E
3A2
2C
2D

Successional trends in the northern range of the juniper *(Juniperus)* ecosystem, where climate extremes limited its distribution, has been little studied. In 1965, permanent plots were located along a Utah juniper *(J. osteosperma)*-black sagebrush *(Artemisia nova)*-bluebunch wheatgrass *(Agropyron spicatum)* ecotone in the foothills of the Bridger Mountains in north central Wyoming to monitor rates of juniper invasion. By 1979 no significant changes in juniper density were observed. The invasion question was then approached by examining size and age structure of stands located along a gradient. All plants were excavated from the soil in order to determine total age. Age structure analysis revealed that significant downslope invasion by Utah juniper had occurred since 1890.

During sampling many trees that appeared from the ground surface to be single, multistemmed individuals, were found to be closely clumped groups of several individuals. Thirty-seven percent of all apparent single, but multistemmed plants were in actuality composed of two or more separate individuals. Over 90 percent of the seedling to sapling age classes (0-40 yrs) were found immediately beneath the canopy of juniper or sagebrush. This growth and seedling establishment characteristic discovery may help explain juniper establishments strategies in the invasion process.

The third and ongoing phase of the study was the evaluation of hypothesized causes of juniper invasion, including a decrease in fire, a decrease in fire frequency, overgrazing and a shift in climate. Climate reconstruction was addressed by correlating annual growth ring structure of older juniper and *(Pinus flexilis)* limber pine in the area with Weather Service data from nearby Thermopolis, Wyoming.

2E 830310 "*Atriplex*/grass and forb relationships under no grazing and shifting precipitation patterns in north central Wyoming." May 6, 1983. Wildland Shrub Symp. The biology of *Atriplex* and related chenopods. USDA Forest Service. Provo, Utah. by H.G. Fisser.

2F2
2A
2C
6A2
1B4
1E1
4A
4B

The Big Horn Basin is a vast, semiarid part of north central Wyoming. Although vegetation and climate are similar to comparable areas of Idaho and Utah, west of the Continental Divide, important differences exist in temperature and moisture characteristics, including delayed initiation, but accelerated spring warm up, and moisture concentrated during the growing season.

Funding and research to combat invasion by halogeton *(Halogeton glomeratus)* were initiated in the mid 1950s. In addition to the establishment of livestock-proof exclosures throughout western Wyoming, the Dry Creek Grazing Intensity Study was initiated in 1956 to determine Gardner saltbush *(Atriplex gardneri)* response to different intensities of its winter utilization by sheep and to determine the related invasion potential of halogeton.

By 1967 it was evident that a winter utilization rate of 35 percent of annual herbage production was acceptable and actually desired in order to stimulate plant and root growth. Halogeton abundance was very specifically disturbance and grazing intensity related. By the late 1960s when the improved condition class of the grazing unit vegetation became evident, there was also a comparable halogeton reduction. Where intensive livestock disturbance continued, halogeton abundance was maintained.

The primary analytic intent of this paper was the identification and the evaluation of long-term herbage production response to precipitation dynamics in conjunction with long-term protection from the stress of livestock grazing. The exclosures were constructed on sites where the plant community appeared to have had many years of continuous overuse. Almost total elimination of perennial grasses and forbs had occurred and reduced vigor of Gardner saltbush was evident. Vegetation from 1955-1971, exhibited population changes suggestive primarily of recovery from many years of continuous overuse. During this period, grass and forb production was minimal, but increasing.

Data from 1972-78, however, portrayed increasing grass and forb production with corresponding decreasing shrub production. By 1978, after 23 years of protection, grass contributed nearly 50 percent of the standing crop. In contrast, it had contributed less than 1 percent after 7 years of protection in 1963.

The results of this study showed that the vegetation composition of three saltbush exclosures changed significantly under protection from grazing. Additionally, these changes occurred during a time when seasonal precipitation was also changing. Spring precipitation was decreasing and winter precipitation increasing. Analyses of the relationship between precipitation and standing crop, based on the entire period would have been misleading. It was demonstrated that the changing plant composition affected the relationship of total standing crop to precipitation. During the earlier period, while the total herbage production was primarily from saltbush, response to moisture in terms of lb./acre/inch of precipitation (kg/ha for 1 cm of precipitation) was minimal. During the later period, with recovery from historical overuse essentially complete and herbaceous species contributing as much as half or more of the total vegetation production, response to precipitation was much more dynamic.

Protection from grazing was only partially responsible for the vegetation trends. The shift of seasonal precipitation, from predominance during spring to greater abundance during winter, became more important through time. Interpretation of climatic vegetation relations based principally on early short-term records, would have suggested a misleading response to nongrazing. The longer term records, however, provided the opportunity to isolate early responses as recovery from historical overuse. In addition, the total data set afforded comprehensive study of vegetation changes as well as greater definitive capability for prediction of climatic/management interrelationships.

4B 830311 "Grazing, vegetation and soils interactions since 1967 in the Powder Rim grazing system."
10D1 May 26, 1983. 12th Wyoming Shrub Ecology Workshop. Baggs, Wyoming. by H.G. Fisser.
4C
1E4 The workshop was sponsored by the Wyoming Section, Society for Range
2F2 Management, and the Wyoming Chapter of The Wildlife Society. Active support of the
7A1 workshop was provided by the Wyoming Agricultural Extension Service, Wyoming
7A3 Game and Fish Department, USDA Forest Service, USDA Soil Conservation Service,
6A3 and USDI Bureau of Land Management.

Forty-nine persons registered for the 1983 workshop to consider the ecology and uses of bitterbrush *(Purshia tridentata)* and other shrub and woodland complexes in the Baggs, Wyoming vicinity. The field chairmen were David S. Moody, Biologist, Wyoming Game and Fish Department and Norman D. Vigil, District Conservationist, Soil Conservation Service. The coordinating committee acknowledged the particular support of the Wyoming Game and Fish Department in staging the 1983 event and the continuing involvement of the other support agencies.

This presentation very briefly described the Baggs grazing system study which was initiated in 1967. Vegetation data were obtained annually through 1971, then during

1975 and 1979. No significant response to grazing system or to protection from grazing was noted. Greater detail is given in items 790504, 800505 and 810603.

7A 830312 “Deer browse impacts on Utah juniper and true mountain mahogany in the Poison
10D1 Butte unit west of Baggs, Wyoming.” May 26, 1983. 12th Wyoming Shrub Ecology
4C Workshop. Baggs, Wyoming. by H.G. Fisser.
1E4

2F2 On the field tour of the 12th Wyoming Shrub Ecology Workshop held in Baggs,
7A1 Wyoming, on May 26, the tour traveled westward from Baggs to the Sand Creek area
7A3 and then southward and eastward through several pastures of the Baggs Grazing
6A3 System study as described previously in item number 830311. The last stop of the grazing system study area was located in the Poison Butte unit a few miles west of Baggs. On this upland area, the tour group traveled to the game-proof combination livestock-proof exclosure in the Utah juniper *(Juniperus osteosperma)*-true mountain mahogany *(Cercocarpus montanus)* type. Observation of the juniper and true mountain mahogany plants in the area revealed intensive use of both shrub species. The juniper bushes had been recently severely highlined in many areas. The mahogany bushes outside the exclosure have continuously been severely overused.

The Wyoming Game and Fish Department personnel indicated that during the past 2-3 years there had been significant increases of mule deer migration from the Colorado summer grazing areas from the southeast and from the Wyoming areas to the east of the Poison Butte winter range. In addition, a severe wintering condition occurred during several years such that tremendous numbers of antelope were forced onto the grazing system units for a considerable length of time. During the past winter of 1982-83, tremendous numbers of mule deer were forced to the juniper-mahogany bluffs adjacent to and to the west of Baggs. The very intensive use of the juniper was extreme and represented, obviously, tremendous overuse of this species. Many of the bushes were highlined as high as 6-7 feet with all of the green materials taken and many of the lower bushes were broken down with twigs as large as a person’s small finger eaten by the deer.

The most surprising aspect of the entire situation, though, was the extreme overuse of the mahogany outside of the exclosure. The exclosure was constructed in 1967 and in the research programs that the University carried on annually through 1971 and then at 4-year intervals in 1975 and 1979, it was readily noted that the mahogany was utilized relatively well. During the past four years the use was extremely excessive and the bushes were essentially on the verge of being completely destroyed.

A surprising response to protection from wildlife use, however, was noted from the appearance of the bushes within the exclosure. At the last time of a previous visit by Fisser to the site four years prior in 1979, there was minimal difference in appearance of the mahogany bushes within the exclosure as compared to those immediately outside. At this trip in May 1983, the shrubs within the exclosure had obviously obtained an ability to develop extensive growth representative of significant response to protection from use. The long-term protection required from 1967-1979, prior to the expression of significant growth by these plants, obviously represented a time period during which the mahogany bushes had to develop new root systems or develop greater amounts of root systems in order to be able to grow significant leaders and leaves on the protected bushes.

The leaves on bushes within the exclosures were quite large and growing rapidly. The leaves of the bushes outside the exclosures, while abundant, were extremely small. The latter situation obviously represented the inherent physiological response of the plants to the stress caused by overuse and represented the ability of the plants to respond to use. The response of the plants within the exclosures obviously represented the ability then, given time to develop adequate new root systems, for the plants to significantly gain in height, abundance of stems and biomass production.

An interesting situation was noted along one of the fence-lines where the deer had been pressing their heads through the wire as far as they could and had completely obliterated branches within their reach and had not been able to reach all of them. The stress upon the fence was significant and was noted to the accompanying Bureau of

Land Management personnel. Hopefully the fence will not be breached and this continuation of significant growth will continue.

8304 Published Proceedings

10D 2E 7A1 2A 2C 1E3 2F2 830401 Fisser, H. G. 1983. Some recent research findings in the true mountain mahogany types. p. 7-10. In H.G. Fisser and K.L. Johnson (eds.) Wyoming Shrublands: Proc. 10th Wyo. Shrub Ecology Workshop. Sybille Canyon, Wyoming. May 28-29, 1981. University of Wyoming, Laramie.

See 810313.

1B1 7A1 2E 2F2 1E2 2A 2C 2D 3A2 3A3 4B 4C 5A1C 5A1D 5A1E 5B4 5C 8C 8D 830402 Fisser, H.G. and K.L. Johnson (eds.) 1983. Wyoming Shrublands: Proc. Ninth Wyoming Shrub Ecology Workshop. June 5-6, 1980. Lander, Wyoming. University of Wyoming, Laramie. 40 p.

The 1980 Shrub Ecology Workshop was held in Lander June 5-6 and addressed the problems associated with the management and use of riparian systems. The workshop was sponsored by the Wyoming Section, Society for Range Management, and the Wyoming Chapter of the Wyoming Wildlife Society. Active support of the workshop was provided by the Wyoming Agricultural Extension Service, Wyoming Game and Fish Department, USDA Forest Service, USDA Soil Conservation Service and USDI Bureau of Land Management.

Forty-nine technicians registered for the workshop and tour to a variety of locations under the field chairmanship of Jack Welch of the Bureau of Land Mangement, Lander. The proceedings of this meeting presented the papers of a number of authors which addressed the riparian stream bank problems, willow (*Salix* spp.) bottom problems, relationships to wildlife and multiple use problems associated with public lands. The tour traveled to several locations of the area south and southeast of Lander along the face of the Wind River mountains. Grazing and wildlife management problems were discussed in some detail.

1B1 2E 7A1 7A2 7A3 8C 8D 2F2 1E2 1E4 2A 2C 2D 3A2 3A3 4B 4C 5A1C 5A1D 5A1E 5B5 5C 8C 8D 830403 Fisser, H.G. and K.L. Johnson (eds.) 1983. Wyoming Shrublands: Proc. 10th Wyoming Shrub Ecology Workshop. May 28-29, 1981. Sybille Canyon, Wyoming. University of Wyoming, Laramie. 40 p.

The workshop was sponsored by the Wyoming Section, Society for Range Management and the Wyoming Chapter of the Wyoming Wildlife Society. Active support of the workshop was provided by the Wyoming Agricultural Extension Service, Wyoming Game and Fish department, USDA Forest Service, USDA Soil Conservation Service and USDI Bureau of Land Management. Twenty-nine technicians registered for the 1981 workshop to consider the ecology, uses and management of true mountain mahogany *(Cercocarpus montanus)* shrublands. The field chairman was Herb Fisser, professor of range management, University of Wyoming. The coordinating committee acknowledged the particular support of the Division of Range Management, U.W., in staging the 1981 event and the continuing involvement of the other support agencies.

This volume reporting the activities of the 10th Wyoming Shrub Ecology Workshop, was the fifth issue of the proceedings. Papers given at this workshop and reports of the field tour activities were presented to further extend the benefits of the Wyoming Shrub Ecology Workshop.

2E 830404 Waugh, W.J. 1983. Stagnation and decadence of *Cercocarpus montanus* in a south-eastern Wyoming big game exclosure. p. 11-14. In H.G. Fisser and K.L. Johnson (eds.). Wyoming Shrublands: Proc. 10th Wyoming Shrub Ecology Workshop, Sybille Canyon, Wyoming, May 28-29, 1981. University of Wyoming, Laramie.

2F2
1E3
1E4
6A
1E2

See 810314.

5A1D 830405 Dean, S.K. and H.G. Fisser. 1983. Long-term vegetation responses to sagebrush control in western Wyoming. abst. 29. In Absts. 36th Ann. Meeting. Soc. Range Manage. Albuquerque, New Mexico. February 14-17, 1983.

3A2
3A3
2A
2C
2E
2F2
5B4
10D

See 830301.

4B 830406 Deboodt, T.L., E.B. Bradley, and H.G. Fisser. 1983. Economic feasibility of resting rangeland to improve forage production in central Wyoming. abst. 215. In Absts. 36th Ann. Meeting. Soc. Range Manage. Albuquerque, New Mexico. February 14-17, 1983.

9A
3A3
3A2
6A
5A1D
11E2

See 830302.

2A 830407 Elwakeel, A.S. and H.G. Fisser. 1983. Perennial grass production dynamics among five vegetation types in north central Wyoming. abst. 208. In Absts. 36th Ann. Meeting. Soc. Range Manage. Albuquerque, New Mexico. February 14-17, 1983.

1E2
2C
2E
2F2
3A2
3A3

See 830303.

1E2 830408 Fisser, H.G. 1983. Phenological description and numerical scoring of sagebrush. abst. 136. In Absts. 36th Ann. Meeting. Soc. Range Manage. Albuquerque, New Mexico. February 14-17, 1983.

2E
2F2
3A2
1E1
11A
11B
11C
11E
2C
2A

See 830304.

3A3 830409 Joyce, L.A. and H.G. Fisser. 1983. Productivity differences of grasslands in semi-desert western Wyoming — A re-examination. abst. 207. In Absts. 36th Ann. Meeting. Soc. Range Manage. Albuquerque, New Mexico. February 14-17, 1983.

1E2
2A
2E
2B

See 830305.

1E2 830410 Kinucan, R.J. and H.G. Fisser. 1983. Dynamics of plant community development for diverse sagebrush sites in western Wyoming. abst. 65. In Absts. 36th Ann. Meeting. Soc. Range Manage. Albuquerque, New Mexico. February 14-17, 1983.

2A
2C
2F2
2E

See 830306.

5A1D830411 Mulholland, J.L. and H.G. Fisser. 1983. Re-establishment rates of sagebrush following
1E1 chemical control on grazed and ungrazed ranges in western Wyoming. abst. 295. In
2A Absts. 36th Ann. Meeting. Soc. Range Manage. Albuquerque, New Mexico.
2B February 14-17, 1983.
2C
2E See 830307.
2F2
11E

2E 830412 Owens, M.K. and H.G. Fisser. 1983. Size and density parameters of western wheatgrass
10D on diverse range sites in western Wyoming. abst. 128. In Absts. 36th Ann. Meeting.
1E2 Soc. Range Manage. Albuquerque, New Mexico. February 14-17, 1983.
11E2
2A See 830308.
2C
2E
2F2

2F2 830413 Waugh, W.J. and H.G. Fisser. 1983. Utah juniper invasion in north central Wyoming and
2E historical climatic and management relationships. abst. 202. In Absts. 36th Ann.
3A2 Meeting. Soc. Range Manage. Albuquerque, New Mexico. February 14-17, 1983.
2C
2D See 830309.

8305 Wyoming Agricultural Experiment Station Publications

830501 Fisser, H.G., S.K. Dean, T.L. Deboodt, R.S. Kinucan and J.L. Mulholland, 1983. Phenology and production studies on semiarid shrub types. 1982 results. Univ. of Wyo. Coop. Res. Rep. to the BLM. Vol. I: Phenology Study, Sections I and II. Wyo. Agr. Exp. Sta. Sci. Rep. 1230. 430 p.

1E2 83050101 Section I - Phenology study.
1E1
3A2 The expanded numerical vegetative and reproductive phenological scales established
2F2 in 1979 were used again during the 1982 field season. The two scales were designated to
2E allow simultaneous scoring of vegetative and reproductive development. The continuous numerical sets, from 1-8, and 8-18, for the vegetative and reproductive stages respectively, facilitated the application of quantitative analysis.

A listing of the phenology study exclosures, exclosure locations by county and the principle phenology species scored in each exclosure was given. An alphabetical listing was given of all the species found in the 12 study exclosures. A phenological inventory was presented by exclosure and sampling date. The prime species intensive phenological survey of each permanently identified plant for each exclosure, including phenological state, plant dimensions, plant age class and percent dead material, was given. A summary of these data, averaged over the 20 sample plants in each prime species by exclosure and sampling date were presented. 1973-1982 data on phenological development at each exclosures for three major growth stages were given for the species listed.

1982 phenology curves for the principal species at each exclosure were compared to curves developed from the earliest and lastest dates observed from 1973-1982.

11F2 83050102 Section II - Phenology multiple regression for predictive modeling.
2E
2F2 A multiple linear regression model was used to analyze vegetative and reproductive
3A2 phenology data. The purpose of the analysis was to first identify causal relationships
1E1 between environmental factors and phenodynamics and secondly, to predict phenological
2A development from environmental interactions. Two grass species, western wheatgrass

(Agropyron smithii) and bluebunch wheatgrass *(A. spicatum)*, and two shrub species, black sagebrush *(Artemisia nova)* and big sagebrush *(A. tridentata)*, were analyzed.

Multiple linear regression has been utilized for comparison of environmental relationships to phenodynamics. The initial program utilized step-wise multiple regression with a forward selection process. That model was changed to a backward selection process of the Hocking-Lamotte-Leslie method of multiple regression analysis and was used through 1980. The procedure used for the present report was based upon an all possible subsets model of multiple regression. The criterion for selection of the "best" model was based upon Mallow's Cp. With this procedure, the total squared error term contained a bias component and a random error component. When Cp values were plotted against p (where p was the expected value of Cp) the regressions with little bias tended to fall near the line Cp^2p. The smallest value of Cp was used to search for the "best" set of independent variables.

Fourteen independent variables were analyzed to develop the regression equations. The regression model was expressed in terms of independent variables and their effect on a dependent or response variable. The model takes the form:

$$Y = B0 + B1 \text{ x } i1 + B2 \text{ x } i2 + B3 \text{ x } i3 + \ldots \\ Bn \text{ x } in + ei$$

The Y was the response variable that was defined by the right side of the equation. In this report the response variables were the vegetative and reproductive phenological stages. The B0 was the Y-axis intercept on the regression and was of little interpretive value in an ecological sense. The B1 B2 ...Bn coefficients resulted from the regression analysis. The X1 X2...Xn represented the independent variables in the model. E1 was the random error component.

A simple T-test was calculated to determine the significance of the variables in each partial model. An analysis of variance was then conducted and the model was tested for significance. The variable selection process was terminated when addition of variables to the model resulted in no significant improvement of the model ($P \leq .05$).

An estimate of the Y response variable could be obtained from the regression equation. The adjusted multiple correlation coefficient and the standard error of the regression determined the accuracy of the estimate. An evaluation of the t-value for regression coefficients, the F-value for the regression and the multiple regression coefficient yielded the optimum predictive model for each data set. These models are necessary to forecast future phenological development rates and, of great importance, to supply quantitative data of interactive relationships.

The regression coefficients (B1 B2...Bn) demonstrated the positive or negative dependence and the magnitude of that relationship to Y according to the influence of the other independent variables. A positive coefficient denoted a variable that promoted rate of plant development. Conversely, negative coefficients indicated variables which protracted the plant development time.

Accelerated plant development rates were expressed by environmental conditions such as rapidly increasing and sustained high summer temperatures; limited precipitation; and rapid soil moisture depletion, a condition common to soils with high sand content and low waterholding capacity. Prolonged phenological development was exhibited as a response to cool summer temperatures, abundant precipitation and long lasting soil moisture associated with abundant precipitation and high waterholding capacity of soils.

Greatest production of herbage was usually associated with environmental factors identified by negative coefficients that prolonged or delayed phenological development. Extremes such as excessively cool and moist spring periods or an exceptionally early warm and dry spring could inhibit or modify plant or phenological development rates to such a degree that subsequent herbage production might be restricted as a result. The complexity of biological interactive responses to time and changing environmental stresses required comprehensive data accumulation and analysis, as well as subjective judgement for definition of predictive models to accurately forecast future events.

The analyses for vegetative and reproductive phenodynamics for each of the four species were listed in separate units in the report.

830502 Fisser, H.G., S.K. Dean, T.L. Deboodt, R.S. Kinucan and J.L. Mulholland. 1983. Phenology and production studies on semiarid shrub types. 1982 results. Univ. of Wyo. Coop. Res. Rep. to the BLM. Vol. II: Precipitation, Productivity and Production, Sections III, IV, and V. Wyo. Agr. Exp. Sta. Sci. Rep. 1231. 256 p.

2A 83050203 Section III - Precipitation pattern study.
3A

During 1969 more than 70 gauges were installed at 6-12 mile intervals throughout the Big Horn and Wind River Basins. A total of 282 weather stations consisting of 217 UW-BLM gauges and 65 U.S. Weather Service stations listed by county were reported in this report.

Most UW-BLM gauges were sheet metal cylinders, 12 inches in height and approximately 2.79 inches in diameter. Precipitation was read to inches of rainfall by measuring the water in milliliters. The gauge diameter allowed 100 ml of water to be equal to 1 inch of precipitation. In the Big Horn Basin the BLM utilized a number of commercial gauges as well as the UW gauges, the latter which were available from the Department of Range Management at the University of Wyoming in Laramie. Of the 217 UW-BLM gauges, 62 were read by the Department of Range Management and 155 by BLM personnel.

Seasonal dates for reading the gauges were established in 1960 based partially upon the traditional four seasons, occurrence of significant moisture periods, and average earliest and latest dates when roads would be passable. The dates selected were April 15, June 30, August 30 and October 15. The winter period thus extended from October 16 of the previous year to April 15, the spring period from April 16 to June 30, summer from July 1 to August 30, and fall from September 1 to October 15. In recent years precipitation recordings have been increased to monthly readings during the growing season (May-October) in the Big Horn Basin.

The UW had also established additional gauges at the 12 exclosures where intensive phenology and productivity studies were conducted. These gauges were read approximately every 2 weeks during the growing season in conjunction with phenology and productivity measurements. They were established to provide a separate data set since frequent readings of a single gauge could easily result in an error. The data from the two gauges are being compared for any discrepancy that may have arisen from mishandling.

The total of 65 Weather Service stations noted above represents most operating in each county. Data from many recording stations were often difficult to obtain at the time of report compilation. The Weather Service employed gauges with a larger orifice and were presumably most accurate. Long-term records of the UW gauges and some of the BLM Taylor gauges showed little difference except for snow that sometimes had not been recorded as accurately as with the larger recorders. The difference had seldom been as much as 10 percent of the total and thus was assumed to be negligible for general comparative purposes.

The total complement of 282 rain gauges distributed over the more xeric portions of western Wyoming provided a data set unequalled. Combined with the long-term herbage production records (obtained annually on some forty sites for as long as 21 years) and intensive phenology and productivity measurements (obtained at 12 locations for 10 years), this provided a volume of data for combined evaluation of precipitation dynamics with herbage production, plant development rates, and vegetation productivity.

Considerable generalized evaluation of the kinds, amounts, rates and distributions of precipitation during the year was given in the manuscript. In general, the precipitation during 1982, as in 1981, was about 10 percent greater than normal. Significant patterns in precipitation differences were noted in the Big Horn Basin. Greatest importance of the precipitation year was the uniqueness associated with much greater than normal fall precipitation throughout the state. Summer rains were above average in some areas, especially in the southwest and western areas of the state. The year will be long remembered for the influence of fall moisture coupled with great amounts of snow pack which caused serious mud slides and floods during spring 1983.

1E2 83050204 Section IV - Productivity-phenology interrelation study.
10D1

2E Relationships between plant phenology, productivity and annual weather variations
2F2 are not well understood. There was a need to determine standing crop variation with
3A2 seasonal progression of annual herbage production, as it related to phenological
1E1 development and weather variations. Such information was essential for determining
2A rangeland herbage production potential from data collected at times other than peak
5A1D standing crop. Interrelationships of time with phenological stages and the development of standing crop of annual herbage production, which could normally be expected at each major plant growth stage, has been little investigated because of extensive monetary, manpower and time commitments necessary for such comprehensive research attempts.

Information on long-term continuous study of herbage production and/or cover and composition existed, on some 100 exclosure locations throughout western Wyoming. Soils, precipitation and other ecological-environmental information had also been obtained from many of them. Recent 6-year intensive studies of phenology and production were conducted on 12 of these locations. Continued studies to address phenology-standing crop dynamics and relationship to range site productive potential can be conveniently integrated into the ongoing program.

The objectives of this study were twofold. First was the determination of plant productivity relationships with environmental factors and phenological growth stages. The environmental variables of annual and long-term precipitation, air temperature and soil temperature were studied in conjunction with both plant phenology and productivity. The second objective involved development of a predictive model for analysis of quantitative data which could be used in management decision processes.

Procedures that were conducted to measure variables in the model involved four major activities:

1) determining early and late cover and standing crop,
2) recording phenology of prime species on existing permanent transect lines,
3) selection and clipping of prime species plant culm and bud units,
4) recording environmental variables.

Determination of cover and standing crop followed existing procedures for estimation of total annual growth production. Other procedures relating to the productivity measurements, the phenology transect procedures, productivity clipping procedures and environmental recording procedures have been described.

Productivity clippings were obtained at 2-week intervals in 1982 during the majority of the growth season beginning in late April and ending in latter September. Data were obtained 10 times on most locations.

Productivity clipping weights, phenology, and related data were presented in tabular form. Changes in culm weights were positive through mid-June. But at many locations, culm weight losses for grasses and for buds of sagebrushes were closely linked with warm summer temperatures and reduced effective precipitation.

3A3 83050205 Section V - Exclosure study: production, cover and precipitation.
5A1D

5B Studies of vegetation forage production from several exclosures and relic areas of the
5C Big Horn and Wind River Basins were initiated during the 1962 field season. The
2A studies included data from 19 livestock exclosures and one relic area. Research activities
2C had been continued annually since that time to include 43 different exclosures throughout western Wyoming. The objectives of the program were to determine:

1) the relationship of vegetation production with phenodynamics and environmental data;
2) the relationship of annual herbage production to area cover percentage;
3) the influence of different treatments, the environment and succession on vegetation changes;
4) the influence of some range improvement practices on herbage production and cover.

Area cover and herbage production studies on sagebrush-grass and juniper sites were conducted on transects of 20 quadrats, 1 ft x 1 ft, spaced systematically along a randomly located 100 ft steel tape. On sites dominated by saltbush, a variation of the sampling technique, described in the 1968 results, was used to measure herbage production. Twenty 1 ft x 10 ft plots were located systematically along a randomly located 100 ft steel tape. Each 1 ft x 10 ft plot was subsampled for vegetative cover using 10 1 ft x 1 ft quadrats.

Area cover of all herbaceous, semi-woody and woody species was estimated within each square foot plot. Data of shrub crown cover and basal cover of plains pricklypear *(Opuntia polyacantha)*, Hood's phlox *(Phlox hoodii)* and Hooker sandwort *(Arenaria hookers)* were not combined when comparing area cover to herbaceous production since this group of plants was not clipped.

Herbage production was determined by clipping herbaceous species at ground or crown level. The clipping procedure was conducted on or near the same date as the previous year for each exclosure. Production field procedures were initiated each year shortly after July 4 on sites dominated by Gardner saltbush *(Atriplex gardneri)*, which was treated as a herbaceous species. All new annual growth was clipped at that time for determination of total production. Clipping procedures for measures of grass and forb production were completed by August 22 each year with latest activities at sagebrush *(Artemisia)*-grass exclosures in the Wind River Basin. Clippings were oven-dried at 70 °C for 24 hours prior to weighing.

On areas where animal use had occurred prior to clipping, estimations of utilization were made by species. These estimates were made at the time each plot was clipped. The utilization values were averaged by species for the 20 plots and used to adjust production for herbage lost by utilization.

An estimate of total production was derived for each site by adding the shrubby and mat forming species production to the clipped production. Shrub, mat forming, and pricklypear plant production was obtained by a modified double sampling technique during September and October. Weight units were estimated on 20 randomly located macro-quadrats, 4 ft x 5 ft, subdivided into 6-inch squares. Each macro-quadrat was 6-12 steps apart across the study area. Shrubby species larger than 6 square inches, primarily big sagebrush *(A. tridentata)*, were estimated using the 6-inch square as one weight unit. Mat-forming plants and shrubs smaller than 6 square inches were estimated by weight units of characteristic plant shape and size. Three of these weight units for each species were clipped for weight determination. Each of these units was dried in an oven at 70 °C for 96 hours. The average gram weight for the three weight units, multiplied by the total number of estimated units in 20 quadrats, was converted to lb./acre as an estimate of annual production.

At the same time clipped production data were collected during July and August and photographs of the transects were taken giving a pictorial history from year to year at each study location.

The production section of the 1982 Annual Report closely followed the format of earlier years. The tabular listings gave information including alphabetical listing of study sites, code names, county locations and the treatments at each site, a list of the plant names which occurred at the sites, the utilization information as indicated on the areas where use had been noted and tabular material listing the estimated production and cover values for both the early spring clipping procedures and the later summer production estimation processes.

Each shrub and perennial grass species had these categories given: shrub, perennial grass, annual grass, perennial forbs, annual forbs, annual herbaceous, forage production and total production. Forage production was given in both pounds per acre and kilograms per hectare. Average percent cover was given for each individual species.

2F 830503 Fisser, H.G. 1983. Aridland shrub ecology and phenology research program. Wyo. Agr.
2E Exp. Sta. Mimeo. 3 p.
2F2
3A2 All aspects of this multi-phase research management-based program have been
1E1 directed by Dr. Fisser since its inception in 1961. The Bureau of Land Management has
2A provided the primary continuous annual support. Other fund sources included McIntire-
5A1 Stennis, Hatch, USDA, Soil Conservation Service, Wyoming Game and Fish Department
11E and other incidentals.

The basic intent had been to obtain quantitative data from geographically and physiognomically diverse semiarid rangeland sites in Wyoming. Use of these data can and has resulted in solution of definitive management problems and identification of related ecosystem functions. Biotic, edaphic and climatic relationships could be identified and extrapolated from this long-term intensive monitoring program for comparison to similar rangelands of the U.S. and to other semiarid areas of the world. Desertification was recognized as being essentially a mismanagement induced phenomenon. Interpretation and synthesis of the aridland shrub ecology and phenology records with other regional and world distributed data sets could significantly assist land managers and users. By application of proper management, reversal of rangeland deterioration could be achieved with resultant improved ecological condition class and increased products for human needs.

Research sites of the aridland shrub ecology and phenology project were located over a vast area of western Wyoming; extending from the towns of Cody, Deaver and Lovell in the north 265 miles southward to Baggs and Burnt Fork, and from near Casper on the east 235 miles westward to Evanston, Sage and Cokeville. Research site locations were established in each of five physiographically defined regional areas within the state: Big Horn Basin, Wind River Basin, Powder River Basin-Platte River Drainage, Red Desert-Little Snake River Drainage, Little Colorado Desert-Green and Bear River Drainages.

Intensive monitoring of vegetation, grazing and environmental factors had been conducted on a variety of vegetation types including sagebrush *(Artemisia)*-grass, saltbush *(Atriplex)*, juniper *(Juniperus)*-mahogany *(Cercocarpus)*, wheatgrass *(Agropyron)*, blue grama *(Bouteloua gracilis)*, greasewood *(Sarcobatus vermiculatus)* and mixed-desert-shrub complexes. Fenced exclosures were utilized on most sites and were 1-4 ha in size. Graduate research effort had resulted in completion of 12 doctoral dissertations and 27 master's theses. One dissertation and five theses were expected during 1983. Annual reports, with detailed data assembly requiring two volumes of 4-500 pages each during recent years, have been submitted to the BLM. The most significant monitoring and research segments of the project are listed below with brief characterization comment.

1) Herbage Production and Cover - Annually, 20 sites in 1962 and increased to 45 in 1965; 20 plot transects, cover estimate and clip, 0.3 m x 0.3 m quadrats in grass complexes, 0.3 m x 3.0 m quadrats in saltbush types; treatments included sagebrush control, grazing, pitting and seeding, within site vegetation type differentiation; shrub production all sites by weight estimate procedures.
2) Environmental Data - Quarterly seasonal precipitation recorded annually at 190 sites since 1962 and U.S. Weather Service records obtained at 60 locations. Soil moisture by neutron probe and gravimetric at several locations during the 1960s and 1970s.
3) Soils - Detailed profile description obtained at 90 locations. Intensive soil-vegetation evaluation at 80 locations.
4) Winter Sheep Grazing - Four rates of use on saltbush, eight pastures, 260 ha, three exclosures, 1956-1965.

5) Summer Steer Grazing - Two pastures, one of native saltbush rangeland, the other improved with waterspreader dikes and grass seedings. Each pasture 250 ha, 40 ha seeded dike area, 1960-65.

6) Permanent Quadrats Shrub Mapping - Demographic and cover measures, 37 locations, 10 quadrats in exclosures, 10 outside, also shrub control treatments; .6 m x 6.0 m shrub mapping quadrats and 1.2 m x 1.2 m quadrats for understory point cover data. Data every four years from 1960.

7) Transects of Permanent Quadrats - Vegetation cover and composition, 20 quadrats 0.3 m x 0.3 m per line, 70 locations with combinations of grazing, chemical, pitting, seeding, burning and dozing on many vegetation types. Transects from 1965 and biennially into 1970s, and then as needed for associated research and management application.

8) Erosion Transects - Detailed point measurements to determine soil surface movement with 1.5 m transect on various treatments at 15 locations, from 1964 and irregularly following.

9) Harvester Ant Control - Three levels of Mirex, grazing and nongrazing, 12 treatment sites 485 ha with 10,800 permanent 0.3 m x 0.3 m quadrats. Annual 1964-68 and partial to present.

10) Grazing Systems - Three grazing systems with one, two and four pasture units. Eight exclosures, 110 permanent quadrat transect lines for vegetation cover and composition on sagebrush-grass, juniper-browse, wheatgrass and mixed-desert-shrub types. Browse utilization and density from seven 200 ft permanent point and strip transects in juniper-mahogany-bitterbrush type. Annually 1967-1971, then 1975 and 1979. Remote sense vegetation type mapping with ERTS imagery, Skylab and high altitude photography.

11) Phenology and Productivity - Intensive study initiated in 1973 on prime species western wheatgrass *(Agropyron smithii)*, bluebunch wheatgrass *(A. spicatum)*, black sagebrush *(Artemisia nova)* and big sagebrush *(A. tridentata)* on 12 geographically and environmentally diverse sagebrush-grass sites from central to north central, south central and western Wyoming. Prime species phenology and life form measured on transects of 20 plants permanently located and identified by distance coordinates. Eleven transects of western wheatgrass, six transects of bluebunch wheatgrass, 3 transects of black sagebrush, and 12 transects of big sagebrush. Also phenology of all species. Monitoring biweekly during growing season. Numerical systems established for separate vegetative and reproductive rankings. Plant height, leaf width and age class recorded for all permanent prime species plants. Environmental measures were ambient air temp., max. and min. air temps. since last sample date, soil moisture at 15, 30 and 46 cm, and soil temp. at 38 cm. Productivity clippings of prime species initiated in 1979 and coordinated with entire phenodynamics program. Annual through 1982.

The 20 years of continuous herbage production and precipitation records from 45 extremely diverse locations, with detailed soils information, provides a data set unmatched. The additional intensive phenodynamics, productivity and related environmental measures for 10 years, also provided a data base from which can be obtained definitive evaluation of phenology relationships to environmental dynamics and management practices.

The initial predictive models of herbage production, phenodynamics and environmental relationships which will be the end production of our Wyoming data analyses could be readily expanded for application to other states and regions, especially within the semiarid sagebrush, saltbush, juniper woodland and greasewood ecosystems of western North America. Interactive discussion with other scientists, inspection of long-term data sets from similar ecosystems and on-site verification trips would provide exceptional value. Definition of the parameters characteristic of ecological systems, and relationship to monitoring procedures, and interpretive applications on rangelands could provide updated information for assistance to land managers and users for minimization of land use impacts.

8306 **Graduate Advisee Completions**

5A1D830601 Dean, S.K. 1983. Plant response to livestock grazing and big sagebrush chemical control in western Wyoming 1962-1980. M.S. Thesis. University of Wyoming. 84 p.

3A2
3A3
2A
2C
2E
2F2
5B4
10D
11E3

Long-term effects of livestock exclusion and chemical control of big sagebrush *(Artemisia tridentata)* with 2,4-D were evaluated for three geographic regions in western Wyoming: the Big Horn Basin, the Wind River Basin and the Bear River Divide. Six study areas represented diverse ecological conditions and grazing use histories. Exclosures were constructed at each study area encompassing both sprayed and nonsprayed areas. Annual herbaceous production data from 1962-1980 were analyzed to identify significant differences between the experimental factors of grazing and spraying. Mean annual production in three five-year intervals from 1966-1970, 1971-75 and 1975-1980 were analyzed to determine significant vegetative trends. Relative species composition, based on mean annual production from 1971-75 and from 1976-1980 was used to calculate range condition ratings based on SCS criteria. Vegetative trend evaluations were determined from changes in range condition ratings from the 1971-75 and 1976-1980 time periods.

There were no significant increases in perennial grass production in the absence of livestock grazing. Production values of Sandberg bluegrass *(Poa secunda)* and bottlebrush squirreltail *(Sitanion hystrix)*, however, were somewhat higher in the grazed areas, indicating their status as increaser species.

The transition of big sagebrush dominated areas in semiarid and arid regions of western Wyoming to higher perennial grass composition and production was an extremely slow process, if it occurred at all, when the only management applied was livestock exclusion. These results have also been found by a number of other workers in the western states.

Shrub control through application of 2,4-D decreased competition from big sagebrush, released available resouces and hastened the rate of increased perennial grass production. A positive relationship was shown between site productivity and recovery rate. Livestock grazing use history and degree of shrub control also affected the rate of recovery. Perennial grass seed reserves were a limiting factor in historically heavily grazed regions. Low levels of perennial grasses at the onset of the chemical control application reflected the scarcity of seed sources. As more perennial grasses became established, and vigor of existing plants was improved, seed sources became more abundant and recovery rate continued at an exponential rate. Anderson and Holte (1981) came to the same conclusion. For this reason, deferment of grazing during the application growing season would be desirable to give perennial grasses an initial advantage for seed production. Arid regions of the Big Horn Basin with approximately 10 percent annual plant species composition, primarily cheatgrass brome *(Bromus tectorum)*, showed the potential for annual plant invasion and prolific production in sprayed areas.

Environmental fluctuations and sampling error resulted in a high degree of variability in the production data. Large changes in production were required to be considered significant. Few significant trends could be detected from analysis of the production data in five-year intervals due to the high year-to-year variation. Total annual precipitation was not useful in reducing variability of the long-term and five-year means. Minimum necessary changes in perennial grass production to be considered significant ranged from 85 kg/ha to 305 kg/ha, and corresponded to potential site productivity.

Additions of increaser plants such as Sandberg bluegrass *(Poa secunda)*, bottlebrush squirreltail *(Sitanion hystrix)* and plains pricklypear *(Opuntia polyacantha)* to the SCS site descriptions were necessary to more closely describe study areas in the Big Horn and Wind River basins. Changes of the site descriptions in allowable percent composition of western wheatgrass *(Agropyron smithii)* would have raised the condition ratings at each study area, particularly in the sprayed treatments.

Species composition based on production was a better indicator of vegetative trend than mean production. Species composition did not fluctuate from wet to dry years in the same manner as annual production. However, composition expressed as relative abundance may not have reflected the actual magnitude of change over the 20-year study period. A large decrease in abundance of one species, such as reducing big sagebrush from 30 to 2 percent, resulted in changing the relative abundance of all species whether or not increases of the magnitude indicated actually occurred.

The determination of climax species composition for purposes of judging range condition was a difficult task. The results of this study and research to date did not show that secondary succession would necessarily follow the same pattern as primary succession. The ultimate equilibrium reached after site disturbance such as that caused by grazing stress or shrub control treatments was primarily a function of the species composition at the time of perturbation. Thus, many different assemblages of the same species could form relatively stable communities on a given site (Anderson and Holte 1981). Unidirectional site descriptions could be misleading when attempting to predict vegetative composition following management induced disturbances such as grazing and spraying.

11E2 830602 Kinucan, R.J. 1983. Phenological relationships and patterns of semiarid plant communities in central and western Wyoming. M.S. Thesis. University of Wyoming. 110 p.

11E3
1E2
2A
2C

A long-term phenology study was conducted jointly by the University of Wyoming and the Bureau of Land Management from 1973-1982. Its purpose was to quantify phenologic development of semiarid shrub types in central and western Wyoming. As part of that study, phenological patterns of development and relationships of those patterns were evaluated at 12 geographically diverse sites.

A numerical index consisting of 18 development phases was used for all phenological observations. The index described phenologic progression from growth initiation through dormancy. Vegetative phenology, which employed the first eight phases, was used in the current analysis because it expressed less variability and was more representative of plant development than reproductive phenology.

Nonlinear regression equations depicting long-term phenophase development were derived for selected species at each site for preliminary analysis. The parameters representing the regression were qualitatively compared to screen out combinations of species which obviously had dissimilar phenologic patterns. A full alternative versus reduced null model hypothesis testing procedure was then used to determine whether similarities were present among the remaining species groups.

Similar patterns of phenophase development were found among several species. Big and black sagebrush exhibited essentially identical average phenological development over a 10-year period at the three sites where they grew concurrently. Either one was acceptable for estimating development of the other. Similarities between big or black sagebrush *(Artemisia tridentata* and *A. nova)* and other shrubs were not found, nor were similarities found among any other combinations of shrub species.

All seven perennial grass species investigated exhibited similar patterns of phenological development at a majority of sites. Sandberg bluegrass *(Poa secunda)*, bottlebrush squirreltail *(Sitanion hystrix)*, bluebunch wheatgrass *(Agropyron spicatum)*, western wheatgrass *(A. smithii)*, needleandthread *(Stipa comata)*, Indian ricegrass *(Oryzopsis hymenoides)* and junegrass *(Koeleria cristata)* displayed similar patterns of development at all sites except Mesa Antelope and Sweetwater. Their development usually took place in the order of the sequence of listing as presented above, but small variations occurred among sites.

Four perennial forbs — low fleabane *(Erigeron pumilus)*, scarlet globemallow *(Sphaeralcea coccinea)*, pale bastardtoadflax *(Comandra pallida)* and Nuttall violet *(Viola nuttallii)* — displayed development similar to one another. Additionally, low fleabane was significantly related to the grass needleandthread. This suggested that grass species might be used to indicate development of these forbs. Two other perennial forbs, prairie onion *(Allium textile)* and woody milkvetch *(Astragalus purshii)*, displayed similar development to one another, but were different from all other groups.

Few annuals displayed similarities within their life forms. They also related poorly to perennial species due to the seasonal variability characteristic of annuals.

The phenological patterns of development for each grouping of species were compared among sites using a full versus reduced model hypothesis test. Statistical differences in phenologic patterns were limited to those sites exhibiting the most extreme temperature and moisture regimes. Mean annual temperature differences exhibited a greater effect on phenological pattern changes than did precipitation variability.

The regression equations developed for preliminary screening also predicted phenophase development based on Julian Date. They generally yielded coefficient of determination values from .70 to .90 and confidence levels of +/− 0.5 to +/− 1.5 ($p \leq .05$). The nonlinear equations depicted phenological development more realistically than did standard regression techniques. They were more accurate because they recognized the curvilinear nature of plant development.

Forb, grass and shrub life forms were compared to determine whether relationships existed among their development patterns. A life form separation was evident and the few existing relationships were of little importance. Grass and forb components depicted little temporal separation between them, indicating less niche separation between these than the shrubs.

Perennial grass species had the greatest potential as indicator species. They were easily scored, widely distributed and significantly related to species within their life form as well as to several forbs. Bluebunch and western wheatgrass are recommended as indicators because they have been intensively studied, were easily quantified phenologically and were dominant species in many communities and locations. Most shrub and forb species did not possess many of the characteristics necessary for suitable indicator species.

Recommendations were proposed to further refine the phenological scoring system used in this study.

8401 Research Activities

2A 840101 Precipitation pattern study.
3A
11A

During 1984 all precipitation data were recorded by BLM personnel. The University program for recording precipitation data was discontinued along with the other portions of the aridland shrub ecology field activities. Most of the Bureau of Land Management offices were sending people to record precipitation data at the locations within their various districts and had taken over recording gauges which were previously read by the University of Wyoming. All of these data were being mailed to Fisser at the range management department and these were to be compiled with existing long-term records as time continued. During the past year the Rawlins district purchased more than 100 of the metal cylinders for rain gauges from the supplier in Laramie.

11A 840102 Computer program development for long-term weather record compilation.
2A

During 1984 the range management department acquired additional computer tapes of long-term weather records from the National Weather Service in Ashville, North Carolina. The data were difficult to abstract from the NOOA tapes because of frequent format change and considerable error in the tape materials. A considerable amount of time was required to essentially check the data from almost every individual day from every station. This does not imply, of course, that each individual numerical value was compared against some other quantitative value. It does mean that through each series of data sets from any one station, the general relationship of values, whether precipitation or temperature, were studied in detail to observe the generalized relationship in change and quantitative amount in order to determine whether errors may have occurred. In several instances, significant quantitative values were noted that were obviously incorrect.

For instance, at the Jeffrey City location, the tape indicated a precipitation value for a single day of more than 6 inches. This same value was found in the monthly climatic reports. Upon reviewing the data, and this required contact with individuals both in Jeffrey City, at the weather station in Cheyenne, and also with individuals at the Climatic Center in Ashville, North Carolina, it was determined that the 6-inch rainfall value was incorrect and was closer to 0.6 inches of precipitation. If this one value had been allowed to remain in the data set, it would have caused a significant upward shift of precipitation on an annual basis, on a seasonal basis, on a monthly basis and would have caused a significant error. Similar errors occur in any data set. Therefore, it has been necessary to review these comprehensive data sets in detail to determine whether significant errors did exist and whether we could be assured that the data did represent real world values.

The major output in terms of program development in the categories of this activity topic during 1984 was the completion of the 23-year record of daily precipitation from 10 U.S. Weather Service stations in the Wind River Basin, and a comparable series of daily precipitation values for the eight major University of Wyoming exclosure rain gauge locations in the Wind River Basin, including McGraw Flat, Granite Mountain, Upper Government Draw, Lower Government Draw, Lander Ant, Sweetwater, Shoshoni Ant and Boysen. Each of these exclosures had been in existence for some 20 years and the precipitation had been recorded at those exclosures on a seasonal basis. The process of constructing the records of daily precipitation for each of these was explained in some detail in other portions of the 1984 bibliographic materials.

11A 840103 Computer program development for estimation of missing climatic data.
2A

During 1984 a major effort was required to develop values for estimating missing climatic data. This was primarily with regard to precipitation, temperature, and to some degree, solar radiation which has been recorded at the Lander Weather Service station.

There were three major categories of interest that were addressed with regard to missing climatic data. At the Weather Service stations which recorded precipitation on a daily basis, the requirement for estimating missing data was addressed by comparing the data from sites with data to those that did not. A generalized approach utilized a modification of the reciprocal of the distance squared procedure for determination of these missing daily climatic data. A second category of missing values was that which occurred with the data from stations which were recorded with hourly gauges. This process involved not only daily precipitation values and/or temperature values, but also values especially of precipitation which may have not been recorded during only 1, 2, 3 or 4 hours of a given day. In these cases additional programs were required to develop estimates of the total daily amounts of precipitation. The generalized computer programs for estimating each total daily precipitation value was developed for data from all Weather Service stations, and these were developed specifically with regard to stations which were most similar to one another. By implication those stations with little similarity were not closely related in the estimation process. Third, the climatic data at the exclosures were recorded on a seasonal basis and required a separate program development process to be able to interpolate the seasonal data to daily values. From those sets of daily values related to the estimation of missing data, the programs developed then were utilized to estimate missing data and with that estimation. These data were interpolated to a daily series of precipitation values.

11E3 840104 Compilation of herbage production and cover data from Wind River Basin exclosures.
3A3
11F1 During latter 1983 and early 1984 the herbage production and cover data for the
2E Wind River Basin exclosures were compiled and transmitted to computer tape. For three
2F2 of the exclosures — McGraw Flat, Granite Mountain and Upper Government Draw — the data were compiled by exclosure, by year and by plot basis for cover and for yield.

For the other exclosures, the production and cover data were compiled on an annual basis and were not processed through the plot by plot compilation. Further analyses will be conducted on the plot by plot data to determine relationships of yield and cover of herbaceous and shrubby species.

11C 840105 Summary of soil moisture and temperature data from Granite Mountain Exclosure.
2A
2C During 1984 the soil moisture and temperature data from the Granite Mountain
2E exclosure taken during the late 1960s and early 1970s with neutron probe readings were
2F2 compiled and entered onto computer tape programs. These data are utilized in part of the validation testing of the ERHYM model. This model required data expressing the amount of available soil moisture at the beginning of each growing season. The relationship of precipitation at a site and the soil moisture as influenced by soil water discharge and utilization by vegetation is of primary importance.

11C 840106 Erosion transect data summarization and input to computers.
8B
2C The erosion transect data from the locations at the halogeton pastures, the steer pastures and the Cumberland exclosures was being summarized and entered into the computer system. A process developed by Dr. Clifton Johnson of the Northwest Watershed Research Center in Boise, has helped develop a graphic portrayal of soil level changes which has been expecially useful for visual comparison of temporally distant transects.

2F 840107 Wyoming rangeland reference sites.
3A
3B The Wyoming rangeland reference site program was addressed only briefly during
3C 1984. A considerable amount of effort has been conducted by Gary Blincow of the
11E1 Bureau of Land Management in Worland who developed considerable information
11E2 regarding the more than 60 exclosure locations in the Big Horn Basin. These data will
11E3 be addressed more intensively during 1985.
2F2

11F2 840108 Yield prediction with ERHYM model.
11F1

11E2 During 1984 several tests were conducted for validation of the ERHYM model.
11E3 These data were derived primarily from exclosures in the Wind River Basin. The earliest
2A test was applied to data from the McGraw Flat exclosure with interpolation to the
2C climatic records at Lander. Lander was some distance from the McGraw Flat exclosure
3A2 and exhibited considerable environmental differences. The output of the program, with
2F2 interpolation to a daily basis of the seasonal precipitation recorded at the McGraw Flat exclosure was significantly related to the occurrence of precipitation in the daily records at Lander. These gave adequate results as a starting point for use of the ERHYM model.

It appeared that this model should apply to range sites in Wyoming and that it requires only updated daily precipitation and temperature files for the climatic records to continue analysis in 1985. Additional efforts during 1984 included validation runs of the model on data from several of the exclosures in the Wind River Basin. The constructed precipitation and temperature files appeared to support the fact that these data yielded real world values. In addition, the ERHYM model could be parameterized to closely interrelate with the yield potential of a site as indicated by the kinds of soil materials on the site, and the kinds of precipitation and temperature records which were definitive for an area.

11A 840109 Compilation of U.S. Weather Service precipitation records - Wyoming.
2A

A major accomplishment during 1984 was the compilation of daily precipitation records at Weather Service stations and exclosures in the Wind River Basin. The Weather Service stations for which the daily precipitation data were compiled and/or constructed by the reciprocal of the distance squared procedure were Muddy Gap, Jeffrey City, Gas Hills, Oregon Trail Crossing, Sand Draw, Lost Cabin, Riverton, Shoshoni and Boysen Dam. The weather data for Lander were considered a separate data set because of Lander's difference and location adjacent to the mountains which gave it cloud cover, precipitation and snow especially, in greater amounts than was common at the other stations, or at the exclosures. These exclosures for which the daily precipitation records were compiled from seasonal precipitation were Sweetwater, Granite Mountain, McGraw Flat, Lander Ant, Shoshoni 7, Upper Government Draw, Lower Government Draw and Boysen. This record of daily precipitation values was initiated on the date of January 1, 1960 and carried continuously through December 31, 1982. This 23 year record will be exceptionally important for validation studies with the ERHYM model.

8402 Journal Publications

3A3 840201 Fisser, H.G. 1984. Culm weight dynamics as an index of sagebrush-grass community
1E2 productivity. J. Colo.-Wyo. Acad. Sci. 16(1):25-26.
2A

2E Indices of herbage productivity were obtained for 3 years from test data on
2B sagebrush-grass ranges in Wyoming. Procedures involved spring (May) and summer
11E3 (August) clippings for early and late standing crop values supplemented with biweekly
11F1 clipping of western wheatgrass *(Agropyron smithii)* and bluebunch wheatgrass *(A. spicatum)* and terminal buds and leaders of big sagebrush *(Artemisia tridentata)* and black sagebrush *(A. nova)*. Data were obtained at 12 locations in western Wyoming and related to ongoing intensive phenodynamics/environmental relation studies.

Biweekly culm and bud clipping required less manpower and time than common productivity procedures. The wheatgrasses and sagebrushes were major herbage components, and growth rates could be extrapolated to identify the all-species productivity dynamics for the community populations.

By the last week of May, standing crop was near 80 percent of the maximum achieved by the communities later in the summer. Both the grasses and the sagebrushes responded to weather changes. Rain and cool summer temperatures were responded to with rapid growth rates. Dry and warm weather caused the culm weights to decline and the sagebrush bud and leader growth was strongly inhibited but not to the extent noted for the grasses. Differences among sites were significant.

11A 840202 Germain, L.P. and H.G. Fisser. 1984. Application of kriging to ecological variables. J.
2E Colo.-Wyo. Acad. Sci. 16(1): 13-14.
2F2
2A Kriging is a statistical technique used to determine the best linear unbiased estimate of the value of a given variable either at a point or on an area basis, based on surrounding measurements of that variable. Kriging has been used extensively in the geological sciences for ore reserve estimation but has been little used in the biological sciences.

Kriging may be a useful technique in the biological sciences under certain conditions. Usually this would mean when measurements were taken over a large area and/or restrictions were made on sampling. It is doubtful that this technique would be worth the time and effort on an intensively sampled area.

A study was conducted to estimate missing precipitation values in the Wind River Basin of Wyoming and compared the results to estimates from traditional techniques and subjective estimates. Preliminary results indicated that kriging did not give superior estimates to the traditional techniques and during some seasons, subjective estimates from an experienced investigator gave the best estimates. However, these results were tested on a simplified kriging model and may not reflect final results.

1E2 840203 Kinucan, R.J. and H.G. Fisser. 1984. Dynamics of vegetative plant part phenodynamics
11E2 and related stress adaptations. Bull. Ecol. Soc. Amer. Program Issue. 65(2):253.
11E3
2A This presentation followed development sequence and growth rate of vegetative
2C plant parts which were investigated for 10 years in diverse big sagebrush *(Artemisia tridentata)* associations of western Wyoming with an exponential model incorporating an asymptotic limit and an additive error in the form: $F(ti) = Yt = B0 + B1 \cdot exp(B2 x ti) + Ei$. A modified reduced versus full model hypothesis test was utilized to compare individual plant species and species grouped by life form.

Numerical phenological scores of many similar life form species exhibited comparable sequence and rate of development patterns. Since the environmental constraints of temperature and precipitation caused short growing seasons, the expected minimal phenologic timing variation occurred, especially when similar rooting habit complemented similar area life form characteristics. This phenomenon was noticeable by year, and particularly so with the composite multiple year scores ($p \leq .01$). Species with dissimilar life form varied significantly in their phenologic patterns. These results indicated that mechanisms of drouth and temperature tolerance or resistance varied with life form.

3A3 840204 Fisser, H.G. 1984. Productivity dynamics of sagebrush grass ecosystems as determined
1E2 from biweekly sampling of a single grass and shrub species. Bull. Ecol. Soc. Amer.
2A Program Issue. 65(2):258.
2E
2B Seasonal productivity dynamics of 12 semiarid sagebrush *(Artemisia)*-grass
11E3 ecosystems in western Wyoming were determined from two annual (May and August)
11F1 total yield measurements and biweekly yield measurements of a dominant grass and shrub species in each plant community. Total productivity was inferred by fitting the biweekly biomass dynamics of western wheatgrass *(Agropyron smithii)* and big sagebrush *(Artemisia tridentata)* to the May and August total standing crop values. Community standing crop was estimated by quadrat clipping herbaceous species and by weight unit estimate double sampling for woody, mat-form and succulents. Biweekly sampling involved clipping individual culms and leaders of the grass and shrub. This

method saved considerable time and labor compared to traditional methods for estimating the seasonal productivity of plant communities. Peak standing crop usually occurred near mid-August, but was subject to annual climatic variation with significant site, year and interaction differences.

8403 Presentations

1E2 840301 "Detailed numerical phenology of sagebrush with visual description statements."
2E February 14, 1984. 37th Ann. Meeting Soc. Range Manage. Rapid City, South
3A2 Dakota. by N.E. Hargis and H.G. Fisser.
1E1

2F2 This poster display provided photographic illustrations of phenophase development
11A research by the University of Wyoming since 1973 under the sponsorship of the Bureau
11B of Land Management, Department of Interior. The Bureau determined it necessary to
11C have a detailed cataloging of phenological phases of the shrubs black sagebrush
11E *(Artemisia nova)* and big sagebrush *(A. tridentata)*, and of the grasses western
2C wheatgrass *(Agropyron smithii)* and bluebunch wheatgrass *(A. spicatum)*. The posters of
2A this display pertained to the shrubs, primarily big sagebrush.
11F1 At the outset of the study in 1973, a general phenological description system was
3A3 developed which would characterize growth phases of the two sagebrush species. It was similar but not identical to some that already had appeared in the literature such as that by Depuit and Caldwell (1983) in the American Journal of Botany, which pertained to sagebrush in Utah.

Historically, phenodynamics had been identified primarily on the basis of plant growth phases such as leaf development, flowering and seed development. The term "stages" was substituted for "phases" in the present research which then allowed additional descriptive statements such as early bloom, mid-bloom and late bloom, stages for the generalized term relating to the progressive development of the flowering phase.

In addition to the first level interpolation from "phases" to "stages," statements of visually identifiable changes within a "stage" were developed. The twig elongation "stage" for example, was interpolated into the following "substages:" internode barely visible between new leaves, internode clearly visible, two or more internodes present, lateral branchlets present on new twigs, and lateral branchlets show internodes. Each stage and substage was assigned a definitive quantitative value for characterization of progressive phenodynamics. The overall purpose of the "stage" and "substage" interpolation was to provide a phenological characterization based on distinctive appearance changes integrated with a numerical scale. With these guidelines a technician with relatively minimal training or education in the natural history of the sagebrushes could quickly become able to identify both the vegetative and reproductive stages in the field. An additional purpose for development of substages and related numerical scoring values was deemed necessary in order to more closely integrate numerically defined phenodynamics with commonly accepted definitive measurement levels of abiotic factors including precipitation, temperature and related soil components. Description of phenological state is admittedly somewhat subjective and, when identified only on the basis of a generalized phase, must be recognized as insensitive to environmental driving variables. Greater sensitivity of phenodynamics to commonly measured abiotic factors could greatly enhance ability to determine biologically significant response levels.

The vegetative elements were scored 1.0 through 8.0 for the general stages. These include winter dormancy (1.0), which technically, for perennial shrubs, is not dormancy but merely a reduced rate of growth. The vegetative bud development stage (2.0) was technically an embryonic cluster, but for descriptive purposes, because of its appearance in the field, was termed a bud. New leaf development (3.0), twig elongation (4.0), completion of twig elongation (5.0), and vegetative twigs begin browning (6.0), identified the progressive stages of seasonal vegetative development. The code 7.0 was available primarily for leaf shed but could, if desired, be utilized to express fall regreening. Winter dormancy with code 8.0 allowed progression to a nonfunctional numerical value to identify the reduced physiological activity associated with the onset of extreme cold winter temperatures. This value was equivalent to the 1.0 score used to identify change from winter dormancy to the increased physiological activity denoted by values of 2.0 or larger. In actuality, if measures of winter energy use and nutrient turnover were identified, then substages from 1.1, 1.2, ... 1.9 could be developed to identify reduced growth activity rates associated with the onset and continuation of cold winter temperatures.

Reproductive phenology was developed similarly to that of the vegetative. The winter dormancy scores (1.0 or 8.0) were generally not recorded in the reproductive phenology column of the data sheet. The earliest entered value was that for the floral bud development stage (9.0). Subsequent stages were the bloom initiation (10.0), mid-bloom (11.0), late bloom, during which fertilization occured (12.0), milk stage (13.0), dough stage (14.0), mature seed (15.0), seeds cast (16.0), fall regreening (17.0) and fall dormancy (18.0). Where necessary these stages, as with the vegetative, were subdivided into substages. For example, additional detail during the stem development period was itemized as reproductive stem development, no buds evident (9.4), buds very small (9.5), buds swelling (9.6), buds enlarged, solid green color (9.7), buds faint yellow at apex (9.8) and bud still closed but very yellow (9.9). The opening of the bud was identified as the beginning of the bloom initiation stage (10.0), which in sagebrushes was soon characterized by the visible presence of exserted stamens.

It was evident that only selected decimal scores had been identified with a substage characterization. To attempt identification of all substages for each stage would be extremely unwieldy. With the present substage statements and scores, the field worker should be able to extrapolate for example from 9.0 through 9.4 in describing the degree of reproductive stem development prior to evidence of reproductive buds. Because of their usually very rapid length increase, the reproductive stems were readily distinguished from those which were identified as vegetative twigs and stems (4.0-5.9).

The primary vegetative photograph on the poster board showed part of a big sagebrush plant that had been characterized as having a phenological vegetative score of 4.1 and a reproductive score of 9.6. The 4.1 value indicated that vegetative twigs were elongating and that an internode was barely visible between clusters of new leaves. The 9.6 score of the reproductive stage indicated presence of reproductive buds and that they had begun to swell.

The development of these substage statements and scores simplifies the process of rangeland monitoring by range technicians. This is important because technicians tend to change from year to year. If we can supply them with the greatest amount of detail possible in terms of visual appearance changes and with data sheets that have typed on them all of the stage and substage identifying statements, there should be minimal error associated with personnel changes because subjectivity should have been limited. In addition, the increased precision of phenological characterization, with potential for detailed phyllochron distinction, provides a sensitivity level much more nearly equivalent to changes of environmental factors such as soil moisture and temperature, soil texture, ambient temperature, precipitation, and others.

Equivalency of biological data with that of the abiotic would surely enhance the ability to integrate field monitoring data with management practices. Also, in the process of ecosystem modeling, the increased relevance of phenodynamics with both empirical and stochastic data manipulation would assist with, and strengthen, the model fit with real world biological/physical interactions.

3A3 840302 "Productivity index procedures for sagebrush-grass rangeland." February 14, 1984. 37th
1E2 Ann. Meeting Soc. Range Manage. Rapid City, South Dakota. by H.G. Fisser.
2A

2E For three years test data had been obtained to derive indices of herbage productivity
2B from sagebrush-grass ranges in Wyoming. Procedures involved spring (May) and summer
11E3 (August) clippings for early and late standing crop values supplemented with biweekly
11F1 clipping of western wheatgrass *(Agropyron smithii)* and bluebunch wheatgrass

(A. spicatum), and terminal buds and leaders of big sagebrush *(Artemisia tridentata)* and black sagebrush *(A. nova)*.

Data were obtained at 12 locations in western Wyoming and related to ongoing intensive phenodynamics/environmental relation studies. Biweekly culm and bud clipping required less manpower and time than common productivity procedures.

The wheatgrasses and sagebrushes were major herbage components and growth rates could be utilized to identify the all-species productivity dynamics of the populations. By the last week of May standing crop was near 80 percent of the maximum achieved by the communities later in the summer.

Both the grasses and the sagebrushes responded to weather changes. Rain and cool temperatures was responded to with rapid growth rates. Dry and warm weather caused culm weights to decline and sagebrush bud and leader growth was strongly inhibited but not to the extent noted for the grasses.

11A 840303 "Estimation of missing precipitation data in Wyoming using traditional, kriging, and best guess estimate." February 15, 1984. 37th Ann. Meeting Soc. Range Manage. Rapid City, South Dakota. by L.P. Germain.
2E
2F2
2A

This study was conducted to determine if any one of five precipitation estimation techniques were more accurate at estimating precipitation. The techniques included the reciprocal of the distance squared, the long term average, multiple regression, kriging and a subjective technique. Preliminary results from data collected in the Wind River Basin of Wyoming showed that no significant differences existed between any technique ($p \geq .90$) during the winter, spring and summer, and only the subjective estimate was significantly better ($p \geq .90$) during the fall.

Some preliminary conclusions that were made from this study included:

1) The regression technique was found inadequate because of the spatial correlation structure of the data.

2) The reciprocal of the distance squared technique would not be suggested because of the anisotropic nature of the data.

3) The long-term average techniques was inadequate during the fall because of limited data and high variability between gauges from year to year.

4) The kriging technique may have been feasible but only if an extremely accurate model was used based on the data. Some suggestions for improvement included:

    a) use of more accurate regression model or alternate procedure to remove trend.

    b) determine the best variogram model to characterize the data. (eg. spherical vs. logarithmic)

    c) model the zone of influence based on the geometric shape observed from the data.

    d) model the variogram as a function of space and time.

5) The subjective estimate in this study was as good or better than any other technique. However, it should be noted that the estimator in this study had 23 years of experience.

6) Further analysis using an accurate kriging model is needed to make final conclusions concerning estimation accuracy.

1E2 840304 "Detailed numerical phenology of sagebrush with visual description statements." February 16, 1984. 37th Ann. Meeting Soc. Range Manage. Rapid City, South Dakota. by H.G. Fisser.
2E
2F2
3A2
1E1 See 840301.
11A
11B
11C
11E1
11F2
2A
2C

3A3 840305 "Culm weight dynamics as an index of sagebrush-grass community productivity." April 27, 1984. 55th Ann. Meeting Colo.-Wyo. Acad. Sci. Gunnison, Colorado. by H.G. Fisser.
1E2
2A
2E
2B See 840201.
11E3
11F1

11A 840306 "Application of kriging to ecological variables." April 27, 1984. 55th Ann. Meeting Colo.-Wyo. Acad. Sci. Gunnison, Colorado. by L.P. Germain.
2E
2F2
2A See 840202.

1E2 840307 "Detailed numerical phenology of sagebrush with visual description statements." May 1, 1984. USDA-ARS/USDI-BLM Regional Workshop on Rangeland Hydrology, Modeling and Monitoring. Boise, Idaho. by H.G. Fisser.
2E
2F2
3A2
1E1 See 840301.
11A
11B
11C
11E
2C
2A
11F1

5A1D840308 "Grazing management longevity and benefits of sagebrush control in western Wyoming." May 1, 1984. USDA-ARS/USDI-BLM Regional Workshop on Rangeland Hydrology, Modeling and Monitoring. Boise, Idaho. by H.G. Fisser.
4B
3A3
2E
2F2
1E1
1E2

In Wyoming, long-term research studies of the value of big sagebrush *(Artemisia tridentata)* control and its relationships to grazing management had been studied for over 20 years. Sagebrush has been controlled at a number of locations and those considered of greatest interest for this report were the following exclosures: Bud Kimball, Smilo, and Cochran in the Big Horn Basin, and Lower Government Draw, Upper Government Draw and Granite Mountain in the Wind River Basin. In the Bear River Divide area south and west of Kemmerer the exclosures identified as Cumberland -1 greasewood *(Sarcobatus vermiculatus)*, Cumberland -2 low sagebrush *(A. arbuscula)* and Cumberland -3 big sagebrush.

Wyoming *(A. t. Wyomingensis)* and mountain big sagebrush*(A. t. vaseyana)* had both been successfully controlled in these research site locations. Little sagebrush reinvasion had occurred on the sprayed areas within the exclosures, but on some of the sites, on the outside of the exclosures, sagebrush reinvasion had, during the last few years, become rather noticeable. Livestock concentrating on the sprayed areas adjacent to exclosures had contributed to a situation which allowed significant sagebrush re-establishment. Overgrazing reduced competition from grasses and forbs, making it easier for sagebrush seedlings to compete and survive. Intense competition from cheatgrass *(Bromus tectorum)* also was noted to provide ample re-establishment environmental potential at the sprayed site at the Smilo exclosure. At the Granite Mountain site, big sagebrush dominated. It was located near sites where A.C. Hull and others had studied early chemical control of big sagebrush and in many instances they too had difficulty in obtaining a good kill of the Wyoming sagebrush.

The harsh environmental conditions at Granite Mountain limited growth potential under the natural community conditions, and as such, the ability for the chemical to be utilized and function in its most optimum manner. Because some shrubs were removed and some were not, the sagebrush plants which did remain exhibited enhanced growth because of reduced competition, and secondarily, because of reduced competition, more sagebrush seedlings were able to become established.

Sagebrush control offered opportunity for increased grass production. Many believe that the control of sagebrush provides an opportunity to envision specifically an increase in potential herbage production of those sites. These data seemed to be in opposition to this concept since it was well shown that the sagebrush and other species which were present prior to the sagebrush spray and which perhaps are taken out by the sagebrush spray, these many species often produced total herbage production closely approximating that of the total production achieved after the spray. The major difference, of course, was that the species composition following control of sagebrush was significantly modified toward species producing forage. In some instances, the response in the Wyoming rangelands had been modified by an excessive increase in annual grasses or forbs. When this situation arose, the annuals provided significant competition to the perennial herbaceous species and as such, the perennial grasses were seldom able to completely overcome the impact of the presence of these annual species within less than 10-15 years.

Perennial grass production on the sites sprayed with 2,4-D for the most part averaged from two to four times as much as that found within the non-treated areas within the exclosures. The protection from grazing also was of some interest in that the areas which had not been sprayed were then able to be monitored for determination of forage production composition changes. In most cases, that resulted in little change. The grazing treatment as such seldom if ever, in the studies carried out in the aridlands of Wyoming, exhibited any significant impact upon the composition of sagebrush dominated rangeland herbage production.

The herbage production increases and lifespan of the sagebrush control projects were impacted and often reduced by the livestock grazing. The sites where sagebrush were strip-sprayed were small portions of large pastures. The sprayed areas tended to be overgrazed because forage was more accessible and livestock movement easier after sagebrush was removed. The fact that this happened is supported by comparison of the data collected outside the exclosures with data collected inside the exclosures. Perennial grass increased on all sprayed sites, but the extent of increase was much greater inside the exclosure. Under proper range management practices, the production increase outside the exclosure should have been as good or better than inside the exclosure.

Estimation of specific life expectancy of values and impact of the sagebrush control treatment is difficult. Variations due to site, climate, soils, initial vegetation composition of understory species, dominance of sagebrush and degree of sagebrush control are among the major factors which affect the treatment of lands for vegetation with chemical control. For the most part it has been noted that life expectancy of sagebrush control sites in the semiarid portions of western Wyoming can be expected to exhibit significant response for at least 15 years.

Even on the Granite Mountain site where sagebrush control was poor and re-establishment earlier than at most other sites, after 20 years, the sagebrush control treatment was still exhibiting significantly greater forage production of herbaceous species and total herbage production of all species than either the adjacent non-grazed native range or the adjacent sites outside the exclosure, both non-chemically treated and chemically treated. All sites, given the chemical control treatment of sagebrush and other broadleaf species still, after 17-24 years, exhibited greater herbage production within the non-grazed exclosures than did the non-grazed protected areas within those sites. All sites whether grazed or non-grazed exhibited greater production when sprayed than when not sprayed. All grazed sites exhibit greater production on the sites given the sagebrush control treatment than those which retained the native vegetation.

3A1 840309 "Long-term Wyoming rangeland monitoring and research cooperative studies of the
3A2 USDI-BLM & University of Wyoming, Department of Range Management." May 1,
3A3 1984. USDA-ARS/USDI-BLM Regional Workshop on Rangeland Hydrology,
2F2 Modeling and Monitoring. Boise, Idaho. by H.G. Fisser.
2E

2C In 1959, the intensive monitoring of vegetation grazing and environmental factors
2A on rangeland areas of the semiarid regions of western Wyoming was initiated by
11E2 cooperative agreement of the University of Wyoming and the Bureau of Land
11E3 Management. These data were continued through the years; and at the present time,
11F1 several site locations offer annual records of herbage production and precipitation. The
11F2 research site locations were established in each of five physiographically defined regional
10D areas within the state: 1) Big Horn Basin, 2) Wind River Basin, 3) Powder River Basin -
8B Platte River Drainage, 4) Red Desert - Little Snake River Drainage, 5) Little Colorado Desert - Green and Bear River Drainages. The initial studies were applied to some 40 sites and involved annual forage production information obtained at near peak standing crop each year. Through the years, additional studies utilized permanent transects and quadrats at a considerable number of additional exclosures to determine the long-term changes and demographic factors related to the vegetation of the semiarid vegetation types.

The general research activities were identified under 13 major categories: 1) Herbage production and cover, 2) Environmental data, 3) Soils, 4) Winter sheep grazing, 5) Summer steer grazing, 6) Permanent quadrats shrub mapping, 7) Transects of permanent quadrats, 8) Erosion transects, 9) Harvester ant control, 10) Grazing systems, 11) Phenology and productivity, 12) Density and 13) Modeling.

The initial predictive models of herbage production, phenodynamics and environmental relationships could readily be applied to other states and regions, especially within the semiarid sagebrush (*Artemisia* spp), saltbush (*Atriplex* spp), juniper (*Juniperus* spp) woodland and greasewood *(Sarcobatus vermiculatus)* ecosystems of western North America. Interactive discussions with the scientists, inspection of long-term data sets from similar ecosystems, and on-site verification trips would provide exceptional value. Definition of parameters characteristic of ecological systems and relationships to monitoring procedures and interpretative applications on rangelands could provide updated information for assistance to land managers throughout the semiarid regions of the world.

2F2 840310 "Herbage production and community dynamics." July 12, 1984. Shrub Research
3A3 Consortium Third Ann. Wildland Shrub Symposium: the biology of *Artemisia* and
2E *Chrysothamnus*. Provo, Utah. by H.G. Fisser.
2A

11F1 Herbage production and community dynamics studies of semiarid rangelands
10D specifically dominated by species of sagebrush (*Artemisia* spp) and rabbitbrush (*Chrysothamnus* spp) were initiated in 1960 in five major areas of Wyoming including the Big Horn Basin, the Wind River Basin and the Red Desert - Little Snake River Drainage, Little Colorado - Green and Bear River drainages. Herbage production was obtained annually at each location at about peak standing crop. At the Gardner saltbush *(Atriplex gardneri)* sites, data were obtained early in July and the field procedures continued into August on the sites with highest elevation and longest growing season

being sampled last. Some 20 locations of rangeland dominated by sagebrush have been monitored continually from 1960-1982.

At each of these sites, weather data, including seasonal precipitation, were recorded as well as temperature and soil moisture data which were obtained at a few locations. Soils information with excavation for profile analysis were conducted on all locations with cooperative effort by the Soil Conservation Service, Bureau of Land Management and University of Wyoming. During the sampling procedures, basal cover of herbaceous species and crown cover of shrubs, half shrubs and mat-form species, including a succulent, Plains pricklypear *(Opuntia polyacantha)* was recorded. Area cover and herbage production studies on the sagebrush-grass sites were conducted with transects of 20 quadrats, 1 ft x 1 ft, spaced systematically along a randomly located 100 ft steel tape. Cover of all herbaceous semi-woody and woody species was estimated within each square foot plot. Data of shrub crown cover and basal cover of Plains pricklypear, Hoods phlox *(Phlox hoodii)* and Hooker sandwort *(Arenaria hookeri)* were not combined when comparing area cover to herbaceous production since this group of plants was not clipped. Forage production was determined by clipping herbaceous species at ground or crown level. Exclosures were clipped on or near the same date as the previous years. Clippings were oven dried at 70 °C for 12 hours prior to weighing. Beginning in 1971, shrub production was obtained by a modified double sampling procedure during September and October annually. The technique involved a weight unit estimation procedure which was rapid and relatively accurate. Estimates were made on transects consisting of 20 quadrats, 4 ft x 5 ft, spaced from 6 to 12 steps apart. The number of weight units in each plot was estimated individually by two men and then checked for accuracy. Weight units were determined prior to estimation procedures. The objective of this and the following comparison papers is to provide a broad characterization of sagebrush related research in Wyoming.

5A1D840311 "Reinvasion following sagebrush control." July 12, 1984. Shrub Research Consortium
1E1 Third Ann. Wildland Shrub Symposium: the biology of *Artemisia* and
1E2 *Chrysothamnus*. Provo, Utah. by H.G. Fisser.
2E
2F2 During the late 1950s and early 1960s, trials to test the ability and requirements for
2C control of big sagebrush *(Artemisia tridentata)* with the chemical 2,4-D were initiated at
2A a number of locations in the Big Horn Basin, Wind River Basin and areas near
3A Kemmerer, Wyoming. Long-term studies with annual records of precipitation and herbage production and cover showed changes and response to the chemical treatment in combination with responses to grazing and protection from grazing. These long-term records have provided a very broad aspect in terms of the factors which may often be detrimental to the effectiveness of chemical control or which may be very beneficial for the control of the shrub.

The presence and importance of sagebrush on a site is identified in terms of the canopy cover of the shrubs, the density and general age of the individual plants, and the development or initiation of sagebrush seedlings. During the more than 20 years of study at the sagebrush controlled test sites of western Wyoming, it was noted that effectiveness of control was most specifically related to 1) optimum moisture and temperature conditions for effective growth of the shrubs and 2) to accurate placement of the chemical.

Poor growing conditions with either cold weather and/or dry spring conditions often resulted in restricted vegetation growth. The rates of sagebrush re-establishment were found to be strongly correllated with degree of sagebrush control, level of grazing intensity, and inherent site productivity. Increased grazing intensity was followed with excessive sagebrush re-establishment. Poor range condition prior to the chemical control site was noted to be followed with a rapid increase in sagebrush following treatment.

The most striking similarity among sites receiving effective control of sagebrush was the relatively slow rate of re-establishment for several years. This contrasted sharply with others who have noted the invasion by this shrub within the first 2-3 years following treatment. The delayed response can be attributed primarily to the semiarid rangeland condition characteristic of these sites and the climatic relationships present on these sites primarily east of the Continental Divide.

At these sites, the long moist spring period during April and May provided excellent opportunity for grass establishment which in general was able to prove competitive to the maintenance of growth of individual sagebrush seedlings. The climatic change from cold winter, often dry, to the spring growing season was much more beneficial and useful for perennial grasses than it was for the shrubs.

A contrasting kind of climate present on the areas west of the Continental Divide with the long, mild winter/spring period allows initiation of growth as early as January or February and provides an excellent growth media for sagebrush and not nearly as positive a situation for perennial grass growth. On the areas west of the Continental Divide, summer precipitation was essentially non-existent. In Wyoming, summer precipitation, while considerably lower than during other periods of the year, could be expected to occur relatively commonly. During years when moisture occurred in great abundance during summer, sagebrush re-establishment benefited.

The areas with greatest potential for sagebrush control were those with high production potential, but within the limits of the semiarid vegetation complex in the basins and foothill areas of the state; not the high elevation and precipitation ($>$ 15 inches precipitation) noted in areas of mountainous rangelands found in the Big Horn Mountains, the Absoraka Mountains, Wind River Mountains and the Wyoming Range to the west. In those areas it has often been noted that the re-establishment of sagebrush was rapid and occurred partially because of the excessive snowfall and cover and because of the tremendously extended moisture periods. Sites at lower elevations were sites in which snow cover was reduced from that of the mountainous regions and the summer temperatures exceeded those of the mountainous regions so that potential sagebrush establishment was restricted.

The sagebrush sprayed areas received greater use by livestock than the adjacent non-sprayed areas, whether conducted by strip spraying or block spraying. The attraction of the areas sprayed by 2,4-D for the livestock was a result of both forage availability and the actual growth of herbaceous plant materials because of the reduced water use by shrubs on the sprayed sites which then resulted in greatly increased herbage species growth.

Chemical control of sagebrush in areas with heavy concentrations of livestock should be avoided unless the grazing management and livestock distribution can be changed. Removal of perennial grass cover through heavy grazing has been shown to favor the re-establishment of sagebrush.

In contrast, however, protection from grazing on all sites has been shown to do little with regard to reduction of that shrub. The sagebrush plants are obviously long-lived and protection from grazing was not a means of reducing their number or density. The understory species associated with the existing sagebrush populations was strongly controlled by the shrubs. In areas with very great shrub cover, understory production was low. The rate of response on the non-grazed and non-chemically treated sites of understory species was only marginally greater than on the areas grazed by livestock and not treated with the chemical unless of course the grazing use in the area was extremely excessive both on the sprayed and non-sprayed sites, in which case the production of herbaceous species was considerably less than that in the non-grazed area.

1E2 840312 "Phenology." July 12, 1984. Shrub Research Consortium Third Ann. Wildland Shrub
2F2 Symposium: the biology of *Artemisia* and *Chrysothamnus*. Provo, Utah. by H.G.
2E Fisser.
3A2
1E1 Intensive studies of plant phenology were initiated in 1973 on 12 locations in
11A western Wyoming. These locations represented the Big Horn Basin, the Wind River
11B Basin, the Little Colorado Desert and Bear River drainages. The sites were also on big
11C sagebrush *(Artemisia tridentata)* dominated rangelands identified in general as semiarid
11E2 in terms of climate, although categorization of sites was done to characterize those
11F2 which were xeric, moderate and mesic in terms of total annual precipitation. The areas of

study expressed a broad ecological amplitude. Seventeen recognizable phenophases were devised for identification of plant development. All flowering species in each study were included in the inventory. Phenological data and environmental parameters (precipitation, maximum and minimum temperatures, soil temperature and percent soil moisture) were measured during the growing season at approximately 2-week intervals. Variations in time and amount of plant growth occurred within the same species the same year in the same location. Soil temperature was one of the major primary driving factors in rate of phenological development. Significant changes in plant growth occurred with only 0.5 °C change in soil temperature. Relationships between soil moisture and phenology were less obvious. Seed dispersal occurred about the same time each year, whereas initiation of growth at bloom times varied. Other important environmental phenomena to driving and regulating plant growth were precipitation from fall through spring, soil moisture at various depths, and maximum and minimum air temperatures. Multiple regression and step-wise selection analysis procedures were used to substantiate or reject relationships observed from the interpretation of field data which included phenological information and 15 environmental variables.

(For greater detail, see 840301.)

11E3 840313 "Validation of a rangeland production model (ERHYM) for sagebrush sites." July 12, 1984. Shrub Research Consortium Third Ann. Wildland Shrub Symposium: the biology of *Artemisia* and *Chrysothamnus*. Provo, Utah. by J.R. Wight.
11F2
11F1
11E2
2A
2C
3A2
2F2

The ERHYM model was developed in eastern Montana with research conducted on shortgrass-midgrass rangeland areas. The acronym has been taken from the title of the research project which was Ekalaka Rangeland Hydrology and Yield Model. The model was developed on the basis of the ratio of potential evapotranspiration to actual transpiration. The water balance portion of the model, as developed in eastern Montana for the current presentation, was developed as a validation test of the parameters identified from the eastern Montana site to those developed in western Wyoming.

The major components of the ERHYM model required maximum and minimum temperatures, daily precipitation and daily solar radiation climatic values. Soil information included profile values of the soils, the water holding capacity and infiltration rates. Daily update of soil water content was carried on by the model, and this related to the precipitation budget and evapotranspiration and plant growth characteristics of the site. The vegetation was identified by the crop coefficient and transpiration coefficient values as well as relative root distribution and a relative growth curve factor. Growth was developed on the basis of the ratio of potential evapotranspiration to actual transpiration. The model integrated several aspects of range environment including amount of precipitation and daily distribution of precipitation. Evaporative demands of the vegetation were also taken into account. Soil moisture by horizon was a major factor and this gave excellent advantage for predicting and forecasting rangeland yield.

Climate and yield data from two locations in the Wind River Basin, namely the Granite Mountain and McGraw Flats exclosures were used in this study. These sites were big sagebrush *(Artemisia tridentata)*-grass complexes with western wheatgrass *(Agropyron smithii)* a prominent species on all. Soil water and soil temperature data for 1963-68 from the Granite Mountain site were used to develop the model's soil water balance and soil temperature components. Precipitation was collected in simple aluminum rain gauges installed at the exclosures at each study site. A daily precipitation record was developed by prorating the site precipitation seasonal records with nearby U.S. Weather Bureau Stations using a distance squared procedure. Daily precipitation records obtained in this manner were generally satisfactory for predicting herbage production, but could induce a few day to day discrepancies between simulated and field measured soil water content values.

Daily solar radiation and maximum and minimum air temperatures were obtained from the climate generating model. For comparison purposes, the model was also run using the maximum and minimum air temperature from the U.S. Weather Bureau Station at Lander, Wyoming.

Soil water content and soil temperature were measured at the Granite Mountain site periodically during the summer months in 1963, 1965, 1967 and 1968. Soil water was measured by the neutron scatter method at soil depths of 6, 12, 18, 24, 36, 48 and 60 inches. Soil temperatures were measured at depths of 1, 8, 15 and 22 inches.

Herbage yields were determined both inside and outside an exclosure at each study site by clipping of herbaceous vegetation by species at ground or crown level with 20 1 ft x 1 ft quadrats spaced systematically along a randomly located 100 ft steel tape. 20 quadrats were clipped at each site each year. The vegetation was clipped within one month following peak standing crop. Beginning in 1971, the yield of shrubs and mat-forming species was obtained using a modified double sampling technique during late September and October. Weight units were estimated on 20 4 ft x 5 ft quadrats, both inside and outside the exclosure. Three of these weight units for each species were clipped and weighed. On areas outside the exclosure, where grazing had occurred prior to clipping, utilization estimates by species were used to adjust the yield estimates.

The effectiveness of ERHYM depended upon its ability to simulate an accurate soil water balance. Because there was little summer runoff at these sites, the soil water components of the model were tested using single growing season runs in which actual field measured soil water values were used to initialize the model. Soil water measurements for the beginning of the growing season were only available in 1963 and 1965. Continuous runs, 1963-67, were utilized to test the models ability to predict over winter recharge.

For yield predictions, the model was run in a continuous mode from 1965-1980 for the Granite Mountain study site and from 1965-1980 for the McGraw Flats study site to match the yield data which was available at the time the model was run. The model's ability to simulate soil temperature was evaluted by comparing model predicted soil temperatures at the Granite Mountain site with field measured values for 1963-68.

Overall, the ERHYM model performed reasonably well on the two sagebrush range sites in central Wyoming. With better quantification of the state variables and model parameters, such as crop coefficient and transpiration coefficient, model accuracy would likely be improved. The results of this study indicated that the ERHYM model had a wide range of potential applications in researching and managing sagebrush-grass rangelands.

1E2 840314 "Herbage yield dynamics and seasonal precipitation relationships." July 12, 1984. Shrub
2A Research Consortium Third Ann. Wildland Shrub Symposium: the biology of
11F1 *Artemisia* and *Chrysothamnus*. Provo, Utah. by C.L. Hanson.
11E3
3A3
2F2

Herbage production and precipitation data have been obtained since 1963 at seven rangeland sagebrush-grass sites in central Wyoming. The intent of this investigation was to determine the linear relationship between seasonal precipitation and total herbage production at these rangeland sites.

In many studies, simple or multiple regression analysis, using herbage yield as the dependent variable and precipitation as the independent variable, have developed predictive equations. However, attempts to correlate annual herbage yields with annual precipitation have generally been unsuccessful. This has been due primarily to the variable distribution of precipitation added to the fact that range plants generally have the greatest rate of growth during the growing season and little, if any, during fall and early winter. Use of seasonal or combinations of monthly precipitation have helped account for the distribution effects, and in many situations, have provided reasonably accurate herbage yield estimates. In Canada, May and June precipitation have provided good estimates of yield. In North Dakota, April through July was closely correlated with annual herbage yields, and in the Northern Great Plains in the area of eastern Montana and western North Dakota, the May precipitation was the best single month to index annual herbage production.

Since precipitation herbage yield relationships often exhibit a very site specific dependancy, the work and effort of this report addressed the herbage yield crop year precipitation as ratios of long-term values. In this study were investigated the yield and precipitation index model for herbage yield adjusting and forecasting. The seven central

Wyoming study sites occurred in the Wind River Basin, which was both a structural and a drainage basin. Its northern boundaries contacted the southern portions of the Owl Creek and Big Horn Mountains. The Wind River Range formed the west and southwesterly boundaries while the southeast and eastern limits were the Sweetwater escarpment and the Rattlesnake Mountains. The basin was asymmetrical in shape with lowest elevations near the Owl Creek Mountains. Elevations ranged from 4,000 to 7,200 ft.

The Wind River Basin has a semiarid climate. At Riverton the mean annual precipitation was 8.54 in and mean annual temperature 43.2 °F. The seven central Wyoming study sites used in this study were the following: McGraw Flat, Granite Mountain, Upper Government, Lower Government, Lander Ant, Sweetwater, Shoshoni -7. Site elevation ranged from 5,220 ft at the Shoshoni -7 location to 7,100 ft at the Granite Mountain area. Annual precipitation varied from 8.09 in at Shoshoni -7 to almost 10 in at Lower Government. Soils were Aridisols, including mixed, typic and frigid haplargids. They exhibited moderate percolation, were well drained, of moderate alkalinity, and contained a zone of increased effervescense in the lower "B" horizon. Sand and clay content was variable. Textural class was loamy. The "A" horizon was usually about 3 inches thick and the "B" horizon about 15 inches thick.

The estimate of total production was derived at each site by adding the shrubby and mat-forming species production to that of the clipped herbaceous species production. The herbaceous species were obtained during July and August with clippings within 1 ft x 1 ft quadrats. The production of shrub and associated species was obtained by a modified double sampling technique during September and October. Weight units were estimated on 20 randomly located macro-quadrats, 4 ft x 5 ft, subdivided into 6-inch squares. Each macro-quadrat was 6 to 12 steps apart from the study, across the study area.

Shrubby species larger than 6-inch square, primarily big sagebrush *(Artemisia tridentata)*, were estimated using the 6-inch square as one unit. Mat-forming plants and shrubs smaller than 6 square inches were estimated by weight units of characteristic plant shape and size. Three of these weight units of each species were clipped for weight determination. Each of these units was dried in an oven at 158 °F for 96 hours. Average gram weight for the three weight units multiplied times the total number of estimated units in 20 quadrats was converted to lb/acre in an estimate of annual production.

Precipitation data was recorded from metal rain gauges installed at each exclosure. These gauges were constructed with a diameter so that each 100 ml of water recorded in the gauge was equal to one inch precipitation. Photographs of the production transects were taken each year for a long-term pictoral history.

Regression procedures were used to determine the relationships between annual and seasonal precipitation and total herbage yield by sites with the seasons identified as 1) Fall - September 1 through October 15, 2) Winter - October 16 through April 15, 3) Spring - April 16 through June 30 and 4) Summer - July 1 through August 31. The relationship between the exclosure outside herbage yield and both exclosure and outside combined were investigated. The total winter and spring precipitation by site was used to develop the precipitation indices in the Sneva and Hyder (1962, J.R.M.) model. Indices were developed from the average of the exclosure and outside herbage yield at each site.

The initial analyses showed that the winter-spring season gave the best overall prediction results. These preliminary analyses also suggested that combining the herbage yield from both inside the exclosures and outside the exclosures gave the most consistent results.

The slopes of the regression lines varied from 52-64 pounds per inch of winter plus spring precipitation with a mean value of 58 pounds per inch. The slope values were all within 10 percent of the mean slope. The mean slope value indicated that one inch of winter-spring precipitation produced about 58 pounds of herbage. The 58 pounds of herbage per inch of seasonal precipitation was less than that estimated at 90 pounds for the intermountain area in the Sneva and Britton (1983, O.S.U.) model. On the other hand, Rogler and Haas (1947, J. Agron.) reported 74 pounds per inch of seasonal precipitation for central North Dakota. Those values, although different from those in Wyoming, were of little specific interest because they were essentially site specific, dependent upon the moisture soil potential of their areas.

The differences in the regression equations developed for the other regions in the western United States and southern Canada strongly suggested that the Wyoming area was significantly different in terms of climatic forage production relationship than the other parts of the western United States. The crop year precipitation utilized by others was about the same season as that utilized in our study, so it appeared important that soils and/or climatic regime of the central Wyoming area had a significant effect upon the herbage production potential relationship of the central Wind River Basin area.

3A3 840315 "Productivity dynamics of sagebrush ecosystems as determined from biweekly sampling of a single grass and shrub species." August 9, 1984. Annual Meeting Ecological Society. Fort Collins, Colorado. by H.G. Fisser.
1E2
2A
2E
2B
11E3
11F1

See 840204.

1E2 840316 "Dynamics of vegetative plant part phenodynamics and related stress adaptations." August 9, 1984. Annual Meeting Ecological Society. Fort Collins, Colorado. by H.G. Fisser.
11E2
11E3
2A
2C

See 840203.

8404 Published Proceedings

3A3 840401 Fisser, H.G. 1984. Productivity index procedures for sagebrush-grass rangeland. abst. 010. In Abst. 37th Ann. Meeting. Soc. Range Manage. Rapid City, South Dakota. February 12-17, 1984.
1E2
2A
2E
2B
11E3
11F1

See 840302.

1E2 840402 Fisser, H.G. and N.E. Hargis. 1984. Detailed numerical phenology of sagebrush with visual description of statements. p. 44 . In Proc 37th Ann. Meeting Soc. Range Manage. Rapid City, South Dakota. February 12-17, 1984.
2E
3A2
1E1
2F2
11A
11B
11C
11E
2C
2A
11F1

See 840301.

11A 840403 Germain, L.P. and H.G. Fisser. 1984. Estimation of missing precipitation data using traditional, kriging, and subjective techniques. abst. 160. In Abst. 37th Ann. Meeting Soc. Range Manage. Rapid City, South Dakota. February 12-17, 1984.
2E
2F2
2A

See 840303.

2E 840404 Fisser, H.G. and W.J. Waugh. 1984. Population structure anomalies of *Juniperus osteosperma* invasion in north central Wyoming, U.S.A. Submitted to 2nd Intern. Rangeland Congress. Adelaide, Australia. May 13-18, 1984. 4 p.
2F2
3A2
2C
2D

The phenomenon of juniper and pinyon increase in the intermountain and southwest regions of the U.S. involves both increased tree density within stands and encroachment onto adjacent grassland or shrub savannah rangelands. This report presented data from the Big Horn Basin of Wyoming, a northeastern geographical limit of the juniper complex for a savannah, shrub and grass rangeland of Utah juniper *(Juniperus osteosperma)*, black sagebrush *(Artemisia nova)* and bluebunch wheatgrass *(Agropyron spicatum)*. This site was near the ecological limit of the ecosystem. Juniper appeared in decreasing density downslope through the ecotone. Specific objectives of this study were: 1) to reconstruct successional history from the present age structure of the juniper trees and 2) to identify patterns or cycles of tree encroachment.

Pattern of tree encroachment was investigated by study of age structures, ground surface features and relative locations of adjacent trees and shrubs. Annual growth ring count was determined after adjustment for false and locally absent rings. Periods of abundant tree establishment were characterized with age class frequency cycles by analysis with an 11-year centered moving average. Regression was conducted to assess climate pattern chronology and tree ring width relationships.

Six time periods with accelerated establishment of Utah juniper were identified. The years separating periods were times of reduced initiation rates. The earliest, extended from approximately 1530, with establishment of the oldest tree, for about 200 years, to 1730. The second, for 87 years to 1817, the third for 75 (around 1892), the fourth for 48 (1940), and the fifth for 27 (1967). The sixth and most recent establishment cycle represents the 13-year period from 1967-1980. Tree establishment was commonplace during the third period to 1892 and involved both expansion within the population and invasion into the adjacent sagebrush community. The fourth period, 48 years to 1940, was striking in terms of significant invasion advancement, while the two most recent periods were characterized by increased tree populations within study communities, but little invasion into adjacent areas.

Possible causes for the observed Utah juniper invasion and population dynamics include: 1) a favorable change in climate, 2) fire suppression and 3) response to overgrazing by livestock. Periods of better than average growing conditions followed by drouth for several years, which caused decay of associated species, and suppression of fire through mechanical means or understory fuel reduction by chemical or other procedures, were reported to enhance its establishment. Livestock overgrazing was considered to favor the tree by weakening forage species and opening a previously closed community.

Since 1890 Utah juniper had increased in density within the stands and had encroached downslope into the shrub-grass savannah community. Seedlings, which established in the favorable micro-habitat under shrub and tree canopies, often occurred in clumps of two or more trees.

Climatic reconstruction efforts indicated no significant change in climate, although periodic localized very short-term temperature extremes had occurred which may have limited communities to a foothill thermal belt. There was little evidence that fire acted as a suppressant to juniper increases. Yet encroachment into the adjacent *Artemisia nova - Agroypyron spicatum* community appeared to be related primarily to the advent of livestock grazing.

See 830413.

5A1D840405 Dean, S.K. and H.G. Fisser. 1984. Grazing management, longevity and benefits of
3A2 sagebrush control on semiarid rangelands. Submitted to Second Intern. Rangeland
3A3 Congress. Adelaide, Australia. May 13-18, 1984. 4 p.
2A
2C The objectives of this study were 1) to evaluate 20-year forage production response
2E to chemical shrub control and grazing exclusion and 2) to define and interpret range
2F2 condition changes. The effects of long-term livestock exclusion and chemical control of
5B4 big sagebrush *(Artemisia tridentata)* with 2,4-D were evaluated for three semiarid regions
10D in western Wyoming; the Big Horn Basin, the Wind River Basin and the Bear River Divide. The six study sites represented diverse ecological conditions and grazing use histories. Exclosures were constructed at each study site between 1958 and 1963

encompassing both chemically treated and nontreated areas. Annual herbaceous production data from 1962-1980 were investigated to identify responses to grazing and spraying treatments. Mean annual production for five-year intervals from 1966-1970, 1971-75, and 1975-1980 was analyzed to determine significant trends. Relative species composition, based on mean annual production from 1971-75 and from 1976-1980 was used to calculate range condition ratings and community rating trend.

No measurable change ($p \leq 0.05$) or ($p \geq 0.05$) of perennial grass composition and production occurred when the only management applied was livestock exclusion. Production of sandberg bluegrass *(Poa secunda)* and bottlebrush squirreltail *(Sitanion hystrix)*, however, were somewhat higher in the grazed areas, indicating their status as increaser species. Perennial forb production was not affected by light and moderate grazing use. Grazing reduced production of annual plants only in the sprayed treatments in the Big Horn Basin.

Chemical shrub control through application of 2,4-D decreased composition from shrubs, and hastened the rate of increased perennial grass production. Site productivity potential was positively related to recovery rate. Significant changes, ($p \leq 0.05$), in mean perennial grass production over the 20-year period ranged from 458 kg/ha (300 percent) to 201 kg/ha (90 percent) in the sprayed areas. western wheatgrass *(Agropyron smithii)* accounted for the major increase at every study area. Annual production of muttongrass *(Poa fendleriana)*, bluebunch wheatgrass *(Agropyron spicatum)* and needleandthread *(Stipa comata)* also increased in the sprayed areas at each site where they occurred.

Livestock grazing use, past and present, and degree of shrub control affected the rate and longevity of production increases. Perennial grass seed reserves appeared limiting at the onset of the chemical control. As more perennial grasses became established, and vigor of existing plants was improved, seed sources became more abundant and recovery continued at an increasing rate. Deferment of grazing during the growing season of the application year thus would be desirable to give perennial grasses an initial advantage for seed production.

When either the shrub control treatment was ineffective, or the area was heavily grazed, maximum benefits of the sprayed treatments were not realized. When less than 30 percent of the shrubs were controlled in the spray treatment, annual production increases of perennial grasses were significantly limited after the first five years. Even when a chemical control treatment was not completely effective, however, a significant perennial grass production increase was evident. Grazing also decreased the longevity of an increased grass production period in a site with incomplete shrub control. The production of perennial grass was significantly decreased between the 5-year period and the last 10 years in the grazed-nonsprayed treatment area.

Few significant trends ($p \leq 0.05$) were detected from analysis of production data in 5-year intervals. Nor was total annual precipitation useful as a covariant to reduce variability of long-term and 5-year means. Minimum perennial grass production changes necessary to denote significance ranged from 85 to 305 kg/ha, and corresponded to potential site productivity, that is, the sites with greater potential productivity require greater amounts of grass production changes necessary to denote significance.

A better indicator of community trend than mean production was species composition based on annual herbage production. These data did not fluctuate from wet to dry years in the same manner as did annual production. The range condition rating analysis indicated definite upward trends at study areas in the Big Horn Basin and in the Bear River Divide. Those in the Wind River Basin showed little or no change. Improvement in range condition in the Big Horn Basin was most dynamic in the sprayed treatments. This may be considered evidence of the seral status of the sprayed compared to the more stable, nonsprayed areas.

The determination of climax species composition for purposes of judging range condition is a difficult task. The results of this study and research have shown that secondary succession does not necessarily follow the same pattern as primary succession. The ultimate equilibrium reached after site disturbance such as that caused by grazing stress or shrub control treatments was primarily a function of the species composition at the time of the perturbation. Thus, many different assemblages of the same species could form relatively stable communities on a given site. Unidirectional site descriptions could be misleading when attempting to predict species composition following management induced disturbances such as grazing and spraying.

See 830601.

1E2 840406 Kinucan, R.J. and H.G. Fisser. 1984. Vegetative phenology patterns of semiarid plant communities in Wyoming, U.S.A. Submitted to Second Intern. Rangeland Congress. Adelaide, Australia. May 13-18, 1984. 4 p.

2A
2C
2F2
2E

Many species of arid and semiarid regions do not regularly develop reproductive components, the commonly utilized characteristics of plant phenological progression. During the long-term joint University of Wyoming/Bureau of Land Management, USDI phenology study in a numerical scoring system to characterize vegetative phenologic progression was developed distinct from that of the floristically based. The objectives of this report are: 1) to characterize non-reproductive plant part phenological developmental differences by species and life forms and 2) to identify from vegetative phenological data best climate variation indicator species.

Nonlinear regression of vegetative phenology scores over time identified distinctive life form related differences of development rates. Best climate variation indicators were western wheatgrass *(Agropyron smithii)* and bluebunch wheatgrass *(A. spicatum)*. Variation among sites was caused primarily by large mean annual temperature differences.

Within semiarid rangeland communities, various groups of species existed which expressed similar patterns of phenological development. These groups tended to be found among species with similar life forms or rooting habits. This was apparently the result of limited growing season, similarities in moisture acquisition strategies, and effects of temperature on those species. Within these groups best key indicator species useful for phenological characterization of one or more associated community members were western wheatgrass and bluebunch wheatgrass.

Within sites, moisture availability appeared to have the greatest effect on developmental pattern differences among the life forms. Among sites, however, mean annual temperature differences exhibited a greater effect on phenological patterns than did precipitation variability. However, differences among sites were only noted when mean annual temperatures varied by 5.0°C or more.

See 830602.

2E 840407 Fisser, H.G. and L.A. Joyce. 1984. Saltbush community changes with and without grazing over a 20-year period in north central Wyoming. Submitted to Second Intern. Rangeland Congress. Adelaide, Australia. May 13-18, 1984. 4 p.

2F2
2A
2C
6A2
1B4
1E1
4A
4B

Vegetation successional changes in the arid western U.S. occur very slowly and seldom are species presence/absence composition changes or extended climatic variability observed in short-term research. This study was undertaken to examine 21 years of annually recorded data from exclosed and grazed sites of seven Gardner saltbush *(Atriplex gardneri)* communities and to determine the relative influences of climatic variability and grazing pressure upon species presence/absence composition and herbage yield changes.

In a joint effort between the University of Wyoming and the Bureau of Land Management, seven exclosures and adjacent grazed sites in the Big Horn Basin were monitored over a 21-year period, 1962-1982. Basal cover of grasses and forbs by species and aerial cover of semi-woody and woody species were recorded prior to clipping annually at approximately peak standing crop. Precipitation was measured on-site and recorded quarterly.

This study involved summarization and analyses of long-term annual records of peak record yield, species presence/absence composition and precipitation. The dominant species at a site were identified as those which had occurred during at least 60 percent of the years sampled. Similarity indexes among sites were calculated from the species presence/absence annual data. Correlation and regression analyses were used to evaluate the precipitation to standing crop relationships. In the early years of this study, shrub standing crop constituted greater than 50 percent of the mean total over all sites. Grass and forbs constituted as little as 10 percent at one site. In the 1963-65 period, spring precipitation was much greater than the long-term average and after 1966, remained low. Winter precipitation increased in amount and proportion of annual total. Herbage yield dynamics shifted over time inside the exclosures. Shrub standing crop decreased and grass and forb increased. These dynamics coincided with and continued after the seasonal precipitation variations were noted.

While species composition was not noticeably different inside versus outside the exclosure, total herbage production was always greater inside. Shifting dynamics of standing crop were clearly visible only inside the exclosure. Whether the patterns seen outside the exclosure were a result of the variation in climatic pattern or of grazing pressure was difficult to determine. It seemed apparent, however, that protection from grazing alone would not have changed the standing crop relationships to the extent now seen if the precipitation patterns had not shifted also.

The ungrazed exclosure communities express reasonable stability; were in at least good condition relative to inherent site potential; and exhibited considerably greater dynamic response to climatic variation than during the early years of the study. Whether re-establishment of historically decimated herbaceous species populations of rangelands could occur while subjected to grazing stress would be partially dependent upon degree of utilization. However, prevalence of long-term historical overuse within the region may have limited herbaceous species seed source adequacy for broad scale improvment. In addition, limited drouth occurrence and continued favorable fall, winter and early spring moisture would be necessary to enhance herbaceous species presence/absence composition as well as herbage yield.

See 840413 and 830310.

5A1D840408 Mulholland, J.L. and H.G. Fisser. 1984. Big sagebrush response and re-establishment following chemical control in Wyoming, U.S.A. Submitted to Second Intern. Rangeland Congress. Adelaide, Australia. May 13-18, 1984. 4 p.

1E1
2A
2B
2C
2E
2F2
11E2
11E3

Big sagebrush *(Artemisia tridentata)* occupies approximately 15 million hectares of rangeland in Wyoming. Control programs with herbicides were initiated during the late 1940s on western U.S. rangelands to increase forage production. Between 1952 and 1976, almost a million hectares were treated on federal and private lands in Wyoming. In most cases, sagebrush re-establishment occurred. The objectives of this study were: 1) to evaluate chemical shrub control effectiveness and duration and 2) to evaluate management usefulness of livestock exclusion.

Five chemically sprayed areas were located in western Wyoming on sites ranging in productivity from 400-1,200 kg/ha and in mean annual precipitation from 18-35 cm. Livestock exclosures were constructed at each study location to include one half in a shrub control area and one half in a native, nonsprayed sagebrush-grass vegetation area. The chemical herbicide used 2,4-D butyl ester.

Canopy cover differences of big sagebrush in 1982 still reflected influence of the chemical shrub control treatment, except for one site where observed wind caused drift may have reduced the actual application and subsequent mortality rate, such that mean shrub canopy cover of the no-spray sites was almost five times greater than on those subjected to herbicide control. Differences over all sites were significant. The grazing treatment caused no significant ($p \leq 0.05$) canopy cover differences. Interactive trends were discernible, in that cover values of the grazed-spray sites tended to be greater than those of the no graze-spray area. While on the non-chemical treatment sites, cover of the graze-no spray tended to be less than the no graze-no spray.

The most striking similarity among sites receiving effective control of shrubs, was the relatively slow rate of re-establishment during the first 10 years following chemical application. Where shrub control was least effective, however, re-establishment began immediately. Concomitantly with effective herbicide treatment, the abundant perennial western wheatgrass *(Agropyron smithii)*, because of its rhizomatous nature, in large part was a deterrent to sagebrush re-establishment even though subjected to intensive herbivory. Additionally, forage production on all sites, whether shrubs were effectively controlled or not, was greatly enhanced by the chemical treatment. During the latter 10-15 years, sagebrush re-establishment increased prominently on both grazed and ungrazed sites. Although some interactive location variation was present, similar response patterns further indicated rate and magnitude of the shrub reinvasion was subject to environmental factors as well. The effective shrub control reduced re-establishment rates of sagebrush for at least 10 years and shrub seedling establishment was inhibited by the rapid establishment of the rhizomatous western wheatgrass in the sprayed areas. Protection from grazing during the year of chemical treatment enhanced early grass establishment and sage production.

See 840601.

11A 840409 Germain, L.P. and H.G. Fisser. 1984. Application of kriging to ecological variables. p. 12. In Proc. 55th Ann. Meeting Colo.-Wyo. Acad. Sci., Western State College, Gunnison, Colorado. April 27-28, 1984.
2E
2F2
2A

See 840202.

3A3 840410 Fisser, H.G. 1984. Culm weight dynamics as an index of sagebrush-grass community productivity. p. 17. In Proc. 55th Ann. Meeting Colo.-Wyo. Acad. Sci., Western State College, Gunnison, Colorado. April 27-28, 1984.
1E2
2A
2E
2B
11E3
11F1

See 840201.

3A3 840411 Fisser, H.G. 1984. Detailed numerical phenology of sagebrush with visual description statements. Posterboard display at USDA-ARS/USDI-BLM Regional Workshop on Rangeland Hydrology, Modeling and Monitoring. Boise, Idaho. April 30 - May 2, 1984.
1E2
2E
3A2
2F2
11A
11B
11C
11E
2C
2A
11F1

See 840301.

5A1D 840412 Dean, S.K. and H.G. Fisser. 1984. Grazing management longevity and benefits of sagebrush control in western Wyoming, U.S.A. Posterboard display at USDA-ARS/USDI-BLM Regional Workshop on Rangeland Hydrology, Modeling and Monitoring. Boise, Idaho, April 30 - May 2, 1984.
3A2
3A3
2A
2C
2E
2F2
5B4
10D

See 840405.

2E 840413 Fisser, H.G. and L.A. Joyce. 1984. *Atriplex*/grass and forb relationships under no grazing and shifting precipitation patterns in north central Wyoming. p. 87-96. In Tiedeman, A.R., E.D. McArthur, H.C. Stutz, R. Stevens and K.L. Johnson (eds.). Proc. Wildland Shrub Symp. "Biology of *Atriplex* and related chenopods." Provo, Utah. May 2-6, 1983. USDA Forest Service General Technical Report 172.

2F2
2A
2C
6A2
1B4
1E1
4A
4B
11A
11E2
11E3
11F1

See 830310 and 840407.

2F2 840414 Fisser, H.G. 1984. Biology and ecology of sagebrush in Wyoming: I. Herbage production and community dynamics. p. 21. In Abst. Shrub Research Consortium Third Annual Wildland Shrub Symposium: the biology of *Artemisia* and *Chrysothamnus*. Provo, Utah. July 9-13, 1984.

3A3
2E
2A
11F1
10D
2F2

See 840310.

5A1D 840415 Fisser, H.G. 1984. Biology and ecology of sagebrush in Wyoming: II. Reinvasion following sagebrush control. p.21. In Abst. Shrub Research Consortium Third Annual Wildland Shrub Symposium: the biology of *Artemisia* and *Chrysothamnus*. Provo, Utah. July 9-13, 1984.

1E1
1E2
2E
2F2
2C
2A
3A3

See 840311.

1E2 840416 Fisser, H.G. 1984. Biology and ecology of sagebrush in Wyoming: III. Phenology. p. 22. In Abst. Shrub Research Consortium Third Annual Wildland Shrub Symposium: the biology of *Artemisia* and *Chrysothamnus*. Provo, Utah. July 9-13, 1984.

2F2
2E
3A2
1E1
11A
11B
11C
11E2
11F2

See 840312.

11E3 840417 Wight, J.R., H.G. Fisser and C.W. Hanson. 1984. Biology and ecology of sagebrush in Wyoming: IV. Validation of a rangeland production model (ERHYM) for sagebrush sites. p. 13. In Abst. Shrub Research Consortium Third Annual Wildland Shrub Symposium: the biology of *Artemisia* and *Chrysothamnus* Provo, Utah. July 9-13, 1984.

11F2
11F1
11E2
2A
2C
3A2
2F2

See 840413.

1E2 840418 Hanson, C.W., H.G. Fisser and J.R. Wight. 1984. Biology and ecology of sagebrush in Wyoming: V. Herbage yield dynamics and seasonal precipitation relationships. p. 13. In Abst. Shrub Research Consortium Third Annual Wildland Shrub Symposium: the biology of *Artemisia* and *Chrysothamnus*. Provo, Utah. July 9-13, 1984.

2A
11F1
11E3
3A3
2F2

See 840314.

3A2 840419 Fisser, H.G. and K.L. Johnson, eds. 1984. Wyoming Shrublands Proc. 11th Wyoming
1E1 Shrub Ecology Workshop. Lander, Wyoming, May 27-28, 1982. University of
1E2 Wyoming, Laramie.
2A

2E Forty-six technicians registered for the 1982 workshop in Lander, Wyoming, to
2F2 consider sagebrush control, reinvasion and management relations. The formal session
5A1D was developed by Herb Fisser, Professor of Range Management. Kendall Johnson of
10D Utah State University, Logan, was program moderator and discussion leader. The field chairman was John Likens, BLM, Lander, for the BLM tour stops and David Reeder, FS, Lander, for the Forest Service stop. The coordinating committee acknowledged the particular support of the Department of Range Management and the Wyoming Agricultural Extension Service, UW, for assistance in staging the 1982 event and the continuing involvement of other support agencies.

8405 Wyoming Agricultural Experiment Station Publications

1E2 840501 Fisser, H.G. and L.P. Germain. 1984. Phenology and production studies on semiarid shrub
11A types: Statistical tests and computer programs. Wyo. Agr. Exp. Sta. Sci. Rep. 1331.
2A

11C Almost all of the professional activities of the authors during 1983 and 1984 had
11E been directed primarily toward the compilation and analyses of data from the phenology
11E and production studies on semiarid shrub types; the program funded by the Bureau of
11E Land Management with the University of Wyoming. The efforts at this time had
11F progressed well, but final analyses and computer programs for forage prediction and
3A forecasting had not yet been finalized.

Fisser was on sabbatical at the Northwest Watershed Research Center in Boise, Idaho, during the 1983-84 academic year. While working with the scientists at that research center, he became associated with the ERHYM model and compiled production data and cover data coordinated with the seasonal precipitation data obtained at eight exclosures in the Wind River Basin. This effort required an extensive cross-check of field data sheets, annual report data sheets, and close inspection of annual herbage production transect slides to remove errors which had cropped up but which had never been completely removed from the 23-year data set. These data were compiled by species and integrated into the computer system at the Watershed Research Center. The data were subsequently transferred to computer tapes which are in operation at the University system in Laramie.

The use of ERHYM was preliminary, but exhibited excellent promise as a means of predicting forage yield from rangeland sites. The validation tests, which were conducted with the Wyoming data compared to the data obtained from the Montana site, from which the model had been developed, showed that there was excellent potential for extrapolating the model capability to the kind of environmental situations extent in Wyoming. Other sites were to be studied as data were compiled.

During the 1983-84 period, Germain was located at the University of Wyoming, and conducted intensive programming development with the climatic data from the Weather Service stations and with those obtained at the various exclosures associated with the phenology and production study contract. The initial effort involved identification of all sites, definition of their locations, and locating these on a map for use in order to interrelate distance relationships as well as other site variables including elevation and precipitation differences.

The kriging process was approached with the intent of utilizing it as a means for estimating missing climatic data elements. This process was objective and exceptionally valuable. But the utilization in terms of the locations and kinds of data available in the present study precluded its active use, especially since kriging analysis required excessive amounts of computer time. As a result, the analysis of climatic data relationship from one location to another was approached with a reciprocal of the distance squared formula to determine the general relationships of one site to another and to extrapolate precipitation values from one location to another.

The ERHYM model is a relatively simple model, but still requires intensive input of climate and yield data. These data, available from the U.S. Weather Service, were purchased on computer tapes. The following manuscript details the problems encountered while extracting the various forms of climate data for utilization in the ERHYM model process. The manuscript is lengthy, but important because it details the unexpected problems found in trying to extract climatic data from long-term records. There are no other records available which indicate the difficulty of obtaining complete data sets.

Daily precipitation values gave input parameters for the ERHYM program developed by Dr. Ross Wight of Boise, Idaho. Because only seasonal (and in some cases monthly) values were recorded at the permanent exclosures in Wyoming, seasonal values required interpolation to obtain a complete daily record for each exclosure. Completing this objective met with problems, barriers and diversions based on the available and obtainable information and the goal of acquiring the best possible estimates from the sum of the total information. In short, what was originally thought to be a relatively direct, simple procedure, became a complex, time consuming, thought provoking process as a result of the information structure, availability and authenticity in addition to the sampling procedures, computer programming and consultation required. The remainder of this discussion will review some of the barriers encountered in the estimation procedure.

Obtaining a complete set of daily data for any exclosure required one prerequisite: before interpolation (or extrapolation) of daily estimates could be accomplished, a complete set of daily data from known locations had to be obtained. To accomplish this task, a set of USWS tapes were purchased from the NOAA in South Carolina containing weather information from daily recording stations across the State. In addition, data tapes were puchased from the Water Resources Center at the University of Wyoming for all hourly recording stations. We were interested in all data between 1960 and 1982, and eventually all historical data prior to 1960. This introduced the first problem. A complete data base for either daily or hourly stations was available only through 1978. This required the purchase of additional tapes to update the data base through 1982. We now had at least five nine-track tapes containing precipitation data. The problems associated with those data tapes may be found at the end of this discussion.

After the problem of reading and extracting the desired data off of each tape was overcome, we found that of the 10 Weather Service stations in the Wind River Basin area, only one, Lander, had a complete data set between the years 1960-1982. Some stations had only a few missing days while some stations such as Jeffrey City had more than 13 years missing. Therefore, before we could interpolate from a seasonal basis to a daily basis at the exclosures, we made daily estimates for missing data at the weather service stations. We could achieve this through the reciprocal of the distance squared technique, however there were additional considerations. First, as the name suggests, an estimate was made from a weighted average based on distance; the Weather Service data tapes did not contain location parameters. Second, the technique assumed that distance was the only factor affecting differences in precipitation, which we know is incorrect. Therefore, we had to make some subjective decisions as to which stations were closely related and run separate analyses on each set. Ironically the only complete data set, Lander, was considered to be independent of the other stations, so an additional check that at least one station within any set had a valid value was needed. A final program which conveyed each set into one data file then had to be accomplished.

Once a complete daily data set of the Weather Service stations was established, attention could be directed to obtaining daily estimates at each exclosure from seasonal data. However, the seasonal data set for the exclosures was also incomplete and estimates of missing seasonal values had to be made. This was accomplished in the same manner as with the daily values for the weather service stations, except that seasonal values from both exclosures and weather stations were used for estimation.

Once this was accomplished, the interpolation from seasonal to daily values could be accomplished. This was done through the use of the reciprocal of the distance squared and percentages of daily values to seasonal values at the Weather Service Station. The results of this procedure showed adequate results and seemed to give the best estimates based on the available information and time constraints.

Review of Problems Associated with Weather Data Tapes

Tape Number:	S02941	USWS STATIONS	0027-3950
	S02942	USWS STATIONS	3951-9925
Tape Structure:	Density - 6250 BPI		
	CODE - ASCII		
	Variable length records		
Contents:	NOAA daily weather data for all stations in Wyoming through 1978		

Primary Steps and Problems:

1) The first step was to convert the tapes purchased from the NOAA into a format and language that the Cyber mainframe computer at the University of Wyoming could understand. Also, because the tapes contained several types of weather records, we wanted to copy only precipitation or temperature records onto separate tapes for efficiency (it took more than 100 seconds of computer time and more than 1 1/2 hours actual time just to read the entire data set on both tapes). This was accomplished by a program that was written by a computer consultant at the University.

2) The second step was taken to construct programs to select the appropriate precipitation or temperature data and write the results in a predetermined format. The results of this procedure revealed significant problems that were not foreseen.
   a) A significant amount of missing data present depended on the station. As a result of the variable length record format of the tapes, if there was an entire month and/or year of missing data (and there were several), then those records simply did not exist on the tape. If there were missing days within a month (a record consisted of 1 month), these also did not exist, the record was simply shortened by the number of days missing (eg. if January had 10 missing days, the record contained 21 days). For these reasons, it was necessary to reconstruct the program to insert missing value indicators when: 1) an entire station was missing, 2) an entire year within a station was missing, 3) an entire month within a year within a station was missing, 4) any day within a month within a year within a station was missing. Following completion of programming for these procedures, additional unforeseen problems were encountered.
   b) Certain stations such as Jeffrey City had an overlap of data. Because the tapes were supposed to be written in sequential order, Station ——> Year ——> Month, the programs were written to utilize this structure to locate and define missing data. However, for Jeffrey City, the data began in January 1962 and proceeded through April 1964. This same sequence was repeated on the tape which was then followed by the 1978 data with everything between May 1964 and December 1979, missing. The program entered an endless loop which required additional constraints to eliminate the problem.

Tape Number:	S03262
Tape Structure:	Density / 6250 BPI
	Code - ASCII
	Fixed length records 402 characters/record
Contents:	NOAA daily weather data for all stations in Wyoming 1980-82

Problems:

1) With the exception of having fixed length records, this tape was believed to be essentially identical to the previous two tapes listed (data through 1978). The presence of fixed length records simply meant that months with missing data or fictitious days (eg. February 30) a false value was assigned to that day. If an entire station, year, or month was missing, again that record simply did not exist on the tape. Restructuring the previous program to accommodate fixed length records proved to be a relatively simple procedure. However, again there were additional differences between the tapes that were not foreseen. The NOAA tapes with data through 1978 were sequentially ordered as follows: Station ——> Year ——> Month. However, the NOAA tape with data 1980-82 was sequentially ordered: Year ——> Station ——> Month. Therefore, the original programs proved to be totally inadequate for the 1980-82 data tape. A new program was constructed for this data.

Tape Number: S03151
Tape Structure: Density: 6250 BPI
Code: Display
Variable and fixed length records
Contents: Wyoming Water Resources data tape containing daily and hourly precipitation records through 1978
Primary Goal: To calculate daily precipitation values from hourly recording stations for a specified range of years

Problems:

1) Because the Wyoming Water Resource tape was set up on a separate data base from the NOAA tapes, it was impossible to use either of the previous programs for generating the desired results. Also, although Water Resources had a program available to calculate daily precipitation from hourly stations, they informed us that it was not possible to use on the data tape they supplied us. Therefore, it was necessary to write a new program specifically for the Water Resources tapes to accommodate the specific characteristics of the tape. Following are some of the specific considerations and problems encountered with the water resources tapes.
   a) Each logical record for the hourly stations varied in length depending upon the amount of rain that occurred each month. One record consisted of one month. The data were recorded as precipitation value and time of month of occurrence. A maximum of 744 hours were possible in any given month. For this reason, it was neccessary to write a program that converted hour of month to day of month.
   b) There were also several flag values associated with the water resources tapes:
      999 = missing
      998 = trace
      997 = fictitious hour (eg. Feb. 30)
      996 = value accumulated in next hour's value

      Therefore, it was also necessary to write the program to identify and account for these flagged values.

Tape Number: S02459
Tape Structure: Density - 1600 BPI
Code - Display
Variable and fixed length records
Contents: Water Resources data tape containing daily precipitation, hourly precipitation, and daily temperatures for most stations in Wyoming for 1978 and 1980.

Problems:

1) Once again the program written for the previous tape (S03151) would not work on this tape even though the data was in the same format. In this case, the problem occurred because Water Resources wrote the tape in what the computer knows as K type blocks. Tape S03151 was written in C type blocks. Unfortunately, there was an important command called "backspace" in the previous program that would work for C type blocks but not K type blocks. The backspace command in the program was important and increased the time efficiency of the program. It therefore became necessary to write another program that would eliminate the backspace command and maintain time efficiency while meeting an additional goal of transportability. To meet the first requirement of efficiency, it was necessary that only those records of interest be read completely. This was the purpose of the backspace command in the previous program. Several new strategies were reviewed to meet the efficiency requirement. Inevitably, they would be specific to Cyber computers and thus not meet the transportability constraint. Finally, after several hours of consultation, an alternative program was constructed for this tape which met both goals. It was also written to work for the 1948—1978 data on the previous tape, which made the previous program obsolete.

2F2 840502 Fisser, H.G. 1984. Statistics and interpretation. March 30, 1984. Wyo. Agr. Exp. Sta. Sci. Rep. 5 p.
3A
11E
11F

Historical and experimental data are two kinds of quantitative scientific information. The two are specifically distinguished by the concepts that experimental quantities are obtained by sampling and thus the data are repeatable, whereas data of a historical nature are usually one of a kind and not subject to confirmation by repeated acquisition and testing procedures.

Scientific research and experimentation are conducted for discovery of new information. The scientific method is a term usually employed to encompass the procedures of systematically collecting and interrelating facts, observation and philosophy. The correct interpretation of new findings should be considered part of the scientific method and assigned as much importance as the proper data acquisition.

Basic precepts to the quantitative nature of any measurable value require an understanding of dispersal characteristics under study. An appreciation of the nature of sampling, deriving a small data set from some vast, often unmeasureable population, is essential. The errors associated with sampling are identified by the terms random sampling "error" and "bias." The former is simply an expression of the sampling process. A sample estimate will seldom, if ever, duplicate the population or universe from which it is derived. Bias, on the other hand, is error associated with the method of data collection and handling. Conceptually, bias is unmeasurable, but it usually assumes minimal importance if sampling design and procedures are acceptable for the needs of the research objectives.

The statistics of research can be viewed in terms of the singular quantitative data record, and of statistical parameters. The latter relates to measures of central tendency, variability and asymmetry. Statistical analyses of data may involve only manipulation of quantitative values to identify means, standard deviation, variance, range, and coefficient of variation. The lack of symmetry of a distribution data set is identified as skewness, a troublesome characteristic commonly associated with biological data.

Whether the research utilized only such data interpretation procedures as those above, or continued with more complex and sophisticated programs and models, it is extremely important to recognize that statistics is no more than a tool for generation of descriptive quantitative values.

When utilized properly, statistical analyses enable more efficient research. Tests of similarity or dissimilarity of various products of statistical procedures at whatever probability level, do not of themselves establish inferences about ecological relationships. These are derived solely through observation, reasoning and interpretation on the basis of prior knowledge and experience. How often have we heard a neophyte state, for example, that his data showed there was no relationship between plant growth and rainfall. Whether his data exhibited this and were accurate is not of importance. That the thoughtless statement was made, implying that plants have no need for water, is important. A researcher must be knowledgeable of biotic-abiotic relationships and be able to interpret statistical results in terms of common characteristics of the measured objects. Failure to do so may lead to grossly misleading findings and assumptions.

Complications in the interpretation of rangeland research occur because natural populations are often dispersed nonrandomly, and unique sampling techniques become necessary. Although data may be obtained by precise methods, the accuracy of estimation may be poor, or the random sampling error may be so large that statistical analyses do not correctly portray real world relationships. It is important, therefore, that sampling methods meet theoretical statistical design and assumption requirements as closely as possible, and also provide biologically acceptable data. In such situations, a reasoned evaluation by a knowledgeable researcher is necessary and must be recognized as a valuable and acceptable interpretive tool.

When addressing the problems of sampling biological systems the ecologist must commonly interpret response in a subjective manner when statistical analysis of data do not appear to reach acceptable levels of sampling measurements. Extreme variability of data because of spatial distribution related to factors such as 1) bare ridges which may occupy less than 5 percent of the surface area in a saltbush (*Atriplex* spp) rangeland, 2) Plains pricklypear *(Opuntia polyacantha)* clumps which may occur in areas as large as 20 feet in diameter but with only one or two clumps per acre, 3) sagebrush (*Artemisia* spp) presence in narrow drainageways of a saltbush rangeland and 4) harvester ant colony pads as large as 40 feet in diameter but with usually only one or two per acre. Random plot distribution often encounters these extreme variants to the general character of rangeland vegetation and causes significant outliers to the data set. Statistical analysis then may appear meaningless in terms of the numerical output; and a statistician's commonly derogatory response to the researcher will imply simply that the data are poorly gathered and/or a result of poor sampling design. The implication is that well-gathered data and acceptable sampling design are an everyday matter of some activity adjustment, and that an inadequately trained nonstatistician really should have become aware of this by earlier consultation with a statistician.

The procedures required to achieve desired statistical significance, however, are often so extreme that time and cost restraints completely preclude implementation. How then does the ecologist attain the data necessary to evaluate such factors as vegetation cover, composition, production or productivity? Too often the statistics of the data [$\bar{X}$, S^2, S, CV, ANOV, Reg. Correl, etc.] are considered the end product of the research effort. An ecologist's view of data are often tempered by knowledge of floral components and relationships to abiotic factors. With a view to statistical analysis simply as a mechanical means of number reduction and as a tool for the research effort the background and experience of an ecologist can often be utilized to derive definitive interpretation of the biological component relationships even though the $P \leq 0.01$ or $P \leq 0.05$ significance is unattainable from the data.

Many criteria of natural vegetation components can often be critically examined with statistical analyses. For those with extreme nonrandom spatial distribution problems, sample adequacy will seldom be possible. Such a status implies a statistically inherent inability to quantify these characteristics because of extreme outlier data elements. Subjective interpretation can however, provide an ability to arrive at a logical quantification of differences or changes due to treatments even though not definable at high probability levels. Such handling of quantitative data with related intuitive interpretive evaluation of relationships can greatly strengthen translation of biological significance to that of statistically based probabilistic significance tests.

Manpower, time and equipment needed to achieve statistically adequate sample data often require prohibitive expenditures. Quantitative biological data from native rangelands will seldom, if ever, be completely statistically adequate under all circumstances of site and climate variability. Because of such sampling problems, data interpretation must be tempered and adjusted to utilize both the numerical and quantitative data derived by well designed statistically accepted procedures, as well as the expertise of knowledgeable and trained scientists. It is crucial, therefore, for users of rangeland research not only to be cognizant of potential data weaknesses; but also to recognize that subjective interpretation may often be more correct than simple numerical probability test values.

The use of statistics is a necessary regimen of quantitative data manipulation; it assists the researcher in minimizing the potential for erroneous conclusion. On the other hand, the use of justifying information from other than a point data set is often necessary for accurate assessment of biological interactions with complex abiotic interrelationships. Minimization of influence caused by outlier data elements has been addressed in centroid cluster analysis (BMDP, P3M) routines and should be considered. Generalization of biotic dynamics, however, still requires data interpretation bolstered by the scientist's knowledge, experience, and intuitive expression.

1E2 840503 Fisser, H.G. and L.P. Germain. 1984. Phenology and production studies on semiarid
11A shrub types. Draft final summary of 1962-1982 period. Wyo. Agr. Exp. Sta. Sci.
2A Rep.
11C
11E1
11E2
11E3
11F1
3A3

8406 Graduate Advisee Completions

5A1D840601 Mulholland, J.L. 1984. Big sagebrush response and re-establishment following chemical
1E1 control in Western Wyoming. M.S. Thesis. University of Wyoming. 72 p.
2A
2B
2C
2E
2F2
11E2
11E3

The long-term effects of grazing, protection from grazing, and chemical control of big sagebrush were evaluated for five areas located in semiarid regions of western Wyoming. The study areas represented ecologically diverse big sagebrush *(Artemisia tridentata)*-grass vegetation types with varying degrees of past and present livestock use. Four geographic regions of western Wyoming were represented; the Big Horn Basin (north central), the Beaver Rim area and Wind River Basin (west central) and Bear River Divide (southwest).

Exclosures were constructed during the late 1950s at each site, with herbicide (2,4-D butyl ester) application coinciding with exclosure construction. The fenced areas included one half in a shrub control area and one half in native sagebrush-grass vegetation. The experimental design consisted of four treatments: grazed-sprayed, ungrazed-sprayed, grazed-nonsprayed and ungrazed-nonsprayed.

Three big sagebrush parameters were measured to evaluate the effects of spraying and grazing and rates of sagebrush re-establishment. These parameters were: sagebrush canopy cover, sagebrush density by age class, and growth ring analyses to determine rates of sagebrush re-establishment.

Sagebrush canopy cover was lowest in the sprayed areas with no differences between grazed and ungrazed areas. The spray treatments, in 1982, were still highly visible with the exception of Granite Mountain where sagebrush control had been least effective. Grazing did not cause differences in sagebrush canopy cover among the four treatments over all sites.

Grazing increased sagebrush density only on the spray plots at three of the five study areas. Juvenile and young sagebrush accounted for the major component of sagebrush plants in these areas. In nonsprayed areas, grazing had no significant effect on sagebrush density. Mature shrubs represented the largest age class of shrubs in non-sprayed areas. The distribution of juvenile and mature shrubs, however, was fairly consistent among sites and indicated continual replacement within the sagebrush population. Thus, changes in the stand structures of big sagebrush in semiarid regions of Wyoming were slow when the only management applied was exclusion of livestock.

Rates of sagebrush re-establishment directly corresponded to degree of sagebrush control, grazing intensity and inherent site productivity. Precipitation influences were noted, but no statistical relationship was found between precipitation fluctuations and differential rates of sagebrush re-establishment. Poor range condition prior to chemical control at the Smilo site caused rapid increases in sagebrush immediately following treatment.

The most striking similarity among sites receiving effective control of sagebrush, was the relatively slow rate of re-establishment during the first 10 years following control. These results were in contrast to Bartolome and Heady (1978, J.R.M.) in southwestern Oregon who found most re-establishment occurred in the first several years after treatment.

Where shrub control was least effective, sagebrush re-establishment began immediately following control. Mature sagebrush remaining after treatment were the seed source for sagebrush of younger age classes. At the Granite Mountain spray an apparent peak in sagebrush re-establishment ten years following control was attributed to recovery of sagebrush partially killed by 2,4-D.

The sprayed areas received noticeably more livestock use than adjacent, nonsprayed areas. The attraction of livestock resulted in greater numbers of sagebrush re-established in grazed than ungrazed areas at Granite Mountain, Lower Government and Cumberland -3. Patterns of re-establishment were also similar between grazed and ungrazed areas which suggested environmental factors were also important in controlling the rate and magnitude of sagebrush re-establishment.

Areas of high production such as Lower Government and Cumberland -3 offered the greatest potential for sagebrush control. Where grazing was excluded on the more productive sites, sagebrush re-establishment was delayed. Competition from non-sagebrush vegetation was an important factor. Observations from this study suggested sagebrush re-establishment would be long delayed on shrub control areas when sites are quickly occupied by perennials.

Chemical control of sagebrush in areas of natural heavy concentrations of livestock such as the Lower Government area should be avoided unless the grazing management and livestock distribution could be changed. Removal of perennial grass cover through heavy grazing use was shown to favor the re-establishment of sagebrush.

4B 840602 Deboodt, T.L. 1984. Vegetation composition and production dynamics in grazed versus non-grazed areas of central Wyoming. M.S. Thesis. University of Wyoming. 87 p.

9A
3A3
3A2
6A
5A1D
11E2

This study examined vegetative dynamics for a big sagebrush *(Artemisia tridentata)*-grassland in central Wyoming. Production, cover and frequency were examined in relation to influences of climatic variations, grazing impacts and harvester ant control.

Twelve study areas, located within the Davison Brothers Allotment 64 kilometers northeast of Lander, Wyoming, were intensively examined for treatment differences. The study areas, each three hectares in size, were divided equally among two grazing treatments, fall-winter-spring use and no use, and three chemical application rates for harvester ant control.

Two study areas were examined to determine which variables best explained production dynamics. Variables analyzed were four precipitation periods and their additive impacts, length of the growing season, percent sagebrush cover recorded on the harvest date, the influence of the grazing management practice and time frame associated with that practice and site differences.

The final model used to help explain the importance of each of the selected variables was highly significant [$F(8,45) = 11.68$] with an R^2 of 0.67. Variables included in the final model were the natural log of winter-spring precipitation, the natural log of the previous summer-fall precipitation, the length of the growing seasons, percent sagebrush cover, the number of years since the initiation of the study and the quadratic specification of that time, the influence of grazing exclusion, and site differences which accounted for soil and overall vegetative composition and production differences between the two areas used for the analysis.

Winter-spring precipitation had the greatest influence on production. Length of the growing season, percent sagebrush cover and time squared all had negative influences on production. Precipitation, time, grazing exclusion and site differences were all positively associated with production.

The natural log transformations of the precipitation data allowed production to increase at decreasing rates as the amount of precipitation increased. The time variable was squared because initial analysis showed a quadratic relationship among the years of the production data.

Cover and frequency data, recorded from permanent plots for the years 1964, 1967 and 1982, were examined to determine changes over time for six plant species. Big sagebrush, western wheatgrass *(Agropyron smithii)*, blue grama *(Bouteloua gracilis)*, Sandberg bluegrass *(Poa secunda)*, needleandthread *(Stipa comata)*, Indian ricegrass *(Oryzopsis hymenoides)* and Plains pricklypear *(Opuntia polyacantha)* were examined for changes in cover on all 12 study areas and changes in frequency for six of the study areas. No species examined showed significant changes in cover or frequency due to harvester ant control.

Big sagebrush, while not increasing significantly in frequency of occurrence, has significantly increased in cover under exclusion of grazing, 7.5 to 10.8 percent. Under fall-winter-spring grazing, sagebrush cover had remained relatively constant for the years examined, 8.01 in 1964 and 8.16 in 1982. Fall-winter-spring rotational grazing had checked sagebrush growth and nonuse had allowed sagebrush to increase.

Plains pricklypear had decreased significantly in cover from 1964 (3.5 percent) to 1982 (0.72 percent). At the same time, frequency of occurrence had remained relatively constant, 0.17 percent in 1964 and 0.15 percent in 1982. Both grazed and nongrazed sites exhibited this pattern, with no significant differences between treatments.

Of the perennial grasses, both western wheatgrass and sandberg bluegrass increased in cover and frequency during the first four years of the study. No significant differences were found between the grazing treatments, but those areas which were grazed had a greater average increase than did the nongrazed areas. Since 1967, both of these grass species have declined significantly in percent basal cover while frequency of occurrence had remained about the same.

Blue grama had decreased in cover (3.88 percent in 1964 and 1.13 percent in 1982) since the beginning of the study with no change in frequency of occurrence over the same time period. This could be the result of the lack of use of this plant during the grazing period. Under no use, blue grama, as shown by the exclosure data, had declined at a faster rate since 1967 when compared to the grazed sites. Again, these differences are not significant but may be indicative that grazing of some sort dampened the dynamics of plant growth associated with weather variations in this sagebrush-grass type.

The needleandthread-Indian ricegrass species complex responded slowly to the changes in grazing management practices. From 1964-67, this species complex declined in both cover and frequency. The decline was greatest in the protected areas. Since 1967, however, this species complex had increased in both cover and frequency. Again in 1982, basal cover and frequency of occurrence were greater in the grazed areas than in the nongrazed areas.

While the difference between grazing treatments was not significant for the perennial grasses in 1982, the greater percent basal cover may be attributed to two things, 1) the increased sagebrush cover in the protected areas reflected an increased competitive advantage for the sagebrush and the disadvantage for nongrazed grasses and/or 2) that grazing can increase perennial grass longevity through improved vigor and nutrient cycling which favored most perennial grasses when compared to extended periods of complete nonuse.

The economic analysis of the production response model for this set of data did not favor complete nonuse or resting of this rangeland beyond what was already called for in the current management plan. Resting this rangeland resulted in a 1.56 kilogram per hectare increase per year. Using the internal rate of return (IRR) as the investment criterion, this increased production would yield an IRR of =9.5 percent based on one year of rest and a 20-year life of the increased production would be needed. A 5-year life span of the increase in production with an investment rate of 10 percent would require a 26 percent or 45 kilogram per hectare increase in production during the year of rest.

It was apparent that, through proper grazing management, trends in range condition could be similar or better than under complete grazing exclusion. The conclusion was that the grazing practices currently being implemented on this site should continue. The only modification in management needed might be to increase the level of use during the late fall-winter periods if a reduction in sagebrush cover was desired.

In the final analysis, protection from grazing, like any other grazing management system, must be implemented only where a need exists and where improvement can be expected. Ecological and economic returns must always be considered and predicted outcomes weighed when implementing any type of grazing management practice.

8502 Journal Publications

3A3 850201 Joyce, L.A. and H.G. Fisser. 1985. Interseasonal dynamics in central Wyoming shortgrass communities. Amer. Mid. Nat. In Press.

8503 Presentations

850301 "Extending the duration of forage yield records through modeling." February 12, 1985. 38th Ann. Meeting Soc. Range Manage. Salt Lake City, Utah. by C.L. Hanson.

See 850402.

850302 "Vegetative recovery with controlled use on low condition salt desert shrub rangeland." August 7, 1985. Shrub Research Consortium. 4th Ann. Wildland Shrub Symposium: Plant-herbivore interactions. Snowbird, Utah. By H.G. Fisser

See 850403.

8504 Published Proceedings/Abstracts

850401 Anderson, S.K. and H.G. Fisser. 1985. Sagebrush control and interaction with climax diversity. 38th Ann. Meeting Soc. Range Manage. Salt Lake City, Utah. February 12, 1985.

Rangelands dominated by the shrub big sagebrush *(Artemisia tridentata)* occupy some 100 million hectares in the western United States. Many methods for increasing forage production have been implemented. Herbage application for shrub control has been among the most successful. The objectives of this study were to 1) evaluate 20-year forage response to chemical shrub control and grazing exclusion and 2) define and interpret range condition and trend.

The effects of long-term livestock exclusion and chemical control of big sagebrush *(Artemisia tridentata)* with 2,4-D were evaluated at six study areas which represented three semiarid range sites in western Wyoming - the Big Horn Basin, the Wind River Basin and Bear River Divide.

Exclosures were constructed at each study site between 1958 and 1963 and encompassed both chemically treated and non-treated areas. Annual herbage production and seasonal precipitation from 1962-1980, sagebrush canopy cover, and ecological condition and trend were investigated.

Range condition ratings as a function of climax species composition were calculated for 5-year intervals; 1970-75 and 1976-1980. Range trend was assessed by comparing the two range condition ratings. No measurable change ($p \leq 0.05$) of perennial grass composition and production occurred over the 20-year period when the only management applied was livestock exclusion. Production of sandberg bluegrass *(Poa secunda)* and bottlebrush squirreltail *(Sitanion hystrix)*, however, were somewhat higher in grazed areas, indicating their status as increaser species. Perennial forb production was not affected by light and moderate grazing use.

Chemical shrub control through application of 2,4-D decreased competition from shrubs and effectively increased perennial grass production. Site productivity potential was positively related to recovery rate. Significant increases ($p \leq 0.05$) in mean perennial grass production over the 20-year period ranged from 458 kg/ha (3 percent) to 201 kg/ha (90 percent) in the sprayed treatments. Western wheatgrass *(Agropyron smithii)* accounted for the major increase at every study area. Annual production of muttongrass *(Poa fendleriana)*, bluebunch wheatgrass *(A. spicatum)* and needleandthread *(Stipa comata)* also increased in the sprayed areas at each site where they occurred. When either the shrub control treatment was ineffective or the area was heavily grazed, maximum benefits of the spray treatment were not realized. When less than 30 percent of the shrubs were controlled with the spray treatment, annual production increases of perennial grasses were significantly limited after the first five years. Even when a chemical control treatment was not completely effective, however, a perennial grass production increase was evident. Intensive grazing also decreased the longevity of an increased grass production period in a site with incomplete shrub control.

The poster board exhibited two photographs showing evidence of short-lived production increase in the Wind River Basin at a location at which only 30 percent of the shrub had been controlled. The chemical had been applied in 1962 and the earlier photograph was taken in 1970 and the later photograph in 1980 and both photographs still showed significant perennial grass response to the chemical treatment even though sagebrush reinvasion had been very strong.

The Big Horn Basin sites showed climatically related potential for annual plant invasion and prolific production in sprayed areas. In this case, the perennial grass production increase in the sprayed treatment occurred very slowly, but accelerated during the last 5 years, up to 1980. Increased perennial grass production coincided with gradual decreases in the annual plant production. Cheatgrass *(Bromus tectorum)* quickly occupied the ecological niche which resulted from the herbicide application. This result suggested that the presence of big sagebrush was the controlling factor in the annual plant population. The sodic and clay nature of the soils at that site were unfavorable to perennial grass seedling emergence. The annual grass presence and abundance for 10-15 years may have modified the site adequately to improve perennial grass seedling establishment potential.

Annual production of perennial grasses exhibited a high degree of variability over the 20-year period. Neither total annual precipitation nor seasonal precipitation was useful as a covariant to reduce variability of the production data. For this reason, few significant trends ($p \leq 0.05$) were detected from analyses of the production data, even when the data was averaged for five-year intervals. Species composition based on annual production was the best indicator of community trend. Range condition analyses based on this measurement indicated definite upward trends at study areas in the Big Horn Basin and in the Bear River Divide for both the sprayed and grazed treatments. Those in the Wind River Basin showed little or no change. Improvement in range condition in the Big Horn Basin was most dynamic in the sprayed treatments.

Trend evaluation was complicated by the annual production dynamics and apparent multiple succession paths. Species composition was characterized by disclimax dominance of grasses typically considered ecologically inferior in terms of climax composition, such as western wheatgrass, Prairie junegrass *(Koeleria cristata)* and mutton bluegrass. The determination of climax species composition for purposes of judging range condition was a difficult task. The results of this study and research have shown that secondary succession did not necessarily follow the same pattern as primary succession. The ultimate equilibrium reached after site disturbance such as that caused by grazing stress or shrub control treatments was primarily a function of the species composition at the time of perturbation.

There was a photograph of the Lower Government Draw exclosure showing the spray strips and the printed items on the 3rd poster board were as follows: Livestock exclusion did not affect: 1) sagebrush canopy cover and 2) annual production of perennial grasses or forbs. The chemical control treatment effects remain quantitatively and individually evident for 20 years. Multiple Secondary Succession Pattern toward climax composition was a function of 1) environmental differences, 2) site specific species interaction and 3) historical use and management. Multiple successional climax pathways must be recognized to effectively interpret range condition and trend responses to management treatment.

2A 850402 Hanson, C.L., H.G. Fisser and J.R. Wight. 1985. Extending the duration of forage yield records through modeling. 38th Ann. Meeting Soc. Range Manage. Salt Lake City, Utah. February 12, 1985.
2C
3A2
3F2

11E2 This presentation addressed a modeling effort to gain insight into data needs for
11F1 range yield prediction and forecasting. The model to be discussed is called ERHYM, and
11F2 was developed by Dr. Ross Wight several years ago. He has been working on it at the

USDA-ARS Northwest Watershed Research Center in Boise, Idaho and now has it developed for use by others. The model will be described in terms of use for extending data through time.

The model input requirements were reviewed briefly. The climatic units are the driving variables and most emphasis was allocated to this section. Other parts of the model were soils, site characteristics, which were strongly hydrology-oriented, and vegetation parameters.

In terms of the climate, this model required daily maximum and minimum temperature, solar radiation, precipitation and, in the climate generation mode, some variables or parameters for generation of quadratic records. There were two modes in terms of the climatic driving variables. One was to use an actual set of data, both in terms of the records that one might have for monitoring environmental factors and, if the factor was a long climatic record, one could use the major values at that particular location to address questions in terms of soil water, soil temperature, evapotranspiration and so forth. In addition, when working with the forecasting mode with long-term records, these could be utilized to generate prediction possibilities in terms of yield forecasting. The model could also be utilized with actual or generated data. Stochastically-generated climatic records could be compared to use of actual data to ask questions regarding expected results from a grazing system program, for example. Also, by extending short-term records for several years, information could be forecasted for three to five years into the future to observe a longer sequence of information.

The maximum/minimum temperatures and solar radiation can either be from actual data or generated. This was noted because in most past range work the concept of generating climatic data has not been stressed. It will be investigated further in the future. This particular model in ERHYM uses actual daily precipitation and from that can generate daily maximum and minimum temperature and total solar radiation. It is conditioned on the fact that there is wet or dry day response. This particular generation scheme came from an earlier effort. In the SPUR model there are tables and graphs for generating the daily amounts of precipitation as well. Both the main flow in terms of using actual data and for generated data mentioned before continues the quadient types of relationships. If we have a short record, how does this fit in a long scheme? If we want to use it in terms of forecasting and we have data up to a certain part of the year, we want to know what the probability is that we'll have some kind of a yield in harvest. From that we can either use actual data sequences or generated.

The ERHYM model has been utilized by Dr. Ross Wight for five years. Several researchers in the western U.S. were also utilizing it. The original request for developing this model was generated by ARS and BLM efforts in Ekalaka, Montana, and the BLM began using this model in management in Montana some three years ago. It was actually used in operation to determine whether stocking rates were necessary as a result of drouth.

To use the model, data to describe soil characteristics including the soil layers, thickness, water content, etc. is necessary. A factor that had been added quite recently was percent rock. This had not been in the modeling previously and was also being used in both the SPUR and ERHYM. Several vegetation components that related to crop transpiration relative to moisture and temperature were noted as important components. Also, the model addressed the growth rates of plants and related these to plant growth initiation and to time of peak growth. The yield of this particular model was computed as the percent that could be expected at the site based on the amount of transpiration versus potential evapotranspiration.

ERHYM was noted to run on a daily time schedule which was maintained similarly in SPUR. This schedule was based on sequential water use from the topsoil increment downward. The root distribution becomes effective and the model integrates water use and soil water content by level downward. It does the same with excess water, so the model is capable of using water from depths. Output of the model was developed by integration of hydrology data subject to soils, evapotranspiration and vegetation components. The model identifies run-off of precipitation as well as melt-off of snow cover.

Data in the report were based on the climatic driving variable, precipitation, with values from 1949-1982, which was a relatively long record not present in most research programs. Vegetation yield from 1967-1982 were entered as basic vegetation parameters. This was the data from Dr. Fisser's work in Wyoming. For total herbage yield prediction a significance level r^2 of 0.68 was obtained. For the perennial grass and forb data with several more years of information an r^2 significance level of 0.78 that is considered acceptable was obtained. The model also keeps track of soil water for generation of soil temperature.

850403 Fisser, H.G. 1985. Vegetative recovery with controlled use on low condition salt desert shrub rangeland. In Abst. Shrub Research Consortium. 4th Ann. Wildland Shrub Symposium: Plant-herbivore interactions. Snowbird, Utah. August 7-9, 1985.

Rotational, deferred and continuous grazing systems with cattle and sheep were initiated in 1967. Data were obtained annually through 1971 and then at four-year intervals to coincide with the deferred four-year cycling on salt desert shrub rangelands of south central Wyoming.

Elevation ranged from 6,200-7,000 ft. Average annual precipitation was 9.20 in recorded at 14 gauges over the 68,000 acre tract. Dominant vegetation types were big sagebrush *(Artemisia tridentata)*-grass, Gardner saltbush *(Atriplex gardneri)*, Utah juniper *(Juniperus osteosperma)* and greasewood *(Sarcobatus vermiculatus)*. Western wheatgrass *(Agropyron smithii)* was the most common perennial grass.

Wildlife included pronghorn antelope, mule deer, coyote, badger, bobcat, prairie dog, rattlesnake, golden eagle, sage grouse and mourning dove. Wild horses were present. Greatest stress from browsing occurred through migratory deer from the mountains to the east and south into Colorado. Significant response to protection from browsing of true mountain mahogany *(Cercocarpus ledifolius)* was not evident until 1982, 15 years following installation of the game proof fence. Cattle grazing did not appear to influence the use of pastures by antelope, sage grouse or the few resident mule deer which occupied the juniper areas during summer. Water developments (ponds and windmills) were used by all wildlife; therefore, once initiated the need to maintain availability was evident. During the study period, adult antelope negotiated the woven wire fences with ever-increasing ease.

The management systems had little influential impact on the rangelands. Western wheatgrass remained stable; but other grasses, including needleandthread *(Stipa comata)*, Indian ricegrass *(Oryzopsis hymenoides)* and bluebunch wheatgrass *(Agropyron spicatum)* made significant gains. Effects of grazing systems on wildlife and wildlife habitat were minimal. Antelope have adapted well to the fences and have extended their range during the course of the study.

The results show continuous grazing with lighter stocking rates can be as effective as rotational or deferred systems in restoring and maintaining a desirable range condition. At a more advanced stage of recovery, results may be different.